Catalyst Characterization

Physical Techniques for
Solid Materials

FUNDAMENTAL AND APPLIED CATALYSIS

Series Editors: M. V. Twigg

Johnson Matthey
Catalytic Systems Division
Royston, Hertfordshire, United Kingdom

M. S. Spencer

School of Chemistry and Applied Chemistry
University of Wales College of Cardiff
Cardiff, United Kingdom

CATALYST CHARACTERIZATION: Physical Techniques
    for Solid Materials
Edited by Boris Imelik and Jacques C. Vedrine

CATALYTIC AMMONIA SYNTHESIS: Fundamentals and Practice
Edited by J. R. Jennings

DYNAMIC PROCESSES ON SOLID SURFACES
Edited by Kenzi Tamaru

ELEMENTARY PHYSICOCHEMICAL PROCESSES
    ON SOLID SURFACES
V. P. Zhdanov

PRINCIPLES OF CATALYST DEVELOPMENT
James T. Richardson

A Continuation Order Plan is available for this series. A continuation order will bring delivery of each new volume immediately upon publication. Volumes are billed only upon actual shipment. For further information please contact the publisher.

Catalyst Characterization
Physical Techniques for Solid Materials

Edited by
Boris Imelik
Late of Institut de Recherches sur la Catalyse
CNRS
Villeurbanne, France

and
Jacques C. Vedrine
Institut de Recherches sur la Catalyse
CNRS
Villeurbanne, France

PLENUM PRESS • NEW YORK AND LONDON

o 6037884
CHEMISTRY

Library of Congress Cataloging-in-Publication Data

Catalyst characterization : physical techniques for solid materials /
  edited by Boris Imelik and Jacques C. Vedrine.
      p.   cm. -- (Fundamental and applied catalysis)
    Includes bibliographical references and index.
    ISBN 0-306-43950-6
    1. Catalysts--Analysis.   I. Imelik, B.   II. Védrine, Jacques C.
  III. Series.
  TP159.C3C37 1993
  660'.2995--dc20
                                                            93-39899
                                                                 CIP

ISBN 0-306-43950-6

© 1994 Plenum Press, New York
A Division of Plenum Publishing Corporation
233 Spring Street, New York, N.Y. 10013-1578

Printed in the United States of America

"Soyez spécialiste en tout"

Boris VIAN

17.02.92

Points
Have no parts or joints,
How then can they combine
To form a line?

— J. A. Lindon —

This book is dedicated to all our friends who are deeply involved with the problem of parts:

Boris

Poem:

Points
Have no parts or joints
How then can they combine
To form a line

        J. A. Lindon

This book is dedicated to all our friends who are deeply involved with the problem of points.

                         B. Imelik
                         February 17, 1992

CONTRIBUTORS

Michel Abon ● Institut de Recherches sur la Catalyse, CNRS, F. 69626 Villeurbanne, France

Aline Auroux ● Institut de Recherches sur la Catalyse, CNRS, F. 69626 Villeurbanne, France

Younès Ben Taarit ● Institut de Recherches sur la Catalyse, CNRS, F. 69626 Villeurbanne, France

Gerard Bergeret ● Institut de Recherches sur la Catalyse, CNRS, F. 69626 Villeurbanne, France

Jean-Claude Bertolini ● Institut de Recherches sur la Catalyse, CNRS, F. 69626 Villeurbanne, France

Laurent Bonneviot ● Département de Chimie, Université Laval, Québec G1K 7P4, Québec, Canada

Paul Bussière ● Institut de Recherches sur la Catalyse, CNRS, F. 69626 Villeurbanne, France

Michel Che ● Laboratoire de Réactivité de Surface et Structure, Université Pierre et Marie Curie, F. 75252 Paris, Cédex 05, France

Gisèle Coudurier ● Institut de Recherches sur la Catalyse, CNRS, F. 69626 Villeurbanne, France

Jean-Alain Dalmon ● Institut de Recherches sur la Catalyse, CNRS, F. 69626 Villeurbanne, France

Jacques Fraissard ● Laboratoire de Chimie des Surfaces, Université Pierre et Marie Curie, F. 75252 Paris, Cédex 05, France

Pierre Gallezot • Institut de Recherches sur la Catalyse, CNRS, F. 69626 Villeurbanne, France

Edouard Garbowski • Institut de Recherches sur la Catalyse, CNRS, F. 69626 Villeurbanne, France

Elio Giamello • Dipartimento di Chimica Inorganica, Chimica Fisica e Chimica dei Materiali, Universitá di Torino, Torino, Italy

Jean Grimblot • Laboratoire de Catalyse Hétérogène et Homogène, URA CNRS, Université des Sciences et Technologies de Lille, F. 59655 Villeneuve D'Ascq, France

Jean-Marie Herrmann • Laboratoire de Photocatalyse, Catalyse et Environnement, URA CNRS Ecole Centrale de Lyon, F. 69131 Ecully, France

Boris Imelik (deceased) • Late Director, Institut de Recherches sur la Catalyse, CNRS, F. 69626 Villeurbanne, France

Hervé Jobic • Institut de Recherches sur la Catalyse, CNRS, F. 69626 Villeurbanne, France

Yvette Jugnet • Institut de Recherches sur la Catalyse, CNRS, F. 69626 Villeurbanne, France

Christiane Leclercq • Institut de Recherches sur la Catalyse, CNRS, F. 69626 Villeurbanne, France

Frédéric Lefebvre • Institut de Recherches sur la Catalyse, CNRS, F. 69626 Villeurbanne, France

Jean Massardier • Institut de Recherches sur la Catalyse, CNRS, F. 69626 Villeurbanne, France

Claude Mirodatos • Institut de Recherches sur la Catalyse, CNRS, F. 69626 Villeurbanne, France

Bernard Moraweck • Institut de Recherches sur la Catalyse, CNRS, F. 69626 Villeurbanne, France

Danièle Olivier • Departement des Sciences Chimiques, CNRS, F. 75016 Paris, France

Hélène Praliaud ● Institut de Recherches sur la Catalyse, CNRS, F. 69626 Villeurbanne, France

Albert J. Renouprez ● Institut de Recherches sur la Catalyse, CNRS, F. 69626 Villeurbanne, France

Jacques C. Vedrine ● Deputy Director, Institut de Recherches sur la Catalyse, CNRS, F. 69626 Villeurbanne, France

FOREWORD
to the
Fundamental and Applied
Catalysis Series

Catalysis is important academically and industrially. It plays an essential role in the manufacture of a wide range of products, from gasoline and plastics to fertilizers and herbicides, which would otherwise be unobtainable or prohibitively expensive. There are few chemical- or oil-based material items in modern society that do not depend in some way on a catalytic stage in their manufacture. Apart from manufacturing processes, catalysis is finding other important and over-increasing uses; for example, successful applications of catalysis in the control of pollution and its use in environmental control are certain to increase in the future.

The commercial importance of catalysis and the diverse intellectual challenges of catalytic phenomena have stimulated study by a broad spectrum of scientists including chemists, physicists, chemical engineers, and material scientists. Increasing research activity over the years has brought deeper levels of understanding, and these have been associated with a continually growing amount of published material. As recently as sixty years ago, Rideal and Taylor could still treat the subject comprehensively in a single volume, but by the 1950s Emmett required six volumes, and no conventional multivolume text could now cover the whole of catalysis in any depth. In view of this situation, we felt there was a need for a collection of monographs, each one of which would deal at an advanced level with a selected topic, so as to build a catalysis reference library. This is the aim of the present series, *Fundamental and Applied Catalysis*. Some books in the series deal with particular techniques used in the study of catalysts and catalysis: these cover the scientific basis of the technique, details of its practical applications, and examples of its usefulness. An industrial process or a class of catalysts forms the basis of other books, with information on the fundamental science of the topic, the use of the process or catalysts, and engineering aspects. Single topics in catalysis are also treated in the series, with books giving the theory of the underlying science, and relating it to catalytic practice. We believe that this approach is providing a collection that is of value to both academic and industrial workers. The series editors welcome comments on the series and suggestions of topics for future volumes.

Martyn Twigg
Michael Spencer

Royston and Cardiff

PREFACE

The rapid expansion of heterogeneous catalysis, begun during the Second World War, has continued in close correlation with industrial growth. Fundamental research and industrial technology have contributed, in a parallel and complementary manner, to the improvement and diversification of contact masses, which tend to become more and more complex.

It was soon evident that whereas a knowledge of the catalytic performance of a material is indispensable for defining the classes of catalysts for each type of reaction, the improvement of known catalysts and the development of new ones imply a knowledge of the physicochemical properties of the catalytic solid (structure, texture, nature of the active sites, gas–solid interface, etc.).

In the early 1950s, the techniques available for the characterization of a catalyst gave access to its chemical composition (chemical analysis), its structure (X-ray diffraction), and its texture (measurement of the specific surface area, pore size distribution). These data proved to be insufficient, not only for understanding the mode of action of the catalyst but also for specifying the conditions required for its preparation. It was the period in which numerous discussions, often acrimonious, had as their sole origin an insufficient definition of the catalytic material. It is therefore not surprising that the development of research in the area of catalysis was subsequently directly related to the evolution of physical and physicochemical techniques.

Since 1960 the physical methods have developed very rapidly, and thanks to a new generation of physical chemists, these techniques have been quickly adapted to the problems of catalysis.

Indeed, there is a world of difference between the materials studied by the physicist and the researcher in catalysis: monocrystals in the former case and polycrystalline or amorphous powders, which are often polyphasic, in the latter. It is therefore important to stress the merits of the researchers who have had the arduous task of giving a new dimension to the physical techniques by modifying the experimental layouts and by contributing to the development of the theoretical basis and to the interpretation of the data obtained.

Today, numerous physical techniques can be used to study the catalysts (homogeneous and heterogeneous), the intermediate species, the adsorbed species, or the interaction of the solid with the reagent or the reaction product. However, as each technique has its own specificity and limits, the use of not one but of several techniques has proved to be indispensable.

The simultaneous use of several "heavy" techniques demands costly investments and the creation of multidisciplinary teams of scientists capable of exploiting not only the physical techniques that they employ, but also the science of catalysis. It is for these reasons that too few laboratories are adequately equipped today.

Over the last decade a limited number of review articles have been devoted to the various physical techniques used in catalysis. It seems, however, that no single volume has grouped all the techniques, analyzed their area of application and validity, and illustrated the respective and complementary contribution of each technique to the knowledge of a particular catalytic system.

The Institut de Recherches sur la Catalyse, which brings together a large number of specialists in physical techniques had, at the outset, organized advanced training courses within the framework of the Stages d'Enseignement des Techniques Avancées de la Recherches (SETAR) of the CNRS. The national and international success of these courses, which provided an initiation to, and an in-depth study of, applied physical techniques in catalysis, together with the insistent demands of many colleagues encouraged us to prepare the present text.

We deemed it essential to present all the techniques used in catalysis and to dwell upon those whose recent developments appear to be promising. It is for this reason that the space reserved for the different methods is unequal and that some, in the absence of significant new information, such as adsorption methods or X-ray diffraction, have not been discussed. In contrast, others like electron microscopy, X-ray radial electronic distribution, XANES (X-ray-absorption near edge structure), STM (scanning tunnel microscopy), AFM (atomic force microscopy), and EXAFS (extended X-ray adsorption fine structure) have been discussed in detail.

The techniques presented are especially adapted to solid materials. In fact any characterization of structures, or crystalline arrangements or of specific active sites requires a rigid framework. Nevertheless selected examples from homogeneous catalysis, in particular in IR and NMR spectroscopies, have been included.

Our intention was not to present an exhaustive list of the results obtained with each technique but rather to describe the technique, to explain its theoretical basis, to define its field of application, and by means of some examples to analyze objectively the validity of the results obtained. Readers should thus find the elements necessary for choosing the best methods for solving their problems. To help them in this choice we have tried, in the final remarks, to give two concrete examples to illustrate the respective contributions of each technique and the complementarity of the data obtained by using several techniques simultaneously.

The articles have been written for the most part by the scientists (or their students) who have participated in the development of the different techniques and who are still actively pursuing research. We wish to thank them sincerely here for their competence, enthusiasm, and devotion.

The impact and originality of the French version prompted M. S. Spencer and M. V. Twigg to suggest the preparation of an English version. We thank them for this initiative and we also thank Technip Editions for granting their rights to Plenum Press.

Each author has consequently redrafted and modified his text in order to extend the fields of application beyond homogeneous and heterogeneous catalysis. Some chapters have been eliminated (PAS, FEM, FIM), and others added (STM, MS) or expanded (EXAFS, XANES). Mike Spencer kindly polished up the manuscripts thereby giving the language a more natural flavor.

We are particularly grateful to Mme Gisèle Coudurier for her invaluable and efficient assistance. R. P. A. Sneeden kindly undertook the critical translation of the preface, introduction, and final chapter. We wish to thank him warmly for this friendly gesture. G. Wicker and Mrs M. T. Gimenez have prepared, as in the case of the French version, all the figures. We wish to thank them most heartily.

B. Imelik[†] and J. C. Vedrine

Villeurbanne

CONTENTS

CHAPTER 3. RAMAN SPECTROSCOPY

E. Garbowski and G. Coudurier

CHAPTER 4. ELECTRONIC SPECTROSCOPY

E. Garbowski and H. Praliaud

CHAPTER 5. NUCLEAR MAGNETIC RESONANCE IN
HETEROGENEOUS CATALYSIS

Y. Ben Taarit and J. Fraissard

CHAPTER 6. ELECTRON PARAMAGETIC RESONANCE:
PRINCIPLES AND APPLICATIONS TO CATALYSIS

M. Che and E. Giamello

CHAPTER 7. FERROMAGNETIC RESONANCE

L. Bonneviot and D. Olivier

CHAPTER 8. MÖSSBAUER SPECTROSCOPY — NUCLEAR GAMMA RESONANCE

P. B. Bussière

CHAPTER 9. AUGER ELECTRON SPECTROSCOPY

J. C. Bertolini and J. Massardier

CHAPTER 10. VIBRATIONAL ELECTRON ENERGY LOSS SPECTROSCOPY

J. C. Bertolini

CHAPTER 11. SECONDARY ION MASS SPECTROMETRY

J. Grimblot and M. Abon

CHAPTER 14. X-RAY ABSORPTION SPECTROSCOPY:
EXAFS AND XANES

B. Moraweck

CHAPTER 15. DETERMINATION OF THE ATOMIC STRUCTURE
OF SOLID CATALYSTS BY X-RAY DIFFRACTION

G. Bergeret and P. Gallezot

CHAPTER 16. SMALL-ANGLE X-RAY SCATTERING

A. J. Renouprez

CHAPTER 17. PHOTOELECTRON SPECTROSCOPIES: XPS AND UPS

J. C. Vedrine

CHAPTER 22. THERMAL METHODS: CALORIMETRY,
DIFFERENTIAL THERMAL ANALYSIS,
AND THERMOGRAVIMETRY

A. Auroux

CHAPTER 23. MASS SPECTROMETRY: PRINCIPLES AND
APPLICATIONS IN CATALYSIS

C. Mirodatos

CHAPTER 24. SCANNING TUNNELING AND ATOMIC FORCE MICROSCOPIES

P. Gallezot

CHAPTER 25. FINAL REMARKS

B. Imelik and J. C. Vedrine

GENERAL INTRODUCTION

B. Imelik[†] and J. C. Vedrine

1.1. INTRODUCTION

Schematically, two principal objectives may be distinguished in the application of physical techniques in the study of catalysis: (1) The characterization of the catalysts; and (2) the acquisition of information relevant to understanding the catalytic phenomenon. The first objective consists of establishing an "identity card" for the catalyst, indicating its structure, morphology, and other fundamental physicochemical data. This information is essential for a meaningful comparison of catalysts prepared in different laboratories. The situation in this area has improved greatly during the last 20 years, but is still not entirely satisfactory. The second objective concerns the catalytic process. Schematically heterogeneous catalytic reactions involve five steps:

1. Transportation of the reagents to the active surface.
2. Chemisorption of at least one reagent when there is more than one.
3. Formation of the reaction intermediate (activated complex).
4. Desorption of the products and regeneration of the catalyst.
5. Transportation of the reaction products.

Steps 1 and 5 follow the same physical laws of diffusion. Porosity is an important factor and kinetic constants as well as catalytic selectivity may be altered by diffusional processes. The laws of diffusion are well established, and the pore size and distribution can be measured with adequate precision.

During Step 2, there is an interaction between the reagents and the catalyst which may involve chemical combination. Several powerful methods are

B. Imelik and J. C. Vedrine • Institut de Recherches sur la Catalyse, CNRS, Villeurbanne, France.

Catalyst Characterization: Physical Techniques for Solid Materials, edited by Boris Imelik and Jacques C. Vedrine, Plenum Press, New York (1994).

1

available for the study of chemisorption. However, the simultaneous study of several adsorbates — coadsorption — is still rare owing to the technical problems involved. Similarly, the study of the chemisorption of very small quantities of an adsorbate also presents certain problems.

Step 3 leads to a lowering of the activation energy of the reaction, that is to say, to a significant increase in the rate of reaction. This is the principal interest of catalysis as chemical reactions are thereby accelerated. However, the energy required for the reaction is governed by the laws of thermodynamics: it is not changed by the presence of the catalyst.

The study of these intermediates in heterogeneous catalysis poses some major problems. If the intermediates are sufficiently stable and if their concentration at the surface of the catalyst is adequate, then interesting information can be acquired by various physicochemical techniques. Nevertheless, it is very important to ascertain that real reaction intermediates are involved as the temperatures at which these studies are usually made are much lower than that of the reaction. If the concentration of the intermediates is low, then serious difficulties are encountered as they can be masked by the dominant precursor phase. Finally, there is no satisfactory technique for the study of short-lived activated complexes.

The reactive properties of a catalyst are expressed in terms of its activity (the degree of conversion of the reagents per unit time) and its selectivity (its capacity to facilitate the formation of one of the desired products in the case of parallel or consecutive reactions). The concept of active sites proposed by Taylor is adopted; the active sites are atoms or groups of atoms directly involved in the reaction, all the others being inactive. If the number of active sites is known, then the turnover frequency, which is the number of times that a site intervenes per unit of time, may be calculated.

The catalysts used in industry today can be extremely complex. They often consist of a mixture of phases including one or more additive/promoter. The rationalization of their activity raises these questions:

1. What are the nature and the real number of active sites?
2. What are the structure and lifetime of the intermediate complexes?
3. How does the presence of the reagents modify the electronic structure and geometric array of the active sites? What modifications do the reagents undergo upon interaction with the catalyst in the course of adsorption? Can the steps leading to the formation of the reaction products be followed?
4. What are the chemical composition and the real atomic arrangement at the catalyst reagent surface? Are they identical at the surface and in the bulk of a solid catalyst and are they modified by the addition of reagents or during the catalytic reaction? What is the importance of the role played by the method of preparation of the catalyst?
5. What is the influence of the coordination state, of the active transition metal ions, of the symmetry of their proximate environment, of their valency state, and of the degree of covalence of the metal–ligand bonds?

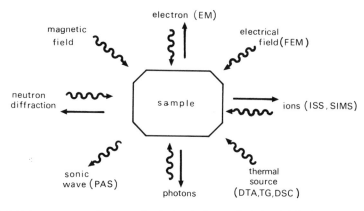

Figure 1.1. Scheme representing any physical technique principle with incident and emitted beams.

TABLE 1.1. Energies Involved in the Various Physical Methods[a]

	Incident probe					
	Radiation ions	X-rays electrons	Photons			
Interaction of probes with:	Nuclei	Electrons inner layer	Electrons outer layer	Molecules	Electron spin	Nuclear spin
Energy (eV)	10^9 10^7	10^5 10^3	10^1	10^{-1}	10^{-3} 10^{-5}	10^{-7}
Wavelength (µm)	10^{-8}	10^{-3}	10^{-2}		10^2 10^4	
Techniques	SIMS-ISS Mössbauer	ESCA, XPS AES, Electron diffraction Electron microscopy Microprobe EXAFS, EELS	UV-visible UPS, LEED APS	IR PAS Raman	Microwave EPR	NMR

[a] Acronyms are listed in Table 1.3.

TABLE 1.2. Presentation of the Principal Methods as a Function of the Nature of the Incident and Emergent Beams[a]

Emitted beams	Incident beams					
	Neutral	Ions	Electrons	Photons	Thermal effect	Magnetic, electric fields or sonic wave
Neutral	INS			Photo-desorption	TPD TG	
Ions	FABMS	SIMS ISS				FIM
Electrons			TEM, Electron diffraction, STEM, SEM AES, EELS, HREELS LEED	ADES, ESCA or XPS, PES PS, UPS		FEM
Photons			APS, EDX, microprobe	ATR, UV-Visible, IR, FTIR, Raman, Mössbauer, (GNR), XRD, EXAFS/XANES		
Thermal effect, magnetic, electric fields or sonic wave	Calorimetry			PAS	DSC DTA	FMR, ENDOR, EPR, NMR, Magnetism, Electric conductivity, STM, AFM

[a] The main acronyms are listed in Table 1.3.

6. What are the causes of catalyst aging? Does it involve a modification of the atomic arrangements, or site blockage by poisons, or a change in the mechanical properties?

7. When a foreign element is added to a material, how is it incorporated? Is there formation of a solid solution or a mixture of phases? Is phase segregation possible, e.g., at the surface? Are the electronic properties of the material affected?

8. When a catalyst is supported on a solid, is there only a dispersion effect which increases the surface-to-volume ratio and thus the number of active sites? Is there a change in the properties of the support (e.g., electron transfer and acidity) and/or in the active phase?

9. What is the role of the transfer of electrons (active phase–support, or for alloys metal–metal), and of the geometric arrangement and size of particles of solid catalysts?

TABLE 1.3. Usual and Widely Used Acronyms

ADES	Angular Distribution Electron Spectroscopy
AES	Auger Electron Spectroscopy
AFM	Atomic Force Microscopy
APS	Appearance Potential Spectroscopy
ARUPS	Angular Resolved UV Photoelectron Spectroscopy
ATR	Attenuated Total Reflection
CEMS	Conversion Electron Mössbauer Spectroscopy
CTEM	Conventional Transmission Electron Microscopy
DRIFTS	Diffuse Reflectance Infrared Fourier Transform Spectroscopy
DRUV	Diffuse Reflectance Ultraviolet Spectroscopy
DSC	Differential Scanning Calorimetry
DTA	Differential Thermal Analysis
EDX	Energy Dispersive X-Ray
EELS	Electron Energy Loss Spectroscopy
ELNES	Energy Loss Near Edge Structure
ENDOR	Electron Nuclear Double Resonance
EPR/ESR	Electron Paramagnetic/Spin Resonance
ESCA	Electron Spectroscopy for Chemical Analysis
ESD	Electron Simulated Desorption
ESEM	Electron Spin Echo Microscopy
EXAFS	Extended X-Ray Absorption Fine Structure
EXELFS	Extended X-Ray Electron Loss Fine Structure
FABMS	Fast Atom Bombardment Mass Spectroscopy
FEM	Field Emission Microscopy
FIM	Field Ion Microscopy
FIR	Far Infrared
FMR	Ferromagnetic Resonance
FTIR	Fourier Transform Infrared
GNR	Gamma Nuclear Resonance
HREELS	High-Resolution Electron Energy Loss Spectroscopy
HREM	High-Resolution Electron Microscopy
INS	Inelastic Neutron Scattering
IR	Infrared
ISS	Ion Scattering Spectroscopy
LAMMA	Laser Microprobe Mass Analysis
LEED	Low-Energy Electron Diffraction
LIF	Laser Induced Fluorescence
LRS	Laser Raman Spectroscopy
MAS-NMR	Magic Angle Spinning Nuclear Magnetic Resonance
MES	Mössbauer Emission Spectroscopy
MS	Mass Spectroscopy
NMR	Nuclear Magnetic Resonance
ND	Neutron Diffraction
NQD	Nuclear Quadrupolar Resonance

(Continued)

TABLE 1.3 (*Continued*)

PAS	Photoacoustic Spectroscopy
PES/PS	Photoelectron Spectroscopy
RBS	Rutherford Backscattering Spectroscopy
RED	Radial Electronic Distribution
SAM/SAEM	Scanning Auger Electron Microscopy
SANS	Small-Angle Neutron Scattering
SAXS	Small-Angle X-Ray Scattering
SEM	Scanning Electron Microscopy
SEXAFS	Surface Extended X-Ray Absorption Fine Structure
SEXELFS	Surface Extended X-Ray Electron Loss Fine Structure
SIMS	Secondary Ion Mass Spectroscopy
STEM	Scanning Transmission Electron Microscopy
STM	Scanning Tunnel Microscopy
TEM	Transmission Electron Microscopy
TG	Thermogravimetry
TPD, TPO, TPR	Temperature Programmed Desorption/Oxidation/Reduction
TDS	Thermal Desorption Spectroscopy
UPS	Ultraviolet Photoelectron Spectroscopy
UV-Vis	Ultraviolet–Visible Spectroscopy
XANES	X-Ray Absorption Near Edge Spectroscopy
XAS	X-Ray Absorption Spectroscopy
XRD	X-Ray Diffraction
XPS	X-Ray Photoelectron Spectroscopy

1.2. GENERAL PRESENTATION OF THE PHYSICAL TECHNIQUES

The general principle of all the techniques is to subject the specimen to the action of an incident beam and to analyze the nature and energy of the beam after interaction with the specimen (Figure 1.1). There are eight fundamental types of probes for the incident beam. These engender four types of particles which leave the sample and carry the information to an appropriate detector. The fundamental probes can be particulate beams composed of electrons, ions, neutrons, or photons, or nonparticulate beams, such as thermal, electric, magnetic, and sonic fields.

The interaction of these probes with a material depends upon the nature of both the probe and the material. For probes involving particles with a nonnegligible mass, the specimen can be profoundly perturbed, and because of the absorption by matter, ultrahigh-vacuum conditions are required. This is a major drawback. In contrast, with nonparticulate or photonic probes, the specimens will be only slightly perturbed, with at times a detrimental overheating; it is thus possible to operate under any atmosphere and to study samples under conditions close to those of the actual catalytic reaction. The energy regions involved in the different physical methods are summarized in Table 1.1.

TABLE 1.4. Comparative Physical Characteristics of the Physical Methods

	Technique							
	Molecular spectroscopy			Resonances				
Characteristic	IR PAS	Raman	UV-visible	EPR FMR	NMR	Mössbauer	Neutrons	Thermal methods
Thickness analyzed	mm	mm	mm	mm	mm	100 μm	0.1 mm	mm
Area analyzed	cm²	μm²	cm²	cm²	cm²	cm²	mm²	mm²
Sample degradation	no	possible	no	no	no	no	possible	possible
Sample preparation	easy	easy	easy	easy	easy	easy	difficult	easy
Quantitative measurements	possible	possible	possible	yes	yes	yes	yes	yes
Gaseous atmosphere	yes	yes	yes	yes	yes	yes	yes	possible
Temperature range (°C)	$\frac{-196}{300}$	$\frac{-196}{300}$	$\frac{-196}{300}$	$\frac{-269}{1000}$	$\frac{-196}{200}$	$\frac{-269}{400}$	$\frac{-269}{800}$	No limit
Information	functional groups; adsorbing species	functional groups; adsorbing species	degree of oxidation; ion symmetry; adsorbing species	para. species; degree of oxidation; symmetry	functional groups	degree of oxidation; symmetry of environment	adsorbing species; atomic structure	energies of formation and adsorption

	Technique							
	Surfaces				X rays			
Characteristic	EELS	XPS	AES	SIMS	Diffr.	EXAFS	Radial distribution	Magnetism	Electrical conductivity
Thickness analyzed	μm	20–50 Å	10–20 Å	2–3 Å	mm	mm	mm	mm	mm
Area analyzed	cm²	cm²	cm²	cm²	mm²	mm²	mm²	cm²	cm²
Sample degradation	very small	possible	possible	no	no	no	no	no	no
Sample preparation	difficult	easy	easy	easy	easy	easy	easy	easy	easy
Quantitative measurements	—	yes	possible	possible	yes	yes	yes	yes	yes
Gaseous atmosphere	no	difficult	difficult	no	yes	yes	yes	possible	possible
Temperature range (°C)	ambient	$\frac{-180}{600}$	$\frac{-180}{600}$	ambient	$\frac{-196}{1000}$	$\frac{-196}{1000}$	$\frac{-196}{1000}$	$\frac{-271}{500}$	$\frac{-196}{800}$
Information	metal-ligand bonds	degree of oxid.; surface composition	surface composition	surface composition	crystallite structure; crystallite size	near environment; number of ligands	near environment	magnetic degree; crystallite size	conduction type; ionosorbed species

TABLE 1.5. Basic Units of the International System of Units
(SI: Système international)

Physical quantity	Usual symbol	Name	Symbol of SI-unit
Length	*l*	Meter	*m*
Mass	*m*	Kilogram	kg
Time	*t*	Second	s
Current Intensity	*I*	Ampere	A
Temperature	*T*	Kelvin	K
Quantity of matter	*n*	Mole	mol

TABLE 1.6. Particular Units Derived from the International System of Units

Physical quantity	Usual Symbol	Name of unit	Symbol of unit	Value in SI-units
Frequency	ν	Hertz	Hz	s^{-1}
Force	*F*	Newton	N	$m\,kg\,s^{-2}$
Pressure	*p*	Pascal	Pa	$N\,m^{-2}$
Quantity of heat	*E.Q.*	Joule	J	$N\,m$
Quantity of electricity	*Q*	Coulomb	C	SA
Electric potential	*E, U*...	Volt	V	$J\,A^{-1}\,s^{-1}$
Magnetic induction	*B*	Tesla	T	$V\,s\,m^{-2}$
Length	*l*	Fermi	F	$10^{-15}\,m$
Length	*l*	Angström	Å	$10^{-10}\,m$
Magnetic induction	*B*	Gauss	Gs	$10^{-4}\,T$
Magnetic field	*H*	Oersted	Oe	$79.57\,Am^{-1}$
Magnetic flux	ϕ	Maxwell	Mx	$10^{-8}\,Wb$
Time	*t*	Minute	min	60 s
Time	*t*	Hour	h	3600 s

In Table 1.2 the various techniques are presented as a function of the incident and emergent probes, and Table 1.3 lists the relevant acronyms.

Table 1.4 summarizes some of the essential characteristics of the various techniques discussed in this book. Thus, to identify functional groups, for example, during a catalytic reaction, IR and NMR spectroscopies are the principal techniques currently used, but in certain cases other techniques such as Raman, UV-visible, and photoacoustic spectroscopies can also be very informative. The environment of a given site can be characterized by diffraction techniques (X-rays, electrons, and neutrons) for crystalline structure, and by UV-visible, Mössbauer, Raman, and EPR spectroscopies for the nature and the number of ligands. For the size and shape of crystallites or metallic particles, electron spectroscopies (TEM, SEM) can be used and, to a lesser extent, X-ray

TABLE 1.7. Interconversion of Energy Units

	Joule	erg	Cal$_{th}$^{a}	eV	cm^{-1}
1 J	1	10^7	0.239	6.242×10^{18}	5.035×10^{22}
1 erg	10^{-7}	1	2.39×10^{-8}	6.42×10^{11}	5.035×10^{15}
1 Cal	4.184	4.184×10^7	1	2.312×10^{19}	2.107×10^{23}
1 eV	1.602×10^{-19}	1.60×10^{-12}	3.829×10^{-20}	1	8065.5
1 cm^{-1}	1.986×10^{-23}	1.986×10^{-16}	4.746×10^{-24}	1.239×10^{-4}	1

^{a} Thermal calorie is such that $1 \text{ cal}_T = 4.1868 \text{ J} = 1.00067 \text{ cal}_{th}$; $1 \text{ cal}_s = 4.1855 \text{ J} = 1.00036 \text{ cal}_{th}$.

TABLE 1.8. Interconversion of Pressure Units

	Pa	bar	atm	Torr (mm Hg)	kgf cm^{-2}
1 Pascal	1	10^{-5}	9.869×10^{-6}	7.5006×10^{-3}	1.0197×10^{-5}
1 Bar	10^5	1	0.9869	750.06	1.0197
1 Normal atmosphere	101.325	1.01325	1	760	1.0332
1 Torr	133.32	1.3332×10^{-3}	1.3158×10^{-3}	1	1.3595×10^{-3}
1 kgf cm^{-2}	98,066.5	0.98066	0.96781	735.55	1

diffraction (line breadth), magnetism, FMR, Mössbauer spectroscopy (GNR), and, for metals, the chemisorption of appropriate molecules (e.g., H_2/O_2 titration). The degrees of oxidation of ions may be determined by spectroscopic techniques, UV-visible, XPS, Mössbauer, magnetism, and EPR. Information about the first surface layers may be obtained by XPS, ADES, ISS, FABMS, SIMS, FIM, and LEED. Only an analysis of the outermost layers of a crystallite is obtained by these methods. A knowledge of the internal surfaces of a porous material can be obtained by the adsorption of appropriate probes, which can be characterized by various bulk physical techniques; these can now be used since it is the adsorbed phase that is being studied.

An important and much sought after aspect is the characterization of a catalytic system under reaction conditions. Many of the methods are inappropriate because an activating probe is needed. Nevertheless, it is possible to get quite close to the reaction conditions especially when the reacting gases and the temperature are not fundamental obstacles to the technique employed. Thus, in principle, X-ray diffraction techniques and IR, UV-visible, Raman, FMR, EPR, NMR, STM, AFM, and Mössbauer spectroscopies and magnetism can be used. For the ultrahigh-vacuum techniques, XPS, UPS, AES, EELS, LEED, and electron microscopies, certain adaptations (gas-bleeding, low residual pressures) can be envisaged but remain exceptions.

TABLE 1.9. Principal Physical Constants Used in the Present Text

Permeability of vacuum:	μ_0	$= 4\pi\, 10^{-7}\,\mathrm{Hm}^{-1}$
Permittivity of vacuum:	$\varepsilon_0 = (\mu_0 c^2)^{-1}$	$= 8.854 \times 10^{-12}\,\mathrm{Fm}^{-1}$
Atomic mass unit:	u	$= 1.661 \times 10^{-24}\,\mathrm{g}$
Plank constant:	$n = h/2\pi$	$= 1.054 \times 10^{-34}\,\mathrm{J\,s}$
Speed of light in vacuum:	c	$= 2.9979 \times 10^8\,\mathrm{ms}^{-1}$
Avogadro's constant:	N	$= 6.022 \times 10^{23}\,\mathrm{mol}^{-1}$
Boltzmann's constant:	k	$= 1.381 \times 10^{-23}\,\mathrm{J\,K}^{-1}$
Elementary charge:	e	$= 1.602 \times 10^{-19}\mathrm{C}$ (Coulomb)
Electron rest energy:	$m_e c^2$	$= 0.511\,\mathrm{MeV}$
Electron rest mass:	m_e	$= 9.109\, 10^{-28}\,\mathrm{g}$
Proton rest energy:	$m_p c^2$	$= 938.7\,\mathrm{MeV}$
Proton rest mass:	m_p	$= 1.673.10^{-24}\,\mathrm{g}$
Molar gas constant:	R	$= 8.314\,\mathrm{J\,mol}^{-1}\mathrm{K}^{-1}$
Bohr radius of hydrogen:	$a_0 = \varepsilon_0 h^2/\pi m_e e^2$	$= 0.529.10^{-10}\,\mathrm{m}$
Ionization energy of hydrogen:	$(\tfrac{1}{2})\alpha^2 mc^2$	$= 13.606\,\mathrm{eV}$
Bohr magneton:	$he/2m_e = \beta_e$	$= 9.274 \times 10^{-24}\,\mathrm{J\,T}^{-1}$
Nuclear magneton:	$\beta_N = he/2m_p$	$= 5.051 \times 10^{-27}\,\mathrm{J\,T}^{-1}$

1.3. UNITS AND CONVERSIONS

One of the problems encountered by researchers stems from the confusion which persists in the systems of units. We tried to use sytematically the international system (SI) of units, Tables 1.5 and 6, but finally decided to keep the units actually used in practice. We include in Tables 1.7 and 1.8 some unit conversions, and in Table 1.9, the values of the principal physical constants used here.

2

INFRARED SPECTROSCOPY

G. Coudurier and F. Lefebvre

2.1. FUNDAMENTALS

The various methods of infrared spectroscopy (IR) are widely used techniques in catalysis laboratories because few physical methods can give so much information so easily on the catalyst structure, its surface properties, the interacions sorbate–sorbent, and the reaction intermediates. Indeed, the IR region $(10,000 > v > 50 \text{ cm}^{-1})$ and particularly the middle region $(4000 > v > 200 \text{ cm}^{-1})$ corresponds to the energies of the vibrations and rotations of the molecules. These movements depend on molecular constants such as the molecular symmetry, the interatomic strength constants, the inertia momentum around some axis.

2.1.1. Origin of the Spectrum[1]

It is well known that if the kinetic energy of a free molecule can vary continuously, the energies of the periodic movements of the molecule, rotations or vibrations, and the internal electronic energy can only have discrete values (quantified energy levels) which characterize stationary states. When it is irradiated by electromagnetic radiation of frequency v, the molecule, initially in the quantum state E can go into the excited state E', if the Bohr condition (resonance law) is satisfied, i.e.,

$$\Delta E = E' - E = hv$$

where h is the Planck constant.

G. Coudurier and F. Lefebvre ● Institut de Recherches sur la Catalyse, CNRS, Villeurbanne, France.

Catalyst Characterization: Physical Techniques for Solid Materials, edited by Boris Imelik and Jacques C. Vedrine, Plenum Press, New York (1994).

If E' is a higher energy state than E, the molecule *absorbs* the radiation of frequency v when it is excited from E to E' and emits at the same frequency when it passes from E' to E. If there is a frequency variation of the electromagnetic radiation, the energy absorption for each frequency corresponds to a transition between rotational, vibrational, or electronic states. In a first approximation, it is assumed that the movements of the electrons do not depend on those of the nuclei and that the nuclei move in a mean constant electronic field. It is also assumed that the vibration and rotation displacements are independent of one another. This formalism (Born–Oppenheimer hypothesis) is vindicated by the fact that the rotation, vibration, and electronic energies are not of the same magnitude and correspond, respectively, to the microwave, IR, and UV regions of the electromagnetic spectrum.

2.1.2. States of Vibrational Energy

A molecule with N nuclei possesses $3N$ degrees of freedom, i.e., $3N$ types of movement. Three of them are the translations of the whole molecule and three others (two if the molecule is linear) its rotations. Thus there remain $3N - 6$ (or $3N - 5$ if the molecule is linear) vibration movements (named vibrational normal modes), which correspond to $3N - 6$ (or $3N - 5$) vibration frequencies (normal or fundamental frequencies). Therefore we can consider the molecule as an assembly of $3N - 6$ (or $3N - 5$) oscillators.

2.1.3. Vibrations Active in IR

If the molecule is subjected to IR radiation whose frequency is equal to the frequency of one of its oscillators, this oscillator will resonate and absorb part of the radiation. The absorption (or emission) intensity is given by the transition probability between the ground and excited states. The rules that specify which transitions are allowed are called selection rules. It can be shown that in absorption or emission, only the transitions corresponding to vibrations with variation of the dipole moment are active in IR, and these are called active modes. In practice the absorption bands of the active fundamental vibrations are the most intense, but weak absorptions due to combinations of identical (overtone) or different (combination) vibrations are often seen. As an example, we describe determination of the vibrating modes and activity in IR spectroscopy of two triatomic molecules.

CO_2 is a triatomic linear molecule, so it has $3N - 5 = 4$ vibration modes (Figure 2.1). As the v_1 vibration does not correspond to a variation of the dipole moment, it is not active in IR. The v_{2a} and v_{2b} vibrations, which are equivalent (doubly degenerate), and the v_3 one correspond to a variation of the dipole moment and are active in IR. The spectrum shows two absorptions at 667 cm^{-1} (v_{2a} and v_{2b} modes) and 2349 cm^{-1} (v_3 mode).[2]

H_2O is a nonlinear triatomic molecule, so it has $3N - 6 = 3$ vibrating modes (Figure 2.2). There is a variation of the dipole moment for each vibration. The IR spectrum of H_2O shows three absorptions at 3657, 1595, and 3756 cm^{-1} corresponding to the v_1, v_2, and v_3 modes, respectively.[3]

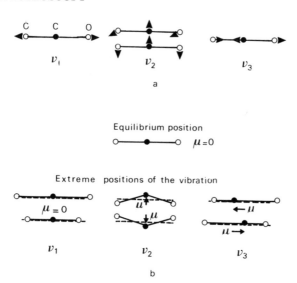

Figure 2.1. (a) Vibrating modes of CO_2. (b) Variation of the dipole moment μ.

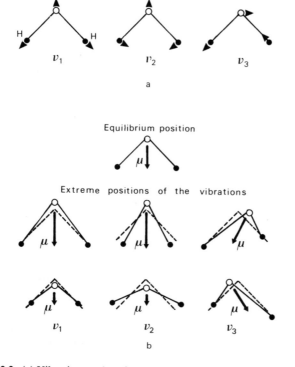

Figure 2.2. (a) Vibrating modes of H_2O. (b) Variation of the dipole moment μ.

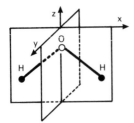

Figure 2.3. Representation of the symmetry elements of the water molecule.

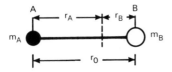

Figure 2.4. The AB diatom molecule (m_A is the A mass, m_B the B mass, and r_0 the A—B length at equilibrium).

2.1.4. Molecular Symmetry

In its equilibrium configuration a molecule can be considered as a geometrical construction defined by interatomic lengths and valence angles. The existence in such a molecule of identical atoms gives the ensemble its symmetry. All the symmetry operations possible with a molecule define its symmetry group. All molecules can be classified into a limited number of symmetry groups which obey the rules of group theory.[3] For example, the water molecule H_2O is in the C_{2v} symmetry group, whose symmetry elements are a rotation axis of order two and two reflection planes xz and yz (Figure 2.3). The knowledge of the symmetry group of a molecule allows the determination, without any other data, of the symmetry class of the $3N - 6$ normal modes of vibration, their description by the symmetry coordinates, their activity in IR and Raman spectroscopy, and the factorization of the secular equation.

2.1.5. Calculation of the Vibration Frequencies

2.1.5.1. General Case

The calculation of molecular vibrations can be considered as a problem of small-movement mechanics. To resolve it, it is necessary to calculate the kinetic and the potential functions of the vibrations as a function of the $3N$ shift coordinates of the N nuclei present in the molecule. If it is assumed that the movements are harmonic, the possible ones and their frequencies are obtained by solving $3N$ Lagrange equations which must be compatible. Finally, by elimination of the rotation and translation movements, one obtains a secular equation which is $(3N - 6)$th order in λ, with $\sqrt{\lambda} = 2\pi\nu$,[3-6] the molecule having $3N - 6$ vibration frequencies, some of which may be identical (doubly or triply degenerate vibrations). The appropriate choice of internal coordinates (variations of the interatomic distances and valence angles) as shift coordinates and the use of symmetry properties allows an easier solution of the secular equation. It is not possible to describe in this chapter all the methods that can be used to solve the secular equation, and it must be pointed out that in catalysis the group frequency notion is often sufficient for the interpretation of the spectra. In these cases, it is only necessary to calculate the vibration frequency of a diatomic molecule.

2.1.5.2. The Diatomic Molecule: The Harmonic Oscillator

The diatomic molecule AB is necessarily linear and has only one vibrating mode ($3N - 5 = 1$), which is the periodic stretching of the AB bond. If one supposes that the interaction type between the two atoms is proportional to the stretching of the bond, this molecule is an harmonic oscillator (Figure 2.4). If $x = r - r_0$ is the variation of the interatomic length, the tensile force is written as

$$f = - kx \quad \text{(Hooke's law)} \tag{2.1}$$

and, by Newton's second law,

$$m_A \cdot d^2 r_A / dt^2 = - kx \quad \text{and} \quad m_B \cdot d^2 r_B / dt^2 = - kx \tag{2.2}$$

If we take into account the relation defining the position of the center of mass, only one equation is obtained:

$$[m_A m_B / (m_A + m_B)] d^2 x / dt^2 = - kx \tag{2.3}$$

which shows that the system is equivalent to a mass $\mu = m_A m_B / (m_A + m_B)$, called the reduced mass. The equation

$$\mu d^2 x / dt^2 = - kx \tag{2.4}$$

is easily solved to give

$$x = x_0 \sin(2\pi v t + \varphi) \tag{2.5}$$

where $v = (1/2\pi)\sqrt{k/\mu}$, with $\bar{v} = Cv$ (C is the velocity of light and $\bar{v}$ is the wave number). The relation

$$\bar{v} = (1/2\pi C)\sqrt{k/\mu} \tag{2.6}$$

is used (a) to calculate the stretching vibration frequencies in a complex molecule (with the hypothesis that the vibrations of an atomic group are relatively independent of those of the rest of the molecule), and (b) to calculate frequencies after isotopic substitution.

TABLE 2.1. Types of Vibrations

1. Bond deformation	
Stretching of one bond	
stretching	ν
Stretching of two bonds	
Symmetric	ν_s
Antisymmetric	ν_a
2. Angle deformations	
Bending (used only when the group is simple, otherwise the following notations are used)	δ
Scissor	δ
Wagging	γ
Twisting	ρ
Rocking	β
3. Cycle deformation	
breathing	

TABLE 2.2. Absorption Frequencies of Some Functional Groups

Stretching vibration functional group	Frequency range (cm^{-1})	Examples (cm^{-1})
O—H organic	3700–3500	Benzyl alcohol: 3618
O—H inorganic	3750–3300	Si—OH: 3740
N—H	3500–3200	CH_3NH_2: 3361 and 3427
C—H	3300–2800	CH_4: 3020
C=C olefinic, aromatic	1690–1500	C_2H_4: 1623
C—O	1430–900	CH_3—O—CH_3: 1103 and 932
Si—O organic	1100–1000	Siloxanes
Si—O inorganic	1100–700	Olivine: 1100–1000 and 830
		Benitoïte: 1035, 930, and 761
		Y zeolite: 1020 and 720

2.1.6. Types of Vibrations

The movements of the atoms of a molecule during vibration can be classified as bond or angle deformations (Table 2.1).

2.1.7. Group Frequencies[1,2,8]

Many tables give correlations between the absorption frequencies and the organic or inorganic functional groups. We shall quote only a few examples showing the influence of the following parameters: mass, force constant, rest of the molecule (Tables 2.2–2.5). It must also be noted that the physical state (solid, liquid, gas) of the molecule and the solvents (variation of molecular interactions) are important.

TABLE 2.3. Influence of the Force Constant

Group	Force constant (N m^{-1})	Wave number (cm^{-1})
C≡C	0.148	1980 (ethyne)
C=C	0.108	1623 (ethene)
C—C	0.046	993 (ethane)

TABLE 2.4. Influence of Mass

	Molecule H$_2$O	Molecule D$_2$O	νH$_2$O/νD$_2$O	$\sqrt{\mu OD/\mu OH}$
ν_1 (cm^{-1})	3657	2671	1.36	1.37
ν_3 (cm^{-1})	3756	2788	1.35	1.37

TABLE 2.5. Influence of the Other Atoms of the Molecule

CH group	Frequencies (cm^{-1})	Examples (cm^{-1})
Alkane	3000–2850	CH$_3$—CH$_3$: 2995 and 2954 CH$_3$: 2960, 2916 and 2852
Alkene	3095–2950	CH: 3012 CH$_2$: 3081
Aromatic	3080–3030	Benzene: 3042
Alkyne	3400–3300	CH≡CH: 3374

2.1.8. Relation between the Absorbed Intensity and Concentration

A sample of l total thickness, containing N molecules, receiving radiation of an intensity I_0 will emit an intensity I such that

$$\ln I_0/I = B_{nm}Nh\nu_{nm}l \qquad (2.7)$$

where B_{nm} is the transition probability, ν_{nm} is the frequency of that transition and h is the Planck constant.[3] For a given transition, $B_{nm}h\nu_{nm}$ is constant and one obtains the Beer–Lambert law (for solutions, liquids, and gases), by observing that N is proportional to the concentration or the pressure:

$$\log I_0/I = \varepsilon Cl \qquad (2.8)$$

where ε is the molar extinction coefficient if the concentration is expressed in mol l^{-1} and the cell length in cm. In heterogeneous catalysis, this relation is used for the determination of relative amounts of sorbed species. Indeed, for a given sample, the length l is a constant and C represents the quantity of sorbed species.[7]

2.2. EXPERIMENTAL CONSIDERATIONS

2.2.1. Introduction

In most spectrometers, electromagnetic radiation passes through the sample under study. The resulting absorptions are analyzed as a function of wavelength. Therefore it is necessary to separate the radiations of different wavelengths. Two methods are used:

1. The wavelengths are separated by a dispersive system, called monochromator, obtained by the use of prisms and diffraction gratings.
2. The periodic signal obtained in the output of a Michelson interferometer is treated mathematically by a Fourier transform.

Two types of spectrometers are made, each of them using one method: the dispersive double-beam spectrometers, which in the new model are connected to a computer, and the Fourier transform spectrometers with single or double beams. Each type of apparatus has advantages and drawbacks. The new dispersive spectrometers are very sturdy, easy both to use and to treat the data. The Fourier transform spectromers have the advantages of short acquisition time, high energy level, and very good signal-to-noise ratio. Both types of spectrometers have the characteristics necessary for the study of catalytic systems:

1. Emitter: highest power because the catalysts are rarely transparent.
2. Resolution: in heterogeneous catalysis, average resolution is often sufficient (1 to 3 cm^{-1}) but in homogeneous catalysis better resolution (0.5 cm^{-1}) is necessary.
3. An electrical zero and an expansion of the scales.
4. Connection with a computer.
5. Sample holder space as big as possible in order to use complex cells.

The methods used (transmission, reflection, or emission spectroscopies), the preparation of the samples (e.g., pressing or dilution) and the choice of the cells are determined by the information expected and the type and the transparency in the IR region of the sample. For example, in heterogeneous catalysis, it is important: (a) to know the catalyst properties (framework vibrations, surface species, acidity, oxidation state of the ion, metal dispersion); and (b) to define the interactions between the sorbents and the catalysts (e.g., properties of the surface groups, adsorption of acid or basic molecules, adsorption of reagents and products, intermediate species, and modification of the oxidation state).

2.2.2. Transmission Spectroscopy

2.2.2.1. Use of Self-Supported Disks

The use of self-supported disks is the most popular method. The solid, as a fine powder, is pressed without any other additive and gives a thin disk ($\phi \simeq 18$ mm, thickness 10^{-2} to 5×10^{-2} mm). The sample obtained under such conditions diffuses little of the radiation, gives a good transmission in IR, and has good mechanical properties. The weight of the disk and the pressure used are dependent on the solid. It may be necessary to use mica disks when the powder sticks to the matrix (Table 2.6). Some solids are not transparent (e.g., oxides with small specific areas) and others cannot be molded as disks and remain as powders. In these cases, the following methods can be used:

2.2.2.1a. Use of a Grate. A gold or nickel electrolytic grate (thickness 15 μm) is used as a support for the disk. This allows the use of less sample (useful for more opaque solids) and enhances the mechanical properties of the disk. The IR absorption of the grate alone is less than 20%. The main disadvantages are the price of the grate and the difficulty of obtaining a homogeneous distribution of powder.

2.2.2.1b. Powder Method. The powder is finely pulverized, sifted, and spread out on a transparent disk (e.g., CaF_2, KBr, or NaCl), which is covered by a second disk. The sample can also be prepared by sedimentation of a solid suspension in a solvent that is easy to remove. However, this method has several disadvantages — diffusion of the radiation (which may be important), existence of a Christiansen filter effect (sudden increase in the transmission for some wave numbers, whose values are determined by the refraction sign of the solid), and finally the interaction of some compounds with the disks (e.g., formation of vanadium oxybromides when V_2O_5 is heated with KBr).

TABLE 2.6. Characteristics of Self-Supported Disks of Some Catalysts

Solid	Mica	Weight (mg)	Pressure 10^{+6} Pa	Transmission at 4000 cm^{-1} (%)
Y zeolite	No	5–15	20	20–40
Silica (Aerosil)	No	20–30	100–300	80
Al_2O_3 (Degussa)	Yes	20–30	100	80
MoO_3	Yes	60–100	100	20–30

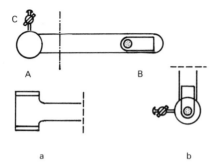

Figure 2.5. IR cell for the study of sorbed phases.

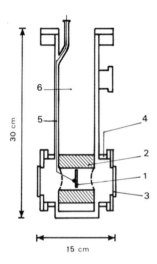

Figure 2.6. High temperature *in situ* reaction cell[9]: (1) sample wafer, (2) copper heater block, (3) windows, (4) water jacket, (5) thermocouple, and (6) Dewar.

2.2.2.1c. Dilution Method. The catalyst is diluted in a compound that is transparent and inert toward the reaction being studied, e.g., SiO_2 or Al_2O_3. In some cases, this dilution gives increased transmission of the disk and better mechanical properties.

TABLE 2.7. Properties of Some Optical Windows

Window	Spectral range (cm^{-1})	Sensitivity toward H_2O	Price
CaF_2	10,000–1000	Slightly soluble	
NaCl	10,000–500	Soluble	
KBr	10,000–300	Very soluble	
CsI	10,000–170	Very soluble	Very expensive
KRS_5 (TlBr, TlI)	10,000–200	Very soluble	Very expensive
Irtran 4	10,000–500	Insoluble	Very expensive

2.2.2.1d. Cells. The disk prepared by one of these methods is put in a sample holder made of quartz, which, when introduced into a cell, allows the treatment of the sample at various temperatures and in a controlled gas phase (Figure 2.5). This cell can be divided in two parts: Part A which is fitted with two optical windows where the sample is put to take the spectra. The window type is determined by the spectral range under study (Table 2.7). The windows are sealed (e.g., epoxy resins, Glyptal resins, wax such as WAX 40, or arabic gum) or held in place by Viton joints. Part (B) made of pyrex or quartz receives the sample for heating. The junction between A and B can be a soldered or a Viton joint, C being a gate (in a greaseless cell) or a vacuum tap. More complex cells have been constructed using the same preparation of the sample: e.g., high- and low-temperature cells (Figure 2.6), high-pressure cell, UV–IR coupled cell. This method allows the study of:

1. The variation of the surface properties of the catalyst during heating and redox treatments.
2. All the possible interactions between the catalyst and the sorbed species.
3. Changes in the catalyst and the adsorbed species during the reaction.
4. In some cases, the evolution of the structure of the catalyst (solids for which the framework vibrations can be observed).

The limitations of this method are due only to the nature of the samples. Other methods are used for the study of opaque solids for IR (e.g., small-area oxides and single crystals) and the highly reflecting solids (metal films, sulfides, etc.).

2.2.2.2. Dispersion Method

The solid is pulverized in KBr (CsI, polyethene) at about 0.1 to 1 mg for 100 and pressed at a pressure of 10^8 Pa. This method is used for the study of the framework vibrations of the solid before and after reaction and in the study of coordination complexes. Quantitative studies are possible.

2.2.2.3. The Powder Coating Method

A few milligrams of the solid under study are finely pulverized and coated in a medium such as nujol, vaseline, hexachlorobutadiene, or fluorolube (decrease of diffusion) and put onto transparent wafers. As in the previous case, this method is suitable for studies of solid catalysts and coordination complexes. Supports purged with argon can be used for air-sensitive compounds.

2.2.3. Reflection Spectroscopy

When none of the various methods described above can be used because the sample is too opaque or too reflecting, it is necessary to resort to reflection techniques. They are more difficult to use and require some additional equipment on the spectrometers as well as optical modifications.

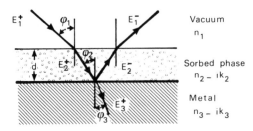

Figure 2.7. Model for the interaction between the IR radiation and the metal-sorbed phase system.

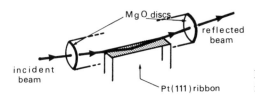

Figure 2.8. Single reflection at grazing incidence.

2.2.3.1. Specular Single or Multiple Reflection: Reflection–Absorption Spectroscopy

Reflection–absorption spectroscopy (RAS) allows the study of metallic films, single crystals, and opaque solids by reflection, and is also often used for sorbed phases on metal surfaces. By use of the following model (Figure 2.7) for the description of the interaction between IR radiation and the system metal-sorbed phase, Greenler[10] has shown that the intensity of absorption in the reflected radiation depends on: (1) the optical constants (n is the refraction index; k is the absorption coefficient; and $n - ik$ is a complex index) of the metal; (2) the thickness of the sorbed phase; (3) the angle of incidence; and (4) the polarization state of the incident radiation.

Theoretically, the absorption factor of radiation polarized parallel to the incident plane should be the highest for an incident angle of 88° and should be about 25 times bigger than that obtained by transmission. As these conclusions have been confirmed experimentally, many workers use this method for the study of catalysts (Figure 2.8).[11,12] They have improved it by (1) increasing the absorption by the use of multiple reflections,[12] and (2) magnifying the signal received by the detector by the use of a Fourier transform apparatus (accumulations),[13] wavelength modulation,[14,15] and polarization state modulation (ellipsometry).[12]

For single crystals, this method can be combined with LEED and Auger techniques (Figure 2.9).[12]

2.2.3.2. Diffuse Reflectance

Diffuse reflectance spectroscopy is largely used for the study of powders in the UV–visible region, but remained unused for a long time in the IR region

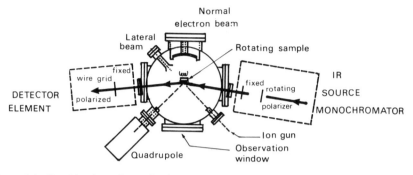

Figure 2.9. Combination of IR reflection spectroscopy and LEED or Auger technique.[12]

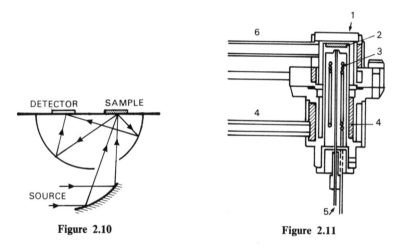

Figure 2.10   **Figure 2.11**

Figure 2.10. Scheme of the IR diffuse reflectance apparatus.

Figure 2.11. Diffuse reflectance cell[18]: (1) KBr window, (2) sample, (3) heater, (4) circulation, (5) thermocouple, (6) vacuum line[18] (reprinted with permission from *Anal. Chem.*, Copyright 1979, American Chemical Society).

because it seemed that the construction of an integration sphere for collecting the diffuse radiation was too difficult. Finally, a spectrometer, whose scheme is shown on Figure 2.10, was designed. The IR light coming from the source is focused on the sample, which is at one focus of an ellipse. The diffuse light is collected by an elliptic mirror and goes to the detector located at the second focus of the ellipse. By using a spectrometer based on this principle, Fuller and Griffiths[16] showed that this method can give good qualitative and quantitative results with many powdered samples. Its application to catalysis is limited but seems promising.[17,18] Figure 2.11 shows a cell which allows *in situ* treatments of samples.

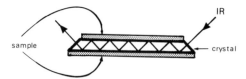

Figure 2.12. Optical way in the KRS5 crystal (internal multiple reflections).

2.2.3.3. Internal Reflectance

This method, also called attenuated total reflectance (ATR) or attenuated multiple internal reflectance, is used for the study of dark solids. The sample, prepared as a fine powder, is dispersed on the two faces of a KRS5 (TlBr + TlI) single crystal, which has a high refractive index. The IR beam enters by the input side of the crystal and has n reflections on the two faces. At each reflection, a small part of the light is absorbed by the sample.[19] Finally, an absorption spectrum is obtained which corresponds to n absorptions (Figure 2.12). The crystal is put into a suitable sample holder, which is placed in a metallic cell. This allows spectra to be obtained under various temperatures, pressures, and gas compositions. This method can be used for the study of some transparent catalysts, but it has disadvantages because the contact between the powder and the crystal is not always very good and the crystal may react chemically at temperatures higher than 250 °C.

2.2.4. Emission Spectroscopy

While transmission and reflection IR spectroscopies study the absorption of radiation by compounds (ground-state–excited state transition), emission spectroscopy studies the reverse phenomenon, i.e., the excited state–ground-state transition. As a consequence of the Boltzmann law, it is necessary to populate the excited state by heating the sample. The emission spectrum is obtained by comparing the emitted radiation of the sample to that of a blackbody heated at the same temperature. From the Kirchhoff laws, the absorbance of a sample at a given temperature is equal to its emittance e. As the radiation is absorbed, reflected, or transmitted in agreement with the relation $a + r + t = 1$ (where r is the reflectance and t the transmittance), when the sample is poorly reflecting, the emission and absorption spectra are similar. The sample is spread as a fine bed on a strongly reflecting support in order to minimize its normal emission. It is heated under vacuum or in a controlled atmosphere in the cell shown in Figure 2.13. This cell is positioned so that the emitted beam is focused on the opening of the spectrometer.[20] This method is not very easy to use because the sample bed must not be too thick (a part of the emitted radiation is absorbed and the measured emission tends to that of a blackbody) and detection of the weak signal can be difficult.[21] The use of Fourier transform spectrometers gives a good signal-to-noise ratio by accumulating the low emitted radiation.[22] However, the application of this method in the study of catalysts is very limited.[20,23,24]

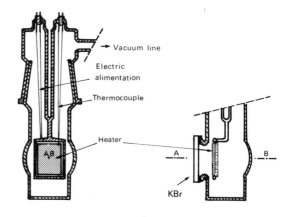

Figure 2.13. Scheme of an IR emission cell.[20]

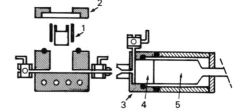

Figure 2.14. Photoacoustic spectroscopy cell[27]: (1) sample holder, (2) cap with optical window, (3) microphone support, (4) microphone, (5) amplifier (reprinted with permission from *Anal. Chem.*, Copyright 1983, American Chemical Society).

2.2.5. Photoacoustic and Deflection Spectroscopies

The principle of the method of photoacoustic spectroscopy (PAS) is based on the photothermic properties of the radiation. The radiation, when it is absorbed by a sample, is degraded into heat and the temperature of the sample increases (photothermic effect). This energy can be transferred to the gaseous phase on the sample. If the sample and the gas are put into a sealed chamber, the heating of the gas corresponds to a pressure increase. Modulation of the radiation will involve a modulation of the gas pressure, which can be detected by a microphone and amplifier. It has been shown that for dark compounds the photoacoustic signal is proportional to the absorption coefficient.[25] The photoacoustic and the absorption spectra are therefore similar. Initially, this method was limited to the UV–visible and near-IR range because there were no IR sources of sufficient energy. The smaller intensity of an IR source can be compensated for by the interferometry advantages.[26] The first results, obtained with the cell shown in Figure 2.14, proved the advantages of this method for the study of surface species without special preparation of the sample.[27,28]

Deflection spectroscopy is also based on the photothermic properties of the IR waves, but it analyzes the refractive index modulations instead of gas pressure changes. These modulations are analyzed by the deflection of a laser beam on the sample surface. The advantage of this method compared to photoacoustic spectroscopy is that the detector is outside the cell.

2.3. EXAMPLES OF THE USE OF IR SPECTROSCOPY IN HETEROGENEOUS CATALYSIS

2.3.1. Catalyst Structure

For these studies, the most used method is the dilution by KBr (when the catalyst is stable in air and does not react with KBr). However, if one wishes to study the structural evolution of the catalyst during the reaction, it is necessary to use the techniques of emission or reflection.

2.3.1.1. Dealumination of Mordenite (Dilution by KBr)

Generally, the IR bands of aluminosilicates are classified into two groups:

1. Bands due to the internal vibrations of the TO_4 tetrahedra ($T = Si$, Al, ...), which establish the primary building unit of the structure. Few of these are sensitive to structural modifications.
2. Bands that correspond to the vibrations of bonds between TO_4 tetrahedra. Naturally their position is highly dependent on the way the tetrahedra are linked. In the zeolites, each oxygen atom of TO_4 bridges two tetrahedra in order to form secondary building units such as rings of four, five or six tetrahedra. These vibrations are characteristic of the framework of the zeolite (external vibrations). It must be pointed out that some bands exist between 650 and 480 cm^{-1} whose number, position, and intensity vary from one structure to another (Table 2.8).

The absorptions of the second group are sensitive to the structure of the zeolite and in some cases give information about the framework, but the bands of the first group have positions that are greatly dependent on the composition of the zeolite.[29] It has been shown that the symmetrical and unsymmetrical stretching frequencies of the TO_4 tetrahedra decrease when the aluminum content of the framework increases. This effect can be explained by a decrease in the mean force constant of the T—O bonds when the aluminum content increases (the Si—O bond is smaller than the Al—O one, and aluminum has a lower electronegativity than silicon). For a given framework, e.g. X and Y faujasites, there is a linear correlation between frequency and aluminum

TABLE 2.8. Characteristic Vibrations of Secondary Building Units in Zeolites

Zeolite	Wave number (cm^{-1})	Intensity	Secondary building unit
Y	570	Medium	Double 6-ring
ZSM-5	550	Strong	5-ring
Mordenite	$\begin{cases} 580 \\ 550 \end{cases}$	Weak	5-ring + 1 TO_4
L	612	Medium	

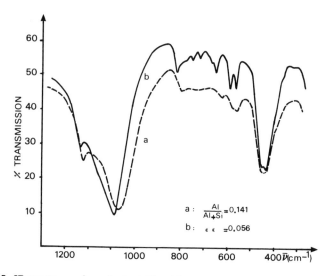

Figure 2.15. IR spectrum of mordenite before (Curve a) and after (Curve b) dealumination.

TABLE 2.9. Influence of Dealumination on the IR Absorptions of Mordenite

Al/(Al + Si)	Wave number (cm^{-1})				
0.147	1054	795	620	580	555
0.141	1054	795	620	580	555
0.131	1063	800	624	580	558
0.056	1078	810	648	588	563
0.008	1087	824	658	593	570

content.[29,30] The same results have been observed in another case[31]: synthesized mordenites (Si/Al ratio = 5.8) were dealuminated by consecutive hydrothermal and acid treatments in order to remove aluminum without destroying the framework. Figure 2.15 shows the IR spectrum of the mordenite before and after dealumination. It can be seen that the number of absorptions is unchanged by dealumination but the wave numbers increase and the linewidths decrease (Table 2.9). If the wave numbers of the vibrations are plotted as a function of the Al/(Al + Si) ratio, linear correlations are observed as for the faujasites (Figure 2.16) and it appears that for the unsymmetrical stretching of the tetrahedra, the relation is independent of the zeolitic framework (Curve 1a).

2.3.1.2. Study of the Reduction of Vanadium Oxide by the Use of Multiple Internal Reflections[31]

The IR spectrum of vanadium oxide V_2O_5 (Figure 2.17) shows a strong absorption at high frequency (1018 cm^{-1}), generally attributed to V=O

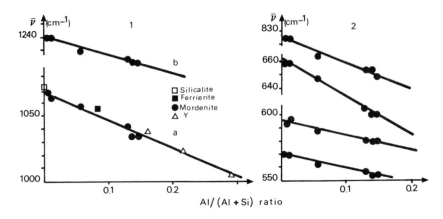

Figure 2.16. Dependence of the wave numbers of zeolites on their aluminum content.

vibrations, and a broad and strong band at 840 cm^{-1}, which corresponds to the V—O vibrations with a bridging oxygen atom. The absorptions at lower frequency correspond to V—O vibrations with oxygen bonded to three vanadium atoms and to bending vibrations. During the reduction process, an IR absorption appears at 970 cm^{-1} while that at 1018 cm^{-1} decreases in intensity and the band at 840 cm^{-1} is shifted until 815 cm^{-1}. Such modifications are caused by the crystallographic shear planes which occur during the reduction process (Figure 2.18): (1) the number of triply bridging oxygen atoms increases while the number of doubly bridging atoms decreases (shift of the IR band at 840 cm^{-1} and lower intensity); and (2) as the vanadium is reduced, the V=O bond is less covalent (shift of the 1018 cm^{-1} band up to 970 cm^{-1}).

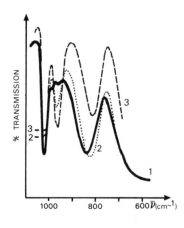

Figure 2.17. Reduction of V_2O_5 by C_3H_6: (1) V_2O_5 activated at 200 °C under oxygen, (2) reduction by C_3H_6 at 220 °C during 16 h, (3) reduction by C_3H_6 at 240 °C during 16 h.

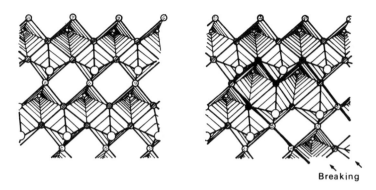

Breaking

Figure 2.18. Structural modifications during the reduction process: ○ terminal oxygen atom alternatively above and below the level; ⊙ oxygen atom linked with two vanadium atoms; ⦻ oxygen atom linked with three vanadium atoms; and ● oxygen atom linked with four vanadium atoms.

2.3.2. Study of the Surface Properties of Catalysts

The surface properties of a catalyst when due to a group whose vibrations are in the 4000–1000 cm^{-1} can be studied directly. The surface hydroxyl group is one example. However, many surface sites can only be seen by the use of a suitable probe molecule, which is sorbed by these sites and has vibrations whose wave number variations will be correlated with the nature of the sites. This indirect method is used for studying, e.g., the Lewis sites of a support and the surface properties of metal particles.

2.3.2.1. Surface Hydroxyl Groups

Most of the supports used in heterogeneous catalysis have IR spectra with one or more absorptions in the 3800–3500 cm^{-1} region. The bands are due to the hydroxyl groups (Brönsted acid sites). For example, the IR spectrum of an Aerosil Degussa silica has two bands at 3742 and 3680 cm^{-1} while γ-alumina gives five ν_{OH} bands at 3800, 3780, 3744, 3733, and 3700 cm^{-1}. Each band can be correlated with a different OH group and Peri[32] has proposed attributions by assuming that all the OH groups are on the (100) crystalline face.

2.3.2.2. Acid Properties

The hydroxyl groups of a catalyst can have acid properties to a greater or lesser extent, depending on their surroundings. If they can react with bases, e.g.,

$$Al-OH + B \rightarrow Al-O^- + HB^+$$

the catalyst shows Brönsted acidity. The intensity of the IR vibrations of HB^+ can be used to calculate the number of acid sites. The variation of the base molecule (various pKa values) and the stability of the HB^+ complex under

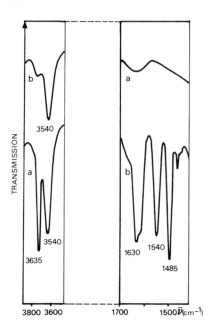

Figure 2.19. Pyridine adsorption on a HY zeolite[33]: (a) zeolite HY activated at 350 °C; (b) adsorption of pyridine at room temperature on the activated sample and evacuation at 150 °C (reprinted with permission from *J. Phys. Chem.*, Copyright 1967, American Chemical Society).

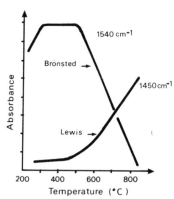

Figure 2.20. Variation of the number of Brönsted and Lewis acid sites in a HY zeolite as a function of the activation temperature[33] (reprinted with permission from *J. Phys. Chem.*, Copyright 1967, American Chemical Society).

heating gives further information on the acid strength. Lewis acid sites (e.g., tricoordinated aluminum atoms) can also be present on the surface of the catalyst. These sites can also react with bases:

$$Al +: B \rightarrow Al \Leftarrow: B$$

If the IR spectra of HB^+ and $Al \Leftarrow: B$ are quite different, the adsorption of only one base can give information on the Lewis and Brönsted acidities. Many nitrogenous bases have that property and are used, e.g., pyridine C_5H_5N, ammonia NH_3, and $C_4H_9NH_2$. As an example, let us consider the adsorption of pyridine on a HY zeolite (Figure 2.19). After activation under oxygen at 350 °C, the IR spectrum of an HY zeolite shows three ν_{OH} bands, due to the terminal hydroxyls: Si—OH (3740 cm^{-1}), the Al—O$_1$H groups in the supercages (3635 cm^{-1}) and the Al—O$_3$H groups in the hexagonal prisms (3540 cm^{-1}). When pyridine (pKa = 5.2) is adsorbed, only the O$_1$H react, yielding to the C$_5$H$_5$NH$^+$ pyridinium ion which has a characteristic IR absorption at 1540 cm^{-1}. On the other hand, by adsorption of stronger bases such as piperidine (pKa = 11.2) not only the O$_1$H but also the O$_3$H groups react.[33] On dehydroxylation of the zeolite, by heating up to 350 °C, Lewis acid sites are generated which also react with pyridine yielding Al $\Leftarrow$: NC$_5$H$_5$ with an absorption at 1450 cm^{-1}. From absorbance measurements of the C$_5$H$_5$NH$^+$ and Al $\Leftarrow$: NC$_5$H$_5$ bands, the changes in the number of Lewis and Brönsted sites can be determined as a function of the activation temperature of the zeolite (Figure 2.20).[33]

2.3.2.3. Properties of Metal Particles

Similar to Lewis acid sites, the surfaces of supported metals cannot be studied directly by IR spectroscopy so it is necessary to use a probe molecule. CO is used most often because of its optical density and for the range of information obtainable from the IR spectra. However, pyridine adsorption has proved that the metal particles have acid properties[34] and NO can be used to corroborate the results obtained with CO.

2.3.2.3a. Nature of the Metal–CO Bond. CO reacts with most of the transition metals to form metal carbonyl complexes. It is often assumed that in these compounds there are a σ and a π (by backdonation) bond between the metal and CO. The σ bond is obtained by the overlapping of the higher occupied orbital of CO (5σ) and an empty orbital of the metal. The π bond is obtained by the overlapping of an occupied orbital of the transition metal of adapted symmetry with the lower empty orbital of CO (2π). The 5σ and 2π orbitals of CO are above all localized on carbon and so we can consider that the metal–carbon bond is multiple. The σ and π components give a synergetic effect: by π donation the metal becomes a better σ acceptor and the CO group becomes a better π donor.[35] The π acceptor properties of CO (or π donor of the metal) can be measured by comparison of the ν_{CO} frequencies, because the filling up of the 2π orbital destabilizes the CO bond and so the vibration frequency. This wave number varies from 2143 cm^{-1} for gaseous CO up to 2000–1800 cm^{-1} for carbonyl complexes.

As the metal particles have many metallic atoms (from about 10 to a few hundred or thousands), several CO molecules can be sorbed on one particle, sometimes with more than one CO for one metal atom. Thus several configurations can be observed with one CO for one metal atom (linear) or one CO for several metal atoms (bridging). It is assumed that, as for the metal carbonyls, there are a σ and a π bond between the metal and the CO group. As a consequence, a decrease in the CO frequency is observed.[36] However, the interpretation is not always so easy because a particle formed with n atoms can sorb n CO molecules and, under these conditions, two properties must be taken

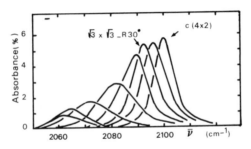

Figure 2.21. Variation of the ν_{CO} band as a function of coverage.

into account: the collective properties of the metallic atoms (electronic or chemical or static effect) and the interactions among the CO molecules (geometric or dynamic effect). The influence of these two effects on the CO vibration frequency is very difficult to ascertain and was the subject of much controversy because the problem is very important: CO can be used as a molecule probe for studying the electronic effect only if the geometric contribution is known. Recently this difficulty was solved by the use of isotopic dilution[37] at a constant coverage. Under these conditions, the CO adsorption gives much information on the interaction between the metal particle and the support.

2.3.2.3b. Influence of Coverage. The ν_{CO} frequency of CO adsorbed on the [111] face of platinum varies between 2063 and 2100 cm^{-1} as a function of the coverage (Figure 2.21).[38] The same effect has been observed for supported platinum[37,39] and many other metals (see Ishi *et al.*[40] and the references therein). However, an inverse effect has been observed on metals such as silver, copper, and gold[40] and two explanations have been proposed:

1. For a given particle, the number of electrons that the metal can give by backdonation has a finite value. When the number of sorbed CO molecules increases, each of them has fewer electrons from metal. Thus the force constant of the M—C bond decreases, that of the C—O bond increases, and the ν_{CO} frequency increases (electronic effect).

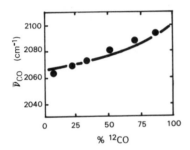

Figure 2.22. Variation of the $\nu^{12}CO$ frequency as a function of the composition of the ^{12}CO–^{13}CO mixture. The points are experimental and the curve is theoretical.

TABLE 2.10. Support Effect on the Vibration
Frequency of CO Adsorbed on Palladium

Support	Pd (%)	Linear CO (cm^{-1})
NaX	2	2035
MgO	2	2065
Al_2O_3	1	2075
NaY	1.7	2075
13% SiO_2-Al_2O_3	1.2	2100
HY	1.2	2105
MgY	1.8	2100
CaY	1.9	2115
LaY	1.8	2105

2. When the number of sorbed CO molecules increases, the average CO—CO distances decrease and the dipole–dipole couplings increase, so that the v_{CO} frequency increases (geometric effect).

The influence of each of these two effects has been determined by the isotopic dilution method with ^{12}CO, ^{13}CO or $C^{16}O$, $C^{18}O$. This method is based on the fact that dipolar coupling is the highest for dipoles vibrating at the same frequency, decreases very rapidly when the difference between frequencies increases, and can be considered as zero when it is bigger than 30 cm^{-1}, which is the case for the ^{12}CO and ^{13}CO molecules or $C^{16}O$ and $C^{18}O$. On the other hand, the π accepting properties of CO are not modified by isotopic exchange. When this method is applied to the previous example,[41] the $v^{12}CO$ frequency varies from 2063 up to 2100 cm^{-1} as a function of ^{12}CO content at saturation coverage (Figure 2.22). Thus the observed shift is due to dipolar coupling.

2.3.2.3c. Support Effect. The vibration frequency of CO adsorbed at saturation coverage on supported palladium is highly dependent on the support (Table 2.10).[42] The shift of v_{CO} toward high frequencies has been explained by a decrease in the electronic density on the metal, due to interaction with acid sites on the support.

2.3.2.3d. Alloying Effect. It is well known that some alloys have catalytic properties (activity, selectivity) quite different from those of their components taken alone. The theoretical explanation takes into account two effects:

1. A geometrical effect: alloying decreases the number of neighboring identical atoms (dilution) and so the selectivity for reactions which require more than one site also decreases.
2. An electronic effect: alloying the metal M_1 with an M_2 metal having more external electrons than M_1 will enrich M_1 electronically and so the bond strength of sorbed species will be modified.

If the ^{12}CO molecule is used as a probe, the geometric and/or electronic effect will decrease the $v^{12}CO-M_1$. This has been observed on the Pt–Cu,[43] Pt–Pb,[44,45] Ni–Cu,[46] and Pd–Ag[47] alloys. Only the isotopic dilution method can differentiate between the two effects. The frequencies $v^{12}CO-M_1$ are plotted against the isotopic dilution ratio $x = {}^{12}CO/({}^{12}CO + {}^{13}CO)$. If the alloying effect is geometric, i.e., if the M_2 effect is only to increase the distances between the M_1-CO dipoles, the frequency of the isolated vibrator must be obtained for $x = 0$ and whatever the alloy composition may be. Thus in this case the curves for M_1 alone and various $M_1 + M_2$ alloys must have a common point at $x = 0$ (Figure 2.23). On the other hand, if the M_2 addition only modifies the electronic density of M_1, the curves must show parallel shifts as the alloy grows rich in M_2 (Figure 2.24). For a Pt–Cu alloy supported on alumina the experimental curves have proved that the alloy effect was purely geometric.[43]

2.3.2.3e. Interaction between the Metal and an Adsorbate. It has been observed that the v_{CO} frequency shifts when another gaseous phase is co-adsorbed with CO.[44,48,49] The size and the sign of the shift are dependent on the coadsorbate (Table 2.11). At constant CO coverage, the shift toward lower frequencies is greater when the ionization potential of the base is small. This effect has been explained by a long-range interaction between the Lewis base (electron donor) and the Pt–CO bond (collective properties of the metal). The following observation is in agreement with this explanation: when pyridine is adsorbed progressively, on platinum covered with 20% CO, the shift increases with the amount of pyridine on the metal (Figure 2.25),[49] i.e., with the electronic density of the particle. On the other hand, by HCl adsorption a shift toward higher frequencies is observed (electronic transfer from the particle to the chlorine atom).

2.3.2.3f. Conclusions. The examples described above show the importance of CO as a probe molecule for the study of metal particles. If the various explanations given by different authors are consistent, it must be pointed out that there are still many controversies.

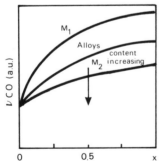

Figure 2.23. Geometric effect on adsorption of CO on alloys.

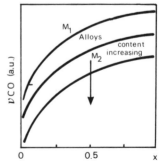

Figure 2.24. Electronic effect on adsorption of CO on alloys.

TABLE 2.11. Variation of ν_{CO} Frequency as a Function of the Co-Adsorbate on Pt–Al$_2$O$_3$ Previously Covered (coverage 20%) with CO[52]

Adsorbate	Ionization potential (eV)	ν_{CO} (cm^{-1})	$\Delta\nu_{CO}$ (cm^{-1})
—	—	2065	—
H$_2$O	12.6	2050	−15
NH$_3$	10.5	2040	−25
Pyridine	9.2	1990	−75
HCl	—	2075	+10

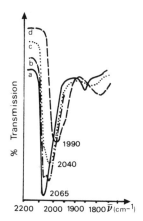

Figure 2.25. Absorption of pyridine on Pt–Al$_2$O$_3$ covered with 20% CO: (a) CO on Pt–Al$_2$O$_3$; (b, c, d) increased amounts of pyridine.

2.3.3. Quantitative and Qualitative Studies of Adsorption

2.3.3.1. Adsorption Isotherms Deduced from Spectroscopic Measurements[50]

Hydrogen adsorption on alumina-supported platinum gives two sorbed species. The stronger bonded species (irreversible H) is not seen by IR spectroscopy. The weak bonded species (reversible H) gives an absorption at 2120 cm^{-1} whose intensity is dependent on hydrogen pressure. From the Beer–Lambert law, its intensity is proportional to the amount of reversible hydrogen adsorbed. It is therefore easy to obtain adsorption isotherms (Figure 2.26). Only the linear transformations corresponding to the dissociative Langmuir model fit in the pressure (0–760 Torr) and temperature (27–95 °C) ranges (Figure 2.27). From this model, we have

$$2Pt + H_2 \rightleftharpoons 2Pt—H$$

If K is the equilibrium constant, D the absorbance (proportional to the adsorbed

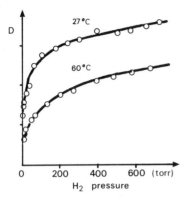

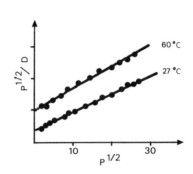

Figure 2.26. Hydrogen adsorption isotherms at 27 and 60 °C on alumina-supported platinum.

Figure 2.27. Linear transformations deduced from the Langmuir dissociative model of adsorption.

amount), D_M the absorbance for the highest coverage, and p the hydrogen pressure, then

$$p^{1/2}/D = p^{1/2}/D_M + 1/(K^{1/2}D_M) \tag{2.9}$$

The equilibrium constant K can be deduced from these curves and the adsorption heat can be determined from

$$[d(\ln K)/dT] = \Delta H/RT^2 \tag{2.10}$$

The value of ΔH obtained from these experiments is -5×10^4 J mol^{-1} and it is in good agreement with that deduced from microcalorimetry (-4.2×10^4 J mol^{-1}).

2.3.3.2. Symmetry of Adsorbed Species

When the adsorbed species has identical groups of atoms, it has symmetry properties that determine the number of active vibrations in IR and their relative intensities. Thus, in some cases, it is possible to determine the geometry of the adsorbed species from its IR spectrum. Often, isotopic exchange is also used to get further information.

2.3.3.2a. Geometry of a Complex When the Number of Ligands is Known. By the reaction of a NaY zeolite, exchanged by Rh(III) ions, with CO at 150–220 °C, a rhodium carbonyl complex is formed, entrapped within the zeolite cavities.[51] Analytical studies of the reactant and product gaseous phases have shown that the carbonylation reaction can be written

$$n\,Rh^{3+} + 3n\,CO + n\,H_2O \rightarrow [Rh(CO)_2^+]_n + n\,CO_2 + 2n\,H^+$$

The IR spectrum of the $[Rh(CO)_2]_n$ species gives two very strong absorptions at 2116 and 2048 cm^{-1} in the ν_{CO} domain. If n is higher than 1, more than

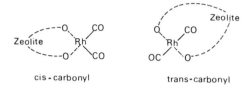

Figure 2.28. *Cis* or *trans* form of $[Rh(CO)_2]^+$ in NaY zeolite.

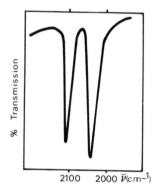

Figure 2.29. IR spectrum of $[Rh(CO)_2]^+$ in NaY zeolite.

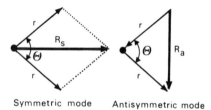

Figure 2.30. Symmetric and antisymmetric modes of the *cis*-carbonyl.

Symmetric mode   Antisymmetric mode

two bands would be expected (more CO linear or bridging CO bonds), so the complex is mononuclear. As all the Rh(I) complexes have a square planar geometry, we can propose such a structure for Rh(CO)$_2$ and two forms can be considered (Figure 2.28).

Two ν_{CO} frequencies are expected for each geometry: symmetric and antisymmetric. However, in the *trans*-configuration, the symmetric mode is inactive by symmetry. The observed spectrum (Figure 2.29) has two ν_{CO} bands and can only be assumed to be a *cis*-structure.

The angle θ between the two CO ligands can be calculated from the intensities of the two vibrations. Each CO molecule is represented by a dipole, and the dipolar vector of each mode is the sum of the two dipoles. As the intensity of a vibration is proportional to the square of the dipolar vector, it is

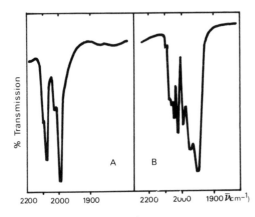

Figure 2.31. IR spectra of the iridium (I) carbonyl complex obtained by reaction with (A) ^{12}CO and (B) $^{12}CO-^{13}CO$.

easy to calculate the ratio of the two intensities as a function of θ (Figure 2.30):

$$R_s = 2r\cos(\theta/2) \qquad R_a = 2r\sin(\theta/2) \qquad (2.11)$$

Thus,

$$I_s/I_a = [4r^2\cos^2(\theta/2)]/[4r^2\sin^2(\theta/2)] = \cot g^2(\theta/2) \qquad (2.12)$$

In this complex, the θ angle has been estimated at 105° and implies a distortion by the zeolitic framework.

2.3.3.2b. Use of $^{12}CO-^{13}CO$ Isotopic Mixtures. Similar to Rh(III) ions, Ir(III) ions exchanged within a NaY zeolite react with CO, giving an Ir(I) carbonyl complex. As in the previous case, its IR spectrum has only two ν_{CO} bands.[52] However, two structures can give two IR ν_{CO} bands — Lx Ir (CO)$_2$ with C_{2v} symmetry and LxIr (CO)$_3$ with C_{3v} symmetry (Lx represents the other ligands). In order to know the structure of this complex $^{12}CO-^{13}CO$ isotopic exchange was used. In the presence of a $^{12}CO-^{13}CO$ mixture, the IR spectrum has seven ν_{CO} bands (Figure 2.31). It is possible to deduce the number of active vibrations for the two structures and ^{12}CO totally or partially substituted by ^{13}CO. It is also easy to calculate their frequencies if the values for the complex with only one isotope are known (Table 2.12).

The experimental spectra are in good agreement with calculation for a tricarbonyl species and so we can conclude that unlike the rhodium complex, the Ir(I) carbonyl complex has the Lx Ir (CO)$_3$ structure.

2.3.4. Detection of Reaction Intermediates

Such experiments are not easy because the system has to be studied during the catalytic reaction. It is necessary to use specially designed IR cells (e.g., to accommodate heating and gas circulation). The intermediates are usually in very low concentrations and have short lifetimes.

TABLE 2.12. Calculated Frequencies for Iridium Di- and Tricarbonyl Species

Dicarbonyl species			Tricarbonyl species		
Species[a]	Symmetry group	ν_{CO} (cm^{-1})	Species[a]	Symmetry group	ν_{CO} (cm^{-1})
$L_xM(^{12}CO)_2$	C_{2v}	2086 2001	$L_xM(^{12}CO)_3$	C_{3v}	2086 2001
$L_xM(^{12}CO)(^{13}CO)$	C_s	2068 2072	$L_xM(^{12}CO)_2(^{13}CO)$	C_s	2075 2001 1966
			$L_xM(^{12}CO)(^{13}CO)_2$	C_s	2060 1956 1979
$L_xM(^{13}CO)_2$	C_{2v}	2036 1956	$L_xM(^{13}CO)_3$	C_{3v}	2036 1956

[a] Force constant: 17.38 mdyne Å^{-1} for ^{12}CO (0.174 N m^{-1}); 17.31 mdyne Å^{-1} for ^{13}CO (0.173 N m^{-1}).

2.3.4.1. Reduction of Methyl Benzoate and Benzoic Acid on Y_2O_3[53]

A dynamic spectroscopic reactor[54] is coupled with a chromatograph and gives both the IR spectra of the catalyst and an analysis of the reaction products. As an example, the IR spectra of Y_2O_3 are shown during the first 60 min of the reduction of methyl benzoate (MB) (Figure 2.32). During the first step, methoxide species and benzoate ions are formed on the surface. The concentration of benzoate ions increases with time until it reaches a constant level while the methoxide reaches a maximum at about 15 min. It can also be seen that the formation of benzoate ions is obtained by reaction of the hydroxyl groups of Y_2O_3. The variation of the reaction products and the surface species, as determined by IR spectroscopy, is shown on Figure 2.33.

Methanol is formed during the first 50 min of the reaction. Subsequently the methanol concentration decreases rapidly as the methoxide species disappear and the benzoate coverage becomes constant. Simultaneously, the production of benzaldehyde, toluene, and benzene begins. We can now describe the following steps of the reaction:

$$MB + S \rightarrow MB_{(s)}$$
$$H_2 + S \rightarrow 2H_{(s)}$$
$$MB_{(s)} + H_{(s)} \rightarrow B_{(s)} + MeOH_{(g)}$$
$$MB_{(s)} \rightarrow B_{(s)} + M_{(s)}$$
$$M_{(s)} + H_{(s)} \rightarrow MeOH_{(g)}$$
$$B_{(s)} + H_{(s)} \rightarrow BA_{(g)}$$

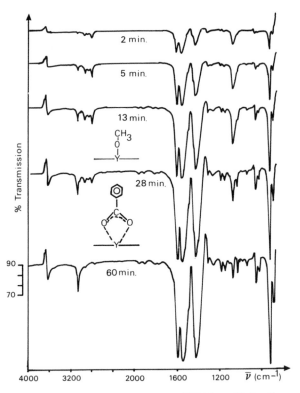

Figure 2.32. Reduction of the methyl benzoate at 270 °C on Y_2O_3. Spectra obtained by substraction of the Y_2O_3 spectrum at $t = 0$.

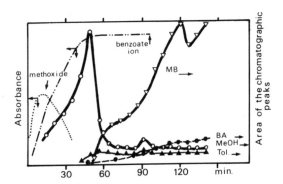

Figure 2.33. Changes in the concentration of gaseous products [methyl benzoate (MB), benzaldehyde (BA), methanol (MeOH), and toluene (Tol.)] and in the optical density of the surface compounds (methoxide and benzoate ion).

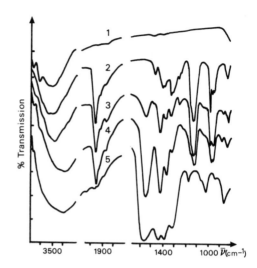

Figure 2.34. CeO$_2$ activated at 400 °C (1) and adsorption of 2-propanol at room temperature (2), 150 °C (3), 250 °C (4), and 350 °C (5).

where (g) is the gaseous species, (s) the adsorbed species, S the surface site, $B_{(s)}$ the benzoate ion, and $M_{(s)}$ the methoxide species.

2.3.4.2. Dehydrogenation and Dehydration of Isopropanol on CeO$_2$[55]

The most common method for characterization of the intermediates is to heat the catalyst with the reactants and to take the IR spectrum of the catalyst after cooling at room temperature. However, this method can only be used for the study of stable intermediates. Figure 2.34 shows the spectrum of 2-propanol adsorbed at ambient temperature and subsequently heated at various temperatures on CeO$_2$. In the same way, the spectrum of the gaseous phase shows the appearance of acetone (150 °C), propene (250 °C and major product at 400 °C), and methane and isobutene (between 250 °C and 400 °C). The spectra of adsorbed species have been interpreted as follows. At room temperature there is hydrogen-bonded species and isopropoxide species are formed. At temperatures above 150 °C, the isopropoxide species loses hydrogen, yielding acetone with an acetonate intermediate:

At temperatures above 250 °C, the formation of propene and isobutene implies a dehydration of isopropanol (propene) and the reaction of an acetone

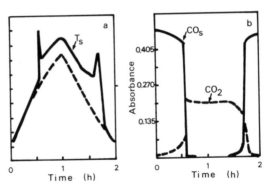

Figure 2.35. (a) Input temperature ramp and surface temperature (T_s) of the catalyst. (b) Infrared spectrograms of the *in situ* CO_2 absorbance response and the linear-bonded CO surface species. Inlet flows N_2: 220 cm³/min; CO: 9 cm³/min; O_2: 5 cm³/min.

molecule with the acetonate species (isobutene) yielding acetate ions on the catalyst:

$$CH_3\text{-}C\text{-}CH_3 \quad + CH_3COCH_3 \xrightarrow{-H_2O} CH_3\text{-}C\text{-}CH\text{-}C(CH_3)_2 \longrightarrow CH_2\text{=}C\text{-}CH_3 + \quad$$

2.3.4.3. Transient Technique Integrating FTIR Spectroscopy with a Temperature-Programmed Reaction and a Concentration-Programmed Reaction[56]

A small-volume IR cell reactor[57] is equipped with a temperature and concentration programming system. The FTIR data (gaseous phase and catalyst) are computed and analyzed by a computer and the software allows the programming of temperature and concentration parameters. The first study was the oxidation of CO on silica-supported platinum because of its apparent relative simplicity. During the temperature-programmed reaction, the dynamic response was simultaneously recorded for surface temperature, CO_2 production, and linear-bonded CO (Figure 2.35). Similarly CO and CO_2 absorbance responses were registered during the concentration-programmed reactions at constant temperature.

These studies have revealed remarkable surface temperature behavior in regions of ignition and quenching (Figure 2.35) and support the previously proposed reaction model.

2.4. CONCLUSIONS

The range of applications of IR spectroscopy in catalysis is so large that the examples treated in this chapter can be considered only as a survey of the possibilities of this method and not as an analysis of all the published work, which would take more than a book to describe. Indeed, the first IR studies of sorbed water molecules were made at the beginning of the century and during the past 30 years IR spectroscopy has been the most commonly used physical method in catalysis laboratories. This spectroscopy is very easy to use and gives much information on the catalyst, the reactants, and the adsorbed reaction intermediates, but it has two limitations: (1) the type of absorbant (but many technical modifications and skills are now used to solve problems), and (2) the low concentrations and/or short lifetimes of the reaction intermediates.

However, as data acquisition is very rapid on the Fourier transform spectrometer one can hope that as transient pulse response methods are developed many studies will be done in the second area. As for all the physical methods, IR spectroscopy alone will rarely solve a catalysis problem. More and more often other techiques are used, simultaneously or not, such as volumetry, gravimetry, and microcalorimetry, and UV-visible, NMR, ESR, and Mössbauer spectroscopies (as described in other chapters).

Technically, the development of the emission, reflection, and photoacoustic spectroscopies and the use of computers show how these old techniques continue to evolve.

REFERENCES

1. W. Brugel, *An Introduction to Infrared Spectroscopy*, Methuen, London (1962); G. Herzberg, *Infrared and Raman Spectra*, D. Van Nostrand, Princeton (1968); B. Mentzen, *Spectroscopies infrarouge et raman*. Monographies du CAST, Masson, Paris (1974); J. Lecomte, *Handbuch des Physik* **26**, Springer-Verlag (1958), p. 244; G. M. Barrow, *Introduction to Molecular Spectroscopy*, McGraw-Hill, New-York (1962).
2. K. Nakamoto, *Infrared and Raman Spectra of Inorganic and Coordination Compounds*, John Wiley and Sons, New-York (1978).
3. F. A. Cotton, *Applications de la théorie des groupes à la chimie*, Dunod, Paris (1968).
4. E. B. Wilson, J. C. Decius, and P. C. Cross, *Molecular Vibrations*, McGraw-Hill, New York (1965).
5. P. Barchewitz, *Spectroscopie atomique et moléculaire*, Tome II, Masson, Paris (1971).
6. J. P. Ferraro and J. S. Ziomek, *Introductory to Group Theory*, Plenum, New York (1969).
7. L. H. Little, *Infrared Spectra of Adsorbed Species*, Academic, New York (1966); M. L. Hair, *Infrared Spectra in Surface Chemistry*, Marcel Dekker Inc., New York (1967).
8. L. J. Bellamy, *Advances in Infrared Group Frequencies*, Methuen, London (1968); *The Infrared Spectra of Complex Molecules*, Chapman and Hall, London (1980); H. A. Szymanski, *Interpreted Infrared Spectra*, Plenum, New York (1964); D. M. Adams, *Metal–Ligand and Related Vibrations*, St. Martin, London (1968); J. R. Ferraro, *Low Frequency Vibration of Inorganic and Coordination Compounds*. Plenum, New York (1971); L. H. Jones, *Inorganic Vibrational Spectroscopy*, Marcel Dekker, New York (1971); C. J. Pouchert, *The Aldrich Library of Infrared Spectra*, Aldrich Chemical (1981); R. A. Nyquist and R. O. Kagel, *Infrared of Inorganic Compounds*, Academic, New York (1971).

9. R. A. Dalla Betta and M. Shelef, *J. Catal*, **48**, 111 (1977).
10. R. G. Greenler, *J. Chem. Phys*. **44**, 310 (1966).
11. R. A. Shigeishi and D. A. King, *Surf. Sci*. **58**, 379 (1976).
12. W. G. Golden, D. S. Dunn, and J. Overend, *J. Catal*. **71**, 395 (1981).
13. A. Ishitani, H. Ishida, F. Soeda, and Y. Nagasawa, *Anal. Chem*. **54**, 682 (1982).
14. K. Horn and J. Pritchard, *Surf. Sci*. **52**, 437 (1975).
15. J. D. Fedyk, P. Mahaffy, and M. J. Dignam, *Surf. Sci*. **89**, 404 (1979).
16. M. P. Fuller and P. R. Griffiths, *Anal. Chem*. **50**, 1906 (1978).
17. L. Kubelkova, H. Hoser, A. Riva, and F. Trifiro, *Zeolites* **3**, 244 (1983).
18. M. Niwa, T. Hattori, M. Takahashi, K. Shirai, M. Watanabe, and Y. Murakami, *Anal. Chem*. **51**, 46 (1979).
19. N. J. Harrick, *Internal Reflectance Spectroscopy*, Interscience, New York (1967).
20. M. Primet, P. Fouilloux, and B. Imelik, *Surf, Sci*. **85**, 457 (1979).
21. D. B. Chase, *Appl. Spectrosc*. **35**, 77 (1981).
22. K. Molt, *Fres. Z. Anal. Chem*. **308**, 321 (1981).
23. P. C. M. Woerkom and R. L. De Groot, *Appl. Opt*. **21**, 3113 (1982).
24. P. C. M. Woerkom, *J. Mol. Struct*. **79**, 31 (1982).
25. R. E. Blank and Th. Wakefield II, Anal. Chem. **51**, 50 (1979).
26. D. W. Vidrine, *Appl. Spectrosc*. **34**, 314 (1980).
27. J. B. Kinney and R. M. Staley, *Anal. Chem*. **55**, 343 (1983).
28. D. A. Saucy, G. E. Cabaniss, and R. W. Linton, *Anal. Chem*. **57**, 876 (1985).
29. E. M. Flaningen, H. Katami, and H. A. Szymanski, *Adv. Chem. Ser*. **101**, 201 (1971); E. M. Flaningen, *idem*, **171**, 80 (1976).
30. P. Pichat, R. Beaumont, and D. Barthomeuf, *C. R. Acad. Sci., Paris* **272**, 612 (1971).
31. G. Coudurier, unpublished results.
32. J. B. Peri, *J. Phys. Chem*. **69**, 220 (1965).
33. T. R. Hughes and H. M. White, *J. Phys. Chem*. **71**, 2192 (1967).
34. M. Primet and Y. Ben Taarit, *J. Phys. Chem*. **81**, 1317 (1977).
35. F. A. Cotton and G. Wilkinson, *Advanced Inorganic Chemistry*, Interscience, New York, (1962), p. 616.
36. G. Blyholder, *J. Phys. Chem*. **68**, 2772 (1964).
37. R. M. Hammaker, S. A. Francis, and R. P. Eischens, *Spectrochim. Acta*. **21**, 1295 (1965).
38. A. Crossley and D. A. King, *Surf. Sci*. **68**, 528 (1977).
39. R. P. Eischens and W. A. Pliskin, *Adv. Catal*. **10**, 1 (1981).
40. S. Ishi, Y. Ohno, and B. Viswanathan, *Surf. Sci*. **161**, 349 (1985).
41. A. M. Bradshaw and F. M. Hoffman, *Surf. Sci*. **72**, 513 (1978).
42. F. Figueras, R. Gomez, and M. Primet, *Adv. Chem. Ser*. **121**, 481 (1973).
43. F. J. C. M. Toolenaar, F. Stoop, and V. Ponec, *J. Catal*. **82**, 1 (1983).
44. A. Palazov, Ch. Bonev, G. Kadinov, and D. Shopov, *J. Catal*. **71**, 1 (1981).
45. A. G. T. M. Bastein, F. J. C. M. Toolenaar, and V. Ponec, *J. Catal*. **73**, 50 (1982).
46. J. A. Dalmon, M. Primet, G. A. Martin, and B. Imelik, *Surf. Sci*. **50**, 95 (1975).
47. M. Primet, M. V. Mathieu, and W. H. M. Sachtler, *J. Catal*. **44**, 324 (1976).
48. R. Queau and R. Poilblanc, *J. Catal*. **27**, 200 (1972).
49. M. Primet, J. M. Basset, M. V. Mathieu, and M. Prettre, *J. Catal*. **29**, 213 (1973).
50. M. Primet, *J. Chem. Soc., Faraday Trans. I* **70**, 293 (1974).
51. F. Lefebvre and Y. Ben Taarit, *Nouv. J. Chim. Fr*. **6**, 387 (1984).
52. P. Gelin, G. Coudurier, Y. Ben Taarit, and C. Naccache, *J. Catal*. **70**, 32 (1981).
53. S. T. King and E. J. Stronjy, *J. Catal*. **76**, 274 (1982).
54. S. T. King, *Appl. Spectrosc*. **34**, 632 (1980).
55. M. I. Zaki and N. Sheppard, *J. Catal*. **80**, 114 (1983).
56. D. J. Kaul and E. E. Wolf, *J. Catal*. **89**, 348 (1984).
57. R. F. Hicks, C. S. Kellner, B. J. Savotsky, W. C. Hecker, and A. T. Bell, *J. Catal*. **71**, 216 (1981).

RAMAN SPECTROSCOPY

E. Garbowski and G. Coudurier

3.1. INTRODUCTION

Raman spectroscopy gives information about rotational and vibrational energy levels of a molecule, like IR spectroscopy. The technique has improved, especially during the past two decades, owing to the development of better technology, i.e., laser sources, solid state detectors, photon counting, charge coupled device camera, signal sampling and averaging and noise filtering with computers, ultralow stray light monochromators, etc. For catalysis and adsorption, this spectroscopy is complementary to classical IR transmission because the selection rules are different, but in some cases it is the only spectroscopy able to detect adsorbent vibrations in a wave number range where the support itself is not transparent to IR radiation.

3.2. THEORY OF RAMAN DIFFUSION[1-3]

First, consider a molecule subjected to an electromagnetic field of radiation of frequency v. This frequency is defined as the exciting radiation, and can be chosen in a range from near-UV to near-IR. The photons can be absorbed by the molecules if their energy $E_0 = hv_0$ corresponds exactly to the difference between the ground state energy level and an excited one. This is the absorption process. After absorption the photons are re-emitted in all directions, without a change in frequency. This process is elastic scattering or the Rayleigh effect. However, in some cases, scattering occurs with a frequency shift corresponding to an

E. Garbowski and G. Coudurier ● Institut de Recherches sur la Catalyse, CNRS, Villeurbanne, France.

Catalyst Characterization: Physical Techniques for Solid Materials, edited by Boris Imelik and Jacques C. Vedrine, Plenum Press, New York (1994).

inelastic collision between the molecule and the photon. This is the Raman effect, which can give much information about the energy levels of the molecule.

3.2.1. Quantum Mechanical Theory

According to the first law of thermodynamics, when a photon whose energy is $h\nu_0$, collides with a molecule whose energy is E_0, the total energy must be constant whatever the process. If the molecule recovers its energy E_0 after the collision process, the emerging photon must also have its former energy, say $h\nu_0$, but its direction, or angular momentum, can be changed. This collision occurs elastically giving the Rayleigh effect. However, if the energy of the molecule has changed to E_i corresponding to a new quantum state, the collision occurs inelastically and the emerging photon now has a different frequency ν_d so that

$$h\nu_d = h\nu_0 - (E_i - E_0) = h\nu_0 - h\nu_i \qquad (3.1)$$

In this expression ν_i is the frequency of a vibrational or rotational energy level E_i. This process is called Raman Stokes diffusion and the molecule has received extra energy from the photon, whose frequency has decreased. In some cases (not very favorable) the molecule can be in an excited state, according to the Boltzmann distribution. During the collision process, the excited molecule can lose energy and return to the ground state. The emerging photon then has more energy, i.e.,

$$h\nu_d = h\nu_0 + h\nu_i \qquad (3.2)$$

This process is called Raman anti-Stokes diffusion. It should be noted that the difference in energy between the colliding photon ($h\nu_0$) and emerging one ($h\nu_d$) is independent of the exciting radiation, and depends only on the vibrational and rotational energy levels of the molecule.

A Raman spectrum consists of a strong scattered radiation where the frequency is ν_0 (Rayleigh diffusion) and a collection of weak sharp lines appearing symmetrically on both sides of the exciting line, but with unequal intensities. It is easy to understand that the number of molecules in an excited state is very low, so the intensity of the anti-Stokes lines is lower than the corresponding Stokes lines, but the number of lines is exactly the same.

3.2.2. Classical Theory

The colliding photon ν_0 consists of an electromagnetic field $\mathbf{E} + \mathbf{B}$. Only the electric field will be considered here. The radiation induces a periodic perturbation of the local field $\mathbf{E} = \mathbf{E}_0 \cos 2\pi\nu t$. The electrical field $\mathbf{E}$ generates a dipole moment μ in the molecule that can be supposed to be an array of positive and negative charges. The dipole moment depends on the intensity of $\mathbf{E}$ and on a constant α. This parameter is the polarizability tensor of the molecule, which depends on the geometry, on the orientation relative to the **E** field. Although the molecule is totally neutral, the instantaneous distribution of charges is not isotropic, so the induced dipole moment μ is not aligned with the field **E**, and its amplitude is dependent on the direction:

$$\mu_i = \Sigma \alpha_{ij} E_j \qquad i, j = x, y, z \tag{3.3}$$

The polarizability α appears as a second-order tensor, i.e., a 3×3 matrix, where the components are the α_{ij} coefficients.

If the molecule undergoes deformations (vibrations, rotations) without α modification, μ oscillates with exactly the same frequency as **E**. The emitted radiation, due to dipole modification, has exactly the same frequency v_0 as the electric field. This is the Rayleigh scattering process, and the intensity is given by

$$I = (16\pi^4/3C^3)v_0^4\mu^2 = (16\pi^4/3C^3)v_0^4\alpha^2E^2 \tag{3.4}$$

according to the theory of electromagnetic fields. However, if the molecule undergoes modification of its geometry due to vibration or rotation, its polarizability varies periodically:

$$\alpha(t) = \alpha_0 + \alpha \cos{(2\pi v_i t + \phi)} \tag{3.5}$$

where v_i is the frequency of the periodic modification (vibration or rotation) and ϕ is an arbitrary phase angle associated with molecule's movement.

According to equation (3.5) the induced dipole is now modulated:

$$\mu = \alpha(t)E = [\alpha_0 + \Delta\alpha \cdot \cos{(2\pi v_i t + \phi)}]E_0 \cos 2\pi v_0 t \tag{3.6}$$

This can be arranged very easily as

$$\mu = \alpha_0 E_0 \cos 2\pi v_0 t + (\Delta\alpha/2)E_0 \cos{[2\pi(v_0 - v_i)t - \phi]}$$
$$+ (\alpha/2)E_0 \cos{[2\pi(v_0 - v_i)t + \phi]} \tag{3.7}$$

The excited molecule can then emit three different photons, one of frequency v_0 (Rayleigh diffusion), the second of frequency $v_0 - v_i$ (Raman Stokes diffusion), and the last of frequency $v_0 + v_i$ (anti-Stokes diffusion). The $v_0 - v_i$ and $v_0 + v_i$ radiations are incoherent due to the presence of the phase factor ϕ.

3.2.3. Selection Rules and Intensity

According to the classical theory of the Raman effect, a vibration can be observed in a Raman spectrum only when there is a variation in the polarizability tensor, that is, only when at least one of the nine components $\Delta\alpha_{ij}$ is different from zero. The $\Delta\alpha_{ij}$ components correspond to operations of a group where the binary products x, y, z, are related to symmetry operations. It follows then that the fundamental vibrations of a molecule are Raman-active when they belong to the same symmetry operation as the binary products x, y, z.

The intensity of the Raman lines depends on the radiation frequency; owing to the scattering laws the diffusion process is proportional to the fourth power of the frequency, v^4, to the square of the polarizability variation, $\Delta\alpha^2$, and to the relative populations of both the E_i and E_0 energy levels. It is then possible to calculate the Stokes/anti-Stokes ratio but this is time-consuming and calculations are relatively cumbersome. The final result is then

$$I_{AS}/I_S = [(v_0 + v_i)/(v_0 - v_i)] \exp [(E\psi - E\psi_i)/kT] \qquad (3.8)$$

where $E\psi_0 = h(v_0 - v_i)$ and $E\psi_i = h(v_0 + v_i)$. When $v_0 \gg v_i$ this can be simplified to

$$I_{AS}/I_S \approx \exp (2hv_i/kT)$$

It then follows that the anti-Stokes lines are much weaker than the corresponding Stokes lines, and for higher v_i values they cannot be observed. It should be remembered that the intensity of the Raman lines depends on $(v_0 - v_i)^4$, i.e., v_0^4. Thus it is worthwhile choosing a high-frequency exciting radiation, but this is not always possible when only a limited number of exciting lines are available (laser). The intensity also depends on $\Delta\alpha^2$, i.e., on the properties of the molecule, and especially on the strength of the bonding. It can be shown that covalent bonding vibrations give intense Raman diffusion, whereas ionic bonding vibrations give very faint lines. This can restrict the area of study, especially in catalysis.

3.3. EXPERIMENTAL CONSIDERATIONS[5-7]

A Raman spectrometer is built with a source, a sample holder, a double monochromator, and a detector (Figure 3.1).

3.3.1. Source

Owing to the low intensity of the diffused radiation, the exciting radiation must be very intense. The development of laser sources for Raman spectroscopy has been the most important advance in the technique; the use of a mercury lamp, as in Raman's original method, is now obsolete. Lasers (gas lasers) give very intense and highly monochromatic lines. They also offer several advantages, such as stability of the output power, nondiverging radiation , which can be focused on a surface area as small as $1\mu m^2$, and the possibility of tuning frequencies with a dye laser.

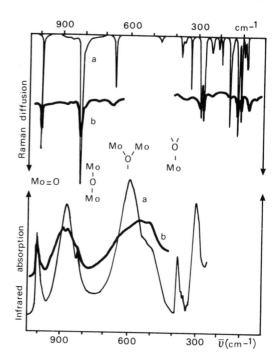

Figure 3.2. IR and Raman spectra of MoO₃ (a) and reduced MoO₃ (b).

Gilson[11] concerning MoO_3 monocrystals. Such an analysis is possible only when the structure of the compound has been characterized completely by X-ray diffraction. It becomes very difficult if there is a decrease in the symmetry of the crystals, or if the number of atoms in the unit cell increases.

In the present case, a comparison of the spectra reported in Figure 3.2 with those of Beattie and Gilson[11] shows clearly that MoO_3 is a pure, polycrystalline oxide. Hence, IR absorption bands and Raman diffusion lines can be assigned unambiguously to single Mo—O and double Mo=O stretching vibrations (corresponding to Mo—O bonding lengths of 1.73 and 1.95 Å, and to Mo=O of 1.67 Å). Reduction by propene (400°C – 20 Torr) gives a semiconducting solid that is almost black. All the Mo—O stretching vibrations are modified, whereas the Mo=O vibration is not. For bismuth molybdates, a similar simplified study has shown that IR and Raman lines of α and β structures can be assigned to vibrations of MoO_4 tetrahedra of the solid. Some hypotheses have been proposed concerning the molybdenum environment in α and β molybdate structures.[12]

3.4.1.2. Supported Oxide

When an oxide is strongly covalent, the Raman diffusion is important and a study can be done easily, e.g., with MoO_3, V_2O_5, or WO_3. Ionic oxides on the other hand, such as SiO_2, Al_2O_3, and zeolites, give a low diffusion and it is not

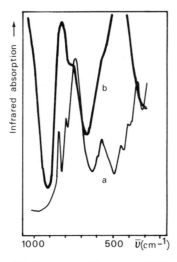 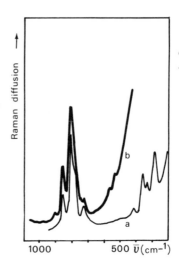

Figure 3.3. Raman and IR spectra of Bi_2MoO_6 (a), and of silica-supported Bi_2MoO_6 (5%) (b).

always possible to obtain Raman spectra. This is due to the fact that Raman lines are dependent upon the bond strength. This support absorbs strongly at some IR frequencies.

This is the case with hydrodesulfurization catalysts, which consist mainly of CoMoO deposited on alumina. The CoMoO-containing phase cannot be studied by IR transmission spectroscopy when the catalyst is in the form of an oxide because alumina absorbs strongly from 900 to 600 cm^{-1}. The situation is worse when the catalyst is sulfided[13] and in that case no transmission spectroscopy can be done. On the contrary, Raman spectroscopy is able to detect the catalyst vibrations because the support does not give any lines, or very few. In this field much work has been done by E. Payen, J. Grimblot et al.[14–19] with CoMo, NiMo, and WMo deposited on alumina. They have observed the modification of the oxidic precursors during the deposition, calcination, reduction, sulfidization steps, and in some cases they have done in situ studies. Raman spectroscopy can detect the metal–oxygen vibrations; the observed wave number shifts are attributed to different oxidation states of molybdenum ions, coordination states, and ligand effects.[14–19]

IR and Raman spectroscopies have been used to characterize a catalyst consisting of 5wt.% Bi_2MoO_6 deposited onto silica.[30] The spectra of the supported phase have been compared to those of pure phase having the same composition and structure (Figure 3.3). The Raman spectrum shows a moderate background due to silica fluorescence, the lines due to the γ-Bi_2MoO_6 phase, and some low-intensity lines corresponding to the α-$Bi_2Mo_3O_{12}$ phase. The IR spectrum shows three strong absorption bands due to SiO_4 tetrahedra vibrations. The silica absorption completely hides the molybdate bands; no details can be observed assignable to Bi_2MoO_6. Thus Raman spectroscopy is the only technique able to characterize the supported molybdate phase and prove that: (1) it is the γ form with traces of α form; and (2) there is no interaction between the silica and the molybdenum phase.

3.4.2. Superficial Properties of a Catalyst

Most studies relating to the acidic properties use pyridine adsorption on catalysts such as SiO_2, Al_2O_3, and zeolites. The strongest Raman lines of pyridine are located at 991 and 1030 cm^{-1}. They correspond to ν_1 symmetrical breathing and ν_2 symmetrical deformation of the cycle. The ν_1 frequency is strongly perturbed by adsorption.[8,20] In particular it has been shown that in the case of X zeolites exchanged by different cations, the ν_1 mode is very sensitive to the nature of the cation, i.e., to the local field around the cation.[21] Two

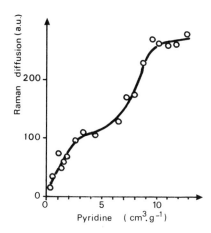

Figure 3.4. Evolution of the intensity of the ν_1 pyridine Raman line as a function of the amount of adsorbed pyridine onto a sodium zeolite.

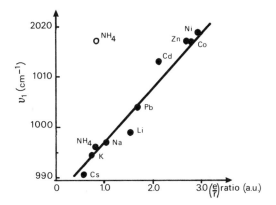

Figure 3.5. Variation of the frequency ν_1 of the adsorbed pyridine Raman line *vs.* the charge/radius ratio of the exchanged cation.

important facts are deduced from these studies:

1. The adsorption can be followed quantitatively. The intensity of the v_1 Raman line can be correlated with the amount of pyridine adsorbed onto the NaX zeolite. Then the adsorption isotherm can be recorded (Figure 3.4).
2. Some energetic aspects of the adsorption can be deduced. The pyridine molecule is coordinated with the cation via its electron pair located on the nitrogen atom. During adsorption, there is an increase in the v_1 frequency. Consistent with the hypothesis that the electrostatic potential is strictly proportional to the charge/cation radius ratio, a very good correlation has been obtained between the electrostatic potential and the v_1 frequency shift (Figure 3.5).

3.4.3. Characterization of Metallic Catalysts

3.4.3.1. Adsorption of Probe Molecules

This technique has been widely used successfully with IR adsorption spectroscopy (e.g., CO and NO). Unfortunately it has not been utilized in Raman spectroscopy owing to the very low intensity of diffusion and the fact that lines are hidden by noise. However, when a spectrum can be recorded, all the vibrations corresponding to a carbonyl species (in the case of CO adsorption) are observed, and they can be used to calculate the force constant of the system (v_{C-O} stretching, v_{Ni-C} stretching, δ_{Ni-CO} bending). Krasser *et al.*[22] studied the Raman spectrum of CO adsorbed onto silica-supported nickel. In the 2100–150 cm^{-1} range the observed Raman lines have been assigned to three species: two linearly adsorbed CO (two different sites) and a bridge adsorbed

TABLE 3.1. Raman Spectra of Chemisorbed CO and Force Constants of Linear and Bridge-Bonded Species with C_{4v} Symmetry

Raman lines (cm^{-1})	Irreducible representation	Force constants (mdyne Å^{-1})
2086 v_{CO} linear	A_1	$f_{CO_l} = 17.41$
2070 v_{CO} linear	A_1	
1804 v_{CO} bridged	A_1	$f_{CO_b} = 12.93$
640 v_{CO} bridged	E	
535 δ_{NiCO}-bridged		
δ_{NiCNi}-bridged	A_1	$f_{OCNi_l} = 0.77$
δ_{OCNi}-linear		$f_{NiC_b} = 2.91$
362 m_{NiC} bridged	A_1	$f_{NiC_l} = 1.43$
355 m_{NiC}-linear	A_1	$f_{OCNi_b} = 1.26$
345 m_{NiC}-bridged	B_1	$f_{NiCNi} \approx 0.46$
307 δ_{NiCO}-bridged	E	
226 δ_{NiCNi}-bridged	B_2	
173 δ_{OCNi}-bridged	B_1	
164 δ_{NiCNi}-bridged	E	

CO giving a species having C_{4v} symmetry. A complete force constant calculation has been performed with the help of the G.F. matrix (Table 3.1).

3.4.3.2. Metal–Metal Vibration

When an organometallic cluster is deposited onto a support, it is important to know whether the complete structure of the cluster is preserved during adsorption. The presence in the Raman spectrum of lines assigned to metal–metal vibrations can help to answer this question. In the case of the $Os_3(CO)_{12}$ cluster deposited onto α-Al_2O_3,[23] the spectrum shows an intense line at 190 cm^{-1} and another broad one having two maxima at 80 and 119 cm^{-1} (Figure 3.6). By comparison with the Raman spectrum of $Os_3(CO)_{12}$, whose symmetry group is D_{3h}, the following facts can be deduced (Table 3.2):

1. When the cluster is deposited onto alumina, the original structure of $Os_3(CO)_{12}$ is preserved, despite a lowering of symmetry (shown by a double E' mode vibration).
2. When the catalyst is heated under CO at 400°C the cluster undergoes transformation giving an atomically dispersed monometallic compound, and the Raman lines at 160 cm^{-1} and 119–80 cm^{-1} disappear.

3.4.4. Adsorbed Molecules and Catalysis

3.4.4.1. Benzene Adsorption on Ni/SiO$_2$ Catalyst[24]

Benzene is a highly symmetrical molecule, so it is useful for theoretical studies in different spectroscopies such as IR absorption, EELS, and Raman. The free molecule belongs to the D_{6h} symmetry group and has 30 vibration modes, which can be classified as (a) IR-active: A_{2u} and E_{1u} modes, and (b) Raman-active: A_{1g}, E_{1g}, and E_{2g} modes.

It was shown above that the number of IR- and Raman-active vibration modes depends on the symmetry of the molecule (Chapter 2). Krasser et al.[24] have shown that C_6H_6 or C_6D_6 adsorbed on nickel supported on silica give vibrations corresponding to B_{2g}, A_{2u}, E_{1u}, B_{2u}, and E_{2u}, which become Raman-active. This symmetry change could be a result of the geometry of the adsorption site. This lowering of symmetry to C_{3v} for adsorbed benzene has also been shown by IR adsorption spectroscopy and by EELS.

TABLE 3.2. ν_{Os-Os} Vibrations of Triosmium Carbonyl Cluster Deposited onto Al_2O_3

$Os_3(CO)_{12}$ (D_3h)	$Os_3(CO)_{12}/Al_2O_3$ (Cs)
A_1 160 cm^{-1}	A′ 160 cm^{-1}
E 119 cm^{-1}	A′ 119 cm^{-1}
A'' 80 cm^{-1}	

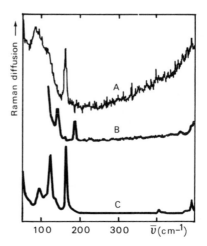

Figure 3.6. Raman spectrum of the triosmium cluster supported on alumina (A) and of $H_2Os_3(CO)_{10}$ (B) and $Os_3(CO)_{12}$ (C) clusters.

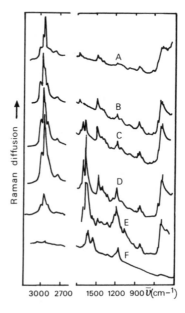

Figure 3.7. Raman spectra of acetaldhyde adsorbed on NaOH-modified silica gel. (A) adsorbed for 4 h at 20°C; (B) 49 h at 20°C; (C) 16 h at 20°C followed by 1 h at 50°C; (D) after an additional 26 h at 50°C; (E) after 91 h at 50°C; (F) after an additional 5 h at 133°C.

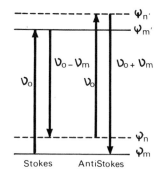

Figure 3.8. Diagram of the Raman resonance effect.

3.4.4.2. Condensation of Acetaldehyde with Modified Silica[25]

The condensation reaction of CH_3CHO on a NaOH-modified silica was studied *in situ* (Figure 3.7). When acetaldehyde is adsorbed at room temperature (A) only the vibrations due to CH_3CHO, i.e., ν_{CH_3} at 2921 cm^{-1}, ν_{CO} at 1719 cm^{-1}, and δ_{CH_3} at 1452 cm^{-1} are observed. The $\nu_{C=O}$ line is relatively weak and can be related to activated chemisorption. When the contact time increases new lines appear at 816, 535, and 489 cm^{-1} (B), which are due to product formation related to dioxane or trioxane. When the temperature increases to 50°C, extra lines appear at 1680 and 1640 cm^{-1}, which can be assigned to crotonaldehyde formed from B species by dehydration. When the reaction runs longer at 50°C (D, E) or at 130°C (F) new lines appear at 1536 and 1146 cm^{-1}, which can be attributed to polyene species having conjugated carbon double bonds.

3.5. NEW DEVELOPMENTS

3.5.1. Raman Resonance Spectroscopy

When the exciting line ν_0 has a frequency close to an electronic absorption frequency ν_{abs}, the intensity of Raman diffusion is strongly enhanced. In this case the population of the upper electronic level is noticeably increased by absorption of the ν_0 frequency. According to the law governing the intensity, the probability of transition increases sharply $P = k\,(\nu_0 - \nu_{abs})^4$, and then Raman emission also increases (Figure 3.8). However, at the same time fluorescence also greatly increases and the Raman spectrum can be completely hidden by this fluorescence. It is then necessary to use a pulse mode laser with synchronous detection; only the Raman diffusion is detected during the pulse (10^{-14} s), and when the delayed fluorescence appears (10^{-9}–10^{-6} s) the detector is switched off electronically.

An example, $WOSe^{2-}$, is shown in Figure 3.9. The Raman emission due to the W—Se vibration is greatly enhanced when the energy of the laser line $h\nu_0$ comes close to the energy of the $\pi(SE) \rightarrow d(W)$ charge transfer transition.

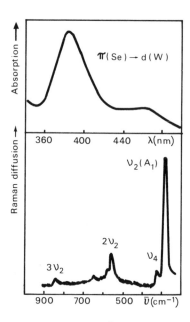

Figure 3.9. Raman spectra of $WOSe^{2-}$, showing the resonance effect.

In some cases, typically of molecules like pyridine adsorbed on a metallic surface, the intensity of diffuse Raman radiation is increased by 10^3 to 10^6. This phenomenon called SERS (strongly enhanced Raman spectroscopy) was first reported for pyridine adsorbed on silver electrodes and has since been observed for many molecules such as CO, pyridine, and benzene adsorbed on metallic surfaces (e.g., Ag, Cu, and Ni). In fact, in the previous example of C_6H_6 adsorbed on nickel supported on silica (see Section 3.4.4.1), the intensity of the v_1 Raman line is nearly 10^3 times higher than that corresponding to liquid benzene.[24] Several theories have been proposed to explain this phenomenon. Some of these take resonance into account[26-29] but, as yet there is no satisfactory explanation for enhanced Raman diffusion.

3.5.2. Coherent Anti-Stokes Raman Spectroscopy (CARS)

The anti-Stokes Raman lines are normally much weaker than the Stokes lines, but in some cases the opposite can be true. Consider two lasers emitting two frequencies v_0 and $v_0 - v_m$. These are used in order to have an absorption and a stimulated emission. Consequently molecules are pumped to the $v = 1$ vibrational energy level (the population between $v = 0$ and $v = 1$ is now inverted). A third laser where the frequency is v_0 excites molecules from the $v = 1$ level and induces an emission of frequency $v_0 + v_m$ because the $v = 0$ level is now sparsely populated, and the anti-Stokes lines appear with a much greater intensity (Figure 3.10).

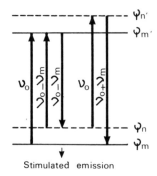

Figure 3.10. Diagram of coherent anti-Stokes Raman spectroscopy.

3.5.3. Fourier Transform Raman Spectroscopy

Laser sources are not strictly monochromatic; all emit some lines in the visible, near-IR, or IR ranges. Thus the neodynium-doped YAG laser gives a strong line at $1.064\,\mu m$ or $9395\ cm^{-1}$, which can be used for excitation. This has been done for some years following the ideas and pioneering work of Hirschfeld and Chase.[31-33]

As the exciting line is in the IR region, Fourier transformation is possible with the advantages of high gain, precision, fast recording, and accumulation. These are the reasons why all FTIR manufacturers now offer Raman accessories in order to obtain both FTIR and FT Raman spectra, within a few minutes, on the same sample.

3.5.4. The Future of Raman Spectroscopy in Catalysis

Catalysis applied laser Raman spectroscopy (CALRS) is still in its infancy due to difficulties arising from the use of lasers, of which the main problem is the fluorescence. When commercially available ultrafast detectors and noise filtering appear, only the instantaneous Raman diffusion will be recorded and delayed fluorescence will not interfere.

Secondly, with ultrafast laser shots at a relatively low rate (e.g. 10 shots per second) excessive heating will be avoided, and Raman spectroscopy could then be used without the danger of desorbing species or of solid state transformations. Moreover the short duration could be valuable for detecting short-lifetime excited species.

The examples cited above show that, despite its newness, Raman spectroscopy can be used in various fields in connection with catalysis, such as structure characterization, superficial site determination, acidic and basic properties, and intermediate species during a reaction.

REFERENCES

1. R. S. Tobias, *J. Chem. Educ.* **44**, 1 (1967).
2. J. A. Koningstein, *Introduction to the Theory of the Raman Effect*, Reidel, Dordrecht (1972), Ch. 1.
3. L. A. Woodward, *Raman Spectroscopy*, Vol. I (M. A. Szymanski, ed.), Plenum, New York (1967), Ch. 1.
4. G. Herzberg, *Molecular Spectra and Molecular Structure, II. Infrared and Raman Spectra of Polyatomic Molecules*, Van Nostrand, New York (1950).
5. M. C. Tobin, *Laser Raman Spectroscopy*, John Wiley and Sons, New York (1971).
6. C. E. Hathaway, *The Raman Effect*, Vol. I (A. Anderson, ed.), Marcel Dekker, New York (1971), Ch. 4.
7. P. J. Hendra, *Spex Speaker* **19**, 1 (1974).
8. T. A. Egerton and A. H. Hardin, *Cat. Rev. Sci. Eng.* **11**, 71 (1975).
9. R. P. Cooney, G. Curthoys, and N. T. Tam, *Adv. Cat.* **24**, 293 (1975).
10. W. H. Delgass, G. L. Haller, R. Kellerman, and J. H. Lunsford, *Spectroscopy in Heterogeneous Catalysis*, Academic, New York (1979), Ch. 3.
11. I. R. Beattie and T. R. Gilson, *Proc. Roy. Soc.* **A. 307**, 407 (1968).
12. I. Matsuura, R. Schuit, and K. Kirakawa, *J. Catal.* **63**, 152 (1980).
13. J. Medema, C. Van Stam, V. H. J. De Beer, A. J. A. Konings, and D. C. Koningsberger, *J. Catal.* **53**, 386 (1978).
14. E. Payen, J. Grimblot, and S. Kasztelan, *J. Phys. Chem.* **91**, 6642 (1987).
15. E. Payen, M. C. Dhamelincourt, P. Dhamelincourt, J. Grimblot, and J. P. Bonnelle, *Appl. Spectrosc.* **36**, 30 (1982).
16. E. Payen, S. Kasztelan, J. Grimblot, and J. P. Bonnelle, *J. Mol. Struct.* **143**, 259 (1986).
17. E. Payen, S. Kasztelan, J. Grimblot, and J. P. Bonnelle, *J. Mol. Struct.* **174**, 71 (1988).
18. E. Payen, S. Kasztelan, J. Grimblot, and J. P. Bonnelle, *Catalysis Today* **4**, 57 (1988).
19. S. Kasztelan, E. Payen, H. Toulhoat, J. Grimblot, and J. P. Bonnelle, *Polyhedron* **5**, 157 (1986).
20. T. A. Egerton, A. H. Hardin, and N. Sheppard, *Can. J. Chem.* **54**, 586 (1976).
21. A. H. Hardin, M. Klemes, and B. A. Morrow, *J. Catal.* **62**, 316 (1980).
22. W. Krasser, A. Fadini, and A. Renouprez, *J. Catal.* **62**, 94 (1980).
23. M. Deeba, B. J. Strensand, G. L. Schrader, and B. C. Gates, *J. Catal.* **69**, 218 (1981).
24. W. Krasser, H. Ervens, A. Fadini, A. J. Renouprez, *J. Raman Spectrosc.* **9**, 80 (1980).
25. A. W. Klaasen and C. G. Hill, Jr., *J. Catal.* **69**, 299 (1981).
26. D. P. Dilella, A. Gonin, R. H. Lipson, P. Mc Breen, and M. Moskovits, *J. Chem. Phys.* **73**, 4282 (1980).
27. M. Ueba and S. Ichimura, *J. Chem. Phys.* **74**, 3070 (1981).
28. T. E. Furtak and J. Reyes, *Surf. Sci.* **93**, 351 (1980).
29. S. Efrima and J. Metiu, *J. Chem. Phys.* **70**, 1602 (1979).
30. G. Coudurier, unpublished results.
31. T. Hirschfeld, *Appl. Spectrosc.* **30**, 68 (1976).
32. T. Hirschfeld and B. Chase, *Appl. Spectrosc.* **40**, 133 (1986).
33. B. Chase, *J. Am. Chem. Soc.* **108**, 7485 (1986).

ELECTRONIC SPECTROSCOPY

E. Garbowski and H. Praliaud

4.1. INTRODUCTION

The absorption bands occurring in the visible and near-UV (and eventually near-IR) regions are used to obtain information on the electronic structure of homogeneous or heterogeneous catalyst supports. The observed bands are related to transitions between the electronic levels of the atoms, ions, complexes, or molecules (organic or inorganic), and various theories have been developed to help the interpretations.[1–18] Furthermore, the spectra of many compounds are known.[4–6,10–13,18–20] Thus electronic spectroscopy can give the valence state and the stereochemistry of transition metal ions used as catalysts as well as their modification during adsorption or reactions.[21] It also gives information on the interactions between adsorbed molecules and solid catalysts or on those among the various components of a liquid mixture.[22–25] The formation of reaction intermediates and reaction products, as well as kinetics of formation can also be determined. For homogeneous solutions, the transmitted light is analyzed. In the case of solid catalysts or turbid solutions, this technique is replaced by diffuse reflectance spectroscopy.[26–31] This nonconventional spectroscopy is widely used in the study of powders (e.g., catalysts, oxides, pigments) or of surfaces (e.g., pigments, papers, glasses, polymers, ceramics).

4.2. PRINCIPLES AND APPARATUS

4.2.1. Principles

Electronic spectroscopy is concerned with the valence electronic transitions between molecular orbitals. The values of the energy of the electronic levels of a

E. Garbowski and H. Praliaud ● Institut de Recherches sur la Catalyse, CNRS Villeurbanne, France.

Catalyst Characterization: Physical Techniques for Solid Materials, edited by Boris Imelik and Jacques C. Vedrine, Plenum Press, New York (1994).

given chemical species A are quantified as solutions of the Schrödinger equations: $H|\psi\rangle = E|\psi\rangle$(4, 5, 14), and make an ordered series $E_0, E_1, E_2, E_3 \ldots E_n$. The species A absorbs (or emits) energy to go from the E_n state to the E_{n+1} state (or E_{n-1}). The emission (or absorption) gives an electromagnetic energy $h\nu = \Delta E = E_{n-1} - E_n$. A photon of frequency ν (in the case of the absorption) will induce the change of the energy state of the chemical species. The total energy of a species includes several factors: E(total) $= E$(translation) $+ E$(rotation) $+ E$(vibration) $+ E$(electronic). The electronic levels are located in the UV–visible range of the electromagnetic spectrum (50,000–3000 cm^{-1}), so electronic spectroscopy is also named UV-visible spectroscopy. It is linked to the repartition of the valency electrons, i.e., to the electron jumps between molecular orbitals. It is not affected by the electrons of the inner shells, which are strongly bonded and which require very large energies (far-UV or X-rays) to modify their configurations. However, the energies associated with the electronic jumps are large enough to provoke vibrations of the molecule [E(electronic) $> E$(vibration)] so the transitions are broadened.

In the literature, most bands are characterized by their wavelength λ, which can be measured in units of mμ [or nanometer (nm)], where 1 mμ $= 10^{-7}$ cm or 10 Å (1 angström unit $= 10^{-8}$ cm). The wave number (reciprocal of the wavelength) $\nu = 1/\lambda$, which is measured in cm^{-1}, is also used.

4.2.2. Apparatus

The absorption spectra are obtained from the analysis of the light either transmitted (solutions or transparent crystals) or reflected (turbid solutions, suspensions, solids) by the absorbent substance, which is positioned between the light source and the detector. The UV-visible spectrophotometers consist of the following components: (1) one or two polychromatic sources, a monochromator to separate the various wavelengths (prism or grating or both kinds), (2) a sample container, (3) a light path or an integrating component such as a sphere or hemisphere, and (4) a detector (photomultiplier tube).

As the light emitted by the source and the response of the detector are functions of the wavelength, the transmitted (or reflected) light intensity I is compared to the incident light intensity I_0 for each wavelength; a chopped monochromatic light enters both the sample and a reference (double-beam spectrophotometer).

4.2.2.1. Transmision

For liquids, gases, or transparent solids, the transmitted light of intensity I is related to the incident light intensity I_0 by the transmittance $T = I/I_0$ ($0 < T < 1$) or by the percentage transmittance $100T$. The optical density (O.D.) or absorbance A [$A = \log 10(I/I_0)$] is also used frequently. The concentration of the absorbent [Abs] is related to the absorbance by the relation $A = \varepsilon$(Abs), where ε is the molar extinction coefficient. For a solution of the path length l (measured in cm) containing a solute of molar concentration c, we have $A = \varepsilon\, cl$, where ε is measured in units of $l\,cm^{-1}\,mol^{-1}$.

4.2.2.2. Reflection and Diffuse Reflectance Spectra

Most heterogeneous catalysts are opaque powdered samples from which light is reflected and not transmitted. The radiation reflected from a powdered crystalline surface consists of two components: the specular component that is reflected from the surface without transmission (mirror reflection) and the diffuse component that is absorbed into the material (penetration of a portion of the incident flux into the interior of the sample) and reappears at the surface after multiple scattering.[29] Commercial spectrophotometers are designed to minimize the specular component and the term "reflectance" is used for diffusely reflected radiation. As the sample absorbs a part of the radiation the diffused intensity I is weaker than the indirect intensity I_0. Furthermore, as the detector receives only a part of the diffuse radiation I, the measurement of I is difficult and an integrating sphere is used. This sphere (or hemisphere) coated on the inside with a highly reflecting layer (e.g. MgO or $BaSO_4$) reflects the diffuse light from the sample and increases the percentage of I reaching the detector (30 to 50%). Relative measurements of reflectance are made against reference materials (nonabsorbing materials as MgO), with the incident light entering the sample or the reference surfaces, alternately.

The reflectance or reflecting power is given by $R = I/I_0$ ($0 < R < 1$). The percentage reflecting power $100R$ is also used. For a nontransparent thick layer, R is called R_∞ and the relative quantity $R'_\infty = R_\infty(sample)/R_\infty(standard)$ is defined as relative measurements are made. The quantity $\log(1/R'_\infty)$ is called the apparent absorbance. Unlike the case of transmitted light (through solutions for instance), this apparent absorbance is not strictly proportional to the concentration of the absorbing species. However it has been possible to relate the reflectance to the absorption of the sample by taking account of various parameters, e.g., particle size, grain size, absorption coefficient, mean depth of penetration of light (very often 1000 layers), wavelength, and possibility of regular reflection.

The measurable reflectance R'_∞ can be used to calculate the Kubelka–Munk function defined as $F(R'_\infty)$:

$$F(R'_\infty) = (1 - R'_\infty)^2/2R'_\infty \qquad (4.1)$$

This quantity, for a layer of infinite thickness, has been related to the parameters k and S, $F(R'_\infty) = k/S$, parameters linked respectively to the absorption and the reflection-diffusion (S is the scattering coefficient). However, the relationship between k and S and the properties of absorption or diffusion of the particles is not simple. The conditions of application of the Kubelka–Munk function have been discussed previously.[31] Sometimes $\log F(R'_\infty)$ is plotted against the wavelength or the wave number and it is possible to find some concentration dependence of the Kubelka–Munk function.

TABLE 4.1. Some Electronic Transitions: Characteristics and Applications

Nature of the transitions	Organic chemistry	Inorganic chemistry
d–d transitions		Transitions metal complexes
Transitions between molecular orbitals mainly localized on the central atom		d orbitals partially empty ionic or covalent solids UV–visible, near IR ranges wide, broad, weak bands $\varepsilon^{(a)}$ range $(10^{-2}\ldots 100)$ (usually $1\ldots 100$) Example: $\mathrm{Ni}\,(H_2O)_6^{2+}$; $\mathrm{Cr}\,(NH_3)_6^{3+}$
Charge transfers	Between acceptor and donors	In the coordination sphere of the metal $M \to L$ or
Transitions between molecular orbitals mainly localized on different atoms	Molecular complexes UV–visible range intense bands ε range $10^3 \ldots 10^6$	$L \to M$ Ionic or covalent solids UV–visible range ε range $10^3 \ldots 10^4$
Electronic transitions across the forbidden energy gap		Covalent solids UV range, continuous band Example TiO_2, ZnO
$e^-(\pi)$ and $e^-(n)$ systems	$^*\pi \to \pi^*$ UV range, ε range $10^4 \ldots 10^6$ ex. aromatics C=C (190 nm) C=C—C=C (214 nm) $^*n \to \pi^*$ UV–visible range ε range 10 ex. C=O (280 nm)	

[a] ε is the molar extinction coefficient in $l\ mol^{-1}\ cm^{-1}$.

4.2.3. Nature of Electronic Transitions: Energy and Intensity

4.2.3.1. Nature of the Transitions

The electronic transitions found in organic or inorganic chemistry are classified into several groups, according to the nature of the electronic jumps. Table 4.1 gives the main possibilities and their usual ranges of energy and intensity.

4.2.3.1a. The d–d Transitions of Transition Metal Complexes. [1,4–6,8–10,12,14–17]
The transition metal ions are characterized by a configuration $3d^n$ or $4d^n$ or $5d^n$. The free metal ion has five degenerate d orbitals (with the same energy). In a

complex the degeneracy is removed and the five d orbitals can have different energies according to the symmetry of the complex and the nature of the ligands. The degeneracy is removed by the field created by the ligands. The incident radiation carries away transitions of the electrons from one d orbital to another d orbital when the d level is not totally occupied ($0 < n < 10$). The energy of the transition is a function of the symmetry of the complex, of the valency of the central metal, of the nature of the ligand (strength of the field created by the ligands, electron accepting or electron donating properties). These transitions are found in the UV range and more often in the visible part of the spectrum, and are responsible for the colors of many transition metal complexes.

4.2.3.1b. Charge Transfers. These involve electron transfers from an occupied orbital localized on a donor to an unoccupied orbital localized on an acceptor, and are related to the acceptor and donor properties of the ligands. Like molecular orbitals, the charge transfers are functions of the geometry of the complex. When the electron transfers involve a change in the oxidation state the terms "spectra of electron transfers" or "oxidoreduction bands" may be used.

In inorganic chemistry,[13] the charge transfer processes occurring in the coordination sphere of the metal are of two types: transfer of an electron from an orbital mainly localized on the metal to an orbital mainly localized on the ligand (abbreviated $M \rightarrow L$) or transfer in the opposite direction ($L \rightarrow M$). The absorption bands which result from such transitions usually (but not always) occur at higher energies than the d–d bands, so they are found mainly in the UV regions of the spectrum. In organic chemistry[2,7,11,18] these transitions arise between electron donors and electron acceptors, and give absorption bands in the UV and visible ranges of the spectrum.

4.2.3.1c. The π–π^* and n–π^* transitions. These involve jumps of π electrons or n electrons between the molecular orbitals of organic molecules (Figure 4.1). It is possible to observe the following transitions: (1) π–π^*, the transition of a π electron from a bonding π orbital to an antibonding π^* orbital; and (2) n–π^*, the transition of an n electron from a nonbonding n orbital to an antibonding π^* orbital.

These transitions occur in the UV range. Other transitions, e.g., $(n \rightarrow \sigma^*)(\sigma \rightarrow \sigma^*)$, are found only in the far UV. For carbocations or unsaturated radicals the particular distribution of π electrons gives strong absorptions in the near-UV and the visible ranges.

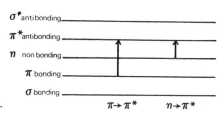

Figure 4.1. Electronic energy levels of a molecule.

4.2.3.2. Selection Rules: Intensity of the Transitions

The value of the extinction coefficient is related to the oscillator strength of the electronic transition and is a measure of the probability of transition between two states defined by the wave functions ψ_1 and ψ_2. This probability has to be greater than zero, which involves selection rules: (a) transitions between wave functions of different spin are forbidden (spin selection rule); (b) transitions between states of the same parity are forbidden (Laporte's rule), for instance $d-d$ transitions are forbidden.

In comparison with IR spectroscopy the symmetry conditions are more severe and the electronic spin has to be invariant. Fully allowed electronic transitions such as the $\pi-\pi^*$ transitions of aromatic molecules have molar extinction coefficients of some 10^4-10^6 l cm^{-1} mol^{-1}. Many charge transfer transitions are also fully allowed and the resulting absorption bands are very intense. At the other extreme spin and Laporte forbidden transitions, such as the crystal field spectra of octahedral manganese (II) complexes, may have intensities as low as $10^{-2}-10^{-4}$ l cm^{-1} mol^{-1}. Although Laporte's rule forbids $d-d$ transitions, it has been shown that such transitions do occur as a consequence of vibration coupling or of the lack of an inversion center in the molecule or of a hybridization $d-p$... The molar extinction coefficients thus range from 1 to 1000 l cm^{-1} mol^{-1}. The intensity depends on the geometry, and the molar extinction coefficient of $d-d$ transitions varies from 1 to 150 l cm^{-1} mol^{-1} for an octahedral complex, from 5 to 250 l cm^{-1} mol^{-1} for a square planar complex, and from 50 to 500 l cm^{-1} mol^{-1} for a tetrahedral complex.

4.3. APPLICATION TO CATALYSIS

Electronic spectroscopy has been widely used for studying catalysts themselves, for example, the valency and coordination state of the transition metal ions (in solutions as well as in solids) and the changes during adsorption or reaction.[21-23,31] Moreover, it can be used to provide information on the nature of interactions between adsorbed molecules and catalyst surfaces, or on the nature of the components present in a liquid mixture.[24,25] Intermediate species (provided their lifetimes are sufficiently long) as well as products of reaction can be detected; some reaction kinetics may also be determined during catalysis.

Reflectance spectroscopy applied to solid state catalysts is somewhat restricted. First, the compounds must be not black at all, and the condition $F(R_\infty) < 1$ has to hold, i.e., the species eventually present must not absorb the light too much. For solids such as ions dispersed in zeolites or on supports, the condition is fulfilled provided the support itself does not absorb too much in the frequency range studied. When the absorbing compound is not diluted, it has to be mixed with a white and inert solid. When quantitative measurements are planned, particle diameters have to be taken into account because they have an

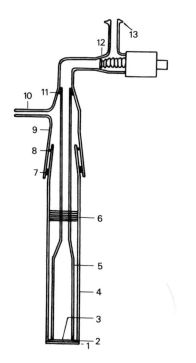

Figure 4.2. Cell for recording reflectance spectra: (1) optical window, (2) sample, (3) quartz frit, (4) body cell, (5) piston, (6) quartz-pyrex connections, (7) Teflon ring, (8) Viton O-ring, (9) conical ground joint, (10) exhaust tube, (11) Viton O-ring, (12) greaseless valve, and (13) greaseless ground joint.

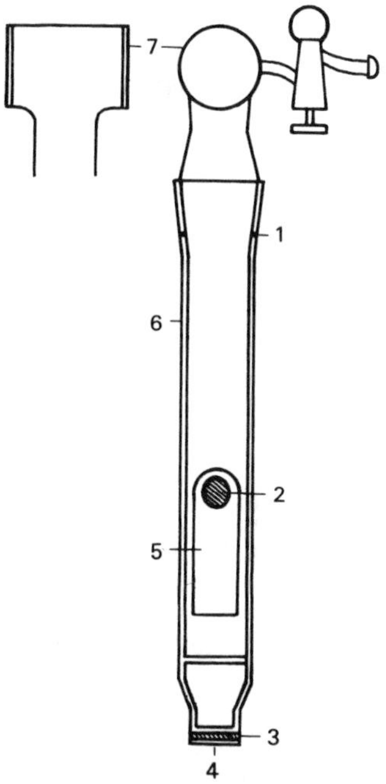

Figure 4.3. Cell for recording reflectance spectra (UV–visible and NIR) and transmission spectra (IR)[32]: (1) Viton O-ring, (2) pellet (for IR measurement), (3) sample (for UV–visible NIR measurement), (4) optical window, (5) IR sample holder, (6) guide, and (7) CaF$_2$ or KBr windows.

effect upon the $F(R)$ value. Generally speaking, it is not necessary to work with pressed powders, but sometimes it is more practical. In this case pressure may change the intensity of the observed bands. For adsorption studies, specific surface area must not be too low (sensitivity). Two techniques may be used: either the adsorbent is in the gaseous phase, or the adsorbent is in solution in a nonpolar solvent. The bands shapes are generally changed and broadened due to vibrational molecular excitation. The spectra obtained have to be deconvoluted into components that have a Gaussian shape before they can be discussed. The interpretation must be completed with other complementary techniques, in particular, IR spectroscopy or ESR, giving information concerning the nature of the ligands or the energy levels of the molecular orbitals.

In practice many cells have been designed and built, and they all have to allow *in situ* treatment and adsorption. The sample is generally activated or

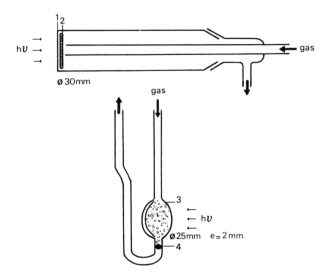

Figure 4.4. Flowing cells acting as catalytic reactors: (1) optical window, (2) pellet, (3) powder, (4) glass frit.

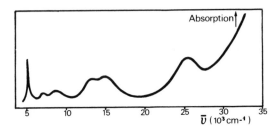

Figure 4.5. Absorption spectrum of a hydrated NiNaY.

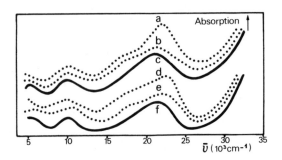

Figure 4.6. Absorption spectra of various zeolites dehydrated at 500°C: (a) NiNaA, (b) NiNaX, (c) NiNaY, (d) NiHA, (e) NiHX, and (f) NiHY.

treated by external heating. Some cells are specially built for recording both reflectance spectra (UV–visible, near IR) and IR spectra (absorption). Others have been designed especially for making *in situ* catalytic reactions and dynamic treatments. Some special cells are used for low-temperature measurements (77K). Several are described in Figures 4.2–4.4.

Some particular examples related to heterogeneous catalysis will now be presented. They are devoted to:

1. The coordination state and the valency state of transition metal ions, and their changes during adsorption or reaction: (a) nickel ions encaged in zeolites — coordination and reducibility; (b) molybdenum ions deposited on various supports — modification during the reaction (propylene oxidation) and relationship with selectivity.

2. The study of molecules adsorbed on solids and their reactivity: (a) adsorption and hydrogenation of benzene on zeolites, (b) the nature of the acidic sites, (c) the strength of the acidic sites and acidity indicators, (d) acidic and ionizing properties — coke formation during ethylene adsorption on HY zeolite.

4.3.1. Nickel Ions in Zeolites: Coordination State and Reduction

4.3.1.1. Ni^{2+} Coordination State: Application of the Crystal Field Theory

Transition metal ions have different coordination states. The coordination is deduced from the $d–d$ electronic transitions whose number, position, and intensity depend on the nature of the ligands and the symmetry of the complex. Ni^{2+} ions are easily introduced into zeolites by exchange. The crystal field theory can explain the values of the observed transitions (Figures 4.5 and 4.6).[33] The model of the crystal field theory (electrostatic model) is developed in Appendix I. The various spectroscopic terms of the free ion ($3d^8$) and of the coordinated ion are also presented (octahedral complex). Provided D or $10Dq$ is the crystal field unit, triplet states have the following energies (singlet states are not observed because the transitions are too weak).

$$
{}^3F \Bigg\langle
\begin{array}{l}
{}^3A_{2g} = -12Dq \\
{}^3T_{2g} = -2Dq \\
{}^3T_{1g} = +8Dq
\end{array}
\qquad {}^3P \rightarrow {}^3T_{1g} = 0
$$

The four triplet states vary in the order (see Figure 4.7)

$$
{}^3A_{2g} < {}^3T_{2g} < {}^3T_{1g}({}^3F) < {}^3T_{1g}({}^3P)
$$

The separation between the two ${}^3F$ and ${}^3P$ spectroscopic terms is a function of

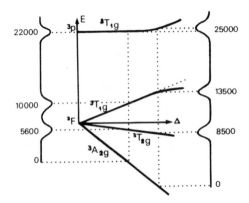

Figure 4.7. Splitting of the ^3P, ^3F levels for an octahedral field.

the Racah parameter and for a free ion

$$E(^3F) - E(^3P) = 15B \qquad (4.2)$$

The observed transitions start from the $^3A_{2g}$ ground state. So three transitions may arise whose energies are:

$$\nu_1 = {}^3A_{2g} \rightarrow {}^3T_{2g} = 10Dq$$
$$\nu_2 = {}^3A_{2g} \rightarrow {}^3T_{1g}(^3F) = 18Dq$$
$$\nu_3 = {}^3A_{1g} \rightarrow {}^3T_{1g}(^3P) = 12Dq + 15B \qquad (4.3)$$

$10Dq$ and B are obtained as soon as ν_1, ν_2, and ν_3 have been determined experimentally. However there is an interaction between the two $^3T_{1g}$ states because they have the same symmetry properties, and they repel each other. So the configuration interaction has to be taken into account (noncrossing rule).

For hydrated zeolites, the spectrum is strictly similar to the spectrum of $Ni(H_2O)_6^{2+}$ in solution (octahedral aquocomplex in the supercage). The experimental values are $\nu_1 = 8500$ cm^{-1}, $\nu_2 = 13,500$ cm^{-1}, and $\nu_3 = 25,300$ cm^{-1}. This leads to $10Dq = 8500$ cm^{-1} and $B = 890$ cm^{-1}. The value B_0 for the free ion is 1040 cm^{-1}. This decrease is related to the repulsion of d electrons and is a measure of the covalency of the ion–ligand bond. There is a partial delocalization of the d electrons on the ligands and so the repulsion of the d electrons is weakened. During dehydration, Ni^{2+} ions migrate toward S_I sites, which are at the centers of hexagonal prisms. The ions are octahedrally coordinated with a Ni—O bond of 2.37 Å.

The crystal field theory specifies that in an octahedral environment, the field is proportional to a^{-5}, a being the ion–ligand distance. For the hexaquocomplex the Ni–O distance is 2.05 Å whereas it reaches 2.37 Å in a zeolite.

TABLE 4.2. Experimental Values of v_1, v_2, and v_3 and Calculated Values of Δ and B (in cm^{-3})[33]

Zeolites	wt. % Ni	v_1	v_2	v_3	$\Delta_{calc.}$	B
NiNaA	1.4	–	10,200	21,900	5900	945
NiNaX	2.7	–	10,000	21,500	5820	935
NiNaY	1.8	5500	10,000	21,200	5740	932
NiHA	1.8	6300	11,500	22,800	6600	980
NiHX	4.5	5600	10,700	22,400	6230	970
NiHY	5.2		10,300	21,800	5820	975
$Ni(H_2O)_6$		8500	13,500	25,300	8500	890

Thus, the field experienced by the ion in a zeolite is considerably lower, as is observed experimentally (Table 4.2). If one considers the $(2.05/2.37)^5$ ratio, the field in zeolites is reduced by half compared to the field created by $6H_2O$. In fact the observed value is quite satisfactory. So the simple electrostatic model may explain the observed spectrum.

It is then concluded that:

1. The crystal field inside the zeolites is weaker than that produced by classical ligands, such as oxygenated molecules.
2. The crystal field varies with the type of zeolite and increases in the order $\Delta Y < \Delta X < \Delta A$.
3. For a given type of zeolite, the field of the protonated structure is stronger than that in the sodium form.
4. The Ni—O bond is almost 100% ionic as revealed by the B values, which are close to the B_0 value of the free ion and far from the B value of the aquocomplex. The rather long Ni–O distance shows that the bond is not covalent.
5. Consequently the crystal field stabilization energy (CFSE) is rather low (thermodynamic aspect). Owing to the Ni–O distance the ion will be very mobile (kinetics aspect), so it must be reactive.

4.3.1.2. Ni^{2+} Ion Reducibility

Ni^{2+} ions are difficult to reduce, and a very high temperature is needed to achieve complete reduction. Metallic nickel is then obtained and the form of the metallic particle is dependent upon the catalyst preparation method. In the zeolite, the field inside the cavities influences the ion properties. It was hoped that new species would be obtained when the reduction conditions were appropriate. All of the six nickel-loaded zeolites shown in Table 4.2 were dehydrated and then put into contact with hydrogen (100 Torr pressure) at 200 and 300°C for several hours (Figure 4.8).[34,35] Two new bands appeared at 13,500 and 29,000 cm^{-1}, but the original bands due to Ni^{2+} located in S_I sites

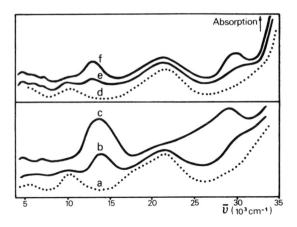

Figure 4.8. UV–visible–NIR reflectance spectra of various zeolites. (1) NiNaY: (a) desorbed at 600°C, (b) treated under H_2 at 200°C, and (c) at 300°C under H_2. (2) NiHY: (d) desorbed at 600°C, (e) treated under H_2 at 200°C, and (f) at 300°C under H_2.

were not diminished. These bands could not be attributed either to any Ni^{2+} complex, or to atomic Ni^0. On the contrary the Ni^+ ion gives an atomic absorption spectrum in which lines lie in the frequency range where a new absorption was observed.

The Ni^+ ion has the $3d^9$ electronic configuration, whose spectroscopic term is ^2D [like Cu(II)]. The crystal field experienced by this ion (octahedral S_I or tetrahedral $S_{I'}$ sites) leads to the splitting of the ^2D level ($^2D \rightarrow {}^2E + {}^2T_2$). A d–d transition is expected in the near-IR range, and experimentally a small absorption is observed at 4700 cm^{-1}. The two absorption bands at 13,500 and 29,000 cm^{-1} cannot be attributed to the $3d^9$ electronic configuration in any case. On the contrary, when a $3d$ electron is promoted toward the $4s$ and $4p$ orbitals, many spectroscopic terms are obtained:

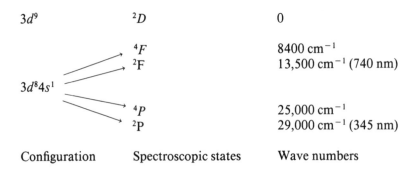

$3d^9$	^2D	0
	^4F	8400 cm^{-1}
	^2F	13,500 cm^{-1} (740 nm)
$3d^84s^1$		
	^4P	25,000 cm^{-1}
	^2P	29,000 cm^{-1} (345 nm)
Configuration	Spectroscopic states	Wave numbers

In the presence of a tetrahedral field, some transitions may arise from the lower groundstate $^2T_2$ to upper spectroscopic states belonging to the excited

TABLE 4.3. Relative Concentration of the Ni^+ Species with the Nature of the Zeolite[35]

Zeolites	% Ni^{2+} (weight)	Δ (in cm^{-1}) for Ni^{2+}	Concentration Ni^+
NiHA	1.8	6600	0
NiHX	4.5	6200	ε
NiHT	5.2	5800	+
NiNaA	1.4	5900	ε
NiNaX	2.7	5800	+ +
NiNaY	1.8	5700	+ +

$3d^8 4s^1$ and $3d^8 4p^1$ electronic configurations. Moreover there will be a splitting of all the spectroscopic states:

$$3d^8 4s^1 \begin{cases} {}^2F \to {}^2T_1 + {}^2T_2 + {}^2A_2 \\ {}^2P \to {}^2T_1 \end{cases}$$

However in this case the $4s$ electron may shield the d electrons. Thus there will not be much splitting of the excited spectroscopic states and it may not be observed. Only a broadening of the band will be observed, which may be asymmetrical. Moreover the ${}^2T_2(3d^9) \to 2F(3d^8 4s^1)$ transition has to be more intense than the ${}^2T_2(3d^9) \to 2P(3d^8 4s^1)$ transition because the number of spectroscopic states is higher for the $3d^8 4s^1$ configuration, as is observed experimentally. The intensity of the bands at 740 and 350 nm (related to the Ni^+ concentration) varies with the type of zeolite and with its acidity, and it is not at all related to the Ni^{2+} initial concentration, as noted in Table 4.3.

It is thus concluded that Ni^{2+} is not well stabilized in the zeolite matrix (low CFSE), and the lower the CFSE, the higher the Ni^+ concentration. This is related to the structure of the zeolite. Ni^+ is isoelectronic with Cu(II), so it should occupy the same crystallographic positions in the zeolite. It is known that Cu(II) prefers the $S_{I'}$ sites, so Ni^+ might well be located at the same site, where the $Ni^+ - O^{2-}$ bond is rather short.

4.3.2. Valence State and Coordination Number of Molybdenum Ions on Various Supports, Changes during the Oxidation of Propylene, and Their Influence on Selectivity

Transition metal ions supported on inert matrixes offer a means of investigating the influence of the symmetry of the catalytic site on the reactivity. For instance molybdenum ions can assume various symmetries depending on the support and the method of preparation. These solids have been tested for partial oxidation of olefins while diffuse reflectance spectroscopy has given information on the valence state and the coordination number of the initial solids and on the changes in these properties during reduction (by propylene, for example) or during the catalytic reaction. The reactions were carried out in the cell used for the spectroscopic measurements. For this purpose, the catalysts were activated *in situ* in an oxygen stream at 500°C and cooled under helium at the reaction temperature (400 or 450°C). The reactants (a mixture of propylene, oxygen, and nitrogen) were then passed over the catalysts and the gaseous effluent was analyzed by gas chromatography. Immediately after the injection required for the chromatographic analysis of the effluent products, changes in the electronic spectra were examined as follows: The reactants were removed through a valve by helium flow and, after rapid cooling, diffuse reflectance spectra were recorded at 25°C. It was possible to obtain spectra for various times of reaction or at the steady state.

Molybdenum ions were deposited on various supports (SiO_2, Al_2O_3, or MgO)[36,37] by impregnation of an aqueous solution of ammonium heptamolybdate (impregnated solids) or by reaction of the support with the molybdenum pentachloride (grafted solids). The charge transfer transition $O^{2-} \rightarrow Mo^{6+}$ of the initial solids (after calcination under oxygen) gave information on the Mo^{6+} coordination, Mo^{6+} ions in tetrahedral coordination absorbing at shorter wavelengths (250–280 nm) than Mo^{6+} ions in octahedral coordination (290–330 nm). Upon reduction the formation of molybdenum ions in various valence states and coordination numbers was also found.[36] Mo^{6+} ions (d^0) have no d–d transition, while Mo^{5+} or Mo^{4+} ions lead to strong absorptions in the visible spectrum owing to d–d transitions, the number and the position of which depend on the symmetry of the cations. A band at 400 nm suggests the presence of Mo^{5+} ions in octahedral symmetry (*Oh*). A distortion of the octahedron leads to additional bands between 600 and 750 nm, the exact position being a function of the distortion. Mo^{5+} in tetrahedral monomeric form is shown by a 500-nm band. The simultaneous occurrence of several transitions may indicate particular surface structures such as molybdenyl MoO^{3+}, *bis*-molybdenyl with dimeric Mo^{5+}, or mixed compounds Mo^{6+}–Mo^{5+}.

It was shown that the stereochemistry of Mo^{6+} depends on the support and on the method of preparation. The Mo/SiO_2 grafted solids prepared with $MoCl_5$ and having a low molybdenum content (between 0.15 and 1 wt.%) possess only tetrahedral Mo^{6+} (coordination four). When the molybdenum content increases (and particularly with the impregnated solids) the coordination six (or five) appears. For the Mo/Al_2O_3 solids two molybdenum coordinations are present, whatever the method of preparation, but the percentage of octahedral coordinated Mo^{6+} on alumina clearly increases with the molybdenum content. Mo–MgO (grafted) solids show only Mo^{6+} in sixfold coordination.

The studies of the reduction (by propylene, for instance) confirmed the interpretations of the spectra of the initial solids, the reduction leading to the formation of Mo^{5+} octahedral or tetrahedral, according to the initial Mo^{6+}. The presence of fourfold coordinated unsaturated Mo^{5+} (Td) is corroborated by the chemisorption of water, which, at 25°C, on partially reduced grafted $Mo–SiO_2$

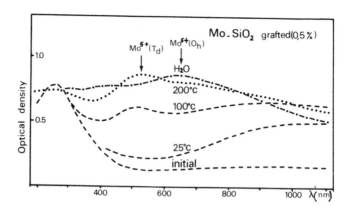

Figure 4.9. Reduction by propylene (100 Torr, 10 h) at various temperatures of a grafted Mo–SiO$_2$ solid (0.5% Mo) and action of water at 25°C on a solid previously reduced at 200°C.

(reduction at 200°C) converts the tetrahedral Mo^{5+} to Mo^{5+} in an octahedral configuration (Figure 4.9)[36] It appears that the chemical reaction of MoCl$_5$ with the support (grafted catalysts) is suitable for the preparation of supported molybdenum oxides of low molybdenum content in which the Mo^{6+} (and Mo^{5+}) coordination can be predominantly tetrahedral.

In further experiments the modifications of the valence states and of the coordination number of the molybdenum was followed during the catalytic reaction itself (oxidation of propylene). The results showed a parallel variation in the selectivity for acraldehyde and the concentration of tetrahedral molybdenum ions on silica and alumina. Both catalysts are more selective at low molybdenum contents when tetrahedral species are predominant over octahedral ones. In contrast, the Mo/MgO solids which possess only octahedral molybdenum have very poor selectivity (Table 4.4).[38]

During the reaction, the increase in the optical density in the visible range arises from the d–d transitions of Mo^{5+} (or Mo^{4+}) due to a reduction of the catalysts. This reduction occurs whatever the C$_3$H$_6$/O$_2$ ratio (even for excess oxygen mixtures); it increases clearly with this ratio and takes place in the first minutes of contact with the reaction mixture. Generally the constant level of optical density is reached more quickly for high C$_3$H$_6$/O$_2$ ratios. The selectivity toward acraldehyde as well as the reactivity is increased when Mo^{6+} and Mo^{5+} coexist. However, to maintain good reactivity, the reduction has to remain partial and even weak. These results show clearly that an oxide catalyst for selective oxidation works in a partially reduced state.

4.3.3. Study of Molecules Adsorbed on Zeolites

Two techniques can be employed for these studies: either the adsorbent is in the gaseous phase, or it is in solution in a nonpolar solvent. If the adsorber is white and has a large specific surface area, very good spectra can be obtained.

TABLE 4.4. Molybdenum Coordination and
Selectivity for Acraldehyde[a]

Solids	Molybdenum coordination[a]	Selectivity for acraldehyde (%)
Mo/SiO$_2$ 0.5% (grafted)	4	37
Mo/SiO$_2$ 1.34% (grafted)	4 + 6	21
Mo/SiO$_2$ 2.0% (impregnated)	4 + 6	1
Mo/MgO 2.0% (grafted)	6	4
Mo/Al$_2$O$_3$ 0.33% (grafted)	4 + 6	15
Mo/Al$_2$O$_3$ 5.3%	4 + 6	7

[a] Tetrahedral or octahedral molybdenum does not take account of the distributions of the tetrahedrons and octahedrons; the coordination is 4 or 6 (or 5 if the distortion of the octahedron is so strong that a ligand along the C$_4$ axis is no longer in the coordination sphere).

Adsorbed species or intermediate species can then be identified. Acidic and redox properties of the surfaces of solid state catalysts can also be determined. The adsorption of various selected molecules facilitates the detection of coke precursors and the measurement of the kinetics of ligand exchange. This is illustrated below.

4.3.3.1. Adsorption and Hydrogenation of Benzene in Zeolites

Benzene was adsorbed on NaX and NaY zeolites exchanged by protons or by Ca^{2+} ions, as well as on zeolites containing platinum (metallic particles). The benzene fine structure was observed, revealing significant differences resulting from adsorption (Figure 4.10).[39] The main conclusions deduced from the spectra are as follows:

1. The 0–0 electronic transition appears. This transition is normally forbidden for symmetry reasons. In effect this transition corresponds to an energy difference between the two $^1A_{2g}$ and $^1B_{2u}$ electronic levels, without changing the vibrational level ($v = 0$ for both states, excited and ground) (see Appendix II). When the molecule is adsorbed it is perturbed, and the symmetry of the skeleton is decreased. Selection rules then change. The transition 0–0, which is normally forbidden for liquid or gaseous

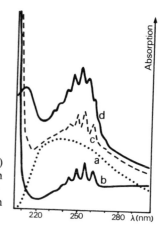

Figure 4.10. Spectra of benzene adsorbed on NaY zeolite: (a) NaY reflectance spectrum (support itself), (b) absorption spectrum of C_6H_6 diluted in hexane, (c) spectrum a + spectrum b, and (d) reflectance spectrum of C_6H_6 adsorbed on NaY zeolite.

benzene, is now allowed, at least partially. Conversely, the appearance of this band is proof that the molecule has lost its D_{6h} symmetry and belongs to another symmetry group, such as C_{3v} or D_{3h}, which in turn means that it may be in a chairlike conformation. In any case either the C—C—C angles are not $2\pi/3$, or the molecule is no longer flat. An extreme case will be an umbrella shape, in which none of the hydrogen atoms is in the carbon atom plane.

2. The energy differences in the fine-structure component are lowered to a mean value of 850 cm^{-1}, while spacing between the 0–0 band and the fine-structure first band gives the value of the E_{2g} vibrational mode making the electronic transition allowed. Calculation leads to a value of 580 cm^{-1}, a little less than the 606 cm^{-1} normal vibration (v_{18} vibration).

3. A new absorption band appears at 215 nm which is very intense for the CaY zeolite, decreases for NaY or NaX zeolites, and is weak for HY or dehydroxylated zeolites (Lewis form). This transition may be explained via a benzene–benzene interaction. The adsorbed molecule is π-electron deficient and behaves like a Lewis acidic molecule (electrophile character).

In general the benzene adsorbs on zeolites, and the greater the number of accessible cations, the stronger the adsorption.

Some zeolites, such as NaY, HY, or CaY, were loaded with platinum (0.5 to 3wt.%) and treated under oxygen at 350°C *in vacuo*. They were then reduced by hydrogen at 350°C and desorbed at the same temperature. C_6H_6 was adsorbed at room temperature. The excess of benzene was removed by a slight pumping and hydrogen was introduced into the cell. The pressure was variable but sufficiently constant within a run. While C_6H_6 was hydrogenated into C_6H_{12} the hydrogenation was followed spectroscopically. The loading of platinum is rather low for two reasons: (a) kinetic — the rate of reaction is low and is easily followed by spectroscopy (during a run the amount of benzene is nearly constant), and (b) spectroscopic — the catalyst is not too black to allow detection of the adsorbed benzene.

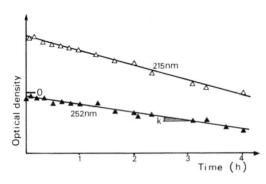

Figure 4.11. Variation of the optical density of C_6H_6 UV bands during a run (catalyst 1% Pt–NaY).

The concentration of adsorbed benzene was followed by recording the UV spectra as a function of time (Figure 4.11). This concentration decreased linearly with time for all supports (NaY or HY), metal loadings, and pressures. Thus the partial order in C_6H_6 is zero (the solid is always saturated with C_6H_6). For a given catalyst, the rate of the reaction is proportional to the pressure, so the order with respect to hydrogen is then equal to 1. The experimental law is

$$v = kP_{H_2^1}[C_6H_6]^0_{ads} \qquad (4.4)$$

On the Pt–CaY catalyst, the reaction is identical to that on Pt–NaY at the very beginning. However, the rate decreases rapidly, and comes close to zero with time. The zero order with respect to C_6H_6 is then no longer observed. A zero order with respect to C_6H_6 means that the platinum particles are always saturated in benzene, whatever the extent of the reaction. So it is concluded that benzene migrates from the support toward the metal. If the opposite were true, the hydrogen would migrate toward C_6H_6 adsorbed on cations located in the big cavities of the zeolites, the rate decreasing with the progress of the reaction. Ca^{2+} ions strongly adsorb C_6H_6, which cannot migrate easily toward metal particles. C_6H_6 is adsorbed on platinum up to 300°C, and to 120°C on Na^+ or H^+ ions present in zeolites. So the equilibrium $C_6H_6/Na^+ \rightarrow C_6H_6/Pt$ is shifted to the right. When Ca^{2+} ions are introduced, the strengths of the $Ca^{2+}-C_6H_6$ and $Pt-C_6H_6$ bonds are similar. Thus C_6H_6 is shared between cations and metal particles, and as soon as the reaction is in progress the metal is free from adsorbed benzene and the rate decreases. It can be concluded that zeolite behaves like a reservoir of reactants which supplies these reactants to the catalytic centers.

4.3.3.2. Nature of the Acidic Centers

When polycyclic aromatic compounds are adsorbed on acidic surfaces, two groups of absorption bands can be identified in the visible range. Some are due to the protonated MH^+ complex, which is formed during adsorption on

Brönsted acidic centers; the others are due to the $M^{\cdot +}$ cation radicals that are formed by adsorption on electron deficient centers (Lewis acidic centers). For example, the perylene cation MH^+ absorbs at 400, 460, and 600 nm, while the cation radical absorbs at 540 nm. For NN-dimethylaniline, the MH^+ cation gives absorption at 260 nm, whereas the $M^{\cdot +}$ species absorbs at 430–470 nm. It is also possible to use pyridine. Gaseous pyridine has a π–π^* transition at 250 nm. When it is protonated, or H-bonded, the vibrational fine structure is preserved, but the maximum is now at 256 nm. When pyridine is adsorbed on a Lewis center, one observes only a broad band without structure whose maximum is at 261 nm. The extinction coefficient is very strong ($\varepsilon \sim 4.10^4$) and so detection of adsorbed pyridine is very easy; conversely the shift in frequency is not as important as in the case of IR spectroscopy.

4.3.3.3. The Strength of Acidic Centers: Acidity Indicators

The acidic strength of a center located on a solid surface is a measure of its ability to convert a base into its conjugated acid, and can be expressed via the Hammett acidity function

$$H_0 = pKa + \log\frac{[A]}{[AH^+]} \tag{4.5}$$

when the reaction takes place on Brönsted acidic centers $(A + BH^+ \rightarrow AH^+ + B)$. An analogous equation for Lewis centers may be set up when the reaction is concerned with an electron transfer from the adsorbate to the adsorbent $(A + :L \rightarrow A: + L)$.

Formally the color of the base changes provided its pKa is greater than the surface H_0 function. In Table 4.5 some values are given which can be used as a scale for an acidic strength measurement. A spectroscopic method can be used by choosing various indicators for which the H_0 function is different, and for which the basic and acidic spectra are different.

For instance, the phenylazonaphthylamine was adsorbed on silica–alumina from solutions (in heptane or benzene) and the spectrum gave only the acidic form of this base, revealing the acidic properties of this solid.[21]

TABLE 4.5. Some Basic Indicators

Indicators	pKa
Phenylazonaphthylamine	+ 4.0
Aminobenzene	+ 2.8
O-Nitroaniline	− 0.2
p-Nitroazobenzene	− 3.3
Benzal acetophenone	− 5.6
Anthraquinone	− 8.1

4.3.3.4. Redox and Acidic Properties: Coke Precursors Formation and Ethylene Adsorption on HY

The protonated HY zeolite is obtained very easily by exchanging Na^+ by NH_4^+, followed by thermal treatment. Its acidity is very strong, as revealed by IR spectroscopy (Brönsted acidity).

Ethylene has only a strong $\pi-\pi^*$ transition in the far UV (165 nm). When it is adsorbed on HY an absorption band appears at 310 nm. This maximum is well known and is due to the π-allylic $C=C-C^+$—cation. This implies that C_2H_4 has been dimerized on a Brönsted acidic center according to the mechanism

$$C_2H_4 + H^+ \rightarrow C_2H_5^+ \tag{4.6}$$

$$C_2H_4 + C_2H_5^+ \rightarrow CH_3-CH_2-CH_2-CH_2^+ \rightarrow CH_3-CH_2-CH=CH_2 + H^+ \tag{4.7}$$

The butene product is immediately ionized via a hydride abstraction on a Lewis center according to the following formula:

$$CH_2=CH-CH_2-CH_3 + Lewis \rightarrow CH_2=CH-CH^+-CH_3 + H^- \tag{4.8}$$

By a subsequent heating at 100°C, extra absorption bands appear at 230, 385, 460, and 560 nm (Figure 4.12)[41] and are assigned to: (a) dienes — $\pi-\pi^*$ transition at 230 nm, (b) π-allylic cations $C=C-C=C-C^+$ at 385 nm, and (c) condensed aromatics that are ionized (460–560 nm), which may be pyrene (460 nm) or perylene (560 nm). Thus coke precursors appear readily on zeolites as soon as olefins are adsorbed on acidic centers which are responsible for such catalytic activity such as butene isomerization.[41]

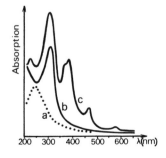

Figure 4.12. UV–visible absorption spectrum of HY zeolite: (a) calcined under O_2 at 350°C then desorbed at 350°C, (b) after C_2H_4 adsorption at room temperature, and (c) after desorption at 100°C.

APPENDIX I. THE d–d TRANSITIONS: CRYSTAL FIELD THEORY (ELECTROSTATIC MODEL)

The d–d transitions of transition metal ions arise from electronic transitions between d orbitals mainly localized on the central metal atom. In a free ion in the gaseous phase (without a surrounding field), the five d orbitals are degenerate. When the ion is placed into a molecular environment (formation of a complex) the degeneracy is lifted, at least partially. The relative energies of the d orbitals and therefore the electronic spectrum depend upon the stereochemistry involved.

Qualitative Analysis

The degeneracy is removed by the electrostatic field created by the ligands. The crystal field theory assumes that the ligands are point charges or charge dipoles and that the d orbitals are entirely nonbonding. The central ion M^{n+} is surrounded by negative ions L^- or by dipoles so arranged that the negative end is pointed toward the central ion. A compound is considered as an aggregate of ions or molecules which interact with each other electrostatically (the electrons of the central ion are repelled by the electrons of the ligands) but which do not exchange electrons, so covalent bonding is neglected completely.

For instance, consider the influence of the octahedron of nearest-neighbor negative ions or oriented dipoles on the d orbitals of a metal ion. The d_{z^2} and $d_{x^2-y^2}$ orbitals point along the x and y axes and so directly toward the ligands. The d_{xy}, d_{yz} orbitals point between the ligands. Thus the electrons in the $d_{x^2-y^2}$ and d_{z^2} orbitals will suffer increased repulsion since they point to negative charges. The d_{xy}, d_{yz} orbitals (t_{2g} set) will be stabilized with respect to the $d_{x^2-y^2}$, d_{z^2} orbitals (e_g set). The d orbitals in an octahedral environment will split into two groups: the triply degenerate t_{2g} and doubly degenerate e_g levels, the total splitting being called Δ or $10Dq$ (Dq is the ligand field parameter). The energy separation between t_{2g} and e_g must preserve the center of gravity; that is, the triply degenerate set of levels is lowered in energy by $4Dq$ and the doubly degenerate set is raised by $6Dq$

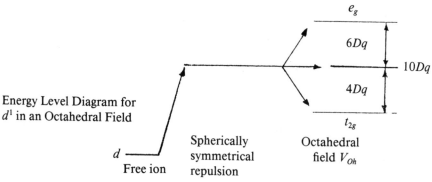

Energy Level Diagram for d^1 in an Octahedral Field

In the case of a single electron (for instance the $3d^1$ ion, Ti^{3+}) this electron will be, in the ground state, located on the lower (stabilized) orbitals (t_{2g}). Upon being influenced by a radiation of energy $hv = \Delta$ (or $10Dq$) the electron will jump to the e_g level (excited state). The radiation with the frequency $v = 10Dq/h$ will be absorbed. For instance, the aqueous solutions of Ti^{3+} absorb at 20,300 cm^- (= 493 nm) and this band is assigned to the transition $^2E_g \rightarrow {}^2T_{2g}$. (Note: small letters are used to describe the symmetry representation of a single electron wave function while capital letters are used for the total electronic wave function of the atom or molecule).

For a multielectron atom or molecule, the electron distribution among the various orbitals is according to the following rules:

1. Pauli exclusion principle — no two electrons can have the same set of four quantum numbers, so each orbital has only two electrons with opposite spins.
2. Aufbau rule — the orbitals with the lowest energies are filled first.
3. Hund's rule of maximum multiplicity — in the various orbitals the electrons with parallel spins are arranged first (maximum number of unpaired spins).

There is a contradiction between the last two rules. The electron distribution is thus a function of the crystal field strength and it is possible to have:

1. High-spin complexes for weak ligand fields with a maximum number of electrons having parallel spins (unpaired electrons) (Orgel's energy level diagrams).
2. Low-spin complexes for strong ligand fields — the field is strong enough to induce spin-pairing of the electrons (Tanabe and Sugano's energy level diagrams).

The ligand field parameter Δ (or $10Dq$) is a function of:

1. The stereochemistry — Δ increases according to the sequence Δ (square planar) $> \Delta$ (octahedral) $> \Delta$ (tetrahedral).
2. The transition metal — Δ increases, often by about 30%, from the first transition series to the second, and by a similar amount from the second to the third series.
3. The valency — for the complexes of a given ligand Δ increases rather rapidly with the valency of the metal ions.
4. The electron–donor or electron–acceptor properties of the ligand (see later on the theory of the molecular orbitals).

The common ligands can be arranged in the order of increasing ligand field strength such that Δ increases along the series in a manner more or less independent of the particular metal ion studied. The order of increasing Δ is found empirically to be I^-, Br^-, Cl^-, F^-, C_2H_5OH, H_2O, NH_3, ethylenediamine, NO_2^-, and CN^- (spectrochemical series).

Quantitative Analysis, Values of the Energies, and the Weak-Field Approach

Free-Ion Terms

It is possible to build up a configuration positioning the electrons on the atomic orbitals. Ti^{2+}, for instance, has the configuration $1s^2 2s^2 2p^2 3s^2 3p^6 3d^2$. However, each configuration gives rise to several electron arrangements and not all have the same energy. For instance, for the Ti^{2+} configuration, there are 45 possible arrangements because there are two electrons each with two magnetic spin quantum numbers [$\pm \frac{1}{2}$ to be permutated in five d orbitals ($10 \times \frac{9}{2}$) arrangements]. The 45 arrangements of the d^2 configuration break up, after consideration of the Pauli principle (no two electrons in the same atom can have the same set of four quantum numbers) into five different levels or *spectroscopic terms* each with a different energy. These *free ion terms* are denoted by ^{2S+1}L. For the d^2 configuration, this notation gives the terms ^3F, ^3P, ^1D, ^1G, ^1S. The *ground term* is thus the ^3F term. This ground term is determined using Hund's rule: maximum spin multiplicity $2S + 1$ and the highest value of L for the maximum spin multiplicity.

It must be noted that: L is the total orbital angular momentum; and the individual l_i of each electron is coupled vectorially $L = \Sigma\, l_i$ to produce various quantified values of the resultant L ($l_i = 0, 1, 2, \ldots$ or orbital angular momentum quantum number); S is the total spin angular momentum and the spin vectors s_i ($s_i = +\frac{1}{2}, -\frac{1}{2}$) are coupled to give the resultant S ($S = \Sigma\, s_i$); and $2S + 1$ is the spin multiplicity.

The various arrangements are therefore characterized by their resultant L and S values, and the terms are denoted by ^{2S+1}L:

Values of L	0	1	2	3	4
Terms	S	P	D	F	G
Values of S	0	$\frac{1}{2}$	1	$\frac{3}{2}$	2
$2S + 1$	1	2	3	4	5

Atomic absorption spectroscopy gives the energies of these levels. The values of the various spectroscopic terms of all the ions are available (see C.E. Moore, *Atomic Energy Levels*, National Bureau of Standards, Circular No. 467. Washington, 1952). Theoretical expressions giving the energy of each level as a function of various parameters (Racah's method) are also available.

Let us look again at the basic equation of quantum mechanics $H\psi = E\psi$. The Hamiltonian of a many-electron atom or molecule (e.g., a $3d^n$ ion) is as follows:

$$H_0 = H(\text{configuration}) + H_e + H_{so}$$

where $H(\text{configuration})$ is linked to the electron distribution among the orbitals and includes the kinetic energy operators of the electrons and the potential energy of attraction between the electrons and the nuclei, H_e is due to the mutual potential energy of repulsion of the electrons and takes the form $\Sigma_{ij} e^2/r_{ij}$ (Σ_{ij} means a sum over all pairs of electrons and r_{ij} is the distance between i and u electrons. H_{so} is written for the spin–orbit coupling. Generally $H(\text{configuration}) > H_e > H_{so}$. The $3d^n$ configuration is thus split in various terms according to the value of H_0. For instance, for the $3d^8$ ion (Ni^{2+}) the following terms are obtained: $^3F\ ^1D\ ^3P\ ^1G\ ^1S$ (order of increasing energy).

Perturbation of the Free-Ion Terms by an Octahedral Field

The electons couple together to give the various spectroscopic terms of the free ions. If the free ion is placed into a crystal field, the degeneracy of the spectroscopic terms may be partially or wholly lifted, to give new terms which are described by group theoretical representations.

Weak-Field Approach

When the crystal field is fairly weak, it may be considered as a perturbation of the free ion levels. The interelectronic repulsion is considered first and then the crystal field is superimposed on the levels so produced:

$$\text{Interelectronic repulsions} \quad > \quad \text{Effects of the crystal field} \quad > \quad \text{Spin-orbit coupling}$$

or

$$\text{Configuration} \xrightarrow[\text{repulsions}]{\text{Interelectronic}} \text{Terms} \xrightarrow[\text{field}]{\text{Weak}} \text{Splitting}$$

The crystal field produces a splitting of the free-ion terms, with the splitting being a function of the number, nature, and geometrical arrangement of the ligands. For example, in an octahedral field, the spectroscopic terms $^3F\ ^1D\ ^3P\ ^1G\ ^1S$ of the free ion Ni^{2+} ($3d^8$) give the following levels:

$$^3F \rightarrow {}^3A_{2g} + {}^3T_{2g} + {}^3T_{1g}$$

$$^3P \rightarrow {}^3T_{1g}$$

$$^1D \rightarrow {}^1E_g + {}^1T_{2g}$$

$$^1G \rightarrow {}^1A_{1g} + {}^1E_g + {}^1T_{1g} + {}^1T_{2g}$$

$$^1S \rightarrow {}^1A_{1g}$$

The orbital degeneracy of the spectroscopic terms of the free ion is given by $2L + 1$; e.g., for the term $F: 2L + 1 = 7$ and for the term $P: 2L + 1 = 3$. In the presence of the crystal field the levels obtained are triply degenerate (T) or doubly degenerate (E) or simple (A). The spin is not changed by the crystal field

$(2S + 1 = 3$ or $1)$. Group theory gives the number and the nature of the new terms obtained when the free ion is placed in a crystal field. Crystal field theory gives the values of the splittings precisely.

A molecule can exist only in discrete energy states characterized by wave functions ψ which are solutions of $H|\psi\rangle = E|\psi\rangle$, where H is the operator whose eigenvalues E are the energy levels of the molecule. Suppose the equation $H_0|\psi_i\rangle = E_0|\psi_i\rangle$ can be solved for the free, unperturbed ion. For a real ion (solvated or fitted in a matrix) the equation $H|\psi\rangle = E|\psi\rangle$ cannot be solved. Perturbation theory allows calculation of the characteristics of the perturbed ion knowing the unperturbed system and the extent of the perturbation. The electrostatic field of the ligands is treated as a perturbation by adding the expression V_{oct} to H_0:

$$(H_0 + V_{oct})|\psi_j\rangle = E_j|\psi_j\rangle$$

where the ψ_j are linear combinations of ψ_i. The energies associated with the wave functions are given by

$$E = \frac{\int \psi^*H\psi d\tau}{\int \psi\psi^* d\tau}$$

where ψ^* is the complex conjugate of ψ.

APPENDIX II. THE π-ELECTRON SYSTEM

The benzene molecule is related to the D_6h symmetry group. All the angles (C—C—C and H—C—C) are equal to $2\pi/3$. The six carbon atoms are sp^2 hybridized, and for each carbon atom there are three such orbitals. Altogether there are 18 sp^2 molecular orbitals: six are used for the six C—C bonds, six are used for the six C—H bonds and the 5 remaining are used for the σ^* molecular orbitals. Each carbon atom carries another $2p$ orbital, which is not used and which is perpendicular to the carbon atoms skeleton. These form three π C—C and three π^* C—C molecular orbitals. All the aromatic properties come from this unique and particular π structure.

The LCAO-MO (linear combination of atomic orbitals–molecular orbitals) theory is useful for specifying the qualitative and semiquantitative aspects of molecular orbitals. The π MOs are formed from the six $2p$ atomic orbitals, and are called $\pi_1, \pi_2, \ldots, \pi_6$ according to their energies (increasing order). The classification comes from symmetry properties in relation to the number of modal planes. The π_1 MO has no nodal planes (except the plane of the carbon atoms); π_2 and π_3 have one plane (perpendicular to the molecule); π_3 and π_4 have two nodal planes; and π_6 have three. Each carbon atom has four valence

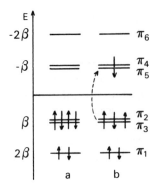

Figure 4.13. Benzene states: (a) electronic ground state, and (b) electronic excited state.

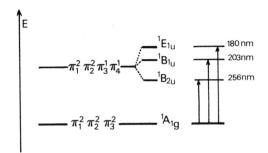

Figure 4.14. Benzene spectroscopic terms.

electrons. Three are used for σ bonds (C—C and C—H). Thus, there is only one electron remaining in the $2p$ atomic orbital. For the six atoms, there are six electrons that are localized in the π molecular orbitals. The difference in energy between one orbital and another is much too great for the application of Hund's rule, so the electrons are coupled together, the lower MOs being filled. The ground state has the $\pi_1^2\pi_2^2\pi_3^2$ electronic configuration. In this case all the π MOs are filled and all the π^* MOs are empty (Figure 4.13). This is a criterion for maximum stability.

The spectroscopic state associated with this configuration is a $^1A_{1g}$ empty according to classical spectroscopic term labeling. During the first excitation, one electron (from π_2 or π_3) is promoted to the π_4^* or π_5^* antibonding molecular orbital. Thus the states related to the configuration $\pi_1^2\pi_2^2\pi_3^1\pi_4^1$ are $^1B_{2u}$, $^1B_{1u}$, and $^1E_{1u}$ (for singlet states) whose energies correspond to wavelengths at 256, 203, and 180 nm (the triplet states are not taken into account) (Figure 4.14). Three spin-allowed transitions can be observed, but only the first is easily detected. The others are located in the far UV and nitrogen purging is required before they can be observed. The transition at 256 nm is rather weak, the integral $\langle\psi_1|r|\psi_1\rangle$ being null for symmetry reasons. For the D_6h symmetry group, the dipole moments are related to $A_{2u}(\mu_z)$ or $E_{1u}(\mu_x,\mu_y)$ irreducible representations.

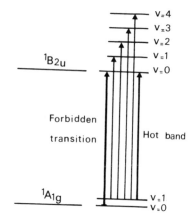

Figure 4.15. Fine-structure band associated with the $^1A_{1g} - {}^1B_{2u}$ transition of benzene: $(0 - 0) = 38,100$ cm^{-1} (262.5 nm), $(1 - 0) = 37,500$ cm^{-1} (266.4 nm), $(1 - 1) = 38,400$ cm^{-1} (260.4 nm), $(1 - 2) = 39,300$ cm^{-1} (254.4 nm), $(1 - 3) = 40,200$ cm^{-1} (248.8 nm), $(1 - 4) = 41,000$ cm^{-1} (243.3 nm).

In this case the direct products

$$A_{1g} \times A_{2u} \times B_{2u} = B_{1g}$$
$$A_{1g} \times E_{2u} \times B_{2u} = E_{2g}$$

do not contain the A_{1g} irreducible representation, and the transition is thus forbidden.

Nevertheless, it is observed at a low intensity, and the reason is a vibronic coupling. The benzene molecule has a lot of vibrational modes, and some have the required symmetry:

$$B_{1g}: \text{ none} \quad E_{2g}: 6a, 6b \text{ at } 606 \text{ cm}^{-1}$$
$$7a, 7b \text{ at } 3048 \text{ cm}^{-1}$$
$$8a, 8b \text{ at } 1595 \text{ cm}^{-1}$$
$$9a, 9b \text{ at } 1178 \text{ cm}^{-1}$$

Thus a E_{2g} vibrational mode can be coupled with the $^1A_{1g}$ fundamental spectroscopic state, which then belongs to the E_{2g} irreducible representation. The first electronic transition issues from the $^1A_{1g}$ state, already excited by E_{2g} vibrational modes. The excited $^1B_{2u}$ state contains a lot of vibrational modes (Figure 4.15). Thus it is easy to observe a fine-structure band whose components are separated by 900 cm^{-1}. This value corresponds to a vibration associated with the excited electronic state, and is related to the totally symmetrical breathing $^1A_{1g}$ mode. Finally the observed fine structure is used to measure the $^1A_{1g}$ vibration, which is called ν_2.

REFERENCES

1. L. E. Orgel, *An Introduction to Transition Metal Chemistry; Ligand Field Theory*, Methuen, London (1961).
2. J. R. Streitweiser, *Molecular Orbital Theory for Organic Chemists*, John Wiley and Sons, London (1962).
3. G. M. Barrow, *Introduction to Molecular Spectroscopy*, Mc Graw-Hill, New York (1962).
4. C. J. Ballhausen, *Introduction to Ligand Field Theory*, Mc Graw-Hill, New York (1962).
5. H. H. Jaffe and M. Orchin, *Theory and Applications of Ultraviolet Spectroscopy*, John Wiley and Sons, New York (1964).
6. C. K. Jorgensen, *Absorption Spectra and Chemical Bonding in Complexes*, Pergamon, London (1964).
7. J. N. Murrell, *The Theory of the Electronic Spectra of Organic Molecules*, John Wiley and Sons, New York (1963).
8. J. S. Griffith, *The Theory of Transition Metal Ions*, Cambridge University Press (1964).
9. T. M. Dunn, D. S. McLure, and R. G. Pearson, *Some Aspects of Crystal Field Theory*, Harper and Row, New York (1965).
10. B. N. Figgis, *An Introduction to Ligand Fields*, Interscience, New York (1966).
11. G. Herzberg, *Molecular Spectra and Molecular Structure III. Electronic Spectra and Electronic Structure of Polyatomic Molecules*, Van Nostrand, New York (1966).
12. H. L. Schlafer and G. Glieman, *Basic Principles of Ligand Field Theory*, Interscience, London (1969).
13. C. K. Jorgensen, *Electron Transfer Spectra: Progress in Inorganic Chemistry* Vol. 12, Interscience, New York (1970), p. 101.
14. M. Orchin and H. H. Jaffe, *Symmetry, Orbitals and Spectra*, Interscience, New York (1971).
15. F. A. Cotton, *Chemical Applications of Group Theory*, John Wiley and Sons, New York (1971).
16. C. K. Jorgensen, *Modern Aspects of Ligand Field Theory*, North-Holland, Amsterdam (1971).
17. M. Gerloch and R. C. Slade, *Ligand Field Parameters*, Cambridge University Press (1978).
18. D. C. Harris and M. D. Bertolucci, *Symmetry and Spectroscopy. An Introduction to Vibrational and Electronic Spectroscopy*, Oxford University Press (1978).
19. A. I. Scott, *Interpretation of the Ultraviolet Spectra of Natural Products*, Monographs Organic Chemistry, Vol. 7, Pergamon, London (1964).
20. H. H. Hershenson, *Ultraviolet and Visible Absorption Spectra*, Academic, London (1966).
21. J. P. Lefti and H. C. Hobson, *Adv. Catal.* **14**, 115 (1963).
22. A. Terenin, *Adv. Catal.* **15**, 227 (1964).
23. K. Tanabe, *Solid Acids and Bases*, Academic, New York (1970).
24. M. Noboru and K. Tanekazu, *Molecular Interactions and Electronic Spectra*, Marcel Dekker, New York (1970).
25. C. N. R. Rao, *Ultraviolet and Visible Spectroscopy: Chemical Applications*, Butterworths, London (1970).
26. W. W. Wendlandt and H. G. Hecht, *Reflectance Spectroscopy*, Interscience, New York (1966).
27. K. Klier, *Catal. Rev.* **1**, 207 (1968).
28. W. W. Wendlandt, *Modern Aspects of Reflectance Spectroscopy*, Plenum, New York (1968).
29. G. Kortum, *Reflectance Spectroscopy: Principles, Methods, and Applications*, Springer Verlag, Berlin (1969).
30. R. W. Frei and J. O. MacNiel, *Diffuse Reflectance Spectroscopy in Environmental Problem Solving*, CRC, Cleveland, OH (1973).

31. W. N. Delgass, G. L. Haller, R. Kellerman, and J. H. Lunsford, *Spectroscopy in Heterogeneous Catalysis,* Academic, New York (1979).
32. H. Praliaud and G. Coudurier, *J. Chem. Soc. Faraday Trans.* I **75**, 2601 (1979).
33. E. Garbowski and M. V. Mathieu, *C. R. Acad. Sci. Paris* **280C**, 1125 (1975).
34. E. Garbowski, M. Primet, and M. V. Mathieu, in: *Molecular Sieves II, ACS Symposium Series* **40**, 281 (1977).
35. E. Garbowski, M. V. Mathieu, and M. Primet, *Chem. Phys. Lett.* **49**, 247 (1977).
36. H. Praliaud, *J. Less Common Metals* **54**, 387 (1977).
37. M. Che, F. Figueras, M. Forissier, J. McAteer, M. Perrin, J. L. Portefaix, and H. Praliaud, *Proc. 6th Int. Congr. on Catalysis,* Vol. 2, The Royal Chemical Society, London, (1977), p. 261.
38. H. Praliaud and M. Forissier, *React. Kin. and Catal. Lett.* **8**, 451 (1978).
39. M. Primet, E. D. Garbowski, M. V. Mathieu, and B. Imelik, *J. Chem. Soc. Faraday Trans.* I **76**, 1942 (1980).
40. M. Primet, E. D. Garbowski, M. V. Mathieu, and B. Imelik, *J. Chem. Soc. Faraday Trans.* I **76**, 1953 (1980).
41. E. D. Garbowski and H. Praliaud, *J. Chim. Phys.* **76**, 687 (1979).

NUCLEAR MAGNETIC RESONANCE IN HETEROGENEOUS CATALYSIS

Y. Ben Taarit and J. Fraissard

5.1. INTRODUCTION TO THE PRINCIPLES AND PRACTICE OF MAGNETIC RESONANCE

5.1.1. Introduction[1]

Magnetic resonance includes both nuclear and electron magnetic resonances. The former applies to nuclei that have nonzero nuclear magnetic moments, the latter to unpaired electrons that all have the same electron magnetic moment. Laws that govern these techniques are quite similar; they differ only on the quantitative scale simply because the largest nuclear magnetic moment (that of the proton) is far smaller than that of the electron (the ratio is 1840). Table 5.1 gives the nuclear spin from which the nuclear magnetic moment derives, as a function of the parity of the mass and atomic number; the angular moment is expressed in $\hbar = h/2\pi$ units. The natural abundance, the relative sensitivity (compared to that of the proton), and the quadrupolar moment (where appropriate) of a limited selection of nuclei often encountered in catalysis are listed in Table 5.2.

Y. Ben Taarit • Institut de Recherches sur la Catalyse, CNRS, Villeurbanne, France. J. Fraissard • Laboratoire de Chimie des Surfaces, Université Pierre et Marie Curie, Paris, France.

Catalyst Characterization: *Physical Techniques for Solid Materials*, edited by Boris Imelik and Jacques C. Vedrine, Plenum Press, New York (1994).

5.1.2. Zeeman Splitting

Nonzero nuclear spin nuclei possess a nuclear magnetic moment μ_N deriving from the spin angular momentum $\mathbf{I}$:

$$\boldsymbol{\mu}_N = \gamma_N \hbar \mathbf{I} = \beta_N \mathbf{I} \tag{5.1}$$

γ_N is the gyromagnetic ratio expressed in radians per second per gauss, g_N is the nuclear g factor, and β_N is the nuclear Bohr magneton given by the following relation:

$$\beta_N = e\hbar/2Mc \tag{5.2}$$

where e and M are the charge and mass of the proton and c the velocity of light. Both g_N and the nuclear angular momentum $\mathbf{I}$ are characteristic of a given nucleus. As noted above, the nuclear magnetic moment may be expressed in Bohr magnetons.

The m_I components of the nuclear angular momentum $\mathbf{I}$ can take only discrete values among $+I, (I-1), \ldots, (-I+1), -I$. These components are also known as the nuclear spin quantum numbers. Spin-$\frac{1}{2}$ nuclei have only $+\frac{1}{2}$ and $-\frac{1}{2}$ as possible spin quantum numbers. The individual nuclear magnetic moments would interact with an external field $\mathbf{H}$ if present and the interaction energy E corresponds to the eigenvalue of the nuclear spin Hamiltonian:

$$\mathcal{H} = \boldsymbol{\mu}_N \mathbf{H} \tag{5.3}$$

Assuming the Oz axis to be parallel to the external magnetic field, the above equation simplifies to:

$$\mathcal{H} = \gamma_N \hbar H I_z = g_N \beta_N H I_z \tag{5.4}$$

In the case of spin-$\frac{1}{2}$ nuclei only two energy levels are possible depending on the spin quantum number: $+\frac{1}{2}$ and $-\frac{1}{2}$ known as the $|\alpha\rangle$ and $|\beta\rangle$ nuclear spin states. As to which is the lowest energy state depends on g_N. The two spin states thus generate two energy levels known as the Zeeman energy levels. The energy separating those two levels is given by

$$\Delta E = g_N \beta_N H = \gamma_N \hbar H \tag{5.5}$$

The energy splitting diagram showing the separation of the nuclear Zeeman levels as a function of the applied external field is shown in Figure 5.1. This energy difference increases linearly with the applied external field.

The absorption of energy may be induced by applying an RF field provided the energy made available by that field is equal to the energy difference between the two Zeeman levels, a condition known as resonance. If this condition is secured then spins of the lower energy level gain enough energy to be promoted to the upper energy level.

TABLE 5.1. Variation of the Nuclear Spin with the Parity of the Mass and Atomic Number

Variable	Symbol	Parity		
Mass number	A	Even	Even	Uneven
Atomic number	Z	Even	Uneven	Even
Spin number	I	Zero	Integer	Half-integer

TABLE 5.2. Some Routine NMR Nuclei and Other Likely Candidates

Isotope	Spin number	Natural abundance	Sensitivity with reference to ^1H	Quadrupolar moment ($\times 10^{-24}$ cm^2)
^1H	$\frac{1}{2}$	99.98	1	0
^{19}F	$\frac{1}{2}$	100	0.83	0
^{13}C	$\frac{1}{2}$	1.108	1.59×10^{-2}	0
^{29}Si	$\frac{1}{2}$	4.7	7.84×10^{-3}	0
^{31}P	$\frac{1}{2}$	100	6.63×10^{-2}	0
^{113}Cd	$\frac{1}{2}$	12.26	1.09×10^{-2}	0
^{15}N	$\frac{1}{2}$	0.37	1.04×10^{-3}	0
^{119}Sn	$\frac{1}{2}$	8.58	5.18×10^{-2}	0
^2H	1	1.5×10^{-2}	9.65×10^{-3}	2.8×10^{-3}
^{14}N	1	99.63	1.01×10^{-3}	2.7×10^{-2}
^{27}Al	$\frac{5}{2}$	100	0.21	0.15
^{11}B	$\frac{3}{2}$	80.42	0.17	3.6×10^{-2}
^7Li	$\frac{3}{2}$	92.58	0.29	$-(1-4)10^{-2}$
^9Be	$\frac{3}{2}$	100	1.39×10^{-2}	2×10^{-2}
^{23}Na	$\frac{3}{2}$	100	9.25×10^{-2}	0.1
^{51}V	$\frac{7}{2}$	99.76	0.38	0.2–0.3
^{93}Nb	$\frac{9}{2}$	100	0.48	$-(0.16-0.4)$
^{133}Cs	$\frac{7}{2}$	100	4.74×10^{-2}	$-(0.3-3) \times 10^{-2}$
^{135}Ba	$\frac{3}{2}$	6.59	4.90×10^{-3}	?
^{137}Ba	$\frac{3}{2}$	11.32	6.86×10^{-3}	0
^{183}W	$\frac{1}{2}$	14.4	7.20×10^{-5}	0
^{203}Tl	$\frac{1}{2}$	29.5	0.18	0
^{205}Tl	$\frac{1}{2}$	70.5	0.19	0
^{207}Pb	$\frac{1}{2}$	22.6	9.16×10^{-3}	0

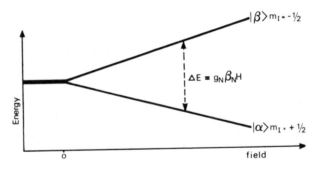

Figure 5.1. Nuclear Zeeman energy splitting as a function of the applied magnetic field.

In order for the resonance condition to be satisfied:

$$h\nu = g_N\beta_N H = \gamma_N \hbar I H$$

or

$$\omega = \gamma_N H \tag{5.6}$$

The population of the two energy levels is, as usual, governed by the Boltzmann distribution law. If N_α and N_β are the respective populations in the $|\alpha\rangle$ and $|\beta\rangle$ spin states then the ratio of the two populations is

$$N_\alpha/N_\beta = \exp(\Delta E/kT) \tag{5.7}$$

where k is the Boltzmann constant.

Equation (5.7) clearly shows how important the magnitude of the applied field is, since the difference in population clearly depends on the strength of the external field. The difference in population will, in turn, determine the intensity of the resonance signal as the latter increases as the probability of transition between the low and high Zeeman levels. The population difference also increases as the temperature decreases as evident from equation (5.7) and therefore samples maintained at the lowest temperatures will give the most intense signal for identical spin populations.

5.1.3. Relaxation, Saturation, and the Bloch Equations[2]

During NMR absorption experiments spins from the lower Zeeman level are continuously being promoted to the upper level. Should this process continue for a long enough period of time the populations of the two levels would eventually equalize and no more absorption should be observed. In fact this is not the case because of the concomitant relaxation process, which causes a given spin that has been promoted to the upper level to lose its extra energy and to fall back to the lower level. This relaxation operates via two distinct mechanisms, one of which involves an energy exchange between the promoted spin and its environment, known as the lattice. This process, termed spin-lattice relaxation, is characterized by a time constant generally labeled as T_1 that measures the period of time necessary for the spin system to reach thermal equilibrium. Upon failure of the relaxation process to operate efficiently, so-called saturation will dominate and little absorption, if any, will be observed. From the macroscopic point of view, as the population of spins in the lower level is slightly higher than that in the upper level, there is a net "magnetization" aligned with the external magnetic field. The effect of the RF field is to tilt the magnetization vector away from the Oz direction, which happens to be the direction of the external magnetic field. T_1 is the period of time necessary for the magnetization vector to return to its initial state after the RF pulse ceases. This is expressed by the Bloch equations, which describe the relaxation process as follows:

$$d\mathbf{M}_z/dt = (\mathbf{M}_0 - \mathbf{M}_z)/T_1 \tag{5.8}$$

$$d\mathbf{M}_x/dt = -\mathbf{M}_x/T_2 \tag{5.9}$$

$$d\mathbf{M}_y/dt = -\mathbf{M}_y/T_2 \tag{5.10}$$

where $\mathbf{M}_x$, $\mathbf{M}_y$ and $\mathbf{M}_z$ are the magnetization components at any instant t along the Ox, Oy, and Oz (that of the external field) axes, respectively. These equations introduce a new relaxation time constant T_2 known as the transverse or spin–spin relaxation time, which relates to the kinetics of the re-establishment of a prevailing equilibrium within a spin system following an RF perturbation. The Bloch equations are important not simply because they describe the time-dependent behavior of a spin system in a given environment accurately but because they appear to be the mathematical basis for modern NMR instrumentation and detection techniques.

5.2. FUNDAMENTAL PARAMETERS IN NMR

5.2.1. The Screening Tensor

The nucleus is by no means an isolated particle, including the proton (H^+). Rather it is in continual interaction with its environment, known as the lattice, which includes electrons, anions, cations, and other small magnetic fields generated by like and unlike spins. The nucleus under observation will thus not only provide information about itself, which is already interesting, but will serve as a probe to monitor its environment, which is far more interesting because this information is much more difficult to acquire by other means.

The sum of the various interactions may be represented by a resulting local magnetic field that adds to or substracts from the applied external field. As is easily understandable, this local field is direction-dependent and therefore most frequently anisotropic. Moreover, its magnitude is directly proportional to the external magnetic field. The effective field may then be written as

$$H = (1 - \sigma)H_0 \qquad (5.11)$$

As the local field is essentially anisotropic, σ must be a 3×3 tensor. It is usually reduced to a diagonal form and the diagonal terms are the principal components of what is termed the screening tensor. Therefore the resonance frequency will not be constant as the effective field is of the tensor type. An infinite number of individual lines corresponding to the various orientations of the individual molecules with respect to the magnetic field will be obtained. The envelope of these lines will be the overall NMR signal, which will be essentially broad. In fact, in the case of liquids the Brownian motion averages out all possible orientations on the time scale of the NMR measurement. Only the isotropic part of the shielding tensor will remain and is therefore dealt with as a scalar in solution NMR, where it is known as the screening constant. In the absence of Brownian motion, rather broad and anisotropic signals are observed. In the solid state, where motion averaging is absent, recently developed devices have made it possible to average out the screening tensor anisotropy.

In fact, in NMR spectroscopy, the screening tensor is seldom used as it essentially refers to the applied external field which varies from one spectrometer to the other. Therefore a more convenient variable has been defined which refers to the resonance of a reference compound under any field. This new parameter, known as the chemical shift, is designated by δ, and is derived from the screening constant, as indicated in the following equation:

$$\delta = (v_s - v_r)/v_r = (\sigma_r - \sigma_s)/(1 - \sigma_r) \qquad (5.12)$$

which is in fact very close to $(\sigma_r - \sigma_s)$; the s and r subscripts indicate sample and reference, respectively. The reference varies from one nucleus to the other. As the most universal example, tetramethylsilane is the reference compound in proton, carbon, and silicon NMR.

In solution the reference is usually an internal one, i.e., one mixed with the sample to be analyzed. Thus the true chemical shift is readily available as the signals of the sample and of the reference compound are obtained simultaneously, and no correction is needed. If for some reason it is not possible to mix the reference with the sample, the former is isolated, usually in a very thin capillary, and termed an external reference. In this case the reference compound and the sample do not experience the same magnetic susceptibility, and a correction for the difference must be made. It can be important for proton NMR and is usually negligible for most nuclei.

In most cases internal references are readily available for solution NMR, but very seldom for solid state NMR. The usual procedure uses an external reference, and magnetic susceptibility corrections are determined with very satisfactory accuracy via NMR measurements for physically adsorbed molecules such as methane, tetramethylsilane, or cyclohexane over the solid under investigation.

5.2.2. Dipolar Interactions

Consider two nuclear spins I_1 and I_2 and the corresponding magnetic moments μ_1 and μ_2 and gyromagnetic ratios γ_1 and γ_2. In the rigid lattice case, the r vector which joins the two magnetic moments has a constant value and an invariable direction. Each of the two spins creates a local field $\pm \varepsilon$ that acts on the neighboring spin, which is clearly independent of the applied external field and usually much less intense. The field experienced by any one spin is therefore the vector sum of the external and local fields $H_r = H \pm \varepsilon$. Thus for a given resonance frequency, the signal will include two components $\pm \varepsilon$ apart from the theoretical resonance value H_0.

The dipolar interaction Hamiltonian is

$$\mathcal{H}_D = g_{N_1} g_{N_2} \beta_N^2 \left[\frac{I_1 \cdot I_2}{r^3} - \frac{3(I_1 \cdot r)(I_2 \cdot r)}{r^5} \right] \tag{5.13}$$

It is easily seen that this interaction decreases drastically as r increases.

The first example to be thoroughly investigated was the one related to water molecules trapped in gypsum. Pake[5] showed that the resulting field prevailing at each ^1H spin of the same water molecule is given by

$$H = H_0 + \tfrac{3}{4} \gamma_N^2 h^2 (1 - 3 \cos^2 \theta) r^{-3} \tag{5.14}$$

where θ is the angle formed by the two magnetic moments. Essentially, the prevailing dipolar interaction is the one involving the nuclei of the same water molecule (two-spin magnetic configuration). However, the dipolar interaction between spins of different molecules is not always negligible and often results in the broadening of the two components specified above. In the case of powder patterns, the signal is even broader owing to the infinite number of θ values. Figure 5.2 compares the theoretical and experimental spectra for this two-spin configuration. Similar studies have been carried out for various magnetic configurations, mainly triangular (equilateral or isosceles triangles).

The dipolar interaction involving nuclear and electronic spins is similar to the one just discussed. Similarly, the dipolar Hamiltonian is of the same type:

$$\mathcal{H}_D = g_e g_N \beta_e \beta_N \left[\frac{I \cdot S}{r^3} - \frac{3(I \cdot r)(S \cdot r)}{r^5} \right] \tag{5.15}$$

where r is again the vector relating the interacting nuclear and electronic spins. This expression is averaged over the probability distribution of the presence of the electron over the space. In the case of a spherical distribution, as is the case for the s orbital of the hydrogen atom, the mean value is averaged out to zero.

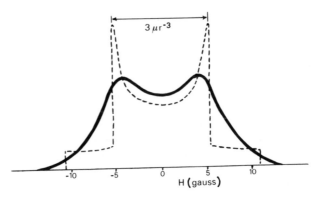

Figure 5.2. An example of a two-spin dipolar interaction spectrum, H_2O in gypsum: broken line, theoretical line shape; solid line, experimental spectrum.

5.2.3. The Fermi Interaction

This contribution arising from the interaction between the nuclear magnetic moment and the electric currents stemming from the nonzero density of the electron at the nucleus was defined by Fermi. The Hamiltonian associated with this isotropic interaction is given by

$$\mathcal{H}_{iso} = a\,\mathbf{IS} \tag{5.16}$$

where a is the coupling constant given by

$$a = (8\pi/3)\gamma_N \hbar g_e \beta \,|\psi(0)|^2 \tag{5.17}$$

where $[\psi(0)]^2$ is the probability of the presence of the electron at the nucleus.

This Fermi interaction arises exclusively from spins in s orbitals, and a is a linear function of the spin density in s orbitals. In the case of an electron spin of a totally ns character, the coupling constant can be written as

$$A = (8\pi/3)\gamma_N \hbar g_e \beta \,|\psi_{ns}(0)|^2 \tag{5.18}$$

where $[\psi_{ns}(0)]^2$ is the spin density at the nucleus, in this particular case of an ns electron.

In general, it is possible to define an electron spin fraction, also termed spin density, at an ns orbital:

$$\rho = a/A \tag{5.19}$$

However, the unpaired spin distribution is most often determined via EPR rather than NMR measurements. In NMR, the Fermi interaction also gives rise

TABLE 5.3. Possible Overall Spin States for the Methylene Group, CH_2

H_a state in CH_aH_b	H_b states	Total spin state	Probability for the spin state relative intensity
α	α	$\alpha\alpha 1$	1
β	α	$\left.\begin{array}{l}\beta\alpha 0 \\ \alpha\beta 0\end{array}\right\}$	2
α	β		
β	β	$\beta\beta - 1$	1

to an additional chemical shift related to the hyperfine coupling constant:

$$\delta_i = a_i(\gamma_e/\gamma_N)[g_e\beta S(S + 1)]/(3kT) \tag{5.20}$$

5.2.4. J Coupling

Experimentally, the 19F NMR spectrum of $POCl_2F$ consists of a doublet of equally intense lines though only one fluorine atom is present within the molecule. Other molecules give rise to doublets, triplets, quartets, quintets, etc. for magnetically equivalent nuclei. The separation between these lines measured on the frequency scale are field-independent and are not affected by changes in temperature. These multiplets arise from the scalar interaction between i and j neighboring spins and has the form $J_{ij} = \mu_i\mu_j$. Obviously the J coupling does not depend on the applied field. J_{ij} is the coupling constant between nuclei i and j. The number and the relative intensities of the multiplet components depend upon the various possible orientations of the interacting nuclear spins. For example, in the case of ethyl chloride, the number of possible overall spin states for the methylene group CH_2 is given in Table 5.3. The same count is possible for the methyl group protons (Table 5.4). Because of the previous distribution of total spin states, the absorptions relative to the methylene group and to the methyl group are in fact split into a quartet and a triplet, respectively. Generally, each line will be split into a multiplet of $2nI + 1$ components, where n is the number of magnetically equivalent spin I nuclei. The relative intensities within a particular multiplet are given by the coefficients of the Pascal triangle.

5.3. APPLICATIONS IN CATALYSIS

5.3.1. Relaxation

The spin–lattice and spin–spin relaxations are influenced by the fluctuating magnetic and electric fields prevailing at the nucleus under investigation. The mechanisms of relaxation vary depending on a number of factors; e.g., transla-

TABLE 5.4. Possible Overall Spin States for Protons in the Methyl Group, CH_3

H_aS state	H_bS state	H_cS state	Total S state	Probability for the spin state relative intensity
α	α	α	$\alpha\alpha\alpha + \frac{3}{2}$	1
α	α	β	$\alpha\alpha\beta + \frac{1}{2}$	
α	β	α	$\alpha\beta\alpha + \frac{1}{2}$	3
β	α	α	$\beta\alpha\alpha + \frac{1}{2}$	
α	β	β	$\alpha\beta\beta - \frac{1}{2}$	
β	α	β	$\beta\alpha\beta - \frac{1}{2}$	3
β	β	α	$\beta\beta\alpha - \frac{1}{2}$	
β	β	β	$\beta\beta\beta - \frac{3}{2}$	1

tional and rotational motion of the molecules in liquids and gases, lattice vibrations in solids, and chemical exchange processes all may determine the relaxation of a given spin system. The following two examples are given to illustrate the use of relaxation phenomena to get information concerning the chemical nature of surface species.

Bermudez[3] has shown that it is possible to determine the concentration of surface hydroxyl groups using T_2' values and taking advantage of the fact that the spin–spin relaxation of surface OH groups is faster than that of adsorbed water molecules. Because of the rapid exchange of the OH and H_2O protons, the observed relaxation time T_2' depends upon the relaxation time constant of the silanol groups and the relative concentration of H_2O and OH in the following way:

$$T_{2R}' = T_{2s} + 2T_{2s}[H_2O]/[OH] \tag{5.21}$$

This relation may be used to determine the concentration of surface OH groups, which is achieved by monitoring the variation of observed T_2' with the concentration of (adsorbed) water.

Frippiat (see Mestdagh et al.[4]) established a relation between the relaxation time constants T_1 and T_2 and the diffusion coefficient D of the protons of HY zeolites, assuming that the relaxation determining mechanism occurs via jumps from one lattice oxide ion to another:

$$T_1^{-1} = k_1 \frac{D}{l^5} \left[\frac{1}{3(\omega_H - \omega_{Al})^2} + \frac{2}{\omega_H^2(\omega_H + \omega_{Al})^2} \right] \tag{5.22}$$

$$T_2^{-1} = k_2(lD)^{-1} \tag{5.23}$$

where $\omega_H/2\pi$ and $\omega_{Al}/2\pi$ are the Larmor frequencies for the proton and aluminum nuclei, respectively, and l is the minimal distance between two spins. These relationships show in a convincing manner that l and D may be determined from the variation of T_1 and T_2 as a function of temperature. Similar studies have been carried out to investigate the motion of various compounds adsorbed on a variety of solids (e.g., benzene and alcohol hydrocarbons).

5.3.2. Chemical Shift

The use of high-resolution NMR to investigate heterogeneous systems is much more complicated than for homogeneous media. This is primarily due to the extreme broadening of the absorption features, which renders the determination of the chemical shift or J coupling inaccurate and unreliable.

The earliest high-resolution NMR experiments, carried out over 25 years ago, involving diamagnetic systems, showed that the resonance frequencies of adsorbed molecules were significantly different from those of free molecules.[6,7] This frequency shift, esentially due to a perturbation in the initial electronic distribution, varies with the nature of the observed functional group and the nature of the surface, particularly when preferential orientation of the molecule with respect to the surface is taking place. It is also true that the molecular motion of adsorbed molecules is severely restricted compared to that of free molecules. Consequently, a broadening of the spectral features is bound to be observed in addition to changes in the chemical shift. The broadening is even more important when these features are associated with groups tightly bound to the surface and less severe for groups at a distance from the surface, which enjoy unrestricted or less restricted freedom of motion.

As promising as it seemed, NMR was very seldom used to invesigate heterogeneous systems for reasons that will be outlined below:

1. Experimental difficulties: among the various difficulties, the most serious was probably the lack of sensitivity of early continuous wave spectrometers. However, since the development of the Fourier transform spectrometer, this is no longer a problem. Thanks to this technological breakthrough, NMR of less sensitive or low-abundance nuclei has become almost routine.
2. Determination of the effective chemical shift: It was not possible to monitor the variation in the chemical shift as a function of a number of parameters, and the determination of peak position relative to conventional references was quite inaccurate, as the substitution method had to be used to measure the chemical shift. In the case of an external reference (see Section 5.2.1), the difference in bulk magnetic susceptibility can be taken into account only if it is known or measurable for a given powder. In fact, direct measurements of χ_v yield very inaccurate figures. This problem was finally solved by measuring the relative chemical shift variation of physically adsorbed molecules as a function of the pressure or coverage.[7]

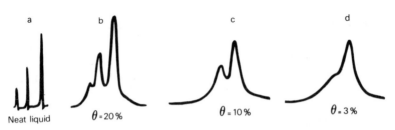

Figure 5.3. ¹H NMR spectrum of ethanol adsorbed on alumina at various surface coverages (θ): (a) pure liquid, (b) $\theta = 20\%$, (c) $\theta = 10\%$, (d) $\theta = 3\%$.

3. Inaccessibility of the chemical shift at zero coverage: To be able to probe the interaction of a given molecule with a particular surface, it is necessary to avoid additional perturbations due to interactions with like molecules. However, the motion of chemisorbed molecules is seriously hindered and consequently the NMR signal is rather broad, which causes two difficulties: broad signals are not easily detectable and if detected, the chemical shift is rather poorly determined. Figure 5.3 shows the evolution of the NMR spectrum of ethanol adsorbed at the surface of alumina as a function of coverage θ.[8]

When the coverage decreases, the more pronounced broadening of the spectrum component relative to the OH groups clearly indicates that the ethanol molecules are bound to the alumina surface through the OH group. By contrast, the methyl groups still possess a significant degree of freedom of motion and at coverages lower than unity, only these groups are in fact detected

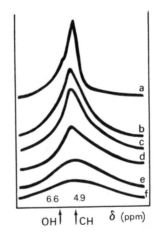

Figure 5.4. ¹H NMR spectrum of formic acid adsorbed on Ketjen silica–alumina pretreated at 300°C at various pressures and temperatures: (a) P/Po > 0.2, 25°C; (b) P/Po > 0.1, 25°C; (c) P/Po > 0.1, 75°C; (d) P/Po > 0.1, 110°C; (e) P/Po > 0.1, 140°C; (f) P/Po > 0.1, 165°C.

at room temperature. Figure 5.4 shows the 1H NMR spectra of formic acid adsorbed on silica–alumina (Ketjen 13%). The observed signal is only associated with the CH group at room temperature and low coverage. This could be easily verified using HCOOD and DCOOH. Lines associated with the solid hydroxyls and the acid OHs may not be detected owing to their significant width with respect to the frequency sweep width of the order of 1 kHz made possible with early continuous wave spectrometers. Yet one may conclude that the formic acid is bound to the surface via its OH group.[9]

At 75°C (Spectrum c), the CH signal is narrowed and a second signal is observed at higher fields, characteristic of exchange-narrowed originally extremely broad signals. This is due to the OH groups of the formic acid at around 11 ppm and the OH groups of the solid at around 4 ppm, via the formation of the intermediate species $HC(OH)_2$, leading to the dehydration of the formic acid over this catalyst. In fact this interpretation of the signal ascribed to the intermediate species became possible only after we were able to identify the very broad signals associated with the molecules interacting strongly with the surface and determine the associated chemical shift in spite of the severe dipolar broadening.

This example illustrates the advantage one may take of mobilizing the adsorbed species to use the NMR technique conveniently. In this way we have been able to detect the signals of two-spin populations that we would not have been able to detect separately.

We will briefly describe three different methods of signal narrowing used in the case of adsorbed phases. These are: (1) exchange narrowing between chemisorbed and physisorbed molecules; (2) magic angle spinning of the sample; and (3) appropriate multipulse sequences.

5.3.3. Equilibrium between Chemisorbed and Physisorbed Molecules

The NMR spectral features of adsorbed phases may be effectively narrowed if the various coexisting species are exchanging rapidly. This is rather interesting when it comes to the investigation of reaction mechanism in heterogeneous catalysis at temperatures close to the reaction temperature.

It is well known that the shape and width of high-resolution NMR signals in solution are critically dependent on time-dependent phenomena such as chemical exchange. For example, when two protons (hydrogen atoms at large) A and B, characterized by chemical shifts δ_A and δ_B, respectively, are rapidly exchanging sites, the observed NMR signal is in fact a single more-or-less narrow line showing at an intermediate chemical shift δ which depends on the previous two δs and on the number of spins A and B, i.e., n_A and n_B as follows:

$$\delta = (n_A\delta_A + n_B\delta_B)/(n_A + n_B) \tag{5.24}$$

Conversely, this relation may be used to determine the chemical shift of a given species (δ_B for instance) assuming that all the other variables are known.

The case of chemically adsorbed molecules (n_{ch}, δ_{ch}) and physically adsorbed molecules (n_ϕ, δ_ϕ) is formally similar to the previous case:

$$\delta = (n_\phi\delta_\phi + n_{ch}\delta_{ch})/(n_\phi + n_{ch}) \qquad (5.25)$$

However, the line widths are usually quite different and the ch component is usually too broad to be detected directly.

Chemical shifts are, in general, measured relative to the gas phase; hence, $\delta_{gas} = 0$. In addition, physical adsorption does not produce any perturbation of the electronic distribution within the adsorbed molecule. Thus $\delta_\phi = \delta_{gas} = 0$ and the equation (5.25) simplifies to

$$\delta = n_{ch}\delta_{ch}/(n_{ch} + n_\phi) \qquad (5.26)$$

or

$$\delta = n_{ch}\delta_{ch}/n \qquad (5.27)$$

where n is the total number of sorbed molecules; thus δ varies as n_{ch}/n. In the case of very strongly chemisorbed molecules, for high enough values of n, δ varies linearly with $1/n$. This is true because n_{ch} remains constant as n increases due to immediate saturation of the strong chemisorption sites. In that case:

$$n_{ch} = S \qquad (5.28)$$

where S is the total number of sites. Determination of this latter number by any appropriate method thus leads to an easy determination of δ_{ch}.

5.3.4. Narrowing via Magic Angle Spinning (MAS)[10]

This method was first discovered by Andrew[10] and extensively used to investigate microcrystalline or amorphous solids. Stejkal (see Schaeffer and Stejkal[11]) and Lippmaa et al.[12] also used this technique. Let us examine again the broadening of the absorption line. Figure 5.5 shows the additional field H_A induced by spin A at spin B. This additional field may be split into two components: one, H_p parallel to H_0, directly adds to (or substracts from) it, and

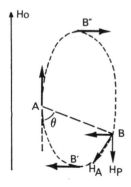

Figure 5.5. Deviation from the Larmor frequency induced by dipolar interaction.

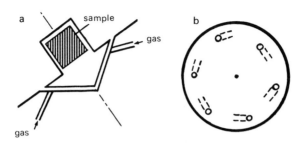

Figure 5.6. MAS sample holder. Andrew type rotor: (a) upper view; (b) cross section.

is responsible for the departure of the resonance frequency from ω_0 given by

$$\omega_0 = \gamma H_0 \tag{5.29}$$

to give a new resonance frequency,

$$\omega = \gamma(H_0 + H_p) \tag{5.30}$$

with

$$\Delta\omega = \omega_0 - \omega = \gamma H_p \tag{5.31}$$

We have given the form of the $\mathbf{H}_p$ component, which is

$$H_p = K_{AB}(1 - 3\cos^2\theta_{AB}) \tag{5.32}$$

where K_{AB} depends on the characteristics of spins A and B and on their distance r_{AB}, and θ is the angle between H_0 and r_{AB}.

In general H_p is nonzero except for two particular positions B' and B'' such that:

$$1 - 3\cos^2\theta_{AB} = 0 \quad \text{or} \quad \theta_0 = 54°44'$$

which is known as the magic angle.

As it is not possible to force all nuclei into such positions, an equivalent process would be the rotation of the whole sample around an axis at an angle θ with respect to the magnetic field. The rotation should be fast enough so that the mean orientation of r_{AB} coincides with that required to cancel out the expression $1 - 3\cos^2\theta_{AB}$. The sample is packed in a rotor which may be of the Andrew type, i.e., a spinning top.[10] The rotor is spun by two gas jets tangential to opposite sides of the rotor (Figure 5.6). The maximum spinning rate varies with the gas pressure and the nature of the gas. The higher the gas pressure, the faster it spins. Differently shaped rotors were also used, including cylindrical rotors, also termed double-bearing rotors.

This technique was first used successfully in ^{13}C NMR. Both chemical shift anisotropy and dipolar broadening were eliminated from the solid state spectrum and the isotropic chemical shift could be easily picked up as the lines obtained were extremely narrow. Later ^{29}Si, ^{31}P, and ^{27}Al NMR were also used under magic angle spinning. Various examples will illustrate the use of this technique to investigate the structure and reactivity of solid catalysts.

5.3.5. Multiple-Pulse Sequences[13]

The evolution of a spin system as a function of time is given by the density matrix. The macroscopic magnetization vector **M**, when tilted off its equilibrium position (aligned with **H**$_0$) under the effect of an RF **H**$_1$ field pulse will return to its initial equilibrium position due to diverse relaxation processes. The total Hamiltonian includes, in the case of rigid lattices, the Zeeman term and additional terms such as the dipolar, the quadrupolar, and various others, the effects of which we wish to average out to zero over a particular pulse sequence.

A first change from the fixed reference axes to a reference system precessing at the Larmor frequency will average out the Zeeman term. Assuming the magnetization vector **M** is aligned with the y axis, 90° pulses around x, $-x$, y or $-y$ will align the magnetization with all the various axes. A cyclic sequence will end up with **M** again aligned with the y axis.

In the absence of interaction, that is, when precessing freely, **M** will not change after a sequence of such manipulations. When various interactions are prevailing (dipolar...) **M** will vary in between pulses. However, it can be shown that for a particular interaction, e.g., dipolar, there exists a particular pulse sequence such that the density matrix $\rho(t)$ at the end of the sequence will be identical to $\rho(0)$ describing the initial state. Therefore measurement of **M** at time t will be devoid of any dipolar interaction contribution. Periodic repetition of such a sequence will yield a series of measurements indicating the evolution of the spin system independent of any dipolar effect or contribution.

This method is not easy to use as the overall sequence duration, which may include as many as 4, 8, 24, or even 48 pulses of 1 μs each and pulse intervals of several μs, must not be too long so as to enable acquision of spectra within a reasonable length of time. More critically, the shape of the pulse must be absolutely rectangular if the efficiency in averaging out various contributions is to be retained. This leads to very difficult technical problems in matching the emitter and transmitter to the NMR probe.

5.4. EXAMPLES IN CATALYSIS

5.4.1. Studies of Acid–Base Interactions at Solid Surfaces via Conventional ^{1}H NMR

Owing to dipolar coupling, the surface protons of a solid acid SXH, where S—X is usually S—O, are characterized by a very broad ^{1}H NMR signal (typically several kHz). Addition of a base molecule AH strong enough to accept a proton according to the equilibrium

$$SXH + AH \leftrightarrows SX^- + AH_2^+ \tag{5.33}$$

will lead, provided there is rapid exchange between the protons of the solid and those of the base, to a single-line ^{1}H NMR spectrum, due to the coalescence of the ^{1}H NMR components δ_{AH}, $\delta_{AH_2^+}$, and δ_{XH}. The position δ_{OBS} of such a line is

$$\delta_{OBS} = P_{XH}\delta_{XH} + P_{AH}\delta_{AH} + P_{AH_2^+} + \delta_{AH_2^+} \tag{5.34}$$

where P_I is the concentration of protons such as XH, AH, and AH_2^+, respectively. Assuming δ_{AH} and $\delta_{AH_2^+}$ are known, the proton concentrations in AH and AH_2^+ and therefore the dissociation coefficient of XSH in the presence of AH may be determined using the following equation:

$$\delta_{OBS}^A = P_{AH}\delta_{AH}^A + P'_{AH_2^+}\delta_{AH_2^+}^A \tag{5.35}$$

where A is a nonzero nuclear spin atom.

Combining equations (5.35) and (5.34) will make it possible to determine δ_{SXH}^H, associated with a line that is too broad to be measured directly.

In fact the above analysis is a rather simplified one as AH was assumed to adsorb exclusively on SXH. At high coverage a sizable fraction of AH is physically adsorbed resulting in the following equilibria:

$$SX - H + AH \rightleftarrows SX^- + AH_2^+ \tag{5.36}$$

$$(AH - H)^+ + AH' \rightleftarrows (AH - H')^+ + AH \tag{5.37}$$

$$SXH \cdots AH + AH' \rightleftarrows SXH \cdots AH' + AH \tag{5.38}$$

Moreover, AH may chemisorb at other surface sites different from SXH (e.g., Lewis sites). This will interfere with the actual quantities of AH interacting with SXH but can be solved chemically. The following are illustrations of this method.

5.4.1.1. NH$_4$Y Zeolites[14]

The ^{1}H NMR spectrum of NH_4^+ ions in Y zeolite is about 32 kHz wide under rigid lattice conditions and about 8 kHz wide at 25°C. Thus the chemical shift of such a proton cannot be measured reliably. Addition of NH_3 results in a single-line ^{1}H spectrum whose position and width vary with the relative concentrations of NH_4^+ and NH_3. In fact this signal is characteristic of the exchange process

$$N_iH_4^+ + N_jH_3 \rightleftarrows N_jH_4^+ + N_iH_3 \tag{5.39}$$

The actual chemical shift $\delta_{NH_4^+}^H$ of the ammonium ion in the zeolite may be determined by extrapolating the following function to zero ammonia concentration:

$$\delta_{OBS}^H = f[NH_3]$$

TABLE 5.5. Variation of the Chemical Shift of the OH Protons
of Various Compounds

Hydroxy compound	EtOH	Silica gel	Silica alumina	HZ zeolite	HCOOH	H$^+$
δ ppm with respect to TMS	0.5	2	4–5	8	11	50

Practically $\delta^{H}_{NH_4^+}$ is found to be $+7.0$ ppm (± 0.1) with respect to gas phase NH$_3$. In solution it was found that

$$\delta^{H}_{NH_4^+} = 6.9 \text{ ppm} \quad \text{and} \quad \delta^{15N}_{NH_4^+} = 43.5 \text{ ppm}$$

with respect to gas phase NH$_3$. These findings indicate that the chemical shift of the ammonium ions does not depend upon the medium (solution or zeolite).

5.4.1.2. HY Zeolite[15]

The chemical shift of the single line observed upon addition of NH$_3$ is a function of the chemical shifts of NH$_3$, NH$_4^+$, and OH of the zeolite and on the dissociation coefficient of the OH groups in the presence of ammonia [equation (5.33)]. The value of δ_{OBS} increases as excess ammonia is removed. Ultimately the number of residual NH$_3$ molecules equals that of the OH groups, which is in agreement with the corresponding value of δ: $\delta^H = 7$ ppm and $\delta(^{15N}) = 43$ ppm. This result proves that the dissociation of OH is total in the presence of NH$_3$.

5.4.2. Brönsted Acidity

The acid strength of an X—H group depends upon the polarization of the X—H bond and therefore upon the electronic environment of the proton; this environment is reflected by the screening tensor σ. Consequently, the chemical shift δ^{H}_{X-H} may be considered as a measure of the Brönsted acid strength. This can be particularly useful in the case of solid Brönsted acids, where no scale similar to the *PK* scale in aqueous media is available. The figures given in Table 5.5 may fit into such a scale for solid acids.

5.5. CATALYSTS AND CHEMISORPTION STUDIES VIA MAS–NMR

Fyfe (see Wasylishen and Fyfe[14]) published a well-documented review on the subject including numerous references. We will limit ourselves to typical examples related to silicate and aluminosilicate zeolite catalysts.

TABLE 5.6. Chemical Shift of ^{29}Si in Si(n Al) Units of an Alumino-Silicate Lattice

Al	Al	Al	Al	Si
O	O	O	O	O
AlOSiOAl	AlOSiOSi	AlOSiOSi	SiOSiOSi	SiOSiOSi
O	O	O	O	O
Al	Al	Si	Si	Si
Si (4Al)	Si (3Al)	Si (2Al)	Si (1Al)	Si (0Al)
4–0	3–1	2–2	1–3	0–4

5.5.1. Catalyst Structural Characterization via Multinuclear NMR

5.5.1.1. Silicon — ^{29}Si NMR

The chemical shift of ^{29}Si in soluble silicates was shown to be very sensitive to the nature and number of substituents present in the second coordination shell of silicon. The relation between the structure of the soluble silicates and the ^{29}Si chemical shift[15] has provided a valuable data base in the NMR investigation of solid silicates and aluminosilicates and more recently metallosilicates. Silicates are characterized by the tetrahedral anion (SiO$_4$)$^{4-}$, labeled Q, with various degrees of condensation. The chemical shift of the central ^{29}Si varies from -60 to -120 ppm relative to TMS with well-defined ranges corresponding to monosilicates Q^0, disilicates Q^1, up to the tridimensional silicates Q^4. In this nomenclature the superscript indicates the number of attached Q units to the central (SiO$_4$)$^{4-}$ tetrahedron. Table 5.6 gives an overview of the chemical shift span for various substitutions.[16]

As an example, Figure 5.7 shows the ^{29}Si NMR spectrum of xonolite. The 2:1 ratio of the observed lines indicates, in agreement with X-ray diffraction, the presence of two distinct sites for silicon: tetrahedra attached to two Q units and tetrahedra attached to three Q units in the relative ratio of 2 to 1.

This technique is also applicable to gels. For example, Figure 5.8 shows the ^{29}Si NMR spectrum of an aqueous gel of silica showing three distinct lines at -90.6, -99.8, and -109.8 ppm assigned to (HO)$_2$Si*Q$_2$, HOSi*Q$_3$, and

Si*Q$_4$, respectively. Upon running the sample under cross-polarization conditions the relative intensities of the former two increase dramatically with respect to that of the latter.

Similarly the replacement of silicon atoms by aluminum atoms that is taking place in aluminosilicate zeolites produces a downfield shift of about 5 ppm per substituted atom (Table 5.6).[17]

For example, the crystal structure of natural natrolite (Na$_2$Al$_2$Si$_3$O$_{10}$2H$_2$O) consists of two types of silicon atoms: those surrounded by three aluminum-centered tetrahedra Si (3Al) and those adjacent to two aluminum-centered tetrahedra Si (2Al) in a 2:1 ratio. The ^{29}Si NMR signal shows two peaks in a 2:1 ratio assigned to these distinct silicon atoms (Figure 5.9).

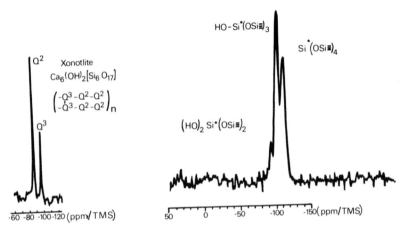

Figure 5.7. ^{29}Si MAS NMR spectrum of xonolite obtained at 39.74 MHz.

Figure 5.8. ^{29}Si MAS NMR spectrum of a silica gel obtained at 11.88 MHz.

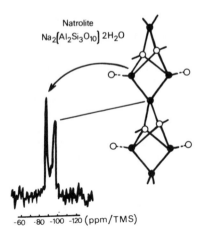

Figure 5.9. ^{29}Si MAS NMR spectrum of hydrated natrolite showing the two crystallographically inequivalent silicon nuclei Si (2Si, 2Al).

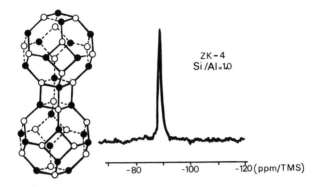

Figure 5.10. ^{29}Si MAS NMR spectrum of ZK-4 of Si/Al = 1.

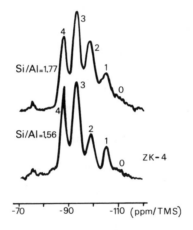

Figure 5.11. ^{29}Si MAS NMR spectrum of ZK-4 of indicated Si/Al ratios.

As an illustration of the sensitivity of the ^{29}Si chemical shift to substitution by aluminum, the case of zeolite A is shown in Figure 5.10. Zeolite A (ZK-4) with a silicon-to-aluminum ratio of 1 is a very simple structure in which each silicon is surrounded by four $(AlO_4)^{5-}$ units. Therefore a single resonance line is observed. At higher silicon-to-aluminum ratios, five distinct cases may be encountered: silicon atoms still adjacent to four $(AlO_4)^{5-}$ tetrahedra, silicon atoms bound to 3, 2, 1, and ultimately zero $(AlO_4)^{5-}$ units giving rise to five distinct lines whose relative intensities vary with increasing Si/Al ratio. The Si (4Al) decreases and simultaneously the Si (0Al) increases (Figure 5.11). The ^{29}Si chemical shift is also sensitive to the T—O—T bond angle with T being either Si or Al. The empirical law[18] governing δ is

$$\delta = 143.03 - 20.34 \sum d_{TT} \qquad (5.40)$$

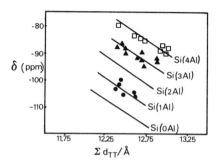

Figure 5.12. Variation of ^{29}Si chemical shift increments as a function of Si—T distance.

where $\Sigma\, d_{TT}$ is the Si $\cdots$ T distance is angstroms calculated on the basis of the T—O—T bond angle; assuming Si—O $= 1.62$ Å and Al—O $= 1.75$ Å:

$$\Sigma\, d_{TT} = 3.37n + 3.24\,(4 - n)\sin\theta/2 \qquad (5.41)$$

Figure 5.12 shows the variation of the chemical shift as a function of the T—O—T bond angle as reflected in $\Sigma\, d_{TT}$ and the substitution of silicon by aluminum. Gross linear plots are observed.

Determination of the framework Si/Al ratio in zeolites: For various reasons, the framework aluminum content may be quite different from the overall aluminum content. As the Lowenstein law prohibits framework Al—O—Al linkages, the intensities of the lines relative to the various Si (nAl) silicon atoms may be used to determine the framework Si/Al ratio. This is given by the relation

$$Si/Al = \sum I_{Si,\, nAl} / \tfrac{1}{4} \sum nI_{Si,\, nAl} \qquad (5.42)$$

In the absence of nonframework aluminum the Si/Al ratios from NMR measurements and X-rays fluorescence coincide thus conforming to the Lowenstein law.[19]

5.5.1.2. Aluminum — ^{27}Al NMR

Nuclei with $I > \tfrac{1}{2}$ are very difficult to deal with and even $+\tfrac{1}{2} \rightarrow -\tfrac{1}{2}$ transitions are dramatically affected by second-order quadrupolar effects. However, valuable information can be obtained provided high fields are used and appropriate corrections made (Figure 5.13).[20]

The ^{27}Al chemical shift, for example, is governed primarily by the aluminum coordination.[21] The octahedral aluminum chemical shift span ranges from 0–22 relative to $[Al(H_2O)_6]^{3+}$, whereas the δ span for tetrahedral aluminum is within 50–80 ppm relative to the same reference. The value of δ also varies with

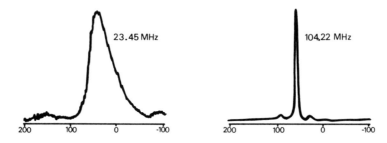

Figure 5.13. Comparison of ^{27}Al MAS NMR spectra at the two indicated frequencies.

TABLE 5.7. Variation of the Chemical Shift of Tetrahedral Aluminum as a Function of the Second-Shell Composition[a]

Al(0Si)	Al(1Si)	Al(2Si)	Al(3Si)
79.5	74.3	69.5	64.2

[a] Relative to $[Al(H_2O)_6]^{3+}$.

the second neighbors (Table 5.7). Hence in aluminosilicates an upfield shift is observed as the silicon second-shell neighbors increase.[22]

The sensitivity of the ^{27}Al signal to quadrupolar effects is not limited to the shape and width of the line; quantitative determination is also affected through relaxation effects. As an example, Figure 5.14 shows the variation of the ^{27}Al NMR signal of framework tetrahedral and nonframework octahedral aluminum of a sodium ZSM-5 as a function of the water content. Whereas no effect was observed in the case of the signal related to the extraframework octahedral species, the line due to tetrahedral framework aluminum was almost not observable in the dehydrated state and slowly increased in intensity as rehydration proceeded. This phenomenon was ascribed to quadrupolar effects arising from significant distortion of the $(AlO_4)^{5-}$ tetrahedra upon dehydration.[23]

5.5.1.3. Phosphorus — ^{31}P NMR

In addition to ^{29}Si and ^{27}Al NMR, widely used to investigate the structure of aluminosilicates, ^{31}P has been used extensively to monitor the structure of various phosphates including aluminum phosphates. PO_4 tetrahedra in various polymeric structures were identified using ^{31}P NMR chemical shift variations as a function of the number of tetrahedra linked to the central phosphorus tetrahedron.

Recently the structure of AlPOs and SAPOs was confirmed using combined ^{27}Al, ^{29}Si, and ^{31}P NMR.[32] The site locations of crystallographically nonequiva-

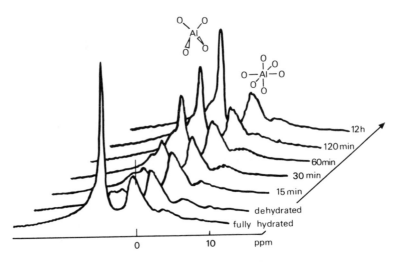

Figure 5.14. Variation of the ^{27}Al MAS NMR spectra for Na-ZSM-5 after various hydration periods.

lent phosphorus atoms were identified in combination with X-ray diffraction studies.[24] Particular phosphorus tetrahedra were shown to interact preferentially with water molecules. Such phosphorus nuclei were also identified on the basis of their chemical shift when the cross-polarization technique was used to obtain the ^{31}P spectra (Figure 5.15).

On the other hand, the sensitivity of ^{31}P NMR to the nature of nearest neighbors was used to monitor the incorporation of paramagnetic cobalt into the structure of molecular sieve aluminum phosphates.[25] The incorporation of such paramagnetic ions induced an important chemical shift anisotropy, which resulted in a very broad pattern of spinning side bands (Figure 5.16).

Postsynthesis isomorphic substitution of aluminum by pentavalent phosphorus was also demonstrated using ^{31}P NMR. Lines with chemical shift similar to those reported for tetrahedrally linked PO_4 tetrahedra at -28 to -32 ppm were indeed observed, indicative of the insertion of PO_4 tetrahedra into the aluminosilicate framework.[26]

5.5.1.4. Boron — ^{11}B NMR

Similarly, ^{11}B NMR was used extensively to probe the location of boron in the first synthesized boron silicates. Tetrahedral boron exhibited a narrow signal whereas trigonal boron was characterized by a broad highly anisotropic signal.[27] These two boron species interconverted upon the addition of water. These findings conclusively demonstrated the incorporation of boron in Boralite and provided useful insight into framework boron subjected to various treatments.

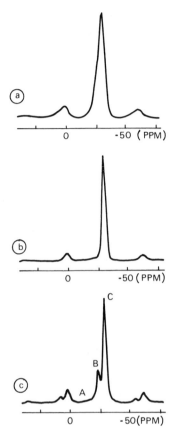

Figure 5.15. ^{31}P MAS NMR of SAPO in: (a) synthesized, (b) calcined, and (c) rehydrated forms.

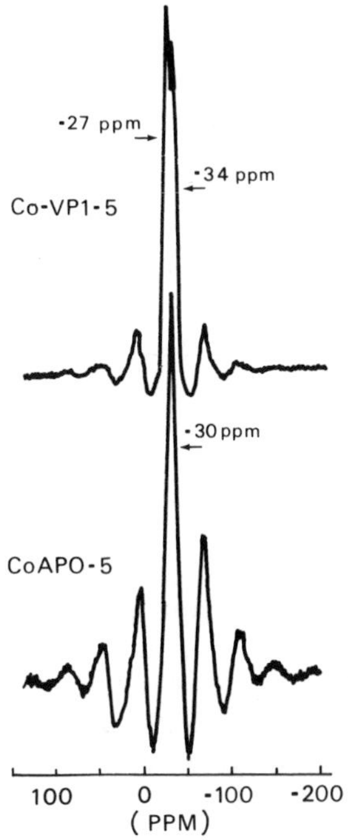

Figure 5.16. ^{31}P MAS NMR of CoAlPO showing the effect of the heteroelement on the ^{31}P chemical shift anisotropy.

5.5.2. Catalyst Structural Characterization via NMR of ^{139}Xe as a Probe Molecule

5.5.2.1. Pure Zeolites

Xenon atoms when adsorbed on porous solids may diffuse from one pore to the other and in the case of zeolites from one channel to another channel and from one cavity to the other. They inevitably collide with channel and cage walls, with the cations, guest molecules, and like atoms. The chemical shift of ^{129}Xe is in fact the result of three terms which reflect the effects experienced by the xenon atoms:

$$\delta = \delta_0 + \delta_s + \delta_E' + \int_0^{\rho_{Xe}} \delta(Xe, Xe)d\rho \qquad (5.43)$$

where δ_0 is the reference (usually xenon gas under quite a low pressure); δ_E is the contribution of the electrostatic field generated by the cations. The term

$$\int_0^{\rho_{Xe}} \delta(Xe, Xe) d\rho_{Xe}$$

where ρ_{Xe} is the density of adsorbed xenon and reflects the Xe–Xe collisions and δ_s is the contribution of the collision of xenon atoms with the zeolite walls and can be formulated as

$$\delta_s = \delta(Xe, S)\rho_s \tag{5.44}$$

where $\delta(Xe, S)$ is characteristic of the wall–xenon collision and depends only on the nature of the wall (essentially oxide ions in the case of zeolites) and ρ_s is characteristic of the shape and size of the cavities or channels.

It is often possible to determine each of the contributions to the overall chemical shift and consequently to monitor: (1) the variation of the electrostatic field with the nature of the cation via δ_E determinations; and (2) the crystallinity of a given zeolite structure via δ_S determinations.

Consider, for instance, a Y zeolite with a given Si/Al ratio. The ^{129}Xe chemical shift, being independent of the Si/Al ratio, does not depend on the number of Na^+ atoms as the number of xenon atoms per cage varies. Therefore the contribution of the δ_E change to the variation of the overall δ may be neglected.[28] At low xenon coverage, δ is primarily determined by the contribution of the xenon–wall collisions. The δ_S values for Y and A zeolites were found to be $\delta_{S(Y)} = 58$ ppm and $\delta_{S(A)} = 82$ ppm, respectively (Figure 5.17), and they reflect the importance of the structure and cage curvature. Consequently, in the case of a mixture of Y and A zeolites the NMR spectrum of adsorbed xenon at low coverage exhibits two distinct components at 58 and 82 ppm whose relative intensities indicate the relative number of Y and A cages in the solid mixture. This can be extended to the determination of the crystallinity of a given structure zeolite compared to a reference zeolite (Figure 5.18). It can also be extended to recognize various cavity types in a single zeolite structure provided they are accessible to xenon.

5.5.2.2. Metal–Y Zeolites and Chemisorption Over Zeolite-Hosted Metals

When guest molecules or particles are distributed within the zeolite cavities their surface is different, in nature, from that of the zeolite. Assuming n_i particles and p types of particles, $0 < i < p$. Each population will give rise to a distinct NMR signal of xenon, with $\delta_{S_i} = \delta_i$ characteristic of the Xe–S_i collisions:

$$\delta_{S_i} = A_{S_i} \times B_{Xe-S_i} \tag{5.45}$$

where A_{S_i} is the contribution term characterizing the chemical nature of S_i whereas B_{Xe-S_i} is characteristic of the Xe–S_i collisions. In the case of slow enough

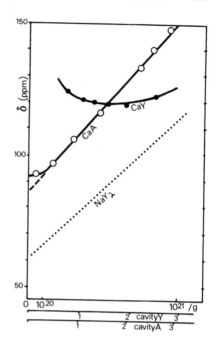

Figure 5.17. Xenon chemical shift variation in CaA, CaY, and NaY as a function of supercages per unit cell.

exchange the ^{129}Xe NMR will include $p + 1$ components to include the xenon–zeolite wall collisions and the p types of particles. The intensity of each component is proportional to the number of collision targets of each particle, n_i. Unfortunately, however, at slow exchange lines are usually too broad to be detected.

By contrast, in the case of rapid exchange, all lines coalesce into a single line whose position will be a probability-weighted average of all the individual contributions:

$$\delta = \sum \alpha_i \delta_i + \alpha_{\text{NaY}} \delta_{\text{NaY}} \tag{5.46}$$

with

$$\sum \alpha_i + \alpha_{\text{NaY}} = 1 \tag{5.47}$$

In addition to the previous two cases, there may be an intermediate situation: assume the number of the distinct and fixed targets S_i, S_j, ... to be smaller that that of the supercages in a particular faujasite sample; in that case there will be up to two to three supercages which do not host such targets. Further assume that the $S_i \cdots S_j$ targets are strong adsorption sites compared to empty supercages, thus favoring maximum concentrations of xenon at the targets rapidly decreasing with distance and particularly at adjacent empty supercages (Figure 5.19).

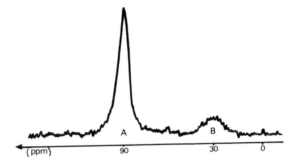

Figure 5.18. Application of xenon NMR to crystallinity determinations: (A) xenon absorption related to the crystalline fraction; (B) xenon absorption related to the amorphous fraction.

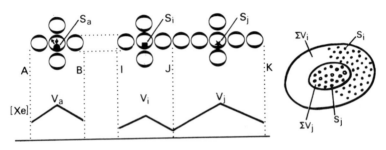

Figure 5.19. Theoretical distribution of the S_i particles in NaY supercages (circles represent the concentration gradient of adsorbed xenon induced by the presence of metal particles.

Let V_i be the volume of samples of low xenon density including a few empty supercages surrounding a target S_i. In each V_i volume a rapid exchange between xenon at the surface of the target and the surrounding volume including xenon colliding with the cage walls thus narrowing the signal of xenon at S_i. If S_i and S_j are remote enough there will not be any exchange of xenon at S_i and S_j. Therefore these will be distinct narrow resonances characteristic of xenon adsorbed on or colliding with S_i and S_j in as much as S_i and S_j are different. If p different $S_1 \cdots S_i S_j \cdots S_p$ targets exist in a given volume we will end up with p separate narrow resonance with S_i chemical shifts given by

$$\delta_i = \lambda_i \delta_{S_i} + \mu_i \delta_{\text{NaY}} \qquad (5.48)$$

where λ_i and μ_i are the xenon collision probabilities with S_i and the zeolite wall, respectively, in a given volume V_i.

Exchange of xenon S_i and S_j is even less probable when these targets are located, depending on their nature, in well separated zones. Consider a particular system $M_x \beta_G$–NaY containing n particles per gram of zeolite, each particle including x atoms on the average, where β is the total chemisorbed gas.

In the case of xenon it has been shown that the prevailing situation is just the one described above, i.e., the xenon NMR spectrum includes as many resonances as there are distinct metallic particles by reference to the number of adsorbed xenon atoms i. The relative intensities are directly related to the populations of each type of particle and the respective chemical shifts are given by the equation (5.48), where S_i is identified with $M_x + iG$.

It is therefore easy to extract the following information from the NMR data:

1. The distribution of the adsorbate G (H_2, O_2, CO, C_2H_4, ...) over the metallic particles for every mean β value. In general, the number of xenon signal components is limited, suggesting that the distribution of the adsorbate over the particles is of a simple nature (integers).
2. The number of metallic particles in a given sample and consequently the average number of atoms per particle. In the case of noble metals this number is usually low $1 < x < 10$ and most of the encaged particles are too small to be directly detected by electron microscopy.

As is obvious from the above this method will also allow the investigation of: (1) competitive chemisorption between gases, (2) surface reactions, and (3) indirectly the metal content at the external surface of zeolite. A few examples will be given below.

5.5.2.2a. Hydrogen Chemisorption.[29] Figure 5.20 shows the chemical shift as a function of the number of xenon atoms per gram, where xenon is colliding

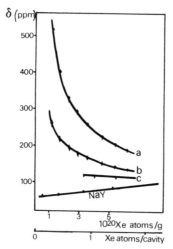

Figure 5.20. Variation of the chemical shift of xenon interacting with platinum particles in Y zeolite with various quantities of preadsorbed hydrogen, as a function of xenon atoms present per cavity: (a) Pt_x–NaY; (b) Pt_x–NaY + 2H; (c) Pt_x–NaY + 4H.

with bare metal particles (a), metal particles which have chemisorbed two hydrogen atoms (b), and four hydrogen atoms (c). Experimentally only two lines are observed at a time and line c only appears as line a has vanished. This indicates that every single particle would chemisorb two hydrogen atoms before any would chemisorb four hydrogen atoms.

5.5.2.2b. Carbon Monoxide Chemisorption.[30]

Chemisorption of CO gives results at variance with the results for hydrogen. Again only two resonances are detected: one is related to bare metal particles and the other to CO-saturated particles with the 1 CO:2 Pt saturation stoichiometry. The added CO molecules saturate the metal particle they encounter before chemisorbing on the next particle. Figure 5.21 shows the variation of the relative intensity of the two possible resonances as total adsorbed CO varies for particles embedded in the supercage.

If for some reason a number of particles were at the external surface, these would be CO-saturated first and this would not affect the xenon resonance colliding with bare particles in the supercages until *all* external particles have been saturated. At that point only, would the second line appear. Therefore this method allows determination of the external particles.

As an illustration, the two solid lines in Figure 5.21 indicate the variations of the relative intensities of the two xenon resonances, where all the particles are within the supercages, whereas the two broken lines indicate similar variations where 18% of the total atoms are at the external surface of the zeolite.

5.5.2.2c. Ethylene Adsorption.[31]

The addition of ethylene to platinum embedded in Y zeolites gives rise, with xenon, to spectrum 2 (Figure 5.22). Comparison with spectrum 1, characteristic of xenon colliding with bare platinum particles, shows that line a is common to the two spectra whereas lines b and c arise from particles holding two hydrogen atoms and one acetylene molecule, respectively. This suggests that ethylene adsorption is dissociative. Upon heating adsorbed ethylene spectrum 3 is observed evidencing an increase in the adsorption and a downfield shift of line c. This is interpreted as being due

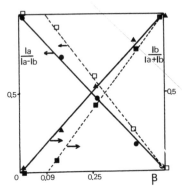

Figure 5.21. Xenon chemical shift as a function of preadsorbed CO quantities over Pt–NaY zeolite; solid lines, all metal particles are within the cavities; broken lines, 19% of the metal is located at the external surface.

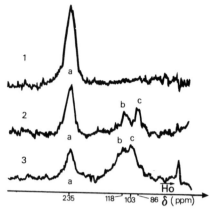

Figure 5.22. Xenon NMR spectra of xenon adsorbed on $Pt-NaY$ zeolite: (a) bare Pt particles, (b) Pt–NaY + 2H, (c) Pt–NaY + 2H + C_2H_2. (Trace 1) xenon spectrum obtained with bare metal particles; (Trace 2) xenon spectrum with preadsorbed alkylene at 25 °C; (Trace 3) xenon spectrum with preadsorbed ethylene treated at 60 °C.

to a reorientation of the adsorbed acetylene from parallel to perpendicular to the surface to form the vinylidene adsorbed complex.[31]

5.5.3. NMR of Probe Molecules for Acidity Determinations

5.5.3.1. Carbon — ^{13}C NMR

The most used probe for acidity determination, apart from ammonia is perhaps pyridine. This molecule, widely used in IR spectroscopy since the pioneer work of Basila and Kantner[32] followed immediately by that of Imelik and his co-workers (see Pichat et al.[33]), has provided the appropriate grounds to discriminate between Lewis- and Brönsted-type acid centers and also to probe the acidity distribution quantitatively and qualitatively via gradual desorption.

Several years ago, even before magic angle spinning became routine, Ian Gay[34] demonstrated the ability of ^{13}C NMR of adsorbed pyridine and a variety of substituted anilines to probe the nature and strength of acid sites at solid surfaces. The ^{13}C resonances of the ortho- and paracarbons relative to nitrogen were shown to vary in opposite directions to each other and the variation of the chemical shift was related to the acid strength of the Lewis center. Pyridium ions, however, were shown to exhibit invariable chemical shifts, not related to the strength of the sites involved.

More recently Dawson et al.[35] investigated the chemical shift of pyridine adsorbed on alumina using both cross polarization (CP) and magic angle spinning at low coverage (0.5% on BET area basis). Surprisingly they observed no shift of the carbon atom resonances with reference to neat pyridine.

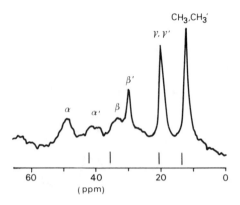

Figure 5.23. ^{13}C CP MAS NMR spectrum of n butylamine held by alumina acid centers: vertical bars indicate resonance frequencies for the liquid base.

However, they indicated a broadening of all resonances that was attributed to the neighborhood of ^{14}N nuclei. In addition, precession and/or wagging of the C_2 axis of the probe molecule were evidenced from the ^{13}C NMR spectra obtained at various temperatures. The authors concluded that there might be a rotation around the C_2 axis. In contrast, the ^{13}C resonances of n-butylamine over the same surface were reported by these authors to differ from those of the neat base. They concluded that the amine was firmly anchored to the surface via the nitrogen atom. It was also observed that in fact two resonances were observed for each single carbon on the amine chain (Figure 5.23).

5.5.3.2. Nitrogen — ^{15}N NMR

^{15}N was thought by Ripmeester[36] to be a more suitable nucleus to probe the surface sites via pyridine adsorption. Alumina and sodium-mordenite did not form pyridinium ions with pyridine in the absence of moisture, while pyridinium ions were formed upon adsorption of pyridine onto an acid-leached mordenite. Similarly Maciel et al.[37] used both ^{13}C and ^{15}N CP/MAS experiments to investigate the nature of the silica–alumina surface-to-pyridine bond and the dynamic behavior of the interacting base molecule. At low coverage, virtually all the pyridine was bound to Lewis centers and only at coverages above the 0.65 monolayer did some anisotropic motion take place. Essentially ^{13}C NMR results consistently paralleled the IR spectroscopy data.

^{15}N chemical shifts were also collected and compared to those measured for pyridine interacting with a variety of compounds such as H_2O, CH_3OH, $Al(CH_3)_3$, and HCl. At intermediate coverages of pyridine over silica–alumina (0.28), the ^{15}N chemical shift was halfway between that of neat pyridine and that of pyridine complexed to $Al(CH_3)_3$. At higher coverages (0.82) the ^{15}N chemical shift was interpreted in terms of dominant hydrogen bonding (pyridine + H_2O, pyridine + CH_3OH).

5.5.3.3. Phosphorus — ^{31}P NMR

Lunsford *et al.*[38] used ^{31}P NMR trimethyl phosphine to probe the nature of the acid sites of decationated Y zeolite. This molecule was preferred over pyridine in view of the abundance of ^{31}P (100% natural) and the large scale of the chemical shifts. Phosphonium ions were found to form over HY zeolite in addition to simply physisorbed trimethyl phosphine. The P—H coupling could be resolved providing additional grounds for the identification of the phosphonium ion. Two such ions, one rather mobile and the other immobile, on the NMR time scale were evidenced. Dehydroxylated Y zeolite equilibrated with TMP showed additional resonances related to TMP interacting with Lewis acid centers. Noteworthy was the suggestion that some molecules interacted with Brönsted and Lewis sites simultaneously. Other trialkylphosphines were used by Baltusis *et al.*[39] to quantify the surface acidity of amorphous silica–alumina. Attempts at absolute determination of Lewis sites failed because of the faint chemical shift and binding constant differences of Lewis-bound and physisorbed phosphines.

5.6. NMR STUDIES OF METALS

5.6.1. The Knight Shift

In the general principles of NMR we have restricted ourselves to "dilute" nuclei. For example we have considered the NMR of ^{1}H in various compounds but not in the hydrogen molecule. Similarly ^{13}C nuclei were considered in various organic or organometallic compounds but not in carbon, and ^{27}Al or ^{29}Si were considered in various aluminum or silicon compounds but not in plain aluminum or silicon. This is to say that the electronic structure of a given element in the condensed homophase is quite different from what it is in heterocompounds.

Indeed in metal and other condensed homomatter (carbon, silicon, etc. ....) the conduction electrons are polarized by the external magnetic field and give rise to a shift of the nominal resonance frequency in addition to the well-known chemical shift. This additional shift is known as the Knight shift and obeys the same relation as the chemical shift, that is, it is directly proportional to the applied magnetic field.

However, in general, the Knight shift is much larger than the chemical shift. Therefore bulk metal pieces or metal small particles will exhibit shifts considerably different from those encountered for compounds of the same elements where electrons are distributed within discrete energy levels and where no conduction electrons are present.

Furthermore, metal pieces or particles may contain an even or odd number of atoms and therefore an even or odd number of electrons per particle. Knight predicted that the Knight shift will be different depending on whether the particle contains an odd or even number of nuclei. Therefore the Knight shift is expected to also vary with the particle size. The following will exemplify the advantages of metal-NMR and the inherent experimental difficulties.

5.6.2. ^{195}Pt NMR of Platinum Catalysts

5.6.2.1. General Considerations

Although many other bulk metals have been extensively studied by NMR including copper and lithium, perhaps the most meaningful example in the field of catalysis is that of platinum. Indeed, platinum is, on the one hand, one of the most widely used and most efficient catalysts, and, on the other hand, one of the most comfortable for NMR use. ^{195}Pt has a rather satisfactory (from the NMR point of view) natural abundance (33.7%) and a reasonably high gyromagnetic ratio, which confers on this nucleus a sensitivity of a little less than 10^{-2} relative to ^{1}H. Most important is the fact that this nucleus is a spin $\frac{1}{2}$, which eliminates complexities brought about by electric quadrupole effects that severely complicate the NMR of non-spin-$\frac{1}{2}$ nuclei.

Owing these favorable circumstances, various platinum compounds have been investigated using ^{195}Pt NMR. There is really nothing to distinguish these studies from other spin-$\frac{1}{2}$ NMR investigations. They have been mostly restricted to simple platinum inorganic compounds including a few polynuclear Pt(0) carbonyls. Figure 5.24 shows the chemical shift scale with the reference compound H_2PtCl_6. Most chemical shifts, as can be seen from Figure 5.24, are positive, i.e., to lower fields, and the overall range is rather limited. The Knight shift, on the other hand, expands over a much larger range and is negative.

As already indicated the Knight shift is expected to vary with even and odd particles and it was therefore predicted that it would vary with the particle size. This, in addition to other phenomena, invariably results in rather broad lines as metal particles are seldom of the same size. Therefore the shape of the experimental signals is a broad envelope, which encompasses signals of platinum atoms in various environments with reference to their coordination

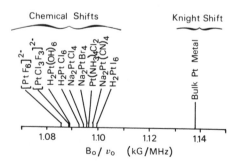

Figure 5.24. The chemical and Knight shift range for platinum diamagnetic compounds and metallic platinum.

number and to the nature of the surrounding ligands. These may be like elements in the bulk but not at the surface. Surface atoms bind to a variety of molecules of the prevailing atmosphere. From this latter consideration, the ^{195}Pt NMR signal also depends significantly upon the particle size, which determines the surface-to-bulk ratio of nuclei.

5.6.2.2. The Experimental NMR Technique

The rather large variation of the Knight shift among metallic particles, owing to environment peculiarities and the large inhomogeneous broadening resulting from the inhomogeneous nature of the sample (conducting metal particles unevenly distributed over an insulator), makes it unrealistic to use the conventional FID technique to obtain the NMR spectra of supported metallic particles. The technique used instead is known as the spin-echo technique.

High-field pulse spectrometers operating at B_0 within the 80–85 kG. range for ^{195}Pt are used. An RF pulse at the appropriate ν_0 frequency and amplitude H_1 is applied for a short duration t_p at time zero ($t = 0$). This pulse is followed by an identical pulse with a double duration ($2t_p$) at time τ after $t = 0$. A spontaneous NMR echo is then observed with an increasing amplitude at $t = 2\tau$. The total duration of this echo is within $1/H_1$. The total NMR signal is then obtained stepwise varying B_0 and maintaining ν_0 fixed. The amplitude of the signal is represented by the area of the spin echo. As the echo is still too weak to be observed, at each B_0 (in the useful range) signal averaging of at least 50,000 scans is necessary to obtain a meaningful envelope.

5.6.2.3. Examples

Figure 5.25 exemplifies the shape and width of the ^{195}Pt NMR spectra obtained under the conditions described above. These signals were all obtained at 77 K and they relate to various platinum-over-alumina samples of various dispersions. All the samples have been reduced and ultimately exposed to air. The number n following Pt in Pt-n-R indicates the average dispersion obtained by the conventional chemisorption method.

All signal areas have been normalized to unity. It can be seen that as the dispersion increases, the low-field feature closest to the "diamagnetic" shift centered at $B_0/\nu_0 \approx 1.089$ increased steadily to be overwhelming at high dispersion.

The peak is naturally ascribed to *surface platinum* nuclei as their relative number increases with the dispersion. Conversely the overriding peak at B_0/ν_0 at 1.138 for the least dispersed sample decreased in intensity as the dispersion increased and was naturally ascribed to the resonance of bulk platinum nuclei. While the identification of this peak is straightforward, the position of the former peak, ascribed to surface platinum atoms, in the "diamagnetic shift" range suggests that such platinum atoms *are not metallic in nature*. The absence of a strong Knight shift was interpreted as an indication of the conduction electrons of the surface platinum atoms being engaged in localized chemical

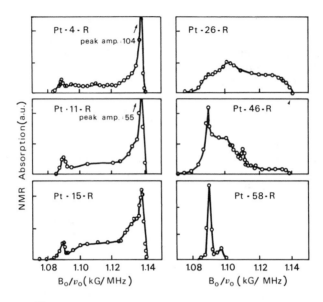

Figure 5.25. The ^{195}Pt NMR spectra of alumina-supported platinum particles as a function of the mean dispersion (note peak shape and position changes).

bonding with oxide ions and the like. The broad intermediate feature was assigned to metallic platinum atoms very close to the surface layer. These metallic subsurface atoms experience different enough environments to end up with a large span of Knight shifts. Both the innermost and subsurface platinum atoms vanish as the particle size decreases; thus the corresponding features disappear for very small metal aggregates.[40]

These assignments have been confirmed upon subjecting the Pt–Al$_2$O$_3$ sample characterized by an average dispersion of 46% (Pt-46-R) to oxygen then to hydrogen at 583 K and finally to vacuum at the same temperature. The relevant signal can be seen in Figure 5.26b to be compared with the signal of the sample exposed to air. Exposure to air again resulted in the signal of Figure 5.26c. Maintaining a residual hydrogen pressure resulted in yet a different signal shown in Figure 5.26d, corresponding to the cleaned sample with presumably no adsorbed species, which lacks the resonance assigned to "diamagnetic" surface platinum atoms bound to oxide ions and similar anions.

Instead the surface atoms now experience an important Knight shift centered around $B_0/v_0 \approx 1.10$. In addition the subsurface layer atoms experiencing intermediate Knight shifts (B_0/v_0 ranging from 1.12 to 1.14) are comparatively more abundant. Re-exposure to air or hydrogen restored the peak at $B_0/v_0 \approx 1.093$ (air) or 1.095 (hydrogen), indicating the presence of surface atoms bound to nonplatinum atoms, and thinned down the subsurface platinum atoms.[40,41]

In a general way the line shape for metallic particles of a given size is roughly as given in Figure 5.27. Clean metal particles do not exhibit a

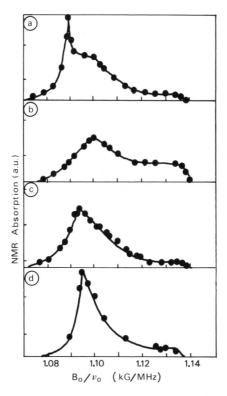

Figure 5.26. The ^{195}Pt NMR spectra of the 46% dispersed platinum over alumina: (a) reduced sample exposed to air, (b) cleaned sample maintained *in vacuo*, (c) re-exposed to air after cleaning, and (d) cleaned and left under a residual pressure of hydrogen.

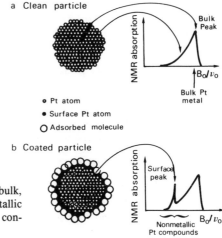

Figure 5.27. Schematic representation of bulk, surface, and subsurface metal atoms in metallic particles in the presence or absence of a contaminant.

symmetrical line because of the differentiated environment of surface atoms; the smaller the particle size the larger the asymmetry.

Contaminated particles exhibit an additional feature apart from the asymmetric line, which corresponds to the surface atoms that are bound to like metallic atoms and to contaminant atoms.

5.7. CONCLUSIONS

The development of a myriad of sophisticated NMR pulse sequences involving one or more nuclei is likely to make it possible to obtain even more information as to the structure and dynamics of catalytic intermediates and/or catalytic precursors. Examples of such techniques include the so-called SEDOR or spin-echo double resonance method, which has already been applied to elucidate the nature of interaction between highly dispersed ^{195}Pt and adsorbed ^{13}CO. The contribution of surface platinum atoms to the SEDOR signal can be isolated.

Spin-echo methods also indicate that the ^{13}C nucleus in CO has acquired a metallic character characterized by an important Knight shift. Many more low-sensitivity nuclei will be routinely investigated considering the tremendously increasing gain in signal-to-noise ratios thanks to improved electronics and higher fields. Multinuclear NMRs run under controlled atmospheres to parallel catalytic conditions will probably be the next breakthrough in the field of physical techniques applied to catalysis.

REFERENCES

1. T. H. Farrar and E. D. Becker, *Pulse and Fourier Transform NMR: Introduction to Theory and Methods,* Academic, New York (1977); L. M. Jackman and S. Sternhell, *Applications of Nuclear Magnetic Resonance Spectroscopy in Organic Chemistry,* Pergamon, New York (1969); A. Carrington and A. D. McLachlan, *Introduction to Magnetic Resonance,* Harper and Row, New York (1967).
2. F. Bloch, *Phys. Rev.* **70**, 460 (1946).
3. V. M. Bermudez, *J. Phys. Chem.* **74**, 4160 (1970).
4. M. M. Mestdagh, W. E. E. Stone, and J. J. Fripiat, *J. Phys. Chem.* **76**, 122 (1972).
5. G. E. Pake, *J. Chem. Phys.* **16**, 327 (1948); G. N. La Mar, W. D. Horrocks, and R. H. Holm, *NMR of Paramagnetic Molecules,* Academic, New York (1973).
6. J. L. Bonardet, A. Snobert, and J. Fraissard, *C. R. Acad. Sci.* **272**, 1836 (1971).
7. J. L. Bonardet and J. Fraissard, *J. Magn. Reson.* **22**, 1 (1976).
8. E. G. Derouane, J. Fraissard, J. J. Fripiat, and W. E. E. Stone, *Catal. Rev.* **7**, 121 (1972).
9. J. Fraissard, S. Bielikoff, and B. Imelik, *Proc. of the ωth Congr. on Catalysis Symposium, No. ϑ, Moscow* (J. Hightower, ed.), The Catalysis Society, Houston, (1968), p. 1577.
10. E. R. Andrew, *Progr. Nuclear Magn. Reson. Spectrosc.* **8**, 1 (1971).
11. J. Schaeffer and O. Stejkal, *J. Am. Chem. Soc.* **98**, 1031 (1976).
12. E. Lippmaa, M. Alla, and T. Tuherm, *Proc. of the 19th Colloque Ampère* (H. Brunner, K. H. Hausser, and D. Schweitzer, eds.), Heidelberg-Geneva (1976).
13. U. Haebelen, *High Resolution NMR in Solids-Selective Averaging,* Academic, New York (1976).

14. R. E. Wasylishen and C. A. Fyfe, *Ann. Rep. NMR Spectrosc.* **12**, 1–80 and 287–290 (1982).
15. E. Lippmaa, M. Alla, T. J. Pehk, and G. Engelhardt, *J. Am. Chem. Soc.* **100**, 1929 (1978).
16. M. Magi, E. Lippmaa, A. Sampson, G. Engelhardt, and A. R. Grimmer, *J. Phys. Chem.* **88**, 1518 (1984).
17. E. Lippmaa, M. Magi, A. Sampson, G. Engelhardt, and A. R. Grimmer, *J. Am. Chem. Soc.* **102**, 4889 (1980).
18. S. Ramdas and J. Klinowbki, *Nature* **303**, 521 (1984).
19. J. Klinowbki, S. Ramdas, J. M. Thomas, C. A. Fyfe, and J. S. Hartman, *J. Chem. Soc., Faraday Trans. II,* **78**, 1025 (1982).
20. D. Freude and H. J. Behreus, *Cryst. Res. Technol.* **16**, K36 (1981).
21. D. Muller, W. Gessner, H. J. Behreus, and G. Scheler, *Chem. Phys. Lett.* **79**, 59 (1981).
22. D. Muller, D. Hoebbel, and W. Gessner, *Chem. Phys. Lett.* **84**, 25 (1981).
23. A. P. M. Kentgens, K. F. M. Scholle, and W. S. Veeman, *J. Phys. Chem.* **87**, 4357 (1983).
24. R. Khouzami, G. Coudurier, F. Lefebvre, B. Mentzen, and J. C. Vedrine, *Zeolites* **10**, 183 (1990).
25. M. I. Davis, C. Montes, P. E. Hathaway, and J. M. B. Garces, *Stud. Surf. Sci. Catal.* **49A**, 199 (1989).
26. M. Kojima, F. Lefebvre, and Y. Ben Taarit, *J. Chem. Soc.* **86**, 757 (1990).
27. Z. Gabelica, J. Bnagy, P. Bodart, and G. Debras, *Chem. Lett.,* 1059 (1984).
28. E. Dempsey, *J. Catal.* **39**, 155 (1975).
29. T. Ito and J. Fraissard, *Chem. Phys.* **76**, 1 (1982).
30. L. C. Menorval and J. Fraissard, *J. Chem. Soc. Faraday Trans. I* **78**, 403 (1982).
31. J. Fraissard, T. Ito, L. C., Menorval, M. A. Springuel-Huet, in: *Metal Microstructures in Zeolites* (P. A. Jacobs, ed.), Elsevier, Amsterdam (1982), p. 179.
32. M. R. Basila and T. R. Kantner, *J. Phys. Chem.* **70**, 1681 (1966).
33. P. Pichat, M. V. Mathieu, and B. Imelik, *Bull. Soc. Chim. France* **00**, 2611 (1969).
34. I. D. Gay and S. Liang, *J. Catal.* **44**, 306 (1976).
35. W. H. Dawson, S. W. Kaiser, P. D. Ellis, and R. R. Inner, *J. Amer. Chem. Soc.* **103**, 6780 (1981).
36. J. A. Ripmeester, *J. Amer. Chem. Soc.* **105**, 2925 (1983).
37. G. E. Maciel, J. F. Haw, I. S. Chliang, B. L. Hawkins, T. A. Early, D. L. McKay, and L. Petrakis, *J. Amer. Chem. Soc.* **105**, 5529 (1983).
38. J. H. Lunsford, W. P. Rothwell, and W. Shen, *J. Amer. Chem. Soc.* **107**, 1540 (1985).
39. L. Baltusis, J. S. Frye, and G. E. Maciel, *J. Amer. Chem. Soc.* **109**, 40 (1987).
40. H. E. Rhodes, P. K. Wang, H. T. Stokes, C. P. Slichter, and J. H. Sinfelt, *Phys. Rev.* **B26**, 3559 (1982).
41. C. P. Slichter, *Surf. Sci.* **106**, 382 (1981).

ELECTRON PARAMAGNETIC RESONANCE: PRINCIPLES AND APPLICATIONS TO CATALYSIS

M. Che and E. Giamello

6.1. INTRODUCTION

Electron paramagnetic resonance (EPR) techniques have been widely employed in catalysis to investigate paramagnetic species which, by definition, contain one or more unpaired electrons. These species, whether catalytically active sites or intermediates, can be located both on the catalyst surface and in the gas phase. In some instances, bulk species have been studied owing to their relevance to catalysis. These techniques are at present routinely used in research laboratories, particularly those concerned with catalysis.[1,2] The nature of the information deduced from EPR may vary from the simple confirmation of the presence of a given paramagnetic species in a catalytic system to the more sophisticated and detailed description of an intermediate or of the coordination sphere of a particular paramagnetic ion supported on a carrier.

The extreme sensitivity of EPR, as compared to those of other spectroscopic techniques, is certainly its most noticeable advantage and has often been used to investigate and characterize low-abundance active sites at catalyst surfaces. In addition, anomalous oxidation states have been observed[3,4] and a number of hypothetical paramagnetic intermediates in catalytic reactions have been detected and identified by means of EPR. As an example, the O^- and O_2^- species often postulated as a result of mechanistic studies and

M. Che • Laboratoire de Réactivité de Surface et Structure, Université Pierre et Marie Curie, Paris, France.   E. Giamello • Dipartimento di Chimica Inorganica, Chimica Fisica e Chimica dei Materiali, Università di Torino, Torino, Italy.

Catalyst Characterization: Physical Techniques for Solid Materials, edited by Boris Imelik and Jacques C. Vedrine, Plenum Press, New York (1994).

conductivity measurements in oxidation reactions have been characterized unambiguously using EPR spectroscopy.[5-7]

6.2. ELECTRON PARAMAGNETIC RESONANCE PRINCIPLE

6.2.1. The Electron Magnetic Moment

A free electron has a spin angular momentum (or simply spin) S which, in a given direction, can only assume the two values $M_S = \pm \frac{1}{2}$ in units $\hbar = h/2\pi$. An electron carries a magnetic moment μ_S which is collinear and antiparallel to the spin itself and given by the expression

$$\mu_S = - g_e \mu_B S \tag{6.1}$$

where g_e is the free electron g value ($g_e = 2.0023$), μ_B is the Bohr magneton $\mu_B = eh/4\pi mc$, with e and m being the electron charge and mass, respectively, and c the velocity of light. The value of the Bohr magneton is $\mu_B = 9.27 \ 10^{-21}$ erg G^{-1}.

6.2.2. The Zeeman Effect

The interaction energy of the electron magnetic moment with an external applied magnetic field $\mathbf{B}$ is classically given by

$$E = - \mu_S \cdot \mathbf{B} \tag{6.2}$$

where $\mathbf{B}$ is the magnetic flux density, i.e., the effect of a magnetic field strength in the matter, measured in Tesla (T) or in Gauss ($1 \ T = 10^4 \ G$).

In quantum mechanics, the μ vector is replaced by the corresponding operator leading to the following Hamiltonian, i.e., the energy operator:

$$\mathcal{H} = g_e \mu_B \mathbf{B} \cdot \mathbf{S} \tag{6.3}$$

If it is assumed that $\mathbf{B}$ lies in the z direction (B_x and B_y are zero and $B = B_z$), the interaction energy corresponds to

$$\mathcal{H} = g_e \mu_B B S_z \tag{6.4}$$

which is the simplest example of a spin Hamiltonian. The energies corresponding to the two allowed orientations of the spin are therefore

$$E = (\pm \tfrac{1}{2}) g_e \mu_B B \tag{6.5}$$

These two energy levels are often referred to as Zeeman levels. The lower energy level corresponds to $M_s = -\frac{1}{2}$, the case in which $\mathbf{B}$ and $\boldsymbol{\mu}_S$ are parallel. The latter two are antiparallel when $M_s = +\frac{1}{2}$, which corresponds to the upper energy level. The energy difference between these two levels is

$$\Delta E = g_e \mu_B B \qquad (6.6)$$

At thermal equilibrium, under the influence of the external applied magnetic field, the spin population is split between the two levels according to the Maxwell–Boltzmann law:

$$n_1/n_2 = e^{-\Delta E/kT} \qquad (6.7)$$

where k is Boltzmann's constant, T the absolute temperature, and n_1 and n_2 are the spin populations characterized by the M_s values of $+\frac{1}{2}$ and $-\frac{1}{2}$, respectively. At 77 K, in a field of about 3000 G, n_1 and n_2 differ by less than 0.005.

6.2.3. The Resonance Phenomenon

The transition between the two Zeeman levels can be induced by irradiating the paramagnetic system with suitable electromagnetic radiation, providing its frequency v fulfills the resonance condition:

$$hv = g_e \mu_B B \qquad (6.8)$$

From equation (6.8), it is easily deduced that the frequency required for the transition to occur is about 2.8 MHz per Gauss of applied field. This means that for the magnetic fields usually employed in the laboratory the radiation required belongs to the microwave region. The energetic scheme of the Zeeman levels and of the corresponding transition is shown in Figure 6.1 as well as the absorption and its first derivative, which, for reasons to be explained later, is the usual presentation of EPR spectra.

The energy absorption necessary to promote electrons from the lower to the upper energy level represents the resonance signal. By this absorption process, the populations of the two energy levels n_1 and n_2 tend to equalize. The odd electrons from the upper level give up the hv quantum to return to the lower level and satisfy the equilibrium Maxwell–Boltzmann law. This energy may be dissipated within the lattice as phonons, i.e., vibrational, rotational, and translational energy. The mechanism by which this dissipation occurs is known as the spin–lattice relaxation, and is characterized by an exponential decay of energy as a function of time. The exponential time constant is denoted T_{1e} or spin–lattice relaxation time. The initial equilibrium may also be reached by a different process. There could be an energy exchange between the spins without transfer of energy to the lattice. This phenomenon, known as the spin–spin

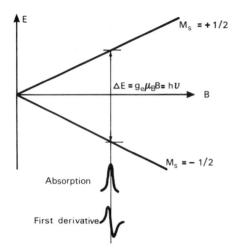

Figure 6.1. The Zeeman energy levels of a free electron in an external applied magnetic field.

relaxation, is characterized by a time constant T_{2e} called the spin–spin relaxation time. When both spin–spin and spin–lattice relaxations contribute to the EPR line, the resonance linewidth can be written as

$$\Delta B \propto 1/T_{1e} + 1/T_{2e} \tag{6.9}$$

In general, $T_{1e} > T_{2e}$ and the linewidth depends mainly on spin–spin interactions: T_{2e} increases on decreasing the spin concentration, i.e., the spin–spin distances in the system. On the other hand, when T_{1e} becomes very short, below roughly 10^{-7} s, its effect on the lifetime of a species in a given energy level makes an important contribution to the linewidth. In some cases, the EPR lines are broadened beyond detection. T_{1e} is inversely proportional to the absolute temperature ($T_{1e} \propto T^{-n}$) with n depending on the precise relaxation mechanism. In such a case, cooling down the sample increases T_{1e} and usually leads to detectable EPR lines. Thus, quite often, EPR experiments are performed at the boiling temperature of liquid nitrogen (77 K) or liquid helium (4.2 K). On the other hand, if the spin–lattice relaxation time T_{1e} is too long, electrons do not have time to return to their initial state. The populations of the two levels (n_1 and n_2) thus tend to equalize and the intensity of the signal decreases, no longer being proportional to the number of spins present in the sample. This effect, known as "saturation," can be avoided by exposing the sample to low microwave power.

The typical shape of EPR lines is Gaussian or Lorentzian. The analytical expressions of the two functions are

$$y = ae^{-bx^2} \qquad \text{(Gaussian)}$$

$$y = a/(1 - bx^2) \qquad \text{(Lorentzian)} \tag{6.10}$$

Figure 6.2 gives the typical features of the two types of lines, in terms of normalized absorption (a) and first-derivative curves (b).

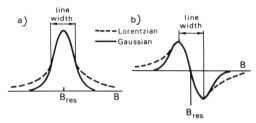

Figure 6.2. Lorentzian and Gaussian line shapes: (a) absorption, (b) first derivative.

6.3. SCHEMATIC DESCRIPTION OF AN EPR SPECTROMETER

A modern EPR spectrometer is designed to detect with high sensitivity the microwave absorption in a sample as a function of the external applied magnetic field. The actual EPR experiment consists of scanning the magnetic field at constant microwave frequency until the resonance condition given by equation (6.8) is fulfilled. Then, a significant amount of energy is absorbed by the sample.

The other approach, i.e., scanning the microwave frequency at a constant magnetic field, is not used as common microwave sources deliver only a limited range of frequencies.

The basic components of the spectrometer (Figure 6.3) are: (a) A microwave source (usually a klystron, but recently Gunn diodes have been introduced) supplying electromagnetic radiation at constant frequency whose power is controlled by an attenuator between the klystron and the sample. The most often employed frequencies are those corresponding to X (about 9.5 GHz) and Q (about 35 GHz) bands. The experimental features of these bands are reported in Table 6.1. (b) A microwave guide system to propagate the microwave radiation from the source to the resonant cavity. (c) A cavity made from a highly conductive metal and having reflecting walls to accumulate power on the sample by multiple reflections of the microwave radiation. The internal dimen-

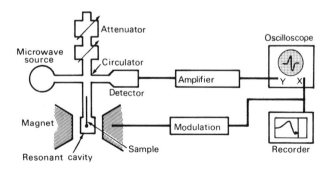

Figure 6.3. Schematic view of an EPR spectrometer.

TABLE 6.1. Typical Conditions for EPR Experiments at
Various Frequencies

Band	v/GHz	Resonance field/T	λ/cm
X	9.5	0.35	2.9
Q	35	1.25	0.8

sions of the cavity are similar to the wavelength of the microwaves. At resonance, standing waves (modes) of various configurations are formed. (d) A powerful electromagnet able to provide a homogeneous field, within the range 0–2 T approximately, which is controlled by a field probe. The range and rate of scanning are adjusted to provide the most suitable conditions for observation of the microwave absorption. (e) A detector diode to measure the energy absorbed by the sample at resonance. (f) A convenient amplifier system, a recorder, and an oscilloscope.

Superimposed on the main magnetic field, an oscillating field is obtained by applying an alternating current (typically 100 kHz) to a set of coils in the cavity walls. This modulating field converts the resonance to an alternating signal which can be separated from random noise using a phase-sensitive detection system. This method of detection has a very high sensitivity and leads to an output whose amplitude is proportional to the slope A_1A_2, so that the resulting EPR signal now appears as the first-derivative curve of the absorption when the magnetic field is scanned (Figure 6.4). The experimentalist selects the amplitude of the modulating field, which should be large enough to obtain a good signal-to-noise ratio but small enough to obtain a good first-derivative curve, i.e., to prevent distortions which occur if the value of the amplitude approaches the resonance linewidth.

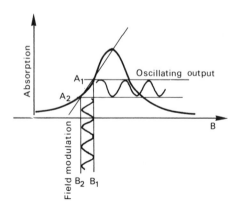

Figure 6.4. Modulation of the magnetic field and the oscillating output.

The samples employed in surface chemistry and catalysis studies are usually polycrystalline solids. Solid samples are usually placed in quartz tubes which may be connected to gas-handling or vacuum lines so as to activate samples or adsorb the desired reagent at the desired temperature and pressure. The sample tubes are then placed in the cavity at the center of the electromagnet gap. The EPR tubes (5 mm in diameter for the most common X band measurements) are filled with the investigated powder, up to an average height of 10–20 mm, which corresponds to a sample weight of 50 and 8 mg for X and Q bands, respectively. Spectra are often obtained at 77 K using a special Dewar fitting into the cavity. The temperature of the sample can, however, be easily altered within the range 90 to 500 K using a flow of cooled dry nitrogen, which is particularly useful when structural evolution of the paramagnetic species with temperature has to be followed.

The digital acquisition of spectra with on-line computers is now common and allows improvement of the signal-to-noise ratio both by spectra accumulation and spectral curve smoothing. Mathematical data treatments can also be used to obtain second and third derivatives of the spectra, which are very useful to detect overlapping signals and more importantly for measuring spin concentrations. This is done by double integration of the first-derivative EPR signal and comparison of the intensity of the integrated signal (which is proportional to the number of spins present in the system) with that of standard samples such as DPPH or copper sulfate with a known number of spins.[8]

6.4. INTERACTIONS BETWEEN THE ODD ELECTRON AND ITS SURROUNDING MEDIUM

6.4.1. The g Factor[9]

6.4.1.1. Theoretical Expression and Physical Meaning of the g Factor

In a "real" chemical system, the odd electron is not free but belongs to an orbital associated with the paramagnetic species. The first interaction is thus defined as occurring between the spin S and the angular orbital momentum L. The latter is associated with a magnetic momentum given by

$$\mu_L = -g_L \mu_B L \tag{6.11}$$

where g_L is the orbital g factor. The coupling between these two momenta, known as the Russell–Saunders coupling, generates a resultant angular momentum:

$$J = L + S \tag{6.12}$$

associated with the magnetic moment

$$\boldsymbol{\mu}_J = -g_J \mu_B \mathbf{J} \tag{6.13}$$

where g_J is called the Landé g factor and is given by

$$g_J = \frac{J(J+1) + S(S+1) - L(L+1)}{2J(J+1)} + 1 \tag{6.14}$$

So far, no interaction between the paramagnetic species and its close neighbors has been considered. In fact, in a solid, for instance, it is subjected to various electrostatic fields which generate the so-called "crystal field." This field quenches the L–S coupling more or less effectively. In fact, the crystal field acts upon the electron motion within the orbital but has no influence on the spin momentum itself; in other words, it acts upon **L** but not upon **S**.

Several cases may be distinguished, depending upon the strength of the crystal field:

1. Strong crystal fields: **L** must align itself with the crystal field and only **S** may orient itself with respect to the external magnetic field. **J** has no meaning and $g = g_e$. A number of organic radicals experience this situation. Yet, g values close to 2 may also be observed whenever the odd electron is in an orbital which is delocalized over the whole molecule. Such an electron would be a quasi-free electron and $g \approx g_e$.

2. Weak crystal fields: **L** is no longer under the constraint of this weak field and the L–S coupling can take place; the g_J values are then given by equation (6.14). This is the situation prevailing in the case of rare earth elements, for which the $4f$ orbitals are very well shielded from their environments by filled $5s$ and $5p$ electron shells.

3. Medium crystal fields: **L** is partly blocked by the crystal field, as in the case of transition metal ions or of inorganic radicals. The vector-coupling model no longer holds. The coupling energy is of the type $\lambda \mathbf{LS}$, where λ, the spin–orbit coupling constant, is characteristic of a given element at a given valence state. Then g_J, which is no longer relevant, is replaced by a spectroscopic g factor, simply called "g factor," which can be calculated on theoretical grounds as shown in Section 6.4.1.4a.

Let us consider a system with a doublet ($S = \frac{1}{2}$) nondegenerate electronic ground state. To begin with, we will consider a system with zero nuclear magnetic moment ($\mu_n = 0$) nuclei exclusively. For such a system, the interaction with the external magnetic field can be expressed in terms of a perturbation of the general Hamiltonian by the following three terms:

$$\mathscr{H}_{\text{pert}} = g_e \mu_B \mathbf{BS} + \mu_B \mathbf{BL} + \lambda \mathbf{LS} \tag{6.15}$$

The first and second terms correspond respectively to the electron Zeeman and orbital Zeeman energies (note the orbital g factor g_L of equation (6.11) is taken here as unity, as expected from equation (6.14), since $S = 0$ and $J = L$ for the orbital only). The third term represents the energy of the spin–orbit coupling. The spin–orbit coupling constant λ mixes the ground state wave function with the excited states. Through the effect of the spin–orbit coupling, the electron can acquire some orbital angular momentum. Standard λ values for various atoms have been obtained from atomic spectra. The previous Hamiltonian, called the spin Hamiltonian as it is concerned only with the spin and its motion within the orbit, can be expressed more simply as follows:

$$\mathscr{H} = \mu_B \mathbf{B} \mathbf{g} \mathbf{S} \tag{6.16}$$

analogous to that given in equation (6.3). S is now the fictitious spin and $\mathbf{g}$ a second-rank tensor (or a symmetric 3×3 matrix) representing the anisotropy of the interaction between the unpaired electron and the external magnetic field. For reasons related to matrix notation, $\mathbf{g}$ is written in between the vectors $\mathbf{B}$ and $\mathbf{S}$. Equation (6.16) also describes the fact that the orbital contribution to the electronic magnetic moment may be different along different molecular axes. In other words, the magnetic moment of the odd electron in a real paramagnetic system is not exactly antiparallel to the spin and its magnitude is not that of a free electron, but depends on the orientation of the system in the applied magnetic field. This concept can be summarized by

$$\mu_S = -\mu_B \mathbf{g} \mathbf{S} \tag{6.17}$$

which is the analogue of equation (6.1).

The $\mathbf{g}$ tensor may be depicted as an ellipsoid whose characteristic principal values (g_{xx}, g_{yy}, g_{zz}) depend upon the orientation of the symmetry axes of the paramagnetic entity with respect to the applied magnetic field (Figure 6.5). The most general consequence of the anisotropy of $\mathbf{g}$, from an experimental point of view, is thus that the resonance field of a paramagnetic species for a given frequency depends on the orientation of the paramagnetic center in the

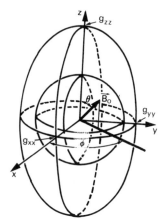

Figure 6.5. Definition of the characteristic angles θ and ϕ to describe the orientation of the magnetic field B_0 with respect to the g ellipsoid in the x, y, z axis system.

magnetic field. This tensor is analogous to the refractive index tensor, which could be represented by an ellipsoid whose principal values depend on the orientation of the incident light with respect to the refringent substance.

The g value for a given orientation depends on θ and ϕ values (Figure 6.5) according to the following relation:

$$g^2 = (g_{xx}^2 \cos^2 \phi \sin^2 \theta + g_{yy}^2 \sin^2 \phi \sin^2 \theta + g_{zz}^2 \cos^2 \theta) \qquad (6.18)$$

Accordingly, the Zeeman resonance will occur at field values given by

$$B_{res} = (h\nu/\mu_B)(g_{xx}^2 \cos^2 \phi \sin^2 \theta + g_{yy}^2 \sin^2 \phi \sin^2 \theta + g_{zz}^2 \cos^2 \theta)^{-1/2} \quad (6.19)$$

In the most general case, the resonance observed for a paramagnetic center in a single crystal is obtained at distinct field values B_x, B_y, or B_z when the magnetic field is parallel to the x ($\theta = \pi/2$, $\phi = 0$), y ($\theta = \pi/2$, $\phi = \pi/2$), or z ($\theta = 0$) crystal axis, respectively. The g values corresponding to these three orientations (g_{xx}, g_{yy}, g_{zz}) are the principal (diagonal) elements of the **g** tensor and are usually distinct. This is the case of orthorhombic symmetry. However, in the case of axial symmetry, the ellipsoid is also axially symmetric as two particular components (e.g., $g_{xx} = g_{yy}$) are equal and different from the third one, g_{zz}. The identical components are usually denoted $g_{zz} = g_{yy} = g_\perp$ and the single component $g_{zz} = g_{//}$, if the Oz axis is the principal symmetery axis. In this case, two particular resonance field values are observed; $B_\perp$ for $\theta = \pi/2$ independent of ϕ and $B_{//}$ for $\theta = 0$. In the case of a spherical symmetry, the ellipsoid changes to a sphere as the three principal components g_{xx}, g_{yy}, and g_{zz} are now equal, and one can denote the g components as g_{iso} ($g_{xx} = g_{yy} = g_{zz} = g_{iso}$). A single resonance field is observed at B_{iso} independent of both θ and ϕ.

6.4.1.2. Experimental Measurement of the g Factor

Absolute determinations of the g values may, in principle, be carried out by independent and simultaneous measurements of B and ν using a gaussmeter and a frequency meter, respectively, according to the equation

$$g = h\nu/\mu_B B \qquad (6.20)$$

which is the equivalent of equation (6.8) for the free electron.

In practice, the g value is often determined by comparing the field values at resonance for the sample investigated and that of a reference sample. As one can write

$$h\nu = g_{ref}\mu_B B_{ref} = g\mu_B B \qquad (6.21)$$

provided ν is invariant through the whole experiment, g is given by

$$g = g_{ref} B_{ref}/B \qquad (6.22)$$

The usual reference samples are diphenyl-picryl-hydrazyl (DPPH, $g =$ 2.0036), Varian Pitch ($g = 2.0029$), and Cr^{3+} in a MgO matrix ($g = 1.9797$). These reference samples give rise to the narrow lines necessary for accurate determinations. The reference sample can be placed in one of the two compartments of a dual cavity or stuck to the investigated sample quartz tube in the case of a single cavity.

6.4.1.3. Experimental Determination of the Principal g Values in Powder EPR Spectra

A powder sample is made of numerous microcrystals randomly oriented in space. The corresponding EPR spectrum is thus the envelope of elementary resonance lines corresponding to the various orientations in space assumed by the symmetry axes of the microcrystals with respect to the magnetic field.

6.4.1.3a. Axial Symmetry. Assuming Oz to be the principal symmetry axis, θ, as defined previously, is then the angle of this principal axis with respect to the magnetic field B. In this case of axial symmetry

$$g_{xx} = g_{yy} = g_\perp \quad \text{and} \quad g_{zz} = g_{//} \qquad (6.23)$$

Equation (6.18) can be written as

$$g = (g_\perp^2 \sin^2 \theta + g_{//}^2 \cos^2 \theta)^{1/2} \qquad (6.24)$$

independent of ϕ as we are dealing with an axial symmetry. The resonance will then occur at a field value given by

$$B_{res} = (h\nu/\mu_B)(g_\perp^2 \sin^2 \theta + g_{//}^2 \cos^2 \theta)^{-1/2} \qquad (6.25)$$

where $g_{//}$ and $g_\perp$ are measured with the axis of the paramagnetic species respectively parallel and perpendicular to the applied magnetic field. The powder spectrum is the envelope of the individual lines corresponding to all possible orientations in the whole range of θ. Assuming that the microcrystals are randomly distributed and that the individual lines have a Dirac delta function line shape, i.e., zero linewidth, simple considerations show that the absorption intensity, proportional to the number of microcrystals at resonance for a given θ value, is a maximum when $\theta = \pi/2$ ($B_\perp$) and minimum for $\theta = 0$ ($B_{//}$). This allows the extraction of the $g_{//}$ and $g_\perp$ values, which correspond to the turning points of the spectrum. Figure 6.6 is a schematic representation of the absorption curve and its first derivative as a function of B for polycrystalline samples with paramagnetic species in an axially symmetric environment. The $B_{//}$ and $B_\perp$ values are taken at the extremes of the first-derivative curve, and $g_{//}$ and $g_\perp$ are determined as indicated in Figure 6.6, which corresponds to $g_{//} > g_\perp$. One can also obtain the reverse case, $g_{//} < g_\perp$.

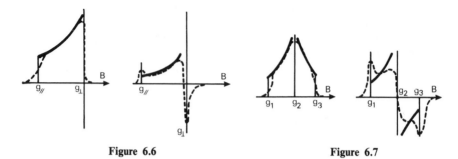

Figure 6.6   Figure 6.7

Figure 6.6. Absorption and first-derivative powder EPR signals in the case of an axial symmetry. Solid lines are theoretical lines assuming a zero width for the individual line corresponding to each microcrystal. Dotted lines are those obtained for a finite individual linewidth.

Figure 6.7. The orthorhombic symmetry case. Dotted lines and solid lines are defined as in Figure 6.6.

In fact, the resolution of each EPR signal is defined by an anisotropy factor:

$$\delta = \Delta B_{an}/\Delta B_i \qquad (6.26)$$

with

$$\Delta B_{an} = B_{/\!/} - B_\perp \qquad (6.27)$$

where $B_{/\!/}$ and $B_\perp$ are the precise field values taken to determine $g_{/\!/}$ and $g_\perp$, and ΔB_i is the individual linewidth characteristic of every microcrystallite. The larger δ, the higher the resolution, and this occurs for narrow individual lines. Thus, determination of the actual width and shape of the individual line is essential, as this gives access to such parameters as the spin distribution and relaxation times. A number of methods have been proposed; some are graphical[10,11] while others need computer methods.[12]

6.4.1.3b. Orthorhombic Symmetry. Similarly, the absorption curve and its first derivative exhibit singular points which allow determination of g_1, g_2, and g_3 as indicated in Figure 6.7. Indeed, three principal components are expected for the **g** tensor for such a symmetry. The extracted values may be identified as the components g_{xx}, g_{yy}, and g_{zz} on the basis of theoretical considerations.

6.4.1.3c. High Symmetry. In this case, the **g** tensor is isotropic with $g_{xx} = g_{yy} = g_{zz} = g_{iso}$. This occurs for perfect cubic, octahedral, tetrahedral, and spherical symmetries. A single line is obtained, which should be symmetrical. This is the simplest case.

6.4.1.3d. Particular Cases. Determination of the **g** tensor components is seldom simple and extreme care must be taken. There are cases where the

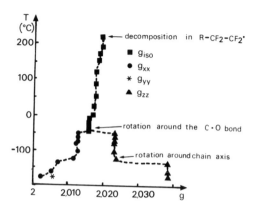

Figure 6.8. The **g** tensor components variation of peroxy radicals in polytetrafluoroethylene (R—CF$_2$—CF$_2$OO·) as a function of recording temperature.[13] The symbols •, * and ▲ are the three g components in the frozen state.

paramagnetic species give rise to unexpected signals. Those most frequently encountered are discussed below.

(i) *Mobility effect.* When a paramagnetic species is undergoing a rapid reorientation, a mean value of the principal components is observed upon averaging out the g anisotropy:

$$g_m = (g_{xx} + g_{yy} + g_{zz})/3 \tag{6.28}$$

In the case of a rapid rotation around a particular axis, two g values are observed, one of which is the average of two components:

$$g_{\perp m} = (g_{xx} + g_{yy})/2, \qquad g_{//} = g_{zz} \tag{6.29}$$

In general, with the assumption that motion is the only averaging process, the averaging of the g components must depend on the temperature of the EPR experiment and must be reversible,[13] provided there is no chemical change in the temperature domain where the rotation takes place (Figure 6.8).

(ii) *Poor resolution.* In a number of cases, two g components may be close enough so that only one line is observed in practice. It is possible to resolve the two close lines by running the EPR experiment at different frequencies but at the same temperature (Figure 6.9).[14] Quite often the Q band (35 GHz) is used, the X band being the usual routine EPR radiation.

(iii) *Concentration effect.* When the concentration of the paramagnetic species is too large, a dipolar broadening is observed, resulting from spin–spin interactions described by the following Hamiltonian:

$$\mathcal{H}_D = g_e^2 \mu_B^2 \left[\frac{S_1 \cdot S_2}{r^3} - \frac{3(S_1 \cdot r)(S_2 \cdot r)}{r^5} \right] \tag{6.30}$$

where **r** is the vector relating the two spins. This interaction depends both on the distance r and the angle between **r** and **B**, and results in resonances which expand around an average position. Thus, the experimental line will be the

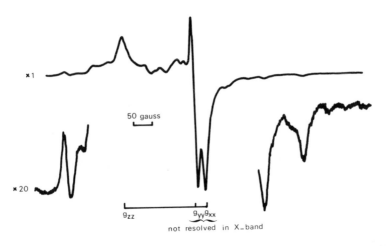

Figure 6.9. The resolution of the two components g_{xx} and g_{yy} in the EPR signal of peroxy radicals in polytetrafluoroethylene ($R—CF_2—CF_2OO^{\cdot}$) at 77 K and in the Q band.[14]

envelope of all individual resonances for various **r** and angles. Equation (6.30) shows that this interaction increases rapidly as r decreases, i.e., as the spin concentration increases. Therefore, a meaningful EPR spectrum of a given paramagnetic species requires an extreme dilution of the paramagnetic substance in a diamagnetic matrix.

This dipolar broadening is often used advantageously to distinguish between bulk and surface species. For this purpose, molecular oxygen (which is a triplet state molecule with two unpaired electrons) is allowed to interact with the paramagnetic substance. Surface species give rise to dipolar broadening in the presence of oxygen

Paradoxically, an exchange narrowing may be observed when the spin concentration increases beyond a certain limit which favors the overlap of atomic orbitals, thus allowing rapid passage or exchange of spins from one radical to another.

(iv) *Saturation effect*. When the relaxation time of a paramagnetic species is long, downward spin transitions become slow with respect to upward induced transitions. The populations of the two energy levels tend to equalize and no further absorption of quanta can occur. This phenomenon is known as the saturation phenomenon and must be avoided by the use of low microwave power. This power must be limited to those values for which the signal intensity remains proportional to the square root of the microwave power.

(v) *Spin–lattice relaxation effect*. As predicted by equation (6.9), short spin–lattice relaxation times (T_{1e}) result in important linewidths ΔB which may preclude g value determinations. It would then be necessary to cool the sample temperature down so as to decrease the linewidth ΔB as predicted by equation (6.7), thus allowing the observation of well-resolved signals.

TABLE 6.2. The **g** Tensors for Some Surface or Bulk Inorganic Radicals

Radical		g_1	g_2	g_3	Reference
17 e$^-$	CO_2^- on MgO	2.0029	2.0017	1.9974	16
	CO_2^- in $CaCO_3$	2.0032	2.0016	1.9973	17
19 e$^-$	SO_2^- on MgO	2.0097	2.0052	2.0028	18
	SO_2^- in $K_2S_2O_5$	2.0103	2.0055	2.0018	19

6.4.1.4. The Use of g Values

6.4.1.4a. Identification of Paramagnetic Species. Radicals fall into two categories: (1) those which may be identified on the basis of the sole **g** tensor; and (2) those which cannot be identified on the basis of the sole **g** tensor.

In the first class,[15] the angular orbital momentum **L** of the unpaired electron is aligned along a particular axis under the influence of an intramolecular field. To a first approximation, **L** does not contribute to the paramagnetism of the molecule involved. The *g* shift from g_e may be ascribed to the spin–orbit coupling which slightly quenches the coupling between the angular orbital momentum and the intramolecular electric field. In this class of radicals, the ground state is nondegenerate and the energy levels are well separated and insensitive to the influence of local crystal fields. In that case, the **g** tensor may be considered as a fingerprint of the radical. Several radicals fall into this class: CO_2^- and SO_2^- are typical examples (Table 6.2).

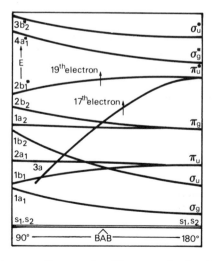

Figure 6.10. The Walsh energy diagram of an AB_2 type molecule with 17 and 19 electrons.[20]

All AB_2 radicals are nonlinear because of the stabilization energy relative to the neutral AB_2 molecule[20] (Figure 6.10); this in turn accounts for well-separated energy levels. Whenever two distinct radicals correspond to the same EPR signal, comparison of the experimental g values with those available in the literature may remove the ambiguity. It must be noted that isoelectronic species give rise to very similar g values.

In the case of the second class of radicals, the ground state is usually degenerate. The presence of the crystal field removes the orbital degeneracy and can drastically modify the g values, which can no longer be used reliably for identification purposes. Identification requires a qualitative agreement between observed and theoretical g values. With the spin Hamiltonian defined in equation (6.15) taken as a perturbation of the total Hamiltonian, perturbation theory gives to first-order the following relation:

$$g_{ii} = g_e - 2\lambda \sum_{(n \neq 0)} \frac{\langle 0 | L_i | n \rangle \langle n | L_i | 0 \rangle}{E_n - E_0} \tag{6.31}$$

where L_i is the angular orbital momentum operator in the i direction, $|0\rangle$ and $|n\rangle$ are the ground- and excited-state wave functions, respectively, and E_0 and E_n the corresponding energies. The g values are thus very dependent upon the mixing in of the excited state by spin–orbit coupling. The spin–orbit coupling constant λ for many-electron ions is related to the positive one-electron spin–orbit coupling ξ by the relation

$$\lambda = \pm \xi/2S \tag{6.31a}$$

where the sign $+$ holds for a shell less than half full and the sign $-$ for a more-than-half-full shell of electrons. When the shell is less than half-filled (for example Ti^{3+}), spin–orbit effects reduce g from 2.0023, while for a shell more than half-filled (for example Cu^{2+}), spin–orbit effects increase g above 2.0023.

Evaluation of the matrix elements $\langle 0 | L_i | n \rangle$ and $\langle n | L_i | 0 \rangle$ in equation (6.31) allows the derivation of g values using perturbation theory. The results can be summarized by writing the following general expression for the g values of $S = \frac{1}{2}$ systems with orbitally nondegenerate ground states:

$$g_{ii} = g_e - n\lambda/(E_n - E_0) \tag{6.32}$$

where the constant n is obtained after evaluating the matrix elements $\langle 0 | L_i | n \rangle$ and $\langle n | L_i | 0 \rangle$. The calculations show that the values of n depend on the nature of the ground state orbital.

In view of equation (6.31), the following observations can be made:

1. The g shift:

$$\Delta g_{ii} = g_{ii} - g_e \tag{6.32a}$$

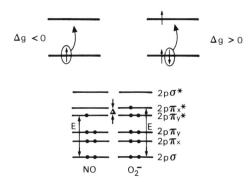

Figure 6.11. NO and O_2^- molecular orbital energy diagram, Δ is the energy splitting due to the surface crystal field.

of the **g** tensor components depends upon the orientation of the paramagnetic species with respect to the magnetic field. Indeed, the L_i components have different abilities to mix ground and excited states and, therefore, generate distinct Δg_{ii}s.

2. The magnitude of Δg_{ii} depends on the ratio of the spin–orbit coupling constant λ to the excitation energy $E_n - E_0$.

3. As shown above, the sign of the second right-hand term of equation (6.31) depends on whether the shell is less than half-filled or more than half-filled. In the first case, it is negative so that Δg_{ii} is negative, while it is positive in the opposite case leading to positive Δg_{ii}.

These two situations are illustrated by the O_2^- and NO radicals whose molecular orbital energy diagrams are given in Figure 6.11.

The separation Δ is the effect of the crystal field, and the Δ and E transitions are due to the L_i operator acting on the ground state [equation (6.31)]. When B is parallel to the internuclear axis taken as the Oz axis, the excited states are those corresponding to the electron jump from the $2p\pi_y^*$ and the $2p\pi_x^*$ orbital for both NO and O_2^-. However, while for O_2^- a positive g shift is expected, the opposite is predicted for NO. These expectations are in agreement with observed g values for both NO and O_2^- adsorbed on MgO (Table 6.3).

An alternative interpretation may be put forward. One might consider the O_2^- and NO energy diagrams as symmetrical and assume that an electron hole exists in the $2p\pi^*$ orbitals for O_2^-. A hole and an electron have

TABLE 6.3. The **g** Tensors for NO and O_2^- Radicals Adsorbed on MgO

Radical	$g_1 = g_{zz}$	$g_2 = g_{xx}$	$g_3 = g_{yy}$	Reference
NO/MgO	1.89	1.996	1.996	21
O_2^-/MgO	2.077	2.001	2.007	22

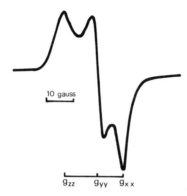

Figure 6.12. X-band EPR signal of O_2^- ions adsorbed at Mo^{6+} ions of silica-supported molybdenum oxide.[24]

opposite behaviors from an electrical point of view which are reflected by opposite g shifts.

6.4.1.4b. Surface Crystal Field Determinations. The accurate calculation of the **g** tensor values has been performed in the case of O_2^-. The following expressions have been obtained[23]:

$$g_{xx} = g_e - \lambda^2/\Delta^2 + \lambda^2/\Delta E \qquad (6.33)$$

$$g_{yy} = g_e + 2\lambda/E - \lambda^2/\Delta^2 - \lambda^2/\Delta E \qquad (6.34)$$

$$g_{zz} = g_e + 2\lambda/\Delta \qquad (6.35)$$

where λ is the spin–orbit coupling constant for oxygen, Δ and E the energy transitions indicated in Figure 6.11 and $g_e = 2.0023$. The calculation method, valid for $\lambda \ll \Delta < E$, clearly indicates that g_{xx} and, to a lesser extent, g_{yy} will be fairly close to g_e while g_{zz} will depend significantly on the $2p\pi^*$ energy splitting due to the crystal field.

Hence, when O_2^- is stabilized on the surface of a properly activated oxide, determination of the surface crystal field strength Δ at the adsorption site will be possible through g_{zz} measurements, using equation (6.35). A reliable pattern of data obtained for a number of oxides has unveiled the g_{zz} variation as a function of the oxidation number of the adsorption site.[6] Figure 6.13 gives the EPR spectrum of O_2^- adsorbed on Mo^{6+} of molybdenum oxide.[24]

6.4.1.4c. Mobility. As already noted in Section 6.4.1.3d, information about rotational and oscillating motion may be gained through examination of the **g**-tensor dependence on the sample temperature.

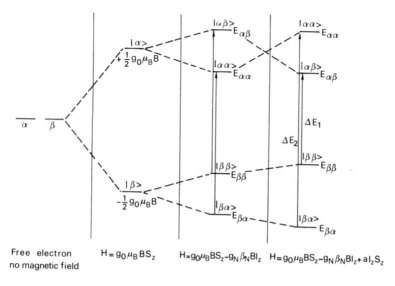

Free electron
no magnetic field

$H = g_0\mu_B B S_z$

$H = g_0\mu_B B S_z - g_N \beta_N B I_z$

$H = g_0\mu_B B S_z - g_N \beta_N B I_z + a I_z S_z$

Figure 6.13. The influence of various terms on the energy of a hydrogen atom in an external magnetic field. For the definition of various symbols, see text.

6.4.2. The Hyperfine Structure

6.4.2.1. The Nuclear Magnetic Moment and the Nuclear Zeeman Effect

An odd electron can be trapped in an anion vacancy of an ionic solid and experience the crystal field prevailing in this location. Such a defect is called an F center. However, the odd electron is often part of an atom, an ion, or a neutral molecule. In the case in which any of these entities has one or more nuclei with a nonzero nuclear spin ($I \neq 0$), the electron spin interaction with such nuclear spins gives rise to the hyperfine structure. Such nonzero-spin nuclei possess a magnetic moment μ_n associated with their nuclear spin $\mathbf{I}$ by the following relation:

$$\mu_n = g_N \beta_N \mathbf{I} \tag{6.36}$$

where g_N is the nuclear g factor and β_N the nuclear magneton, which is smaller than the eletronic Bohr magneton by a factor 1838, i.e., the ratio of the proton to electron mass. Equation (6.36) is the analogue of equation (6.1) with the difference that, here, μ_n and $\mathbf{I}$ are collinear and parallel. For a nuclear spin $I = \frac{1}{2}$, the nuclear Zeeman effect, i.e., the interaction of the nuclear magnetic moment with an external applied magnetic field, can be described by the following Hamiltonian:

$$\mathcal{H} = - g_N \beta_N \mathbf{B} \mathbf{I} \tag{6.37}$$

If it is assumed that **B** lies in the z direction (B_x and B_y and zero and $B = B_z$), the interaction energy corresponds to

$$\mathcal{H} = -g_N\beta_N B I_z \tag{6.38}$$

It is essentially similar to equation (6.4) for the electron spin, except that the sign of the right-hand term is now *negative*. The energies corresponding to the two allowed orientations of the nuclear spin are therefore

$$E = (\mp \tfrac{1}{2})g_N\beta_N B \tag{6.39}$$

These two energy levels are often referred to as nuclear Zeeman levels. The lower energy level corresponds to $M_I = +\tfrac{1}{2}$, the situation where **B** and μ_n are antiparallel. The latter two are parallel when $M_I = -\tfrac{1}{2}$, which corresponds to the upper energy level The hyperfine interaction may be isotropic or anisotropic. We shall start by considering the first case.

6.4.2.2. EPR Spectra with Isotropic g and Hyperfine Tensors

This case is illustrated by organic radicals in solution which possess nuclei with nonzero spins, particularly hydrogen ($I = \tfrac{1}{2}$). The corresponding EPR spectra exhibit a number of hyperfine lines, separated by hyperfine coupling constants and characterized by their relative intensities. The latter three parameters define the so-called hyperfine structure. By virtue of motional averaging in solution or because the radical under consideration has spherical or cubic symmetry, anisotropic effects are absent and the energy levels can be described by means of the following spin Hamiltonian:

$$\mathcal{H} = g_0\mu_B \mathbf{BS} - g_N\beta_N \mathbf{BI} + a\mathbf{IS} \tag{6.40}$$

The terms of the Hamiltonian are the electronic Zeeman term, the nuclear Zeeman term, and the hyperfine electron–nuclear coupling term. In the electronic Zeeman term, the usual **g** tensor is replaced by g_0, a scalar, since **g** is isotropic in solution. In equation (6.40), a is the hyperfine coupling constant since the hyperfine tensor is also isotropic; **I** is the nuclear spin; g_N and β_N are the usual constants, i.e., the nuclear g factor and the nuclear magneton. The hyperfine coupling constant a is given by

$$a = (8\pi/3)g_e g_N \mu_B \beta_N |\Psi(0)|^2 \tag{6.41}$$

where $|\Psi(0)|^2$ is the odd-electron density at the nucleus concerned. Consequently, the hyperfine interaction defined by

$$\mathcal{H} = a\mathbf{IS} \tag{6.42}$$

is, in principle, possible when the odd electron is located in an orbital with some s character. This interaction is in fact of the quantum type, and represents the interaction energy of the nuclear magnetic moment with magnetic fields induced by electric currents generated by the odd-electron spin at the nucleus.

The hydrogen atom (in free space) is a simple example to discuss because, by virtue of its spherical symmetry, anisotropic effects are absent. In this system, the electron spin is interacting with the proton which possesses a nuclear spin $I = \frac{1}{2}$. Assuming **B** lies in the z direction (B_x and B_y are zero and $B = B_z$), we can write the Hamiltonian of equation (6.40) as

$$\mathcal{H} = g_0 \mu_B B S_z - g_N \beta_N B I_z + a I_z S_z \qquad (6.43)$$

neglecting the second-order effects due to $a I_x S_x + a I_y S_y$ normally associated with the third term of the Hamiltonian.

The spin states can be conveniently represented by the notation $\alpha\alpha$, $\alpha\beta$, etc., where α refers to a spin state of $+\frac{1}{2}$ and β to a spin state of $-\frac{1}{2}$. The first term represents the electron spin state and the second term the nuclear spin state. Thus, $\alpha\alpha$ indicates $M_s = \frac{1}{2}$ and $M_I = \frac{1}{2}$, $\alpha\beta$ $M_s = \frac{1}{2}$ and $M_I = -\frac{1}{2}$, etc. The state with lowest energy will be the $\beta\alpha$ state, i.e., that corresponding to $M_s = -\frac{1}{2}$ (see Section 6.2.2) and $M_I = +\frac{1}{2}$ (see Section 6.4.2.1). The energies corresponding to the various spin states and derived from equation (6.43) are thus:

$$E_{\alpha\alpha} = +\tfrac{1}{2} g_0 \mu_B B - \tfrac{1}{2} g_N \beta_N B + \tfrac{1}{4} a \qquad (6.44)$$

$$E_{\alpha\beta} = +\tfrac{1}{2} g_0 \mu_B B + \tfrac{1}{2} g_N \beta_N B - \tfrac{1}{4} a \qquad (6.45)$$

$$E_{\beta\beta} = -\tfrac{1}{2} g_0 \mu_B B + \tfrac{1}{2} g_N \beta_N B + \tfrac{1}{4} a \qquad (6.46)$$

$$E_{\beta\alpha} = -\tfrac{1}{2} g_0 \mu_B B - \tfrac{1}{2} g_N \beta_N B - \tfrac{1}{4} a \qquad (6.47)$$

The EPR transitions correspond to $\Delta M_s = \pm 1$ and $\Delta M_I = 0$ and the NMR ones to $\Delta M_I = \pm 1$ and $\Delta M_s = 0$. The spectrometer described in Section 6.3 is set up to detect only EPR transitions, i.e., those from $\beta\beta$ to $\alpha\beta$ (transition 1) and from $\beta\alpha$ to $\alpha\alpha$ (transition 2). From equations (6.44) to (6.47) the energy differences associated with those transitions are derived as

$$\Delta E_1 = E_{\alpha\beta} - E_{\beta\beta} = g_0 \mu_B B - \tfrac{1}{2} a \qquad (6.48)$$

$$\Delta E_2 = E_{\alpha\alpha} - E_{\beta\alpha} = g_0 \mu_B B + \tfrac{1}{2} a \qquad (6.49)$$

The two expressions resemble equation (6.6), which led to a single EPR line. Owing to the presence of the proton which carries a nuclear spin $I = \frac{1}{2}$, the EPR spectrum of the hydrogen atom consists of two lines of equal intensity, separated by about 506 G, the hyperfine coupling constant a, and are symmetrically disposed around the position expected if there had been no nuclear spin in interaction with the unpaired electron.

The value of a is very often reported in units of G, MHz, or cm^{-1}. It should be stressed that the line separation in a spectrum expressed in G is given by $a/g_0\mu_B$, where a is in ergs and $g_0\mu_B$ in ergs G^{-1}. When g departs from the value of 2, it is incorrect to give a in G. One has to multiply the line separation by $g_0\mu_B$ and then divide by $g_e\mu_B$, where g_e is the free electron g value of 2.0023, to report a correct value of a in G. Since a is an energy, it is suitable to give its value as an energy (i.e. in Joules). However, the value of a is also commonly reported in cm^{-1}. This unit is obtained by multiplying the line separation in G by $g_0\mu_B$, with μ_B in cm^{-1} G^{-1}. There is no g-value dependence for this unit. The value of a in MHz is obtained by multiplying a (cm^{-1}) by c (3×10^{10} cm s^{-1}) and dividing by 10^6.

When several nuclei with nonzero nuclear spins ($I \neq 0$) interact with the unpaired electron with the radical, the EPR spectrum is more complex. It is easily shown that the total number of hyperfine lines is given by the general relation

$$N = (2n_1 I_1 + 1)(2n_2 I_2 + 1)(2n_3 I_3 + 1) \cdots \qquad (6.50)$$

where $n_1, n_2, n_3, \ldots$ are the numbers of equivalent nuclei interacting with the odd electron with nuclear spin, respectively $I_1, I_2, I_3, \ldots$ and giving rise to the hyperfine coupling constants $a_1, a_2, a_3, \ldots$. In equation (6.50), there are as many right-hand terms as groups of equivalent nuclei interacting with the odd electron.

The last important parameter is the distribution of the relative intensities. When the radical only contains nuclei with $I = \frac{1}{2}$ nuclear spins, such as hydrogen or carbon-13, the relative intensities are given by the coefficients of the binome $(1 + x)^n$, where n represents the number of equivalent nuclei. The determination of the relative intensities, though not complicated, becomes increasingly more tedious, when the nuclei with nonzero nuclear spins ($I \neq 0$) are not of the same nature and increase in number.

From the knowledge of the three parameters involved in the hyperfine structure, namely the number of hyperfine lines, their relative intensities and the hyperfine splitting constant(s), it is possible to deduce the nature of the radical involved. Conversely, knowing the radical nature, it is also possible to derive the expected hyperfine structure.

6.4.2.2a. Spin Polarization.[25] Isotropic hyperfine structures are observed in the case of aromatic radicals such as $\cdot C_6H_6^-$. In this case, the unpaired electron is located in a π molecular orbital (MO) which expands to all the carbon structure of the molecule. This MO, formed from the overlap of atomic $2p_z$ orbitals (where z is the direction perpendicular to the carbon ring) has a node in the molecular plane containing the protons which give rise to the hyperfine structure. The question arises as to how the s orbitals of the hydrogen atoms can share a finite spin density to account for the observed hydrogen hyperfine structure.

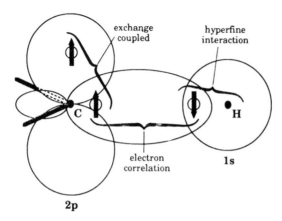

Figure 6.14. Possible electron pairing in a C(sp²)–H(1s) fragment. The $2p$ orbital is the nonhybridized carbon $2p_z$ orbital perpendicular to the plane determined by the three sp^2 orbitals principal axes.[25]

Consider the isolated —C—H fragment with a π electron located in the carbon $2p_z$ orbital lying perpendicular to the plane formed by the three trigonal bonds depicted by this fragment. Two structures are possible. The structure where the two carbon electrons have parallel spin (Figure 6.14) is energetically favored with respect to the opposite structure, as expected from Hund's rule. The electron in the hydrogen 1s orbital must thus be antiparallel to (paired with) the sp^2 hybridized carbon electron of the σ(C—H) bond. The exchange forces couple the (C—H)σ bond electron and the π electron of the carbon. Assuming the odd electron is in an α state ($M_s = +\frac{1}{2}$), there will be excess spin density in the σ bond originating from the spin polarization imposed by the coupling. Note that an α state of the odd π electron will induce a β state in the hydrogen 1s orbital. Theoretically, a single π electron generates a negative spin density of (–0.05) in the hydrogen 1s orbital, corresponding to a coupling constant of 20–25 G.

6.4.2.2b. The McConnell Relation.[26] In the aromatic radicals, the unpaired π electron is delocalized over all carbons. Therefore, all (C—H)σ bonds are polarized in proportion to the π electron spin density ρ_π, which is expressed by

$$a_H = Q\rho_\pi \qquad (6.51)$$

a_H is the hyperfine coupling constant due to hydrogen nuclei and Q is the McConnell constant of –22.5 G for aromatic hydrocarbons and –25 G for aliphatic hydrocarbons. Table 6.4 illustrates this relation.

TABLE 6.4. Hyperfine Coupling Constant, Spin Density and Effective
Q Value in the Cyclic Polyene Radicals $^{\cdot}C_nH_n$

Radical	Spin density ρ_π	Hyperfine coupling constant a_H (G)	Effective Q value (G)
$^{\cdot}CH_3$	1	-23.04	-23.04
$^{\cdot}C_5H_5$	$\frac{1}{5}$	-5.98	-29.9
$^{\cdot}C_6H_6^-$	$\frac{1}{6}$	-3.75	-22.5
$^{\cdot}C_7H_7$	$\frac{1}{7}$	-3.91	-27.4
$^{\cdot}C_8H_8^-$	$\frac{1}{8}$	-3.21	-25.7

6.4.2.3. EPR Spectra with Anisotropic g and Hyperfine Tensors

By contrast with the isotropic hyperfine interaction, anisotropic electron–nuclear interactions are frequent since they arise from the dipolar interaction between the electron magnetic moment μ_s and the nuclear magnetic moment μ_n. The corresponding energy is given by

$$E = \mu_s \mu_n/r^3 - 3(\mu_s \cdot r)(\mu_n \cdot r)/r^5 \tag{6.52}$$

where r is the vector relating the electron and nuclear magnetic moments μ_s and μ_n and r is the distance between these two moments. The quantum mechanics analogue, obtained by expressing μ_s and μ_n as a function of I, S, g_e, g_N, μ_B and β_N, is known as the anisotropic hyperfine Hamiltonian:

$$\mathscr{H}_{an} = -g_e\mu_B g_N\beta_N\{I \cdot S/r^3 - [3(I \cdot r)(S \cdot r)]/r^5\} \tag{6.53}$$

Equation (6.53) must be averaged over the entire probability of the spin distribution $|\Psi(r)|^2$. $\mathscr{H}_{an}$ is averaged out to zero when the electron cloud is spherical (s orbital) and comes to a finite value for axially symmetric orbitals (e.g., the p orbital). Thus, the hyperfine interaction due to protons is necessarily isotropic.

In practice, the experimental hyperfine tensor arises most frequently from the contribution of both isotropic and anisotropic hyperfine interactions. If the total hyperfine tensor is denoted as $\mathbf{A}$, it could be suitably represented by a diagonal matrix, provided adequate axes are chosen. This tensor may be split into an isotropic and an anisotropic part as follows:

$$\mathbf{A} = \begin{vmatrix} A_1 & 0 & 0 \\ 0 & A_2 & 0 \\ 0 & 0 & A_3 \end{vmatrix} = a + \begin{vmatrix} b_1 & 0 & 0 \\ 0 & b_2 & 0 \\ 0 & 0 & b_3 \end{vmatrix} \tag{6.54}$$

with

$$a = (A_1 + A_2 + A_3)/3 \qquad (6.55)$$

In a number of cases, the second-term matrix of equation (6.54), is traceless $(b_1 + b_2 + b_3 = 0)$ and of the form $(-b, -b, +2b)$. The anisotropic part of the **A** tensor corresponds to the dipolar interaction as expressed by the Hamiltonian of equation (6.53).

When both the **g** and **A** tensors are anisotropic, the total spin Hamiltonian is written as

$$\mathcal{H}_{an} = \mu_B \mathbf{BgS} - g_N \beta_N \mathbf{BI} + \mathbf{IAS} \qquad (6.56)$$

It can be seen that equations (6.40) and (6.56) are similar. In equation (6.56), however, **g** and **A** are now tensors and owing to matrix notation placed in between vectors **B** and **S**, and **I** and **S**, respectively. The first term has already been dealt with as equation (6.16) in Section 6.4.1.1, the middle term is the nuclear Zeeman energy [see equation (6.37) in Section 4.2.1] and the last one expresses the anisotropic electron–nucleus hyperfine coupling interaction.

The s and p characters of the orbital hosting the unpaired electron are given by the following relations:

$$C_s^2 = a/A_0 \qquad (6.57)$$

and

$$C_p^2 = b/B_0 \qquad (6.58)$$

where A_0 and B_0 are theoretically calculated coupling constants, assuming pure s and p atomic orbitals for the element under consideration, and

$$C_s^2 + C_p^2 + \cdots = 1 \qquad (6.59)$$

In a few cases, it is possible to determine particular bond angles ϕ from the hybridization ratio[24]:

$$\lambda = (C_p^2/C_s^2)^{1/2} \qquad (6.60)$$

In the case of C_{2v} symmetry:

$$\phi = 2 \cos^{-1}(\lambda^2 + 2)^{-1/2} \qquad (6.61)$$

while in a C_{3v} symmetry:

$$\phi = \cos^{-1}[1.5/(2\lambda^2 + 3) - \tfrac{1}{2}] \qquad (6.62)$$

6.4.2.4. *Experimental Determination of the* **g** *and* **A** *Tensors in Powder EPR Spectra*

The **g** and **A** tensors of the Hamiltonian in equation (6.56) may be isotropic, axial, or orthorhombic. The hyperfine interaction will result in the periodic repetition of the pattern observed with no hyperfine interaction, arising simply from the principal **g** tensor components. The periodicity will depend upon the nuclear spin value(s). One should observe $2I + 1$ lines for a single nucleus of nuclear spin I or $2nI + 1$ lines for n magnetically equivalent nuclei of the same spin I. Figure 6.15 illustrates a few cases where $I = 1$.

It is sometimes difficult to decide whether a particular pattern arises from the **g** tensor anisotropy or from a hyperfine splitting. The following method may help in the determination.

6.4.2.4a. Recording the Signal at Two Different Microwave Frequencies. The X and Q bands are the ones most often employed. Assuming both **A** and **g** to be axially symmetric, one can write for the X-band spectrum:

$$hv_X = g_\perp \mu_B B_{\perp_X}$$
$$hv_X = g_{//} \mu_B B_{//_X} \tag{6.63}$$

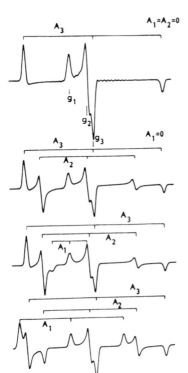

Figure 6.15. Powder spectra including a hyperfine structure due to one nucleus with $I = 1$.[28] The g components are the same in all spectra, the A components are given in each case.

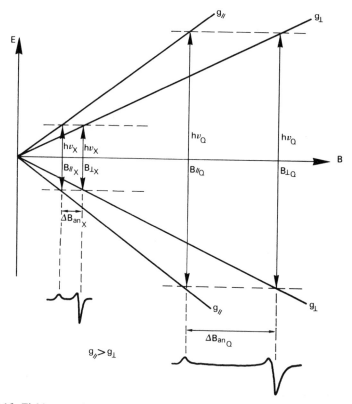

Figure 6.16. Field separation between the principal g values in the X and Q bands in the case of axial symmetry.

and for the Q-band spectrum:

$$hv_Q = g_\perp \mu_B B_{\perp_Q}$$

$$hv_Q = g_{//}\mu_B B_{//_Q} \tag{6.64}$$

The $g_\perp$ and $g_{//}$ values are measured at the inflection points of the spectra as shown earlier in Figure 6.6, i.e., at field values obeying the following relations:
For the X band

$$(B_{//} - B_\perp)_X = (hv_X/\mu_B)(1/g_{//} - 1/g_\perp) \tag{6.65}$$

For the Q band

$$(B_{//} - B_\perp)_Q = (hv_Q/\mu_B)(1/g_{//} - 1/g_\perp) \tag{6.66}$$

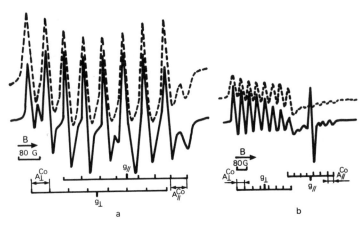

Figure 6.17. Experimental (solid lines) and simulated (dotted lines) spectra of $[Co(CH_3N^{12}C)_6]^{2+}$ in the X-band (a) and Q-band (b) modes. The isotropic line at the center of the $A_{//}$ pattern is due to another species.[30]

With the values of ν_X and ν_Q shown in Table 6.1, it then follows that

$$(B_{//} - B_\perp)_X/(B_{//} - B_\perp)_Q = \nu_X/\nu_Q \cong 1/3.7 \qquad (6.67)$$

Equation (6.67) indicates that the field separation between the principal g value positions will vary linearly with the resonance frequency (Figure 6.16), while patterns arising from hyperfine interaction are frequency-independent to a first approximation,[29] as is apparent from equation (6.41) and (6.53). The example of Figure 6.17 illustrates the use of two microwave frequencies in order to resolve **g** and **A** anisotropies.[30]

Clearly, the splitting constant observed in the perpendicular pattern centered around $g_\perp$ and that observed in the parallel pattern centered around $g_{//}$ did not vary significantly as the microwave frequency changed, while the spacing between $g_{//}$ and $g_\perp$ increased by the ratio of the two frequencies. This resulted in the appearance of additional lines in the parallel pattern which overlapped in the X-band spectrum with the more intense lines of the perpendicular pattern.

There are cases, however, where the use of the Q band does not improve the resolution of the spectrum when important site inhomogeneities exist, resulting in a broad distribution of g values which then overlap to give a broad featureless spectrum (Figure 6.18).[31]

6.4.2.4b. Isotopic Labeling. (i) *Use of isotopes having different nuclear spins.* Figure 6.19 illustrates such a procedure.[21] The odd electron of NO adsorbed on MgO is interacting with the ^{14}N nuclear spin ($I = 1$). Three lines with a splitting constant $A_{xx} = A_{yy} = A_\perp$ are observed around the perpendicular component $g_{xx} = g_{yy} = g_\perp$ while the parallel component was not resolved and appeared as a single line at $g_{zz} = g_{//}$. When ^{15}NO is adsorbed ($I = \frac{1}{2}$), a doublet is

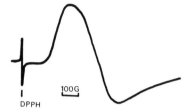

Figure 6.18. EPR spectrum (Q band, 300 K) of a ^{95}Mo enriched MoO₃/SiO₂ catalyst reduced at 500 °C.[31]

observed around $g_\perp$, thus confirming that the electron is actually interacting with a single nitrogen nucleus.

It can be seen that: (1) the hyperfine splitting varies as one goes from ^{14}NO to ^{15}NO and the splitting constants are in the ratio: $(g_N\beta_N)_{15}/(g_N\beta_N)_{14}$ [equations (6.41) and (6.53)]; (2) the $g_{//}$ component is broadened with ^{15}NO showing that $A_{//}$ is not zero but rather that the hyperfine splitting is simply poorly resolved. This method is frequently used for ^{14}N/^{15}N and ^{1}H/^{2}H isotopes or among the

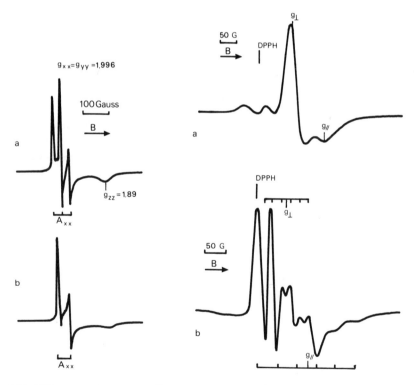

Figure 6.19. EPR spectra (93 K) of (a) ^{14}NO and (b) ^{15}NO adsorbed on MgO evacuated at 800 °C.[21]

Figure 6.20. EPR spectrum (X band, 300 K) of MoO₃/SiO₂ catalyst reduced by hydrogen at 500 °C: (a) molybdenum in natural abundance; (b) ^{95}Mo-enriched. MoO₃/SiO₂ catalyst.[31]

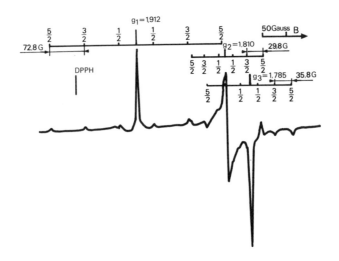

Figure 6.21. EPR spectrum (X band, 77 K) of a sample with 0.51 wt.% Mo in TiO_2 (rutile) after calcination to 800 °C in air.[32]

$^{19}F(I = \frac{1}{2})$, $^{35,37}Cl(I = \frac{3}{2})$, $^{78,81}Br(I = \frac{3}{2})$ and $^{127}I(I = \frac{5}{2})$ series in analogue compounds.

(*ii*) *Selective isotopic enrichment.* Figure 6.20 shows an example of the use of isotopic enrichment.[31] Natural molybdenum isotopes have zero or $\frac{5}{2}$ nuclear spins. Enrichment in the $I = \frac{5}{2}$ isotope gives an easy separation of hyperfine lines from features due only to the **g** tensor. Under such circumstances, it is possible to recognize an axial **g** tensor for Mo^{5+} species and to extract all the **A** tensor components.[31]

When the spectrum is well resolved, the relative intensity of the lines must be in accordance with the relative abundances of isotopes with $I = 0$ and $I \neq 0$. This is the case of Mo^{5+} ions dispersed in a rutile matrix (Figure 6.21).[32] Molybdenum has two isotopes ^{95}Mo and ^{97}Mo ($I = \frac{5}{2}$) with natural abundances of 15.78 and 9.60%, respectively, and with about the same magnetic moment. Thus, to a good approximation 25.38% isotopes have an $I = \frac{5}{2}$ nuclear spin giving rise to six hyperfine lines, while the remaining 74.62% have no nuclear spin and thus give only **g** components. The final powder spectrum will depend on the symmetry of the environment around each Mo^{5+} ion. The important point to stress is that each **g** tensor component will concentrate 74.62% intensity and appear in the middle of the hyperfine sextet, where each line contributes for 5.7% (i.e., the ratio 25.38/6).

6.4.2.4c. Spectra Simulation. Once the **A** and **g** principal components are obtained, the EPR spectrum can be simulated. The fit between the experimental and the computed spectrum confirms the extracted g and A values and even permits improvement of the accuracy of these values.[30]

6.4.2.4.4. Particular Cases. (i) *Mobility effect.* The hyperfine tensor will be subject to the same averaging effects as the **g** tensor and similar relations are used:

$$A_m = (A_{xx} + A_{yy} + A_{zz})/3 \tag{6.68}$$

or in the case of a rotation around the z axis:

$$A_{\perp_m} = (A_{xx} + A_{yy})/2$$
$$A_{//_m} = A_{zz} \tag{6.69}$$

(ii) *Concentration effect.* As already noted, a spin–spin dipolar interaction when the spin concentration increases results in an important broadening which undermines the nuclear spin coupling. Further concentration increases beyond a certain limit might result in a line narrowing via spin–spin exchange, yet no hyperfine splitting will be observed. The DPPH is a spectacular illustration of such behavior.[33]

In the polycrystalline form, a single line 3 G wide is observed. In a dilute solution, the same radical will give rise to five lines arising from the interaction of the odd electron with the two ^{14}N nuclei ($I = 1$). Further dilution and oxygen removal from the solution (which eliminates dipolar broadening) give rise to a hyperfine pattern of over a hundred hyperfine lines.[33]

(iii) *Saturation effects.* See Section 6.4.1.3d (iv).

(iv) *Spin–lattice relaxation effect.* See Section 6.4.1.3d (v).

6.4.2.5. The Use of the Hyperfine Tensor

The information obtained from the hyperfine tensor can be summarized as follows: (1) s and p characters of the unpaired electron orbital; (2) spin density distribution over the nuclei interacting with the odd electron; (3) bond angle determinations; (4) identification of the paramagnetic species (nature and number of nuclei in the radical, equivalence of nuclei) deduced from the number and relative intensity of the hyperfine lines, and (5) motion of radicals.

6.4.3. The Superhyperfine Structure

The unpaired electron may be coupled not only with the nucleus of the atom or ion to which it belongs, but also with more remote nuclei. When one or more such remote nuclei have a nonzero spin, the problem can be treated as in the previous case, and a superhyperfine structure will be observed. The same rules that apply to the hyperfine structure are obeyed. The superhyperfine structure can be superimposed on the hyperfine lines. A Hamiltonian similar to that given in equation (6.56) will account for this situation; the difference would reside in the fact that I will represent nuclear spins of nuclei not directly involved in the paramagnetic species. Determination of the magnetic parameters will follow the procedure given in Section 6.4.2.4.

Valuable information may be gathered from the superhyperfine parameters. In particular, the nature of the adsorption sites in catalysis and the ground state of transition metal ions can be accurately identified.[30,31,34]

6.5. APPLICATIONS OF EPR IN CATALYSIS

We have so far introduced the parameters relevant to EPR spectroscopy (g factor, hyperfine and superhyperfine tensors) as well as the information derived from these parameters in order to characterize a given paramagnetic center and its environment. As a catalyst is very often in a powdered polycrystalline form, we have also illustrated the different types of powder spectra that can be observed as well as the way to extract the EPR parameters from experimental powder spectra. The illustration of the case of more than one unpaired electron originating the fine structure of the EPR spectra has been omitted for the sake of brevity and because it is less relevant to catalytic studies. This is because the inhomogeneities and distortions of the crystal field at a solid surface give rise to very broad signals that are difficult to interpret. In the following, we will illustrate some applications of EPR techniques to catalytic phenomena.

6.5.1. Redox Properties of Surfaces[35]

Metal oxides are often involved in catalytic systems, either as catalysts or as supports. In many cases, their surfaces possess redox sites whose nature, number, and strength often must be known in order to understand their catalytic behavior. These values can be ascertained using charge transfer reactions, i.e., reactions involving the transfer of a single electron in solid–liquid or solid–gas systems. Charge transfer complexes are generally produced between the solid surface (S) and electron acceptor (A) or donor (D) organic molecules usually dissolved in benzene. The reactions that lead to the formation of radical ions can be written as

$$S + A \rightarrow S^+ + A^-$$

$$S + D \rightarrow S^- + D^+$$

The ability of a surface to enable such reactions depends respectively on the ionization potential (IP) or electron affinity (EA) of the organic molecule. The

IP (or EA) threshold corresponding to the electron transfer characterizes the site strength. Furthermore, on the basis of the intensity of the EPR signal, the number of donor (acceptor) sites can be determined by means of a double integration of the first-derivative EPR signal (see Section 6.3). Typical molecules with low IP are perylene, anthracene, and naphthalene (π bases) while commonly used high EA compounds are tetracyanoethylene, dinitrobenzene, and trinitrobenzene.

6.5.1.1. Oxidizing Properties

When perylene (IP = 6.8 eV) adsorbed on silica–alumina activated *in vacuo*, the color of the solid changes from white to purple and an EPR signal appears (Figure 6.22).[36] The hyperfine coupling constant (3.4 G) and the relative intensities of the hyperfine lines are in good agreement with those obtained in the case of Pe^+ formed by electron transfer between perylene and an acid solution of H_2SO_4. Anthracene and naphthalene, which also exhibit a low ionization energy, are transformed into positive ions as well by contact with

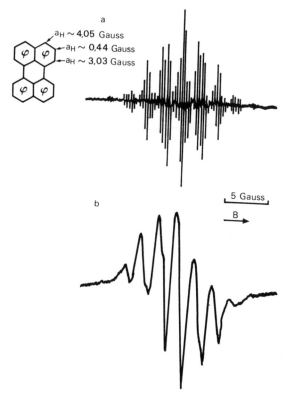

Figure 6.22. EPR spectra of the cation radical Pe^+ prepared (a) in concentrated sulfuric acid; (b) on silica–alumina ($\sim$ 13 wt.% Al_2O_3) at 293 K.[35]

amorphous silica–alumina samples. The oxidative properties of the silica–alumina system depend largely on the temperature to which the solid has been activated. Hydroxylated samples, in fact, are completely inactive toward electron transfer, whereas the maximum activity is observed for samples activated at 800 °C

The oxidation of aromatic hydrocarbons on silica–alumina very probably takes place on Lewis acid sites (vide infra). The method described above allows a distinction between the role of Brönsted and Lewis sites in heterogeneous catalysis as the aromatic hydrocarbons such as perylene interact with these latter sites only, leaving the former sites free.

Similar studies have been done on zeolites which are crystalline aluminosilicates exhibiting cages of different sizes. In the case of NaY zeolites, with natural faujasite crystal structure, the sodium ions can be substituted by ammonium ions by the ion exchange technique in solution. In the resulting zeolite (NH_4Y), the ammonium ions are associated with the tetrahedral AlO_4^- ions. By heating the NH_4Y zeolite the ammonium ions decompose evolving ammonia and leaving H^+ ions which form hydroxyl groups with lattice O^{2-} ions. The system obtained by this method is called HY zeolite. By increasing the temperature of the thermal treatment a second fraction of water molecules is eliminated in parallel with dehydroxylation and formation of tricoordinated aluminum ions according to the process:

The agreement between the weight loss of the zeolite and the appearance of oxidative properties measured by the concentration of adsorbed anthracene cations is fairly good (Figure 6.23). This confirms the existence and the nature of the electron acceptor sites generated by dehydroxylation of HY zeolite.[37]

The transition metal ion exchanged zeolites exhibit oxidative properties as well. Naccache and Ben Taarit[38] were able to show that in the case of Cu^{2+} exchanged zeolites, the copper ions were reduced by adsorption of anthracene with a parallel appearance of the EPR signal of the radical anthracene cation. Conversely, no EPR signal due to this cation was observed after preliminary reduction of Cu^{2+} to Cu^+ by CO. These results indicate that Cu^{2+} ions are indeed the electron acceptor sites. Similar conclusions were reached in the case of Ce^{3+} exchanged zeolites,[39] where no oxidative properties were observed except in the case of oxidation of Ce^{3+} to Ce^{4+}. In this case, the adsorption of anthracene or perylene causes the appearance of strong EPR signals, thus suggesting that the electron transfer with formation of the radical cation does indeed involve the Ce^{4+} ions.

Electron donor molecules were also used to measure the accessibility of oxidative sites. This can be illustrated by considering three different probe molecules such as benzene (IP = 9.2 eV), naphthalene (2 benzene rings,

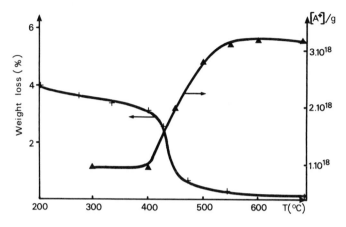

Figure 6.23. The relationship between the differential weight loss ($\Delta m/\Delta T$) and the concentration of anthracene cation radicals (A^+) adsorbed on a NH_4Y zeolite at a function of pretreatment temperature.[37]

TABLE 6.5. Influence of the Zeolite Nature on the Intensity of the EPR Signal of Radical Cations Formed from Different Electron Donor Molecules

Zeolite	Benzene	Naphthalene	Anthracene
ZSM-5	Strong signal	No signal	No signal
NH_4 mordenite	Weak signal	Strong signal	Weak signal
NH_4Y	No signal	Weak signal	Strong signal

IP = 8.2 eV), anthracene (3 benzene rings, IP = 7.2 eV). When these molecules are adsorbed on various dehydrated zeolites the formation of an EPR signal due to the adsorbed radical cation is observed in some cases. Table 6.5, deduced from the results in references Ben Taarit[40] and Wierzchowski *et al.*,[41] shows the main results obtained.

Oxidative sites were identifed that corresponded to the Lewis acid sites present in the system. It was observed that the ZSM-5 zeolite, which exhibits the highest acidity, gave rise to an intense signal with benzene but no signal with naphthalene and anthracene, which have an even lower ionization energy. These two large molecules cannot penetrate the very small channels typical of the ZSM-5 structure, so there is no oxidation of these hydrocarbons. Mordenite (Table 6.5) is accessible to benzene, naphthalene and, very weakly, to anthracene. This zeolite, however, shows a peculiar oxidative ability toward naphthalene (medium IP) indicating the presence of sites in the solid with medium acidic properties. Finally, the Y zeolite, with large cavities accessible to all three molecular probes, is characterized by a large number of weak acid sites, indicated by the high intensity of the signal observed in the case of anthrancene adsorption.

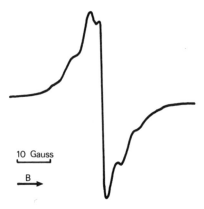

10 Gauss

B

Figure 6.24. EPR spectrum of the anion radical TCNE$^-$ adsorbed on alumina.[42]

6.5.1.2. Reducing Properties

Tetracyanoethylene adsorption $[(CN)_2C{=}C(CN)_2$, TCNE] on activated alumina causes the color of the solid to turn from white to red. At the same time, a nine-line EPR signal appears (Figure 6.24) due to the interaction of an unpaired electron with the four equivalent nitrogen nuclei. This signal is due to the formation of the TCNE$^-$ radical anion. If the activation temperature is lower than 300 °C, the donor sites are surface OH$^-$ groups whereas at higher temperatures the donor sites are low-coordination O^{2-} ions generated upon dehydroxylation of the surface.[42,43] Similar results have been obtained with magnesium oxide and titanium oxide[44] using trinitrobenzene and TCNE. Baird and Lunsford[45] showed that the rate of isomerization of 1-butene on MgO can be related to the concentration of surface donor sites as determined by TCNE adsorption.

As in the case of oxidative properties, the transition metal ions play a role in some reactions involving electron transfer toward an organic molecule. Mo^{5+}, for instance, can give TCNE anion radicals by direct electron transfer to TCNE.[46]

Pink et al.[47–49] have shown that some oxides (aluminas, silica–alumina, zeolites) are able to stabilize simultaneously positive (Pe$^+$) and negative (nitrobenzene) ions in adjacent positions. These findings gave deeper knowledge of the active sites present at the surface of oxides. The redox properties of alumina, for instance, have been used to explain the catalytic properties of this oxide.

6.5.2. Identification of Active Sites in Catalysis[50]

The adsorption of natural nitric oxide ^{14}NO on a dehydrated γ-alumina leads to the EPR spectrum shown in Figure 6.25a. The spectrum is due to the interaction of the unpaired electron of ^{14}NO with both ^{14}N and ^{27}Al exposed at the surface of the solid giving rise to the hyperfine and superhyperfine struc-

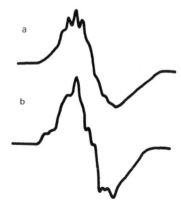

Figure 6.25. EPR spectrum of NO radicals adsorbed on γ-Al$_2$O$_3$: (a) after adsorption of 2×10^{12} NO molecules cm^{-2}; (b) spectrum simulated with the parameters indicated in the text.[50]

tures, respectively. A simulation of the experimental spectrum performed on the basis of the following parameters is shown in Figure 6.25b:

$$g_{xx} = 1.99, \quad g_{yy} = 1.9845, \quad g_{zz} = 1.945$$

$$A_{xx}^{N} = 30 \text{ G}, \quad A_{yy}^{N} = 5 \text{ G}, \quad A_{zz}^{N} = 10 \text{ G}$$

$$A_{xx}^{Al} = 12.5 \text{ G}, \quad A_{yy}^{Al} = 10 \text{ G}, \quad A_{zz}^{Al} = 12.5 \text{ G}$$

The effect of H$_2$S on the catalytic activity of γ-Al$_2$O$_3$ in the isomerization of 1-butene has been followed by the effect on the nitric oxide EPR spectrum, as a function of increasing H$_2$S doses. The results given in Figure 6.26 show that the intensity of the EPR signal of adsorbed NO and the catalytic activity both decrease with increasing doses of H$_2$S. This trend indicates that the exposed Al^{3+} ions capable of coordinating the NO molecule are indeed the active sites

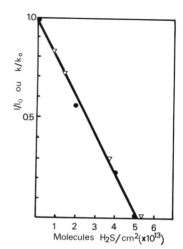

Figure 6.26. Variation of the relative rate constant k/k_0 of the 1-butene isomerization (triangles) and of the relative spin concentration of NO radicals (black circles) as a function of the doses of H$_2$S adsorbed on γ-Al$_2$O$_3$.[50]

for the catalytic isomerization reaction and that H_2S is an effective poison of both the adsorption of NO and the catalytic reaction. The high sensitivity of the EPR technique is particularly relevant to this application as less than 1% of the surface atoms are involved as active sites in the catalytic process.

6.5.3. Identification of Intermediates in Catalysis[51]

An important reaction in catalysis and in pollution problems is that between hydrogen sulfide and sulfur dioxide according to the following scheme (Claus reaction):

$$2\,H_2S + SO_2 \Leftrightarrow 2\,H_2O + (3/x)S_x$$

The usual catalyst is alumina but basic systems such as silica impregnated with NaOH have also been investigated. In the latter case, the adsorption of SO_2 causes the formation of SO_2^- radicals whose concentration has been monitored by EPR as a function of the amount of NaOH present on the sample (Figure 6.27). The maximum radical concentration is attained for a composition of 1.4 wt.% NaOH.

The SO_2^- radicals must react with H_2S because their EPR signal disappears on contact with the latter gas. The highest observed initial rate of the Claus reaction at 180 °C was that measured for a sample with 1.4 wt.% NaOH (Figure

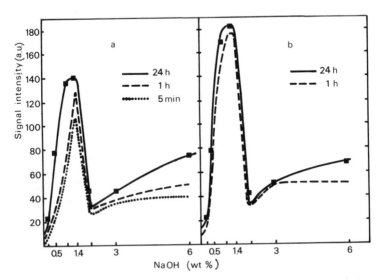

Figure 6.27. Effect of the concentration of NaOH deposited on a silica gel and of the SO_2 adsorption time on the concentration of the anion radicals SO_2^- (adsorption pressure 20 Torr): (a) at 0 °C, (b) at 26 °C.[51]

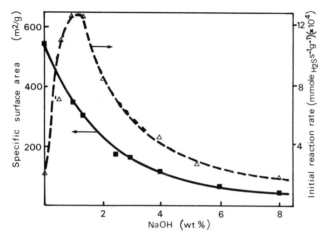

Figure 6.28. Specific surface area and initial Claus reaction rate at 180 °C as a function of NaOH deposited. Black squares — specific surface area. Open triangles — initial rate.[51]

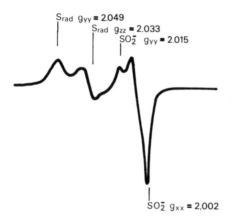

Figure 6.29. EPR spectrum during the Claus reaction carried out under dynamic conditions. Silica gel with 1.4 wt.% NaOH, 180 °C, 4 Torr H_2S and 2 Torr SO_2 in helium. Spectrum obtained after 5 min of reaction.[51]

6.28), so it was concluded that the SO_2^- radical is an intermediate of the catalytic reaction. To confirm this hypothesis, the EPR spectrum of the catalyst was recorded under the same conditions as the catalytic reaction (Figure 6.29). The spectrum shows the simultaneous presence of SO_2^- and of S_x radicals, in agreement with the scheme of the Claus reaction given above. Further, the surface SO_2^- radical can also be used to measure the electron donor properties of the solid.

6.5.4. Coordination Chemistry and Catalytic Properties of Supported Ni$^+$ Ions

Olefin dimerization in homogeneous media can be catalyzed by nickel complexes, probably with nickel in the Ni$^+$ oxidation state. Immobilization of isolated Ni$^+$ ions on a support (CaX zeolite, SiO$_2$) allows the same process to occur by heterogeneous catalysis.[52,53] Ni$^+$ is a paramagnetic ion with a $3d^9$ configuration. The coordination of the supported ions has been investigated by EPR following the adsorption of CO in both natural and ^{13}C-enriched forms. The results obtained in the case of Ni$^+$/SiO$_2$ are collected in Table 6.6 and correspond to the spectra shown in Figure 6.30.

Conclusions about the symmetry of the surface carbonyl species obtained by CO adsorption are based on the coherent analysis of the following four parameters:

1. The number of the **g** tensor components present in the spectrum which is related to the symmetry of the Ni$^+$ ion (isotropic, axial, orthorhombic).
2. The relative order of the **g** tensor components which allows the identification of the ion ground state.
3. The superhyperfine structure (number and relative intensities of the superhyperfine lines, superhyperfine splitting constant) due to the ^{13}C of the ligand, which indicates the structure and the number of ligands in the coordination sphere.
4. The evolution of the coordination sphere by reversible elimination–addition of CO molecules.

Finally, it is quite reasonable to suppose that the symmetries of the carbonyl species have to be in agreement with those observed for Ni$^+$ homogeneous complexes.[55]

In order to illustrate the method, an example can be considered from Table 6.6. The tetracarbonyl species 4 is characterized by an axial **g** tensor with the

TABLE 6.6. EPR Parameters of the Ni^+ Carbonyl Species Formed by Adsorption of CO at Various Pressures on the Ni^+/SiO_2 System

CO pressure (Torr)	Species	g tensor		Hyperfine ^{13}C A tensor (G)	
<1	1	$g_1 = 2.392$ $g_2 = 2.35$ $g_3 = 2.020$	$g_\perp$ $g_\parallel$	Not resolved	
1–40	2	$g_1 = 2.191$ $g_2 = 2.086$ $g_3 = 2.066$	$g_\parallel$ $g_\perp$	$A_1 = 30$ $A_2 = 32.5$ $A_3 = 32.5$	(2 equivalent CO)
20–400	3	$g_1 = 2.200$ $g_2 = 2.162$	$g_\perp$	A_1 not resolved A_2 not resolved	
		$g_3 = 2.005$	$g_\parallel$	$A_3 = \begin{cases} 37 \\ 18.5 \end{cases}$	(1 CO) (2 equivalent CO)
100–600	4	$g_\perp = 2.130$		$A_\perp = \begin{cases} 51.5 \\ 25 \end{cases}$	(1 CO) (3 equivalent CO)
		$g_\parallel = 2.009$		$A_\parallel = \begin{cases} 55 \\ 25 \end{cases}$	(1 CO) (3 equivalent CO)

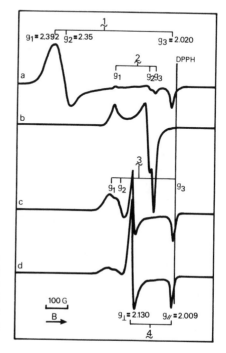

Figure 6.30. EPR spectra (X band, 77 K) obtained after adsorption at 293 K of ^{12}CO under a pressure of: (a) 10 Torr followed by evacuation for 15 min at 340 K, (b) 10 Torr, (c) 100 Torr, and (d) 400 Torr. Sample: 1.7 wt.% Ni/SiO_2 pretreated at 973 K and reduced by H_2 (400 Torr, 77 K, 15 h).[54]

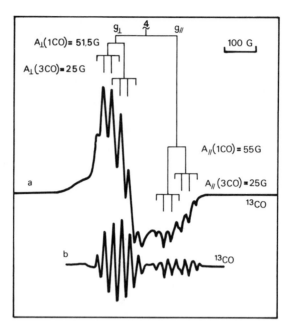

Figure 6.31. EPR spectra (X band, 77 K) obtained after adsorption at 293 K of ^{13}CO under a pressure of 400 Torr: (a) first derivative, (b) third derivative. Sample: 1.7 wt.% Ni/SiO$_2$ pretreated at 973 K and reduced by H$_2$ (400 Torr, 77 K, 15 h).[54]

following order of the various components $g_\perp > g_{//} \approx g_e$. This order is observed when the unpaired electron is located into the d_{z^2} orbital. In this case, the g values calculated according to McConnell and Chestnut[26] are

$$g_{//} = g_e$$
$$g_\perp = g_e + 6\lambda/\Delta \tag{6.70}$$

where λ is the spin–orbit coupling constant of Ni$^+$, g_e the free electron g value (2.0023), and Δ the difference between the energy of the d_{xz}, d_{yz} doublet and that of the d_{z^2} orbital. The fact that the ground state is d_{z^2} indicates that the ligand located along the z axis of the complex exhibits a stronger interaction with the Ni$^+$ ion than those along the x and y axes. This is the case for both a compressed octahedron (D_{4h} or C_{4v}) and a compressed trigonal bipyramid (D_{3h} or C_{3v}).

The ^{13}C superhyperfine structure allows the determination of the most probable ligand configuration. The spectrum in Figure 6.31 indicates the presence of four CO ligands, one of which has a larger superhyperfine coupling constant than the other three. This ligand is thus located in an axial position, along the z axis, where the interaction with the d_{z^2} orbital which contains the unpaired electron is larger than in the other directions. The remaining three CO molecules give rise to a four-line superhyperfine structure with a smaller coupling constant and with relative intensities 1:3:3:1, indicating the equivalence of the three ligands that must therefore be located in the xy plane in an equatorial position.

As the signal is clearly axial, no other ligand in an equatorial position can be considered apart from the three CO molecules, thus eliminating the other hypothesis of a compressed octahedron. The only possibility is therefore the pentacoordination of the trigonal bipyramid. The fifth ligand along the z axis is thus a surface oxygen of the support oxide.

Among the four carbonyl species in Table 6.6, only species 2 shows the relative order of the g components: $g_{//}(g_1) > g_{\perp}(g_2 \approx g_3) > g_e$. This order indicates a $d_{x^2-y^2}$ ground state for a d^9 ion. Thus, the degree of electron–ligand interaction taking place in the xy plane is larger than that along the z axis. This applies to symmetries such as an elongated octahedron (D_{4h} or C_{4v}), an elongated trigonal bipyramid (D_{3h} or C_{3v}), and a square planar arrangement (D_{4h}).

The ^{13}C superhyperfine structure indicates the presence of two equivalent CO molecules in the coordination sphere of Ni^+, as confirmed by IR spectroscopy.[56] Thus, species 2 is probably a tetracoordinated Ni^+ ion with the four ligands (two CO molecules and two surface oxygens) in the xy plane (Table 6.6). This symmetry is supported by the observation that a further CO molecule can be incorporated in the complex, transforming species 2 into the tricarbonyl

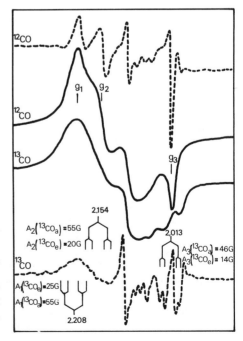

Figure 6.32. EPR spectra (X band, 77 K) obtained after reaction of species 2, $[Ni(CO)_2]^+$ in NiCaX zeolite with 250 Torr of C_2H_4 for 1 min at 298 K: (a) first derivative, (b) third derivative, a = axial, b = equatorial.[53]

species 3. The reversible transformation of species 1 into species 2, 3, and 4 is easily achieved by increasing the CO pressure on the system. The carbonyl species described above show the same symmetries as those of Ni^+ complexes in solution.

As Ni^+ ions are the active sites in the olefin dimerization reaction,[52,53] the coordination of ethene onto supported Ni^+ ions has also been investigated. In this case, however, the superhyperfine structure due to the ^{13}C atoms of the enriched ethene was not observed, indicating that ethene is attached side-on to the Ni^+ center. The number of ethene molecules that are required in the coordination sphere during the catalytic process could therefore not be determined directly. An indirect method, based on the use of both C_2H_4 and ^{13}C-enriched CO and on the knowledge of the usual symmetries of surface Ni^+ complexes, has been adopted. The dicarbonyl species 2 reacts progressively with ethene giving rise to a complex EPR spectrum (Figure 6.32). This was interpreted by employing ^{13}C-enriched CO and analyzing the third-derivative spectrum, which in several cases contains more information than the conventional first-derivative presentation.[57]

A novel species was identified in the zeolite NiCaX[53,54] with the following parameters: $g_1 = 2.208$, $g_2 = 2.154$, $g_3 = 2.013$, $A_1(^{13}CO_a) = 55$ G, $A_1(^{13}CO_e) = 25$ G, $A_2(^{13}CO_a) = 55$ G, $A_2(^{13}CO_e) = 20$ G, $A_3(^{13}CO_a) = 46$ G, and $A_3(^{13}CO_e) = 14$ G. This spectrum is assigned to the $[Ni(CO)_2(C_2H_4)]^+$ species 5, with two inequivalent CO ligands. The similarity between the **g** tensors of species 5 and 3 suggests that the structure of the complex 5 is probably a trigonal bipyramid:

where the axial ligand CO_a interacts to a larger extent than the equatorial one CO_e with the unpaired electron in the d_{z^2} orbital and is therefore responsible for the larger superhyperfine splitting constant. Thus the ethene molecule is in an equatorial position and species 5 is the result of the addition of one ethene molecule to species 2.

Increasing the ethene pressure up to 600 Torr caused a further change in the EPR spectrum. Using ^{13}C-enriched CO and the third-derivative presentation, a new species, 6, can be identified with the following parameters: $g_1 = 2.29$, $g_2 = 2.19$, $g_3 = 2.007$, and $A_1 = A_2 = A_3 = 45$ G.[53,54] Only one CO molecule is in the coordination sphere of this complex and the relative order of the **g** tensor components indicates again a trigonal bipyramid structure. The coupling constant values (45 G) can be compared with those found for species 5, suggesting that the CO molecule is in an axial position. The two ethene molecules should thus be in equatorial positions as shown in the following structure:

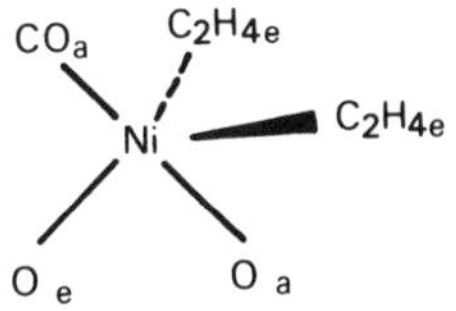

$$CO_a \quad C_2H_{4e}$$

(structure diagram showing Ni complex with CO_a, C_2H_{4e}, C_2H_{4e}, O_e, O_a ligands, labeled 6)

Species 6, $[Ni(CO)(C_2H_4)_2]^+$, is obtained from species 5 by a ligand exchange reaction with C_2H_{4e} replacing CO_e in the complex.

Catalytic activity tests were performed on systems containing the two species 5 and 6. With species 5, no appreciable dimerization reaction was observed after 24 h, whereas with species 6, reaction took place. Product distribution was very close to that observed in the catalytic tests performed without CO. It is therefore necessary that two ethene molecules be bonded to the same Ni^+ ion for the dimerization reaction to take place.

These results indicate that EPR spectroscopy can be a very powerful tool in elucidating the nature and structure of the active sites in catalysts.

6.5.5. Oxidative Coupling of Methane

The evidence of the oxidative dimerization of methane at high temperatures on metal oxides was first reported by Keller and Bashin.[58] Since then, the reaction, leading to the formation of small amounts of ethane and ethene, has been the subject of a great deal of work. Among the various oxide systems active in the catalytic reaction, magnesium oxide and, in particular, lithium-doped magnesium oxide (Li–MgO), are the most thoroughly investigated. Because of the presence of radicals during the catalytic reaction, EPR has played a major role in detecting them both in the gas phase (reaction intermediates) and on the surface (active sites).

The formation of methyl radicals $(CH_3\cdot)$ at 773 K in a gas phase composed of methane and oxygen in the presence of MgO was first observed by Lunsford. The radicals were detected by the matrix isolation ESR (MIESR) technique at 14 K in an argon matrix.[59,60] The $CH_3\cdot$ radical was observed with low oxygen partial pressure and exhibits the well-known quartet of hyperfine lines with a 1:3:3:1 relative intensity ratio. At higher oxygen partial pressures, the formation of increasing amounts of the methyl peroxy radicals $(CH_3OO\cdot)$ is observed exhibiting an axial signal without any hyperfine structure (Figure 6.33).

The state of the MgO surface during the catalytic reaction has been explored by quenching a catalyst working at the reaction temperature in liquid oxygen and comparing the EPR spectrum so obtained (Figure 6.34b) with that registered after slow cooling of the solid to 298 K (Figure 6.34a).[61] In the former case, various paramagnetic oxygen species are observed such as O_2^- (superoxide ions), O_3^- (ozonide ions) and Li^+O^- pairs (Figure 6.34b).

The O^- site (i.e., a positive hole trapped on an oxide O^{2-} ion) is believed to be responsible for the initial activation of the methane molecule with formation

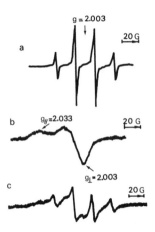

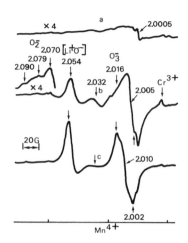

Figure 6.33. Radical EPR spectra: (a) pure methyl radical (CH$_3$·) spectrum, (b) pure methyl peroxy radical (CH$_3$O$_2$·) spectrum; (c) mixture of methyl and methyl peroxy radicals spectrum.[70]

Figure 6.34. EPR spectra of 7% Li/MgO after heating in 192 Torr at 923 K for 1 h: (a) sample cooled slowly to 298 K; (b) sample quenched in liquid O$_2$ at 77 K; and (c) sample (a) irradiated in 15 Torr of O$_2$ for 30 min.[61]

of the methyl radical according to the reaction

$$CH_{4(g)} + O^-_{(sur)} \rightarrow CH^{·}_{3(g)} + OH^-_{(sur)}$$

According to this hypothesis, the formation of ethane (C$_2$H$_6$) occurs by simple radical coupling, whereas ethene would be produced by further more complex reactions involving ethane and a surface O$^-$ or, in the gas phase, ethane and methyl radicals.[62]

The EPR techniques appear to be very powerful ways of improving our understanding of heterogeneous–homogeneous catalytic reactions.[63–66]

6.6. CONCLUSIONS

In this chapter, we have outlined the importance of magnetic parameters available from EPR spectroscopy (*g* factor, hyperfine and superhyperfine tensors) and the information which could be deduced to identify and characterize the paramagnetic centers together with their environment. As in most catalytic and surface studies, polycrystalline samples, rather than single crystals, are usually available. We have also presented the various powder EPR patterns generally observed and the analysis method to extract the EPR magnetic parameters of interest. A series of examples has been given in relation to catalysis.

EPR is a fast and extremely sensitive technique (10^{15} spins/G are commonly detected) that requires only very small amounts of sample for obtaining spectra. Further, EPR is a nondestructive technique because the energy involved in the experiment (about 10^{-3} kJ mol^{-1}) is too small to cause chemical changes in the sample.[66]

A major disadvantage of the technique is that it is limited to paramagnetic substances, but this can be useful when change in a paramagnetic species in a complex medium needs be followed. Another serious disadvantage of EPR, unfortunately common to some other spectroscopies, is the dramatic increase of linewidth in the conditions of temperature and pressure usually encountered for catalytic reactions.

Few techniques give a clear distinction between surface and bulk species. The best criterion is to observe the effect of adsorbed gases. If the EPR signal of the species is destroyed or changes while previously stable in vacuo at the same temperature, then it is likely that the signal corresponds to a species within a few angstroms from the surface. If no reaction occurs, the signal may be broadened by addition of a paramagnetic gas. A dipolar or exchange broadening by O_2 is commonly used because of its reversibility and pressure dependence. It arises from the interaction of the two unpaired electrons of the dioxygen molecule with that of the surface paramagnetic species. We have also given examples of the determination of both the ligand vacancies and the symmetry of surface transition metal ions by means of probe molecules such as ^{12}CO and ^{13}CO. As with all other spectroscopic techniques, EPR has a certain number of limitations. Paradoxically one of the most serious for EPR is its extremely high sensitivity. All the information obtained by EPR about a given system should therefore be set in the quantitative scale obtained by means of other spectroscopic (IR, UV, etc.) and nonspectroscopic (gravimetric, structural, etc.) techniques. The combined use of various experimental techniques is the only one that allows a deeper and more general understanding of the catalytic phenomena.

REFERENCES

1. J. H. Lunsford, *Adv. Catal.* **22**, 265 (1972).
2. J. C. Védrine, in: *Characterisation of Heterogeneous Catalysts* (F. Delannay, ed.), Marcel Dekker, New York (1984), p. 161.
3. C. Naccache, J.-F. Dutel, and M. Che, *J. Catal.* **29**, 179 (1973).
4. L. Bonneviot, D. Olivier, and M. Che, *Chem. Comm.* (1982), p. 952.
5. M. Che and A. J. Tench, *Adv. Catal.* **31**, 77 (1982).
6. M. Che and A. J. Tench, *Adv. Catal.* **32**, 1 (1983).
7. M. Che, E. Giamello, and A. J. Tench, *Colloids Surf.* **13**, 231 (1985).
8. K. Dyrck, A. Madej, E. Mazur, and A. Rokosz, *Colloids Surf.* **45**, 135 (1990).
9. M. Che, J. Védrine, and C. Naccache, *J. Chim. Phys.* **66**, 579 (1969).
10. M. Che, J. Demarquay, and C. Naccache, *J. Chem. Phys.* **51**, 5177 (1969).
11. Ya S. Lebedev, *Zh. Strukt. Khim.* **4**, 19 (1963).
12. J. A. Ibers and J. D. Swalen, *Phys. Rev.* **127**, 1914 (1962).
13. D. Olivier, C. Marachi, and M. Che, *J. Chem. Phys.* **71**, 4688 (1979).
14. M. Che and A. J. Tench, *J. Chem. Phys.* **64**, 2370 (1976).

15. F. J. Adrian, *J. Colloid Interface Sci.* **26**, 317 (1968).
16. J. H. Lunsford and J. P. Jayne, *J. Phys. Chem.* **69**, 2182 (1965).
17. S. A. Marshall, A. R. Reinterg, R. A. Serway, and J. A. Hodges, *Mol. Phys.* **8**, 223 (1964).
18. R. A. Schoonheydt and J. H. Lunsford, *J. Phys. Chem.* **76**, 323 (1972).
19. P. W. Atkins, A. Horsfield, and M. C. R. Symons, *J. Chem. Soc.* (1964), p. 5220.
20. A. D. Walsh, *J. Chem. Soc.* (1953), p. 266.
21. J. H. Lunsford, *J. Chem. Phys.* **46**, 4347 (1967).
22. J. H. Lunsford and J. P. Jayne, *J. Chem. Phys.* **44**, 1487 (1966).
23. W. Känzig and M. H. Cohen, *Phys. Rev. Lett.* **3**, 509 (1959).
24. M. Che, C. Naccache, and A. J. Tench, *J. Chem. Soc., Faraday Trans.* I **70**, 263 (1974).
25. A. Carrington and A. D. McLachlan, in: *Introduction to Magnetic Resonance*, Harper, London (1967), p. 81.
26. H. McConnell and S. Chestnut, *J. Chem. Phys.* **28**, 107 (1958).
27. P. W. Atkins and M. C. R. Symons, in: *The Structure of Inorganic Radicals*, Elsevier, Amsterdam (1967), p. 257.
28. M. Che and E. Giamello, in: *Spectroscopic Analysis of Heterogeneous Catalysts*, Vol. B (J. L. G. Fierro, ed.), Elsevier, Amsterdam (1990), p. 165. (Note that the captions of Figures 5.9 and 5.10 in this reference have been reversed.)
29. G. Bleaney, *Phil. Mag.* **42**, 441 (1951).
30. J. H. Lunsford and E. F. Vansant, *J. Chem. Soc., Faraday Trans.* II **69**, 1028 (1973).
31. M. Che, J. C. McAteer, and A. J. Tench, *J. Chem. Soc., Faraday Trans.* I **74**, 2378 (1978).
32. M. Che, G. Fichelle, and P. Mériaudeau, *Chem. Phys. Lett.* **17**, 66 (1972).
33. J. Turkevich, *Physics Today* **18**, 26 (1968).
34. Ph. de Montgolfier, P. Mériaudeau, Y. Boudeville, and M. Che. *Phys. Rev.* B **14**, 1788 (1976).
35. B. D. Flockhart, in: *Surface and Defect Properties of Solids*, Specialist Periodical Report, Vol. 2, The Chemical Society, London (1973), p. 69.
36. G. M. Muha, *J. Phys. Chem.* **71**, 633 (1967).
37. C. Naccache and Y. Ben Taarit, *VI Seminario Latino Americano de Quimica*, Chili, (1974), p. 8.
38. C. Naccache and Y. Ben Taarit, *J. Catal.* **22**, 171 (1971).
39. Y. Ben Taarit, M. V. Mathieu, and C. Naccache, *Adv. Chem. Series* **102**, 362 (1971).
40. Y. Ben Taarit, C. Naccache, and B. Imelik, *J. Chim. Phys.* **67**, 389 (1970).
41. P. Wierzchowski, E. Garbowski, and J. C. Védrine, *J. Chim. Phys.* **78**, 41 (1981).
42. B. D. Flockhart, C. Naccache, J. A. N. Scott, and R. C. Pink, *Chem. Comm.* (1985), p. 238.
43. C. Naccache, Y. Kodratoff, R. C. Pink, and B. Imelik, *J. Chim. Phys.* **63**, 341 (1966).
44. M. Che, C. Naccache, and B. Imelik, *J. Catal.* **24**, 328 (1972).
45. M. J. Baird, and J. H. Lunsford, *J. Catal.* **26**, 440 (1972).
46. M. Dufaux, M. Che, and C. Naccache, *J. Chim. Phys.* **67**, 527 (1970).
47. B. D. Flockhart, I. R. Leith, and R. C. Pink, *Trans. Faraday Soc.* **66**, 469 (1970).
48. B. D. Flockhart, I. R. Leith, and R. C. Pink, *Trans. Faraday Soc.* **65**, 542 (1969).
49. B. D. Flockhart and L. McLoughlin, *J. Catal.* **25**, 305 (1972).
50. J. H. Lunsford, L. W. Zingery, and M. P. Rosynek, *J. Catal.* **38**, 179 (1975).
51. Z. Dudzig and Z. M. George, *J. Catal.* **63**, 72 (1980).
52. L. Bonneviot, D. Olivier, M. Che, and M. Cottin, *Proc. 5th Int. Conf. on Zeolites*, Recent Progress Reports (R. Sersale, ed.), Giannini, Naples (1980), p. 168.
53. L. Bonneviot, D. Olivier, and M. Che, *J. Mol. Catal.* **21**, 415 (1983).
54. L. Bonneviot, *Thèse Doctorat ès Sciences*, Paris (1983).
55. K. Nag and A. Chakravorty, *Coord. Chem. Rev.* **33**, 87 (1980).
56. M. Kermarec, D. Delafosse, and M. Che, *Chem. Comm* (1983), p. 411.
57. M. Che, B. Canosa and A. R. Gonzalez-Elipe, *J. Phys. Chem.* **90**, 618 (1986).
58. G. E. Keller and M. M. Bashin, *J. Catal.* **73**, 9 (1982).
59. D. J. Driscoll, W. Martir, J. X. Wang, and J. H. Lunsford, *J. Am. Chem. Soc.* **107**, 58 (1985).

60. D. J. Driscoll, W. Martir, J. X. Wang, and J. H. Lunsford, in: *Adsorption and Catalysis on Oxide Surfaces* (M. Che and G. C. Bond, eds.), Elsevier, Amsterdam (1985), p. 403.
61. J. X. Wang and J. H. Lunsford, *J. Phys. Chem.* **90**, 5883 (1986).
62. T. Ito, J. X. Wang, C. H. Lin, and J. H. Lunsford, *J. Am. Chem. Soc.* **107**, 5062 (1985).
63. D. J. Driscoll, K. D. Campbell, and J. H. Lunsford, *Adv. Catal.* **35**, 139 (1987).
64. J. H. Lunsford, *Langmuir* **5**, 12 (1989).
65. T. A. Garabyan and L. Ya. Margolis, *Catal. Rev.* **31**, 355 (1989).
66. M. Che, in: *Magnetic Resonance in Colloid and Interface Science* (J. P. Fraissard and H. A. Resing, eds.), Reidel, Dordrecht (1980), p. 79.

FERROMAGNETIC RESONANCE

L. Bonneviot and D. Olivier

7.1. INTRODUCTION

Ferromagnetic resonance (FMR) is a magnetic resonance comparable to nuclear magnetic resonance (NMR) and electron paramagnetic resonance (EPR); in these techniques the effect of a microwave irradiation is to flip the magnetic moments oriented in a magnetic field. Unlike EPR or NMR resonances, which operate on nuclear or electron spin, FMR concerns magnetic domains of a ferromagnetic material, i.e., the so-called Weiss domains. The first observation of FMR was reported by Griffiths in 1946[1] for electrolytically deposited films of iron, cobalt, and nickel. The first application to catalysis was made by Hollis and Selwood in 1961 on nickel-supported catalysts.[2] This technique can be used not only for ferromagnetic materials (metals and their alloys)[3,4] but also for ferrimagnetic materials (oxides such as garnets).[5] Usually, the magnetic moment, whose intensity depends on the Weiss domain volume as we shall see later on, is about three orders of magnitude greater than the magnetic moment of an electron. As a consequence, a quantum mechanical description of FMR phenomena is not necessary as it is for EPR or NMR spectroscopies. Furthermore, the temperature-dependent magnetization of ferromagnetic materials is at least three orders of magnitude more intense than in the case of paramagnetic materials. Therefore, though the FMR linewidths are usually very broad, this is the most sensitive spectroscopy for characterization (about one hundred times more sensitive than EPR).

Nevertheless, this technique was essentially developed by physicists and it is little used by chemists, mainly due to the complexity of the theory and the

L. Bonneviot ● Département de Chimie, Université Laval, Québec, Canada. D. Olivier ● Institut de Recherches sur la Catalyse, CNRS Villeurbanne, France. Present address: Departement des Sciences Chimiques, CNRS, F-75016 Paris, France.

Catalyst Characterization: *Physical Techniques for Solid Materials*, edited by Boris Imelik and Jacques C. Vedrine, Plenum Press, New York (1994).

difficulty of interpreting the results from experimental powder pattern spectra. Some elements of the theory are treated in textbooks in French by Herpin[3] and in English by Morrish.[4] A more complete description of the physical basis of FMR can be found in a chapter in *Magnetic Oxides* by Patton[5] and in *Microwave Ferrites and Ferrimagnetics* by Lax and Button.[6] Finally, the reader will find useful information on the properties of ferromagnetic materials in a recent series edited by Wohlfarth.[7]

In this chapter, elements of the theory are introduced when necessary and illustrated with examples from oxide-supported nickel catalysts. Indeed, most of the FMR studies of catalysts have been performed on Raney nickel and supported nickel particles. The experimental observations on nickel can usually be extrapolated to iron, cobalt, and alloys. Nevertheless, some substantial differences among these systems can arise from the different Curie temperatures and the relative importance of the various magnetic anisotropies reviewed here for nickel. The only reviews on FMR in the field of catalysis are this chapter and that of Slinkin.[8]

7.2. BACKGROUND

7.2.1. Theoretical Principles

7.2.1.1. Resonance Conditions

The classical model tells us that a magnetic moment (μ) placed in a uniform magnetic field **H** is oriented parallel to the field direction if the field is large enough. After slight disturbance from this equilibrium, μ precesses about the field direction to return eventually via this indirect motion to its original position. The loss of energy associated with this so-called precession motion leads to a damping or a relaxation effect. One can stimulate this precession by applying a sinusoidal field with frequency v equal to the precession frequency v_0 of the magnetic moment. At the resonance ($v = v_0$) the magnetic moment is flipped antiparallel to the field direction. For a free electron, the resonance frequency v_0 is given by

$$h v_0 = g_e \beta \mathbf{H}$$

or

$$2\pi v_0 = \gamma_e \mathbf{H}$$

where g_e is the electron g factor, and γ_e ($= g_e/2m$) is the gyromagnetic ratio and β_e ($= h\gamma_e/2\pi g_e$) the Bohr magneton for the electron.

In a ferromagnetic material, the resonance concerns electrons that are strongly coupled to one another within a Weiss domain. The resonance conditions of these electrons no longer coincide with the resonance conditions of a free electron or localized and unpaired electrons. The natural g factor (called g_e for a free electron and g_J, the Landé factor, for an electron in a complex) as well as the natural γ gyromagnetic ratio of the metal characterizes the magnetic properties of the polarized electrons of the conduction band at the Fermi level.

Furthermore, magnetic anisotropies due to the crystalline structure, the shape of the sample, and the mechanical strain undergone by the material are very strong in ferromagnetic materials. They are so intense ($\sim 10^6$ G) that they impose the directions of easy magnetization along crystal directions (such as 111 or 100 in Ni) corresponding to a spontaneous orientation of μ without the need of an external field. These anisotropies are conveniently represented in the Weiss molecular field theory[9] by an internal magnetic field or anisotropy field $\mathbf{H}_a$ aligned in the direction of easy magnetization. In this representation, the local or effective field $\mathbf{H}_{\text{eff}}$ results from the composition of the external and internal applied fields:

$$\mathbf{H}_{\text{eff}} = \mathbf{H} + \mathbf{H}_a$$

The resonance is achieved when the effective field is equal to the intrinsic resonance field $\mathbf{H}_0^0$ of the material under investigation, following the equation

$$\mathbf{H}_{\text{eff}} = \mathbf{H}_0^0$$

and

$$h\nu_0 = g\beta \mathbf{H}_0^0 \quad \text{or} \quad 2\pi\nu_0 = \gamma \mathbf{H}_0^0 \tag{7.1}$$

This definition implies that, in the absence of anisotropy, the resonance occurs when the external field equals the intrinsic field. The resonance field $\mathbf{H}_{\text{res}}$ is indeed altered by the presence of the anisotropy field such that

$$\mathbf{H}_{\text{res}} = \mathbf{H}_0^0 - \mathbf{H}_a \tag{7.2}$$

When the applied field and the anisotropy field point in the same direction, the resonance shifts to lower fields.

In the following text, the resonance fields $\mathbf{H}_{\text{res}}$ are discussed as scalar entities as $H_{//}$, $H_{\perp}$, $H_{[100]}$, $H_{[110]}$, and $H_{[111]}$, depending on the nature and the symmetry of the anisotropy which dominates. The anisotropy fields are put at $\mathbf{H}_a$ by default but depending on the case, the following notation used, $\mathbf{H}_d$, $\mathbf{H}_s$, $\mathbf{H}_{s+d}$, $\mathbf{H}_{mc}$, ... (see Glossary). Finally, three experimental (external) fields, H_0, H_+, and H_-, are defined from the powder pattern FMR spectra (*vide infra*, Section 7.4.1.2b and Glossary).

The magnetization is another property of ferromagnetics that contributes to their resonance. The difference from paramagnetics comes from the nature of the magnetic moments and their ability to align with the external applied field. In ferromagnetics, one deals with the magnetic moments of Weiss domains μ, whose intensity depends on the number of polarized electrons per unit volume and on the volume v of the domains:

$$\mu = \mathbf{M}_s v \tag{7.3}$$

where M_s is the spontaneous magnetization of the materials.[10] The spontaneous magnetization determined by the magnetic coupling via the exchange integral depends on the temperature and vanishes at the Curie temperature T_C. This temperature dependence is conveniently aproximated by the Weiss law:

$$\mathbf{M}_s = \mathbf{M}_{so} th[(M_s/M_{so})(T_C/T)] \qquad (7.4)$$

where $\mathbf{M}_{so}$ is the spontaneous magnetization at 0 K. Below $0.5 T_C$, $M_s \approx M_{so}$ and above $0.8 T_C$, M_s decreases drastically and equals zero at the Curie temperature. The spontaneous magnetization is measured in a field strong enough to overcome the internal field and align all the magnetic moments of the Weiss domains of the sample. In the case of interest (small single domain particle, no particle–particle magnetic interaction, and weak anisotropy, *vide infra*), M_s and M_{so} are equal to the experimental magnetization measured at saturation in a strong applied field for a given temperature T or at 0 K, respectively. From the measurement at saturation, we can calculate the number of polarized electrons per atom, the so-called Bohr magneton number, $\beta(X)$. For instance, $\beta(Ni) = 0.6$, $\beta(Co) = 1.7$, and $\beta(Fe) = 2.22$. The Curie temperature depends on the number of atom neighbors and the exchange integral between coupled spins in the material. T_C is therefore a characteristic of a ferromagnetic material and varies in a wide range, from 77 K for EuO to 1403 K for Co with intermediate values of 289, 631, and 1043 K for Gd, Ni, and Fe, respectively.[5-7]

For multidomain systems, the magnetization depends on the way the domains are forced to align their magnetic moments along the magnetic field. This is attributed to the displacement of the frontiers between domains and leads to the hysteresis cycle. In supported metal catalysts, the particles are usually small enough to be considered as single domains (< 20 nm) and no hysteresis is observed.

7.2.1.3. Relaxation

The FMR lineshape of a crystallite is usually Lorentzian. The intrinsic peak-to-peak linewidth $\Delta \Gamma_{pp}$ determined on the first derivative curves (*vide infra*) characterizes the relaxation process undergone by the spins which have been flipped. A classical model taken from mechanics[11,12] explains the relaxation by damping forces which operate on the spins returning to equilibrium. The Bloch theory of NMR[13] extrapolated to FMR by Bloembergen[14] introduces the longitudinal and the transversal relaxation times T_1 and T_2. T_1 is the time required for the magnetization to realign in the direction of the magnetic field applied in the absence of the microwave field and describes the spin–lattice relaxation by which the energy accumulated in the spin system during the absorption is dissipated in the bulk of the sample. During the resonance, the microwave field generates a phasing of the resonant spins. T_2 is the time necessary for the dephasing of these spins when the microwave action has been suppressed. There is a relation between T_1 and T_2:

$$1/T_2 = 1/2T_1 + 1/T_2^* \tag{7.5}$$

where the relaxation time T_2^* describes the spin–spin interactions. The intrinsic linewidth $\Delta\Gamma_{pp}$ is written in the Bloch–Bloembergen theory as [γ is defined in equation (7.1)]

$$\Delta\Gamma_{pp} = 2/\gamma T_2 \tag{7.6}$$

and in the Landau–Lifschitz theory as

$$\Delta\Gamma_{pp} = (4\pi\nu/M_s\gamma^2)\lambda_L \tag{7.7}$$

where λ_L is the damping constant. The latter shows the linewidth explicitly as proportional to the resonance frequency, but this model does not take into account the spin–spin relaxation mechanism. For our purpose, the experimental FMR spectra of catalysts are taken from powders and, in most cases, heterogeneous samples where the linewidths are constituted by a combination of intrinsic linewidths and powder patterns due to magnetic anisotropies. This point is developed in Section 7.4.1.

7.2.2. Experimental FMR Data

The experimental FMR data are collected on an EPR spectrometer described in the preceding chapter. The microwave irradiation is kept constant while the magnetic field is swept to find the resonance conditions. The FMR spectrum is the first derivative of the absorption curve plotted against the external applied field, as well as for EPR. The experimental g factor is conveniently measured using the DPPH, with $g = 2.0036$ taken as a reference. The calculations are performed from equation (7.1) knowing that the measured resonance field H_{res} is the external field and not the intrinsic resonance field H_0^0 [equation (7.2)]. Therefore, the experimental g factor characterizes the resonance in the presence of anisotropies, and does not occur at the intrinsic resonance field. The situation is more complicated when dealing with a powder pattern (see Section 7.4.1).

The intrinsic resonance for metallic nickel ($g = 2.22$ in the absence of anisotropy) is expected at 3000 G for a frequency of 9.28 GHz. This FMR resonance can be observed with a conventional X-band EPR spectrometer, without modification of the apparatus. Nevertheless, one has to be careful about the quality of the resonance in the presence of a ferromagnetic material. The use of low-microwave power, a small quantity of product, or a diluted sample is recommended because of the very large magnetization of these materials.

Ferromagnetic or paramagnetic signals are collected under very similar experimental conditions; both are first-derivative curves (see Section 7.4.1.2b). In a spectrum, superimposition of both types of signal may occur and cause some problems of assignment. Fortunately, a ferromagnetic signal is in most cases broader (100 to 1000 G) than a paramagnetic signal (< 100 G).[15] This first qualitative approach is consolidated by a careful analysis of the thermal magnetic behavior of each species. It is, therefore, recommended that the variation in the signal be plotted *vs.* temperature to give the so-called *thermomagnetic curve*. A $1/T$ dependence is characteristic of paramagnetic behavior while other types of dependence can be attributed to antiferro-, ferri-, or ferromagnetism.[2-7] In the case of ferromagnetism, the metal particle size in a nickel-supported oxide catalyst can be calculated from the thermomagnetic curves. From the analysis of the FMR lineshape and its temperature dependence, one can derive the nature and the intensity of the *magnetic anisotropies*. The shape of the particles, the effect of surface coverage by an adsorbate, and the mechanical strain undergone by the supported particles can be deduced from these data.

7.3. THERMOMAGNETIC CURVES

7.3.1. Superparamagnetism and Particle Size

The magnetic moment μ is aligned along a direction of easy magnetization. Reorientation of μ can easily occur for small anisotropy. For large particles, the crystalline anisotropy is large and the magnetization is blocked. As a consequence, there is a residual magnetization, M_r, which is maintained for a longer time the larger the particles. For a uniaxial anisotropy, the anisotropy energy E_a can be considered as an energy barrier to be overcome for the reorientation of the magnetization. To explain the domain size effect on the residual magnetization, Néel[16] proposed that E_a depends on the volume of the particle and the anisotropy constant K as follows:

$$E_a = Kv \tag{7.8}$$

and showed that the probability of reorientation occurring at a correlation time τ is given by the relation

$$1/\tau = f_0 e^{-\alpha} \tag{7.9}$$

with $\alpha = Kv/kT$, where f_0 is the frequency factor of the order of $10^9 \, \text{s}^{-1}$ and k the Boltzmann constant. As a consequence, after the field is removed the residual magnetization $\mathbf{M}_r$ will vanish with the time t as

$$\mathbf{M}_r = \mathbf{M}_s \exp - (t/\tau) \tag{7.10}$$

For a measurement time t long enough ($t > \tau$), no remanence is observed. Magnetic domains, in such conditions, behave as single-domain particles. An assembly of such domains resembles that of paramagnetic centers. The main difference lies in the nature of the associated magnetic moment, which is several orders of magnitude larger for a domain [depending on its volume, see equation (7.3)] than for a paramagnetic center. This magnetic state where the thermal fluctuation of the magnetic moment cancels the residual magnetization ($M_r = 0$) is called superparamagnetism.[17]

The transition between superparamagnetism and ferromagnetism occurs over a small range of sizes as shown in the following examples. If it is assumed that the crystalline anisotropy dominates all the other anisotropies, an iron particle 23 nm in diameter will have a correlation time of 10^{-1} s at room temperature and will rapidly reach thermal equilibrium; hence no residual magnetization will result. For a slightly larger particle of 30 nm diameter, τ will be 10^9 s and the magnetization will remain very stable; the particle will behave as a permanent magnet.[17] Indeed, an increase of 15% in volume ($\sim 5\%$ in diameter) is enough to increase τ by a factor of a hundred.

For magnetic measurements, $t = 10^2$ s, the magnetic transition ($t = \tau$) occurs for an energy barrier of about $25\,kT$. Below a given temperature called the blocking temperature T_b the magnetic moment will be blocked. For a given temperature, there is a critical volume v_c or a critical diameter (d_c for spherical particles) above which the freezing occurs. At room temperature ($T_b = 25$ °C), the critical diameter is 8 nm for hcp cobalt, 28 nm for fcc cobalt, and 25 nm for iron. The superparamagnetism transition occurs at smaller sizes for stronger anisotropies; an elongated iron particle with an optimum shape anisotropy (rod shape, *vide infra*) has a critical volume equal to that of a sphere 6 nm in diameter.[17]

7.3.2. FMR of Superparamagnetic Particles

7.3.2.1. Superparamagnetic Conditions in FMR Experiments

Equation (7.10) clearly shows that the remanence vanishes in all cases, but the process takes longer for larger particles. Therefore, the conditions for observing the superparamagnetism depend on the measurement time. For FMR, the measurement time t ($\approx 1/v$) is very much shorter than for magnetic susceptibility (10^{-1} to 10 s) or Mössbauer measurements (10^{-8} s) and equals about 10^{-10} s. Equation (7.9) proposed by Néel is not valid for such a short time, i.e., for very small particles and weak anisotropies.[18] The more appropriate relation for FMR is the low-field and small-domain approximation of Aharoni's calculations[19,20]:

$$\tau = (M_s v / \gamma kT) \tag{7.11}$$

for

$$\alpha = (M_s H_a v / 2kT) < 1 \tag{7.12}$$

where all the parameters have been defined previously ($\gamma \approx 2 \times 10^7$ G^{-1} s^{-1}). For spherical nickel particles with a cubic crystalline anisotropy only, the barrier is $Kv/12$ ($K < 0$, [111] easy direction of magnetization) and α equals to $Kv/12kT$. The blocking temperature of small nickel particles is calculated for various sizes from equation (7.11) and α is given to justify the validity of equation (7.12).

TABLE 7.1. Blocking Temperature of Nickel Spherical Particles for X-band FMR Measurements[18,21] ($t = 1.09 \times 10^{10}$ s) and $\alpha = Kv/12kT$

d/nm	1	2	3	4	5	6	7
α	0.39	0.39	0.41	0.32	0.21	0.08	—
T_b/K	1.0	7.6	26	61	120	204	310

For instance, in Table 7.1 one notes that particles larger than 7 nm are ferromagnetic up to 310 K while particles smaller than 4 nm are superparamagnetic at a temperature as low as 61 K. However, the table only gives a rough indication since other and stronger anisotropies than the crystalline anisotropy could also contribute to the spin freezing and increase the blocking temperature or decrease the critical volume of the particles.

7.3.2.2. Intensity of the FMR Signal of Superparamagnetic Particles

Under superparamagnetic conditions, the thermal energy is high enough to allow the magnetization of each particle to overcome the anisotropy energy; the anisotropies can be considered as small as in the case of a paramagnetic system. Furthermore, if the particles are far enough apart so that it can be assumed that there is no mutual interaction between them, the overall magnetization of the system follows the Langevin law. Since the magnetization is aligned along the applied external field, the magnitude alone suffices to fully describe the phenomena using a scalar entity:

$$M = M_s L(x) \tag{7.13}$$

with

$$x = (\mu H/kT) \tag{7.14}$$

where $L(x) = \coth(x) - x^{-1}$ and the magnetic moment depends on the particle volume following equation (7.3) such that

$$M = M_s[\coth(M_s v H/kT) - kT/M_s v H] \tag{7.15}$$

with a simplified form in the low-field, high-temperature or small-volume domain approximations,

$$M = M_s^2 v H/3kT \tag{7.16}$$

If the resonance is obtained under the appropriate conditions indicated above, the FMR response is proportional to the magnetization of the sample or, more precisely, to the magnetization of the ensemble of particles involved in the

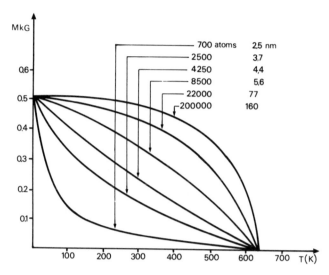

Figure 7.1. Magnetization *vs.* temperature for spherical nickel particles in an applied field of 3000 G, assuming that only crystalline anisotropy is operative.[22]

resonance phenomenon. Therefore, for a given particle size the FMR signal intensity plotted *vs.* temperature will follow a characteristic curve, which can be calculated from equations (7.15) or (7.16) (Figure 7.1). Practically, one estimates that the Langevin approximation is valid when the magnetic anisotropy field H_a is half the external applied field. This condition will be easier to fulfill in Q-band ($\nu \approx 36$ GHz, $H_{res} = 12,000$ G) than in X-band ($\nu \approx 9$ GHz, $H_{res} \approx 3000$ G) experiments.

In real catalysts, there is always a distribution of sizes which should be taken into account in equation (7.15) but for narrow distributions this can be neglected.[18,21]

7.3.3. Experimental Parameters

The comparison of the theoretical thermomagnetic curves and the experimental ones is a delicate operation since FMR is not the appropriate technique to get an absolute measurement of magnetization. Nevertheless, it is possible to plot a relative thermomagnetic curve with the intensity of the FMR signal taken at the lowest temperature T_0 as a reference,[18]

$$[M(T)/M(T_0)] = f(T/T_0) \tag{7.17}$$

where $M(T)$ and $M(T_0)$ are the magnetization at T and T_0, respectively. These theoretical curves relative to T_0 are easily compared with the experimental curves,[18,21]

$$[I(T)/I(T_0)] = f(T/T_0) \qquad (7.18)$$

where $I(T)$ and $I(T_0)$ are the FMR signal intensity at T and T_0, respectively.

Practically, the FMR signal is a first derivative which has to be integrated once to get the adsorption curve and twice to get the signal intensity. The integration program takes into account (i) a baseline correction and (ii) a field correction. The baseline is taken on the derivative curves and on the adsorption curves at high fields, where there is no signal for nickel. As the magnetic field is swept during the FMR experiments, the magnetization of the superparamagnetic particles varies proportionally to the field following equation (7.16). A simple correction for the magnetization consists of dividing by the external field value and the magnetization can be calculated at the intrinsic resonance H_0^0 following the relation

$$I(T) = H_0^0 \int_0^\infty \left[\int_0^H Y'(H, T)dH \right] dH/H \qquad (7.19)$$

where $Y'(H, T)$ is the first-derivative FMR curve recorded on the spectrometer. This calculation only applies when all the particles of the sample experience superparamagnetic behavior. An indication that this is not the case for some of the particles is revealed on the FMR spectra by a resonance at zero field attributed to a remanent magnetization. From equation (7.15), which relates the particle size to the magnetization and equation (7.17), which assumes a superparamagnetic behavior, it appears that it is impossible to evaluate the fraction of reduced nickel and the particle size from a thermomagnetic curve for a heterogeneous particle size distribution.

7.3.4. Data Analysis

7.3.4.1. Homogeneity of the Particle Size Distribution

The homogeneity of the particle size distribution can be deduced from the regularity of the thermomagnetic curves. In this case, the thermomagnetic curve can be fitted by a single size of particle taken as an average. Figure 7.2 represents the experimental thermomagnetic curve of a NiCaX sample which consists of a X-type zeolite containing ion exchanged Ni^{2+} ions reduced at 0 °C by hydrogen atoms generated in a cold plasma. Electron micrograph pictures of this sample reveals that the nickel particles are located in the supercages (1.3 nm inside diameter) of the zeolite structure. The thermomagnetic curve of such particles does indeed exhibits a smooth monotonic shape. By contrast, two irregular thermomagnetic curves are shown in Figure 7.3. These curves have been obtained from silica-supported Ni–Cu bimetallic particles. The inhomogeneity is introduced in this case by the presence of copper which modifies the Curie temperature. There are two populations of particles, one with low T_C due to alloyed particles, and the other, to copper-free particles with the natural T_C of nickel.

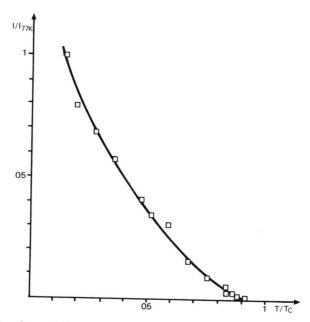

Figure 7.2. Experimental thermomagnetic curve of nickel particles in NiCaX zeolites; the average particle diameter is 1 nm.[18,23]

7.3.4.2. Curie Temperature

At the Curie temperature T_C the ferromagnetic order disappears, the magnetization drastically decreases and, consequently, the FMR signal vanishes. The Curie temperature is obtained experimentally by extrapolation of the thermomagnetic curves in the vicinity of T_C ($0.9 < T/T_C < 1$). For convenience, one takes advantage of a simplified analytical equation of the magnetization close to T_C: for a ferromagnetic ensemble of particles [from equation (7.4)],

$$M^2 = 3M_{so}^2 (1 - T/T_C) \qquad (7.20)$$

and for a superparamagnetic ensemble of particles [from equations (7.16) and (7.20)],

$$M = M_{so}^2 vH/k[(1/T) - (1/T_C)] \qquad (7.21)$$

The equations give two different experimental plots for the extrapolation of T_C depending on the magnetic state of the particles under investigation: I^2 vs. T or I vs. $1/T$ for ferromagnetic or superparamagnetic particles, respectively. An interesting example is provided by the case of nickel particles obtained by thermal reduction in hydrogen of nickel-exchanged Y-type zeolites.[22,24] The

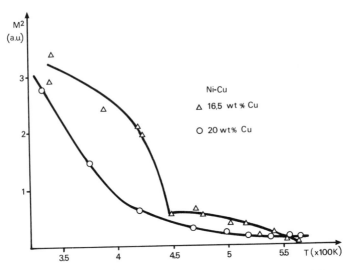

Figure 7.3. Experimental thermomagnetic curves of Ni–Cu bimetallic particles supported on silica.[22]

authors show that a fraction of the nickel particles located in the zeolite porous structure migrates outside the zeolite. This fraction increases when the reduction temperature increased. The internal nickel Ni_{int} located in the supercages of the zeolite structure is characterized by a low Curie temperature T_{C_1} due to an incomplete reduction (*vide infra*). The external nickel Ni_{ext}, obtained by migration, forms large and completely reduced particles characterized by a T_{C_2} which coincides with the natural T_C of nickel. After reoxidation, the external nickel is easier to reduce than the internal nickel. Therefore, it was possible to determine the fraction of external nickel by thermoprogrammed reduction, TPR (see Table 7.2).

A quantitative measurement by FMR is also proposed by the authors. Based on equation (7.20), assuming that the particles are still ferromagnetic at this temperature, it appears that the slope of the curves is proportional to M_{so}^2/T_C. Therefore, they solve the problem with the help of the following equation:

$$Ni_{int}/(Ni_{ext} + Ni_{int}) = \alpha_1 T_{C_1}(x_1 T_{C_1} + \alpha_2 T_{C_2}) \qquad (7.22)$$

where α_1 and α_2 are the slopes for internal and external particles, respectively (see Figure 7.4). The values obtained by FMR correlates the TPR. Nevertheless, one has to be careful in such calculations as most of the particles of these investigated samples should be superparamagnetic close to T_C (at about 600 K) (see Table 7.1, which indicates that particles as large as 7 nm are expected to be superparamagnetic even at room temperature). Equation (7.21) should be used instead of equation (7.20) and an I vs. $1/T$ plot instead of the plots of Figure 7.4.

TABLE 7.2. FMR and TPR Characteristics of Bidispersed
Nickel Particles in a Y-Type Zeolite[22,24]

Sample[a]	T_{C_1} (K)	T_{C_2} (K)	% Ni_{ext}[b] FMR	TPR
NiY-678	619	—	0	4
NiY-723	593	634	38	30
NiY-773	602	637	48	54
NiY-823	592	638	57	53
NiY-873	626	641	65	64.5

[a] The number following the dash in the sample name is the reduction temperature.
[b] Ni_{ext} is the external nickel, see text.

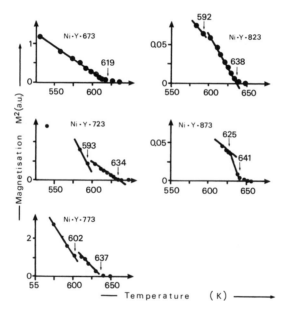

Figure 7.4. Curie temperature determination for nickel particles in nickel containing a Y-type zeolite reduced at various temperatures.[22,24]

Furthermore, the validity of the empirical Weiss law in the vicinity of T_C is not experimentally verified.[3-9] Equation (7.22) should be considered as an empirical fit of the results rather than as a formula that can be applied in every case.

The decrease in the Curie temperature is an intriguing point already noted above (Figures 7.3 and 7.4). This has been noted either for very small particles or in presence of the formation of an alloy with, e.g., copper.[18,21,24,25] The Curie temperature depends on the magnetic environment of each atom of a ferromagnetic material that can be altered in one way or another in the cases cited above. The Weiss molecular field W is related to the number of nearest neighbors Z and to the Heisenberg exchange integral J[3]:

$$W = (ZJ/4N\beta^2)$$

where N is the number of atoms per cubic centimeter ($\propto$ density), and related to T_C by

$$T_C = CW$$

where C is the classical Curie constant. It follows that T_C is proportional to Z and J or, more precisely, to their average value, following equation (7.23),[2,25]

$$T_C \propto \langle Z \rangle \langle J \rangle \tag{7.23}$$

The coordination number may vary with the structure of the particles or with size when they are small enough. The exchange integral decreases when the interatomic distance lengthens or cancels for "nonmagnetic neighbors." Practically, this means that T_C for nickel is expected to decrease in Ni–Cu alloys depending on the copper fraction and the homogeneity of the mixture.[22] One would also expect a decrease for very small particles of pure nickel (< 6 nm).[2,21,25] For such a size range, the structure of the particles evolves owing to a relaxation phenomenon (the surface curvature increases as well as the surface tension) and, as a consequence, the interatomic distance decreases. Therefore, it is impossible to predict how T_C should vary. Curie temperatures higher than the natural value for nickel have been observed for supported nickel catalysts.[18,21,24,25] In presence of an adsorbed gas, the surface tension drastically decreases and the relaxation effect suppressed. In this case, one expects a simple correlation between the size and T_C.

The effect of gas adsorption on nickel particles supported on silica is shown on Figure 7.5, which depicts the shift of the thermomagnetic curve toward low temperature in the presence of oxygen. One notes that oxygen has a double action on the surface: a surface tension suppression and an oxidation of nickel atoms, the latter being no longer ferromagnetic. As a consequence, the oxidized surface atoms do not belong to the ferromagnetic volume of the particle.

7.3.4.3. Particle Size Determination

In the superparamagnetic state, the size is directly related to the magnetization of the sample [equations (7.15) and (7.16)]. Comparison of the experimental and theoretical thermomagnetic curves allows the characterization of the range of particle sizes (Figures 7.1, 7.2, and 7.6). Nevertheless, several parameters render this method inaccurate: (i) strong anisotropy fields drastically distort the signal, whose integration becomes very delicate, (ii) the presence of ad-

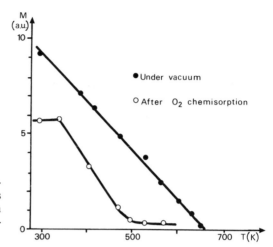

Figure 7.5. Effect of oxygen chemisorption on the thermomagnetic curves of nickel particles having a 1–5.5 nm range of diameters in a Ni/SiO$_2$ catalyst.[22.25]

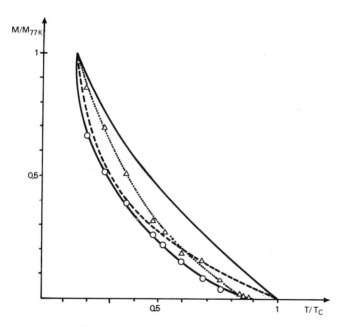

Figure 7.6. Comparison of experimental and theoretical thermomagnetic curves: $M_T M_{77K} = f(T/T_C)$, with $T_C = 631$ K; (——) and (----) for theoretical curves of 4- and 2-nm particles, respectively, and (O) and (△) for 1.7 and 4.3 wt.% nickel in Ni/SiO$_2$ catalysts.[18,21]

sorbed gas may vary the superparamagnetic volume of small particles considerably with a high surface/volume ratio (see text above), (iii) large size distribution, and (iv) the presence of ferromagnetic particles. Therefore, the size determination by FMR should wisely be considered as approximate.

7.4. MAGNETIC ANISOTROPIES

7.4.1. Powder Pattern and Lineshape

For simplification, the effect of a powder pattern is only treated, in this section, for uniaxial anisotropies for which it is possible to define the sign of the anisotropy field H_a.

7.4.1.1. Sign of Uniaxial Anisotropies

The sign of uniaxial anisotropies can be defined from a consideration of the internal field H_a, and the applied field H in a direction parallel to the anisotropy axis. H_a is positive when the local field increases along the anisotropy axis. Therefore, the parallel component of the FMR resonance occurs at a lower field for a positive anisotropy field or, in short, a positive anisotropy [equation (7.2)]. Conversely, a negative anisotropy leads to a high-field resonance for the parallel component of the FMR response. Of course, these definitions do not apply for multiaxial anisotropies, such as crystalline anisotropy (*vide infra*). In the case of nickel, cobalt, and iron, crystalline anisotropy is usually overcome by uniaxial anisotropies that are discussed in Sections 7.4.2, 7.4.3, and 7.4.4.

7.4.1.2. Determination of the Anisotropy Sign

The powder pattern of an EPR signal is discussed in Chapter 6 (on EPR) and the main features apply to an FMR signal. The differences lie in the parameters used to describe the signal ($H_{//}$, $H_{\perp}$, instead of the g and A factors, see Section 7.2). The FMR signal is a superimposition of the intrinsic peak-to-peak linewidth $\Delta\Gamma_{pp}$ (see Section 7.2.1.3) of all the particles randomly oriented. As $\Delta\Gamma_{pp}$ and the anisotropy field are of the same order of magnitude, several hundred gauss, the FMR signal is weakly asymmetric. The analysis of such signals requires a different approach to that used for EPR signals. A and B, the amplitude asymmetry parameters, and A' and B', the width asymmetry parameters, are conveniently defined[22] as depicted in Figure 7.7.

The signal crosses the baseline at an apparent resonance field, H_0, different from the intrinsic resonance H_0^0. There is no direct measure of $H_{//}$ and $H_{\perp}$ but H_+ and H_-, which are the fields for the maximum and the minimum of the derivative curve, can be obtained. A' and B' are defined as $(H_0 - H_+)$ and $(H_- - H_0)$, respectively. Based on the fact that a parallel component in a powder pattern has a smaller intensity than a perpendicular component, a positive anisotropy should be expected when $A/B < 1$. Furthermore, the shape,

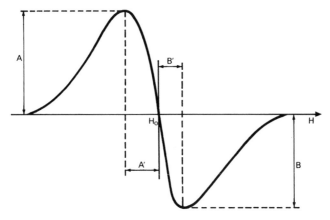

Figure 7.7. Definitions of the asymmetry parameters for the analysis of anisotropy field (from Simoens[$^{(2)}$]; see also Jacobs *et al*.[(24)] and Derouane[(25)]).

surface, or magnetostriction axial anisotropies have a zero trace tensor, so the perpendicular component occurs at a field twice as close to H_0^0 than the parallel component. Taking into account that A' and B' are defined with H_0, and that H_0 is close to H_0^0 for small anisotropies, $A'/B' > 1$ is expected for positive anisotropies. These parameters can also be used as indicators for the sign of strong anisotropies. On A and B there are, of course, no changes but differences appear on A' and B', as, in this case, the magnetization aligns preferentially along the facile magnetization direction defined by the anisotropy field instead of along the external field. This contributes to a drastic increase in the low-field component of the resonance.[(26)] The A'/B' ratio increases regardless of the anisotropy sign, as depicted on Figure 7.8, which summarizes the different situations encountered.

7.4.1.3. Determination of the Anisotropy Intensity

The calculation depends on the sign and on the strength of the anisotropy. There is no simple calculation for weak anisotropies. The best description would be provided by a complete simulation of the data including the powder pattern associated with each anisotropy and the intrinsic linewidth. Very few of these calculations have been attempted. The main difficulty lies in the separation of the different components of a broad signal. The peak-to-peak linewidth $\Delta H_{pp}(= H_- - H_+)$ indeed depends on the $\Delta\Gamma_{pp}$ and H_a.

In the case of nickel, $\Delta\Gamma_{pp}$ is constant over a wide range of temperatures (130–600 K), so in this range the temperature dependence of ΔH_{pp} is essentially due to anisotropies. At high temperatures or for very weak anisotropies, the effect of the anisotropies is negligible and the lineshape is Lorentzian. At low temperatures or for stronger anisotropies the Lorentzian shape is distorted. In practice, the intrinsic lineshape is progressively transformed from a Lorentzian

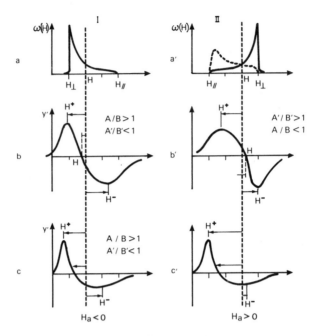

Figure 7.8. Effects of the strength and the sign of an axial anisotropy on the FMR lineshape [from Simoens[22]]; I and II for negative and positive anisotropies, respectively, a and a' for resonance distribution of randomly oriented magnetic domains or particles (the dashed line represents the case of a strong anisotropy), b and b' FMR line shapes for weak anisotropies, and c and c' FMR lineshapes for strong anisotropies.

to a Gaussian type[18,27] and H_0 shifts away from the natural resonance at H_0^0. The experimental determination of H_a is based on the linear dependence of $\Delta\Gamma_{pp}$ [equation (7.7), Section 7.2.1.3] and the invariability of the internal field anisotropy $vs.$ the frequency v applied for the resonance. For the range of small anisotropies in X and Q bands:

$$\Delta H_a = \alpha'[\Delta H_{pp}(X) - \Delta\Gamma_{pp}(X)]$$
$$\Delta H_a = \alpha'[\Delta H_{pp}(Q) - \Delta\Gamma_{pp}(Q)] \qquad (7.24)$$

with

$$\Delta\Gamma_{pp}(Q) = (v_Q/v_X)\Delta\Gamma_{pp}(X) \qquad (7.25)$$

and $\Delta H_a = |H_{//} - H_\perp|$ for an uniaxial anisotropy,[29,30] v_Q and v_X are the microwave frequencies for Q and X bands and, α' equals 1.732 or 1.776 for Lorentzian or Gaussian shapes, respectively, for the intrinsic resonance lines. Equations (7.24) and (7.25) give[18]:

$$\Delta H_a = \alpha'\{[v_Q\Delta H_{pp}(X) - v_X\Delta H_{pp}(Q)]/(v_Q - v_X)\} \qquad (7.26)$$

$$\Delta\Gamma_{pp}(X) = v_X\{[\Delta H_{pp}(Q) - \Delta H_{pp}(X)]/(v_Q - v_X)\} \qquad (7.27)$$

Then equations (7.26) and (7.27) can be used to calculate the anisotropy intensity from the linewidth of the FMR signal taken at two different frequencies.

For a strong anistropy, we can make a convenient approximation and neglect the effect of the intrinsic linewidth, assuming that the extremes of the derivative curve correspond to $H_{//}$ and $H_\perp$. When H_a is positive, the parallel component is on the weak-field side, with the approximation $H_+ \approx H_{//}$ and $H_a = H_0^0 - H_{//}$, and we have

$$H_a \approx H_0^0 - H_+ \tag{7.28}$$

When H_a is negative, the perpendicular component is located on the weak-field side where the signal is most intense. The approximation is now $H_+ \approx H_\perp$, and since $H_a = 2(H_\perp - H_0^0)$, we have

$$H_a \approx 2(H_+ - H_0^0). \tag{7.29}$$

7.4.1.4. Apparent Anisotropy for Superparamagnetic Domains

In the superparamagnetic regime, the anisotropies are averaged during the reorientation of the magnetic moments of the particles or magnetic domains. As a consequence, the anisotropies appear smaller than expected. This is, of course, dependent on the temperature. The apparent anisotropy $\langle H_a \rangle$ is given by the following relation for uniaxial symmetry[31]:

$$\langle H_a \rangle / H_a = [1 - 3x^{-1}L(x)]/L(x). \tag{7.30}$$

For a cubic anisotropy, the relation is different[33]:

$$(\langle H_a \rangle / H_a) = [1 - 10x^{-1}L(x) + 35x^{-2} - 105x^{-3}L(x)]/L(x) \tag{7.31}$$

where x is defined by equation (7.14), μ by equation (7.3), and $L(x)$ by equation (7.13). The thermal averaging effect is more pronounced for cubic anisotropies to the point that uniaxial anisotropies usually dominate for small superparamagnetic particles (< 4 nm for Ni).[18]

7.4.2. Shape Anisotropy

7.4.2.1. Demagnetizing Field and Shape Anisotropy[3,4]

The demagnetizing field $\mathbf{H}_d$ was first proposed by Kittel to explain the disparity of the FMR responses of samples of the same conducting materials but of different shapes. He suggested a relation between the internal field and the induced current due to the motion of the conduction electrons in the bulk of the samples. These induced current loops or Foulcault currents generate internal magnetic fields in the bulk of the sample. Locally, these magnetic fields are opposed to the applied field and reduce the magnetization, and are the so-called demagnetizing fields. This is effectively observed for a ferromagnetic plate placed perpendicularly to the direction of the magnetic field. However, the overall effect depends on the size of the loops and their relative situations to one another (mutual magnetic interactions). As a consequence, the resulting demagnetizing field is a function of the shape and the orientation of the sample leading to an effective field that can be either larger or smaller than the external applied field.

The demagnetizing field $\mathbf{H}_d$ is also called the shape anisotropy field. For negligible anisotropy fields other than the shape anisotropy field, we have

$$\mathbf{H}_d = \mathbf{H}_{\text{eff}} - \mathbf{H}_{\text{res}} \tag{7.32}$$

The demagnetizing field depends on the shape coefficients N_{ij} (elements of a tensor) which are reduced to only three for simple shapes, such as ellipsoids (diagonal tensor). These three coefficients, N_x, N_y, and N_z, are calculated along the Ox, Oy, and Oz axis, so the resonance occurs following equation (7.33), at

$$H_0^0 = \{[H_{\text{res}} + (N_y - N_z)M_s][H_{\text{res}} + (N_x - N_z)M_s]\}^{1/2} \tag{7.33}$$

with $N_x + N_y + N_z = 4\pi$.

For a spherical magnetic domain, N_x, N_y, and N_z are all equal to one another and H_d is cancelled. For an ellipsoid with a rotational axis, the calculation of the resonance field parallel and perpendicular to the axis gives

$$\begin{aligned} H_{//} &= H_0^0 - \Delta N M_s \\ H_\perp &= \tfrac{1}{2}[\Delta N M_s + (\Delta N^2 M_s^2 + 4H_0^{02})^{1/2}] \end{aligned} \tag{7.34}$$

where $\Delta N = N_b - N_a$, with N_a the shape coefficient along axis a, and $N_b = N_c$ the coefficients for axes b and c, respectively. From equation (7.33) we have

$$\Delta N = \tfrac{1}{2}(4\pi - 3N_a) \tag{7.35}$$

and N_a calculated by Osborn[28] is given by the following equations:

$$N_a = \frac{4\pi m^2}{m^2 - 1}\left\{1 - \frac{\text{arc sin}[(m^2 - 1)^{1/2}/m]}{(m^2 - 1)^{1/2}}\right\} \tag{7.36}$$

for an oblate ellipsoid with $m = b/a$ and,

$$N_a = \frac{4\pi}{m^2 - 1}\left[\frac{m}{2(m^2 - 1)^{1/2}}\,ln\left(\frac{m + (m^2 - 1)^{1/2}}{m - (m^2 - 1)^{1/2}}\right) - 1\right] \tag{7.37}$$

for a prolate ellipsoid with $m = a/b$.

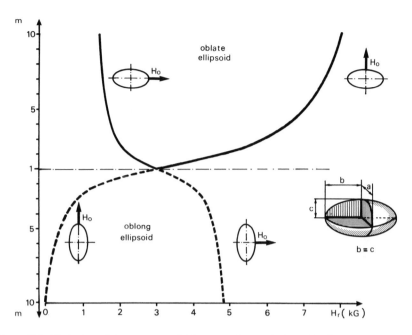

Figure 7.9. Variation of the resonance field $vs.$ m, where m is the ratio of the small to the large axis of an oblate ellipsoid (top) and prolate ellipsoid (bottom).[22]

The variation of m with the resonance field is shown in Figure 7.9. There are two limiting cases for these ellipsoids: the rod for the prolate ellipsoid and the plate for the oblate ellipsoid. For a rod, we have $N_a = 0$ with $N_b = N_c = 2\pi$, and $\Delta N = 2\pi$ leading to

$$H_{//} = H_0^0 - 2\pi M_s \tag{7.38}$$

For a plate, we have $N_a = 4\pi$ with $N_b = N_c = 0$, and $\Delta N = 4\pi$ leading to

$$H_{//} = H_0^0 + 4\pi M_s \tag{7.39}$$

7.4.2.2. Case of Ellipsoids, and Plate, and Rod Shapes

As H_a parallel to the axis of the anisotropy is $\Delta N M_s$ [equation (7.34)], the sign of the anisotropy only depends on ΔN:

1. For an oblong ellipsoid, $N_a < N_b$ and H_a is positive (case II in Figure 7.8).
2. For an oblate ellipsoid, $N_a > N_b$ and H_a is negative (case I in Figure 7.8).

For the particular cases of the rod and the plate, H_a is equal to $+2\pi M_s$ and $-4\pi M_s$, i.e., positive and negative, respectively, in accordance with the general case of ellipsoids treated above. It is important to note that for very small domains (<6 nm) the shape anisotropy is less important than the surface anisotropy (*vide infra*), and that, as a consequence, shapes and signs have the opposite relationship to that given here for larger particles.

7.4.2.3. Surface Anisotropy in Very Small Particles

Néel[33] has introduced the surface anisotropy to describe the situation encountered by surface atoms with a different environment than those of the bulk. This anisotropy is based on the dipolar magnetic coupling between two "magnetic neighbors." The overall effect of this anisotropy is a summation of the local effect on each surface atom and depends as well as on the shape of the magnetic domain and on the presence of adsorbates. For a spherical domain, this summation leads to a cancellation of local effect, and there is no resultant surface anisotropy.

A film with a thickness d or an ellipsoid with an eccentricity e and a small axis length b has a surface anisotropy varying in $1/d$ and e^2/b, respectively; the surface anisotropy becomes very important for very thin films and small particles. Néel shows that the surface anisotropy becomes more important than the shape anisotropy at a critical diameter d_c:

$$d_c = 4\,|K_s|/\pi M_s^2 \tag{7.40}$$

where K_s is a surface anisotropy constant. For nickel, d_c equals 6 nm with a $K_s = -0.12 \times 10^{-3}$ J m^{-2}. The surface anisotropy field of ellipsoids is related to the demagnetization field so that the overall anisotropy (shape + surface), $\mathbf{H}_{s+d}$, is given by equation (7.41):

$$\mathbf{H}_{s+d} = \mathbf{H}_d(1 - d/d_c) \tag{7.41}$$

where d is the thickness of a film or the smallest axis b of an ellipsoid, i.e., for practical purposes, the smallest dimension of the particle. For instance, for particles of 3 and 1 nm the overall anisotropy field equals to -1 and -5 times $\mathbf{H}_d$. This shows that a clear predominance of the surface anisotropy for small particles is to be expected, as found experimentally.[34]

7.4.3. Magnetostriction Anisotropy

7.4.3.1. Mechanical Strain and Anisotropy

The magnetostriction anisotropy is attributed to the distortion of the lattice under mechanical stress applied to the magnetic material. A magnetic anisotropy field H_σ is related to a (3×3) stress tensor, σ, by

$$H_\sigma = 3\lambda\sigma/M_s \qquad (7.42)$$

where λ is a magnetostriction coefficient having different values depending on the crystal directions. For nickel, along the $\langle 100 \rangle$ or the $\langle 111 \rangle$ directions, these coefficients are $\lambda_{100} = -66 \times 10^{-6}$ and $\lambda_{111} = -29 \times 10^{-6}$.[7] For polycrystalline samples, it is convenient to take an average value of λ ($\lambda_{av} = -37 \times 10^{-6}$)[35] and σ as a scalar entity; the direction of the stress is a function of the symmetry of the system (see below). In the absence of external mechanical strain, this anisotropy is typically the result of interfacial interactions related to the nature of those interfaces developed around a supported particle: metal–gas or metal–vacuum interfaces, on one hand, and metal–support, on the other. This results in an anisotropic stress tensor and leads to a lattice distortion. This problem has been widely treated in thin-film studies, where mechanical stability and strong adhesion to the substrate are essential qualities.[36]

7.4.3.2. Magnetostriction in Supported Small Particles

The concept developed in this field can be applied to supported particle catalysts.[18,21,22] The stress consists of two major components, a "thermal" component arising from the difference in the thermal expansion coefficients of the film (particle) and the substrate (support), and an "intrinsic" component depending on the structure and the growth of films (particles). The thermal component stress, well understood, is due to the constraint imposed by the film–substrate or the particle–support bonding and is given by

$$\sigma_{th} = (\alpha_m - \alpha_s)\Delta T E_{met} \qquad (7.43)$$

where σ_{th} is the thermal stress tensor, α_m and α_s are the average coefficients of expansion of the metal and the support, respectively, ΔT is the temperature of the substrate during film deposition minus its temperature of measurement, and E_{met} is Young's modulus of the film (particle). For nickel films deposited on NaCl[37,38] or aluminum,[23] a large thermal strain was found to be associated with a large lattice misfit, while this was not the case for MgO, SiO$_2$,[38] and CaF$_2$[23] substrates. In all cases, even with a copper or a Cu$_2$O support,[39] for which no mismatch is expected, the thermal stress in nickel films is strongly dependent on the temperature of the substrate during vapodeposition.

For small particles, ΔT cannot be calculated simply from the temperature of reduction at which the particles are formed. A relaxation of the structure of the deposited metal cluster is likely to occur, leading to an equilibrium temperature T_e such as ΔT is probably smaller than expected. For nickel particles supported on silica,[21] the overall anisotropies were negative below 473 K, and positive above. This change of sign is attributed to the thermal component of the magnetostriction anisotropy, which overcomes the other negative anisotropies at high temperatures, indicating that T_e was smaller than 473 K.

The intrinsic stress is a much more complicated situation in thin films than the thermal stress.[36] For very thin films, there is a strong dependence of the magnetization on the film thickness d. This dependence can be attributed to surface tension effects such as the average intrinsic stress and can be written[35]:

$$\sigma_{int} = -(\gamma_1 + \gamma_2)/d \qquad (7.44)$$

where γ_1 and γ_2 are the surface tension for the film–vacuum and the film–support interfaces. For nickel at room temperature, we have[35]

$$H_{\sigma_{int}} = -1.13 \times 10^4/d \qquad \text{(in gauss, with } d \text{ in nm)} \qquad (7.45)$$

This relation was verified for nickel deposited on NaCl. Any gas chemisorption leads to a complete cancellation of this intrinsic anisotropy.[35] In this situation and for supported nickel particles in the absence of chemisorbed gas, the intrinsic anisotropy is uniaxial and negative with an axis perpendicular to the substrate (support) surface.

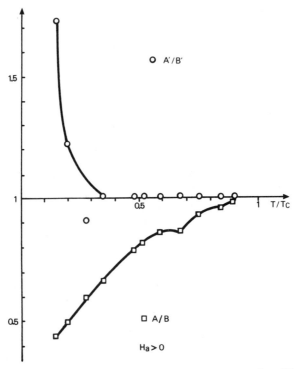

Figure 7.10. Temperature dependence of the asymmetry parameters for a $Ni_{10}CaX$ sample reduced at 273 K by hydrogen atoms.[18]

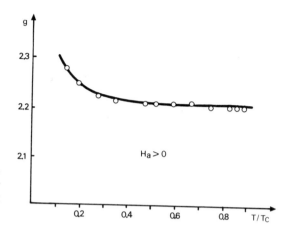

Figure 7.11. Temperature dependence of the experimental g factor for a $Ni_{10}CaX$ sample reduced at 273 K by hydrogen atoms.[18]

7.4.3.3. Experimental Spectra and Magnetostriction Anisotropy[21,18]

In most of the experimental work on powders and some on films the magnetostriction anisotropy is considered to be uniaxial. This is mainly because the magnetostriction coefficient λ and expansion coefficient α do not vary sufficiently from one direction to another for such a subtlety to be significant in powder pattern spectra. Therefore such anisotropies can be considered as uniaxial with the principal axis perpendicular to the substrate or the oxide support.

The origin of the magnetostriction is obtained experimentally from the thermal behavior and the sensitivity of the anisotropy to the chemisorption of molecules such as CO or hydrogen. The intrinsic magnetostriction varies monotonously as $\lambda\sigma/M_s$, and though less intense, persists up to T_C. Gas adsorption can be expected to cancel this intrinsic component while leaving the thermal component unchanged. The latter varies as ΔT, and $\lambda\sigma/M_s$ cancels and changes in sign at T_e.

The sign of the anisotropy is related to the state of the strain. A compression in the principal axis leads to a positive anisotropy and a low-field shift of the parallel component of the anisotropy in the case of nickel ($\lambda < 0$). Such a

TABLE 7.3. Determination of the Anisotropy Field Intensity and Intrinsic Linewidth in the NiCaX Sample Reduced by Hydrogen Atoms at 0 °C[18]

T/K	$\Delta H_{pp}/G$	$\langle \Delta H_a \rangle/G$	$\Delta \Gamma_{pp}/G$
93	1600	1200	650
223	700	900	260
293	600	570	310

compression is expected for a particle free of adsorbates, i.e., for the intrinsic component of magnetostriction. The thermal component leads to the same shift at temperatures lower than T_e, where the support forces the elongation of the particle in all the directions of the surface plane, i.e., a compression perpendicular to the surface (note: the stress tensor at equilibrium is characterized by a zero trace matrix). Quantitative determination of the anisotropy should be treated as discussed in Section 7.4.4, taking into account the size of the particles and the averaging effect.

7.4.3.4. Magnetostriction in Nickel Particles Stabilized in an X-Type Zeolite Framework

The reduction of a nickel-containing X-type zeolite, $Ni_{10}CaX$, at 0 °C by hydrogen atoms leads to small nickel particles of 1 nm average diameter (calculated from susceptibility measurements).[40] The internal and external particle distributions have been calculated directly from electron micrograph pictures of slices of zeolite particles prepared on a microtome: in the zeolite structure, the particles are smaller than the cages ($d < 1$ nm) and, on the external part of the crystallites, the sizes range from 1 to 4 nm.[18] The sign and the intensity of the anisotropy for such samples have been studied in detail.[18] The ratio of the amplitude asymmetry parameters A/B stays less than 1 over the whole range of temperature while the ratio of the width asymmetry parameters A'/B' is greater than or equal to 1; the anisotropy is positive (Figure 7.10). The g factor is consistently larger than the natural g value for nickel (Figure 7.11). Above 300 K, the signal progressively tends to be symmetrical ($A/B \rightarrow 1$ and $A'/B' \rightarrow 1$) when the temperature increases as a result of the averaging effect expected for very small particles.

The linewidth is insensitive to hydrogen or oxygen chemisorption, which indicates the existence of a chemisorbed phase that cancels any intrinsic constraint. Furthermore, no significant sign modification can be detected even close to T_C; this also excludes the possibility of a thermal component of magnetostriction. Therefore, a shape anisotropy or rather surface anisotropy is dominant for such small particles. A positive surface (+ shape) anisotropy H_{s+d} is attributed to oblate ellipsoids or flat nickel particles. A more quantitative estimation of the distortion from a purely spherical shape necessitates a measure of the anisotropy strength. The intensity of the anisotropy has been estimated by the multifrequency method, Q and X bands [(equations (7.26) and (7.27)], and the results are given in Table 7.3.

A strong anisotropy field of 1200 G dominates the signal linewidth at 93 K, decreases down to 570 G at 293 K, and vanishes close to T_C. Such anisotropy is attributed to the largest particles of the sample that experience the weakest averaging effect, i.e., to those located in the external part of the zeolite particle (1–3 nm from electron micrograph observations). Indeed, the analysis of the lineshape at 77 K shows that the signal is the superimposition of a large Gaussian resonance which accounts for the signal shift (large particles) and a narrower Lorentzian resonance with a negligible shift (small particles). From these sizes and the averaged field intensity, a length-over-width ratio for the particle shape was estimated to be between 3 and 4.

7.4.4. Magnetocrystalline Anisotropy

The magnetocrystalline anisotropy arises from the spin–orbit coupling experienced by atoms in a nonspherical symmetry. In ionic systems or complexes, the crystal field symmetry is often used to explain how the orbitals are no longer degenerate (different associated energies). This leads to an anisotropic distribution of the angular momentum and the spin momentum as well, via the spin–orbit coupling. The crystal generates local asymmetric environments depending on its structure and imposes this structural anisotropy. In a ferromagnetic material, this anisotropy is very intense because of the strong ferromagnetic coupling between neighboring spins and it is called the magnetocrystalline anisotropy.

7.4.4.1. Crystal Structure and Anisotropy

The magnetocrystalline anisotropy can also be considered as a strong local fiel H_{mc} or magnetocrystalline anisotropy field in which the magnetic moment of Weiss domain is oriented in directions of facile magnetization in the absence of applied external field. The direction of facile magnetization depends on the nature of the metal and its structure (Table 7.4). For nickel and iron the cubic structure imposes a cubic anisotropy with facile magnetization in the [111], [110], and [100] directions.

The resonance fields H_{res}, arise at $H_{[100]}$, $H_{[111]}$, and $H_{[110]}$ when the external field is parallel to the [100], [111], and [110] crystalline directions such as, for instance, [equation (7.2)],

$$H_0^0 = H_{[111]} + H_{mc} \quad \text{with } H_{ext} \text{ parallel to } [111]$$

In these situations, the resonance is conveniently calculated with the help of one anisotropy constant K_1, and equations (7.46)–(7.48):

$$H_0^0 = H_{[100]} + 2K_1/M_s \tag{7.46}$$

$$H_0^0 = H_{[111]} - 4K_1/3M_s \tag{7.47}$$

$$H_0^0 = [(H_{[110]} + K_1/M_s)(H_{[110]} - 2K_1/M_s)]^{1/2} \tag{7.48}$$

In reality there are two other anisotropy constants, K_2 and K_3, but these are small enough to be neglected.[7] K_1 is a temperature-dependent parameter. For nickel, the constant K_1 and, of course, the anisotropy cancel at about 400 K. A negative K_1 implies that the resonance occurs at low field in the [111] direction

TABLE 7.4. Characteristics of the Magnetocrystalline Anisotropy for Iron, Nickel, and Cobalt[3,7]

Metal	Direction of facile magnetization			$K_1/\text{erg cm}^{-3}$ at 298 K	H_{mc}/G at 298 K
Fe	[100]	[110]	[111]	$+4.81 \times 10^5$	538
Ni	[111]	[110]	[100]	-5.48×10^4	-131
Co	[0001]			$+4.12 \times 10^6$	3060

and at high field for the [110] and [100] directions for nickel. Conversely, iron exhibits a positive K_1 with a low-field resonance along the [100] direction and a high-field shift along the [110] and [111] directions. Cobalt has a hexagonal compact structure leading to a uniaxial magnetocrystalline anisotropy. When the external field is parallel to this direction, the resonance occurs at a field $H_{//}$ given by

$$H_0^0 = H_{//} + (K_1/M_s) \tag{7.49}$$

This anisotropy is more intense for cobalt (Table 7.4) than other metals. K_1 is positive up to 536 K and becomes negative above this temperature. In the whole range of temperature (below T_C) K_2 stays positive[7]; the temperature dependence of the magnetocrystalline anisotropy of cobalt is more complicated than that of iron or nickel.

7.4.4.2. Powder Pattern and Magnetocrystalline Anisotropy

7.4.4.2a. Cubic Anisotropy (Iron and Nickel). A cubic anisotropy is characterized by three directions; this situation is similar to an orthorhombic g factor treated for a paramagnetic case. Nonetheless, there is a very important difference due to the intensity of the anisotropy, which forces the magnetic moments of the Weiss domain to orient themselves along these three directions of facile magnetization. It is no longer possible to consider a random distribution of the directions of the magnetic moment. As a consequence, the resonance is enhanced along the facile magnetization directions and the low-field component is enhanced in comparison with the two high-field components, as shown in Figure 7.12.

In practice, on the derivative absorption curve for nickel, only $H_{[111]}$ and $H_{[110]}$ can be detected with sufficient accuracy to estimate the anisotropy strength, $\Delta H_a = |H_{[111]} - H_{[110]}|$, which is equal to $10K_1/3M_s$, i.e., 314 G at 298 K. As the intrinsic linewidth is already about 300 G at this temperature and the averaging effect will decrease the apparent anisotropy, the magnetocrystalline anisotropy can be neglected at room temperature for small particles.

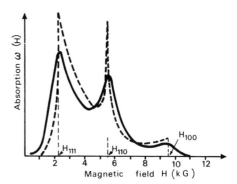

Figure 7.12. Effect of the magnetocrystalline anisotropy on the FMR resonance absorption curve: (---) theoretical curve and (——) experimental curve.[26]

7.4.4.2b. Uniaxial Anisotropy (Cobalt). Below 536 K, K_1 is positive and the uniaxial anisotropy is positive. The resonance in a direction parallel to this axis $H_{//}$ arises at low field and the powder pattern spectrum is typically represented by case II, a' and b' in Figure 7.8. The magnetocrystalline anisotropy field is much larger (≈ 3000 G) than the intrinsic linewidth (≈ 100 G), allowing the observation of a well-resolved powder pattern unless the particles are very small and the averaging effect highly efficient. Nonetheless, the low-field component of the resonance $H_{//}$ is expected to be more intense than the perpendicular component $H_\perp$.

7.4.5. Relative Intensity of the Various Anisotropies

It is important to identify and discriminate the various types of anisotropies which can operate simultaneously on supported particles; otherwise the random information obtained is misleading about the structure, shape, and mechanical strains. For magnetocrystalline anisotropy (Sections 7.4.3.3 and 7.4.3.4), the difference between thermal and intrinsic components can be determined by the use of chemisorption to cancel the latter. More generally, temperature is the only parameter available to identify the nature of the anisotropies via the selection of the right range of temperatures where one anisotropy dominates the others. Indeed, each anisotropy follows a characteristic temperature dependence depicted in Figure 7.13: the magnetocrystalline in K/M_s, the magnetostriction in $\lambda\sigma/M_s$, and the shape-related anisotropies in M_s.

The thermal component of magnetostriction follows the temperature dependence modified from case c in Figure 7.13 by a factor ΔT, or more exactly, $\lambda\Delta T/M_s$, with a change of sign at T_e [equation (7.43)]. An additional difficulty arises from the averaging effect that modifies the shape of the curves of Figure 7.13. The apparent anisotropy is smaller than expected for smaller particles and higher temperatures [equations (7.30) and (7.31), Section 7.4.1.4]. The averaging effect is much more pronounced for cubic than for uniaxial anisotropies. It follows that for small particles the cubic anisotropies, such as magnetocrystalline for iron and nickel, are much more affected by the temperature than shape

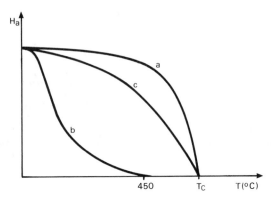

Figure 7.13. Temperature dependence of various anisotropies: (a) shape, (b) magnetocrystalline, and (c) magnetostriction (intrinsic component σ being constant) (from Bonneviot et al.[18,22]).

or magnetostriction, which are more axial by nature. For example, for 4-nm nickel particles, the magnetocrystalline anisotropy is negligible above 77 K. On the other hand, at low temperatures, magnetocrystalline anisotropy may dominate and its partial averaging may be used as a measure of the particle size.[41]

7.5. CONCLUSION

Ferromagnetic resonance is a very powerful technique for the study of oxide-supported particles of ferromagnetic materials such as iron, cobalt, and nickel and their alloys. *In situ* experiments are possible as with EPR and NMR and like these techniques, FMR is a nondestructive photon–photon spectroscopy (nonsurface specific). Its sensitivity is greater than most other spectroscopies, including EPR.

The diversity of information that can be obtained from this technique is of special interest for supported particles: particle size and shape and mechanical constraints undergone at the metal–support and the vacuum (or gas)–metal interfaces. Nevertheless, the information contained in one single FMR spectrum is not directly available. This difficulty arises from the impossibility of determining several independent parameters from a broad powder pattern signal. Fortunately, intrinsic linewidth and magnetic anisotropy follow different temperature-dependent behaviors for each parameter. A study of the intensity and linewidth temperature dependencies is necessary to characterize (i) the homogeneity of the particle size, (ii) their sizes in case of homogeneous dispersion, and (iii) their magnetic anisotropies (shape and constraint). It is important to note that the point (i) concerning the homogeneity is crucial since most of the subsequent information is related to the particle size and therefore meaningless for inhomogeneous samples.

For a narrow size distribution, the average particle size is easily estimated from the shape of the thermomagnetic curves. A typical accuracy for such measurements is about ± 0.5 nm. Better accuracies are obtained in studies over larger temperature ranges and narrower size distributions. Furthermore, these estimates are affected, particularly for very small particles, by the presence of an adsorbed phase which may demagnetize the surface atoms. The same situation occurs in size determination by magnetic susceptibility measurements although this technique is more accurate for size determination than FMR measurements.

The various types of anisotropies can be distinguished by careful analysis of the temperature dependence of the signal asymmetry. For very small particles, the magnetocrystalline anisotropy is usually negligible (case of Ni) and the remaining averaged anisotropies can be treated in the uniaxial approximation. Uniaxial anisotropies are characterized by a sign related to the shape (flat or elongated) or to the constraint (compression or elongation). This sign is determined from the experimental spectrum by signal shape parameters, H_0, H_+, H_-, A, B, A', and B' (Figure 7.7), and when the anisotropies are strong, their intensity is conveniently estimated from these parameters. For small anisotropies or highly averaged anisotropies of very small particles, the only available experimental approach is the analysis of the signals acquired at different frequencies, typically X and Q band.

Though it appears that FMR is neither a straightforward nor an universal technique as are vibrational, elecron, or X-ray absorption spectroscopies, this resonance has some very specific advantages. It is the only technique we can use to analyze particle stress, to give shape characterization in a very small size range (< 2 nm) not easily attainable by transmission electron microscopy, and (because of its sensitivity) to analyze traces (≈ 100 ppm) and metal compositions for bimetallic particles (see the example on Ni–Cu/SiO$_2$).

GLOSSARY

$\left.\begin{array}{l}A\\A'\end{array}\right\}$	asymmetry parameters (Figures 7.7 and 7.8)
α'	shape coefficient for Lorentzian and Gaussian curves [equations (7.24) and (7.26)]
α	exponent in relaxation time [equations (7.9), (7.12), and Table 7.1]
α_x	expansion coefficient [equation (7.43)]
$\left.\begin{array}{l}B\\B'\end{array}\right\}$	asymmetry parameters (Figures 7.7 and 7.8)
$\beta(x)$	Bohr magneton number for a metal (x = Fe, Co, Ni, . . .)
β	Bohr magneton [equations (7.1) and (7.23)]
d	spherical domain diameter or film thickness [Table 7.1, equations (7.40), (7.41), (7.44) and (7.45)], see also v
d_c	critical diameter [equations (7.40) and (7.41)]

$\Delta\Gamma_{pp}$	intrinsic peak-to-peak linewidth [equations (7.6), (7.7), (7.24), (7.25), and (7.27)]
ΔH_a	see H_a [equations (7.24), (7.26), and Table 7.3)
ΔH_{pp}	peak-to-peak linewidth [equations (7.24), (7.26), and (7.27), Figure 7.7, and Table 7.3]
ΔN	see N_m [equations (7.34), (7.35), (7.38), and (7.39)]
ΔT	temperature splitting between T_e and T (see T_e, E_{met} and α_x)
E_a	anisotropy energy [equation (7.8)]
E_{met}	Young's modulus of the supported metal [equation (7.43)]
f_0	frequency factor [equation (7.9)]
γ	gyromagnetic ratio [equations (7.1), (7.6), (7.7), and (7.11)]
γ_1 γ_2	surface tension at film–vaccum and film–support interfaces [equation (7.44)]
γ_e	electron gyromagnetic ratio
g_e	electron g factor
H_+ H_-	experimental field for the maximum and the minimum of the first-derivative curve [Figure 7.7, equations (7.28) and (7.29)]
$H_{//}$ $H_\perp$	resonance field in specific orientations [equations (7.34), (7.38), (7.39), and (7.49)], see ΔH_a
$H_{[111]}$ $H_{[110]}$ $H_{[100]}$	resonance field in specific orientations [equations (7.46)–(7.48) and Figure 7.13]
H_0	apparent resonance field taken from the first-derivative curve (Figure 7.7)
$\mathbf{H}_{s+d}$	sum of the shape and surface anisotropy fields [equation (7.41)]
$\mathbf{H}_a$	anisotropy field [equations (7.2), (7.12), and (7.28)–(7.31)], see ΔH_a
$\mathbf{H}_d$	demagnetizing field [equations (7.32) and (7.41)]
$\mathbf{H}_{eff}$	effective or local field [equations (7.1) and (7.32)]
$\mathbf{H}_{mc}$	magnetocrystalline anisotropy field [equations (7.46)–(7.49) and Table 7.4]
$\mathbf{H}_0^0$	intrinsic resonance field [equations (7.1), (7.2), (7.19), (7.33), (7.34), (7.38), (7.39), and (7.46)–(7.49)]
$\mathbf{H}_{res}$	experimental resonance field [equations (7.2), (7.32), and (7.33)], see $H_{[111]}$, $H_{[110]}$ and $H_{[100]}$
$\mathbf{H}_\sigma$	magnetostriction anisotropy field [equations (7.42) and (7.45)]
$\mathbf{H}_s$	surface anisotropy field, see H_{s+d}
$I(T)$	FMR signal intensity at temperature T [equation (7.19)]
K	anisotropy constant, K_m with $m = s, a, 1, 2, 3$ (equations (7.8), (7.9), (7.41), and (7.46)–(7.49)]
k	Boltzmann constant
$L(x)$	Langevin function, $\coth(x) - x^{-1}$ [equations (7.13)–(7.16), and (7.31)]
λ	magnetostriction coefficent [equation (7.42)]
λ_L	damping constant [equation (7.7)]
$M(T)$	magnetization of the sample [equations (7.13), (7.15)–(7.17), (7.20), and (7.21)]
M_r	residual magnetization [equation (7.10)]

M_{so} spontaneous magnetization at 0 K [equations (7.4), (7.20), and (7.21)]

μ magnetic moment of Weiss domains or small particles [equations (7.3) and (7.14)]

M_s spontaneous magnetization [equations (7.3), (7.4), (7.7), (7.10)–(7.13), (7.15), (7.16), (7.33), (7.34), (7.38)–(7.40), (7.42), and (7.46)–(7.49)]

ν frequency delivered by the klystron, X or Q bands [equations (7.24)–(7.27)]

N_m shape anisotropy constants, $m = x, y, z, a$ or b [equations (7.33)–(7.39)]

ν_0 precession frequency [equation (7.1)]

σ strain tensor [equations (7.42)–(7.44)]

T temperature of the sample

τ correlation time [equations (7.9)–(7.11)]

t measurement time [equation (7.10) and Table (7.1)], see blocking temperature T_b

T_1 longitudinal relaxation time [equation (7.5)]

T_2^* spin–spin relaxation time [equation (7.5)]

T_2 transversal relaxation time [equations (7.5) and (7.6)]

T_b blocking temperature (Table 7.1)

T_C Curie temperature [equations (7.4) and (7.20)–(7.23) and Table 7.2]

T_e equilibrium temperature between a particle and the support [equation (7.43), see ΔT

T_0 lowest temperature of measurements used for thermomagnetic curves [equations (7.17) and (7.18) and Figure 2)

v volume of the particle or Weiss domain for monodomain particles [equations (7.3), (7.8), (7.9), (7.11), (7.12), (7.15), (7.16), and (7.21)], see also d

v_c critical volume for superparamagnetic transition, see d_c and T_b

x variable $= \mu H/kT$, see $L(x)$

$Y'(H, T)$ first derivative of the FMR absorption curve [equation (7.19)]

REFERENCES

1. A. H. E. Griffiths, *Nature* **158**, 670 (1946).
2. D. Hollis and P. W. Selwood, *J. Chem. Phys.* **35**, 378 (1961).
3. B. Herpin, *Théorie du Magnétisme*, PUF, Paris (1968).
4. C. H. Morrish, *The Physical Principle of Magnetism*, John Wiley and Sons, New York (1965).
5. D. E. Patton, in: *Magnetic Oxides* (D. J. Craik, ed.) John Wiley and Sons, New York (1975), p. 575.
6. E. Lax and K. J. Button, *Microwave Ferrites and Ferrimagnetics*, McGraw-Hill, New York (1962), p. 145.
7. F. P. Wohlfarth, *Ferromagnetic Materials*, North-Holland, Amsterdam (1980).
8. A. A. Slinkin, *Russ. Chem. Rev.* **37**, 642 (1968).
9. P. Weiss, *J. Phys., Ser. 4*, **6**, 661 (1907).
10. P. W. Selwood, *Chemisorption and Magnetization*, Academic, New York (1975).
11. L. D. Landau and E. M. Lifschitz, *Phys. Z; Soviet.* **8**, 153 (1935).

12. T. L. Gilbert, *Phys. Rev.* **100**, 1243 (1955).
13. F. Bloch, *Phys. Rev.* **70**, 460 (1946).
14. N. Bloembergen, *Phys. Rev.* **78**, 572 (1950).
15. L. Bonneviot, F. X. Cai, M. Che, M. Kermarec, O. Legendre, C. Lepetit, and D. Olivier, *J. Phys. Chem.* **91**, 5912 (1987).
16. L. Néel, *Ann. Geophys.* **5**, 99 (1949).
17. C. P. Bean and J. D. Livingston, *J. Appl. Phys.* **30**, 120-S (1959).
18. L. Bonneviot, Thesis, Université P. et M. Curie, Paris (1983).
19. A. Aharoni, *Phys. Rev.* **B 7**, 1103 (1973).
20. A. Aharoni, *Phys. Rev.* **117**, 793 (1969).
21. L. Bonneviot, M. Che, D. Olivier, G. A. Martin, and E. Freund, *J. Phys. Chem.* **90**, 2112 (1986).
22. A. J. Simoens, Thesis, Faculté Universitaire Notre-Dame de la Paix, Namur, Belgium, (1980).
23. D. Fargues, F. Vergand, E. Belin, C. Bonnelle, D. Olivier, L. Bonneviot, and M. Che, *Surf. Sci.* **106**, 239 (1981).
24. P. A. Jacobs, H. Nijs, J. Verdonck, E. G. Derouane, J. P. Gilson, and A. J. Simoens, *J. Chem. Soc. Faraday Trans. I* **75**, 1195 (1979).
25. E. G. Derouane, A. J. Simoens, C. Colin, G. A. Martin, J. A. Dalmon, and J. C. Védrine, *J. Catal.* **52**, 50 (1978).
26. E. Schlömann, *J. Phys. Chem. Solids* **6**, 257 (1958).
27. S. Bagdonat and M. J. Patni, *J. Mag. Res.* **15**, 359 (1974); C. M. Srivastava and M. J. Patni, *J. Phys.* **38**, Cl, 267 (1977).
28. J. A. Osborn, *Phys. Rev.* **67**, 351 (1945).
29. C. P. Poole, *Electron Spin Resonance,* John Wiley and Sons, New York (1967), p. 525.
30. M. Che, J. C. Védrine and C. Naccache, *J. Chim. Phys.* **66**, 579 (1969).
31. J. D. Livingston and C. P. Bean, *J. Appl. Phys.* **30**, 318-S (1959).
32. R. S. De Biasi and T. C. Devezas, *Phys. Lett.* **50 A**, 137 (1974); *Phys. Lett.* **B 87**, 1425 (1977).
33. L. Néel, *J. Phys. Radium* **15**, 225 (1954).
34. W. Göpel and B. Wiechmann, *J. Vac. Sci. Technol.* **20**, 219 (1982).
35. M. M. P. Janssen, *J. Appl. Phys.* **41**, 384 (1970).
36. K. L. Chopra, *Thin Film Phenomena,* McGraw-Hill, New York (1969), p. 266.
37. J. F. Freedman, *J. Appl. Phys.* **33**, 1148 (1962).
38. S. Chikazumi, *J. Appl. Phys.* **32**, 81-S (1961).
39. S. Kuriki, *J. Appl. Phys.* **48**, 2992 (1977).
40. M. Che, M. Richard, and D. Olivier, *J. Chem. Soc. Faraday Trans. I* **76**, 1526 (1980).
41. V. K. Sharma and A. Baiker, *J. Chem. Phys.* **75**, 5596 (1981).

MÖSSBAUER SPECTROSCOPY — NUCLEAR GAMMA RESONANCE

P. Bussière

8.1. INTRODUCTION

The chemical and structural properties of an atom can be studied from the shifts and degeneracy of its nuclear energy levels. These changes in the nuclear state are easily observed by using the resonance spectroscopy of the gamma rays emitted in a nuclear transition toward the ground state for an isotope of the element in question. Nuclear gamma resonance (NGR), is possible only for solids, as was shown by Mössbauer in 1957. It is illustrated in Figure 8.1, which shows in particular the mean energy transition E_0 within a Lorentzian distribution with a linewidth W_0 (or Γ_0).

Throughout this chapter we shall be dealing mainly with iron and tin, of which the isotopes ^{57}Fe and ^{119}Sn, respectively, are especially suitable for Mössbauer spectroscopy. The corresponding sources are ^{57}Co and ^{119m}Sn, for which the decay schemes are given in Figure 8.2 (thick arrow for the Mössbauer transition). They provide all the theoretical and technical features of NGR:

The branching ratios and the conversion coefficients determine the intensities of radiation (including electrons and X-rays from internal conversions).

The mean energy E_0 available for NGR is in the approximate range from 5 to 150 keV. Linewidth W_0, according to the relationship $W_0\tau = \hbar$ (τ is the mean life of the excited state), generally falls between 10^{-5} and 10^{-9} eV, i.e., a fraction of E_0 from 10^{-10} to 10^{-13}.

The spin I and the parities $+$ or $-$ govern:

1. The degeneracies: $(2I + 1)$ values of the quantum number m_I.

P. Bussière ● Institut de Recherches sur la Catalyse, CNRS, Villeurbanne, France.

Catalyst Characterization: *Physical Techniques for Solid Materials*, edited by Boris Imelik and Jacques C. Vedrine, Plenum Press, New York (1994).

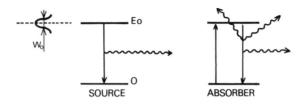

Figure 8.1. Nuclear gamma resonance.

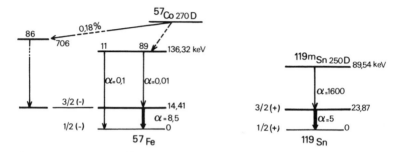

Figure 8.2. Decay schemes of ^{57}Co and ^{119m}Sn, with the Mössbauer transitions of ^{57}Fe and ^{119}Sn.

2. Selection rules: in our example Δm_I may only be either 0 or ± 1.
3. The corresponding relative intensities through Clebsch–Gordan coefficients.

The shifts of the nuclear states studied by NGR are either of the same order of magnitude as W_0 or from a few times to some tens of times larger or smaller. It follows that:

1. No detector has the resolution required in NGR spectroscopy. The solution of this problem consists of moving either the radioactive source or the absorber so that the energy offered to the resonance has a value given by the Doppler effect. In our two examples the velocities range from 0 to 1 cm s^{-1}. Figure 8.3 illustrates the principle of this technique. Mostly the sample to be studied is kept immobile as the absorber and the source are moved. This is the case throughout this chapter, except in Section 8.7.1, which describes emission Mössbauer spectroscopy.
2. The phenomenon is quite specific; no interference is possible from any other isotope.
3. Small perturbations are sufficient to escape the suitable energy domain. A strong perturbation is recoil energy E_R ($E_0^2/2Mc^2$, i.e., $E_\gamma^2/2Mc^2$), which is $1.96 \; 10^{-3}$ eV in the case of ^{57}Fe. NGR is therefore impossible with isolated atoms. This difficulty disappears when the atom is inserted

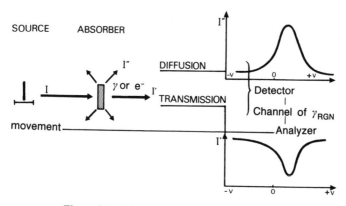

Figure 8.3. Scheme of Mössbauer spectroscopy.

in a solid lattice as the mass involved in recoil is very large, resulting in a negligible E_R. There remains only one important perturbation — phonon excitation, which generally implies an energy consumption of some 10^{-3} eV.

Finally the possibility of NGR is governed by the probability of emission and absorption of the photon gamma without recoil and without phonon excitation. The probability f is named the Lamb–Mössbauer factor, the Debye–Waller factor, or the recoil-free fraction. Many books describe Mössbauer spectroscopy.[1-6] The application of NGR to heterogeneous catalysis is described in exhaustive reviews[7-9] and in short and critical ones such as Bussière.[10]

8.2. THE f FACTOR AND RESONANT ABSORPTION

8.2.1. Elements of the Theory

The f factor can be analyzed by considering the system (nucleus, photon) where the nucleus is a charged particle moving in the electromagnetic field carried by the photon. The probability w for a transition is proportional to the square of the matrix element of the interaction Hamiltonian which leads the initial state i to the final state f:

$$w = \text{Const} \left| \int \psi_f^H \text{ int } w_i dt \right|^2 \tag{8.1}$$

State functions contain contributions of the γ photon of the nucleus and of the gravity center G of the nucleus. Due to the short range of nuclear forces, and given that each term of the Hamiltonian operates only on the part of the wave function containing like coordinates, the contributions of the photon and the nucleus appear as constants in the matrix element, which can be expressed as

$$w = \text{Const} \left| \int \psi_f^G \exp(ikx)\psi_i^G dt \right|^2 \tag{8.2}$$

where the exponential term originates from the potential characterizing the waves whose superimposition describes the electromagnetic field of the photon.

We obtain f when the lack of phonon excitation results in the same state of the solid lattice after the interaction as before, i.e.,

$$f = \text{Const} \left| \int \psi_i^G \exp(ikx)\psi_i^G dt \right|^2 \tag{8.3}$$

As wave functions sum to unity, normalization gives

$$f = \exp(-K^2 \langle x^2 \rangle) = \exp \frac{-E^2 \langle x^2 \rangle}{(\hbar c)^2} \tag{8.4}$$

where K is the modulus of the wave vector, E the energy of the photon, and x the vibrational amplitude of the atoms in the solid lattice. We can derive $\langle x^2 \rangle$ from the Debye model of the vibrations, so that

$$f = \exp \left\{ -\frac{3E_R}{2kT_D} \left[1 + 4 \left(\frac{T}{T_D} \right)^2 \int_0^{T_D/T} \frac{y dy}{e^y - 1} \right] \right\} \tag{8.5}$$

where E_R is the recoil energy, T the temperature, and T_D the Debye temperature with $y = h\nu/kT$. Tables have been compiled for the integral.

It is interesting to consider the following approximations:

When $T \ll T_D$:

$$f_{LT} = \exp \left[-\frac{3E_R}{2kT_D} \left| 1 + \frac{2\pi^2}{3} \left(\frac{T}{T_D} \right)^2 \right| \right] \tag{8.6}$$

which therefore never reaches the value of 1.

When $T \gg T_D$:

$$f_{HT} = \exp - \left(\frac{6E_R}{kT_D} \frac{T}{T_D} \right) \tag{8.7}$$

From these equations it is possible to deduce the conditions for sufficiently large values of f: M not too small ($\gtrsim 40$); E_0 not too large ($\lesssim 150$ keV); T_D not too small (rather strong chemical bonds); T not too high [at room temperature and above (such as the usual conditions of catalysis), only the following nuclei are available inside a metallic lattice or an ionic compound: ^{57}Fe, ^{119}Sn, ^{151}Eu, ^{149}Sm, ^{161}Dy, and ^{181}Ta].

Compensation for undesirable effects can be obtained by decreasing the temperature. In the two first cases (effects of M and E_0) both the source and the absorber, therefore the moving holder itself, must be cooled down to 77 or 4 K.

8.2.2. Particular Cases

8.2.2.1. Surface Atoms

Mössbauer spectroscopy using γ-rays appears as a technique for bulk studies, where of course the surface layers contribute part of the spectrum dependent on dispersion. The detection of conversion electrons (CEMS) favors this contribution, which becomes the major one in the case of DCEMS (see Section 8.7.3). In all systems the environment of Mössbauer atoms in top layers can result in an increase in their vibrational amplitude perpendicular to surface and, consequently, in a decrease in their Lamb–Mössbauer factor according to equation (8.4). Examples of such a situation are given in Sections 8.5.2.3 and 8.8.2.2.

8.2.2.2. Fine Powders

When the grains are loosely bound together, the experimental resonant absorption is often very much smaller than expected from the recoil-free fraction f_{lat} corresponding to the lattice vibrations.[11] This could be due to a large probability for the excitation of a movement of the whole particle that contains the absorbing atom, which results in a very low probability f_{part} for nonexcitation. Now an experimental factor f_{exp} should be considered:

$$f_{exp} = f_{lat} * f_{part}$$

Values of f_{part} can be determined by the preparation procedure: SnO_2 aerosol particles of about 5 nm exhibit an f_{part} as 0.01, while SnO_2 particles of similar size but prepared by precipitation from an aqueous solution and subsequent thermolysis behave with $f_{part} \approx 1$. Sampling methodology is also important since $f_{part} = 0.6$ for the same SnO_2 aerosol in a wafer pressed at 10^8 Pa.

8.2.3. Resonant Absorption (Nonresonant Absorption Negligible)

The transmitted line (Figure 8.4) is still Lorentzian but its theoretical width W is twice W_0. At each point corresponding to a velocity v the absorption is: $r(v) = [I'_\infty - I'(v)]/I'_\infty$, i.e., at the minimum (generally at $v \neq 0$): $r_{max} = (I'_\infty - I'_{min})/I'_\infty$.

(i) $r_{max} \lesssim 0.1$

$$r_{max} = f_s t/2 \tag{8.8}$$

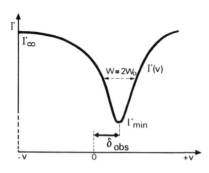

Figure 8.4. Mössbauer absorption line.

where f_s is the f for the source and t is the effective thickness equal to $\beta n d \sigma_0 f_a$, a dimensionless parameter, where β is the isotopic relative abundance, n the number of atoms of the chemical element in one unit volume, d the linear thickness of the sample, σ_0 the cross section for resonant absorption, and f_a the recoil-free fraction of the absorbant. The effective thickness is thus proportional to n and f_a.

(ii) $r_{\max} \gtrsim 0.1$

$$r_{\max} = f_s[1 - J_0(it/2)\exp(-t/2)] \tag{8.9}$$

inasmuch as the line stays nearly Lorentzian, which can be assumed when

$$t < 5 \tag{8.10}$$

Moreover,

$$W_{\exp} = W_s + W_a(1 + 0.27t) \tag{8.11}$$

(iii) It is also interesting to consider the absorption area

$$A = \int_{-\infty}^{+\infty} r(v)dv = CWf_s tP(t) \tag{8.12}$$

where C is a constant and $P(t) \approx 1$ when t is small.

8.2.4. Applications

8.2.4.1. Line Width of the Absorber W_a

W_a is calculated from equation (8.11) and the value of W_s and f_a. Its value generally exceeds the theoretical one, and this gives information on the solid

since there are several possible causes of the increase in W_a: (1) superimposition of neighboring lines, (2) nonresolved hyperfine splittings, (3) solid state defects, (4) electronic or magnetic relaxation phenomena, and (5) atomic diffusion, for which the dependence of W on temperature gives a means for determining a diffusion coefficient.

8.2.4.2. Effective Thickness t

This parameter can be derived from equations (8.8), (8.9), and (8.12). Knowledge of f_a follows from the knowledge of n or n from that of f_a. The latter case enables one to obtain a quantitative analysis of the different chemical species of the element, by allotting each one its own f_a.

An absolute determination of f_a or n generally gives a value of low accuracy. It requires a good selection of the resonant γ-rays from the other radiations, which can seldom be achieved with high precision, the value of f_s, and, for determining n, that of f_a, which are not often known precisely.

The situation is better when using a standard. In relative determinations, e.g., the evolution of a phase in a catalyst during an activation, a precision within a few percent can be achieved when it is certain that f_a does not change during the experiment.

8.3. EXPERIMENTAL TECHNIQUES

8.3.1. The Mössbauer Elements and the Sources

The conditions given in Section 8.2.1. ($M > 40$, $E_0 < 150$ keV) define the requirements for a useful source. The elements in which at least one isotope has a transition interesting in transmission Mössbauer spectroscopy (more or less easily) are: Am, Sb, Ba, Cs, Dy, Er, Eu, Ga, Ge, Au, Hf, Ho, I, Ir, Fe, K, La, Lu, Hg, Nd, Np, Ni, Os, Pt, K, Pr, Re, Ru, Sm, Ag, Ta, Te, Tr, Th, Tl, Sn, W, U, Xe, Yb, and Zn.

Emission Mössbauer spectroscopy can be an alternative, although caution is required in assessing the results (see Section 8.7.1). The following elements can be used in this technique: Cd, Cs, Co, Cu, Ga, Pt, Rh, and Rb.

Making the sources requires a delicate technology in order to emit the resonant γ-rays as a single line (see Sections 8.5 and 8.6) with a line width as close as possible to W_0 (see Section 8.2.3.1). Specialized laboratories sell sources with long half-lives, in particular ^{57}Co (diffused into cubic nonmagnetic metals such as Cr, Cu, Pd, and Rh) and ^{119m}Sn (Ca^{119m}SnO$_3$).

8.3.2. The Samples

Sampling is determined mainly by the requirements of equations (8.8) and (8.12). First the thickness x of the matrix including the Mössbauer element must not exhibit a high nonresonant absorption (from photoelectric and Compton effects). More precisely, if $x_{1/2}$ transmits half the intensity of the beam, a valuable condition is

$$x < x_{1/2} \qquad (8.13)$$

For instance, some values of $x_{1/2}$ in NGR of ^{57}Fe are 10, 70, and 1000 mg cm^{-2} of Fe, Al, and C, respectively.

Secondly, the intensity of the resonant absorption must be considered. When it is rather small [see equation (8.8)], the Mössbauer lines can be distorted or even hidden by the statistical fluctuations of the baseline. When it is rather large, saturation effects occur, resulting in a decrease in the height of the lines and even in reaching a deviation of the line shape from the Lorentzian profile [see equation (8.10)].

Finally a good compromise can be found as

$$0.1 < t < 5 \qquad (8.14)$$

For instance, at 295 K for 1 mg cm^{-2} of natural iron metal $t = 0.40$ and for 1 mg cm^{-2} of natural SnO$_2$, $t = 0.27$.

When a small amount of the absorbing element is present in the samples together with a low content in the resonant isotope, acceptable values of t can be obtained by using isotopic enrichment (if it is possible at not too great a cost). Powders should preferably be pressed into self-supporting wafers: (1) to avoid absorption by a holder, (2) to avoid any lowering of apparent f factor in the case of very small particles (see Section 8.2.2.2), and (3) to allow the use of a horizontal bench.

8.3.3. The Cells

Special cells are needed to study pellets under conditions other than in air at room temperature. For this it is not difficult to design (or buy) a furnace to heat the sample up to 1100 K in any atmosphere at 10^5 Pa or below, down to a typical secondary vacuum. A furnace allowing for rapid heating and quenching has been described.[12]

Since even the soft γ-rays from ^{57}Fe are easily transmitted by gases at pressures of several MPa as well as by convenient thicknesses of liquids made of light atoms, NGR studies in catalysis at high pressures are possible and appropriate cells have been made.[13,14]

It can be difficult to study a sample prepared at high temperature at low temperatures without contact with traces of oxygen. A cell has been designed for *in situ* Mössbauer spectroscopy from 750 °C down to 77 K[15]; another one has been described recently, going down to 4 K and less.[16] Without such devices, the transfer of a sample from a furnace to a cryostat needs the use of a glove-box or of some ingenious device.[17] Cells have to be small and their windows thin so

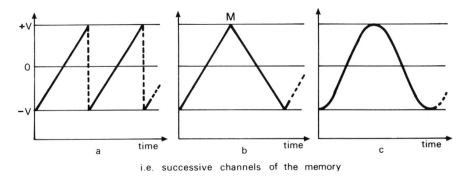

i.e. successive channels of the memory

Figure 8.5. The movement supplied by the drive. Modes: (a) saw-tooth, (b) triangular, (c) sinusoidal.

that they do not diminish the detection efficiency too much. Moreover, they often have to be small enough to fit into the gap of a magnet.

8.3.4. The Spectrometer

8.3.4.1. The Driving Unit

The most typical component of the device shown in outline in Figure 8.3 is a constant acceleration oscillator, a kind of very precise loudspeaker, where all the convenient velocities, either positive (when the source approaches the absorber) or negative (reverse sense), are given within a fraction of a second along a length of several millimeters, according to one of the three modes shown in Figure 8.5. Simultaneously the clock of the driving unit, coupled with a multichannel analyzer, successively addresses the countings corresponding to the velocities to the memory channels.

8.3.4.2. The Detection of Radiation

A Mössbauer spectrometer is at first a radiation spectrometer, because of the presence of parasitic rays either from the decay scheme or from diffusion, conversions, etc. The most usual detectors of γ-rays in a transmission mode are a proportional counter and a scintillation probe. The high resolution of solid state detectors is often useful, as is the specificity of some absorbers. Conversion electrons (see Section 8.7.3) need a high vacuum with channeltrons or a well-defined atmosphere with proportional counters.

8.3.4.3. The Assembly of the Components

The main components of the spectrometer (drive, cell, detector, shields, and collimators) stand on an optical bench on an antivibration rack or table.

8.3.5. Calibration

The velocities can be deduced from the channel numbers in two ways: either from the spectrum of a standard or from direct measurement of the velocities by interferometry. The former method is by far the most usual.

8.3.6. Computer Calculations

8.3.6.1. Converting the Raw Spectrum to the Net Spectrum

The angular points in the velocity–time diagrams of Figure 8.5 are not very well known. In particular, in the most often used triangular mode each period of the movement results in two spectra, symmetrical with respect to time, superimposed one on the other by folding around M. This operation can be performed by the multichannel analyzer itself, but generally the accuracy is insufficient so a separate, better computer is preferred.

8.3.6.2. Fitting the Net Spectrum to Single Lines and/or Multiplets

A program taking into account the Lorentzian shape of the lines and the formulas reported in Sections 8.5 and 8.6 below provides the best values of Mössbauer parameters of the components of the spectrum.

8.3.7. Performance

8.3.7.1. Spectrum Accumulation Time

This depends on the detection conditions (e.g., activity of the source, conversion coefficients, distance between the source and the detector, and nonresonant absorption of all kinds), and on the required precision (e.g., complexity of the spectrum or need for high accuracy on some Mössbauer parameters, even for a single line, and intensity of the resonant absorption). Usually a good spectrum is obtained after some hours, but it is not unusual to spend several days or only half an hour. Such times are not consistent with the requirements of a true kinetic study; this becomes possible when recording the intensity of one line of the spectrum, this needs only a few minutes.

8.3.7.2. The Precision on Mössbauer Parameters

As detailed in 8.2.3.2, absolute values of t (effective thickness) and therefore of either f (Lamb–Mössbauer factor) or n (corresponding number of atoms) are obtained only with low accuracy, but relative values can be determined within a few percent. Nuclear level shifts brought about by the electronic and structural environment (see Sections 8.4–8.6) and also the line widths are usually determined from fitting to within ± 0.02 mm s^{-1} in the best cases in iron and tin Mössbauer spectroscopy.

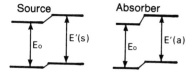

Figure 8.6. Nuclear level shifts in the Mössbauer transition.

The temperature of the laboratory should be constant within 1 or 2 K in order to reach these performance levels.

8.3.8. Radioprotection

In the vicinity of a Mössbauer spectrometer in which the source is shielded, the dose rate keeps below the maximum admissible value. For instance, a nonscreened ^{57}Co source of 3.7 GBq gives 0.1 mSv h^{-1} at 1 m, while the dose rate falls to 0.02 mSv h^{-1} at 10 cm from the same source shielded by 2 cm of lead. According to the proportionality of the dose to the inverse square of distance, it is clear that a direct handling of sources requires safety procedures: small screens, small telehandlers, lead glasses for the eyes, and measurements of doses even on the fingers. Working with samples and cells on the bench also needs caution. It is essential to comply with all the legal regulations covering the use of radioactive substances.

8.4. THE ISOMER SHIFT AND THE SECOND-ORDER DOPPLER SHIFT

8.4.1. Elements of the Theory

The simple scheme of Figure 8.1 must be modified (Figure 8.6) because the eletronic density inside the source nucleus $|\psi(0)|_s^2$ shifts the two levels of the transition, of which the mean energy becomes $E'(s)$, which is slightly different from E_0. Similarly, for the absorbing nucleus $|\psi(0)|_a^2$ changes E_0 to $E'(a) \neq E'(s)$ when the atoms have different electronic configurations in the source and in the absorber, which is the usual case. Then the energy $\delta = E'(a) - E'(s)$ is needed to reach resonance so that the minimum of the transmitted line stands at $v \neq 0$, as seen earlier in Figure 8.4.

Each nuclear level, for which the nucleus containing Z protons can be described as a sphere of radius R, is shifted by the energy ΔE:

$$\Delta E = (2\pi/3)e^2 Z \langle R^2 \rangle |\psi(0)|^2 \times \text{a relativistic correction} \qquad (8.15)$$

For a transition from the excited state (index e) to the ground state (index g), when gathering constant terms into C and C':

$$\Delta E_e - \Delta E_g = C \,|\, \langle R_e^2 \rangle - \langle R_g^2 \rangle \,|\, |\psi(0)|^2 = C'(\Delta R/R)|\psi(0)|^2 \qquad (8.16)$$

Finally the isomer shift δ between the source and the absorber is

$$\delta = C'(\Delta R/R)[\,|\psi(0)|_a^2 - |\psi(0)|_s^2\,] \qquad (8.17)$$

which is independent of temperature.

Now the experimentally observed positions of the Mössbauer lines vary with T. For instance, that of α-iron decreases from 0 at 295 K to about -0.5 mm s^{-1} at 1073 K. This behavior originates in the second-order term in the relativistic effect on the emitted γ photon $h\nu$, produced by the vibrational movement of atoms with the velocity v (c is the velocity of light):

$$v'_{obs} = v\left(1 - \frac{v}{c}\right)\left(1 - \frac{v^2}{c^2}\right)^{1/2} \simeq v\left(1 - \frac{v}{c}\right)\left(1 - \frac{v^2}{2c^2}\right) \qquad (8.18)$$

It will be noted that the v^2/c^2 term can be neglected in the calculation of energy shift in the movement of the source with respect to the absorber.

As the vibrational periods are shorter than the excited state mean life, average v is null, which results in

$$v' = v\left(1 - \frac{\langle v^2 \rangle}{2c^2}\right) \qquad \text{or} \qquad \delta' = -\frac{\langle v^2 \rangle}{2c^2} E_\gamma \qquad (8.19)$$

From Debye's theory:

$$\delta' = -\frac{1}{2c}\frac{9kT_D}{8M}\left[1 + 8\left(\frac{T}{T_D}\right)^4 \int_0^{T_D/T} \frac{y^3 dy}{e^y - 1}\right] \qquad (8.20)$$

with $y = h\nu/kT$. The second-order Doppler term (SOD) introduces a shift δ_{SOD} that is the difference between the two expressions, similar to equation (8.20), given by the source and the absorber, respectively. Finally:

$$\delta_{obs} = \delta + \delta_{SOD}$$

with δ_{SOD} depending on T/T_D and on the nature of the source and the absorber through M and T_D. It is often very small at room temperature and below.

In addition to the general references listed,[1-10] the volume edited by Shenoy and Wagner[18] is particularly important when dealing with isomer shift.

An isomer shift is given with respect to a standard, so appropriate corrections are often necessary in comparisons of results given by different authors. In the case of Fe NGR the usual standard is either metal (α-Fe) or sodium nitroprusside (SNP), in Sn NGR δ values are referred to SnO_2 at room temperature, Sections 8.5 and 8.6 illustrate how δ is found when nuclear levels are split.

TABLE 8.1. Isomer Shifts of ^{57}Fe for Several Typical Chemical States of Iron^a

Iron metal ($4s^x 3d^{8-x}$): 0.25	Alloys: 0.1 to 0.8	
Ionic compounds:		
$Fe^{3+}(4s^x 3d^5)$	$Fe^{2+}(4s^x 3d^6)$	$Fe^+(4s^x 3d^7)$
0.4 to 0.9	1.0 to 1.8	1.6 to 2.2
Covalent and coordination compounds, all oxidation states:		
	0.0 to 0.8	

^{a} δ with respect to NPS, mm/s, at 295 K.

8.4.2. Applications

8.4.2.1. Chemical States

The value of δ is governed by electronic densities (mainly from s-electrons) inside the nucleus, therefore especially by the outer electronic configurations. This ensures a characterization of the oxidation state and of the chemical $|\psi(0)|^2$ which is determined by: (1) the s-electrons themselves, (2) screening by p-, d- or f-electrons, and (3) the ionicity or covalency and length of the chemical bond. When $\Delta R/R$ is positive, δ increases with $|\psi(0)|^2_a$, as observed in ^{119}Sn NGR. When $\Delta R/R$ is negative the effect is reversed, as in the case of ^{57}Fe, illustrated by Table 8.1, which shows that the determination of chemical state can be difficult, in particular at the limit of Fe^{2+} and Fe^{3+} ranges. However, it can usually be resolved by considering the chemistry of a particular system. Unusual values of δ appear in Mössbauer spectroscopy of fine particles owing to several phenomena (see, e.g., Bussière[10]).

8.4.2.2. Vibrational States

The shift δ_{SOD} gives a value for the Debye temperature, which is determined from its variation with temperature.

8.5. ELECTRIC QUADRUPOLE INTERACTION ENERGY AND QUADRUPOLE SPLITTING

8.5.1. Elements of the Theory

8.5.1.1. Interaction Energy E_Q

An electric field gradient (EFG) splits a nuclear level of quadrupole moment Q (spin $I > 1$) into the $2I + 1$ sublevels characterized by the values of the magnetic quantum number m_I. When the EFG does not have axial

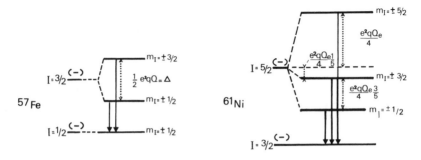

Figure 8.7. Level diagrams in case of quadrupole interactions with $\eta = 0$ and $q_{zz} > 0$.

symmetry each nuclear level is shifted to a sublevel by E_Q:

$$E_Q = e^2 q_{zz} Q \frac{3m_I^2 - I(I+1)}{4I(2I-1)} \left(1 + \frac{\eta^2}{3}\right)^{1/2} \tag{8.21}$$

where eq_{zz} is the main component of EFG and η is the asymmetry parameter $(q_{xx} - q_{yy})/q_{zz}$, which is null in an axial EFG.

The difference between two expressions similar to equation (8.21), one for the excited state (with Q_e) and the other for the ground state (with Q_g), gives the energy change for each allowed transition. Corresponding level diagrams are presented in Figure 8.7. A double degeneracy remains, as two opposite values of m_I provide the same sublevel. The characteristic parameter is the quadrupole coupling constant, often termed ε: $e^2 q_{zz} Q/h$ [in Hz], or $e^2 q_{zz} Q$ [referred to Doppler velocity, mm s^{-1}].

In the simple case of two transitions, e.g., with ^{57}Fe and ^{119}Sn, the spectrum is a doublet (see Figures 8.7 and 8.8). The difference between the two line positions Δ is referred to as quadrupole splitting (QS):

$$\Delta = \tfrac{1}{2} e^2 q_{zz} Q (1 + \eta^2/3)^{1/2} \tag{8.22}$$

8.5.1.2. Isomer Shift

The schemes of Figure 8.7 show the positions of the unsplit nuclear levels, from which it is easy to determine the isomer shift. In case of a doublet, δ is measured at the center (see Figure 8.8).

8.5.1.3. Relative Intensities of the Lines

The relative intensities of the lines are calculated with direction factors which are functions of θ, the angle between the observation direction and the

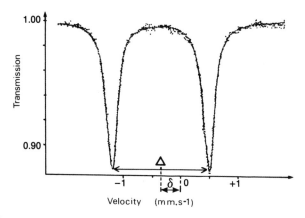

Figure 8.8. Quadrupole doublet of powdered sodium nitroprusside; δ with respect to $\alpha - $ Fe.

nuclear kinetic moment vector, oriented along the main axis of EFG. When all the values of θ are equally probable, e.g., in a well-prepared powder sample, the corresponding averages provide the relative intensities. A doublet is symmetric in such a case, while a single crystal of the same solid oriented with respect to the γ-ray beam so that $\theta = 0$ gives one line three times smaller than the other.

8.5.1.4. Origins of EFG

The electronic configuration of the atom results in an EFG when it corresponds to a nonspherical distribution of the charge density, e.g., in $Fe^{2+}(4s^x 3d^6)$: five $3d$-electrons give a spherical distribution and the sixth $3d$-electron induces an asymmetry. The main component of such an EFG is termed as q_{el} or q_{val}.

When the effective charges of neighboring ions or atoms build a highly symmetrical electrical environment they do not create an EFG. This should be particularly true in the Mössbauer sources, such as $Ca^{119m}SnO_3$ and ^{57}Co diffused into a cubic metal. In the reverse situation, quadrupole interaction originates from the structure, most clearly when the resonant nucleus does not experience an EFG from the surrounding electrons, e.g., in case of $Fe^{3+}(4s^x 3d^5)$. The main component of EFG created by the lattice is termed q_{lat}.

The values of q_{el} and q_{lat} do not simply add because of distortions induced in the atom's electron cloud by these EFG, which result in

$$q = (1 - R)q_{el} + (1 - \gamma_\infty)q_{lat} \qquad (8.23)$$

where R and γ_∞ are screening and antiscreening Sternheimer parameters, respectively. R lies between 0 and 1 and γ_∞ generally lies between -10 and -100. It is possible to distinguish the two components by studying the dependence of q on temperature, which does not significantly influence q_{lat}.

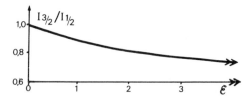

Figure 8.9. Karyagin–Goldanskii effect for a $\frac{3}{2} \rightarrow \frac{1}{2}$ quadrupole doublet.

8.5.1.5. Karyagin–Goldanskii Effect

When the Lamb–Mössbauer factor f is anisotropic, for instance because of a lattice break in the vicinity of an interface and therefore in particular for very small particles, the relative intensities should be calculated by introducing a dependence of f on θ. In the case of a doublet corresponding to the transitions of $|\pm 3/2\rangle$ and $|\pm 1/2\rangle$ excited states to the ground level with the respective intensities $I_{3/2}$ and $I_{1/2}$:

$$\frac{I_{3/2}}{I_{1/2}} = \frac{\int_0^\pi f(\theta)(1 + \cos^2 \theta)\sin \theta \, d\theta}{\int_0^\pi f(\theta)(\frac{5}{3} - \cos^2 \theta)\sin \theta \, d\theta} \tag{8.24}$$

By developing f according to equation (8.4) and using the anisotropy parameter $\mathscr{E} = (\langle z^2 \rangle - \langle x^2 \rangle)/\lambda^2$, it is possible to plot the curve of Figure 8.9. As $\mathscr{E}$ is proportional to E_0^2, it can be understood why KGE, although not strong in ^{57}Fe NGR, often arises when using ^{119}Sn. It increases with temperature.

8.5.2. Applications

8.5.2.1. Electronic States

As noted in 8.5.1.3, Fe^{2+} ions are characterized in Mössbauer spectroscopy not only by their isomer shift but also usually by a high value of q_{el} resulting in a large quadrupole splitting. However, the effect of q_{el} can be reduced or even cancelled in several cases, e.g., by the occurrence of a large value of q_{lat} of opposite sign or by a dynamical Jahn–Teller effect.

8.5.2.2. Crystalline and Molecular Structures

Crystalline and molecular structures can be characterized by q_{lat}. This component of EFG is nil in a highly symmetrical environment. Sodium nitroprusside exhibits a large QS (1.75 mm s^{-1}) (Figure 8.8) because of the asymmetry of the octahedral environment of the Fe(II) brought by the ligand NO.

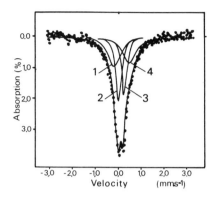

Figure 8.10. Pt–Fe/C at room temperature.[19]

8.5.2.3. Particle Size

A special asymmetry of a local structure is that induced by an interface. As absorption Mössbauer spectroscopy is a bulk technique, it cannot reveal this feature of a solid where dispersion is low. When particle size is small it becomes possible to observe quadrupole splittings which do not exist for the atoms of the bulk (e.g., superparamagnetic γ-Fe_2O_3) and even a correlation of Δ with the mean particle size can be found. Another effect of small particles is the superimposition of two doublets, one for the bulk and the other for surface layers, with relative intensities in a ratio that approaches the value of dispersion.

The latter case was especially clear in a study of carbon-supported Pt–Fe catalysts[19]; one of those spectra is reproduced in Figure 8.10 (lines 2 and 3 form the doublet that is known for the bulk alloy, and lines 1 and 4 that corresponding to surface iron atoms).

Apart from the electronic effect of alloying (δ different from that of α-iron), this experiment showed all the features of Mössbauer spectroscopy with fine particles: asymmetry of the doublets likely due to KGE (see Section 8.5.1.5), f factor five times lower than f_{lat} because of a weak alloy-support bonding (see Section 8.2.2.2), superparamagnetism (see 8.6.2.3). Moreover, hydrogen desorption erased the doublet corresponding to surface iron atoms, probably by allowing them vibrational amplitudes perpendicular to surfaces larger than in the bulk, which resulted in strongly decreasing f (see Section 8.2.2.1).

Figure 8.11 shows a Sn NGR spectrum of a Pt–Sn/Al_2O_3 catalyst,[20] which was fitted to three lines with the parameters reported in Table 8.2. When comparing the isomer shifts with those that are known for the chemical states of tin, it appears that this spectrum must be considered as being the superimposition of two doublets ascribed to Pt–Sn alloys. According to the intensities (A %) of the lines, at least one of these doublets exhibits a large KGE.

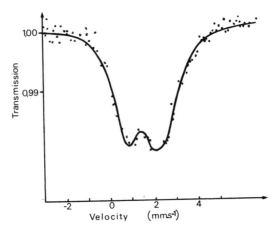

Figure 8.11. Sn NGR spectrum of Pt 0.95%–Sn 0.50%–Al_2O_3.[20]

TABLE 8.2. Decomposition of the Spectrum of Figure 8.11[a]

δ (mm/s)	1.09	1.82	2.12	A/A (SnO_2)
W (mm/s)	1.00	1.40	1.40	0.63
A (%)	12	55	33	
	Pt_3Sn	PtSn		

[a] δ with respect to SnO_2.

8.6. MAGNETIC INTERACTIONS

8.6.1. Elements of the Theory

In an internal or external magnetic field each nuclear level of nonzero spin is split into $2I + 1$ sublevels, reached by shifting it by $-g\mu_N m_I H$, where g is the nuclear g-factor, μ_N the nuclear Bohr magneton, m_I the magnetic quantum number, and H the magnetic field. This results in a spectrum that exhibits as many lines as allowed transitions (six in the case of ^{57}Fe).

The relative intensities of these lines are calculated with direction factors given by the angle θ (magnetic field, direction of the gamma beam). When all the orientations are equally probable (in particular in a polycrystalline sample with an internal magnetic field), averaged values are easily calculated and observed, such as in a sextet from ^{57}Fe: 3, 2, 1, 1, 2, 3 (see e.g., the spectrum of iron foil in Figure 8.14A).

H is proportional to the difference between the positions of two lines of the spectrum, especially the outermost ones, which generally yield a value of H within a few oersteds. In the presence of a simultaneous electric quadrupole

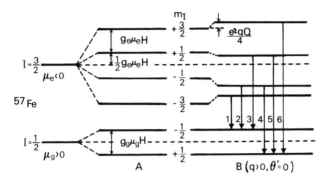

Figure 8.12. Levels diagrams for a magnetic interaction without (A) and with (B) a simultaneous quadrupole interaction.

interaction resulting from an axial EFG and very much smaller than the magnetic interaction, the shift giving a sublevel is

$$- g\mu_N m_I H_+ (-1)^{(|m_I|\,1/2)} (e^2 qQ/4)(3\cos^2\theta' - 1)/2 \qquad (8.25)$$

where θ' is the angle (principal axis of the EFG tensor, direction of the magnetic field).

Figure 8.12 shows the energy levels and transitions of the ^{57}Fe nucleus subjected to a magnetic field without and with quadrupole interaction. The determination of the isomer shifts easily follows, e.g., from the center of a symmetrical sextet.

Figure 8.13 illustrates how the shape of a spectrum can vary with the ratio of the energies of the magnetic and quadrupole interactions in the case of ^{57}Fe and $q > 0$. For $q < 0$ the diagram is reversed, which explains how the sign of an EFG, generating a symmetric doublet, can be determined by using a rather weak external magnetic field. On the other hand, a rather weak quadrupole interaction (and its sign) is exhibited in a sextet, as the difference between the distance from line 1 to line 2 and that from line 5 to line 6.

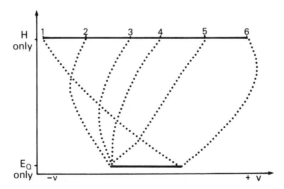

Figure 8.13. Mössbauer levels diagrams for ^{57}Fe in the case of varying relative energies of magnetic and quadrupole interactions.

8.6.2. Applications

8.6.2.1. Determination of Magnetic Transition Temperatures

Iron metal for instance is paramagnetic and gives a single line above its Curie point at 1043 K. At lower temperatures the internal field generated by its ferromagnetism results in a sextet.

8.6.2.2. Characterization of a Magnetic Solid by Its Internal Magnetic Field(s) H_i

Iron metal has an internal magnetic field of 330 kOe at 295 K, which gives a sextet with a distance of 10.65 mm/s between lines 1 and 6, widely used for calibrating Mössbauer spectrometers. This field enables one to distinguish this species from alloys, carbides, ferric oxides, and oxihydroxides. The knowledge of H_i ensures a good characterization inside a group of similar compounds: for instance the Mössbauer parameters of β- and γ-FeOOH doublets at room temperature are nearly the same, δ (NPS) = 0.64 and $\Delta = 0.65$ mm/s, but the Néel point of the former is close to 273 K so that the spectrum turns into a sextet at reduced temperatures (at 77 K, $H_i = 476$ kOe), while a sextet from the latter is observed at very much lower temperatures (at 4 K, $H_i = 460$ kOe).

8.6.2.3. Particle Size and Superparamagnetism

In a magnetic monodomain, the flip of spins from a position of easy magnetization to another one occurs within the relaxation time

$$\tau = \tau_0 \exp(KV/kT) \tag{8.26}$$

where K is the anisotropy constant and V the volume of the grain. If τ is significantly longer than the Larmor precession period of the nucleus [$\sim 10^{-8}$ s for ^{57}Fe, corresponding to a size of about 7 nm for particles of metal iron according to equation (8.26)], a sextet appears. Otherwise the sample behaves like a paramagnetic solid whose NGR spectrum consists of a single line or doublet.

As an example, small particles of Pt–Fe alloys at compositions which result in ferromagnetism exhibited the doublets displayed in Figure 8.10.[19] The case of iron itself is illustrated in Figure 8.14, where A is the spectrum from a usual polycrystalline foil and B the spectrum done without contact with air after the decomposition of MgO-supported $Fe_3(CO)_{12}$, about 1% Fe b.w.; the assignments corresponding to spectrum B are given in Table 8.3.

Particle size distributions can be obtained from Mössbauer spectroscopy alone either by using equation (8.26) and working at several temperatures or by blocking the spins by means of an external magnetic field.[22]

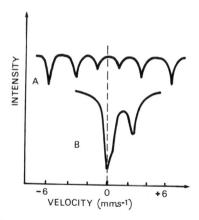

Figure 8.14. Spectra of iron at room temperature: (A) polycrystalline foil, (B) ex-$Fe_3(CO)_{12}$/MgO.[21]

TABLE 8.3. Components of Spectrum B of Figure 8.14

δ (NPS) mm/s	W mm/s	Δ mm/s	A %	Species
0.32	0.29	0.00	36	Fe^0
1.52	0.38	1.54	64	Fe^{2+}

8.6.2.4. An Example of Complex Spectra

An *in situ* study of the evolution of an Fe/Al$_2$O$_3$ catalyst used in Fischer–Tropsch synthesis[23,24] provides a good example of the use of Mössbauer spectra. Figure 8.15 shows a set of spectra together with the bar diagrams of their components, obtained from a sample of 10% Fe/Al$_2$O$_3$ at 523 K in hydrogen and in a mixture 1 CO:1 H$_2$. The first spectrum consists of an iron metal sextet (particles larger than 10 nm) and a doublet from Fe^{2+} strongly bound to alumina. In the presence of CO the former progressively disappears while the latter does not vary. Two progressively growing sextets, with the same isomer shift and the internal fields of 137 kOe (iron atoms Fe(I) with three nearest-neighbor carbon atoms) and 205 kOe (Fe(II) with two neighboring carbons), characterize a hexagonal iron carbide Fe$_x$C ($2 < x < 2.4$). The relative surface areas of these sextets yield the values of x, from about 2.25 at the beginning to about 2.03 at the end of the evolution. Finally the doublet S, with nearly the same isomer shift as the carbide sextets and growing slightly throughout the treatment, can be ascribed to superparamagnetic carbide species. A quantitative study of the evolution of all the species involved in this experiment is given in Figure 8.16.

Some conclusions concerning Fischer–Tropsch synthesis are supported by this study and by the investigation of the behavior of the same catalyst modified

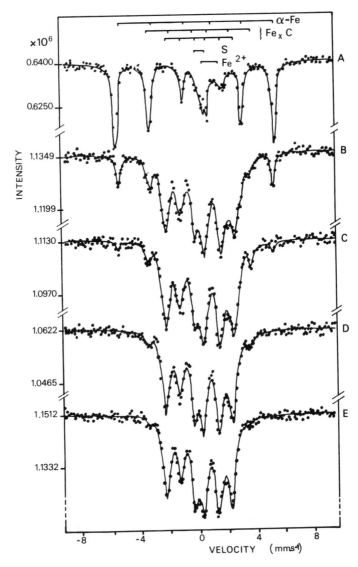

Figure 8.15. The evolution of the Mössbauer spectrum of a catalyst Fe/Al_2O_3 during the reaction $H_2 + CO$ at 523 K: (A) without CO; (B) in the presence of CO after 1 h; (C) 3h; (D) 5h, and (E) 22h.

by another chemical element (Mn, Mo, and Cr at 6 at.% with respect to iron; K at 50 at.%), which resulted in changes in the catalytic properties. In particular:

1. The iron-metal content decreases down to an undetectable value which could still be a significant amount of surface atoms. However, the catalyst in such a state exhibited no activity in ethylene hydrogenation,

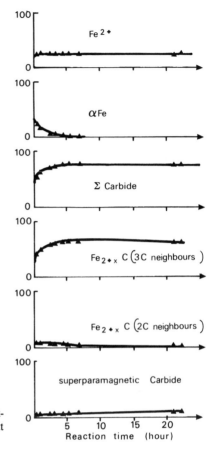

Figure 8.16. Changes in the phases of Fe/Al$_2$O$_3$ catalyst in the course of a Fischer–Tropsch reaction at 523 K.[24c]

so that its activity in Fischer–Tropsch synthesis under these reported conditions could rather be ascribed to carbide species.

2. Carburization increases dispersion since superparamagnetic carbides appear throughout the evolution of the catalyst.

3. After the additions noted above, iron metal appeared with an unchanged isomer shift and internal magnetic field. The observed changes in the catalytic properties, therefore, can be ascribed to modifications of the surface of the support, which consequently can be thought to play a role in the reaction mechanism.

8.7. FURTHER TECHNIQUES

8.7.1. Emission Mössbauer Spectroscopy

The state of a source element can be investigated if the decay mode does not induce chemical and structural modifications. For example, hydrodesulfurization (HDS) catalysts made of supported or nonsupported cobalt and molybdenum have been the subject of many NGR studies after they have been prepared with ^{57}Co. Such work requires knowledge of the emission spectra of all the species involved. Table 8.4 lists the main assignments proposed for such catalysts on several supports.[25]

According to these investigations a mixed cobalt and molybdenum sulfide could explain the improvement of the catalytic properties that appears when cobalt is added to molybdenum. Co–Mo–S is fairly well distinguished from other compounds that could occur in this system, owing to the behavior of iron in NGR after the decay of ^{57}Co[25b]; it exhibits a doublet with an isomer shift of an Fe^{3+} but with a quadrupole splitting depending on temperature, as shown in Figure 8.17, where it corresponds to the compounds Co/MoS_2 (ppm), Co/MoS_2 (0.25), and $Co–Mo–Al_2O_3$. This variation is typical of a large electronic EFG which characterizes a low-spin Fe^{3+} since $q_{el} = 0$ for a high-spin ferric ion (see Section 8.5.1.4). The cobalt atoms therefore lie in a sulfur environment but this structure differs from that of the usual compounds, as is shown in Figure 8.17. Co–Mo–S has not been found by other authors[26] studying carbon-supported catalysts of small surface area after a more severe sulfiding treatment. A doublet with nearly the same Mössbauer parameters at room temperature has been found without molybdenum in studying carbon-supported catalysts.[27]

8.7.2. A Mössbauer Element Used as a Probe

A Mössbauer element can be inserted into a solid if its concentration and/or structural position does not change the properties being investigated. This

TABLE 8.4. Behavior of Co–Mo HDS Catalysts Studied by Emission Mössbauer Spectroscopy[a]

Composition	Co after calcination	Co after treatment by H_2/H_2S or C_4H_4S
Co 0.25%–Al_2O_3	Co in Al_2O_3 spectra exhibit Fe^{3+} and Fe^{2+} due to decay radiation	Co–Mo–S
Co 1–Mo 6–Al_2O_3 More much Co	" Co_3O_4 appears	" Co–Mo − S + Co_9S_8
Co 1–SiO_2	Co_3O_4 + Co in SiO_2	Co_9S_8
CO–1–Mo 6–SiO_2	$CoMoO_4$ poorly crystallized	Co–Mo–S and much Co_9S_8
Co 1–Mo 6–O 900 m^2/g	$CoMoO_4$ poorly crystallized	Co–Mo–S and Co_9S_8

^{a} After Topsoe *et al.*[25]

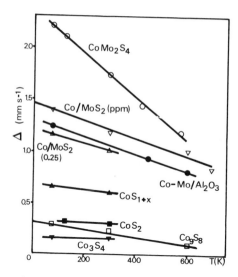

Figure 8.17. Variation of the quadrupole splitting of sulfided compounds of cobalt: dependence on temperature.[23b]

condition is particularly well fulfilled when the addition is carried out by using a carrier-free tracer in order to perform source experiments (see Section 7.1 and the limitations). For example, hydrogen chemisorption on nickel has been studied by using ^{57}Fe containing nickel.[28] ^{57}Fe NGR spectra were given by the hyperfine magnetic field of nickel and their changes in dependence on temperature allowed calculations of the anisotropy constants $K_{vac} = (6 \pm 2) \, 10^4 \, \text{J m}^{-3}$ and $K_H = (3 \pm 1) \, 10^4 \, \text{J m}^{-3}$. From spectra done in an external magnetic field it was possible to conclude that the magnetic moment of the particles decreased when they adsorbed hydrogen.

8.7.3. Conversion Electron Mössbauer Spectroscopy, CEMS

As can be seen from the general experimental scheme shown earlier in Figure 8.3 and from the decay scheme of ^{57}Co presented in Figure 8.2, it is possible to use Mössbauer spectroscopy to study the first surface layers of a solid and the underlying ones: working in the diffusion mode and observing the conversion electrons of which the range in the case of ^{57}Fe, e.g., is only 0.3 μm in iron metal because of their low energy (7.5 keV). Energy selection easily limits the observed depth to some tens of nanometers. In the case of finely dispersed catalysts this does not differ from the transmission mode and in the case of a single crystal the contribution of underlying layers overwhelms that of the first ones. Ultrahigh-vacuum and high-resolution electron spectroscopy enable one to study only 1 nm and to obtain a composition profile along the depth [depth selective CEMS (DSCEMS)].

TABLE 8.5. Mössbauer Parameters and Composition of Reduced Pt–Sn/Al$_2$O$_3$ Catalysts[a]

Composition (% b.w.)		Catalysts (n^0)	Mössbauer parameters (mm/s)		Assignment and relative spectral area	
Pt	Sn		δ	Δ		
0.47	0.47	1	0.58	0	H$_2$SnCl$_6$	30
			2.34	0	PtSn$_4$	60
			1.95	4.2	PtSn	10
0.95	0.95	2a	1.40	0.75	Pt$_3$Sn	30
			1.92	0.3	PtSn	70
			1.97	4.25	PtSn	20
		2b	1.88	0.70	PtSn	40
			2.82	2.60	SnO or Sn(OH)$_2$	40
0.50	0.20	3	0.32	0	SnCl$_4$, nH$_2$O	65
			2.30	0	PtSn$_4$	35

[a] After Bacaud et al.[20] Spectra in the presence of hydrogen at 295 K.

8.8. A SURVEY OF APPLICATIONS TO HETEROGENEOUS CATALYSIS

8.8.1. Explaining Catalytic Properties from Characterization of the Components of Catalysts

The preceding paragraphs gave examples of such investigations, dealing with metallic catalysts. Mössbauer spectroscopy is especially efficient in studying polymetallic supported catalysts which other techniques often fail to characterize satisfactorily. Let us examine in detail the case of Pt–Sn/Al$_2$O$_3$ catalysts.[20] Table 8.5 presents the main results, including those shown in Figure 8.11 and Table 8.2, here referred to as Catalyst 2a. All these results show the possibilities of Mössbauer spectroscopy in such characterizations and lead to information as to the catalytic properties of these solids, of course on the condition that there is no significant change in their composition during the reaction:

1. In all cases alloys have been produced, which accounts for the decrease in the catalytic activity for hydrogenation due to inhibition of hydrogen chemisorption on platinum enriched in electrons. Alloying also results in diminished cracking, which explains the better stability than with platinum alone.

2. Sn^{2+} and Sn^{4+} are present in different amounts in the samples prepared with different concentrations. Tin ions can react with the strong acidic sites of alumina so that the activity for cyclization and isomerization can proceed by using the weak acidity.

3. In catalysts Nos. 1 and 3, not much of platinum is included in the NGR lines ascribed to alloys, although these solids do not exhibit the catalytic properties of platinum alone. It can be thought, therefore, either that a very small amount of tin is present in a very dilute solid solution or that very small particles of alloys have a low f factor. The possibility that nonalloyed platinum is electronically modified by neighboring tin ions cannot be excluded.

Oxide catalysts and oxide precursors of catalysts have been studied extensively using Mössbauer spectroscopy. The case of Sb–Sn oxides prepared by thermolysis of the corresponding mixed hydroxides at different temperatures introduces ^{121}Sb NGR.[29,30] These solids exhibited antimony Mössbauer spectra such as those of the catalyst containing 20 at.% and prepared at 1223 K, shown in Figure 8.18. Both oxidation states clearly appear, a large quadrupole interaction ($e^2qQ = 16.7$ mm/s) outlines the eight-line spectrum of Sb^{3+} while only computer fitting gives the smaller quadrupole interaction of Sb^{5+} (3.8 mm/s). The isomer shift of Sb^{3+} with respect to the Ca^{221}SnO$_3$ source (see the arrow in Figure 8.20) is -14.44 mm/s, quite different from that of Sb$_2$O$_3$ (-11.7), but equal to that of Sb$_2$O$_4$ (-14.36) and close to that of Sb$_6$O$_{13}$ (-14.7). As the latter cannot be present after heating at 1223 K, a first phase is characterized: Sb$_2$O$_4$. After subtracting its contribution to the Sb^{5+} peak, a second Sb^{5+} signal appears which must be ascribed to Sb^{5+} in solid solution in SnO$_2$ since its isomer shift (-1.0 mm/s) is smaller than that of Sb$_2$O$_5$ ($+0.2$) and the latter species is not expected to remain after heating at 1223 K. The concentration of this solid solution can be estimated by assuming that it involves all SnO$_2$ and that the f factor of the corresponding Sb^{5+} is equal to that of Sb$_2$O$_4$: the result is 13 at.%. ^{119}Sn Mössbauer spectroscopy indicates an electronic modification of SnO$_2$ by Sb:$\delta = 0.05$ mm/s instead of 0.00.

For a set of catalysts of various compositions, prepared at various temperatures, this study resulted in the characterization of the antimony oxides (Sb$_6$O$_{13}$ occurred in some cases) and of the SnO$_2$-based solid solution. The latter must contain Sb^{3+} ions, which another similar study claimed to have found.[31] A comparison of these results and of the data from other techniques (XRD, IR) leads to the conclusion that the sites active in the mild oxidation of propene to

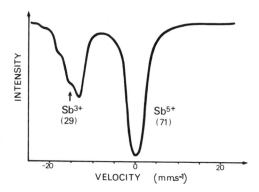

Figure 8.18. ^{121}Sb NGR spectrum at 4.2 K of the catalyst with an overall concentration of 20 at.%, prepared at 1223 K, and percentages of spectral area.[32]

acrolein were $Sb^{5+}-Sb^{3+}$ pairs present at the surface of the SnO_2-based solid solution[33a] and probably correspond to Sb_2O_4 oriented as a (100) face.[33b]

8.8.2. Chemisorption

Let us recall the experiment of desorption carried out with catalysts Pt–Fe/C described in Section 8.5.2.3: determination of the number atoms by the corresponding NGR spectrum component, which vanished in hydrogen desorption because of an increase in vibration amplitudes perpendicular to surface.[19] Such behavior would be of great interest if it were general, but it is not because inorganic oxides probably have few truly superficial cations, and oxidizable metals supported on basic oxides undergo oxidation when subjected to hydrogen desorption, owing to a redox equilibrium with the support[33,34]:

$$Me + nH^+ \rightleftharpoons Me^{n+} + n/2\ H_2 \qquad (8.27)$$

Mössbauer spectroscopy demonstrated this kind of metal-support reactivity in the cases of catalysts Fe/MgO,[33] Fe/Al$_2$O$_3$,[34] and Pt–Sn/Al$_2$O$_3$.[20] The first case is illustrated in Figure 8.19.

Nevertheless a study of carbon monoxide and hydrogen chemisorption could be achieved by using the less reactive support SiO_2.[35] This work is illustrated by the spectra of Figure 8.20 done at 77 K in an external magnetic field of 12.4 kOe. They contain a sextet ascribed to surface atoms, of which both δ and H_i increase during hydrogen or carbon monoxide chemisorption, suggesting an increase in density of iron $3d$-electrons.

Iron-exchanged zeolites have been investigated. For instance adsorption of water, ethanol, t-butylic alcohol, piperidin, and S_2C in Fe(II)Y and Fe(II)A involves mainly ferrous ions inserted in site II (sodalite windows of supercages),

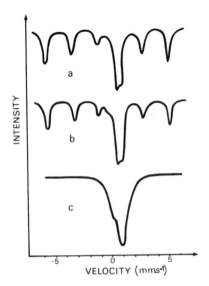

Figure 8.19. (a) Catalyst Fe/MgO in the presence of hydrogen; (b) the same sample after desorption ($\approx 10\%$ oxidized); (c) another sample with a high defect content, fully oxidized by hydrogen desorption (after Bussière et al.[33]).

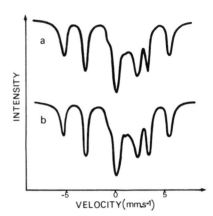

Figure 8.20. Highly dispersed Fe/SiO₂ catalyst: (a) after desorption, (b) after hydrogen chemisorption (after Mørup et al.[35]).

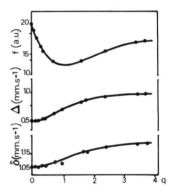

Figure 8.21. Values of f, δ, and Δ for Fe^{2+} in zeolite Fe(II)A after C_2H_4 adsorption (q molecules per Fe^{2+}).[38]

which can even migrate into the intragranular fluid, while ethylene, CO, NO, and pyridine only modify the corresponding doublet, which indicates a weak bonding.[37,38]

The example of ethylene in a Fe(II)A (at 50% exchange) has been the subject of a detailed study.[38] Figure 8.21 shows the variations in the principal Mössbauer parameters as functions on the amount of adsorbed ethylene. In particular f decreases as up to one ethylene molecule is adsorbed for two Fe^{2+}, and the isomer shift and quadrupole splitting undergo small changes. This diagram suggests a weak bonding which results in a slow residence time for the adsorbed molecule; relaxation of ferrous ions between their two states (with and without adsorbate) could increase vibration amplitudes and consequently decrease f. This statement is supported by NGR at temperatures below 296 K (Figure 8.22). Below 200 K a second ferrous doublet, ascribed to the adsorbed complex, appears. An estimate of the relaxation time of iron between its two states yields an activation energy of 23 kJ mol⁻¹ for this chemisorption.

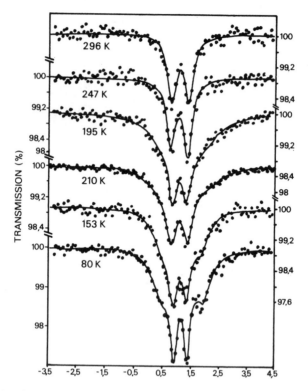

Figure 8.22. NGR spectra at various temperatures with zeolite Fe(II)A after adsorption of 0.8 molecule of C_2H_4 per Fe^{2+}.[38]

8.8.3. Study of Catalysts during Reaction

Section 8.6.2.4 described work done on a Fischer–Tropsch reaction and Section 8.7.1 a similar study on hydrodesulfurization catalysts. This kind of investigation is surely the most promising and justifies the technical effort required to develop the most suitable *in situ* cells (see Section 8.3.3).

8.9. CONCLUSIONS

Some features of NGR limit its wide use in studying catalysts. A few of them are not truly scientific: the rather high cost of sources and isotopically enriched elements, fear of radioactivity, and hesitation about acquiring the necessary knowledge in nuclear physics. Others are relevant to the basic principles of the technique: a rather small number of suitable elements, particularly the impossibility of studying those of low atomic mass, and lack of information about the surface of the solids, except in case of DSCEMS and sometimes of CEMS.

However, at least three important scientific points favor Mössbauer spectroscopy:

1. High specificity as there can be no interference.
2. Information: on one hand, each spectrum is accounted for by the various structural sites occupied by the element, each of them described by several parameters related to the electronic state of the atom, to the structure, and the vibrational state of the lattice, to magnetism, and even to the texture of the solid; on the other hand, surface is characterized either together with volume when dispersion is high or when the species detected in the spectra are prepared or produced only in the top layers of the grains.
3. Possibility of performing *in situ* experiments in heterogeneous catalysis.

Although NGR is the most successful in studying iron, cobalt and tin species, this technique can be used for other elements, and discussions about several of them are found in the bibliography pertinent to catalysis investigation using Mössbauer spectroscopy: Sb, Eu, Au, Ir, Pt, Ru, and Te. This small list should be extended in the future, when the theoretical and/or technical difficulties have been solved by the further collaboration between chemists and physicists.

REFERENCES

1. C. Janot, *L'effet Mössbauer et ses applications à la physique du solide et à la métallurgie physique,* Masson et Cie, Paris (1972).
2. N. N. Greenwood and T. C. Gibb, *Mössbauer Spectroscopy,* Chapman Hall, London (1971).
3. J. M. Friedt, *Techniques de l'ingénieur, section méthodes spectroscopiques,* Academic, New York (1977), p. 2605; J. M. Friedt and J. Danon, *Modern Physics in Chemistry* (E. Fluck and V. I. Goldanskii, eds.), Academic, New York (1979), p. 195.
4. *CRC Handbook of Spectroscopy, Vol. III, Mössbauer Spectroscopy* (J. W. Robinson, ed.) CRC, Boca Raton, Florida (1981), pp. 403–528.
5. R. L. Cohen (ed.), *Applications of Mössbauer Spectroscopy, Vol. I,* Academic, New York (1976); *ibid. Vol. II* (1980).
6. (a) *Mössbauer Effect Data Indexes* [available from Mössbauer Effect Data Center (M.E.D.C.): 1958–1965], A. H. Muir, K. J. Ando, and H. M. Coogan, Interscience, New York (1966); 1966–1976, J. G. and V. E. Stevens *et al.,* IFI/Plenum, New York (various years). (b) *Mössbauer Effect Reference and Data Journal,* Mössbauer Effect Data Center (M.E.D.C.), University of North Carolina, Asheville (1977 *et seq.*). (c) *Catalysts Mössbauer Handbook, ibid.* (1986).
7. J. A. Dumesic and H. Topsø, *Adv. Catal.* **26,** 121 (1976).
8. H. Topsø, J. A. Dumesic, and S. Mørup, *Applications of Mössbauer Spectroscopy, Vol. II,* Academic, New York (1980), p. 55.
9. *Catalysts Mössbauer Handbook,* M.E.D.C. ed. (1986).
10. P. Bussière, *Rev. Phys. Appl.* **16,** 477 (1981).
11. P. Bussière, *Rev. Phys. Appl.* **15,** 1143 (1980).
12. J. M. Dubois and G. Le Caër, *J. Phys. E* **13,** 1002 (1980).
13. W. N. Shen, J. A. Dumesic and C. G. Jr. Hill, *Rev. Sci. Instr.* **52,** 858 (1981).

14. P. A. Montano, A. S. Bommannavar and V. Shah, *Fuel* **60**, 703 (1981).

15. J. A. Dumesic and H. Topsø, *Adv. Catal.* **26**, 59 (1976).

16. M. B. Madsen, L. Nielsen and S. Mørup; Communication to ISIAME, Parma (1988).

17. J. W. Niemantsverdriet, C. F. J. Flipse, and A. M. Van Der Kraan, *Proc. ICAME*, 1981, Indian National Science Academy, New Delhi (1982), 426.

18. G. K. Shenoy, F. E. Wagner (eds.), *Mössbauer Isomer Shifts*, North-Holland, Amsterdam (1978).

19. C. H. Bartholomew and M. Boudart, *J. Catal.* **29**, 278 (1973).

20. R. Bacaud, P. Bussière, and F. Figueras, *J. Catal.* **69**, 399 (1981).

21. F. Hugues, P. Bussière, J. M. Basset, D. Commereuc, Y. Chauvin, L. Bonneviot and D. Olivier, *Proc. 7th Intern. Congr. on Catalysis*, Kodansha Ltd, Tokyo (1981), p. 418.

22. S. Mørup and H. Topsø, *Proc. ICAME*, Institute of Physics Bucharest, Vol. I, (1978), p. 229.

23. N. Nahon, V. Perrichon, P. Turlier, and P. Bussière, *J. Physique Colloq.* **41**, C1–339 (1980); *ibid.*, in: *Magnetic Resonance in Colloid and Interface Science* (J. P. Fraissard and H. A. Resing, eds.), Reidel, Dordrecht (1980), p. 337.

24. (a) M. Pijolat, G. Le Caër, V. Perrichon, and P. Bussière, *Proc. ICAME 1981*, Indian National Science Academy, New Delhi (1982), p. 431. (b) G. Le Caër, J. M. Dubois, M. Pijolat, V. Perrichon, and P. Bussière, *J. Phys. Chem.* **86**, 4799 (1982). (c) M. Pijolat, V. Perrichon, and P. Bussière, *J. Catal.* **107**, 82 (1987).

25. (a) H. Topsø, B. S. Clausen, R. Candia, C. Wivel, and S. Mørup, *Bull. Soc. Chim. Belge*, **90**, 1189 (1981). (b) *ibid.*, *J. Catal.* **68**, 433 (1981).

26. M. Breysse, B. A. Bennett, and D. Chadwick, *J. Catal.* **71**, 430 (1981).

27. A. M. Van Der Kraan, M. W. J. Craje, E. Gerkema, W. L. T. M. Ramselaar, and V. H. J. De Beer, *Hyperfine Interactions*, **46**, 567 (1989).

28. S. Mørup, B. S. Clausen, and H. Topsø, *J. Physique Colloq.* **40**, C2–78 (1979).

29. J. L. Portefaix, P. Bussière, F. Figueras, M. Forissier, J. M. Friedt, F. Theobald, and J. P. Sanchez, *J. Chem. Soc., Faraday Trans. I* **76**, 1652 (1980).

30. (a) B. Benaichouba, P. Bussière, J. M. Friedt, and J. P. Sanchez, *Appl. Catal.* **8**, 237 (1983). (b) J. C. Volta, P. Bussière, G. Coudurier, J. M. Herrmann, and J. C. Vedrine, *Appl. Catal.* **16**, 315 (1985).

31. F. J. Berry, *J. Catal.* **73**, 349 (1982).

32. R. Dutartre, M. Primet, and G. A. Martin, *React. Kin. Catal. Lett.* **3**, 249 (1975).

33. (a) P. Bussière, R. Dutartre, G. A. Martin, and J. P. Mathieu, *C. R. Acad. Sci. Ser. C* **280**, 1133 (1975). (b) R. Dutartre, P. Bussière, J. A. Dalmon, and G. A. Martin, *J. Catal.* **59**, 382 (1979).

34. N. Nahon, V. Perrichon, P. Turlier, and P. Bussière, *React. Kin. Catal. Lett.* **11**, 281 (1979).

35. S. Mørup, B. S. Clausen, and H. Topsø, *Surf. Sci.* **106**, 438 (1981).

36. W. N. Delgass, R. L. Garten and M. Boudart, *J. Phys. Chem.* **73**, 2970 (1969).

37. B. L. Dickson and L. V. C. Rees, *J. Chem. Soc., Faraday Trans. I* **70**, 2038 (1974).

38. B. L. Dickson and L. V. C. Rees, *J. Chem. Soc., Faraday Trans. I* **70**, 2051 (1974).

AUGER ELECTRON SPECTROSCOPY

J. C. Bertolini and J. Massardier

9.1. THE AUGER PROCESS

Different energetic beams can be used for excitation of solids with the creation of holes in electronic levels: photons, ions, electrons. In this chapter only the electron impact excitation will be taken into account, but many of the conclusions can easily be extended to other excitation modes. Incident electrons with primary energy (E_p) larger than the energy of a core electronic level (E_x) of the impacted material can generate a core hole by ionization. The excited atom then relaxes by filling the hole via a transition from an outer level E_y. The excess energy released can be accommodated by the atom in either of two ways: by emitting an X-ray photon at that energy or by giving this excess energy to another electron which is ejected from the atom. A schematic representation of the electron impact Auger process is given in Figure 9.1. The kinetic energy of the ejected electron is characteristic of the electronic levels involved and therefore of the analyzed material. This process was first discovered by Auger[1,2] and the ejected electrons are called (XYZ) Auger electrons, where X indicates the deep level $(K, L_1, \dots)$ on which has been created the hole which is then filled by an electron coming from the Y level $(L_1, L_2, \dots, V)$, the Auger electron being ejected from an outer level Z $(L_2, \dots, V)$; the Auger transition therefore is written as XYZ.

J. C. Bertolini and J. Massardier ● Institut de Recherches sur la Catalyse, CNRS, Villeurbanne, France.

Catalyst Characterization: Physical Techniques for Solid Materials, edited by Boris Imelik and Jacques C. Vedrine, Plenum Press, New York (1994).

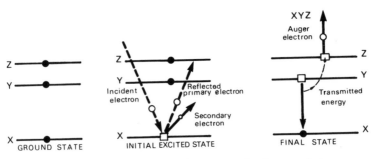

Figure 9.1. Schematic diagram of the process for Auger emission.

9.1.1. Auger Electron Energies

The Auger electron energy is, in principle, determined by the difference in the total energies before and after the transition. In the example in Figure 9.1, the energy of the outgoing electron is given by the relation

$$E_{XYZ} = E_X(A_0) - E_Y(A_0) - E_Z(A^*) \tag{9.1}$$

where $E_X(A_0)$ and $E_Y(A_0)$ are the energies of the X and Y levels of the atom in its ground state A_0. Indeed, the lifetime of the deep hole is very short and it is generally assumed that no significant relaxation occurs during the XY transition. Even if some relaxation does take place, the excited state with one missing electron would be followed by a shift of the same magnitude for both the X and Y levels. Consequently $X(A_0) - Y(A_0)$ represents quite well the energy gained by the system before the ejection of the Auger electron.

The energy of the Z level in the final state corresponds to a doubly ionized state for the atom, A^*, and therefore the kinetic energy of the emergent Auger electron may be represented by the following equation:

$$E_{XYZ} = E_X(A_0) - E_Y(A_0) - E_Z(A_0) - I_{YZ} + R^{2+} \tag{9.2}$$

where I_{YZ} is the coupling energy between the holes in the Y and Z levels and R^{2+} the relaxation energy corresponding to a doubly ionized state. An empirical relation has been proposed which gives the Auger electron energy regardless of the nature of the considered transition:

$$E_{XYZ} = E_X(Z_0) - \tfrac{1}{2}[E_Y(Z_0) + E_Y(Z_0 + 1)] - \tfrac{1}{2}[E_Z(Z_0) + E_Z(Z_0 + 1)] \tag{9.3}$$

where Z_0 is the atomic number of the element. A semiempirical approach for E_{XYZ} in which one uses the experimentally determined binding energies [$E_X(A_0)$, $E_Y(A_0)$ and $E_Z(A_0)$] and the calculated values for the other terms of equation (9.2) [$I_{(YZ)}$, R^{2+}] is sometimes proposed. It is often not very satisfactory and the

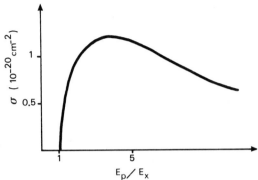

Figure 9.2. Ionization cross section *vs*. the reduced primary energy (K level of aluminum): E_p represents the primary energy and E_x the energy of the core level from which the electron is being ejected.

empirical formula[3] or better the generally tabulated experimental values[3-5] should be used.

As the Auger process involves the participation of three electrons, hydrogen and helium (one and two electrons, respectively) are not detectable. Auger electron energies are very characteristic of the element sample. Moreover, these energy values depend on the chemical state of the element and small shifts can be observed. Correlations can be established with the photoemission XPS shifts for the binding energies from which the relaxation effects can be deduced so giving the true binding energy shifts accurately. The principle of this calculation is given in the Appendix.

9.1.2. Auger Line Shapes

The width of Auger lines Γ is governed primarily by the lifetime τ of the transition ($\Gamma = \hbar/\tau$). The order of magnitude for τ is 10^{-16} s, which leads to a linewidth of about 6–7 eV. Moreover, the energetic width of the valence band induces an additional widening for the Auger XYV or XVV transitions in which the electrons of this band are concerned. Auger fine structures may be induced which reflect the local density of states of the less energetic filled electronic levels. Secondary inelastic processes can also cause the broadening of Auger lines, and can even generate peaked features mainly on the low-energy side of the Auger peak, such as plasmon loss peaks.

9.1.3. Auger Line Intensities

The intensity of Auger lines depends:

1. On the value of the ionization cross section σ_X for the creation of the electronic vacancy in the deep level by the primary electrons; it varies with the energy of the incident electrons. In Figure 9.2, the variation of the σ_X cross section *vs*. the incident energy is given for the aluminum K level. Such a curve is specific of both the level and the element, but the shape remains quite the same whatever they are; the maximum is

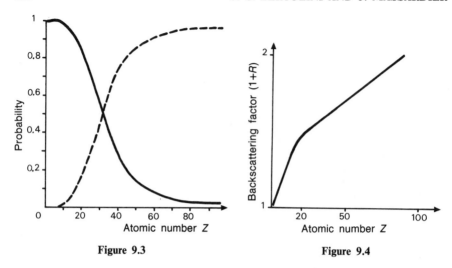

Figure 9.3   **Figure 9.4**

Figure 9.3. Relative probabilities for de-excitation by emission of an X-ray photon (dashed line) and by emission of an Auger electron (solid line) following creation of a core hole in the K shell (the energy increases with the atomic number Z).

Figure 9.4. Backscattering coefficient as a function of the atomic number Z.[6]

       reached when the primary energy is 3 to 5 times higher than the binding energy E_X of the deep level.
2. On the Auger yield (often written $1 - w$, where w is the probability for de-excitation by photon emission), which is near unity for low-energy E_X–E_Y transitions. In contrast X-ray photon emission is dominant for high-energy transitions. Figure 9.3 shows the probability for de-excitation by the two processes of a deep hole created on the K level for the 80 lighter elements of the Periodic Table.
3. On the I_p intensity of the primary beam.
4. On the backscattering factor. The primary beam (I_p current and E_p energy) remains efficient for the Auger process even if it has undergone inelastic events modifying both its energy and trajectory; it remains efficient even for depths much greater than the mean free path of Auger electrons (i.e., several tens or hundreds of angstroms). The electrons backscattered by the underlayers can therefore generate excitations in the upper layers leading to ejection of additional Auger electrons. Consequently, we have to consider not only the primary current I_p but also a mean additional term RI_p ($1 + R$ is the backscattering factor), which takes into account that secondary contribution to the Auger line intensities. R is near zero for light elements and increases with the atomic number Z of the constituent of the matrix as shown in Figure 9.4.
5. On the roughness factor r. Some studies have shown that the Auger current emitted from a rough surface is noticeably different from that of an ideally flat surface. Such a phenomenon can be understood intu-

itively if one considers the shadowing effects induced by the geometric surface defects, which induce an intensity decrease, while the roughness increases the apparent surface area, which may have an opposite influence. The factor r depends therefore upon the surface morphology, and its determination is difficult. However, it is often close to unity and so it is generally neglected.

6. On the depth at which the Auger emitter atom is located, i.e., on the depth of the material from which the electrons are coming. In practice, it is assumed that the number of electrons, ejected without any inelastic interactions, decreases exponentially with the depth according to $e^{-d/\lambda}$, where λ is the mean free path for the emitted electrons and d the emission depth.

The detected intensity for the XYZ Auger electrons characteristic of an element A located in a solid material is therefore given by the following relation:

$$I^A_{XYZ} = \sigma^A_X(1 - w)(1 + R)I_p r T_E \int\int\int N^A(Z)\, f(\lambda, \theta, z)dx\, dy\, dz \qquad (9.4)$$

where σ^A_X is the probability of forming a hole on the X level of the element A for an excitation primary energy E_p; $(1 - w)$ the de-excitation probability by the Auger process; $1 + R$ the backscattering coefficient of the material; I_p the intensity of the primary current; r the roughness factor (≈ 1); T_E the transmission coefficient of the spectrometer at the kinetic energy E; $N^A(z)$ the A concentration as a function of the depth in the material; $f(\lambda, \theta, z)$ an attenuation function which depends on the mean free path λ, on the analysis angle with respect to the normal at the surface, θ, and on the depth z and

$$f(\lambda, \theta, z) = e^{-z/\lambda \cos\theta} \qquad (9.5)$$

9.1.4. Angular Effects

Angular features of the intensities of the Auger spectra have been found on single crystals. They are related to diffraction phenomena, and their effects are generally not very important. Nevertheless, some inaccuracies may appear in the quantitative determination of elemental concentrations when a directional spectrometer is used. However, additional information on the crystallography of the analyzed area of crystalline targets can be obtained in this way. To be more accurate one would have to take into account the incident angle of the primary beam, but such an influence is significant only if the depth for excitation by the primary electrons becomes less than the mean free path of Auger electrons. As the primary excitation depth is much greater than the mean free path, and the primary energy is chosen to be much larger than the Auger electron energies, an angular influence of the primary beam can appear only for very grazing incidence for the primary electron beam. Moreover, it has been

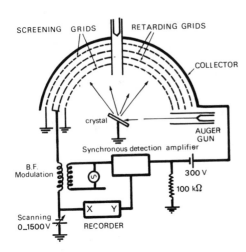

Figure 9.5. Schematic representation of a four grid AES spectrometer.

proposed that the backscattering factor is dependent on the incident angle of the primary beam.[7,8] This increases as the incident angle moves away from normal incidence, which amplifies the contribution to the Auger signal of the outermost layer with respect to the inner layers.

9.2. EXPERIMENTAL CONSIDERATIONS

The Auger electrons are detected by energy analysis, from a few eV to some hundreds of eV, of the electrons emitted from the target. Two main analysis methods have been generally developed: (1) the retarding field method for which one needs only a low-energy electron diffractometer and the appropriate electronics, and (2) the use of speed selectors.

9.2.1. Retarding Field Analyzers

The simplified scheme of the retarding field analyzer (RFA) apparatus with its associated electronics is presented in Figure 9.5. The spectrometer is composed of concentric grids. A variable retarding potential E is applied to at least one of the inner grids. The current $i(E)$ is then given by the expression

$$i(E) = K \int_E^{E_p} N(E)dE \qquad (9.6)$$

where $N(E)$ is the number of electrons at the E energy.

The detection can be greatly improved by an electronic differentiation of the spectrum, which is achieved by the superposition of a frequency modulation $\Delta E = a \sin(\omega t)$ on the retarding voltage applied to the suppressor grid. After amplification, the signal is compared in a lock-in amplifier with either the fundamental or the first harmonic of the modulation frequency, which gives either the energy distribution $N(E)$ or the differential distribution $dN(E)/dE$. The principle is described below.

The intrinsic resolution $\Delta E/E$ of commercial RFA spectrometers is about 0.5 to 1%; it also depends upon the amplitude of the superimposed modulation. RFA apparatus presents a large analyzing angle, but the signal-to-background-noise ratio is not very good due to its large pass-band.

9.2.2. Speed Selector Analyzers

The most often used system for Auger spectrometry is the cylindric mirror analyzer (CMA) (Figure 9.6). The electrons emitted from a source located on the axis pass through an aperture between two cylinders on which is applied a variable potential V. Only the electrons of E energy (with $E = KV$, where K is a constant dependent on the geometry of the particular CMA filter) pass through the output aperture and are focused on the detector. The current $i(E)$ is proportional to the energy according to the expression

$$i(E) = KEN(E) \tag{9.7}$$

As with the RFA in order to obtain better detection a modulation signal $\Delta E = a \sin(\omega t)$ is superimposed on the scanning voltage and this gives $EdN(E)/dE$. The high transmission of the CMA spectrometer is combined with a good resolution (intrinsic $\Delta E/E \approx 0.3\%$). Additionally the narrow pass-band confers a high signal-to-background-noise ratio.

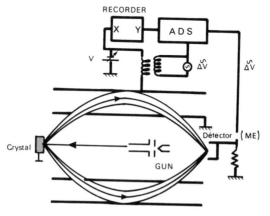

Figure 9.6. Schematic representation of a CMA spectrometer and associated electronics for use in AES.

9.2.3. Frequency Modulation and Synchronous Detection

When a small alternating voltage, $\Delta E = a \sin(\omega t)$ is applied on the retarding grid (RFA) or on the external cylinder of the CMA, the current $i(E + \Delta E)$ can be expressed in a Taylor series:

$$i(E + \Delta E) = i(E) + \frac{di(E)}{dE}\Delta E + \frac{d^2 i(E)}{dE^2}\frac{\Delta E^2}{2} + \frac{d^3 i(E)}{dE^2}\frac{\Delta E^3}{3!} + \cdots \quad (9.8)$$

and taking into account that $\Delta E = a \sin(\omega t)$

$$i(E + \Delta E) = i(E) + \left[a\frac{di(E)}{dE} + \frac{a^3}{8}\frac{d^3 i(E)}{dE^3} + \cdots \right]\sin(\omega t)$$
$$- \left[\frac{a^2}{4}\frac{d^2 i(E)}{(dE)^2} + \frac{a^4}{48}\frac{d^4 i(E)}{dE^4} + \cdots \right]\cos(2\omega t) \quad (9.9)$$

With a small modulation amplitude, the contribution of the third and following derivatives of the signal can be neglected.

The lock-in amplifier integrates the product of $i(E + \Delta E)$ by an alternating signal with an ω (or 2ω) frequency:

$$I = C \times \frac{1}{Ti} \int_0^{Ti} i(E + \Delta E) \sin k\omega t\, dt \quad (9.10)$$

with $k = 1$ or 2. Only the terms of $i(E + \Delta E)$ which have a "$k\omega$" pulsation ($k = 1$ or 2) contribute to the measured signal I (the other terms being zero) as long as the integration time T_i is much greater than $2\pi/k\omega$, a condition which is always filled when the modulation frequency is around or higher than 1000 Hz.

Using an RFA analyzer, where

$$i(E) = K \int_E^{E_p} N(E)dE$$

equation (9.9) can be simplified as follows:

$$i(E + \Delta E) = i(E) + aN(E)\sin(\omega t) - \frac{a^2}{4}\frac{dN(E)}{dE}\cos(2\omega t) \quad (9.11)$$

The synchronous detection adjusted to the frequency ω thus gives the $N(E)$ value (spectral density in energies of electrons). When adjusted to 2ω it gives the first derivative $dN(E)/dE$ of this density, which is the more commonly used mode.

With a CMA spectrometer, $i(E) = KEN(E)$, and

$$i(E + \Delta E) = i(E) + ak\left[E\frac{dN(E)}{dE} + N(E)\right]\sin{(\omega t)} \qquad (9.12)$$

The detection in ω gives

$$ak\left[E\frac{dN(E)}{dE} + N(E)\right]$$

This corresponds to the derivative of the electron energy distribution, whose amplitude varies rapidly when an Auger feature is present. The additional $N(E)$ term presents a rather smooth variation; at low energies it even cancels the slope of the background which is a considerable convenience when compared with measurements using an RFA analyzer.

9.3. APPLICATIONS IN CATALYSIS

As previously specified, the energies of Auger peaks are characteristic of a given element. The intensity of the peaks increases with the elemental concentration in the very first atomic layer, and the sensitivity depends upon the given element. In general, one can detect about 10^{-2} monolayers by this technique. Auger spectrometry was first used as a complement of LEED to control the cleanliness of single crystals. However, the Auger technique is not confined to studies on single crystals. It is now largely used for surface control in metallurgy, in the elaboration processes for electronic components, and to some extent in the analysis of surfaces of supported or unsupported catalysts. By scanning the primary electron beam, it is possible to draw a chart of the elements present in surface. The spatial resolution is generally limited by the dimensions of the primary beam, and reaches some hundreds of angstroms for the best usual electron guns working at rather low energy (some keV). Even if one uses a very small spot for the incoming electron beam, which is the case, for example, in electron microscopes, one cannot expect to have a resolution better than about 200 Å, owing to an intrinsic physical limitation, the "pear effect" associated with the scattered electrons diffusing in the surface (and in the bulk) of the bombarded material. For quantitative analysis, the peak-to-peak height in the derivative (as it is reported in the Table of Pure Elements[2]) is generally used. It would be more accurate to take into account the area under the Auger peaks in the $N(E)$ mode. The peak-to-peak height remains proportional to the Auger intensity as long as the shape of a given Auger line does not vary. This condition holds generally, except for elements in very different chemical states and/or environments, when one considers Auger transitions involving valence electrons.

9.3.1. Bimetallic Alloys and Surface Segregation

The sensitivity of low-energy Auger emitted electrons to the very first atomic layers allows the Auger technique to be used to determine the surface segregation of multicomponent materials. As an example, we have chosen a very well characterized alloy: a $Pt_{10}Ni_{90}$ single-crystal alloy cut along the (111) orientation. The Auger spectra recorded for the same experimental conditions both for the $Pt_{10}Ni_{90}$ (111) alloy and for the pure Pt(111) and Ni(111) single crystals used as standards are shown in Figure 9.7. In the given PtNi sample the

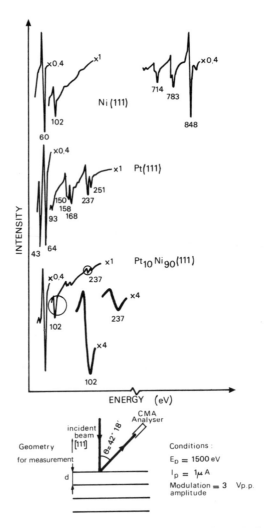

Figure 9.7. AES spectra of pure Ni(111), pure Pt(111), and $Pt_{10}Ni_{90}$(111) alloy samples.

very intense 61 and 64 Auger peaks have not been used since they do not display sufficient separation. The quantitative treatment has therefore been based on the M_1VV (102 eV) and $N_{5,6}N_{6,7}V$ (237 eV) Auger lines, respectively characteristic of nickel and platinum. The general equation (9.4) cannot be used directly as the ionization cross section for the formation of a deep hole σ^A is not well known. Nevertheless, the sensitivity factor for the two components of the alloys can be experimentally determined by using the results relative to the pure metals. Equation (9.4) giving the Auger peak intensity can be then expressed as follows:

$$I_{\text{pure metal}} = I_0 \sum_{n=0}^{\infty} \exp(-nd/\lambda \cos\theta) = I_0/(1 - \exp(-d/\lambda \cos\theta)) \quad (9.13)$$

where I_0 is the contribution of the first layer to the overall intensity, d and θ are defined in Figure 9.7, and n is the number associated with the successive atomic layers.

It can be also expressed by

$$I_{\text{pure metal}} = N i_0(1 + R) \sum_{n=0}^{\infty} \exp(-nd/\lambda \cos\theta) \quad (9.14)$$

where i_0 is then the contribution to the overall intensity of one atom layered on its substrate, N the number of atoms by surface area unit, and $1 + R$ the back-scattering coefficient.

The estimation of inelastic mean free paths λ for the Auger peaks can be deduced either from the universal curve: escape depth $vs.$ the electron energy (given in Chapter 17) or from the relations given by Seah and Dench.[9] The best way, however, is to use experimental determinations when they are available. In the present example, we have chosen to take the values determined by Kuijers on platinum and nickel[6]: $\lambda_{237} = 7.3$ Å and $\lambda_{102} = 4$ Å.

The R values have been deduced from the curve of Figure 9.4[6] ($R_{\text{Pt}} = 0.92$ and $R_{\text{Ni}} = 0.53$) and $d_{\text{Pt(111)}} = 2.261$ Å and $d_{\text{Ni(111)}} = 2.03$ Å are the well-known interplane distances for the (111) orientation of these metals.

For pure platinum:

$$I_{237}^{\text{Pt}} = N^{\text{Pt}} \times i_0^{\text{Pt}} \times 1.92 \times 2.92 \quad (9.15)$$

and for pure nickel:

$$I_{102}^{\text{Ni}} = N^{\text{Ni}} i_0^{\text{Ni}} \times 1.53 \times 2.01 \quad (9.16)$$

In the alloy (al) the corresponding values are

$$I_{237}^{\text{al}} = N^{\text{al}} i_0^{\text{Pt}}(1 + R^{\text{al}}) \sum_{n=0}^{\infty} C_n^{\text{Pt}} \exp(-nd/\lambda_{237} \cos\theta) \quad (9.17)$$

$$I_{102}^{\text{al}} = N^{\text{al}} i_0^{\text{Ni}}(1 + R^{\text{al}}) \sum_{n=0}^{\infty} C_n^{\text{Ni}} \exp(-nd/\lambda_{102} \cos\theta) \quad (9.18)$$

where C_n is the concentration in the nth layer, $C^{Ni} = 1 - C^{Pt}$; and N^{al}, d^{al}, and R^{al} are the parameters for the alloys as previously defined for the pure metals ($d^{al} = 2.062$ and $R^{al} : 0.57$).

In order to simplify these relationships, it is better to consider the Auger peak ratios: $I^{al}_{237}/I^{al}_{102}$ rather than their absolute values. The i^{Pt}_0 and i^{Ni}_0 values are deduced from the measurements of Auger peak heights of the pure metals, which have to be recorded under identical conditions.

If the composition of the alloy is assumed to be homogeneous (i.e., for invariant C_n concentrations) in the analyzed volume (which corresponds to a depth of about 15–20 Å), one gets a platinum concentration of 17.4% with the following experimental values: $I^{Pt}_{237}/I^{Ni}_{102} = 1.09$ and $I^{al}_{237}/I^{al}_{102} = 0.24$. A platinum enrichment is therefore found in the surface region.

Since theoretical predictions show that the outer layer should be the most changed in the segregation process of binary alloys, the simpler assumption is that there is only a single-layer segregation. The formulation can be then expressed as follows:

$$\frac{I^{al}_{237}}{I^{al}_{102}} = \frac{i^{Pt}_0}{i^{Ni}_0} \frac{C^{Pt}_0 + \sum\limits_{n=1}^{\infty} C^{Pt}_b \exp(-nd^{al}/\lambda_{237} \cos \theta)}{(1 - C^{Pt}_0) + \sum\limits_{n=1}^{\infty} C^{Ni}_b \exp(-nd^{al}/\lambda_{102} \cos \theta)} \tag{9.19}$$

with $C^{Pt}_b = 0.1$ and $C^{Ni}_b = 0.9$.

The result so obtained is $C^{Pt}_0 = 0.3$ ($C^{Ni}_0 = 0.7$), which indicates a strong platinum enrichment in the surface. Such a result is in good agreement with those deduced from XPS measurements, or given by ISS experiments.

On metal alloys, the surface segregation can extend to more than one layer, and a better description would use a model with multilayer segregation. This indicates the need to have the possibility of varying the respective weights of the successive layers to the overall Auger signal. In order to do so one can use: (i) many Auger lines at different energies, which correspond to different mean free paths; this method has been applied to the study of AgAu[10] and PtCu[11] polycrystalline alloys; (ii) the changes with the emission angle: the more grazing the emergence, the higher the sensitivity to the surface layer[12]; and (iii) the changes with the incident angle,[13] which method needs measurements up to very grazing incidence.

Consequently, data with high precision and confident values for backscattering and mean free path parameters are needed for a good definition of the surface of alloys.

For a polycrystalline sample, the intensity of an Auger peak characteristic of an A element in a matrix AB, emerging at an energy E can be expressed by

$$I^A_E \propto N_V \times i^A_0 (1 + R^{AB}) \int_0^{\infty} C^A(z) \exp(-z/\lambda \cos \theta) dz \tag{9.20}$$

where N_V represent the total number of atoms per unit volume, and $C^A(z)$ the concentration in A element in the elemental volume located between Z and $Z + dZ$ below the surface.

If one assumes a homogeneous distribution of A and B elements, i.e.,

$$\int_0^\infty C(z)e^{-z/\lambda}\cos\theta = C\lambda\cos\theta$$

and combines the previous equations, one gets the practical formulation:

$$\frac{C^A}{C^B} = \frac{I_{E_A}^{al}}{I_{E_B}^{al}}\frac{I_{E_B}^{\text{pure B}}}{I_{E_A}^{\text{pure A}}}\frac{(1+R^A)}{(1+R^B)}\frac{N_V^A}{N_V^B} \tag{9.21}$$

In practice, for dispersed catalysts having rough surfaces only this last equation can be applied reliably, with the use of I values for pure elements tabulated[4] or, better, determined in its own spectrometer. The backscattering $1 + R$ factors can be taken from Figure 9.4.

The study of two copper-based solid metallic catalysts illustrates the ability of Auger to measure the surface segregation of some elements present in the bulk, even at a very low concentration.[14] The poor performances of some copper catalysts were probably due to a surface enhancement of the lead present in very low quantities in the catalysts.

The study of catalysts deposited on oxide supports is more difficult. Indeed, charge effects and local heating can occur on materials of low conductivity; these points will be discussed in detail later but to a large extent these perturbations can be avoided by using unfocused primary electron beams of low intensity and increasing the length of time for data acquisition. The surface analysis of supported catalysts is never possible if the crystallites are located in the bottom of the pores.

In the previous treatments, we have always considered the same mean free path for a given energy, indepenent of the metallic matrix. Slight variations do exist depending upon the atomic number, especially for high-energy Auger transitions, an experimental way of determining the dependence of λ upon the alloy composition has been proposed.[15]

9.3.2. Depth Profiling

By the combined use of argon ion sputtering and Auger electron spectroscopy, one can get depth profiles of materials. Indeed, medium-energy argon ions remove successive atomic layers; the resolution for sputtering stays in the range of 2 to 3 atomic distances. The following example, of a glassy Ni_2Zr ribbon (Figure 9.8) illustrates the possibilities of such a methodology.[16] The measurements show the presence of three regions: (i) a contamination layer in which carbon and oxygen dominate; (ii) an intermediate part in which oxygen and

J. C. BERTOLINI AND J. MASSARDIER

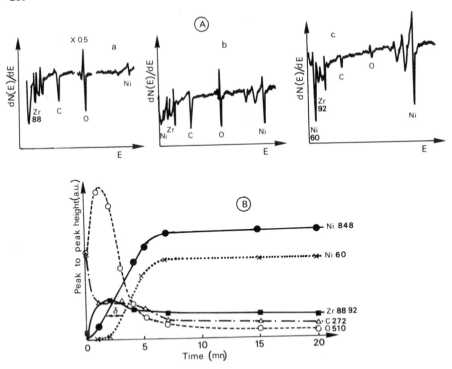

Figure 9.8. (A) AES spectra after different sputtering times: (a) 1 min, (b) 3 min, (c) 7 min ($E_p = 3$ keV, CMA analyzer, modulation amplitude of 3 V_{pp}). (B) AES peak-to-peak heights vs. sputtering time (and depth) of a glassy Ni_2Zr ribbon.

zirconium are the main elements (the intensity of the Auger peaks characteristic of these elements decreases while the 60 and 848 eV Auger transitions associated with nickel are first detected and then increase in intensity); (iii) a final stage associated with the bulk, where the Auger spectrum remains unchanged along the sputtering time.

In order to have a depth profile, one has to calibrate sputtering time vs. eroded depth. This is done in separate experiments in which the delay time necessary to sputter thin films of accurately known thickness is measured under well-controlled conditions. In the present experiment, the correspondence between the sputtering time and the depth was measured on gold thin films; it was 10 Å/min.

A thin layer of zirconium oxide (the mean value of the thickness is about 30 Å) covers the Ni_2Zr alloy. An upward chemical shift is simultaneously observed on the Auger peaks characteristic of zirconium, related to the change from the oxidation state to the metallic state. Moreover, the observation of the two nickel Auger peaks at 848 and 60 eV energy, respectively, verifies the in-depth vs. sputtering-time calibration. Indeed, the very first emergence of an Auger signal associated with an element located below the oxide interface corresponds to a depth of about 2 to 3 apparent mean free paths ($2-3 \lambda \cos \theta$). With $\lambda_{848} \approx 14-17$ Å and $\lambda_{60} \approx 5-6$ Å, the difference between the corresponding mean free paths is about 7 Å, i.e., $2-3 \lambda \cos \theta$ ranges between 14 and 21 Å, a value close to the 15 Å deduced from the time scale difference $\delta = 1.5$ min of the two related variations in Figure 9.8b.

For typical nonplanar and rougher samples, the definition of the sputtering angle and that of the emergence angle of Auger electrons with respect to the surface are less well defined. This lessens the precision needed in the definition of the depth of the interface to be determined. Indeed, the sputtering efficiency depends upon the incident angle of the impinging ions (see Chapter 11 on SIMS and Chapter 12 on ISS for similar problems) and the analyzed depth varies with the emergence angle of the electrons. A theoretical analysis of AES depth profiling in multilayers has been given by Petrakian and Renucci[17]; the results were compared with experimental determinations.

9.3.3. Surface Analysis of Oxides

Surface analysis by electron impact AES must be done cautiously in order to avoid irradiation damage and significant charge effects. These points will be discussed later in Section 9.5. By the use of low primary currents one can generally avoid these perturbations and gain significant results. The surface composition of mixed oxides can be determined by using the procedure for binary alloys (Section 9.3.1).

9.3.3.1. Mixed Oxides

Let us consider as an example a typical catalyst for mild oxidation, the mixed tin–antimony oxide, which behaves differently depending on the antimony content and the thermal activation treatment. Samples were analyzed using, on the one hand, an RFA with a high current for the incident beam (some tenths of 10^{-4} Å) and a large beam size (diameter $D = 500-1000 \,\mu m$) and, on the other hand, a CMA with a primary current of about 10^{-7} Å well focused ($D = 50 \,\mu m$) or not ($D = 600 \,\mu m$). These results can be compared to those gained by the use of XPS photoemission on the same samples and reported in Chapter 17.

The Sb–Sn–O system looks favorable for surface analysis by AES:

1. The samples have a measurable electric conductivity so spectra can be obtained without significant charge effects, i.e., the energetic position of Auger peaks is determined with high accuracy.
2. Tin and antimony are close together in the Periodic Table of the elements. Their characteristic Auger lines therefore correspond to the same transitions, and consequently the kinetic energies are very close together. Thus, for both tin and antimony the parameters to be used for quantitative Auger analysis (ionization cross section σ, probability for de-excitation by the Auger process $1 - w$, retrodiffusion factor R, mean free path λ, and ana-

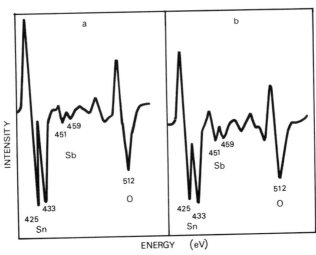

Figure 9.9. AES spectra of SbSnO catalysts treated at 950 °C: (a) 1.5% Sb, (b) 5% Sb (E_p = 1500 eV, RFA spectrometer, modulation amplitude of 10 V_{pp}).

lyzer transmission T) are closely similar. A simple comparison between the parent tin and antimony Auger peaks gives a direct determination of the average surface composition. The spectra corresponding to samples heated at 950 °C and containing, respectively, 1.5 and 5 wt.% of antimony, recorded using an RFA spectrometer are shown in Figure 9.9. The features at 425 and 433 eV (451 and 459 eV) are characteristic of the $N_4N_{4,5}N_{45}$ and $M_5M_{4,5}N_{4,5}$ transitions of tin (and antimony). They are shifted toward lower energy (4 ± 1 eV) if one refers to the known energy values of these transitions on antimony and tin in their metallic state.[4,5] This chemical shift will be discussed below (Section 9.3.5). From the relative intensities of the 425 and 451 eV Auger peaks the content of antimony relative to tin in the surface region was determined (Table 9.1). Other results obtained by using a CMA Auger spectrometer (with a lower intensity for the primary beam) and by XPS photoemission are also given in the table; all the results are in good agreement.

A surface enrichment of antimony appears clearly. It corresponds to the contribution of a surface region involving some atomic layers; the first outer layer does not contribute more than 15–30% to the total signals, depending on the technique used. AES is more sensitive to the surface itself (kinetic energy E_c of the MNN considered transitions at 420–460 eV) than is photoemission ($3d$ photoemission lines at E_c = 950–1000 eV), which could explain the higher Sb/Sb + Sn ratios determined by AES. However, one has to keep in mind that the irradiation damage generated by photons is less important than that from electrons.

TABLE 9.1. Comparative Values of the Antimony Content in the Surface Region of Sb–Sn–O Catalysts Determined from Data Gained by AES (RFA and CMA Spectrometers) and XPS

Sample	Treatment temperature (°C)	% Sb/(Sb + Sn) (surface)		
		Auger RFA	Auger CMA	ESCA (3d levels)
Sb–Sn–O	500		6	3
1.5%Sb	950	8	11	11
Sb–Sn–O	500		8	5
5%Sb	950	17	24	18
Sb–Sn–O	500	23	21	20
20%Sb	950		41	35

9.3.3.2. Highly Nonconducting Oxides

Powdered materials of high purity, used as supports or/and catalysts (SiO_2, Al_2O_3, NaA-zeolite) have been the subject of extended studies by AES surface analysis in many laboratories, within the framework of the American Society for Testing and Materials (ASTM).[18] Wide disparity among the results proves that this kind of analysis by AES needs to be done with care, both as regards the conditioning of the sample and the choice of condition for the analysis. The extent of beam damage during analysis can be assessed by control of the electron dose. In this way Suib et al.[19] studied A, X, Y, ZSM-5 zeolites. By the combined use of AES and 1-keV argon ion sputtering (depth profiling) they found that the Si/Al ratio was unchanged in surface with respect to the bulk in these zeolites. Moreover, these authors showed that a migration of metal and rare-earth ions toward the external surface of the zeolite occurred during the reduction process.[20] The AES depth resolution is limited to the external surface of the zeolite material.

9.3.4. Adsorbates and Deposits

Auger electron spectroscopy is in general not well adapted to analyze molecular adspecies on surfaces. Indeed, desorption and/or dissociation of admolecules frequently occurs due to electron-stimulated bond breakage.

AES is a useful method for the quantitative determination of strongly bonded adatoms. Surface concentration vs. time, partial pressure, and/or temperature leads to the possible determination of important physicochemical parameters such as the sticking coefficient and the heat of adsorption. For a layer-by-layer deposit, the increase of the AES line intensity characteristic of the adsorbate occurs simultaneously with the decrease of the substrate signal. From the analysis of the variations of these Auger signals, information on the growth mechanism of

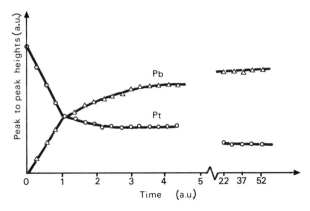

Figure 9.10. Variations of the amplitude of the AES peaks characteristic of lead and platinum, during lead deposition on a platinum substrate.[21]

the adsorbate can be gained. One example, the deposit of lead on platinum, is given in Figure 9.10.[21] At a given deposition time, arbitrarily normalized to unity, a break is clearly visible, which corresponds to the filling of the first adlayer. The mechanism of further growth looks more complicated. Although attenuated, the AES signal relative to the platinum substrate never reaches the zero value. Two hypotheses can explain this observation: (i) the growth of three-dimensional aggregates following the first monolayer filling, the layer remaining sufficiently thin for a part of the surface to allow the possible observation of the platinum substrate; or (ii) the migration of some platinum atoms above the lead film, and such a formation of two-dimensional Pt–Pb binary alloy in the outer layer has been proposed by Praline et al.[21] A simple layer-by-layer growth would have given a succession of straight lines, as observed up to the first monolayer completion.

Even in the case of strongly bonded species, AES must be used carefully to avoid damages induced in the analyzed surface region, i.e., it is better to decrease the primary electron beam to the weakest possible value.

The analysis of deposits, residues, etc. on supported catalysts is obviously more difficult; in addition to any analytical problems, difficulties can arise from their nonhomogeneous distribution. Moreover, the signals associated with the active phase are rather weak on lightly loaded supported catalysts.

9.3.5. Chemical Shifts and Line Shapes in AES

By precise measurements of the Auger transition energies one can get information on the chemical state and the chemical environment of the analyzed element. Indeed, electron transfers associated with the formation of chemical bonds induce energy shifts (chemical shifts) of the deep electronic levels. The Auger electron energies are therefore changed. However, the interpretation of the observed AES chemical shifts is frequently not trivial. Indeed, final state effects may affect the measured AES chemical shift, mainly due to the reaction of the surrounding atoms to the apparent lack of electrons on the emitter atom during the Auger process. Moreover, changes in the electronic density of states in the valence band directly modify the energetic position and the shape for the Auger transitions implying these valence electrons, such as XYV and XVV transitions.

In Section 9.3.3.1, we described how AES can be used to show that antimony segregates in the surface region of the SbSnO catalysts. Moreover, on the spectra shown in Figure 9.9, it can be seen that the $M_4N_{4,5}N_{4,5}$ (and $M_5N_{4,5}N_{4,5}$) Auger transitions for tin (and antimony) are shifted by 4 ± 1 eV toward lower energies relative to their values measured for tin and antimony in their metallic state.[4,5] These AES chemical shifts are more significant than the corresponding core-level shifts measured by XPS (1–2 eV). This observation indicates that final state effects play a role in the observed shifts; by coupling AES and XPS experiments one can deduce both the amplitude of the relaxation (final state) and of the true chemical shift (initial state) following the procedure described in the Appendix.

Another example is given by palladium supported on SiO_2 and Al_2O_3 obtained by vapor deposition of atomic palladium with varying particle size.[22] An important shift downward of the Pd M_5VV Auger electron kinetic energy was observed when the particle size was decreased. Kohiki[22] measured the related Pd $3d^{5/2}$ XPS electron binding energy shift, and by using the procedure described in the Appendix he separated the true chemical shift (initial state effect) from the other physical parameters (final state effect). A true chemical shift does exist, dependent upon the particle size. However, one has to be careful since the Auger transition involves the valence electrons, which makes it difficult to determine the Pd M_5VV Auger electron kinetic energy, especially if the Auger line shape is modified.

When the Auger transition implies more or less discrete valence levels, its Auger line shape can be very strongly modified. Carbon, aluminum, and silicon in different chemical environments give different AES fine structures. The carbon KVV Auger spectrum is a characteristic example.[23] In Figure 9.11, different fine structures are shown; the Auger line shapes of carbon in carbide compounds (with carbon–metal bonds) and in graphite (as in compounds with carbon–carbon bonds) are well differentiated. Slight deviations from these typical spectra can appear depending upon some changes in the carbon chemical environment, as observed for carbon monoxide and hydrocarbons chemisorbed on metal surfaces.[24,25] Consequently, such an analysis can provide interesting information on the nature of carbonaceous residues, such as poisons left on the surface after catalytic reactions. Silicon in different states of matter (semiconductor materials, metal silicides, silica) can be easily differentiated by their Auger line shapes in the low-energy region, as it appears from the spectra presented in Figure 9.12. Theoretical calculations of the Auger line shapes of the $L_{2,3}VV$ Auger transition of silicon have been undertaken in recent years and

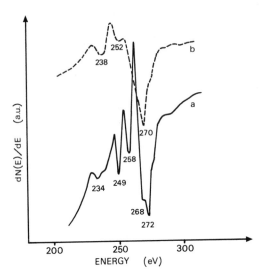

Figure 9.11. Fine structures of the KVV Auger line of carbon: (a) carbidic form; (b) graphite ($E_p = 1500$ eV, RFA spectrometer, modulation amplitude of 3 V_{pp}).

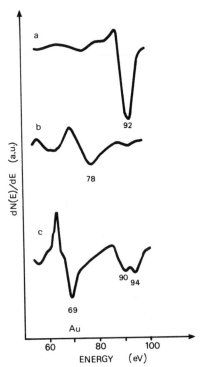

Figure 9.12. AES line shape for: (a) silicon; (b) silica; (c) Si–Au alloy ($E_p = 3$ keV, modulation amplitude of 1 V_{pp}).[26]

tentatively compared to the experimental data for pure silicon.[27] It has been proposed that electronic modifications induced by surface effects have to be taken into account to fit the theoretical and experimental curves.

9.4. IMAGING IN AES

Electron guns allow focusing of the impinging beam on a well-defined and small area of the sample by simple adjustments of the electronic optics of the gun. In a typical AES apparatus, working at primary energies lower than 10 keV, the lateral resolution does not exceed 2000 Å. Only more sophisticated electron guns, such as those included in electron microscopes, allow one to obtain a lateral resolution of about 10 Å. However, the "pear effect" (associated with the fact that the scattered electrons diffusing in surface or in the bulk can generate Auger electrons even if they have lost a part of their initial energy) leads to a physical limitation so that the lateral resolution cannot be less than 200 Å. Scanning Auger electron spectrometers record either secondary electron images (with the same mechanism as in scanning electron microscopes SEM) or lateral variations in the surface composition achieved by recording the peak-to-peak height for a characteristic Auger transition of the element being considered. Thus, the "Auger maps" can be correlated with the corresponding feature in the surface topography.

Such a technique has been used successfully for the surface analysis of promoted catalysts where inhomogeneity in the lateral distributions of different elements was found.[28] More recently, Ertl et al.[29] reported the results of a detailed study of an industrial ammonia synthesis catalyst. On the reduced samples, the completely reduced iron particles are uniformly covered by a "K + O" adlayer acting as an "electronic promoter," which confirms previous hypotheses drawn from indirect methods. The high spatial resolution combined with a high sensitivity to minority elements needs the use of a rather high energy and a high current density for the primary electron beam. Care has to be taken when analyzing the results as there can be damage, as discussed below. Moreover, on insulating materials the charge effect limits the spatial resolution.

9.5. LIMITATIONS ON THE USE OF AES

The review of Pantano and Madey[30] describes the possible damage that can occur during AES experiments. Let us summarize here the main current problems limiting the use of AES.

First, the examination of insulating materials is complicated by charging of the surface. As a consequence, the AES spectra are widened and shifted in energy in an uncontrolled way. Depending upon the secondary electron yield, (ratio between the backscattered electron current and the primary current) the surface is positively or negatively charged. Therefore Coulombic forces between charged particles generate electromigration of the ionic species in the surface.

The secondary electron yield varies with the primary energy and the incidence angle, so by a judicious choice of these parameters it is sometimes possible to minimize the charge effect influence. The charge effect can also be compensated for by bombarding the surface with an additional low-energy electron (or ion) beam. However, the residual conductivity of the materials is often sufficient for AES measurements at a very low primary current density.

Second, irradiation damage has to be considered. This has two distinct origins: (i) the local beam heating increasing with the beam power $E_p \times I_p$ and decreasing with the thermal conductivity of the material and the beam diameter; and (ii) the electron stimulated desorption (or adsorption) subsequent to electronic excitations of surface species inducing desorption (or adsorption) of fragments.

Electron beam damage in AES is then dependent on the nature of the analyzed material, of the primary beam current density and of the electron irradiation dose. Each sample has to be considered with its own properties by the analyst. As an illustration, the decomposition of alumina into metallic aluminum has a threshold estimated to about 10 C cm^{-2} for the primary beam.[31] Oxygen release has also been observed on SiO_2 under electron irradiation during the Auger analysis.[32] For organic compounds the permissible beam current electron dose is drastically lower, by several orders of magnitude compared to Al_2O_3 and SiO_2.

In conclusion, AES is a powerful tool for elementary surface analysis. The low-energy Auger transitions make the AES technique very sensitive to the very first atomic layers of solids. With well-focused primary electron beams, local analysis, and AES maps (surface region concentration of given elements) obtained by scanning of the primary beam can be drawn. Complementary information on the electronic state and the chemical environment can be deduced from the AES chemical shifts and Auger line shapes. Deconvolution techniques may be used for a better determination of the energetic position and/or the line shape, mainly by Fourier transform methods.[33] However, great care has to be taken to avoid charge effects and electron irradiation damage, mainly in the AES analysis of insulating materials.

APPENDIX

The chemical shift measured in Auger experiments between two different states of an element (e.g., more or less ionic state, modification of the structural and chemical environment) obeys the relation

$$\Delta E^{AES}(XYZ) = \Delta E(X) - \Delta E(Y) - \Delta E(Z) - \Delta I(YZ) + \Delta R^{2+} \quad (9.22)$$

In XPS experiments, the chemical shift of a binding energy E_b is expressed as follows:

$$\Delta E^{\text{XPS}} = \Delta E - \Delta R^+ \qquad (9.23)$$

where ΔE is relative to the "true" chemical shift. Theoretical work has shown that

$$\Delta I(XZ) \approx 0$$

$$\Delta E(X) \approx \Delta E(Y) \approx \Delta E(Z) \approx \Delta E$$

$$\Delta R^{2+} \approx 3\Delta R^+ \qquad (9.24)$$

R^{2+} and R^+ being the relaxation energies associated with the doubly or simply ionized states, respectively. Therefore the shift measured by AES is expressed according to the simplified equation

$$\Delta E^{\text{AES}}(XYZ) = -\Delta E + 3\Delta R^+ \qquad (9.25)$$

By combination of the two equations (9.23) and (9.25), R^+ may be obtained

$$\Delta R^+ = (\Delta E^{\text{AES}} + \Delta E^{\text{XPS}})/2 \qquad (9.26)$$

where $(\Delta E^{\text{AES}} + \Delta E^{\text{XPS}})$ represents the variation of the so-called "Auger parameter."

It is possible by this method to differentiate the true chemical shift (characteristic of the electronic state of the element) and the relaxation energy (characteristic of the structural arrangement and the chemical nature of the atoms surrounding the element being considered).

REFERENCES

1. P. Auger, *C. R. Acad. Sci. Paris* **177**, 169 (1923).
2. P. Auger, *Surf. Sci.* **48**, 1 (1975).
3. K. Siegbahn, *et al.*, *Electron Spectroscopy for Chemical Analysis*, Nova Acta Regiae Science, Uppsala (1968).
4. P. W. Palmberg, C. E. Riach, R. E. Weber, and N. C. MacDonald, *Handbook of Auger Electron Spectroscopy*, Physical Electronics Industries, Edina, MN (1972).
5. D. Briggs and M. P. Seah, *Practical Surface Analysis by Auger and X-Ray Photoelectron Spectroscopy*, John Wiley and Sons, New York (1985).
6. F. J. Kuijers, Thesis, Leyden, The Netherlands (1978).
7. S. Ichimura and R. Shimizu, *Surf. Sci.* **112**, 386 (1982).
8. R. Shimizu, *Jpn. J. Appl. Physics* **22**, 1631 (1983).
9. M. P. Seah and W. A. Dench, *Surf. Interf. Anal.* **1**, 2 (1979).
10. T. S. King and R. G. Donnelly, *Surf. Sci.* **151**, 374 (1985).
11. A. O. van Langeveld and V. Ponec, *Appl. Surf. Sci.* **16**, 405 (1983).
12. S. A. Chambers, T. R. Greenlee, C. P. Smith, and J. M. Weaver, *Phys. Rev.* **B 32**, 4245 (1985).

13. J. P. Segaud, E. Blanc, C. Lauroz, and R. Baudoing, *Surf. Sci.* **203**, 297 (1988).
14. M. M. Bhasin, *J. Catal.* **34**, 356 (1974).
15. A. Sulyok and G. Gergely, *Surf. Sci.* **213**, 327 (1989).
16. J. C. Bertolini, J. Brissot, T. Le Mogne, H. Montes, Y. Calvayrac, and J. Bigot, *Appl. Surf. Sci.* **29**, 29 (1987).
17. J. P. Petrakian and P. Renucci, *Surf. Sci.* **186**, 447 (1987).
18. T. E. Madey, C. D. Wagner, A. Joshi, *J. Electron Spectrosc. Rel. Phenom.* **10**, 359 (1977).
19. S. L. Suib, G. D. Stucky, and R. J. Blattner, *J. Catal.* **65**, 174 (1985).
20. S. L. Suib, G. D. Stucky, and R. J. Blattner, *J. Catal.* **65**, 179 (1985).
21. G. Praline, N. Pacia, J. J. Ehrhardt, A. Cassuto, and J. P. Langeron, *Surf. Sci.* **105**, 289 (1981).
22. S. Kohiki, *Appl. Surf. Sci.* **25**, 81 (1986).
23. G. Dalmai, J. C. Bertolini, and J. Rousseau, *Surf. Sci.* **27**, 379 (1971).
24. T. W. Haas, J. T. Grant, and J. C. Dooley, *Adsorption–Desorption Phenomena* (F. Ricca, ed.), Academic, London (1972), p. 359.
25. S. D. Foulias, K. J. Rawlings, and B. J. Hopkins, *Surf. Sci.* **114**, 1 (1982).
26. A. Cros, F. Salvan, M. Commandre, and J. Derrien, *Surf. Sci. Lett.* **103**, 109 (1981).
27. R. Vidal, M. C. G. Passeggi, E. C. Goldberg, and J. Ferron, *Surf. Sci.* **201**, 97 (1988).
28. K. Hangi, H. Shimizu, H. Shindo, T. Onishi, and K. Tamaru, *J. Res. Inst. Catal., Hokkaido Univ.* **28**, 175 (1980).
29. G. Ertl and D. Prigge, *J. Catal.* **79**, 359 (1983).
30. C. G. Pantano and T. E. Madey, *Appl. Surf. Sci.* **7**, 115 (1981).
31. A. Van Oostrom, *Surf. Sci.* **89**, 615 (1979).
32. S. Thomas, *J. Appl. Phys.* **45**, 161 (1974).
33. K. W. Nebesny and N. R. Armstrong, *J. Electron Spectrosc. Rel. Phenom.* **37**, 355 (1986).

VIBRATIONAL ELECTRON ENERGY LOSS SPECTROSCOPY

J. C. Bertolini

10.1. INTRODUCTION

Various techniques can be used to measure surface vibrations on solids: infrared emission or absorption (IR), inelastic neutron scattering (INS), inelastic electron tunneling spectroscopy (IETS), Raman spectroscopy, and electron energy loss spectroscopy (EELS). This last technique is the most recent, and still under development with respect to both the technical and the theoretical points of view. We will not describe here details of the basic principles that govern vibrating systems: group theory. The number of vibration modes depends upon the symmetry of the system and the vibration energies depend upon the force constants (derivative of the harmonic potential). The surface vibrations can be artificially divided into two classes depending upon their energy values: those being closely connected to the substrate phonons (i.e., bulk or surface phonons) and the intramolecular vibrations of adsorbed species having higher energies, which can be considered as decoupled from the substrate.

Much information can be gained by using vibrational spectroscopy:

1. The geometric structure of the adsorbate–substrate couple (e.g., adsorption site and orientation of the adsorbed molecules with respect to the substrate). Often the vibrating system can be considered as decoupled from its neighbor; i.e., there is short-range local order.
2. The force constants both inside the adsorbed molecules and with the substrate, and hence the bond strengths.

J. C. Bertolini ● Institut de Recherches sur la Catalyse, CNRS, Villeurbanne, France.

Catalyst Characterization: *Physical Techniques for Solid Materials*, edited by Boris Imelik and Jacques C. Vedrine, Plenum Press, New York (1994).

3. The chemical nature of the adsorbed species by comparing their vibrating modes in the adsorbed phase with those of their free (gaseous) counterparts.

The first vibrational EELS experiment was reported in 1967,[1] but was extended to several laboratories only after 1975. Its use is to date limited mainly to the observation of flat surfaces. In catalysis, its main impact is in studies of the perturbations induced on adsorbates bonded to single-crystal model catalysts. Such adsorbates can either characterize the chemical and electronic properties of the surface, or be regarded as the intermediates or poisons in catalytic reactions.

10.2. ELECTRON–VIBRATION INTERACTIONS

In an EELS experiment, the energetic distribution of electrons backscattered from a target bombarded with low-energy electrons is measured, and the energy analysis is performed around the primary energy along a given direction with respect to the surface normal (Figure 10.1a). In most experiments, the electron detector is located very near the specular direction ($\theta' \sim \theta$);

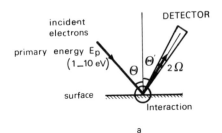

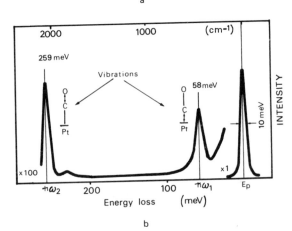

Figure 10.1. (a) Schematic representation of the geometry for an EELS experiment. (b) A typical EELS vibration spectrum: CO on Pt(111) ($E_p = 3$ eV, $\theta = \theta' = 50°$, CO coverage = 0.3).

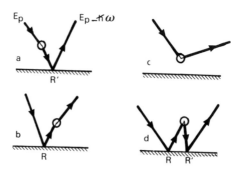

Figure 10.2. Schematic representation of the four contributions to the electron-vibration interaction: O represents the interaction, R and R' are the reflection coefficients of the electrons on the surface at E_p and $E_p - \hbar\omega$ energies, respectively.

but by analysis out of the specular direction one can get additional information, which will be discussed below. A typical EELS spectrum is presented in Figure 10.1b. An intense peak dominates the spectrum at an energy equal to the primary energy E_p, which corresponds to the elastic (no loss) peak. Small structures which can be detected with large amplification are associated with loss peaks. The energy losses correspond to vibration energy quanta $\hbar\omega_n$, i.e., their kinetic energies are $E_p - \hbar\omega_n$. Two electron–vibration interaction processes can appear; the long-range dipolar mechanism is frequently dominant, but a short-range resonant interaction can also be important.

10.2.1. Long-Range Dipolar Interactions

The dipolar electron–vibration interaction mechanism has been fully described by Evans and Mills[2] and Sokcevic et al.[3] The elementary excitations generate electric field fluctuations which extend far beyond the sample. The potential energy that governs the interaction between the primary electron and the electric field generated by a vibrating dipole is

$$V = - e(\mu + \partial\mu)\mathbf{r}/r^3 \tag{10.1}$$

where μ is the static dipolar moment, $\partial\mu$ the dynamic dipolar moment, and $\mathbf{r}$ the vector that measures the position of the interacting electron with respect to the mass center of the vibrator.

Thus, the primary electron can interact with the electric field associated with the vibrating dipole all during its trajectory outside of the solid. Figure 10.2 shows the four contributions[2,3] expected during the interaction process. Most of the interaction occurs at long distances (i.e., at distances r large with respect to the interatomic distances). The a (interaction and then diffusion on the surface) and b (diffusion on the surface and then interaction) processes are largely dominant. Consequently, the dipolar electron–vibration interaction contribution is located near the specular direction as much as the vibration energy mode $\hbar\omega_n$ is small with respect to the primary energy E_p.

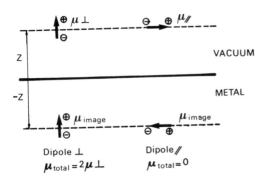

Figure 10.3. Schematic representation of dipoles and of their induced images on a metallic substrate.

10.2.1.1. Vibrating Dipoles on Metal Surfaces

On metal surfaces, the conduction electrons screen the electric field associated with the vibration dipole inside the solid. Such a screening acts instantaneously when the vibration energy $\hbar\omega$ being considered is much lower than the energy associated with the plasmons of the solid, $\hbar\omega_p$. This is the case if one considers vibrating modes having energies $\hbar\omega = 20 - 600$ meV on metals for which $\hbar\omega_p$ ranges between 5 and 30 eV. Therefore, in a first approximation one has to consider not only the electric field generated by the dipole itself, but also its image underneath the metal surface (Figure 10.3). At long range, the electric field of a dipole and its image is null if it is polarized parallel to the surface, but reinforced for a dipole normal to the surface. This is also the case for the field fluctuation associated with the vibration mode of such a dipole. Only the vibrating modes having a component perpendicular to the surface are detectable, i.e., only the IR active modes can contribute. Thus vibrational EELS spectroscopy can be compared to the IR absoption/emission spectroscopies. Nevertheless, cross sections for dipolar interactions are larger in EELS than in IR spectroscopy.

The EELS cross section is often expressed as the intensity ratio between a loss peak $(E_p - \hbar\omega)$ and the elastic peak E_p.

with

$$\frac{I(E_p - \hbar\omega)}{I(E_p)} = \frac{4\Pi}{E_p\hbar\omega}\frac{m}{M}e^2\left(\frac{\partial q}{\partial s}\right)^2\frac{Ns}{\cos\theta}f(\theta,\alpha) \qquad (10.2)$$

with

$$f(\theta,\alpha) = \frac{\sin^2\theta - 2\cos^2\theta}{1+\alpha^2} + (1+\cos^2\theta)\ln(1+(1/\alpha^2)) \qquad (10.3)$$

and

$$\alpha = \frac{\hbar\omega}{2E_p}\frac{1}{\Omega} \qquad (10.4)$$

where E_p and θ are the energy and incident angle (with respect to the surface normal) of the primary beam; m and e the mass and charge of the electron; M, $(\partial q/\partial s)$ and $\hbar\omega$ the reduced mass, normal component of the dynamic dipole moment, and energy of the vibration mode; N_s is the number of vibrators; and Ω the half-value of the aperture solid angle of the electron energy analyzer. This equation is given for well-defined angles for the impinging and outgoing analyzed beams, i.e., for flat surfaces.

10.2.1.2. Vibration Modes of Ionic Crystals

The electric field associated with the surface vibrations of ionic crystals (e.g., oxides) is not screened by the solid. On the contrary, the solid responds to the excitation so that one has to consider not only the N_s^0 vibrating cells per unit surface area, but also the N_v^0 cells per unit volume, as contribution to the electric field, that interacts with the incoming electron. For a large aperture angle of the spectrometer the intensity ratio between a given loss peak $[I(E_p - \hbar\omega)]$ and the elastic peak $[I(E_p)]$ can be expressed as follows[2]:

$$I(E_p - \hbar\omega)/I(E_p) = \pi e^2 \omega_p^2 / \hbar \omega_s^2 V (1 + \varepsilon_0)^2 \qquad (10.5)$$

where V is the normal componnent, with respect to the surface, of the velocity of the primary electrons; ε_0 the dielectric constant of the material; and $\hbar\omega_p$ and $\hbar\omega_s$ are the energies of the bulk and surface plasmons.

10.2.2. Short-Range Interactions

The dipolar theory for electron–vibration interactions on metals remains strictly valid only for dipoles located at a $z = 0$ distance from the surface. For adsorbed molecules ($z \neq 0$), the wavelength λ of low-energy electrons used in EELS spectroscopy (from 2 to 20 eV) is of the same order of magnitude as the adsorbate–substrate distances Z. Vibrational EELS thus differs from IR spectroscopy, for which λ is much larger than $Z (\lambda \gg Z)$. Therefore, impinging electrons might "see" the vibrating dipole and its image independently, which would make possible the observation of both normal and parallel modes. Nevertheless, the length of time for such short-range interactions is very short (with respect to the length of time during which the incoming electron interacts at long range with the electric field associated with the dipole and its image, i.e, during its entire trajectory), and so only a very small contribution is expected from this interaction.

Resonances and/or trapping of electrons in the surface region can greatly increase the duration of short-range interactions. LEED resonances, i.e., a trapping of low-energy electrons close to the surface, inducing a long residence time of these electrons on the surface, are the usual causes of short-range interactions in vibrational EELS spectroscopy. Such electrons can then interact with all surface vibration modes whatever their nature: IR or Raman active, parallel or perpendicular to the surface. The cross section for excitation of vibrations must then be increased when the LEED intensity in the specular beam is decreased; such a phenomenon is quite general for hydrogen adsorbed on metal surfaces and will be illustrated below. Trapping of an electron on antibonding orbitals of adsorbed species, as observed for gases, with the temporary formation of a negative ion, can also induce short-range effects.[4] In this case, one has to consider in addition the vibration modes of the excited state of the adsorbed group that contains $n + 1$ electrons. This has been done only for OH groups[5]; in general, the lifetime of excited states of adsorbed species is too low to allow the observation of electrons interacting in such a way.

For short-range interactions, the selection rules do not apply and all vibration modes can be observed. Electrons which have then interacted are located not only in the vicinity of the specular direction but also in the whole direction space ($\theta' \neq \theta$). Consequently, by analysis in and out of the specular direction one can get further information about the nature of the surface vibration modes of species adsorbed on metals, e.g., the orientation of a dynamic dipole with respect to the surface. In this way very interesting structural information on adsorbed species can be obtained. No satisfactory theory of short-range electron–vibration interactions yet exists. Hence the possibility of occurrences of such processes cannot be calculated.

10.3. EXPERIMENTAL CONSIDERATIONS

Electrons are usually produced from hot filaments. The energy distribution of emitted electrons is then Maxwellian. The full width at half-maximum (FWHM) $\Delta E_{1/2}$ depends on the temperature T of the source:

$$\Delta E_{1/2}[\text{eV}] = 2.45T/11,600 \, [\text{K}] \qquad (10.6)$$

Typically equation (10.6) leads to a FWHM of 420 mV for a 2000 K filament electron emitter. An electron energy monochromator is therefore necessary to study vibrations in the 30–500 meV (250–4000 cm^{-1}) energy loss range. Such a monochromatization is performed with electrostatic electron velocity selectors — 127° cylindrical or hemispherical selectors.[6] The energy analysis of backscattered electrons must also be done with very high resolution which is made possible with the same kind of electron velocity selectors. The characteristics of the different selectors, with respect to their geometry and to the transmitted electron energy, are described in a book by K. D. Sevier[6] and in a review article by D. Roy and J. D. Carette.[7]

Figure 10.4 shows the schematic arrangement of a high-resolution vibrational EELS spectrometer combining an electron monochromator and an electron energy analyzer. Such an apparatus is used in our laboratory to study surfaces vibrations.[8] The energy resolution $\Delta E_{1/2}$ depends upon the size of the selector (mean radius R), the size of the input and/or of the output aperture

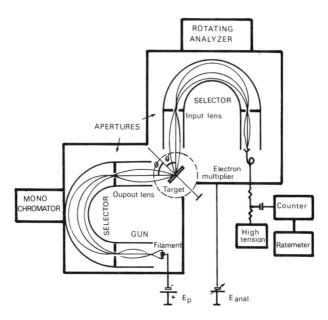

Figure 10.4. Schematic representation of high-resolution EELS spectrometer using hemispherical selectors.

(diameter ϕ), the angular spread of the electron beam at the entrance and/or at the output of the selector (half-value of the solid angle α), and the electron energy transmitted by the selector, E_T, which is defined by the electric field placed between the outer and inner hemispheres:

$$\Delta E_{1/2} = E_T \phi / (2R + k\alpha^2) \qquad (10.7)$$

where k is small, depending upon the selector (k equals 0.25 for a hemispherical selector).

Usually spectrometers work with a total resolution in the 7–10 meV range, at a primary electron current of about 10^{-10} A. Better resolutions are now possible, and experiments with less than 2-MeV resolution have been reported recently.

10.4. SOME EXAMPLES OF INTEREST

Usually, the EELS technique is employed to suty surface vibrations of adsorbates on well-defined single crystal metal surfaces. It is in this field that EELS spectroscopy has had the most impact, by giving new and relevant data on the adsorption site, the "activation," and the surface reaction processes (e.g., dissociation and rearrangements) for small molecules after chemisorption.

Some data are relevant to the surface of oxides. Tentative experiments have also been done on chemisorption properties of very small supported metal particles.

10.4.1. Adsorbates on Metal Single Crystals

10.4.1.1. Unsaturated Hydrocarbons: C_2H_4 and C_2H_2 on $Ni(111)$

10.4.1.1a. Nature of the Adsorbed Species and Surface Reactions. At room temperature, ethylene and acetylene adsorption both lead to the formation of the same surface complex giving the structures shown in the EELS spectra (Figure 10.5).[9,10] On Ni(111), C_2H_2 and C_2H_4 adsorb at 300 K with the formation of the same hydrocarbon surface complex; C_2H_4 is partly dehydrogenated. The surface complex is identified as adsorbed C_2H_2 by the following points:

1. The vibration located at 151 meV (little shifted, at 147 meV, for the deuterated product) attributed to the ν_{CC} stretching mode and proving adsorption without C—C bond breaking.
2. The structure at 365 meV (274 meV for the deuterated complex) attributed to ν_{CH} stretching modes.
3. The lack of vibrations associated with CH_2 groups, expected in the 165–193 meV region and appearing as peaks of large intensity when present.[11]

After heating such a C_2H_2 surface complex, up to 375–425 K the vibrational EELS spectrum is significantly modified (Figure 10.6). The ν_{CC} vibration peak vanishes. The new surface complex is identified as $\equiv$C—H surface groups, with a ν_{CH} stretching mode at 375 meV and a δ_{CH} deformation mode at 97 meV. The C—C bond breaking occurred before the dehydrogenation of the C_2H_2 surface species.

10.4.1.1b. Molecular Bond Strengths in the Chemisorbed State. From the previous experimental data, one can simply estimate the C—C force constant:

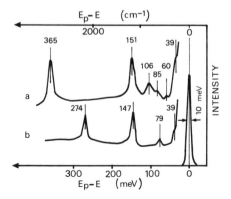

Figure 10.5. EELS spectra obtained after adsorption on Ni(111) at ambient temperature: (a) C_2H_2 (0.75 L) or C_2H_4 (6 L), (b) C_2D_2 (0.75 L) or C_2D_4 (6 L) ($E_p = 3$ eV; $\theta' = \theta = 50°$). [1 langmuir exposure (L) = 10^{-6} Torr·s]

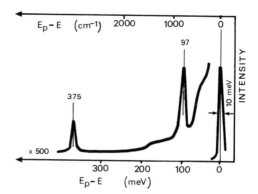

Figure 10.6. Vibration spectrum observed for C_2H_2 adsorbed on Ni(111) after heating at 375–425 K ($E_p = 3$ eV; $\theta = \theta' = 50°$).

$f_{CC} = 5$ mdyn/Å. Such a value can be compared with the f_{CC} force constant of the free molecules C_2H_2 ($f_{CC} = 15.6$ mdyn/Å), C_2H_4 (9.6), and C_2H_6 (4.5). The chemisorption-induced perturbation of the C_2H_2 species is important. In the chemisorbed state, the hybridization of the carbon atoms is near sp^3. This simple estimate gives only a rough value of the f_{CC} force constant. We will consider below a more sophisticated calculation which gives its value with better precision.

10.4.1.1c. Adsorption Site and Geometry of the Surface Complex. A complete determination of the whole vibration system is very difficult. Indeed, one has first to find all the vibration modes. Moreover, it would be necessary to consider all the interactions: not only inside the adsorbed molecule and with the metal substrate, but also between the adsorbed molecules and between the adsorption site and the neighboring metal atoms. A simplification can be introduced by considering only the complex formed by the adsorbed molecule and its adsorption site, isolated from the neighborhood atoms. This approach remains valuable only as long as the energy values of the chosen vibrations are much higher than those of the highest-energy phonons of the substrate (about 20–40 meV for metals). Therefore, intramolecular vibrations of adsorbates can be satisfyingly fitted in such a way, but large uncertainties exist in the determination of molecule–substrate force constants. In order to take into account the coupling of the atoms forming the adsorption site with their neighbors, one can assign a larger mass to these atoms (the rigid surface approximation).

For C_2H_2/Ni(111), a triangular adsorption site as shown in Figure 10.7 is in agreement with the previous conclusions. A mechanical calculation using the FG matrix method[12] gave the energy values of the vibration modes as shown in Figures 10.5a and b with the following force constants ($f_{CC} = 4$ mdyn Å^{-1}; $f_{CH} = 4.7$ mdyn Å^{-1}; $f_{CNi_1} = 0.5$ mdyn Å^{-1}; $f_{CNi_2} = 0.35$ mdyn Å^{-1}), and the geometrical parameters given in Figure 10.7b. Vibrations at 365, 151, 39, 106, 85, and 60 meV are, respectively, associated with CH, CC, and CNi stretching modes and with three CH deformation modes. The force constants so determined look reasonable with respect to the values reported for

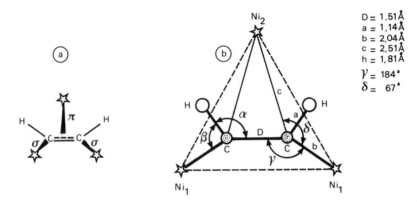

D = 1,51Å
a = 1,14Å
b = 2,04Å
c = 2,51Å
h = 1,81Å
γ = 184°
δ = 67°

Figure 10.7. Schematic representation of the triangular site adsorption of C_2H_2 fragment (h represents the height of the C—C bond with respect to the nickel surface).

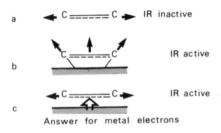

Figure 10.8. Schematic representation of the dipoles associated with a C—C stretching mode (the C—C bond is parallel to the surface): (a) free radical, (b) radical strongly coupled to the surface, and (c) possible reaction of the metal electrons to change in the C—C bond length.

comparable gaseous molecules.[12] In the adsorbed state, the C—C bond length of the C_2H_2 fragment is very near the length of a C—C single bond, i.e., near the sp^3 hybridization state of the carbon atoms. The C—C stretching mode looks largely polarized parallel to the surface, i.e., only slightly active with respect to the selection rules discussed above. However, either the coupling with the surface (Figure 10.8b) or the reaction of metal electrons to a C—C bond length increase (Figure 10.8c) makes possible the observation of such a vibration mode. The larger the coupling to the metal substrate, the greater the polarization will be.

10.4.1.2. Carbon Monoxide on Metals: CO on Ni(111), (100), and (110)

10.4.1.2a. Adsorption Site. Carbon monoxide adsorption has been the subject of many studies. On nickel,[13] it chemisorbs at room temperature into different states depending upon the surface orientation (Figure 10.9). On Ni(111) only one adsorption state is found. The CO stretching vibration shifts from 1815 to 1920 cm^{-1} when the coverage is increased from 0 to 0.5. The symmetric metal–CO v_{MCO}^{sym} mode located at 380 cm^{-1} is characteristic of multiply bonded (probably bridge-bonded) CO adspecies.[11,13] On Ni(100) and (110) two adsorption states coexist, which correspond to linearly bonded and

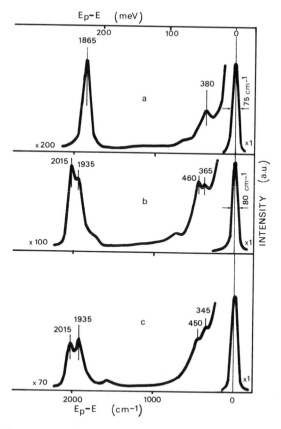

Figure 10.9. Vibration spectra of CO adsorbed at 300 K: (a) on Ni(111), $\theta_{CO} = 0.2$; (b) on Ni(100), $\theta_{CO} = 0.5$; and (c) on Ni(110), $\theta_{CO} = 0.5$ ($E_p = 3$ eV; $\theta = \theta' = 50°$).

bridge-bonded CO. The assignment of the two ν_{CO} stretching modes to linearly adsorbed CO (higher ν_{CO} values) and to multiply adsorbed CO (lower ν_{CO} values), respectively, can be made by analogy with known values of well-defined organometallic carbonyl clusters. Moreover, the energetic value of the symmetric ν_{M-CO}^{sym} stretching mode of CO adsorbed in a top and a bridge position can be related if the adsorption energies of both adsorbed states are very close. That is the case for CO adsorbed on nickel. In the valence force approximation (with the C—Ni bond length equal to 1.90 Å, the known value for polynuclear nickel carbonyls), the energy ratio of ν_{M-CO}^{sym} modes relative to the linear and bridge positions of CO adspecies is 0.76. Such a value is very close to those experimentally measured for CO adsorbed on Ni(100) and Ni(110) (Figure 10.9 and Table 10.1), taking experimental uncertainties into account.

10.4.1.2b. Comparison between Nickel Single Crystals and Ni/SiO$_2$ Supported Catalysts. The EELS vibration spectra of CO adsorbed on Ni(100) and

TABLE 10.1. Energy of the Vibration Modes Associated with CO Adsorbed on Ni(111), (100) and (110) Surfaces at 300 K

	Energy, cm^{-1}					
Surface	Ni(111)		Ni(100)		Ni(110)	
Coverage θ_{CO}	Low	0.5	Low	0.5	Low	0.5
O | v_{CO} C | Ni v_{M-CO}			2000	2015	1990	2015
				460		~ 450
O | v_{CO} C /\ v_{M-CO} Ni Ni	1815	1910	1900	1935	1880	1935
	380	380		365		~ 345

Ni(110) closely resemble the IR spectra of CO adsorbed on a supported Ni/SiO_2 catalyst (Figure 10.10).[14] On Ni(111) the v_{CO} stretching vibration has a lower energy. The results obtained on well-defined surfaces can therefore be used to determine the presence of (111) faces on an unknown nickel catalyst.

10.4.1.2c. The v_{M-CO} Stretching Vibration in Relation to the M–CO Bond Strength. The linear configuration of adsorbed CO has often been observed on various metals. A linear relationship exists (Figure 10.11) between the adsorption energy (measured by a separate method; e.g., thermal desorption spectroscopy, or a technique giving the isosteric heat of adsorption) and the square root of the vibration energy of the v_{M-CO} mode (proportional to the square of the force constant f_{M-CO} and measured by vibrational EELS spectroscopy). This indicates that the shape of the adsorption potential changes very little and that it is only the depth of the potential well that varies with the metal–CO system. The relationship shown in Figure 10.11 can thus be used to estimate the adsorption energy of linearly adsorbed CO on metals from the metal–CO stretching vibration energy. The same correlations could be done for multiple bonded CO adspecies.

10.4.1.3. Hydrogen on Metals: An Example for Short-Range Resonant Interactions

Metal surfaces covered with hydrogen give rise to resonant phenomena with low-energy electrons. This allows observation of all the vibration modes whatever their symmetry and their orientation with respect to the surface. The first

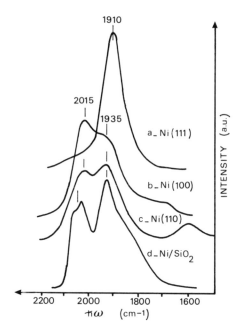

Figure 10.10. Vibration spectra of CO adsorbed at 300 K on nickel single crystals (EELS results) and on a Ni/SiO₂ catalyst (IR spectrum). The coverage is near 0.5.

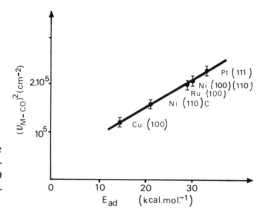

Figure 10.11. Relationship between the square of the v_{M-CO} stretching vibration and the adsorption energy of CO linearly bonded on several metal surfaces (low CO coverage).

experiment reported by Ho *et al.*[15] for hydrogen adsorbed on tungsten is a good illustration of the observation of all vibration modes parallel to the surface under specific experimental conditions. At a primary energy of 9.65 eV an abrupt decrease in the elastic peak intensity is observed on a hydrogen-covered W(100) surface ($\theta_H = 2$); i.e., a resonance exists for an incoming electron beam energy. The vibration spectrum then shows three loss peaks at 80, 130, and 160 meV for measurements out of the specular direction, instead of only one at 130 meV in the specular direction (Figure 10.12). These peaks are associated

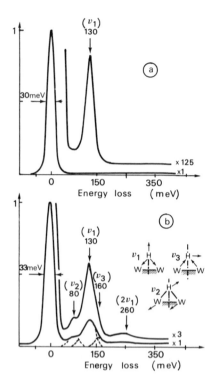

Figure 10.12. Vibration EELS spectra of hydrogen adsorbed on W(100) (hydrogen coverage $\theta_H = 2$; incident angle $\theta = 23°$; primary energy = 9.65 eV): (a) analysis in the specular direction, and (b) analysis out of the specular direction ($\theta' = \theta + 17°$).

with the excitation of two parallel modes (80 and 160 meV) in addition to the normal mode (130 meV) of hydrogen adsorbed in a C_{2v} symmetry corresponding to a bridge adsorption site. In conclusion, by the combined use of a defined primary energy of electrons (resonance condition) and of observation out of the specular direction one can measure both normal and parallel vibration modes. The energies of the vibration modes are all related through the geometry of the hydrogen/ adsorption site couple, which can therefore be determined from EELS experiments.[15]

10.4.2. Adsorption on Supported Metals

Only a few experiments have been reported on vibrational EELS studies of adsorbates on supported metals. In all cases model catalysts prepared by metal deposition on flat supports were used. The first reported study was carbon monoxide chemisorbed on rhenium supported on a thin film of alumina.[16] The rhenium particles were 20–30 Å in diameter. Results are given in Figure 10.13. The structure which emerges at 860 cm^{-1} is associated to the support itself; i.e.,

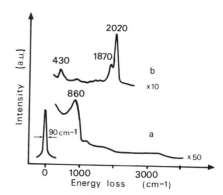

Figure 10.13. EELS vibration spectra of: (a) a clean alumina thin film; and (b) carbon monoxide adsorbed on rhenium particles (20–30 Å diameter) vapodeposited on the alumina support (primary energy = 5 eV).

it corresponds to the energy of the surface phonon of an aluminum oxide. Its intensity vanishes in the presence of the metal deposit. Vibration frequencies of CO adsorbed at 2020 and 1870 cm^{-1} are associated with CO stretching vibrations of linearly and multibonded CO adspecies, respectively; the low-frequency mode at 430 cm^{-1} corresponds to a metal—CO stretch or to a metal—CO deformation mode. Heating of the sample to 250 °C induced a preferential desorption of the multiply bonded state, as indicated by the disappearance of the 1870 cm^{-1} loss peak.

New EELS results have recently been reported on carbon monoxide adsorption on metal particles deposited on various supports, e.g., Pd/SiO$_2$,[17] NiAl$_2$O$_3$,[18] Pt/Al$_2$O$_3$,[19] and Pd/C.[20] However, few data are available on other adsorbates. Hydrocarbon residues have been detected on the surface of palladium particles deposited on amorphous carbon or on graphite, after butadiene 1–3 hydrogenation at room temperature at near atmospheric pressure; these residues were present only on particles larger than about 20 Å in diameter.[20] On smaller particles (i.e., smaller than about 12 Å), which are not active in the hydrogenation reaction, no residues were detected after reaction, and there is no CO chemisorption.[20] Acetylene adsorbs at room temperature on 25 Å diameter particles and the hydrocarbonated residues gave the EELS structures shown in Figure 10.14. The spectrum is very similar to that obtained after room temperature C$_2$H$_2$ adsorption on Pd(100).[21]

These pioneering experiments open interesting possibilities for the use of the EELS technique in catalysis.

10.4.3. Prominence of Surface Phonons of Solid Materials

10.4.3.1. Surface Phonons of Metal Substrates

Ibach and Bruchmann[22] first showed surface phonons on a stepped [6(111) × (111)] platinum sample. The detection of surface phonons is favored because the energies of these phonons are higher than that of the highest bulk phonon of the platinum substrate (Figure 10.15). Moreover, in this case

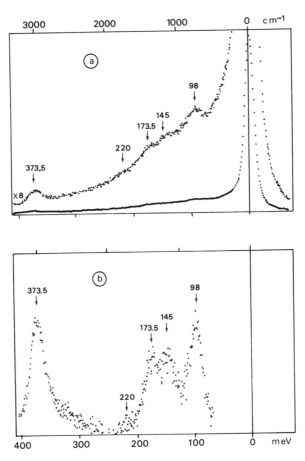

Figure 10.14. EELS spectrum measured after C_2H_2 adsorption (L, 300 K) on palladium particles (25 Å diameter) vapodeposited on a graphite support: (a) as registered, (b) after background subtraction.

interaction with the primary electron beam is possible as, owing to the electron deficit of these sites, localized dipoles appear on the atoms located on the step edges. The peak associated with the surface phonon excitation is very sensitive to the presence of small amounts of adsorbed CO. Indeed, CO adsorbs preferentially in the stepped region and induces big changes in both the energy position and the dipole moment associated with the presence of the steps. From the energy of the observed phonons one can deduce that the force constants between the atoms located on the step side and their neighbors are 1.7 times higher than in the bulk; i.e., surface atoms have fewer neighbors but the individual bonds are stronger. Rearrangements of the electronic configurations have occurred in the step region, with a contraction of the metal–metal bond lengths of these surface atoms with respect to the bulk values.

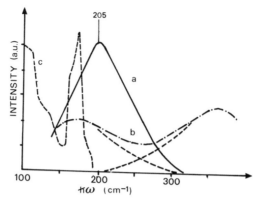

Figure 10.15. Loss peaks induced by the excitation of surface phonons of a [6(111) × (111)], Pt sample[22]: (a) clean surface, (b) surface partially covered with CO ($\theta_{CO} = 2.7 \times 10^{-2}$), and (c) calculated density of states of the substrate bulk phonons.[23]

10.4.3.2. Surface Phonons of Ionic Crystals

The energy of a Fuchs–Kliewer surface phonon can be calculated from the optical constants of the material by the following relationships[24]:

$$\omega_s = \left(\frac{\varepsilon_0 + 1}{\varepsilon_\infty + 1}\right)^{1/2} \omega_{TO} \qquad \omega_{LO} = \left(\frac{\omega_0}{\omega_\infty}\right)^{1/2} \omega_{TO} \qquad (10.8)$$

where ω_s is the surface phonon frequency of a crystal of infinite thickness, ω_{TO} the frequency of the transverse optical mode, ω_{LO} the frequency of the longitudinal optical mode, and ε_0 and ε_∞ are the static and the high-frequency dielectric constants.

For example, the experimental energy value of the Fuchs–Kliewer surface phonon of a ($1\bar{1}00$)ZnO crystal (68.8 ± 0.5 meV) is in close agreement with that calculated from the known optical constants of ZnO (69 meV).[25] On Ni(100), the oxide layer grows with the same orientation as the metal substrate.[26] Surface phonons emerge in EELS spectra with very high intensity (Figure 10.16). The energy value of such phonons is expected to vary with the oxide thickness.[24] In the spectrum in Figure 10.16 there are gain peaks on the high-energy side of the elastic peak, and their intensity depends upon the temperature of the solid by Boltzmann law:

$$I_{gain}/I_{loss} = \exp(-\hbar\omega/kT) \qquad (10.9)$$

As a consequence of the strong coupling of the surface dipoles with the bulk dipoles, the energetic value of the Fuchs-Kliewer surface phonons is no longer characteristic of the outer plane of the solid only but depends on the geometry and the bond strengths in the surface region. With thick oxides and more generally with ionic crystals (nonconducting materials) it may be necessary to use external beams to cancel the charges carried by the material.[27]

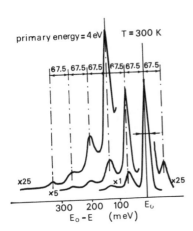

primary energy = 4 eV   T = 300 K

67.5 67.5 67.5 67.5 67.5 67.5

x25   x1   x25
x5

300   200   100 E_0
$E_0 - E$ (meV)

Figure 10.16. EELS spectra of a thin NiO(100) layer grown on a (100) nickel surface.[26]

10.5. COMPARISON WITH OTHER METHODS: CONCLUSIONS

It is difficult to compare the various spectroscopies applied to surface vibration measurements: IR, Raman, INS, and vibrational EELS. Indeed, their application depends upon the nature and morphology of the sample itself and its support. Vibrational EELS can be more closely compared with reflection IR spectroscopy, primarily when considering the dipolar interaction process, which is indeed dominant. EELS presents larger cross sections than IR spectroscopy,[28] which is what makes measurements on small quantities of matter possible; the EELS technique is therefore suitable for the study of samples having a small specific area, such as the surfaces of single crystals. With respect to optical methods, the resolution remains poor but the new generation of experimental setups allows resolutions better than 2 MeV (i.e., 16 cm^{-1}). Moreover, the use of low-energy electrons makes it necessary to carry out EELS measurements under vacuum conditions. The possibility of short-range interactions opens a new field of investigation for the EELS technique with respect to IR but much still has to be done to quantify the phenomena.

REFERENCES

1. F. M. Propst and T. C. Piper, *J. Vac. Sci. Technol.* **4**, 53 (1967).
2. E. Evans and D. L. Mills, *Phys. Rev.* **B5**, 4126 (1972).
3. D. Sokcevic, Z. Lenac, R. Brako, and M. Sunjic, *Z. Physik* **B28**, 273 (1977).
4. J. W. Davenport, W. Ho, and J. R. Schieffer, *Phys. Rev.* **B17**, 8 (1978).
5. S. Andersson and J. W. Davenport, *Solid State Commun.* **28**, 677 (1978).
6. K. D. Sevier, *Low Energy Electron Spectrometry*, Interscience, New York (1972).
7. D. Roy and J. D. Carette, *Electron Spectroscopy for Surface Analysis, Topics in Current Physics* (H. Ibach, ed.), Springer-Verlag, Berlin (1977).
8. J. C. Bertolini, G. Dalmai, and J. Rousseau, *J. Microsc, Spectrosc. Electron* **2**, 575 (1977).
9. J. C. Bertolini and J. Rousseau, *Surf. Sci.* **83**, 531 (1979).

10. J. C. Bertolini, J. Massardier, and G. Dalmai-Imelik, *J. Chem. Soc. Faraday Trans.* **74**, 1720 (1978).
11. H. Ibach, H. Hopster, and B. Sewton, *Appl. Surf. Sci.* **1**, 1 (1968); *Appl. Phys.*, **14**, 21 (1977).
12. E. B. Wilson, J. C. Decius, and P. C. Cross, *Molecular Vibrations: The Theory of Infrared and Raman Vibrational Spectra*. Mc Graw-Hill, New York (1955).
13. J. C. Bertolini and B. Tardy, *Surf. Sci.* **102**, 131 (1981).
14. M. Primet, Private communication.
15. W. Ho, R. F. Willis, and C. W. Plummer, *Phys. Rev. Lett.* **40**, 1463 (1978); *Surface Sci* **80**, 593 (1979).
16. L. H. Dubois, P. K. Hansma, and G. A. Somorjai, *Appl. Surf. Sci.* **6**, 173 (1980).
17. D. Schleich, D. Schmeiser, and W. Gopel, *Surf. Sci.* **191**, 367 (1987).
18. J. G. Chen, J. E. Crowell, and J. T. Yates, Jr., *Surf. Sci.* **187**, 243 (1987).
19. D. Venus, D. A. Hensley, and L. L. Kesmodel, *Surf. Sci.* **199**, 191 (1988).
20. B. Tardy, C. Noupa, C. Leclercq, J. C. Bertolini, A. Hoareau, M. Treilleux, J. P. Faure, and G. Nihoul, *J. Catal.* **129**, 1 (1991).
21. B. Tardy and J. C. Bertolini, *J. Chim. Phys.* **82**, 407 (1985).
22. H. Ibach and D. Bruchmann, *Phys. Rev. Lett.* **41**, 958 (1978).
23. D. H. Dutton, B. N. Brochouse, and P. A. Miller, *Can. J. Phys.* **50**, 2915 (1972).
24. S. R. Fuchs and K. L. Kliewer, *Phys. Rev.* **140**, 2076 (1965); *Phys. Rev.* **153**, 498 (1967).
25. H. Ibach, *Phys. Rev. Lett.* **24**, 416 (1970).
26. G. Dalmai-Imelik, J. C. Bertolini, and J. Rousseau, *Surf. Sci* **63**, 67 (1977).
27. P. A. Thiry, M. Liehr, J. J. Pireaux, and R. Caudano, Vibrations at Surfaces, 1985 Proceedings (D. A. King, N. V. Richardson, and S. Holloway, eds.), Part B, p. 69.
28. H. Ibach, *Surf. Sci.* **66**, 56 (1977).

SECONDARY ION MASS SPECTROMETRY

J. Grimblot and M. Abon

11.1. INTRODUCTION

When a solid target is bombarded by ions having energies of several keV, different complex processes may occur simultaneously. When the ion energy is low (typically less than 2 keV), the target surface can scatter the incident ions by an elastic collision mechanism. Analysis of the energetics of this collision is called ion scattering spectroscopy ISS, which is reviewed in Chapter 12. In addition to this elastic collision, in particular for higher ion energies, a certain depth of the solid is perturbed and excited, which results in emission of secondary electrons and photons and sputtering of neutral or charged fragments. Some of the incident ions may also be trapped or implanted in the solid. The mass analysis of these charged fragments (i.e., negatively or positively charged ions) is called secondary ion mass spectrometry (SIMS). Ions of chemically inert gases are generally used in SIMS as primary exciting particles but, sometimes, for given purposes, more reactive ions such as O_2^+ or Cs^+ are also employed.

The sputtering process, first observed in 1852–1853,[1] was developed rapidly for numerous applications (e.g., sample cleaning, new-materials preparation, reactive etching) or for surface analysis by ionic microprobe and SIMS. References 2 to 19 deal with the first SIMS experiments (on solids, thin films, and surfaces) and present reviews.

SIMS appears to be a powerful analytical technique for detection of surface contamination, for characterization of implanted elements or of concentration

J. Grimblot • Laboratoire de Catalyse Hétérogène et Homogène, URA CNRS, Université des Sciences et Technologies de Lille, Villeneuve D'Ascq, France. M. Abon • Institut de Recherches sur la Catalyse, CNRS, Villeurbanne, France.

Catalyst Characterization: Physical Techniques for Solid Materials, edited by Boris Imelik and Jacques C. Vedrine, Plenum Press, New York (1994).

profiles in a large number of practical materials such as thin films, semi-conductors, alloys, ceramics, and glasses. The analyzed sample may be organic or inorganic. A specific application, the study of surface reactions, adsorption phenomena, surface composition, interactions between active species and solids such as metals, alloys, oxides, supports, and promoters will be emphasized in the last part of this chapter.

11.2. THEORETICAL AND PRACTICAL ASPECTS

11.2.1. The Ion–Solid Collision: The Sputtering Process

Figure 11.1 shows the schematic representation (Werner[12]) of the ion–solid collision process, which results in the emission of neutral, excited, and ionized particles.

The primary ions of keV energy penetrate the target and there are binary collisions with the atoms in the solid. After these initial collisions, the primary ions, having lost their energy, come to rest in the solid (ion implantation in the penetration depth), and the target atoms recoil. After successive recoils by a cascade process,[20-26] some of the target atoms are in motion in the excitation depth. Those that are close enough to the surface with a velocity vector directed outside the solid can be ejected if they overcome the work function of the solid. This sputtering process is efficient only if the kinetic energy of the moving atom is higher than the binding energy within the solid. As a result the kinetic energy of the emitted particles is weak (between 1 to 10 eV on average), with a maximum at about 4 eV on transition metals[27] and a high-energy tail up to 100 eV. This is illustrated

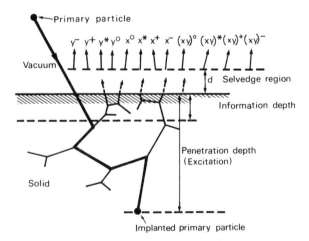

Figure 11.1. Schematic representation of the ion-solid collision process.[12] Emission of the target fragments comes from the information depth. At the selvedge of the solid, complex reorganization of the fragment may occur.

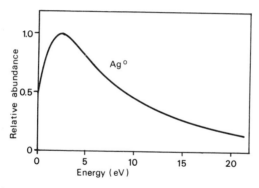

Figure 11.2. Kinetic energy distribution of silver atoms sputtered by bombarding silver using 1-keV Ar^+ ions.[28]

in Figure 11.2 for a silver surface bombarded by 1-keV Ar^+ ions. These energies increase only weakly with primary ion energy.

The information depth from which neutral, excited, and ionized fragments (atoms X, Y or molecules XY) of the target are ejected is smaller than the perturbation or penetration depth (Figure 11.1). This information depth depends on the mass and energy of the primary ions and on the nature of the target. The mean escape depth is generally between 2 and 3 atomic layers. For example, the mean value is 6 Å when silicon is bombarded by 5-keV Ar^+ ions at an incident angle of 60°.[12] Very few particles come from a depth more than 20 Å. For fragments containing more than two atoms (molecular ions and clusters), the escape depth is limited to the first two layers.

Some complex ionic rearrangements and neutralization may occur in a zone (of thickness d) outside of the solid. These phenomena, not always well understood, represent one of the reasons for difficulties in the interpretation of SIMS results.

The sputtering yield variations S (defined by the total number of emitted particles per incident ion) as a function of mass, incident energy, and impact angle between the trajectory of the ions and the normal to the target are shown in Figures 11.3 and 11.4. The optimum angle appears to be 60–70°. The nature and surface structure of the target is also a parameter affecting the sputtering rate (and therefore the emitted ion yield). In the case of copper bombarded by 5-keV Ar^+ ions, the ratio between the sputtering rates of the (111) and (110) planes is close to two.[9] The ejection mechanism itself is dependent on the structure of the exposed surface. In general, sputtering by unique collision between the incident ion with a surface atom is a rather rare phenomenon (recoil sputtering) but there are exceptions.

Figure 11.5 shows the ejection of atoms from the bombardment of copper by 600-eV Ar^+ ions at normal incidence. The target atom of the Cu(001) plane is not emitted since it is moving back into the bulk of the material. On the (110) surface, however, the target atom does not recoil far into the crystal because it is

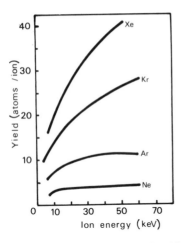

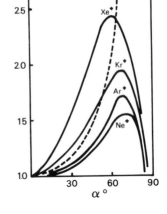

Figure 11.3. Sputtering rate of gold as a function of the ion energy.[29]

Figure 11.4. Sputtering rate variation of copper as a function of the incident angle relative to the normal of the target.[30]

sitting on four atoms of the second layer so that it is more easily ejected. Transfer of energy in that case is through propagation along the atomic rows of the exposed layer.[31] When comparing aluminum surfaces, the sputtering rate S varies between 1 and 4 by impact with Ar^+ ions. Maximum values of S are obtained on compact surface planes, whereas minimum values are observed on open surfaces.[11,32,33] Consequently, on practical surfaces when several surface exposed planes exist together, the exact values of the sputtering rate are not well known. Moreover, surface impurities, defects, and surface roughness may affect the expected values considerably.[9]

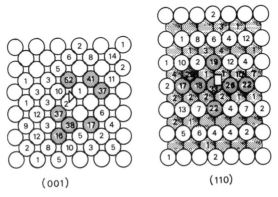

(001)   (110)

Figure 11.5. Ejection of atoms from the bombardment of copper by 600-eV Ar^+ ions at normal incidence. (a) Cu(001), (b) Cu(110). The numbers refer to the percentage of ion impacts in which the particular atom was ejected. The shaded atoms are the ones ejected more frequently. The symmetry zone for the ion impacts is shown on each face.[31]

TABLE 11.1. Sputtering Rate S with 1-keV Ions at the Normal Incidence Angle[34]

Target	He$^+$	Ne$^+$	Ar$^+$	Kr$^+$	Xe$^+$
Be	0.35	0.80	1.1	0.8	0.7
Al		1.13	1.94	1.53	
Si			1.0		
Ti			1.13		
Fe		0.84	1.34	1.44	
Ni	0.22	1.45	1.85	1.89	2.0
Cu	0.65	2.75	3.64	3.62	2.42
Ge			1.55		
Zr			1.06		
Nb			0.98		
Mo		0.43	1.14	1.41	1.63
Pd			3.06		
Ag	1.8	2.4	3.8	4.7	
Cd			11.2		
Ta			0.91		
W			1.10		
Pt	0.08	0.85	2.0	2.35	2.52
Au		1.53	3.08	3.86	

Table 11.1 lists the sputtering rates of several pure materials bombarded by rare gas ions.[34] Note that with Ar$^+$ ions at the normal incident angle, the sputtering rate does not much depend on the chemical nature of the target (most S values range from ~ 1 to ~ 2 with Ar$^+$ ions). The chemical state of the elements at the surface or in the information depth also has a noticeable influence on the sputtering rate. It is well known, for example, that the sputtering rate of a metal in the oxide state is equal to or less than the sputtering rate of the pure metal.[7,35,36] It is shown in Table 11.2, from the study of Kelly and Lam[35] that the ratio S(oxide)/S(metal) is always higher than 1 (except for Al, Mg, or Ti in comparison with Al$_2$O$_3$, MgO or TiO$_2$). However, when the metal molar fraction x in the oxide corrects the S oxide value, it appears that there are fewer sputtered metallic atoms in the oxide than in the pure metal. The only exception in Table 11.2 is V$_2$O$_5$ compared to vanadium.

11.2.2. Preferential Sputtering

When multicomponent materials containing several chemical elements in one or several phases are examined by means of SIMS, the phenomenon of preferential sputtering is observed partly because the different elements do not have the same sputtering rate S (see Table 11.1) as well as because the exposed surface may have a critical effect on the sputtering process, as discussed previously. This results in textural (increase of surface roughness by crater formation) and chemical composition modifications. This change of chemical composition has been seen many times on alloys[9,11,14] and binary compounds for which the actual surface composition has been determined by the use of another technique such as Auger electron spectroscopy.

On Ni–Cu alloys bombarded by 500-eV Ar^+ ions, the observed sputtering rate ratio $S(Cu)/S(Ni)$ is 1.9.[37] The same value is obtained when considering the results given in Table 11.1 (1-keV Ar^+) for the pure elements. However, in general, the experimental sputtering rate depends on the alloy composition and can differ from the calculated ratio with data of the pure elements. In the Ag–Au alloys, $S(Ag)/S(Au)$ varies from 2.3 to 2.9,[38] whereas $S(Ag) = 3.8$ (Table 11.1) and $S(Au) = 3.08$ (Table 11.1) or 3.6 (according to Wehner[9]), which gives a theoretical ratio of 1.15 ± 0.10.

The preferential sputtering rate between two elements in a target is a function of[39,40,41]:

1. The effective cross section between the incident ions and the elements of the target.
2. The bonding energy of the elements in the near-surface region.
3. The mass ratio $M1/M2$ between the mass of the incident ion, $M1$, and the mass of the element in the target, $M2$.

TABLE 11.2. Comparative Study of the Sputtering Rates of Oxides and Metals Bombarded by 10-keV Kr^+ Ions[35]

Oxides	S(oxide) (atoms/ion)	S(metal) (atoms/ion)	$\dfrac{S(\text{oxide})}{S(\text{metal})}$	$\dfrac{S(\text{oxide})}{S(\text{metal})} \cdot x$
Al_2O_3	1.6; 1.4 ± 0.2	3.2 ± 0.5	0.5	0.2
MgO	1.8 ± 0.5	8.1	0.2	0.1
$Mo\,O_3$	9.6 ± 0.4	2.8 ± 1.0	3.4	0.9
Nb_2O_5	3.4 ± 0.5	1.6; 2.0	1.9	0.5
SiO_2	4.2; 3.0 ± 1.5	2.1	1.7	0.6
SnO_2	15.3 ± 1.8	6.7; 6.4 ± 0.6	2.3	0.8
Ta_2O_5	2.5 ± 0.5	1.6 ± 0.3	1.6	0.4
TiO_2	1.9 ± 1.4	2.1 ± 0.8	0.8	0.3
UO_2	3.8 ± 0.5	2.4	1.6	0.5
V_2O_5	12.7 ± 1.7	2.3 ± 0.4	5.5	1.6
WO_3	9.2 ± 1.2	2.6 ± 1.0	3.6	0.9
ZrO_2	2.8 ± 0.1	2.3	1.2	0.4

x is the metal molar fraction in the oxide.

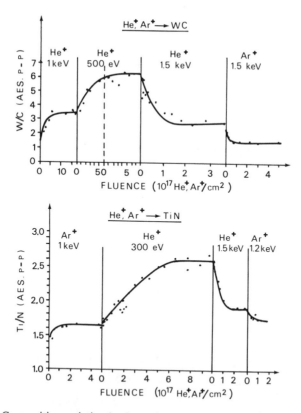

Figure 11.6. Composition variation (as determined by AES) of WC and TiN under sequential bombardment with ions of different masses and energies.[43]

The last aspect is illustrated in Figure 11.6, where the surface compositions of two binary materials (WC and TiN) of promising use in catalysis[42] are modified by interaction with rare gas ions of different energies.[43] Qualitatively, it can be seen that the lighter element is preferentially removed when the mass and energy of the incident ion beam are small. For the heavier ions at highest energy, the metallic element (Ti, W) is removed as well, so that the surface composition (determined by AES) returns to a value close to the bulk composition.

It can be seen in Figure 11.6 that under given experimental conditions, the measured surface composition is constant so that a steady state is obtained (equilibrium sputtering). The concentration difference between the surface C^s and the bulk C^b of the compound is compensated for by the different sputtering rate S_A or S_B of the two elements of the compound AB. This relationship can be written as[37]

$$C_A^b / C_B^b = (C_A^s / C_B^s)(S_A / S_B) \qquad (11.1)$$

From this relationship, it appears that, in principle, SIMS can be a quantitative technique.

When there is initially a concentration gradient between the surface and the bulk on the sample being tested, the concentration difference can be largely attenuated by the collision process itself. Following the collision cascade the target atoms at the surface or in the near-surface region are displaced from their initial positions, and cascade mixing may occur.[14]

11.2.3. The Ionic Yield and the Ionization Coefficient

In the sputtered fragments of the target, only the positive and negative ions are analyzed in the SIMS experiment so that it is necessary to distinguish the

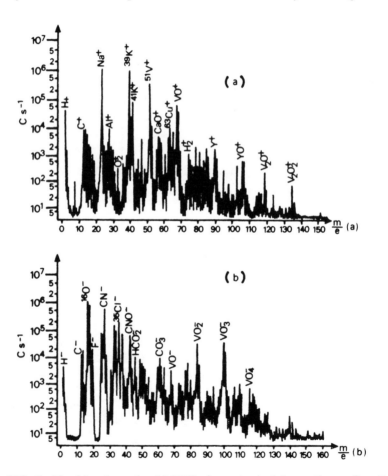

Figure 11.7. Positive (a) and negative (b) SIMS of a contaminated vanadium surface. Bulk concentration of yttrium is $< 5 \times 10^{-4}$. Primary current $I_p = 5 \times 10^{-8}$ A, analyzed sample A = 0.1 cm^2.[6]

TABLE 11.3. Yields of Oxide-Specific Secondary Ions $MeO_n^{\pm}$ for Al, Ti, Cu, and Ge[44]

Metal	Me^+	MeO^+	MeO^-	MeO^{2-}
Al	0.7	6×10^{-4}	2×10^{-2}	2×10^{-2}
Ti	0.4	0.5	—	8×10^{-3}
Cu	7×10^{-3}	—	1×10^{-3}	1.5×10^{-2}
Ge	2×10^{-2}	1×10^{-3}	8×10^{-4}	4.5×10^{-2}

sputtering rate S (total number of emitted particles relative to one primary incident ion) from the positive and negative yields S^+ and S^-. The ionization coefficients or degrees of ionization γ^+ and γ^- are given by[10]

$$\gamma^+ = S^+/S \quad \text{and} \quad \gamma^- = S^-/S \tag{11.2}$$

Ion yield variations may reach four orders of magnitude when comparing various elements, mainly in the metallic state. Such large variations make the SIMS spectra complex.

In Figure 11.7, the complementary aspects of positive and negative SIMS are seen clearly: γ^+ is very high for electropositive elements and, conversely, γ^- is very high for electronegative elements. Therefore, both on practical surfaces and on very clean surfaces (single crystals), ions such as K^+, Na^+, Ca^{2+}, C^-, $Cl^-, \ldots$ are easily detected, often with a high intensity that contrasts with their usually minute concentration. For these reasons, the minimum detectable concentration can be as low as 1 ppb.[12]

It should be noted especially that H^+ and H^- can also be detected (Figure 11.7), which is of particular importance in many systems including catalysts. For this reason as well as its very sensitive character, SIMS is superior to other surface techniques such as XPS or AES in many applications. For some typical elements in oxides which have potential applications in catalysis, negative spectra can be valuable; however, it is the positive spectra that are generally studied. This is the case, for example, for germanium and copper compared to aluminum or titanium (see Table 11.3).

Under the following conditions (3-keV Ar^+, at an incident angle of 70°) the absolute yields S^+ (only Me^+ is considered) were measured on clean metals and on their respective oxides for comparison (Table 11.4).[45] For the calculation of γ^+, S is assumed to be the same for the clean metal and for the oxide, which is probably not true in all cases. The value chosen is about 2, a typical value given in Table 11.1.

The results in Table 11.4 clearly indicate that S^+ of metals is always considerably smaller than S ($\gamma^+ \ll 1$). However, the ion yield increases from clean metals to metallic oxides whatever the nature of the metal. The last column indicates the positive ion yield ratio between the element and the corresponding oxide; the enhancement factor in the positive ion yield can be as

TABLE 11.4. Absolute Yield S^+ and Ionization Coefficient γ^+ of Clean Metallic Surfaces and of the Corresponding Oxides[7,45]

Element	S^+(clean)	γ^+	S^+(oxide)	γ^+	S^+(oxide)/S^+(clean)
Mg	8.5×10^{-3}	4×10^{-3}	1.6×10^{-1}	8×10^{-1}	20
Al	2×10^{-2}	1×10^{-2}	2	1	100
V	1.3×10^{-3}	7×10^{-4}	1.2	6×10^{-1}	10^3
Cr	5×10^{-3}	3×10^{-3}	1.2	6×10^{-1}	200
Fe	1×10^{-3}	5×10^{-4}	3.8×10^{-1}	2×10^{-1}	380
Ni	3×10^{-4}	2×10^{-4}	2×10^{-2}	1×10^{-1}	70
Cu	1.3×10^{-4}	7×10^{-5}	4.5×10^{-3}	2×10^{-3}	30
Sr	2×10^{-4}	1×10^{-5}	1.3×10^{-1}	7×10^{-2}	700

high as 10^3 for vanadium. This rise in positive ion yield S^+ does not occur only when pure metals are compared to their corresponding pure oxides, but is also observed when an oxygen chemisorbed layer or a molecular physisorbed layer is present on the metallic substrate. Three examples will illustrate this point.

The ionic currents corresponding to the SIMS signals of Cr^+, Mn^+, Fe^+, and Ni^+ are shown in Figure 11.8 as a function of the oxygen pressures to which the corresponding metals were exposed. The curves have been normalized to 1 when the oxygen pressure is 10^{-4} Pa.[13] It is assumed that a surface oxide layer is formed progressively on those metals so that the ionic yield S^+ increases. The γ^+ and S^+ values become stable and high when the surface is saturated with oxygen. It is also claimed that upon exposing the clean metal surface to oxygen,

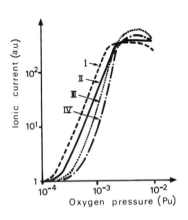

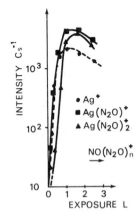

Figure 11.8. Cr, Mn, Fe, and Ni SIMS signal evolutions as a function of oxygen pressure in the chamber.[13]

Figure 11.9. Intensity variation of Ag^+, $Ag(N_2O)^+$, and $Ag(N_2O)_2^+$ SIMS signals as a function of NO exposure on an Ag(111) surface at 25 K (unpublished results).

the electronic work function of the solid increases so that the positive ion emission is favored.

For some systems, it is not even necessary to create chemical bonds between the adsorbed molecule and the substrate to observe enhanced signals from the metallic target. When N_2O is condensed at low temperature (30 K) on an Ag(111) surface, no chemical modifications are detected by XPS.[46] One can only observe the transition between the first layer and the multilayer buildup. Nevertheless, a considerable increase of the Ag^+ signal (and also of the $Ag(N_2O)_2^+$ and $Ag(N_2O)^+$ signals) occurs during formation of the N_2O mono-layer (Figure 11.9). Presence of physisorbed N_2O then considerably enhances the silver ionization yield.

Similar to the modification of the S^+ yield during surface oxidation of metals, changes in the S^- yield evolution have also been observed when molybdenum is progressively exposed to 10^{-7} Torr of oxygen. In contrast, increased yields of negative ions (S^-) are possible by using cesium ions as a primary source of ions[48] or by introducing cesium into the target.[49] Higher yields of Si^-, Au^-, and Al^- are obtained by this technique.

It can be concluded, based on the observed evolutions of S^+, that where O_2 is introduced to react with the surface of a metal, a quantitative analysis is possible as, when surfaces are oxygen-saturated, the γ^+ coefficients are stable and high for many metals.[13] However, physisorbed or chemisorbed states may also considerably modify S^+ of a given target so that, in many cases, the quantitative information is limited by the presence of small levels of (uncontrolled) surface contamination. For example, water, an easily adsorbed polar molecule may greatly affect the ionic yield, even if its surface concentration is very weak (background adsorption in a UHV chamber).

11.2.4. Experimental Basis of SIMS

11.2.4.1. Quantitative Aspects

One of the objectives of a SIMS experiment is to determine the concentration of an element A in a target by measuring the secondary ion current $I_A^\pm$ (which has to be corrected for the isotopic abundance of the element of mass A). In principle, it can be written[12]:

$$I_A^\pm = \eta_A^\pm S_A^\pm I_p C_A \qquad (11.3)$$

or

$$I_A^\pm = \eta_A^\pm S_A \gamma_A^\pm I_p C_A \qquad (\text{as } \gamma_A^\pm = S_A^\pm / S_A)$$

where I_p is the primary ion current; $\eta_A^\pm$ is an instrumental parameter correlated to the transmission factor of the mass analyzer (assumed to be constant for given experimental conditions); $S_A^\pm$ is the number of secondary ions (either positive or negative) per incident primary ion; and C_A is the atomic concentration in the matrix of the element A.

When a uniform and circular ion beam is used, we have

$$I_p = 0.25\pi D_p d_p^2 \qquad (11.4)$$

(in ions s^{-1} or in A as for singly charged ions 1 mA = 6.2×10^{-15} ions s^{-1}) where d_p is the spot diameter (cm) and D_p is the ionic current density (ions $cm^{-2} s^{-1}$).

Typical values of those parameters are the following[10]: $\gamma_A^\pm$: 10^{-5} to 10^{-1}; S_A : 1 to 10; $\eta_A^\pm$: 10^{-5} to 10^{-2}; D_p : 10^{-6} to 10^{-2} mA cm^{-2}, and d_p : 10^{-4} to 10^{-1} cm. The ion current density D_p is most important because the erosion rate Z (as defined by the layer thickness sputtered per unit time) is indeed proportional to D_p. According to Werner [12], there is a relationship between Z and D_p:

$$Z[\mu m\ h^{-1}] = 3.6 \times 10^{-4}\{M[amu]/\rho[g\ cm^{-3}]\}\{D_p[\mu A\ cm^{-2}]\}S \qquad (11.5)$$

where ρ is the density of the target and M is the atomic mass of an atom in the target (the proportionality coefficient is 0.06 if Z is in Å min^{-1}).

In order to estimate the quantitative analytical aspects of SIMS, it is important to point out the relationships among the different parameters noted in equations (11.3)–(11.5). The detection limit can be as low as 10^{-14} g (10^{-6} atomic layer) if the target materials are consumed by sputtering. For example, 10 ppm of aluminum (ionic yield $\simeq 10^{-3}$) can be detected with a precision of 3% with a primary Ar^+ impact of 100 μm (dp). It is necessary, in this case, to remove a layer 130 Å thick. If the chosen dp value is 2 mm, only 0.33 Å of material needs to be sputtered.[50] The lowest detectable concentrations C_{min} (in atomic ppm or ppb) for some elements relevant to catalytic materials under given experimental conditions[12] are shown in Table 11.5. To make quantitative measurements, it is necessary to determine the ionization coefficient $\gamma_A^\pm$ of the element A relative to a reference element in the *same matrix* with, if possible, the same sputtering rate. From equation (11.3), it can be deduced that

$$I_A^\pm C_A^{-1}/I_{ref}^\pm C_{ref}^{-1} = \gamma_A^\pm / \gamma_{ref}^\pm = \delta_A \qquad (11.6)$$

According to McHugh,[10] concentration measurements are accurate to about 10%. It must be noted in equation (11.3) that the secondary ion current $I_A^\pm$ depends on C_A, the parameter to be determined for a quantitative analysis, and also on $S_A^\pm$ (or $\gamma_A^\pm$), which is dependent on the chemical nature of the examined materials. S^+ variations were described in Section 11.2.3 as a function of various parameters (e.g., metal $vs.$ oxide, P_{O_2}, condensed layer, and the presence of electropositive or electronegative surface impurities). As the influencing parameters are not always controllable, it must be clear that quantitative SIMS is not always possible on practical surfaces, such as those of catalysts.

TABLE 11.5. Calculated C_{min} for Cu, Ni and Al (Pure Metals or Oxides) for Two Values of $Ip^{(12)}$ [a]

		C_{min}	
	S^+	ppma	ppba
Elements			
Cu	1.3×10^{-4}	150	15
Ni	3×10^{-3}	7	0.7
Al	2×10^{-2}	1	0.1
Oxides			
Cu	4.5×10^{-3}	5	500
Ni	2×10^{-2}	1	100
Al	2	0.01	1
$I_p[A] \rightarrow$		10^{-8}	10^{-7}

[a] It is assumed that $\eta^+ = 10^{-3}$ and a minimum detectable ion current at the ion collector $I_{min} = 2 \times 10^{-8}$ A. Values of S^+ have been taken as in Table 11.4.

TABLE 11.6. Typical Values for Primary Ion Current I_p, Beam Diameter d_p, Primary Ion Current Density D_p, and Erosion Rate Z (using $S = 3$)[12]

Mode	I_p (A)	d_p	D_p	Z ($S = 3$)
Static	10^{-11}	1 mm	1 nA/cm² ($= 0.01$ nA/mm²)	1 Å/h
Dynamic	3×10^{-8}	1 mm	3 μA/cm² ($= 30$ nA/mm²)	300 Å/h
	10^{-11}	1 μm	1 mA/cm² ($= 10$ μA/mm²)	10 μm/h
	10^{-5}	1 mm	1 mA/cm² ($= 10$ μA/mm²)	10 μm/h

11.2.4.2. Perturbations in the Examined Solid

Depending on the choice of the experimental parameters, the effective ion current density and thus the erosion rate Z [equation (11.5)] can be varied widely. When Z is small (of the order of 1 atomic layer) during the time for recording a SIMS spectrum, the *static* mode is used. When erosion rates are high, the experiment is relevant to *dynamic* SIMS. The two modes are delineated in Table 11.6.[12]

When a high ion current density D_p is used (dynamic SIMS), the bombarded materials are changing constantly as part of the solid is sputtered and simultaneously there are large modifications in the penetration depth (e.g., ion implantation, crystal lattice destruction, chemical reactions, and composition evolution). Therefore dynamic SIMS is not really appropriate for characterizing the surface layers of a solid, but rather is more relevant for observing modifications in the material over significant depths (determination of concentration profiles in, e.g., alloys and semiconductors).

11.2.4.3. Experimental Equipment

In some of the review articles cited previously[12,15,19] detailed information about the components of SIMS equipment is given. Schematically, the spectrometer is composed of an ion gun and a mass analyzer mounted in a vacuum chamber. Generally for surface analysis, the SIMS components are present, in the same vacuum system, with other surface-sensitive techniques such as XPS, AES, UPS, and LEED, so that *all the available information can be cross-checked. Often, this is the only way to interpret results correctly.* An independent system is often required for pumping off the relatively high pressure of rare gases (usually 10^{-5} Torr of argon) in the ion gun. If insulators are examined, it is necessary to neutralize the surface charge, often with low-energy electrons from an electron flood gun. It is also possible to obtain an ionic picture of the secondary ion by scanning the primary ion beam over the examined surface. The similar ionic microprobe developed by Castaing and Slodzian[51] (which has well-documented applications in metallurgy, mineralogy, and biology) will not be discussed here.

11.2.5. Static SIMS

11.2.5.1. General Characteristics

For surface analysis by SIMS in catalysis the static mode is used most often. As shown first by Benninghoven,[6] small doses of ions (typically 10^{-9} A cm^{-1} or less for about 1,000 s, see Table 11.6) on large surfaces ($\simeq 1$ mm^2) give low erosion rates and only a very small portion of the uppermost layer is removed during analysis. Moreover, the mean distance between two ionic impacts is high (~ 100 Å), consequently each incident ion is in collision on a nonperturbed surface. Under these conditions, to avoid surface contamination, the pressure in the chamber should be very low (less than 10^{-10} Torr) as is the rule with other surface-sensitive techniques (XPS, AES, and ISS).

Thus within static SIMS conditions, studies of surface reactions and analysis of catalytic materials is possible. Dynamic SIMS is used for catalysts, only if concentration profiles along different particles are needed; the X-ray microprobe technique can also be used for this analysis. Further, concentration profiles can also be obtained by XPS or AES analysis in conjunction with ion sputtering.

TABLE 11.7. Main Characteristics of Static SIMS[a]

Advantages	Disadvantages
The uppermost layers are analyzed	Large sensitivity differences: function of
Analysis of all the atomic elements,	the surface structure and of presence of
including hydrogen	contaminants (partly hamper quantitative
Detection of molecules	analysis)
(see molecular SIMS)	Difficulties in interpretation of molecular
Lateral resolution	fragmentation patterns
Isotope separation	Surface reactions induced by ions
High sensitivity — Quantitative analysis	Modification of signal intensity by
possible	presence of pores in the bombarded solid
Weak surface destruction	

[a] Adapted from Benninghoven.[6]

11.2.5.2. Limitations

In Table 11.7 are shown the main advantages and disadvantages of the low-damage static SIMS which allows surface characterization. These can be discussed in the light of the characterization of catalytic solids. As already pointed out, detection of $H^{\pm}$ and fragments containing H is of particular importance. In the study of molecules or salts by SIMS (molecular SIMS, see Section 11.2.5.3), the presence of $(M - H)^-$ and $(M + H)^+$ ions has been observed, which easily allows identification of the molecule of mass M. The high sensitivity of SIMS is also very attractive but, unfortunately, when the surface structure or composition is not well established or when H_2O or OH groups are present (as is commonly observed on catalytic surfaces), quantitative interpretation can be difficult. Moreover, the presence of pores in most catalytic materials of large surface area reinforces the problem of quantitative analysis. Is the surface of the pores accessible to the incident ions? If so, what are the effects of the pore walls on the survival probability of the emitted ions? On a smooth surface, there is a zone at the selvedge of the solid of thickness d, where neutralization or ionization occurs (see Figure 11.1). How are these phenomena modified in the pore structure? These questions arise in connection with porous systems if only the external surface is analyzed. In this case, SIMS analysis can only probe the less interesting surface whereas the catalytic processes occur mainly within the pore structure. The influence of the pore structure is far less important in XPS and AES studies as, with these techniques, the mean analyzed depth is some 5 to 50 Å. But in ISS similar questions may be debated as the fundamental process is an elastic collision between the incident ions and atoms of the uppermost layer of the solid.

The lateral resolution (1 to 0.5 μm) imposed by d_p [equation (11.4)] could be useful in some cases but, as with other available microprobe techniques, weak segregation effects at the nanometer level, currently encountered on catalysts, cannot be observed.

11.2.5.3. Molecular SIMS

The static SIMS working mode allows the analysis of organic molecules by ion bombardment without destroying the desorbed molecules in the process. Initially, static SIMS was applied to monomolecular layers adsorbed or condensed on metals,[52] and only several years later to the desorption and analysis of organic molecules.[53] This approach is now called molecular SIMS and is also used to investigate the molecular structure of various kinds of nonvolatile and thermally stable molecules (organic, inorganic, or biological compounds as well as polymeric materials).[54-57]

The interpretation of SIMS spectra of organic compounds can be easy when data on the dissociations of gas phase organic ions are available. SIMS spectra incorporate features due to ion–molecule reactions and to unimolecular dissociations which parallel those observed in other forms of mass spectrometry. The dual characteristics of molecular SIMS combine a new method of ionization (in contrast to classical mass spectrometry) and a surface analysis technique. It is now an important analytical technique and can be used to study working surfaces and catalysts for which one may expect to obtain information about the adsorbed molecule and the surface.

In molecular SIMS, there are three distinct processes of molecular ion formation:

1. Cationization (or anionization). A metal ion Me^+ is attached to the organic molecule M to give a $[Me + M]^+$ cation. In general, the molecule is deposited on a metallic surface in such a way as to get a monolayer-like thin film, but molecular aggregation on the surface cannot always be avoided. It is well known that acid-etched substrates of silver, copper, or nickel give the best results.
2. Electron transfer to form M^- and M^+ ions or alternately $[M - H]^-$ and $[M + H]^+$.
3. Direct sputtering of the organic cations or anions when salts are examined. In addition to these processes, dissociation of the molecular ions formed can lead to several fragments which are a fingerprint of the molecule.

The SIMS spectrum of phenanthrene deposited on silver foil is shown in Figure 11.10.[55] It is easy to recognize the $[M + Ag]^+$ ion, where M represents neutral phenanthrene. There are two peaks due to the isotopic repartition of the silver (^{107}Ag and ^{109}Ag). It is proposed that this ion is formed by an ion (Ag^+)–molecule (M) reaction in the selvedge region outside the solid (represented with a thickness d in Figure 11.1). The M^+ peak is also present on the same spectrum. The high intensity of M^+ is assumed to be due to the great stability of radical ions of polycyclic aromatic hydrocarbons.

This example indicates some very promising applications in catalysis, e.g., the analysis of heavy hydrocarbon cracking reactions or catalyst poisoning by carbon compounds (aromatic hydrocarbons, coke, amorphous carbon, or gra-

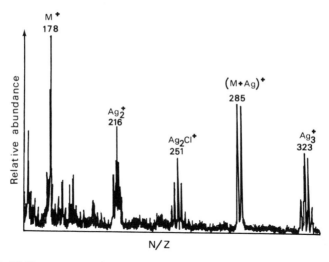

Figure 11.10. SIMS spectrum of a bulk sample of phenanthrene supported on silver foil: Ar^+ bombardment, 5×10^{-10} A, 2 mm² spot size, 4 keV energy.[55]

phite) which block the active sites. To our knowledge, no systematic investigation has yet been pursued in this direction with static SIMS.

When N_2O is progressively adsorbed (condensed) on a silver surface, ions with silver are obtained first (Ag^+, $Ag(N_2O)^+$, and $Ag(N_2O)_2^+$). When the monolayer is reached, fragments without silver are detected (Figure 11.9). This observation gives precise evidence of the transition between a 2D–3D overlayer buildup.

11.3. APPLICATIONS

11.3.1. Metals and Alloys

11.3.1.1. Short-Range Order

Many models have been proposed to explain the secondary ion emission of metallic compounds and several processes can occur simultaneously. It seems that a strong relationship exists between the ionization probability and the electronic structure of the metal or the alloy. As the electronic density at the Fermi level depends on the alloy composition, matrix effects have been detected in SIMS experiments on diluted alloys.[11] SIMS can be then considered as a technique giving information about the electronic properties of metallic systems. However, as described above, the phenomenon of preferential sputtering may hamper this interpretation and cause many problems when a careful and precise surface composition is required. This is particularly true in catalysis by bimetallic materials. The composition of their upper-

TABLE 11.8. Ionic Ratios Ni_2^+/Ni^+ and Ni_3^+/Ni^+
Obtained on Different Exposed Planes of Ni:
Comparison with Theoretical Calculations[58]

	(110)	(100)	(111)
Probability of producing Ni_2^+	2	4	6
Ni_2^+/Ni^+ (experimental)	1	1.8	3.2
Probability of producing Ni_3^+	4	24	48
Ni_3^+/Ni^+ (experimental)	1	6	38

most layer is required because the molecules interact with this layer during their transformation.

An interesting point to note when examining well-defined surfaces of single crystals is the possibility of obtaining information on the local environment or local bond order of the emitted particles. Barber et al.[58] analyzed the ionic fragment patterns of three nickel surfaces (fcc structure) and combined them with the probabilities of producing Ni_2^+ or Ni_3^+ clusters (only the nearest neighbors on the first exposed atomic layer were taken into account) (Table 11.8).

The ionic fragments (Ni^+, Ni_2^+, or Ni_3^+) come mainly from the first exposed layer. The agreement between calculated probabilities and experiment is quite acceptable (except perhaps on the dense (111) face for Ni_3^+ clusters). If the second layer is taken into account, there are large discrepancies in the results. Moreover, the calculations are correct only if the exposed surface is not perturbed. Induced damages by the ion bombardment modified the experimental ratios which then became similar on the three nickel planes at the end of the experiment. In addition, the presence of impurities or adsorbed oxygen (or OH) increased the nickel yield by several orders of magnitude,[59] which may modify the ionic fragment ratios.

Thus, under favorable conditions, SIMS can give information on local bonds of metal and alloy surfaces. However, there is still much debate on the conclusions deduced from SIMS experiments, e.g., whether the fragments come directly from the surface (which would lead to a local order description) or whether the observed fragments are the result of reactions in the selvedge region of the solid (Figure 11.1). This problem must be always kept in mind in the following examples especially in the sections dealing with metal-O_2 interactions and CO adsorption.

11.3.1.2. Analysis of Surface Contaminants

One of the first examples of an application of SIMS to metallic catalysts came from a study by Benninghoven[5,47] in which he showed the evolution of surface contaminants in a silver catalyst after various treatments. Etching eliminated several surface species such as SO_4^-, SO_3^-, SO_2^-, nitrate, cyanide, and organic fatty acids; only Cl^- remained after 20 layers had been removed.

Heating led to diffusion from the bulk to the surface and reaction with the surrounding gas phase.

11.3.1.3. Surface Composition of Bimetallic Catalysts

11.3.1.3a. Ag/W(110). This example nicely illustrates the ability of SIMS to follow the progressive coverage of an exposed surface [the (110) plane of W] by deposition of a second metal (Ag). The Ag^+ or W^+ intensity was plotted as a function of the silver coverage (Figure 11.11). Simultaneously AES measurements were performed.[17,60] Clearly the W^+ SIMS intensity decreases and vanishes at one silver monolayer whereas Ag^+ increases and reaches a plateau for that coverage. In the AES plot, only a slope change is noted. These results in an ideal case indicate the monolayer sensitivity of SIMS and emphasize the superiority of SIMS over AES in this type of analysis.

11.3.1.3b. Cu or Au/Ru(0001). These bimetallic systems have been characterized by a combination of techniques among which we will mention only some static SIMS results. The objective of this work[61] was a better understanding of the influence of copper or gold deposits on the surface reactivity of Ru(0001). This system was chosen because Group I metals have significant influences over Group VIII metals in reforming reactions and also because copper and gold do not alloy with ruthenium. The samples were prepared by evaporation onto a clean Ru(0001) surface, maintained at 540 or 1100 K, of increasing quantities of copper or gold. The Cu–Ru system has been studied extensively by SIMS and a $RuCu^+$ cluster has been detected.

11.3.1.3c. Fe–Ru Fischer-Tropsch Catalysts. These unsupported bimetallic powders of various atomic compositions were observed simultaneously by XPS and SIMS after reduction by hydrogen at typically 400 °C and exposed to air a few minutes before being introduced into the spectrometer.[62,63] This is enough to detect oxygen by XPS and oxygen-containing clusters in positive SIMS (i.e., FeO^+, $FeOH^+$, RuO^+ and $RuOH^+$). However, the most important feature is the

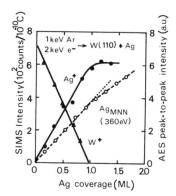

Figure 11.11. Comparison of SIMS and AES analysis of the growth of silver on tungsten in the monolayer range.[17,60]

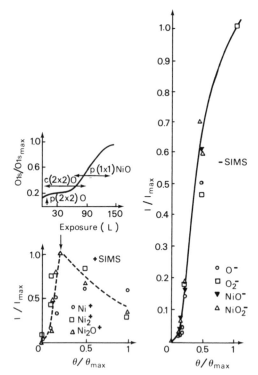

Figure 11.12. Variation of positive and negative SIMS ion intensities as a function of coverage, relative to saturation, for Ni(100) at 300 K. The relationship between relative coverage, as determined from XPS O(1s) intensities, exposure, and the regions of chemisorption and oxidation, are shown in the upper-left-hand part of the figure.[59,64]

presence of the $FeRu^+$ ion (a similarity with the $CuRu^+$ cluster detected in the Cu–Ru system). This suggests the presence of Fe–Ru bonds in the surface layer. A physical mixture of iron and ruthenium powders does not produce such a mixed cluster; Ru^+, Fe^+, Ru_2^+, Fe_2^+ Fe_2H^+, $Fe_2H_2^+$, and $RuNaO^+$ are other positive ions detected in the typical spectrum. The FeO^+/Fe^+ intensity ratio is roughly equal to the RuO^+/Ru^+ ratio whereas XPS shows oxidation of iron and the presence of metallic ruthenium. The system is described by formation of oxidized iron islands at the ruthenium surface, which is still detectable and influenced by an interaction with oxygen.

11.3.1.4. Metal–Oxygen Interaction

There is, as noted above, much controversy about interpretation of the SIMS results obtained by adsorption and reactions at metal surfaces. The reasons for the controversy lie in the high sensitivity of the method: the secondary ion yield variations depend very much on small modifications of the chemical nature of the surface species. The very small fraction of any species that escapes as ions depends strongly on the charge distribution in the bonds being broken and on the work function at the surface.

The systems O/W and O/Ni, two systems relevant to catalysts, have been

TABLE 11.9. Relationship between M_nCo^+ ($n = 1, 2, 3$) Cluster Populations and CO Bonding States[65]

Surface	SSIMS			CO bonding IR/HREELS/LEED assignment
	MCO^+	M_2CO^+	M_3CO^+	
Cu(100)	0.9	0.1	—	Linear only
Ru(001)	0.9	0.1	—	Linear only
Ni(100)	0.8	0.2	—	Linear + bridged
Ni(111)	0.6	0.4	—	Linear + bridged
Pd(100)	0.3	0.6	0.1	Bridged
Pd(111)	0.3	0.4	0.3	Triply-bridged

investigated by surface techniques including SIMS. In Figure 11.12 the variations of the positive and negative SIMS ion intensities are shown as functions of relative coverage of a Ni(100) surface exposed to oxygen at 300 K.[64] The downward arrows indicate the known coverage at which NiO nucleation begins. The positive ion current variation is sensitive mainly to the chemisorbed oxygen coverage but not to the structure of the oxygen overlayer. On the other hand, the main contribution to the intensity of negative ions is from oxide growth. The same general behavior is also found for Ni(110) and Ni(111) surfaces.

In the case of the W(100)/O_2 system, the WO^{3-}, WO^+ and WO^{2+} ions have a similar role to the negative ions detected in the O/Ni system, i.e., they are indicative of oxide nucleation. This study also shows the usefulness of combining SIMS measurements with XPS, LEED (when single crystals are concerned), TPD, and other techniques when available to give a full view of the chemical and structural processes with reactive molecules at the surfaces.

11.3.2. Adsorption Studies

11.3.2.1. CO Adsorption

Static SIMS studies of CO adsorption on different metals of importance in catalysis generally give two types of metallic clusters: the M_n^+ clusters ($n = 1, 2, 3$ when M is nickel, for example) and the M_nCO^+ clusters associating one CO molecule with M_n^+. Absolute and relative intensities of these positive ions vary considerably with the CO coverage. Several authors have tried to correlate the relative populations of the respective MCO^+, M_2CO^+ and M_3CO^+ clusters with identified bonding states of CO at the metal surface, namely as linear, bridged, or triple-bonded sites. Characterization of the bonding modes is available from vibrational spectroscopy. Table 11.9 shows the static SIMS patterns associated with a unique or combined structure of the adsorbed molecule.[65] From this, it was proposed that the linear CO bonding state

M—CO will give a ratio MCO^+/M_2CO^+ close to 9, while a bridge-bonded site will give a ratio MCO^+/M_2CO^+ of about 0.5.

However (see Section 11.2.5.3 on molecular SIMS), it has been argued that the molecular clusters do not leave the surface intact but complex arrangement, dissociation, neutralization, or ionization processes intervene in the selvedge region of the solid after the sputtering step. Consequently, it is not easy to interpret the intensity results from CO (or other molecules) adsorption.

Brundle et al.[64] have argued against some of the conclusions concerning the CO/Ni(100) system, and have insisted that the $Ni_2CO^+/NiCO^+$ ratio appears to be more a function of coverage with no correlation with the linear or bridge-bonded CO adsorption states. At 0.5 monolayer, the $C(2 \times 2)$ overlayer, with only linear-bonded CO gives a ratio $NiCO^+/Ni_2CO^+$ of 10.0 but at 0.68 monolayer, a range of structures between linear and bridged is present and the $NiCO^+/Ni_2CO^+$ ratio is 11.0. Thus it is sometimes difficult to correlate the changes in this ratio with geometric effects because other parameters (e.g., coverage, heat of adsorption, and work function) vary simultaneously.

On the Ru(0001) surface, as the $MCO^+/\Sigma\, MnCO^+$ ratio is constant (0.9) whatever the CO coverage, it is assumed that a linear-bonded CO site structure is present at all coverages.[61] When the surface is precovered by gold ($\theta_{Au} = 0.34$), there is no ratio change. Electron energy loss-spectroscopy confirms that gold has no effect on CO adsorption on ruthenium. With copper deposited on ruthenium, the relative MCO^+ yield is lower at low CO coverage, which implies that a significant proportion of CO has been moved into a bridge-bonding coordination. Concurrently, the presence of a $RuCuCO^+$ ion would indicate the presence of a "mixed"-site bonding.

11.3.2.2. The Adsorption of Other Molecules

To a lesser extent, the adsorption of other molecules has been followed by SIMS. When C_2H_4 is adsorbed at 130 K on Ru(0001) and Cu/Ru(0001) surfaces,[61] the $Ru_2C_2H_4^+$ species present on pure ruthenium is replaced by $CuC_2H_4^+$ ions when copper is added at low coverage. A new $RuCuC_2H_4^+$ peak is also present. At higher coverage the $RuCuC_2H_4^+$ and $Ru_2C_2H_4^+$ ions are absent and only $CuC_2H_4^+$ and $RuC_2H_4^+$ ions remain. These results with annealing studies and other techniques give some insight into the electronic promoter effect of such a system, which has intermediate properties between ruthenuim and copper (or gold). Static SIMS experiments have been performed when hydrogen interacts (dissociative adsorption) with the Ni(100) surface at equilibrium conditions.[66]

SIMS can also help in the understanding of surface kinetics phenomena. In this case it is necessary to verify that the primary ion beam is neither sputtering the surface at a great rate not inducing chemical reactions. The following example relates the interaction of H_2, O_2, and H_2O on a Pt(111) surface. Figure 11.13 shows the H_2O formation profile as a function of substrate temperature when hydrogen from the gas phase is interacting with the oxygen that was adsorbed initially; after an induction period, the fast reaction can be followed easily, thus allowing the determination of kinetic parameters.[67] Combined with other experiments using the ^{18}O isotope and with TPD results, a detailed mechanism of H_2O formation and OH reactions has been proposed. In particular, an oxygen-covered platinum surface is rearranged upon exposure to H_2O to form an intermediate state that is composed of rows of OH groups with H_2O between them. Reactions with alkenes and methoxy formation on Pt(111) are given in the same publication.[67]

11.3.3. Oxide Catalysts

11.3.3.1. Ion Emission in Oxides

The ion yield in oxide or in oxygen-covered metallic surfaces is considerably higher than in pure metals (Table 11.4, Figure 11.8). Therefore much work has been devoted to the interpretation of ion emission in oxidized metals, e.g., the review article of Wittmark.[17]

A situation similar to largely ionic MX halides can be found with SiO_2, where $(SiO_2)_i Si^+$ and $(SiO_2)_i O^-$ ($i = 0, 1, 2, \dots$) have been detected. However, the situation is more complex as $Si_m O_n^{\pm}$ species are also observed. Thus, in general, interpretation of oxide SIMS spectra is much more difficult. Nevertheless a simple relationship plotting the intensity ratio $I(MO^+)/I(M^+)$ as a function of the bond energy dissociation $D(M^+ - O)$ exhibits a rather good correlation.[17]

11.3.3.2. The Selective V—P—O Catalyst

The vanadium–phosphorus mixed oxide is a unique system that is highly efficient in the selective oxidation of butane to maleic anhydride $C_4H_2O_3$. This complex reaction, which requires incorporation of three oxygen atoms in the carbon frame occurs on the vanadyl pyrophosphate active phase. Numerous catalytic and physical characterizations have been developed to give a better understanding of this system. Although the bulk compositions of the $(VO)_2P_2O_7$

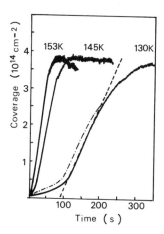

Figure 11.13. Variation of the H_3O^+ SIMS yield with time as a function of substrate temperature. The initial oxygen-coverage was 0.25 monolayer, the hydrogen pressure was $1 \times 1 \ 10^{-7}$ Torr, and the Ar^+ beam current was 0.5 nA. The H_3O^+ signal was normalized to that originating from the 0.25 monolayer of H_2O. The slope of the dashed line depicts the maximum rate of H_2O formation at 130 K. The dot–dash curve represents a correction to the 130 K profile arising from variations of the H_3O^+ sensitivity with H_2O coverage.[67]

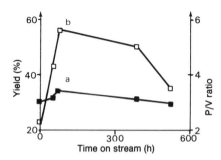

Figure 11.14. Yield of maleic anhydride (a) and P/V surface ratio (b) *vs.* time on stream.[68]

active phase is well defined it is known that a small excess of phosphorus (P/V bulk ratio between 1.1 and 1.2) is necessary to achieve maximum selectivity in butane transformation. SIMS investigations can find the uppermost P/V composition of the catalyst before or after use, and for several types of preparations or, combined with sputtering, a profile from the surface to the interior of the solid can be established. The SIMS results of Hass *et al.*[68] indicate that the selectivity for maleic anhydride is essentially determined by an excess of phosphorus relative to vanadium near the surface and by the P/V gradient profile. They found that on a typical, good catalyst the P/V ratio can reach 3/1 at the surface whereas the bulk ratio value extends to about 2 nm below the surface.

No details were given by Hass *et al.*[68] on the calibration procedure (the real P/V values are therefore subject to discussion) but clearly there is a phosphorus surface enrichment on their catalysts. The P/V surface ratio and the yield in maleic anhydride are shown as functions of the time on stream (Figure 11.14). Clearly the selectivity is strongly related to changes in surface composition. However, the yield changes of maleic anhydride are small while the reported P/V ratio evolution is rather important. Catalytically, it is not possible to explain these small variations in selectivity with the high surface concentration modifications. The surface of the analyzed catalyst was probably perturbed during the SIMS experiment unless the SIMS intensities are incorrectly calibrated.

11.3.3.3. Other Oxide Catalysts

Pomonis and Vickerman[69] have used SIMS to study vanadium-oxide-based catalysts, of the general formula $Sn(1 - x)V_xO_2$, for methanol oxidation. The intensity ratio V^+/Sn^+ as a function of x was proportional to the bulk vanadium content for x below 0.03 or above 0.03 but in a small domain close to $x = 0.03$ there was an apparent fall in the relative sensitivity to vanadium. This might have one of two possible causes: a surface depletion of vanadium or a modification in its electronic state at the surface for this given composition which reduces the vanadium secondary ion emission. This second hypothesis was assumed to be the most reasonable. The relationship between dissociative methanol adsorption and oxygen mobility in these oxides was also discussed in this paper.

In a review paper, Brown and Vickerman[65] give other examples of catalyst characterization by SIMS. The spectra of the FeO_x^- and SbO_x^- ions emitted by the Fe_2O_3–Sb_2O_4 system exposed to $^{18}O_2$ reveals that ^{18}O was incorporated first into SbO_x^-, which shows that antimony is the site of the initial oxidation. Later ^{18}O is also present in FeO_x^- ions, which implies that oxygen is mobile on the surface of these mixed oxides. In the same review, examples using fast atom bombardment mass spectrometry (FABMS) are given. This technique, derived from SIMS, in which neutral particles are used instead of primary ions to eject material fragments, will not be discussed here. Surface segregation of silicon or aluminum in zeolites and the Mo oxide coverage in Ni–Mo–Al_2O_3 catalysts have been studied by FABMS.

To conclude this section, we note a study on adsorption of water, methanol, acetone, and benzene molecules on TiO_2, a solid in wide use in catalysis as a support.[70] At low coverages, it is assumed that specific interactions with the support determine the nature of the emitted ions. Acidic protons are abstracted to produce OH^-, $TiOH^+$, $TiOCH_3^+$ ions, whereas covalent hydrogen bonds are less affected. In view of the discussion above on SIMS interpretations, this conclusion seems ambitious. Further work is necessary to assess the possibility, by using SIMS, of observing acid–base interactions at catalyst surfaces.

The influence of traces of fluorine in the hydroxylation of $TiO_2(100)$ (rutile structure)[71] is also illustrative of the potential of SIMS in the study of OH groups present on catalyst supports.

11.3.4. Supported Catalysts

11.3.4.1. Hydrotreating Catalysts

These catalysts, used to remove heteroatoms (sulfur, nitrogen or vanadyl or nickel present as petroporphyrins) of petroleum fractions, are generally composed of cobalt or nickel (the promoter) associated with molybdenum on a support. The examples given here deal with the problems of cobalt migration into γ-Al_2O_3,[72] with the changes in molybdenum dispersion following calcination and water treatments[73] and with the state of the vanadium or nickel present in used cracking catalysts.[74,75]

Two catalysts, with 4 and 30 wt.% of cobalt supported on γ-Al_2O_3, are compared with Co_3O_4, $CoAl_2O_4$, and γ-Al_2O_3. XPS, ISS, and SIMS are used as complementary techniques.[72] On the 4% cobalt catalyst (calcined at 600 °C), Al_3^+ and $CoAl^+$ ions were detected. On the 30% cobalt catalyst (calcined at 400 °C), the previous peaks were absent but the CoO_2^+ ion was present, also detected on bulk Co_3O_4. The emerging picture is that the 4% catalyst is almost identical to $CoAl_2O_4$, whereas the 30% Co catalyst resembles Co_3O_4 with a few strong peaks of Al_2O_3. SIMS helped to find different cobalt environments in these supported catalysts.

On the silica-supported and alumina-supported MoO_3 catalysts a large number of positive or negative secondary ions can be detected whatever the molybdenum loading.[73] Calcination–hydration treatments do not induce changes in the nature of the ions observed by SIMS, but the relative intensities are sensitive to these treatments. A reversible variation of the MoO^+/Mo^+ and MoO_2^+/Mo^+ ratios was seen for the Mo–SiO_2 catalysts as a function of calcination. H_2O treatment indicated that static SIMS is sensitive to the variations in the structure of surface molybdates. On the Mo–Al_2O_3 system, the structure of supported molybdenum is stable in contrast to the Mo–SiO_2 system. Imaging SIMS also locates vanadium and nickeld deposited on catalysts by an artificial impregnation procedure[74] or after practical use in a fluid cracking catalyst unit.[75]

11.3.4.2. Other Supported Systems

In the area of supported catalysts, SIMS has been used to characterize the state of the supported species. The structure of rhenium-oxide aluminum metathesis catalysts resembles Re(VII) oxide.[76] The vanadium-oxide promoted Ru/Al_2O_3 catalyst is used in CO hydrogenation to produce liquid fuels. In the Ru/Al_2O_3 system Ru^+ and Al^+ were detected in a SIMS experiment.[77] With RuV/Al_2O_3, Ru^+, RuO^+, VO^+, and V^+ are detected whereas no RuO^+ is detected when a mechanical mixture of Ru/Al_2O_3 with vanadium oxide is examined. The promoter effect is thought to be due to intimate contact between vanadium oxide and ruthenium, which permits the formation of RuO^+ ions in SIMS experiments.

11.4. CONCLUSIONS

Through the numerous examples cited in this review, it appears that SIMS is a technique of high potential for surface analysis which may be applied to characterizations of a great variety of solid materials including heterogeneous catalysts. As with other surface science techniques, the positive aspects are sometimes counterbalanced by some disadvantages. We note again the main features we have encountered:

1. SIMS is a destructive technique and only ionized particles at a relatively low level of concentration are detected. In static SIMS, the destructive character is considerably attenuated compared to dynamic SIMS.
2. The spectra are generally complex, which often limits interpretations, but in molecular SIMS, identification of molecules or of ionic compounds is sometimes easy.
3. The mechanism of SIMS (including excitation, fragment ejection, ionization, neutralization, recombination phenomena outside the solid, influence of the solid work function) is far from fully elucidated, which makes the quantification of SIMS difficult.
4. The ejected ions do not come directly from the initial impact with the primary ion beam but generally from a depth of two of three atomic

layers. In static SIMS, the information depth is limited to the upper-most layer. This can be considered as a useful quality.

5. Analyzing insulators is sometimes a problem as in Auger electron spectroscopy. Neutralization by low-energy electrons is frequently used to reduce these charge effects.

6. The analyzed surface is not well defined in the case of porous materials. SIMS is the most sensitive technique as less than 1 ppm of monolayer can be detected.

7. All the elements are detected. This gives SIMS an advantage over other surface-sensitive techniques (AES-XPS and ISS).

8. Isotope differentiation opens the possibility of examining complex mechanisms (e.g., oxygen mobility in oxides).

9. SIMS is sometimes described as a technique which offers (controversial) information on local bond order and on sites and modes of adsorption.

10. Imaging solid surface by scanning SIMS can be achieved with a fairly good lateral resolution (0.5 µm) so that SIMS can be used as a local microprobe. Thus complementary information from HREM is necessary.

Despite its high potential in adsorption studies for surface composition and local order determination, SIMS has not yet been used to a large extent for catalyst characterization. As a general comment, complementary analyses with other techniques (e.g., XPS, AES, TDS) are needed to confirm the SIMS interpretation.

REFERENCES

1. W. R. Grove, *Phil, Mag,* **5**, 203 (1853); *Phil, Trans. Soc.* **142**, 87 (1852).
2. F. L. Arnot and J. C. Milligan, *Proc. Roy. Soc. Ser.* A **156**, 538 (1936).
3. R. F. K. Herzog and F. P. Viehbock, *Phys. Rev.* **76**, 855 (1949).
4. H. W. Werner and H. A. M. De Grefte, *Surf. Sci.* **35**, 438 (1973).
5. A. Benninghoven, *Surf. Sci.* **35**, 427 (1973).
6. A. Benninghoven, *Surf. Sci.* **53**, 596 (1975).
7. H. W. Werner, *Surf. Sci.* **47**, 301 (1975).
8. M. Kaminsky, *Atomic and Ionic Impact Phenomena on Metal Surfaces.* Springer-Verlag, Berlin (1965).
9. G. K. Wehner, *Methods of Surface Analysis* (A. W. Czanderna, ed.), Elsevier, Amsterdam, (1975), p. 5.
10. J. A. Mc Hugh, *Methods of Surface Analysis* (A. W. Czanderna, ed.), Elsevier, Amsterdam, (1975), p. 223.
11. G. Blaise, *Materials Characterization Using Ion Beams* (J. P. Thomas and A. Cachard, eds.), Plenum Press, London, NATO Advanced Study Institute Series, Volume **28**, (1978), p. 143.
12. H. W. Werner, *Electron and Ion Spectroscopy of Solids,* Plenum Press, New York, NATO Advanced Study Institute Series, Volume **32**, Series B Physics (1978), p. 324; *Developments in Applied Spectroscopy, 7A* (E. L. Grove and A. J. Perkins, eds.), Plenum, New York (1969), p. 239.

13. B. Blanchard, *Analyse par émission ionique secondaire SIMS,* Techniques de l'ingénieur, (1981), p. 2618.
14. J. C. Pivin and C. Roques Carmes, *Le Vide, les couches minces,* numéro spécial mars (1979), p. 221.
15. W. Katz and J. G. Newman, *Mater. Res. Bull.* **40**, Aug.-Sept. (1987).
16. D. Schuetzle, T. L. Riley, J. E. Devries, and T. J. Prater, *Mass Spectrom. Rev.* **3**, 527 (1984).
17. K. Wittmaack, *Surf. Sci.* **89**, 668 (1979).
18. R. E. Honig, *Intern. J. Mass Spectros. and Ion Proc.* **66**, 31 (1985).
19. J. C. Vickerman, *Spectroscopy of Surfaces* (R. J. H. Clark and R. E. Hester, eds.), Advances in Spectroscopy, Vol. 16, John Wiley and Sons (1988), p. 155.
20. J. Lindhart, V. Nielsen, and M. Scharff, *Kgl. Danska, Videnskab, Selskab, Mat. Phys. Medd.* **33**, 14 (1963); **36**, 10 (1968).
21. W. K. Chu, *Material Characterization Using Ions Beams* (J. P. Thomas, A. Cachard, eds.), Plenum Press, London, NATO Advanced Study Institute Series Vol. 28, (1978), p. 3.
22. J. B. Sanders, *Can. J. Phys.* **46**, 455 (1968).
23. P. Sigmund, *Sputtering Yield of Ion Bombarded Solids,* Proc. 3rd Nat. Conf. on Atomic Collisions in Solids, Kiev (1974).
24. P. Sigmund, *Radiation Damage Processes in Materials.* Proc. 1973 NATO Advanced Study Institute, Aleria, Noordhoff, Leiden (1975), p. 3.
25. D. E. Harrison, W. L. Moore, and H. T. Holcombe, *Radiat, Eff.* **17**, 167 (1973).
26. P. Sigmund, *Phys. Rev.* **184**, 383 (1969); **187**, 768 (1969).
27. G. Blaise and G. Slodzian, *Rev, Appl. Phys.* **8**, 105 (1973).
28. F. Bernhardt, H. Dechsner, and E. Stumpel, *Nucl, Instrum. Meth.* **132**, 329 (1976).
29. P. K. Rol. D. Onderlinden, and J. Kistemaker, *Proc. 3rd. Intern. Vacuum Congr. Vol. 1,* Pergamon, Oxford (1966), p. 75.
30. H. Oechsner, *Z. Phys.* **261**, 37 (1973).
31. N. Winograd and B. J. Garrison, *Acc. Chem. Res.* **13**, 406 (1980).
32. G. Slodzian, *Surf. Sci.* **48**, 161 (1965).
33. M. Bernheim, *Radiat, Eff.* **18**, 231 (1973).
34. H. Oechsner, *Z. Phys.* **238**, 433 (1970).
35. R. Kelly and N. Q. Lam, *Radiat. Eff.* **19**, 39 (1973).
36. G. Blaise and M. Bernheim, *Surf. Sci.* **47**, 324 (1975).
37. M. Shimizu, M. Ono, and K. Nakayama, *Surf. Sci.* **36**, 817 (1973).
38. M. Yabumoto, K. Watanabe, and T. Yamashina, *Surf. Sci.* **77**, 615 (1978).
39. J. W. Coburn, *Thin Solid Films* **64**, 371 (1979).
40. G. Betz, *Surf. Sci.* **92**, 283 (1980).
41. R. Kelly, *Surf. Sci.* **100**, 85 (1980).
42. L. Leclercq, *Surface Properties and Catalysis by Non-Metals* (J. P. Bonnelle, B. Delmon and E. Derouane, eds.), Reidel, Dordrecht (1983), p. 433.
43. E. Taglauer, *Appl. Surf. Sci.* **13**, 80 (1982).
44. A. Benninghoven, *Chemistry and Physics of Solid Surfaces* (R. Vanselow and S. Y. Tong, eds.), CRC Press, Cleveland, OH (1977), p. 207.
45. A. Benninghoven and A. Mueller, *Phys. Lett.* **A 40**, 169 (1972).
46. J. Grimblot, P. Alnot, R. J. Behm, and C. R. Brundle, *J. Electron Spectrosc. Rel. Phenom.* **52**, 175 (1990).
47. A. Benninghoven, *Surf. Sci.* **28**, 541 (1971).
48. P. Williams, R. K. Lewis, C. A. Evans, and P. R. Hanley, *Anal. Chem.* **49**, 2023 (1977).
49. M. Bernheim and G. Slodzian, *J. Phys.* **38**, L325 (1977).
50. M. Morabito and R. K. Lewis, *Anal. Chem.* **45**, 869 (1973).
51. R. Castaing and G. Slodzian, *J. Microsc.* **1**, 395 (1962).
52. A. Benninghoven, *Z. Phys.* **230**, 403 (1970).
53. A. Benninghoven, D. Jaspers, and W. Silktermann, *Appl. Phys.* **11**, 35 (1976).

54. R. J. Colton, *J. Vac. Sci. Technol.* **18**, 737 (1981).
55. S. J. Pachuta and R. G. Cooks, *Chem. Rev.* **87**, 647 (1987).
56. G. M. Lancaster, F. Honda, Y. Fukuda, and J. W. Rabalais, *J. Am. Chem. Soc.* **101**, 1951 (1979).
57. R. J. Day, S. E. Unger and R. G. Cooks, *Anal. Chem.* **52**, 557A (1980).
58. M. Barber, R. W. S. Bordoli, J. C. Vickerman, and J. Wolstenholme, *Proc. 3rd Intern. Conf. on solid Surfaces* (P. Dobrozensky *et al.*, eds.), (1977), p. 983.
59. H. Hopster and C. R. Brundle, *J. Vac. Sci. Technol.* **16**, 548 (1979).
60. H. Niehus and E. G. Bauer, *Electron Fisc. Apli.* **17**, 53 (1974).
61. B. Sakakini, A. J. Swift, J. C. Vickerman, C. Harendt, and K. Christmann, *J. Chem. Soc., Faraday. Trans.* **I 83**, 1975 (1987).
62. A. Shepard, R. W. Hewitt, W. E. Baitinger, G. J. Slusser, N. Winograd, G. L. Ott, and W. N. Delgass, *Quantitative Surface Analysis of Materials* (N. S. Mc Intyre, ed.), American Society for Testing and Materials (1978), p. 187.
63. W. N. Delgass, L. L. Lauderback, and D. G. Taylor, *Chemistry and Physics of Solid Surfaces IV*, Springer Series in Chemical Physics, Vol. 20, 51 (1981).
64. C. R. Brundle, R. J. Behm, P. Alnot, J. Grimblot, G. Polzonetti, H. Hopster, and K. Wandelt, *Catalyst Characterization Science, ACS Symposium Series* (M. L. Deviney and J. L. Gland, eds.), Vol. 288 (1985), p. 317, and references therein.
65. A. Brown and J. C. Vickerman, *Surf. Interface Anal.* **6**, 1 (1984).
66. X. Y. Zhu, S. Akhter, M. E. Castro, and J. M. White, *Surf. Sci.* **195**, L145 (1988).
67. (a) K. M. Ogle and J. M. White, *Surf. Sci.* **139**, 43 (1984). (b) P. L. Radloff and J. M. White, *Acc. Chem. Res.* **19**, 287 (1986).
68. J. Hass, C. Plog, W. Maunz, K. Mittag, K. D. Gollmer, and B. Klopries, *Proc 9th Intern. Cong. on Catalysis, Vol. 4*, (M. J. Phillips and M. Terman eds.), (1988), p. 1632.
69. P. Pomonis and J. C. Vickerman, *J. Chem. Soc. Faraday Disc.* **72**, 247 (1981).
70. E. De Paw and J. Marien, *J. Phys. Chem.* **85**, 3551 (1981); *Intern. J. Mass Spectrom.* **46**, 519 (1982).
71. S. Bourgeois, L. Gitton, and M. Perdereau, *J. Chim. Physique* **85**, 413 (1988).
72. R. L. Chin and D. M. Hercules, *J. Phys. Chem.* **86**, 360 (1982).
73. L. Rodriguo, A. Adnot, P. C. Roberge, and S. Kaliaguine, *J. Catal.* **105**, 175 (1987).
74. S. Jaras, *Appl. Catal.* **2**, 207 (1982).
75. E. L. Kluger and D. P. Leta. *J. Catal.* **109**, 387 (1988).
76. A. K. Coverdale, P. F. Dearing, and A. Ellison, *J. Chem. Soc., Chem. Comm.*, 567 (1983).
77. N. Takahashi, T. Mori; A. Furuta, S. Komai, A. Miyamoto, T. Hattori, and Y. Murakami, *J. Catal.* **110**, 410 (1988).

ION SCATTERING SPECTROSCOPY

J. Grimblot and M. Abon

12.1. INTRODUCTION

Similarly to SIMS, ion scattering spectroscopy (ISS) is a surface-sensitive technique that is the result of the interactions between an ion beam and a solid target. More precisely, the ions coming from a monoenergetic ion beam scattered by the surface of the material to be investigated are energetically analyzed in a given direction. Generally rare gas ions are used. This technique is also called LEIS or low-energy ion scattering because the energy of the primary ions is generally small (200 to 2000 eV) to avoid large target surface modifications.

ISS is a relatively new technique even though the first paper by Smith[1] in 1967 mentions the possibility of surface analysis using ion beam scattering. Smith also showed that the interaction between the rare gas ion (He^+, Ne^+, Ar^+) and the atom of the target can be described by an elastic binary collision process. Later papers by Taglauer and Heiland,[2-5] and Brongersma et al.,[6-9] and review articles listed as references 10–18 are illustrative of the development of this technique.

The most important characteristic of ISS is its high sensitivity to the uppermost layer of the solid target. In principle, this gives ISS an advantage over the other surface science techniques which are concerned with analysis of several surface layers depending on the mean free path of electrons. The incident ion interacts by scattering with only one surface atom of the target and the elastic collision process can be calculated by way of classical mechnanics. As a consequence, the observations (number and energy of the scattered ions) are

J. Grimblot ● Laboratoire de Catalyse Hétérogène et Homogène URA CNRS, Université des Sciences et Technologies de Lille, Villeneuve d'Asq, France. M. Abon ● Institut de Recherches sur la Catalyse, CNRS, Villeurbanne, France.

Catalyst Characterization: Physical Techniques for Solid Materials, edited by Boris Imelik and Jacques C. Vedrine, Plenum Press, New York (1994).

apparently easy to interpret. However, as already noted in Chapter 11 (the principles and uses of SIMS), whenever ions are involved, there are complications due to phenomena that occur at the surface or in the selvedge region, such as particle neutralization and sputtering. This latter phenomenon is less pronounced in ISS then in SIMS experiments, as the energy of the incident ion rarely exceeds 2 keV, but it is always present (see e.g., Figure 11.6). The main field of ISS applications is surface analysis. The progressive destruction of the solid target upon exposure to the ion beam can be used with benefit to follow the sample composition layer by layer starting from the uppermost one. In the case of investigations of single-crystal surfaces, with or without adsorbed species, the information may also be of a structural nature because some of the surface atoms, depending on their sizes, may mask the adjacent atoms (in the same top layer or in the underlying layer) from the ion beam (shielding and shadowing effects). In that way ISS is a complementary technique to LEED (low-energy electron diffraction).

In this chapter, after a description of the principles and main properties of ISS, we will focus on applications, in particular those relevant to catalytic materials. For this aspect, we will use the recent article of Horell and Cocke[18] as well as our personal experience. The problems related to the induced effects caused by ion beams, such as preferential sputtering, will not be developed further as they have already been detailed in Chapter 11.

12.2. THEORETICAL AND PRACTICAL ASPECTS

12.2.1. Mechanics of a Binary Collision

12.2.1.1. Principles and the Fundamental Relationship

In the ISS experiment, an incident beam of low-energy ions (typically less than 10 keV but generally in the 200–2000 eV range) interacts with the solid target to be analyzed and the scattered ions are energy-analyzed by using an electrostatic analyzer. Figure 12.1 is a diagram of the collision between the incident ions of mass M_i and energy E_0 with an atom of the target of mass M_a, where θ is the angle between the direction of the primary ions and that of the scattered ions. The primary ions, correctly mass and energy filtered, are noble gas ions, typically $^3He^+$, $^4He^+$, $^{20}Ne^+$, or more rarely $^{40}Ar^+$. Owing to the electron neutrality of the solid target, there are no long-distance electrostatic interactions between the incident ion and the atoms of the target.

Those incident ions have a corresponding very short wavelength, typically 4×10^{-3} Å for 1-keV He^+ ions or 10^{-3} Å for Ne^+. For these values, far less than the interatomic distances, there are no diffraction effects on the ions from the target structure. For an ion energy of 1 keV, 5×10^{-16} s (He^+) to 10^{-15} s (Ne^+) is the time necessary to cover 10^{-10} m. As a consequence the interaction time between the incident ion and the atom of the target at rest is 10^{-15} to 10^{-16} s, which is very short compared to the characteristic vibration periods of the solid

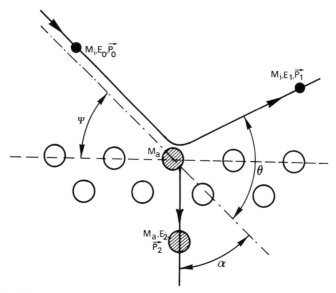

Figure 12.1. Schematic representation of the collision between the incident ions (E_0, M_i) with an atom of mass M_a of the solid target; θ is the scattering angle (laboratory coordinate system).

phonons (10^{-12} to 10^{-13} s). Therefore each atom of the target can be considered by the incident ion as an isolated free atom.[12]

The noble gas ions are chemically very unstable so that the neutralization probability upon the first collision is very high. For He$^+$, only 1 ion over 1000 survives after the collision with the surface atom of a metallic sample. Heavier ions have higher survival probabilities and two successive collisions may occur before they are neutralized. The high neutralization probability of rare gas ions facilitates the interpretation of the ion–atom interaction as only one collision generally needs to be considered. This also explains why only the uppermost layer is analyzed by ISS: monolayer adsorption may hide the substrate completely.[6]

All these general aspects indicate that the binary elastic collision between the primary ion (Mi, E_0, $\mathbf{p}_0$) with the surface atom of mass M_a can be calculated according to the fundamental principles of mechanics, i.e., energy E and momentum $\mathbf{p}$ conservation. Therefore the fundamental ISS relationship of a scattered ion in the direction of angle θ in its initial trajectory is

$$\frac{E_1}{E_0} = \left[\frac{\cos \theta + (\gamma^2 - \sin^2 \theta)^{1/2}}{1 + \gamma} \right]^2 \tag{12.1}$$

with $\gamma = M_a/M_i \geq 1$.

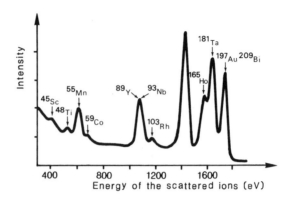

Figure 12.2. ISS spectrum of a target containing several heavy elements. The intensity is plotted as a function of the energy E_1 of the scattered Ne^+ ions. $E_0 = 2,500$ eV and $\theta = 142°$.[14]

On the other hand, the target atom receives a recoil energy E_2 with

$$E_2/E_0 = 4\gamma \cos^2 \theta/(1 + \gamma)^2 \qquad (12.2)$$

However, this second relationship is of no real importance in using ISS as a surface spectroscopic technique. When the scattering angle θ is 90°, equation (12.1) becomes simpler:

$$E_1/E_0 = (M_a - M_i)/(M_a + M_i) \qquad (12.3)$$

In a given experiment, E_0, M_i, and θ are chosen and maintained as constant parameters so that the measure of E_1 gives the mass M_a of the surface atom of the target: ISS is therefore considered as a *surface mass spectroscopy*. Typically, Figure 12.2, shows the mass analysis of a surface target containing several heavy elements.

As in practice H^+ ions cannot be used as primary ions and with the condition $\gamma \geq 1$, all the elements of the Periodic Table can be detected by He^+ ions except hydrogen, which is not detected by itself. Indirectly, however, surface hydrogen species often can be indicated by the shadowing effect. Typical examples will be presented in Section 12.3. We can also note that deuterium atoms have been detected on a tungsten surface by use of He^+ ions.[19] Sometimes, for improved resolution (see Section 12.2.1.2), Ne^+ or even heavier noble gas ions can be used instead of the He^+ source. In that case, of course, the elements situated before the primary ion in the Periodic Table will not be detected; using Ne^+ for example prevents, in particular, detection of carbon, nitrogen, and oxygen, elements frequently present in a large variety of materials. If both the maximum number of elements and the best resolution for the heavier atoms are required, experiments must use He^+, Ne^+, and even Ar^+ alternately.

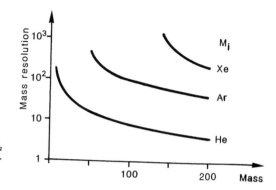

Figure 12.3. Mass resolution $M_a/\Delta M_a$ as a function of M_a for different primary ions.[4]

12.2.1.2. Mass Resolution

Starting with the equation (12.1), the mass resolution $M_a/\Delta M_a$ can be deduced by the following equation:

$$\frac{M_a}{\Delta M_a} = \frac{E_1}{\Delta E_1} \frac{2\gamma}{1+\gamma} \frac{\gamma + \sin^2\theta - \cos\theta(\gamma^2 - \sin^2\theta)^{1/2}}{\gamma^2 - \sin^2\theta + \cos\theta(\gamma^2 - \sin^2\theta)^{1/2}} \quad (12.4)$$

with $\gamma = M_a/M_i \geq 1$. When $\theta = 90°$ equation (12.4) is simplifed into

$$M_a/\Delta M_a = (E_1/\Delta E_1)[2\gamma/(\gamma^2 - 1)] \quad (12.5)$$

The mass resolution of the ISS spectrum depends on both γ and θ ánd decreases when the mass of the target atoms increases. Optimal mass resolution is obtained when M_a is equal to M_i (Figure 12.3). As a consequence, when heavy atoms have to be distinguished in an ISS experiment, heavier ions such as Ne^+, Ar^+, or even Kr^+ or Xe^+ must be used. However, in these cases the perturbation of the surface is considerably increased. On the other hand, for a given value of $\gamma = M_a/M_i$, the mass resolution increases with the scattered angle θ. Below 90° (Figure 12.4) the mass resolution is very poor so that in practical equipment θ is always equal to or higher than 90°. Heiland and Taglauer[20] have shown that with a scattering angle of 90°, 0.8 keV-He^+ ions cannot resolve the components present on a Cu–Nb surface whereas Ar^+ ions of the same energy can: the copper and niobium peaks are well separated. For the same scattering angle, by using equation (12.3), one can show that the ^{63}Cu and ^{65}Cu isotopes are resolved (energy separation $\Delta E_1 = 15$ eV) when Ar^+ (1 keV) ions are used whereas they are not separated ($\Delta E_1 = 3$ eV) if He^+ (1 keV) ions are chosen. As a whole, the mass resolution obtained in ISS experiments is rather poor,[4] which limits its potential applications. In catalysis, distinction between the two metals in bimetallic catalysts is sometimes very difficult.

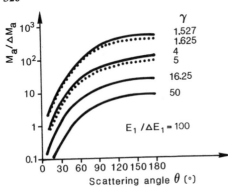

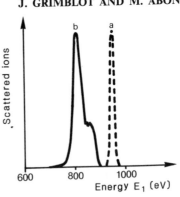

Figure 12.4. Mass resolution $M_a/\Delta M_a$ as a function of the scattering angle for different values of γ.[4]

Figure 12.5. Energy spectra of He$^+$ (a) and Ne$^+$ (b) ions scattered by a nickel target. $E_0 = 1,000$ eV and $\theta = 45°$.[12]

12.2.1.3. Comparison of Practical Spectra with Theory

Figure 12.5, shows the ISS spectra obtained from the scattering of He$^+$ or Ne$^+$ ions on a nickel target. With He$^+$, only one peak is observed in agreement with collision theory as the probability of the ion being scattered by only one binary collision is very high. With Ne$^+$, two peaks are detected: the more intense peak corresponds to one binary collision while the shoulder is the result of multiple successive collisions (in practice two collisions occur) between the incident ion and several atoms of the target. Therefore, to have only a binary collision and to use the basic ISS relationship [equation (12.1) or (12.3)], it is necessary to use light ion sources (He$^+$, Ne$^+$) and high scattering angles.[5] Figure 12.6 gives a clear guide for the choice of ion. The noble gas ions have to be used instead of more stable ions such as H$^+$ because, in that case, H$^+$ may undergo several collisions before being neutralized. Even when light noble gas ions are used, there is still some difference between the idealized system as deduced by using the fundamental relationship and the

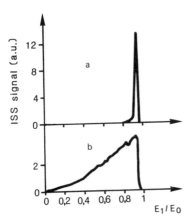

Figure 12.6. Comparison between the spectra obtained with He$^+$ (a) and H$^+$ (b) ions scattered by a gold target. $E_0 = 1,800$ eV.[12]

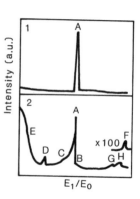

Figure 12.7. Comparison of an idealized ISS spectrum (1) that corresponds to an elastic binary collision with a spectrum (2) in which some additional features are practically detected on some materials: (A) scattered peak from an elastic binary collision; (B) double collision-scattered peak; (C) tail due to inelastic collision; (D) scattered peak of an impurity on the sample; (E) background of low-energy ions coming from sputtering of the sample; (F) scattered peak due to doubly ionized species (Mi^{2+}); and (G) memory effects due to traces of gas previously introduced in the ISS equipment.[13]

real observed spectrum, which can be much more complex and less easy to interpret. This comparison is shown schematically in Figure 12.7. One can see that the low-energy background [E feature in Figure 12.7(2)] is the result of the sputtering effect of the primary ion on the target. The SIMS technique described in Chapter 11 is devoted to the analysis of these positively or negatively charged emitted species. On the other hand, by the impact of positively charged ions, the sample also emits secondary electrons and then the analyzed sample can be positively charged.[21] For nonconducting samples this charging effect may be a problem which prevents the detection of any reliable spectrum, and a neutralization procedure using low-energy electron beams (flood gun) is sometimes required simultaneously with the ion scattering experiment. A similar procedure can also be used in Auger analysis of insulators.

12.2.2. Scattered Ion Yield

12.2.2.1. General Relationship

The factors contributing to the spectral intensity I_m or the number of scattered and detected ions per second resulting from a surface layer containing m atoms or of density N_m atoms per unit area is given by

$$I_m = T(E)I^0 A N_m [d\sigma_m(E)/d\Omega]\Delta\Omega P^+(E) \qquad (12.6)$$

where $T(E)$ is a combined instrument factor that includes analyzer transmission, detector sensitivity in the energy range of interest and any appropriate conversion factors; I^0 is the primary ion current density or number of primary ions per second arriving at the target per unit area, A is the overlap between the bombarded area and the area sampled by the analyzer; $d\sigma_m(E)/d\Omega$ is the differential scattering cross section into a unit solid angle; $\Delta\Omega$ is the acceptance angle of the analyzer; and $P^+(E)$ is the survival probability of the scattered ions. Use of equation (12.6) for quantitative surface analysis is limited owing to uncertainties in $d\sigma_m(E)/d\Omega$ and P. Despite this problem, it is possible to use ISS for semi-quantitative analysis by using pure reference samples.

12.2.2.2. The Differential Scattering Cross Section

The differential cross section $d\sigma/d\Omega$ can be calculated by using the interaction potential $V(r)$ between the incident ion and the atom at rest in the surface target for a given scattering angle θ. The repulsion potential between the charged nucleus Z_i of the ion and the charged nucleus Z_a of the atom belonging to the target is, according to the Coulomb interaction:

$$V(r) = Z_i Z_a e^2/r \qquad (12.7)$$

where r represents the distance between the nuclei. However, due to screening effects of the electronic clouds around the two nuclei, the more appropriate potential to consider is given by the following equation based on the Firsov or Thomas–Fermi–Moliere approximations[22,23]:

$$V(r) = (Z_i Z_a e^2/r) \cdot f(r/a) \qquad (12.8)$$

The term $f(r/a)$ takes the screening function into account,[22,23] where a is a screening length of the form

$$a = \alpha a_0/(Z_i^{1/2} + Z_a^{1/2})^{2/3}$$

with α an adjustable parameter, and a_0 the Bohr radius.

By using the Rutherford scattering approximation,* the different cross sections can be calculated. Theeten and Brongersma[12] have calculated $d\sigma/d\Omega$ for the $H^+ \rightarrow Ni$ interaction with a scattering angle of 60°. When a screening function was included in the interatomic potential, the values they obtained were about 200 times weaker than those deduced when only the Colomb interaction was considered (Figure 12.8). Indeed, the screening function is an important factor in calculating reliable scattering cross sections. As was pointed out,[12] all the electrons of the interacting species (the incident ion and the surface atom of the target) have an influence but the valence electrons do not have any particular effect. As a consequence, the scattering cross sections, and therefore the ISS signal intensities, are not very sensitive to the chemical state of the elements in the target. No information on chemical bonds is available in an ISS experiment, unlike with other surface-sensitive techniques XPS, AES, and even SIMS for which a matrix effect exists because the secondary ion yield is dependent on the local bonds of the emitted particles or clusters. Thus, in principle, by appropriate calibration, ISS can be a quantitative method; in adsorption studies, changes in surface coverage may be followed by ISS. This particular property is often used when exploring supported catalysts with various metal loadings. In addition to the choice of the interaction potential, the other main parameters affecting the scattering cross sections are the energy E_0, the scattering angle θ, and the mass M_a of the atoms in the target. In Figure 12.8

---

* For the basic principles of atomic and molecular collisions, see, e.g., Child.[24]

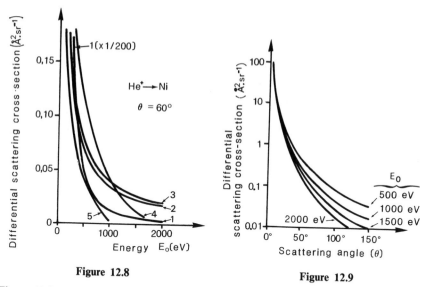

Figure 12.8

Figure 12.9

Figure 12.8. Variation of the scattering cross section of He^+ ions as a function of the incident energy E_0 for different interaction potentials: (1) Coulomb potential, (2) Fersov, (3) Moliere, (4 and 5) Born–Mayer with different parameters.[12]

Figure 12.9. Variation of the scattering cross section as a function of the scattering angle θ for the $Ne^+ \rightarrow Ni$ system, calculated using the Born–Haber interaction potential.[4]

it can be seen that the cross section decreases with E_0; on the other hand, the scattering cross section also decreases when the scattering angle θ increases (Figure 12.9). The inverse evolution is obtained for the mass resolution (Figure 12.4), therefore a compromise must be found for θ (generally $90° \leq \theta \leq 140°$). Figure 12.10 shows the evolution of the scattering cross section with the mass M_a of the surface atom of the target.

We will see in Section 12.3 that hydrotreatment catalysts have been well characterized by ISS. Hence, to interpret the data on this type of catalyst as completely as possible values of the scattering cross sections have been calculated by Kasztelan,[25]* and Jeziorowski et al.[26] for typical elements (i.e., O, Mo, Al, S, Ni, Co, W) present in such systems.

12.2.2.3. The Survival Probability

The survival probability $P^+(E)$ or, in other words, the ion neutralization probability $(1 - P^+)$, one of the parameters of equation (12.6), causes many problems in ISS quantification as it is the least predictable one. Even if the

---

* In that thesis the scattering cross sections have been calculated for all the elements up to uranium and for four values of incident energies E_0 (0.5, 1.0, 1.5, and 2.0 keV).

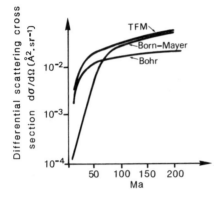

Figure 12.10. Variation of the scattering cross section as a function of the mass M_a of the surface atoms of the target. Three different interatomic potentials have been used for the calculations: Thomas–Fermi–Moliere, Born–Mayer and Bohr. The presented evolution concerns 1 keV He$^+$ ions.[4]

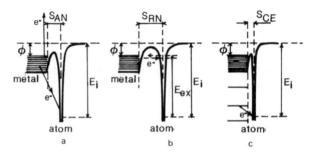

Figure 12.11. The different neutralization processes of a rare gas ion at the metallic surface for $E_i > \Phi$ (E_i is the ionization energy of the rare gas and Φ is the work function of the metal): (a) Auger neutralization AN, (b) resonant neutralization RN, (c) quasi-resonant charge exchange (CE).[5]

scattering cross sections are calculated with empirical interatomic potentials, comparisons among elements are significant. On the other hand, several complex processes, depending on the nature of the solid target, may need to be taken into account to arrive at the survival probability P^+.

At solid surfaces, in particular metallic ones, the work function is rarely higher than 5 to 6 eV, a value far smaller than the ionization energy E_i of the rare gas (24.4, 21.6, and 15.6 eV for, respectively, He, Ne, and Ar). As a consequence, a valence electron present in the conduction or valence hand of the solid is near a deep potential well created by the rare gas ion.[4] Therefore, the neutralization probability is high, always higher than 90% and sometimes even higher than 99%: less than 1 incident ion out of 100 may escape in the ionized state upon scattering. Heiland and Taglauer (5) created potential diagrams showing different neutralization processes of a rare gas ion at a metallic surface (Figure 12.11).

According to Smith[27] and Hagstrum,[28] the ion survival probability P^+ has a velocity dependence of the form

$$P^+ = \exp(-v_c/v_n) \tag{12.9}$$

where v_n is the component of velocity normal to the surface and v_c is a material constant.

As a consequence, by combining equations (12.6) and (12.9), the plot of $\ln[I_m/(d\sigma/d\Omega)]$ vs. $E_0^{-1/2}$ should be a linear relationship as the ion velocity is proportional to $E_0^{0.5}$. This has been checked experimentally for $H^+ \rightarrow Ni$ or $He^+ \rightarrow Cu$ interactions.[12,27] In the $He^+ \rightarrow Pb$ interaction, dramatic oscillations in the ion yield as a function of the ion energy were reported.[29] This effect was attributed to resonant electron exchange which results from coupling of the atomic core levels of similar energies in the collision partners (see Figure 12.11). It was found that these oscillations may depend on the chemical state of the atom in the target; indium compounds show this behavior.[30] However, these oscillations are generally of weak amplitude and the $(d\sigma/d\Omega)P^+$ product does not change much with E_0.[12] With this argument, it is therefore reasonable to consider ISS as a quantitative technique if a correct calibration is performed. On binary ionic compounds, the P^+ parameter is less understood and also less studied. Nevertheless, in the ion–insulator surface collision with compounds like NaF or NaCl, the neutralization process tends to be localized on the F^- (or Cl^-) ion by a nonresonant charge exchange and the intensity ratio I(anions)/I(sodium) varies linearly with the inverse velocity V_0^{-1} (Figure 12.12).

When the MoS_2 basal plane surface (with a sulfur and molybdenum hexagonal arrangement) was examined by ISS, it was observed (Figure 12.13) that the ion escape probabilities for first (sulfur) and second (molybdenum) layers differ drastically. The escape probabilities for He^+ scattering at molybdenum sites vary (log scale) with the inverse normal velocity of the incident ion as predicted by equation (12.9) for metals whereas at the first sulfur layer, the scattering, and therefore the survival probability does not display this simple dependence.[32]

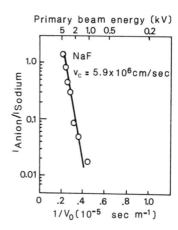

Figure 12.12. Plot of I_F/I_{Na} (log scale) vs. V_0^{-1} indicating inverse exponential initial velocity dependence on ion survival from F collisions, assuming unity survival probability for $He^+ \rightarrow Na$.[31]

12.2.2.4. Shielding and Shadowing Effects

Experimentally, it has been shown that the presence of surface impurities or adsorbed species generally causes variations in the scattered ion intensity. This phenomenon is mainly explained by geometrical and interaction considerations depending on the direction of the incident and scattered ions relative to the analyzed surface orientation; adsorbate species in the top layer may "protect" the atoms underneath from collision. Theeten and Brongersma[12] have shown that, from the Moliere interaction potential, an apparent size or impact parameter can be calculated for a given scattering angle. For example, the impact parameter of ^{58}Ni changes from 0.2 to 0.9 Å when exposed to $^4He^+$ ions (1 keV) at $\theta = 60°$ or 5°, respectively. These apparent atomic size differences induce shielding of atoms, a phenomenon that can be described by the shadowing effect. Figure 12.14a is a laboratory diagram showing the possible trajectories of an ion scattered by a target surface. The position of the initial target atom is indicated by a dot. A shadow cone is formed behind the initial position of the target atom and an atom located in this cone cannot contribute to ion scattering. Conversely, in Figure 12.14b, an ion initially scattered by an atom can be scattered by a second atom, shown by a dot. The shadow formed behind the position of the second atom is called the blocking cone.[33] As the interatomic potential used for scattering calculations is a screened Coulomb potential, the radii of the shadowing and blocking cones are comparable to the atomic radius or impact parameters. For example, the shadowing cone for the 1 keV He^+/O system has been calculated.[34]

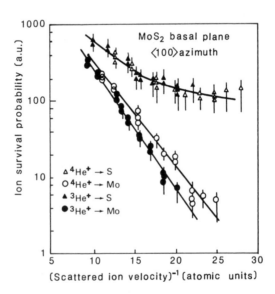

Figure 12.13. Relative escape probabilities for $^3He^+$ and $^4He^+$ scattering at sulfur and molybdenum sites along the $\langle 100 \rangle$ direction of the MoS_2 basal plane surface as a function of the outgoing ion velocity (1 atomic unit = 2.187×10^8 cm s^{-1}).[32]

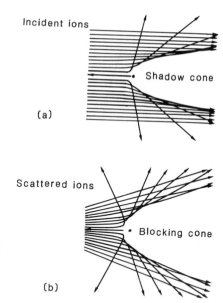

Figure 12.14. Schematic figures depicting (a) the shadow cone of an atom in an incident ion beam, and (b) the blocking cone of an atom in a scattered ion flux.[33]

As a consequence, the general equation (12.6) giving the intensity or the scattered ion yield I_m has to be modified to take into account the effect of adsorbed species n on a surface layer having an initial density N_m of m atoms[3,31]:

$$I_m = TI_0 \cdot A(N_m - N_n\sigma_m\beta)(d\sigma_m/d\Omega)\Delta\Omega P^+ \qquad (12.10)$$

where σ_n is the total scattering cross section for species n; N_n is the density of the adsorbate species n; and β is a parameter depending on E that determines the shielding effects. This relationship will be used when a substrate surface is progressively covered by an adsorbed species: adsorption of molecules on clean metal surfaces or adsorption of salts or ionic species on oxide surfaces in the preparation of supported catalysts.

Shadowing effects are also detected in the examination of binary compounds, where the atomic or ionic radii of the species are quite different. Similar to the shadowing parameter β, this effect is energy-dependent. To illustrate the influence of the incident ion energy E_0, Figure 12.15 shows the oxygen and silicon ISS spectra of SiO_2 in which the silicon is located in a tetrahedral oxygen environment; at low energy (0.5 keV), the big oxygen surrounding species are almost completely shielding the small Si^{4+} cation while at high energy (typically 4 keV) the silicon intensity is higher than the oxygen peak. For such a high energy, the shielding effect is very small. As a consequence a low incident ion energy is required to take advantage of these shadowing effects.

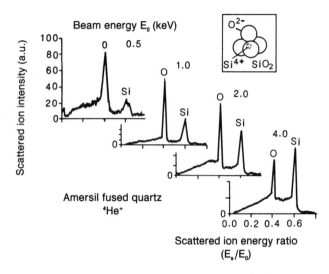

Figure 12.15. Effect of variation in E_0 on the appearance of $^4He^+$ ISS spectra for SiO_2 (fused quartz) and interstitial location of Si^{4+} ion (insert).[31,35]

12.2.2.5. Surface Sensitivity

The principle of ISS based on the mechanics of binary collisions as previously described gives this technique the ability to analyze the uppermost layer of a solid target. There was an attempt to demonstrate this surface-sensitive property by analyzing CdS. The (111) surface of this solid is made of cadmium atoms only, whereas the $(\bar{1}\bar{1}\bar{1})$ surface is composed only of sulfur atoms. Owing to differential sputtering by the incident ions during an ISS experiment, the first investigations were not conclusive. However, Brongersma and Mul[36] have shown a decisive difference between those exposed surfaces when weak incident ion beam currents are used. The surface sensitivity has also been demonstrated by adsorption of bromine on the (111) surface of silicon[37] or by comparing the (100) and (111) surfaces of silicon: the different detected intensities vary with the actual atomic density.[38] It has been claimed that the sensitivity limit is between 10^{-4} and 10^{-3} surface monolayer for Au/Si[39] and S/Ni or O/Ni systems.[3,40]

The ISS signal is generally proportional to the surface concentration, which implies that there is no matrix effect and also that the survival probability is independent of the coverage.[3,4,40] For example, the scattered ion (He^+ or Ne^+) intensity varies linearly with the Auger intensity ratio of the S/Ni (111) system,[40] the slope being larger when Ne^+ incident ions are used. However, deviation from linearity has sometimes been observed.[41] According to Taglauer and Heiland,[4] the scattered ionic current relative to the primary current has been estimated to be about 10^{-3} when the scattering cross section $d\sigma_m/d\Omega$ is 1 $Å^2$ sr^{-1}; the surface density of a monolayer, N_m, is 10^{15} atoms cm^{-2};

and the acceptance angle of the analyzer, Ω, is 10^{-3} sr. However, assuming that the survival probability is about 10^{-3}, the I/I_0 ratio is of the order of 10^{-6} for a complete monolayer.

12.2.3. Experimental Considerations

12.2.3.1. The Experimental Setup

In a typical ISS experimental setup the target is mounted on a rotatable sample manipulator in a UHV system. The ion gun is generally pumped separately from the main chamber in order to get at least 10^{-9} Torr in the analytical chamber. This is a prerequisite to getting reliable analyses, as ISS is surface-sensitive and contamination must be avoided. In general, the electrostatic analyzer is cylindrical or hemispherical with good resolution (a few eV at 1000 eV) and associated with an electron multiplier (channeltron type) to obtain good amplification as the scattered ion current is weak. It is common to have a multitechnique experimental setup in which AES, XPS, LEED, TDS, UPS, and SIMS facilities are available simultaneously so that all the data obtained on the same chemical surface state of a sample can be cross-checked.

12.2.3.2. Analysis of Single Crystals

By considering that one of the potentialities of ISS is the acquisition of structural information by taking advantage of the shielding effects, clean and covered surfaces must be examined under different orientations by changing the incident, polar, or azimuthal angles. Sometimes it is interesting to modify the scattering angle θ, so the analyzer must be rotatable. At the least, to have good alignment of the ion gun, the target, and the analyzer, translation movements of the target holder are required. In general, a very good manipulator is necessary in order to examine single-crystal surfaces.

13.2.3.3. Analysis of Powder Catalysts

Most laboratory-prepared catalysts are powders. Industrial catalysts have generally well-defined shapes (spheres, cylinders, etc.) made of small grain aggregates. To examine these samples by surface-sensitive techniques, i.e., ISS, the materials should be finely ground in a mortar. The resulting power can be pressed to make self-supported pellets, supported on a metallic holder (foil, grid) (e.g., by evaporation of the liquid of the powder suspension) or pressed on a soft metal foil (indium). In these cases, angle considerations are not important as each particle, in which several domains or elemental grains may exist, has various orientations. No structural information is available on such practical catalysts. The way to conduct the experiment and the sample conditioning appear very important. This has been discussed in an ISS analysis of Ni–Mo–Al_2O_3 catalysts.[42] The same sample, as either self-supported pellets or deposited as a thin film on gold foil, does not show the same evolution as a function of time during examination with 1-keV $^4\mathrm{He}^+$ ions. Another parameter examined was the evacuation time in a preparation chamber before the ISS analysis. Among the observations, it was significant that the nickel atoms were always shielded by contaminants which were not removed quickly enough from the surface by the sputtering effect. It is suggested that this sample acts as a reservoir of contaminants, probably OH species, highly mobile at room temperature, and there is a competition between diffusion from the bulk toward the surface and sputtering. This problem will probably be encountered in the examination of large-surface-area porous materials with OH groups such as alumina, zeolithe, and silica, all of which are used directly as catalysts or as supports.

12.2.3.4. Experimental Difficulties

The difficulties of examining powder samples by ISS were described in the preceding section. More generally, ISS is considered to be a difficult technique for several reasons. For precise structural determinations, alignment and orientation should be correct. As already pointed out in Chapter 11, the use of ions perturbs the surface by ion etching, which can be selective when several elements are present; inducing desorption of adsorbed species, and causing roughness increase and layer mixing. The way to minimize induced surface modifications due to the ion bombardment is to use light ions (He^+) of weak energy and low ionic current density. However, with a sputtering yield of 1 atom/ion, the ionic current must be smaller than 10^{15} ion cm^{-2} as one atomic layer corresponds roughly to 10^{15} atoms cm^{-2}. During 100 s, with an ion density of 10^{11} ions cm^{-2}, the ion dose is weak (10^{13}) compared to 10^{15}, but the scattered ion current will be 10^6 weaker (i.e. 10^5 ion cm^{-2} or 1.6×10^{-14} A). This is really a small value and it is not always possible to reduce the primary ion current sufficiently. The damage caused by exposure to the ion beam can be eliminated by annealing the sample. This has been checked, for example, in the case of MoS_2 in which the S/Mo stoichiometry is restored after annealing to 870 K.[32] For nonconducting samples, it is sometimes necessary to use a flood gun to get reliable spectra.

12.3. APPLICATIONS

The following applications focus on the surface structure and composition determinations of materials relevant to catalysis by means of ISS. The structure and nature of adsorbed species, ion-induced desorption, and practical catalysts are examined as well with typical examples.

12.3.1. Atom and Molecule Desorption Induced by Low-Energy Ions

The atom or molecule desorption induced by the ion beam in the 0.2–2.0 keV energy range can be of benefit for studying the reactivity and bonding of the adsorbed phases. This effect is attractive in the exploration of the H_2–W(100) and CO–Ni(110) interactions.[43,44] As equation (12.6) is always valid, there is a proportionality between the scattered ion intensity I for a given element and the superficial density N. Using the same subscripts as in equations (12.6) and (12.10) (m for the substrate and n for the adsorbed species) and assuming that the induced desorption rate is first order, we can deduce the following relationships:

$$I_n(t) = I_n^0 \exp(-I_0 \sigma_d t) \qquad (12.11)$$

and

$$I_m(t) = I_m^0 [1 - K \exp(-I_0 \sigma_d t)] \qquad (12.12)$$

where I^0 is the incident ion current density; σ_d the desorption cross–section; K a constant; I_n and I_n^0 are, respectively, the scattered ion density relative to the adsorbed species at time t or initially; and I_m and I_m^0 have the same meaning but they concern scattering on the substrate.

Equations (12.11) and (12.12) are only valid for coverages of less than 1 monolayer and it is assumed that no readsorption occurs during the ISS measurements; σ_d is dependent on the mass of the primary ions and also on the angle of incidence between the ion beam and the target. In the case of hydrogen adsorption, equation (12.11) cannot be used as hydrogen atoms are not detected by ISS, and equation (12.12) must be used instead. Figure 12.16a shows two ISS spectra of the clean and hydrogen-covered (100) surface of tungsten, and in Figure 12.16b, confirmation of the validity of equation (12.12) has been established: the desorption cross section σ_d was found to be equal to 1.2×10^{-16} cm^2.

Similarly, the desorption of CO adsorbed on Ni(110) has been followed as a function of the exposure time under He$^+$ (600 eV) either by using the oxygen signal [equation (12.11)] or the nickel signal evolutions [equation (12.12)]. The results give similar cross-section values (7×10^{-15} cm^2). This suggests that induced desorption does not provoke CO dissociation, which should lead, in that case, to a carbon residue on the nickel surface. This residue should affect the nickel ISS intensity variations.

12.3.2. Analysis of Practical Catalysts by ISS

12.3.2.1. General Comments

The use of ISS as a surface-sensitive technique to characterize practical catalysts appeared belatedly in the literature by comparison with the other applications of ISS described above. The first report was probably that of Shelef et al.[45] who examined aluminate-based catalysts and took advantage of the induced ion sputtering to obtain concentration profiles. In the recent review article by Horrell and Cocke,[18] many examples dealing with ISS characterization of different kinds of catalytic materials, such as bimetallic reforming catalysts, zeolites, intermetallic systems, oxidation catalysts, and many others

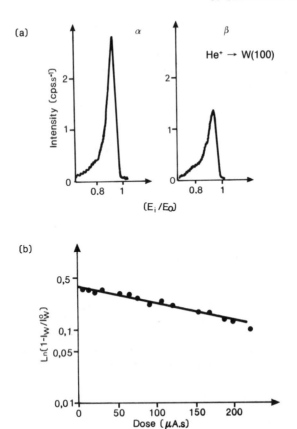

Figure 12.16. (a) ISS spectra of clean (α) and hydrogen-covered (β) W(100) surfaces (from Taglauer and Heiland[15]) (b) Variation of $\ln (1 - I_W/I_W^0)$ as a function of the incident ion dose ($I_0 t$) received by the sample during hydrogen desorption from the W(100) surface.[43]

are presented. It is not our intention here to analyze all these examples again. We will focus mainly on hydrotreating based catalysts, generally composed of a supported phase which associates molybdenum or tungsten with a promoter nickel or cobalt, all of them as oxides in both the laboratory preparations and the commerical catalysts. The support generally used is alumina. On those supported catalysts, the intensity ratio of a signal of a supported species relative to the signal of the support (aluminum) is often analyzed and the discussion is based on the shielding effects of one component by the other, as already noted for adsorption studies. In some cases, bulk unsupported sulfides have also been characterized. Some other typical examples of the use of ISS to characterize the surface of noble metal-TiO_2 relevant to SMSI, not reviewed by Horell and Cocke, are also presented.

12.3.2.2. Hydrotreating and Related Catalysts

ISS has contributed largely to a better knowledge of these molybdenum- or tungsten-based catalysts in the oxide form. However, in almost all the papers, ISS is used in conjunction with other techniques (AES, XPS and laser Raman spectroscopy) so that it is sometimes difficult to estimate accurately the actual impact of ISS on the results.

The $NiO-MoO_3-Al_2O_3$ catalysts are the most widely studied systems[25,26,42,46-50] but the cobalt (or other promoters such as titanium) – molybdenum interactions have also been examined.[51-54] For comparative purposes, catalysts with only the promoter cobalt or nickel have also been investigated,[55,56] and some work also deals with supported tungsten systems.[57-61] Many parameters, such as the supported metal content, the order of impregnation when two metals are deposited, the preparation parameters including the spreading of one phase over the support oxide by mixing and calcining the pure oxides,[62,63] and the duration of the calcination, among others, have been studied. There have been some discrepancies between results of different groups, which are probably due to the various origins of the samples and the fact that ISS experiments are not always performed under identical conditions. In particular, the ion incident energy may have a considerable effect on the sputtering rate, the scattering cross sections, and the shielding effects. Figure 12.17 shows the spectra obtained on an $NiO-MoO_3-Al_2O_3$ catalyst bombarded with He^+ ions of different energies.[42] Clearly at 1 or 2 keV, the aluminum signal is greatly enhanced, which implies that the supported phase and/or the oxygen surrounding atoms do not play the shielding role as expected. To avoid surface modifications and

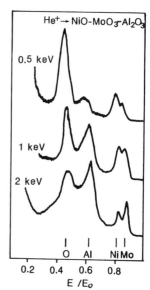

Figure 12.17. ISS spectra of a $NiO-MoO_3-Al_2O_3$ sample after 5 min of etching with $^4He^+$ at 2,000 eV.[42]

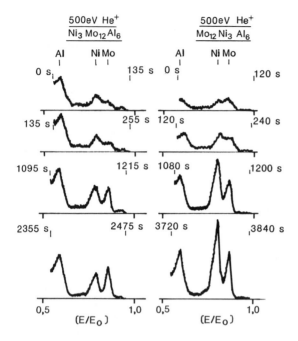

Figure 12.18. Ion scattering spectra of $Ni_3Mo_{12}Al_6$ and $Mo_{12}Ni_3 Al_6$ samples. (3wt.% NiO and 12 wt.% MoO_3). The first metal in the formula is impregnated after the metal in the middle. The last number indicates the pH of the molybdate solution.[46]

retain the potential of ISS for surface analysis, energies of less than 1 keV are often required for such systems.

However, to investigate the repartition of the supported species in terms of monolayer dispersion, Ni–Mo interaction, and Ni diffusion within the alumina support, it is interesting to record the ISS spectra as functions of time after the progressive removal of the outer layers. This has been achieved on Ni–Mo catalysts prepared by the pore filling method with molybdate solutions at different pH. A two-step procedure was followed to deposit molybdenum first or vice-versa.[46] The effect of the final calcination temperature was also considered.[26] Typical evolutions are shown in Figure 12.18. Nickel is always detected, and as it is proposed that a molybdate layer is present at the surface of these catalysts, the nickel is interacting with the molybdate monolayer and with the support to form a spinelle structure. The sequence of impregnation affects the nickel distribution between these two locations: when molybdenum is deposited first, it prevents Ni^{2+} from interacting with Al_2O_3.[46,48]

Calcination at 870 K modifies the molybdate distribution on the support: small molybdate patches not entirely covering the alumina are present at lower temperatures of calcination. At 870 K or higher, they are spread over the alumina surface in a bidimensional monolayer. Nickel location is also affected by these different calcination temperatures[26] and times.[47] Similarly, the Ni–Mo or Co–Mo interactions, evident by a strong shielding effect of the molybdenum element by the promoter nickel or cobalt, have been explained by formation of supported isopolymolybdate salts of cobalt or nickel.[50] The

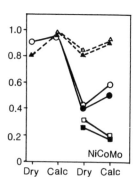

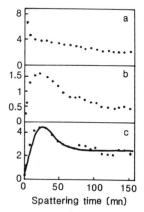

Figure 12.19. Evolution of the ISS intensity ratio (He$^+$, 1000 eV) *vs.* the preparation step of the samples. Solid and open triangles, circles, and squares represent the addition of cobalt and nickel, respectively: $\bigcirc$ and $\bullet$: $(I_{Mo}/I_{Al})_{ISS}$; $\square$ and $\blacksquare$: $(I_{Ni}(I_{Co})/I_{Al})_{ISS}$; $\triangle$ and $\blacktriangle$: $(I_{Mo}/I_{Al})_{XPS}$.[50]

Figure 12.20. Time dependence of the ISS intensity ratios: (a) I_{Mo}/I_{Al}, (b) I_{Co}/I_{Al}, and (c) I_{Co}/I_{Mo}. The full curve has been drawn from a mathematical model of the erosion process.[51]

(I_{Mo}/I_{Al}) ISS intensity ratio is strongly perturbed during the preparation sequence (Figure 12.19). By impregnation with cobalt or nickel nitrate, this ratio clearly decreases whereas the subsequent calcination (500 °C for 2h) does not affect it noticeably. This evolution had been already shown by XPS but to a lesser extent (dashed line in Figure 12.19), which again demonstrates the higher surface sensitivity of ISS over XPS if, of course, all the precautions have been taken for avoiding surface destruction. ISS can also be used to detect limits in monolayer-like coverage of the deposited species by using relationships similar to equations (12.6) and (12.10).[50]

Investigations of other systems, e.g., Co–Mo or Ni–W, emphasize the problem of location of the promoter (nickel or cobalt) relative to the supported molybdate phase. A bilayer model has been proposed by Delannay *et al.*,[51] Chin and Hercules,[53] and Horell *et al.*[61] whereas the previous cited papers dealing with the Ni–Mo supported catalysts suggest direct Ni–Mo interaction. Indeed, on the Co–Mo oxidic precursor, the results of the erosion profiles (Figure 12.20) are interpreted by the presence of a Co–Mo double layer as: (i) no cobalt is detected in the outer layers while the Mo signal is the highest, and (ii) the cobalt signal is a maximum after a certain erosion time. ISS spectra have been

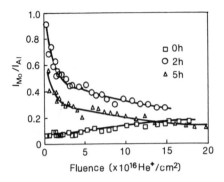

Figure 12.21. ISS molybdenum and aluminum peak height ratios as functions of He$^+$ ion flux. Data for the 7.6% Mo O$_3$/Al$_2$O$_3$ physical mixture without heat treatment (0 h) and after 2 and 5 h of calcination in O$_2$ in the presence of water vapor.[63]

obtained on sulfided Co–Mo–Al$_2$O$_3$ catalysts.[52] A decrease of the cobalt and molybdenum contributions is detected upon increasing the sulfidation temperature. Comparison with bulk MoS$_2$ and Co$_9$S$_8$ sulfides helped to interpret the general ISS features of the supported sample. It was concluded that a sulfidation temperature provokes increasing MoS$_2$ sintering.

An interesting way to study the surface mobility of the supported phase (MoO$_3$-like compound) in such catalytic systems is to prepare the catalysts by mixing of the pure oxide powders followed by a thermal treatment.[62,63] Some typical ISS evolutions are presented in Figure 12.21. The I_{Mo}/I_{Al} intensity ratio of the noncalcined MoO$_3$–Al$_2$O$_3$ mixture does not change very much with the He$^+$ ion dose used to obtain the ISS spectra, which implies that there exists a random distribution of molybdenum and aluminum. On the contrary, after mixing and calcination, the profile evolution appears similar to that observed on classically prepared MoO$_3$–Al$_2$O$_3$ samples where the spreading phenomenon has occurred. Analogous results have been obtained with the MoO$_3$–TiO$_2$ system.[62]

It is also worth noting results obtained on bulk Ni–Mo mixed sulfides.[64] For an atomic Ni/(Ni + Mo) ratio between 0.2 to 0.8, there is a particular surface composition resulting from the existence of a phase of composition Ni/(Ni + Mo) between 0.2 to 0.3. This identified phase, called the NiMoS phase, in which nickel is interacting with the edges of the small MoS$_2$ platelets, is of primary importance as the general catalytic synergy between MoS$_2$ and promoters (here nickel) results from the presence of that phase.

12.3.2.3. Contribution of ISS in Characterizing the SMSI Effect

The strong-metal-support-interaction (SMSI) effect, which modifies the adsorption and catalytic properties of a Group VIII metal (e.g., Pt, Rh) supported on easily reducible oxides (e.g., TiO$_2$, ZnO, V$_2$O$_5$) upon reduction at relatively high temperatures, is a fascinating phenomenon. It deserves great interest but the interpretations, still contradictory, are a matter of debate. ISS has sometimes been used, in conjunction with other surface-sensitive tech-

TABLE 12.1. Surface Composition of Rh–TiO$_2$ and
Pt–TiO$_2$ Samples

Rh–TiO$_2$ (5% Rh)	(Rh/Ti)$_{XPS}$	$(I_{Rh}/I_{Ti})_{ISS}$
R200^a	0.27	0.08
R300	0.23	0.05
R500	0.18	0.03
Pt–TiO$_2$ (1% Pt)	**(Pt/Ti)$_{XPS}$**	**$(I_{Pt}/I_{Ti})_{ISS}$**
R200	0.14	0.14
R500	0.10	0.12

^{a} Indicates the temperature of reduction.[65]

niques and adsorption or catalytic measurements, to explain the behavior of such systems. The following two investigations, among others, are really significant as they lead to opposite conclusions. In the study of the SMSI effect observed on practical Me–TiO$_2$ catalysts (Me = Rh or Pt) reduced at different temperatures, ISS, in tandem with AES or XPS, indicates a decrease in the Me/Ti ratio when the reduction temperature increases (Table 12.1).[65] Other information leads the authors to conclude that SMSI can be attributed to metal-support electronic interactions whereas metal encapsulation by the support fits the results. On the other hand, Hoflund *et al.*[66] claimed that, upon reduction (Figure 12.22), TiO$_2$ is partly reduced into TiO, which moves over platinum and covers it. Note, however, that the sample described in that second example was prepared quite differently from the former one.

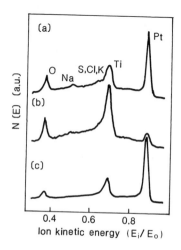

Figure 12.22. ISS spectra taken from a platinized titania sample after: (a) cleaning and annealing the sample in 10^{-6} Torr O$_2$, (b) annealing the sample in 10^{-5} Torr of H$_2$ for 1 h at 500 °C, (c) lightly sputtering the sample with 1-keV Ar$^+$.[66]

12.4. CONCLUSIONS

Low-energy ion scattering spectroscopy (LEISS or ISS) is a relatively new technique which has interesting applications in catalysis because of its extreme surface sensitivity. Noble gas ion interactions with atoms of the target, i.e., the scattering process, can be described by a binary elastic collision. The kinetics of this scattering interaction provides information on the masses of the top surface atoms of the target. As a consequence, this "surface mass spectroscopy" allows detection of all of the atoms of the Periodic Table except hydrogen, whose presence can be deduced indirectly from shielding effects.

Quantitative analysis is often difficult, in particular with practical catalysts, owing to the interference of several factors that affect the scattered ion intensity. The determination of scattering cross sections is dependent on the choice of the interaction potential. The survival probability is a function of the normal component of ion velocity and also of a matrix parameter which has to be determined. As a whole, this factor is the least understood as several neutralization processes may occur simultaneously, but in well-defined systems it can be estimated.

Shielding effects are potentially interesting as they can be used to deduce, from intensity measurements, the presence and also the location of adsorbed species as well as the local environment of the surface scattering atoms of the target. This has often been used to complement LEED measurements on adsorbed molecules on well-defined surfaces. Nevertheless, direct information on chemical bonds is not available.

As is usually observed in the use of ions interacting with solid targets, ion-induced sputtering is also encountered in ISS experiments. By using light ions (He^+, Ne^+) with low primary current densities, the destruction of the target surface is minimized. This progressive surface etching is beneficial to the investigation of profile concentrations, but preferential sputtering effects must be evaluated to determine the actual surface composition evolution. The use of low ion currents may shift the limits of detection to quite sensitive values: typically in the best cases 10^{-3} to 10^{-4} monolayer has been claimed to be detected. This limit is probably higher in practical surfaces due, in particular, to sensitivity attenuation by surface roughness.

ISS has a promising future in the characterization of catalyst surfaces but systems such as multicomponent catalysts are probably too complex and the resolutions too low when adjacent elements in the Periodic Table are present simultaneously to be characterized by this technique alone. Presence of pores in the examined solid and diffusion of mobile species to the external surface may be also a handicap (an argument also pointed out for SIMS) in the use of such a technique. Complementary measurements with other surface-sensitive techniques may in part overcome these difficulties.

REFERENCES

1. D. P. Smith, *J. Appl. Phys.* **38**, 340 (1967).
2. W. Heiland, H. G. Schaffler, and E. Taglauer, *Surf. Sci.* **35**, 381 (1973).
3. E. Taglauer and W. Heiland, *Surf. Sci.* **47**, 234 (1975).
4. E. Taglauer and W. Heiland, *Appl. Phys.* **9**, 261 (1976).
5. W. Heiland and E. Taglauer, *Nucl. Inst. Meth.* **132**, 535 (1976).
6. H. H. Brongersma, *J. Vac. Sci. Technol.* **11**, 231 (1974).
7. H. H. Brongersma and P. M. Mul, *Surf. Sci.* **35**, 393 (1973).
8. H. H. Brongersma, F. Meijer, and H. W. Werner, *Phil. Tech. Rev.* **34**, 362 (1974).
9. H. H. Brongersma and T. M. Buck, *Nucl. Instr. Meth.* **132**, 559 (1976); *Surf. Sci.* **53**, 649 (1975).
10. R. E. Honig and W. L. Harrington, *Thin Solid Films* **19**, 43 (1973).
11. J. B. Theeten, *La recherche* **70**, 770 (1976).
12. J. B. Theeten and H. H. Brongersma, *Rev. Phys. Appl.* **11**, 57 (1976).
13. W. L. Baun, *Appl. Surf. Sci.* **1**, 81 (1977).
14. H. H. Brongersma, L. C. M. Berens, and G. C. J. Van Der Ligt, Electron and Ion Spectroscopy of Solids, Lecture Notes for NATO Advanced Study Institute, Ghent (1977).
15. E. Taglauer and W. Heiland, *Proc. 3rd Inter. Conf. on Solid Surfaces* (R. Dobrozemsky *et al.*, eds.), Wien (1977), p. 2495.
16. T. M. Buck. *Methods of Surface Analysis* (A. W. Czanderna, ed.), Elsevier, Amsterdam (1985).
17. J. M. Poate and T. M. Buck, in: *Experimental Methods in Catalytic Research,* Vol. 3 (R. B. Anderson and P. T. Dawson, eds.), Academic, New York (1976).
18. B. A. Horrell and D. L. Cocke, *Catal. Rev., Sci. Eng.* **29**, 447 (1987).
19. H. H. Brongersma, *Proc. 2nd Intern. Conf. on Solid Surfaces,* Kyoto (1974).
20. W. Heiland and E. Taglauer, *J. Vac. Sci. Technol.* **12**, 352 (1975).
21. W. Englert, W. Heiland, E. Taglauer, and D. Menzel, *Surf. Sci.,* **83**, 243 (1979).
22. I. M. Thorrens, in: *Interatomic Potentials,* Academic, New York (1972).
23. G. Moliere, *Z. Natur.* **2a**, 133 (1947); O. B. Firsov, *Sov. Phys. JEPT* **36**, 1076 (1959).
24. M. S. Child in: *Molecular Collision Theory,* Academic, London and New York (1974).
25. S. Kasztelan, Doctoral Thesis, Lille, France (1984).
26. H. Jeziorowski, H. Knözinger, E. Taglauer, and C. Vogdt, *J. Catal.* **80**, 286 (1983).
27. D. P. Smith, *Surf. Sci.* **25**, 171 (1971).
28. H. D. Hagstrum, *Phys. Rev.* **96**, 336 (1954).
29. R. L. Erickson and D. P. Smith, *Phys. Rev. Lett.* **34**, 297 (1975).
30. D. L. Christensen, V. G. Mossotti, T. W. Rusch, and R. L. Erickson, *Chem. Phys. Lett.* **44**, 8 (1976); *Nucl. Insrum. Meth.* **149**, 587 (1978).
31. R. C. Mc Cune, *J. Vac. Sci. Technol.* **18**, 700 (1981).
32. S. M. Davis, J. C. Carver, and A. Wold, *Surf. Sci,* **124**, L12 (1983).
33. M. Aono and R. Souda, *Jap. J. Appl. Phys.* **24**, 1249 (1985).
34. D. J. Godfrey and D. P. Woodruff, *Surf. Sci.* **105**, 438 (1981).
35. H. Naguib and R. Kelly, *Radiat, Eff.* **25**, 1 (1975).
36. H. H. Brongersma and P. M. Mul, *Chem. Phys. Lett.* **19**, 217 (1973).
37. H. J. Brongersma and P. M. Mul, *Chem. Phys. Lett.* **14**, 380 (1972).
38. W. L. Harrington and R. E. Honig, *22nd Annual ASMS Conf.,* Philadelphia (1974).
39. D. J. Ball, T. M. Buck, D. Mc Nair and G. H. Wheatley, *Surf. Sci.* **30**, 69, 1972.
40. E. Taglauer and W. Heiland, *Appl. Phys. Lett.* **24**, 437 (1974).
41. H. Niehus and E. Bauer, *Surf. Sci.* **47**, 222 (1975).
42. S. Kasztelan, J. Grimblot, and J. P. Bonnelle, *J. Chim. Phys.* **80**, 793 (1983).
43. E. Taglauer, U. Beitat, and W. Heiland, *Nucl. Inst. Meth.* **149**, 605 (1978).
44. E. Taglauer, G. Marin, and W. Heiland, *Appl. Phys.* **13**, 47 (1977).

45. M. Shelef, M. A. Z. Wheeler, and H. C. Yao, *Surf. Sci.* **47**, 697 (1975).
46. H. Knözinger, H. Jeziorowski, and E. Taglauer, *Proc. 7th Inter. Congr. on Catalysis.* Tokyo (Seiyama, T., and Tanabe K., eds.), Elsevier, Amsterdam (1980), p. 604.
47. B. Canosa Rodrigo, H. Jeziorowski, H. Knözinger, N. Thiele, X. Z. Wang, and E. Taglauer, *Bull. Soc. Chim. Belg.* **90**, 1339 (1981).
48. J. Abart, E. Delgado, G. Ertl, H. Jeziorowski, H. Knözinger, N. Thiele, X. Z. Wang, and E. Taglauer, *Appl. Catal.* **2**, 155 (1982).
49. S. Kasztelan, E. Payen, H. Toulhoat, J. Grimblot, and J. P. Bonnelle, *Polyhedron* **5**, 157 (1986).
50. S. Kasztelan, J. Grimblot, and J. P. Bonnelle, *J. Phys. Chem.* **91**, 1503 (1987).
51. F. Delannay, E. N. Haeussler, and B. Delmon, *Bull. Soc. Chem. Belg.* **89**, 255 (1980); *J. Catal.* **66**, 649, (1980).
52. J. M. Beuken and P. Bertrand, *Surf. Sci.* **162**, 329 (1985).
53. R. L. Chin and D. M. Hercules, *J. Phys. Chem.* **86**, 3079 (1982).
54. M. Houalla, C. L. Kibby, E. L. Eddy, L. Petrakis, and D. M. Hercules, *Proc. 8th Intern. Cong. on Catalysis, Berlin,* IV Verlag Chemie, Berlin (1984), p. 415.
55. M. Wu, R. Chin and D. M. Hercules, *Spectrosc. Lett.* **11**, 615 (1978).
56. M. Wu and D. M. Hercules, *J. Phys. Chem.* **83**, 2003 (1979).
57. L. Salvati, L. E. Makovsky, J. M. Stencel, F. R. Brown, and D. M. Hercules, *J. Phys. Chem.* **85**, 3700 (1981).
58. J. C. Carver, I. E. Wachs, and L. L. Murrell, *J. Catal.* **100**, 500 (1986).
59. D. Ouafi, F. Mauge, J. C. Lavalley, E. Payen, S. Kasztelan, M. Houari, J. Grimblot, and J. P. Bonnelle, *Catal. Today* **4**, 23 (1988).
60. P. J. C. Chappell, M. H. Kibel, and B. G. Baker, *J. Catal.* **110**, 139 (1988).
61. B. Horell, D. L. Cocke, D. L. Sparrow, and G. Murray, *J. Catal.* **95**, 309 (1985).
62. R. Margraf, J. Leyrer, E. Taglauer, and H. Knözinger, *React. Kinet, Catal.* **35**, 261 (1987).
63. R. Margraf, J. Leyrer, E. Taglauer, and H. Knözinger, *Surf. Sci.* **189/190**, 842 (1987).
64. S. Houssenbay, S. Kasztelan, H. Toulhoat, J. P. Bonnelle, and J. Grimblot, *J. Phys. Chem.* **93**, 7176 (1989).
65. E. S. Shpiro, B. B. Dysenbina, O. P. Tkachenko, G. V. Antoshin, and K. M. Minachev, *J. Catal.* **110**, 262 (1988).
66. G. B. Hoflund, A. L. Grogan, Jr, and D. A. Asbury, *J. Catal.* **109**, 226 (1988).

APPLICATIONS OF NEUTRON SCATTERING TO CATALYSIS

H. Jobic

13.1. INTRODUCTION

With the construction of intense neutron sources, such as at the Institut Laue–Langevin in Grenoble, the number of studies of condensed matter with neutron scattering has increased tremendously. The range of applications includes crystal and magnetic structures, molecular dynamics, disordered materials, polymers, structural excitations, and biology.

It may seem surprising that neutrons, which are weakly scattered by matter, can be used to study surface chemistry. In fact, only catalysts that have a large surface area and scatter neutrons much less than adsorbed species can be studied. The importance of neutron scattering for surface work is that it can characterize some properties of chemical systems that cannot be characterized by other techniques.

13.2. INTRODUCTION TO THE THEORY

13.2.1 Properties of Neutrons

The neutron wavelength is given by the de Broglie relation

$$\lambda = h/mv \tag{13.1}$$

where h is Plank's constant, m the neutron mass, and v its velocity. The wave

H. Jobic • Institut de Recherches sur la Catalyse, CNRS, Villeurbanne, France.

Catalyst Characterization: Physical Techniques for Solid Materials, edited by Boris Imelik and Jacques C. Vedrine, Plenum Press, New York (1994).

vector k associated with the neutron has the magnitude

$$k = 2\pi/\lambda \qquad (13.2)$$

Some useful neutron properties are: a mass of 1.675×10^{-24} g (at rest); spin equal to $\frac{1}{2}$ and a charge of 0. The magnetic moment $\mu = -1.913\ \mu_N$ (nuclear magnetons) and the momentum $\mathbf{p}$ and kinetic energy E are, respectively,

$$\mathbf{p} = m\mathbf{v} = \hbar\mathbf{k}$$
$$E = \tfrac{1}{2}mv^2 = \hbar^2 k^2/2m = h^2/2m\lambda^2 \qquad (13.3)$$

The unit most used in neutron scattering is meV. Some important conversion relationships are:

$$1\ \text{meV} = 8.07\ \text{cm}^{-1} = 11.6\quad \text{K} = 0.24 \times 10^{12}\quad \text{Hz} = 23.1\ \text{cal}$$

A useful property of neutrons is that the wavelength is of the order of magnitude of interatomic distances so that the structure of the sample can be studied. Further, the energies of neutrons produced from cold, thermal, or hot sources cover a wide range so that most of the molecular motions can be studied.

13.2.2. Scattering Cross Sections

The interaction of neutrons with matter occurs via nuclear forces (we leave aside the interaction of the magnetic moment of the neutron with the magnetic moment of unpaired electrons within the sample). The exact form of the interaction potential $V(r)$ is not known with precision but since the neutron-nucleus interaction occurs at a much shorter range ($\approx 10^{-4}$Å) than the neutron wavelength (≈ 1Å), the scattered wave will be isotropic so that one can use a pseudopotential defined as

$$V(\mathbf{r}) = (2\pi\hbar^2/m)b\delta(\mathbf{r} - \mathbf{R}) \qquad (13.4)$$

where b is the scattering length for a particular nucleus, m the neutron mass, and $\mathbf{r}$ and $\mathbf{R}$ are the neutron and nucleus positions. The scattering length of the nucleus, b, can be real or complex. The real part is usually positive and the imaginary part is related to the absorption of neutrons (it is small for most nuclei).

In neutron inelastic experiments, the quantity which is actually measured is the partial differential cross section $d^2\sigma/d\Omega dE$, which represents the number of neutrons scattered into a solid angle $d\Omega$ with energy in the range dE. This quantity can be calculated from perturbation theory[1] and it is related to matrix elements of the form: $\langle \lambda' | \exp(i\mathbf{Q} . \mathbf{R}) | \lambda \rangle$, where $|\lambda\rangle$ and $|\lambda'\rangle$ are the initial and final states and $\mathbf{Q}$ the neutron momentum transfer defined by

$$\hbar\mathbf{Q} = \hbar(\mathbf{k}_0 - \mathbf{k}') \qquad (13.5)$$

where $\mathbf{k}_0$ and $\mathbf{k}'$ are, respectively, the incident and scattered wave vectors. It therefore appears that in neutron scattering both the spatial and temporal properties of the sample can be investigated.

The total cross section is given by

$$\sigma_{tot} = \int \int \frac{d^2\sigma}{d\Omega dE} \, d\Omega dE = 4\pi b^2 \qquad (13.6)$$

This quantity is measured in barns ($1 \text{ barn} = 10^{-28}\text{m}^2$) and represents the size of the nucleus during the neutron–nucleus interaction (note that this quantity is indeed small).

13.2.3. Coherent and Incoherent Scattering

The scattering lengths vary irregularly from one atom to another, or even from one isotope to another, in contrast to X-ray scattering, where the atomic scattering factors vary linearly with the number of electrons in the atoms.

The coherent scattering takes into account interference effects between the waves scattered from each nucleus. The mean scattering potential which gives rise to coherent scattering is proportional to the mean scattering length $\langle b \rangle$. For a sample containing only one type of atom (one isotope) with zero spin, then $\langle b \rangle = b$ and the scattering will be totally coherent (e.g., ^{12}C and ^{16}O).

The incoherent scattering corresponds to the mean-square deviation from the mean potential and is proportional to $(\langle b^2 \rangle - \langle b \rangle^2)^{1/2}$. The deviations from the average are due to isotopic or spin effects.

The values of $\langle b^2 \rangle$ and $\langle b \rangle^2$ can be calculated for each natural element by an average over all its stable isotopes and their spin distributions. It is usual to display the results under the form of the coherent and incoherent cross sections, which are defined as

$$\sigma_{coh} = 4\pi \langle b \rangle^2$$
$$\sigma_{inc} = 4\pi(\langle b^2 \rangle - \langle b \rangle^2) \qquad (13.7)$$

In Table 13.1, these cross sections are given for some common elements, together with their mean scattering lengths.

For diffraction studies where the scattering can be considered as elastic (Bragg), the incoherent scattering is troublesome as it gives an intense background. It is thus better to use deuterated compounds, because the incoherent cross section of deuterium is 40 times weaker than that of hydrogen.

For inelastic studies, the energy of the scattered neutrons is analyzed and both coherent and incoherent scattering can be studied. For single crystals,

TABLE 13.1. Scattering Lengths and Coherent, Incoherent, and Absorption Cross Sections (in barns) for Some Common Elements[a]

Element	$\langle b \rangle$ $\times 10^{-14}$ m	σ_{coh}	σ_{inc}	σ_{abs}
H	-0.37	1.8	80	0.19
D	0.67	5.6	2	0
B	$0.53-0.02i$	3.54	1.7	426
C	0.66	5.6	0	0
N	0.94	11.01	0.49	1.1
O	0.58	4.23	0	0
Al	0.35	1.5	0	0.13
Si	0.42	2.2	0	0.1
V	-0.04	0.02	5.18	2.82
Ni	1.03	13.3	5.2	2.5
Cd	$0.51-0.70i$	3.3	2.4	1400
Pt	0.95	11.6	0.13	5.7

[a] σ_{abs} is proportional to λ, and it corresponds here to $\lambda = 1$ Å.

coherent scattering allows a determination of the phonon dispersion curves. For molecules adsorbed on surfaces, primarily incoherent scattering is studied, as it can be seen from Table 13.1 that in order to have a good contrast, the adsorbate must contain hydrogen atoms and the substrate must as far as possible be free from residual hydrogen.

13.3. EXPERIMENTAL TECHNIQUES

All the experimental results which will be given in the sections below have been obtained at the Institut Laue–Langevin, in Grenoble. The high-flux reactor of the ILL operates at a power of 57 MW, giving a maximum thermal (300 K) neutron flux in the D_2O moderator of 1.2×10^{15} n cm^{-2} s^{-1}. These thermal neutrons have a Maxwell distribution whose wavelength at the maximum of the flux is $\lambda_m = 1.2$ Å. This distribution is modified for some beam tubes by using different moderator temperatures. There is a hot source which is a block of graphite of 10 dm^3 and which reaches a temperature of 2000 K by nuclear heating after a few hours of reactor operation. There is also a cold source, which is a vessel containing 25 dm^3 of liquid deuterium at 25 K. For these sources, the maximum of the flux distribution is shifted to $\lambda_m = 0.5$ Å for hot neutrons and to $\lambda_m = 4$ Å for cold neutrons. At present, there are 35 instruments operational at the ILL.[2] Each has its own characteristics and is subject to continuous improvement.

A characteristic of neutron scattering is that large sample environments can be accommodated at the sample position, e.g., cryostats, furnaces, superconducting magnets, high-pressure cells, dilution refrigerators. The beam size at the sample position is typically 5×2 cm^2. As most of the results presented below were obtained on a few instruments at the ILL, we will describe some of these briefly. Other steady-state reactors (e.g., HFBR at Brookhaven or HFIR at Oak Ridge) have similar spectrometers.

13.3.1. Beryllium Filter Detector Spectrometer INFB

This spectrometer, which is facing the hot source, is used for measuring high-energy transfers, between 100 and 5000 cm^{-1}. A general sketch of this spectrometer is shown in Figure 13.1.

On this instrument, the incident energy is selected by Bragg scattering from copper monochromators. The monochromator support carries three different monochromators: Cu(200) Cu(220), and Cu(331). The changing of the mono-chromators is automatic and controlled by the program. For each copper plane, there is a variable vertical focusing which increases the incident flux. Thus the flux and the resolution can be adapted to the different regions of energy transfer.

To obtain a spectrum, the incident energy is varied stepwise by rotating the monochromator. The scattered neutrons are counted for a constant number of incident neutrons detected by a monitor placed in front of the sample. For each energy step, only those neutrons which have lost almost all their energy, due to a vibrational transition of the sample, pass through the beryllium filter and are detected. Indeed, only neutrons whose energy is smaller than the cut-off energy of beryllium, $E_c = 42$ cm^{-1} ($\equiv 4$ Å), are transmitted; neutrons which have

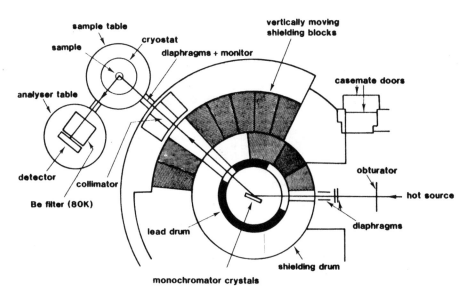

Figure 13.1. Schematic view of the beryllium filter detector spectrometer INFB (ILL).

higher energies are Bragg-scattered out of the beam and are absorbed by cadmium plates placed around the beryllium filter. The beryllium filter is cooled to 77 K to decrease the inelastic scattering and thus increase the transmission.

The observed peaks, corresponding to the vibrational transitions, are shifted to high energies by about 30 cm^{-1} because of the beryllium filter response function. The measured peak widths depend on the intrinsic widths of the modes and on the instrumental resolution. This instrumental resolution is dominated at low-energy transfers by the filter transmission function, and at high-energy transfers by the mosaic spread of the monochromator and by the collimation. It is of the order of $\Delta E/E = 5\%$. The resolution of the analyzer has recently been improved by using a beryllium–graphite combination filter.[3] Provided that the other contributions to the overall resolution (monochromator, collimation) are smaller, the resolution can be improved by using a filter with a lower cut-off; graphite has a cut-off energy of $E_c = 14$ cm^{-1} so that the resolution was improved by a factor 2.5.[3] Other recently built spectrometers in the world have similar performances.[4,5]

The intrinsic width of the peaks depends on the effective mass of the molecule and it varies from a δ function[3] to a Gaussian with a full width at half maximum (FWHM) of a few hundred cm^{-1}.[6]

13.3.2. Time-of-Flight Spectrometer IN6

IN6 is a time-of-flight (TOF) spectrometer used for quasi-elastic scattering (see Figure 13.2). This instrument is designed for intermediate elastic resolu-

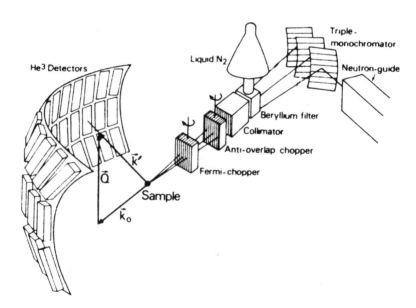

Figure 13.2. Schematic diagram of the time-of-flight spectrometer IN6 (ILL).

tion: 70–200 μeV (0.56–1.61 cm^{-1}) but the incident flux is very high due to vertical and horizontal focusing. The incident wavelength is in the range 4–6 Å (cold neutrons).

In the horizontal plane, time focusing is obtained by chopping sequentially three neutron beams of slightly different wavelengths produced from three identical monochromators having different orientations. These monochromators are in fact made up of composite pyrolytic graphite using the full height (20 cm) of the neutron guide and focusing the beam at the sample position.

On this spectrometer, a cooled beryllium filter is placed in front of the sample to remove the second-order reflection of the graphite monochromator. After scattering from the sample, the neutrons pass through a box filled with helium in order to reduce the background (natural helium scatters neutrons much less than air). The detector bank covers a range of scattering angles $10° < 2\theta < 115°$ and houses 337 elliptical ^{3}He detectors. The distance between the sample and the detectors is 2.47 m (flight path). The detectors can be grouped by software in several Q domains to get better statistics and to avoid the Bragg peaks of the sample. The neutrons scattered by the sample are analyzed: (i) in energy, by measuring their time-of-flight, and (ii) in momentum transfer, according to the scattering angle.

Other TOF spectrometers with different characteristics are also available at the ILL. For example, IN5 has a lower incident flux but a larger range of incident wavelengths: 2–15 Å. The backscattering spectrometer IN10 is well-suited for high-resolution work (0.3 μeV or 0.002 cm^{-1}). These spectrometers are chosen for each study according to the desired resolution and momentum transfer region.

13.3.3. Small-Angle Scattering

The instrument D11, used for small-angle neutron scattering (SANS), is installed at the exit of a cold neutron guide and the range of available wavelengths is $4.5 < \lambda < 20$ Å. A schematic view of D11 is shown in Figure 13.3.

The incoming beam of neutrons is monochromated by a helical slit velocity selector whose rotation speed determines the transmitted neutron wavelength.

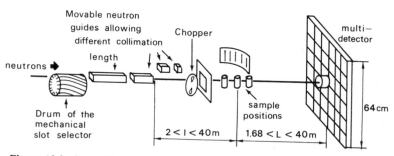

Figure 13.3. General layout of the small-angle scattering instrument D11 (ILL).

The collimation of the incident beam can be varied by shifting movable guides into the beam. The detector is a multiwire proportional chamber. The 128 wires form a 64×64 cm^2 grid containing 3808 active elements (the corner regions are not active). The detector can be positioned at any distance between 1.1 and 37 m from the sample; this modifies the resolution and the Q range ($4\pi \sin \theta/\lambda$), which varies from 10^{-3} to 0.5 Å^{-1}.

Other instruments such as D16 or D17, also equipped with multidetectors, can cover a different Q range.

13.3.4. Two-Axis Powder Spectrometers

The diffractometer D1B, shown in Figure 13.4, gives both a high flux and a good resolution at low angles (FWHM down to 0.2°).

The instrument is installed near a thermal guide tube and the incoming beam of neutrons is shared with another diffractometer. The monochromator is

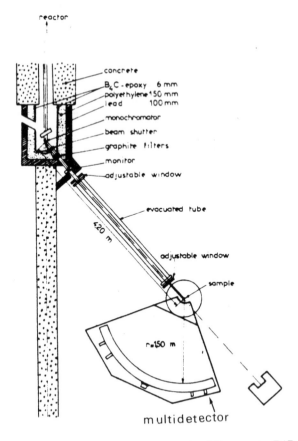

Figure 13.4. Schematic view of the two-axis powder diffractometer D1B (ILL).

pyrolytic graphite (002) or germanium (311) so that the wavelength can be taken as 2.52 or 1.28 Å, respectively. The scattered intensity is recorded as a function of scattering angle by sweeping the multidetector through the entire 2θ range and recording the intensity at 2θ intervals of typically 0.05 Å. This detector is a curved multidetector which contains 400 active elements covering $2\theta = 80°$. For kinetic experiments, the typical time to obtain a good pattern is less than 1 h.

There are also high-resolution diffractometers ($\Delta d/d$ down to 5×10^{-4}), where a wide choice of wavelengths is available (1.2 to 5.7 Å).

13.4. NEUTRON DIFFRACTION

The neutron diffraction phenomenon or coherent scattering is analogous to the one observed with X-rays for a crystal. The expressions giving the scattered intensities are, however, simpler. The amplitude scattered by a single nucleus p at the position $\mathbf{R}_p$ is b_p, the phase of the scattered radiation is $\exp(i\mathbf{Q}\cdot\mathbf{R}_p)$, so that the coherent cross section is

$$d\sigma_{coh}/d\Omega = \left| \sum_p \langle b_p \rangle \exp(i\mathbf{Q}\cdot\mathbf{R}_p) \right|^2 \tag{13.8}$$

If we consider that in a crystal the atoms lie in a periodic three-dimensional array, we can rewrite the position vector $\mathbf{R}_p$ as

$$\mathbf{R}_p = \mathbf{l} + \mathbf{d}_p \tag{13.9}$$

where $\mathbf{l}$ is a lattice vector referring to a particular cell and $\mathbf{d}_p$ gives the position of atom p within the cell, so that the cross section can be written as

$$d\sigma_{coh}/d\Omega = \left| \sum_{\mathbf{l}} \exp(i\mathbf{Q}\cdot\mathbf{l}) \right|^2 \times \left| \sum_p \langle b_p \rangle \exp(i\mathbf{Q}\cdot\mathbf{d}_p) \right|^2 \tag{13.10}$$

If we define the structure factor of the unit cell as

$$F = \sum_p \langle b_p \rangle \exp(i\mathbf{Q}\cdot\mathbf{d}_p) \tag{13.11}$$

the coherent scattering cross section becomes

$$d\sigma_{coh}/d\Omega = |F|^2 \left| \sum_{\mathbf{l}} \exp(i\mathbf{Q}\cdot\mathbf{l}) \right|^2 \tag{13.12}$$

The second factor in equation (13.12) ensures that the scattering is limited to certain values of the scattering vector $\mathbf{Q}$ (related to θ by the expression $Q = 4\pi \sin \theta/\lambda$), which are the Bragg peaks; the structure factor F governs the intensities of the reflections.

In the case of a layer of atoms q adsorbed on a substrate made of atoms p, the scattering amplitude can be divided into two terms:

$$F_p \sum_p \exp(i\mathbf{Q}\cdot\mathbf{l}_p) + F_q \sum_q \exp(i\mathbf{Q}\cdot\mathbf{l}_q) \qquad (13.13)$$

The coherent cross section, which is proportional to the squared modulus of the amplitude, will then include four different terms which will be proportional to $|F_p|^2$ for the substrate and to $|F_q|^2$ for the adsorbate, and there will be two terms containing $F_p F_q^*$ and $F_p^* F_q$ representing cross interference between the adsorbate and the substrate. The term containing $|F_q|^2$ corresponds to the diffraction from a two-dimensional layer. The size of the crystalline domains will modify the shape of the diffraction peaks. The cross-interference terms can be used to obtain information about the position of the adsorbate relative to the substrate, both normal and parallel to the surface.

Most of the studies published so far in this field deal with small molecules physisorbed on graphite (e.g., inert gases, CD_4, ND_3, . . .); see Thomas[7] for a good review.

In a more recent study of deuterium chemisorbed on Co_3O_4,[8] it was shown by diffraction that deuterium is dissociated at room temperature but not at 40 K, and that this chemisorption involves electron transfer toward the solid (presumably toward Co^{3+} ions in tetrahedral sites).

Another research domain of growing activity is the use of neutron diffraction to locate molecules adsorbed in zeolites.[9-11] The adsorption of methane and acetylene in Na–A has been studied in this way.[9] At room temperature for C_2D_2 and at 40 K for CD_4, the molecules are found to be trapped in front of the NaII and NaIII cations. The fact that the axis of the acetylene molecule is not parallel to the axis of the window was proposed as a possible explanation for the very long time needed to reach equilibrium of C_2H_2 adsorption in Na–A. (This molecule can jump into neighboring cavities only when its axis is parallel to the axis of a window.) In the case of methane, the molecule becomes more and more delocalized as the temperature is increased.

The adsorption of benzene in Na–Y has recently been studied at 4 and 300 K by neutron powder diffraction.[10] At 4 K, two different benzene sites are found, and at 300 K the molecules occupy the same sites but their disorder is very high. The first benzene molecule (see Figure 13.5) is in the supercage and is bonded via its π electrons to a SII sodium ion (four such sites are available per supercage). The second molecule is centered in the window consisting of a 12-membered oxygen ring, between adjacent supercages (there are, on average, two of these sites per supercage).

The two-site occupancy is a function of benzene loading. At lower coverage, the benzene molecules are found to be mainly bonded to the SII sodium ions. At

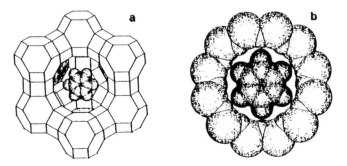

Figure 13.5. Diagram showing the two benzene molecule species in Na–Y zeolite: (a) within the supercage, at a distance of 2.7 Å from the Na(2) ion, (b) in a window between two supercages.[10]

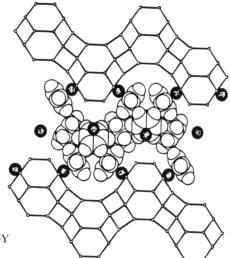

Figure 13.6. Cross section through Na–Y showing clustering of benzene molecules.[48]

high coverage these molecules move toward the center of the supercage and form clusters through the molecules in the windows (see Figure 13.6). This clustering of benzene molecules in Na–Y was also suggested from a small-angle neutron scattering study.[12]

13.5. VIBRATIONAL SPECTROSCOPY

13.5.1. Calculation of the Differential Cross Section Corresponding to the Vibrational Motions

The neutron can exchange energy with the sample and excite transitions between vibrational levels. This corresponds to inelastic scattering. The energy transfer is given by

$$\hbar\omega = E - E' = (\hbar^2/2m)(k_0^2 - k'^2) \tag{13.14}$$

Just as the scattering cross section can be split into coherent and incoherent contributions, the differential cross section can be written as

$$\frac{d^2\sigma}{d\Omega dE} = \frac{k'}{k}\frac{N}{4\pi}\left[\sigma_{\text{inc}}S_{\text{inc}}(\mathbf{Q}, \omega) + \sigma_{\text{coh}}S_{\text{coh}}(\mathbf{Q}, \omega)\right] \tag{13.15}$$

As explained in Section 13.2.3, hydrogenated compounds are often studied in chemisorption so that the main term in equation (13.15) will be the incoherent one. The scattering law, $S(\mathbf{Q}, \omega)$, then depends only on the self-motion of an atom:

$$S_{\text{inc}}(\mathbf{Q}, \omega) = \frac{1}{N}\sum_{\lambda\lambda'}P_\lambda\sum_l |\langle\lambda'|\exp(i\mathbf{Q}\cdot\mathbf{R}_l)|\lambda\rangle|^2\delta(\hbar\omega + E - E') \tag{13.16}$$

where P_λ is the probability of occupation of the initial state λ:

$$P_\lambda = \frac{\exp(-E_\lambda/kT)}{\sum_{\lambda'}\exp(-E_\lambda/kT)} \tag{13.17}$$

This formalism is found in other spectroscopies, the only difference lying in the matrix elements in equation (13.16). In IR, the matrix elements are $\langle\lambda'|\mathbf{e}\cdot\mathbf{\mu}|\lambda\rangle$, where $\mathbf{e}$ is a unit vector of the incident electromagnetic field and $\mathbf{\mu}$ the electric dipole moment. In Raman spectroscopy, the matrix elements are $\langle\lambda'|e_0\cdot\tilde{\alpha}\cdot e'|\lambda\rangle$, where $\tilde{\alpha}$ is the polarizability tensor. Therefore in these optical spectroscopies a vibration will be active only if there is a change in $\mathbf{\mu}$ or $\tilde{\alpha}$ during the vibration. For simple systems it is possible in neutron scattering to calculate the matrix elements in equation (13.16) directly, but for polyatomic molecules it is more convenient to expand the position vector $\mathbf{R}_l$ into translational, rotational, and vibrational contributions, as they all occur at different time scales:

$$\mathbf{R}_l = \mathbf{d} + \mathbf{b}_l + \mathbf{u}_l \tag{13.18}$$

where $\mathbf{d}$ gives the position of the molecular center of mass, $\mathbf{b}_l$ the position of the atom l relative to the center of mass, and $\mathbf{u}_l$ the displacement of the atom from the equilibrium position under the effect of the vibrations. Assuming that these motions are uncoupled, we can write equation (13.16) in the form of a convolution product:

$$S(\mathbf{Q}, \omega) = S_{\text{trans}}(\mathbf{Q}, \omega)\otimes S_{\text{rot}}(\mathbf{Q}, \omega)\otimes S_{\text{vib}}(\mathbf{Q}, \omega) \tag{13.19}$$

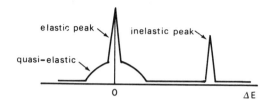

Figure 13.7. Schematic spectrum corresponding to quasi-elastic, elastic and inelastic peaks.

The translational and rotational motions will be discussed in Section 13.6. These motions which imply small energy transfers (≤ 1 meV) occur near the elastic peak, in an energy domain that is called quasi-elastic (Figure 13.7).

The vibrational motions involve higher energy transfer (a few to 500 meV), and they produce inelastic peaks well separated from the elastic peak, so that the method is called neutron inelastic scattering (NIS).

For an harmonic system, the vibrations of a molecule can be separated into normal modes of motion. The normal modes are dynamically independent so that the displacement vector for an atom l, $\mathbf{u}_l$, can be expressed in terms of the normal coordinates $\mathbf{q}$:

$$\mathbf{u} = \mathbf{Cq} \tag{13.20}$$

where the $\mathbf{C}$ matrix describes the atomic displacements in mass-weighted Cartesian coordinates. It has been shown[13] that the differential cross section for the intramolecular modes is given by

$$\frac{d^2\sigma}{d\Omega dE} = \frac{k'}{k}\frac{\sigma_H}{4\pi} \sum_l \prod_k \left\{ \exp(-2W_l) \exp\left(\frac{n_k \hbar\omega}{2kT}\right) \frac{1}{|n_k|!} \right.$$
$$\left. \times \left[\frac{\hbar(\mathbf{Q}\cdot\mathbf{C}_l^k)^2}{4m_H\omega_k \sin h(\hbar\omega/kT)} \right]^{|n_k|} \delta\left(\omega - \sum_k |n_k|\omega_k\right) \right\} \tag{13.21}$$

The observed spectrum consists of a series of δ functions corresponding to transitions of n quanta $\hbar\omega$ ($n > 0$ for neutron energy loss). The Debye–Waller factor, $\exp(-2W_l)$, only includes the mean-square displacement $\langle u^2 \rangle$ due to the intramolecular modes:

$$\exp(-2W_l) = \exp(-Q^2 \langle u^2 \rangle_{\text{int}}) \tag{13.22}$$

Harmonics or simultaneous transfer of single quanta to different modes can occur because of the product $\prod_k$ in equation (13.21). In general, the fundamentals will be the most intense features of the NIS spectrum, with their intensity given by ($n_k = 1$)

$$\frac{d^2\sigma}{d\Omega dE} = \frac{k'}{k}\frac{\sigma_H}{4\pi} \sum_l \exp(-2W_l)\frac{1}{1 - \exp(-\hbar\omega/kT)} \frac{(\mathbf{Q}\cdot\mathbf{C}_l^k)^2}{2m_H\omega_k} \delta(\omega - \omega_k)$$

$$\tag{13.23}$$

This is the basic equation in NIS and several conclusions can be drawn from it:

1 There are no "optical selection rules" as in IR or Raman spectroscopies, i.e., all the modes can be observed.
2. Neutrons are very sensitive to hydrogen because of its large incoherent cross section (σ_H) and its low mass (m_H) so that for hydrogenous compounds, the sum over the atoms Σ_l can be restricted to the hydrogen atoms.
3. The modes involving large amplitudes (C^k) for the protons will have high intensities. Polarization effects can be measured for crystalline compounds.[14]
4. The scattering from other atoms can be measured. For example by selective deuteration one can not only shift the vibrational frequencies but also vary the intensity of the modes involved because deuterium has a smaller cross section (cf. Table 13.1).
5. It is possible to compute, from a normal coordinate analysis, both the frequencies and the intensities of the vibrational transitions; examples will be given below.

It must be noted that the measurement of the internal modes for a molecular crystal is complicated by the fact that they are convoluted with the low-frequency external modes.[15–18] With the new spectrometers which are now in use across the world,[3–5] it is possible to resolve the external modes around the internal modes. The intensity which is taken from the fundamentals and redistributed to the external modes and to multiphonon processes depends on the Debye–Waller factor from the external modes, $\exp(-Q^2\langle u^2\rangle_{ext})$. This quantity varies with temperature and neutron momentum transfer; in order to sharpen the fundamentals, it is useful to work at very low temperatures.[18]

13.5.2. Application of NIS to Adsorbed Species

It was noted in the general introduction that large-area substrates are needed for surface work. Therefore the sample, usually in powder form, is porous and tends to have an inhomogeneous surface, e.g., silica, alumina, and Raney metals. On the other hand, zeolites are well-crystallized materials with a framework enclosing channels or cavities thereby offering large surface area. All these powders are widely used in catalysis and they are not easily studied by other spectroscopic techniques, i.e., IR, Raman, or EELS.

1. These materials are usually optically opaque or they have only limited frequency windows; they may also fluoresce in a laser beam. On the other hand, they can be almost transparent to neutrons if they include a small quantity of hydrogen. Therefore the whole energy range, from 1 to 5000 cm^{-1}, can be studied.
2. The assignment is made easier by the fact that there are no selection rules and that the NIS vibrational intensities can be calculated, whereas

in IR, Raman, or EELS, the calculation of the intensities of the molecular vibrations is impossible in most cases.
3. For neutron experiments, containers are made of aluminum, stainless steel, or quartz (pyrex cannot be used since it contains boron, an element which absorbs neutrons strongly). Cells can be made for vacuum or high-pressure work and placed in cryostats or furnaces.

13.5.2.1. Hydrogen Chemisorbed on Raney Metals

Raney nickel catalysts are widely used in industry for hydrogenation reactions. Other metals (e.g., Pd, Pt, and Ru) can be prepared in that form and are suitable for neutron studies. Raney nickel is made by leaching aluminum out of an Ni–Al alloy with boiling sodium hydroxide. The catalyst appears as nickel crystallites supported on Ni–Al cores (the residual quantity of aluminum is 2–5 wt.%). The surface area of these materials depends on the temperature of pretreatment, and is of the order of 30 to 80 m^2g^{-1}.

Hydrogen adsorption on nickel has been studied by different spectroscopic methods, i.e., IR, EELS, and NIS. The IR experiments were performed on nickel particles supported on alumina.[19] EELS data have been obtained on single crystals[20-25] and since only a limited number of sites are available on well-defined crystal planes, these results are useful for assigning powder spectra. Most of the NIS results that have been reported have been obtained with Raney nickel.[26-30] Although the NIS spectra are interpreted in terms of multiply bonded protons on the surface, the assignment of the vibrational peaks to hydrogen bonded to two, three, or four nickel atoms is controversial.[26-32] The proportion of terminal (or linear) hydrogen is also difficult to estimate.

It has been shown in Section 13.5.1 that the intensities of the peaks are proportional to the mean-square displacements of the hydrogen atoms. For a simple case such as hydrogen adsorbed on metals, the metal atoms can be considered as fixed during the local modes of hydrogen (in the range 300–3000 cm^{-1}). Therefore the nondegenerate modes of hydrogen will have roughly the same NIS intensity and an E mode will be, with a good approximation, twice as intense as an A mode. A measure of the integrated intensities of the peaks can then yield the populations of the various sites.

13.5.2.1a. One type of spectrum, which is obtained after evacuating the Raney catalyst to 10^{-4} Pa at 473 K and then heating at 573 K for 1 h, is shown in Figure 13.8a. The main features of the spectrum obtained under these experimental conditions are a peak at 940 cm^{-1} and a shoulder at 1100 cm^{-1}. Hydrogen adsorption (corresponding to $\theta = 1$) was performed after recording the spectrum of bare nickel (the spectrum shown in Figure 13.8a is a difference spectrum between the gas-loaded sample and the bare sample). It can be noted that the spectra corresponding to $\theta = 1$ and $\theta = 0.42$ are very similar.[26]

On the basis of the frequency range and the relative intensities of the bands, the NIS spectrum of Figure 13.8a can be easily assigned. The peak at 1100 cm^{-1} is attributed to the symmetric stretch (ν_s, A_1 symmetry) and the peak at 940 cm^{-1} to the antisymmetric stretch (ν_{as}, E symmetry) of a C_{3v} symmetry species.

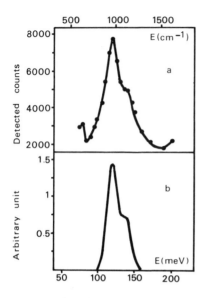

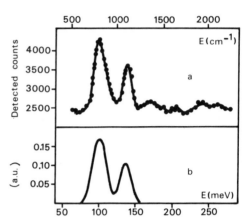

Figure 13.8. NIS spectra of hydrogen adsorbed at 300 K on Raney nickel and corresponding to a C_{3v} species: (a) experimental, (b) calculated.[29]

Figure 13.9. NIS spectra of hydrogen adsorbed on Raney nickel and corresponding to a nearly C_{3v} species: (a) experimental, (b) calculated.[29]

This spectrum can be simulated by taking the mode at 940 cm^{-1} to be twice as intense (because it is doubly degenerate) as the one at 1100 cm^{-1}. The calculated spectrum is shown in Figure 13.8b. The widths of the calculated peaks are fitted to the experimental one; this takes into account the instrumental resolution and the intrinsic width of the modes. Since there is a good agreement between the experimental and calculated spectra, it is found that this type of spectrum can be simulated with only one hydrogen species of threefold symmetry, as on Ni(111). If a significant proportion of terminal hydrogen were adsorbed, the bending mode of this species would modify the NIS profile between 700 and 1200 cm^{-1}. The small band at about 600 cm^{-1} in Figure 13.8a is assigned to the symmetric stretching mode of a hydrogen atom bonded in a fourfold site.[20] According to its intensity this species, as well as terminal hydrogen, represents a small proportion of the adsorbed hydrogen ($< 10\%$).

For the threefold symmetry sites, the values of v_{as} and v_S can be used to calculate the Ni–H bond length.[33] A value of 1.87 ± 0.03 Å is derived, in good agreement with the distance obtained from LEED on Ni(111), which was 1.84 Å.[34]

13.5.2.1b. For other experimental conditions, e.g., Raney nickel covered with hydrogen at saturation for 10 days and then partially degassed at 600 K just before the neutron experiment, there are different peak positions, as can be seen in Figure 13.9.

Two peaks are observed at 780 and 1080 cm^{-1} which are assigned to hydrogen adsorbed on nearly threefold symmetry sites. These sites are found on the (110) unreconstructed surface (taking into account the second nickel layer), or on the reconstructed surface that is obtained above 220 K in the presence of hydrogen.[35]

The peak at 1080 cm^{-1} is assigned to the symmetric stretch and the one at 780 cm^{-1}, which is twice as intense, to the antisymmetric stretch. The width of the peak at 780 cm^{-1}, which is larger than expected, suggests that ν_{as} is not of E symmetry and that a lifting of degeneracy has occurred. This spectrum can be simulated (Figure 13.9b) by taking two modes at 750 and 815 cm^{-1} (± 10 cm^{-1}) and a third one at 1080 cm^{-1} (all these modes having the same intensity).

13.5.2.1c. Another type of spectrum, such as the one reported in Figure 13.10a, can also be obtained by performing the hydrogen adsorption several days before the experiment or by using a higher treatment temperature.

In this case, the two species previously described are found on the surface. From the simulation (Figure 13.10b), it can be deduced that about 60% of the hydrogen atoms are on the C_{3v} sites and 40% on the nearly C_{3v} sites. The broad band centered at 1900 cm^{-1} is due mainly to the overtones and combinations between the fundamentals.

13.5.2.1d. To conclude, it is clear that the NIS spectrum of hydrogen adsorbed on Raney nickel reflects the different experimental conditions of sample preparation. It is possible to follow the surface reconstruction by measuring variations in the populations of the various sites. Contrary to some authors who still claim that after dissociation one hydrogen atom is bonded to one nickel atom,[36] it is now clear that the main species adsorbed on nickel are multiply bonded protons. The experimental results obtained on single crystals[20-25] and recent theoretical calculations[37-39] also favor highly coordinated species.

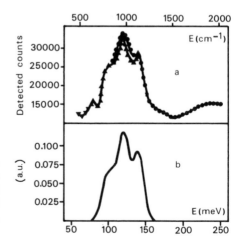

Figure 13.10. NIS spectra of hydrogen adsorbed on Raney nickel and corresponding to a mixture of two species: (a) experimental [points (▲) and (●) obtained with the (200) and (220) copper planes, respectively], (b) calculated for the main peaks.[29]

In another study performed on Raney nickel as a function of hydrogen pressure (up to 80 bars), the progressive filling of the adsorption sites could be followed.[30] The saturation of the Raney nickel surface with hydrogen is obtained at about 1 bar and it corresponds to 1.3 to 1.4 monolayers. The hydrogen atoms are still mainly bonded to three nickel atoms.

NIS results have also been obtained for hydrogen adsorbed on MoS_2,[40] or on palladium and platinum.[41-42] On metals, the main hydrogen species on the surface is always multiply bonded, even if recent results obtained on some Raney nickels indicate that the proportion of terminal hydrogen may reach approximately 20%. The formation of hydrides in small particles of palladium supported on Y zeolite has also been studied by NIS.[53] It can be noted that several neutron scattering techniques have been used to study hydrides: (i) neutron diffraction, to localize the adsorption sites, (ii) NIS, to study the potential wells, and (iii) quasi-elastic neutron scattering, to follow the jump dynamics of hydrogen between the different sites.

13.5.2.2. Benzene Adsorption on Raney Nickel and Platinum

Catalytic hydrogenation of benzene on transition metals is of great interest as a model reaction.[43] However, the reaction intermediates leading to the formation of cyclohexane are not yet known. Various spectroscopic methods have been used to determine the structure of adsorbed benzene, but because of the difficulties described in Section 13.5.2, only the results obtained from EELS and from NIS are of significance from a vibrational point of view. The structure and the perturbation of the adsorbed molecule can be studied by NIS. The frequencies and the intensities of the vibrational transitions can be calculated from a normal coordinate analysis.[46] The validity of the model and the assignment can therefore be tested more thoroughly. An example of this type of computation is shown in Figure 13.11. The experimental spectrum was obtained at 77 K on the old version of the INFB spectrometer at the ILL. The spikes in the calculated spectrum were derived from a force field determined by Favrot et al.[47]; they were then folded with the instrumental resolution in order to obtain the continuous profile. All the vibrational modes of benzene, except the C—H stretching modes situated around 3000 cm^{-1}, are shown and have a more-or-less intense contribution in the NIS spectrum. With the new version of INFB[3] as well as for other recently built spectrometers,[4,5] the vibrational modes and the combinations with the lattice modes at 480 and 775 cm^{-1} can be better resolved.[48] It can be noted that the NIS spectra calculated from other force fields published for benzene are very similar, because the assignment is the same for all authors. However, the NIS spectrum is in general very sensitive to changes in the assignment.

The NIS spectrum of benzene adsorbed on Raney platinum (Figure 13.12a) is significantly different from the one obtained for the free molecule. Some vibrations are shifted to higher frequencies, others to lower frequencies, the shifts being as high as several tens of cm^{-1}. New peaks are observed below 500 cm^{-1}, and these correspond to the vibrational modes of benzene relative to the

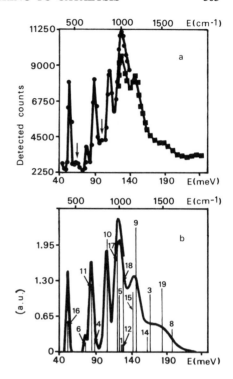

Figure 13.11. NIS spectra of solid benzene[44]: (a) experimental (77 K), combination peaks are indicated by arrows, (b) calculated, the numbering system of Wilson[45] is used for the frequencies.

surface. However, no adsorbed hydrogen is detected as this would give an intense peak around 600 cm^{-1}[42] in a region where the scattered intensity is the weakest in Figure 13.12a. It can therefore be deduced that the molecules are adsorbed parallel to the surface and that no dehydrogenation has occurred.

In order to assign the vibrational modes of adsorbed benzene, the force field of a model molecule, $(\eta^6 - C_6H_6)Cr(CO)_3$ was transferred to the case of adsorbed benzene by changing a few force constants. The adsorbed molecule is adsorbed flat on the platinum surface (estimated distance 2 ± 0.5 Å) and bound to one metal atom. The comparison between the force constants of the surface complex and those of the free molecule gives information on the perturbation of the benzene ring and on the strength of interaction with the platinum surface.[49] The agreement between the experimental and calculated spectra (Figure 13.12) shows that this model is satisfactory.

The NIS spectrum of benzene adsorbed on Raney nickel (Figure 13.13a) is very different. In order to interpret the spectrum, it is necessary to include a contribution from adsorbed hydrogen coming from a small proportion ($\approx 15\%$) of benzene totally dehydrogenated and firmly bound to the surface (Figure 13b). The adsorption geometry used in the normal coordinate analysis[44] is the same as on platinum; the benzene molecule is taken to be parallel to the surface and bound to one nickel atom because the EELS spectra obtained on the (111) and (100) planes were almost identical.[50] However, benzene could keep an effective

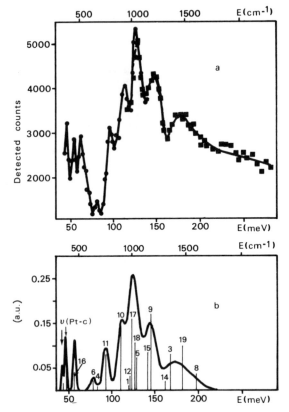

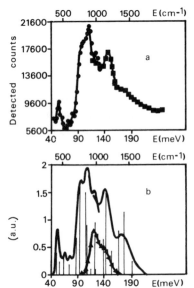

Figure 13.12. (a) NIS spectrum of benzene adsorbed at 300 K on Raney platinum and recorded at 5 K, (b) calculated spectrum.[49]

Figure 13.13. (a) NIS spectrum of benzene adsorbed at 300 K on Raney nickel and recorded at 77 K, (b) calculated spectrum with added contribution from adsorbed hydrogen.[44]

sixfold local symmetry while the site group would depend on the crystalline plane.[54]

When the surface of Raney nickel is precovered with hydrogen ($\theta_H = 0.5$), a weaker perturbation of the benzene ring is observed compared with the bare surface.[51] The adsorbed benzene molecules stay in the form of benzene and are not hydrogenated to cyclohexane. The vibrational frequencies of adsorbed benzene are closer to those of the free molecule and the Ni—C force constants are weaker than on the bare nickel surface. These observations are in agreement with other results, such as magnetism, which indicate a variation of the saturation magnetization of 4.6 BM (Bohr magneton) per adsorbed benzene molecule on bare nickel and only 1.5 BM on hydrogen-covered surface.[52]

The adsorption of other hydrocarbons such as acetylene, ethylene, or cyclohexane on Raney nickel has also been studied by NIS.[55-57] Some recent applications of NIS concern molecules adsorbed in zeolites (e.g., see Jobic et al.[48] and Howard et al.[64]).

13.6. QUASI-ELASTIC SCATTERING STUDY OF DIFFUSIVE MOTIONS

In order to study very small energy transfers ($< kT$) that result from unquantified diffusive motions, such as translations, it is better to use an alternative representation of the matrix element formula, (equation 13.16), in terms of time-dependent atomic coordinates, i.e., correlation functions. The theoretical expression for the cross section, (equation 13.16), can be developed into a more useful form using the Fourier representation of the δ function:

$$S_{\text{inc}}(\mathbf{Q}, \omega) = (1/2\pi\hbar) \int_{-\infty}^{+\infty} dt \exp(-i\omega t)I_s(\mathbf{Q}, t) \qquad (13.24)$$

where $I_s(\mathbf{Q}, t)$ is called the intermediate scattering function. Introducing the space-time correlation function $G_s(\mathbf{R}, t)$,[58] we can express the scattering law as the double Fourier transform of this function:

$$S_{\text{inc}}(\mathbf{Q}, \omega) = (1/2\pi\hbar) \int\int drdt \exp[i(\mathbf{Q} \cdot \mathbf{R} - \omega t)]G_s(\mathbf{R}, t) \qquad (13.25)$$

For a classical system, $G_s(\mathbf{R}, t)$ is the probability that if a particle is at the origin at time zero, the same particle will be at position $\mathbf{R}$ at time t (hence the subscripts which stands for self). If the system contains protons, only the incoherent scattering has to be considered, as in inelastic scattering. As the range of Q and ω which can be covered by neutron instruments is not sufficient to perform the Fourier transform of the experimental scattering law, the approach that is used is to determine $G_s(R, t)$ for model systems, calculate the theoretical $S(Q, \omega)$, and compare it with the experimental results. For a molecular system, the usual approximation is to consider that the different molecular motions are uncoupled and can be treated separately. The vibrational term was discussed in the previous section; the translational and rotational contributions will now be considered.

13.6.1. Translational Motion

For a liquid, or for a system of particles undergoing continuous diffusion, the diffusion equation takes the form

$$D\nabla^2 G_s(\mathbf{R}, t) = \partial G_s(\mathbf{R}, t)/\partial t \tag{13.26}$$

where D is the self-diffusion coefficient. The solution of this equation, which can be recognized as the usual Fick's law, is a Gaussian expression:

$$G_s(\mathbf{R}, t) = [1/(4\pi Dt)^{3/2}] \exp(-R^2/4Dt) \tag{13.27}$$

This solution has the form $\delta(R)$ as $t \to 0$ and when integrated over the volume, it is equal to unity. The space Fourier transform gives the intermediate scattering law:

$$I_s(\mathbf{Q}, t) = \exp(-DQ^2 t) \tag{13.28}$$

and the spectrum is the time Fourier transform of $I_s(\mathbf{Q}, t)$:

$$S_{\text{inc}}^{\text{trans}}(\mathbf{Q}, \omega) = (1/\pi)DQ^2/[\omega^2 + (DQ^2)^2] \tag{13.29}$$

The shape is a Lorentzian with a half-width at half-maximum DQ^2. As a large number of spectra are measured simultaneously at different Q values (see Section 13.3.2), a graph of the width as a function of Q^2 should be a straight line of slope D. This is one test of the correctness of the model. It should be noted that this model is essentially valid at small Q values corresponding to large distances (i.e., for $QR \ll 1$, where R is the molecular radius).

In heterogeneous systems, other diffusion models can apply. In particular, the molecule can jump between sites. Several models for jump diffusion have been proposed.[59-61] At small Q values, i.e., large distances, the details of the jumps are lost and the scattering laws reduce to equation (13.29). The jump distances can be derived from the experimental results whereas this quantity has to be estimated for NMR measurements. The mean time between jumps, τ, is related to the jump distance L and to the self-diffusion coefficient by Einstein's equation

$$\tau = L^2/6D \tag{13.30}$$

There is a complication when the system is not isotropic.[7,62] In zeolites for example, the diffusion of molecules may take place in one or two dimensions; in this case it is necessary to do a powder average of the scattering law.[63]

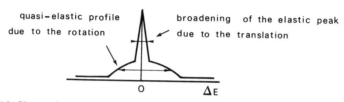

Figure 13.14. Shape of spectra showing separation of translation and rotation contributions.

13.6.2. Rotational Motion

For some systems, the motions of translation and rotation occur on the same time scale, and it is difficult to separate the two contributions in the quasi-elastic profile. When the broadenings of the two peaks are different in magnitude, both motions can be characterized with precision; the spectra then have the shape shown in Figure 13.14.

As a general rule, the incoherent scattering law for a rotational motion can be written as the sum of an elastic term and of a quasi-elastic component:

$$S_{inc}^{rot}(Q, \omega) = A_0(Q)\delta(\omega) + \sum_l A_l(Q)\mathscr{L}(\omega, \Gamma_l) \qquad (13.31)$$

The elastic peak intensity is governed by $A_0(Q)$, which is known as the elastic incoherent structure factor (EISF). The existence of an elastic peak originates from the localized character of the rotational motion (i.e., it occurs within a restricted volume). The EISF is defined by

$$S_{inc}^{el}(\mathbf{Q}) = \int dR \exp(-i\mathbf{Q}\cdot\mathbf{R})G_s^\infty(\mathbf{R}) = \left| \int dR \exp(-i\mathbf{Q}\cdot\mathbf{R})p(\mathbf{R}) \right|^2 \quad (13.32)$$

$A_0(Q)$ is therefore the space Fourier transform of the proton trajectory averaged over long times, $p(R)$. The experimental EISF is determined by measuring the elastic intensity over the total intensity, elastic + quasi-elastic, as the Lorentzian functions are normalized to unity and the structure factors obey the relation

$$\sum_{l=0}^{N} A_l(Q) = 1 \qquad (13.33)$$

The variation of the EISF with Q is a direct measure of the geometry of rotation. The type of reorientational motion can be determined from the comparison of the experimental EISF with the values calculated for different models, e.g., isotropic rotational diffusion, uniaxial rotation, and fluctuations around equilibrium positions.

On the other hand, the dynamic behavior of the molecules is contained in the quasi-elastic contribution [equation (13.31)], which is expressed as a sum of Lorentzians $\mathscr{L}(\omega, \Gamma_l)$, whose number l and width Γ_l depend on the model.

13.6.3. Applications of Quasi-Elastic Neutron Scattering (QENS) to the Study of Molecules Adsorbed in Zeolites

The dynamics of molecules adsorbed in zeolites can be studied on the microscopic scale by NMR and QENS. Neutron scattering is well suited because sufficient quantities of hydrogenated molecules can be adsorbed in these materials. The only problem is to avoid Bragg scattering from the zeolite because the coherent scattering from the zeolite is modified upon hydrocarbon adsorption.

Various neutron spectrometers are available at the ILL, and in other research centers, to measure slow diffusive motions. The lower limit is given by the resolution which is at present about 0.01 μeV (0.001 cm^{-1}), which corresponds to diffusion coefficients of the order of 10^{-9} cm^2 s^{-1}. However, the counting rates corresponding to this resolution are too small for surface work, where the signal of the substrate has to be subtracted. The typical range of diffusion coefficients that can be covered for heterogeneous systems is 10^{-7} to 10^{-4} cm^2 s^{-1}. Correlation times for rotational motions can be investigated in the range from 10^{-8} to 10^{-12} s.

13.6.3.1. Benzene Adsorbed in Sodium Mordenite

For this system, the rotational and translational dynamics of benzene have been investigated at two loadings in the temperature range 300–450 K.[65] Mordenite has a one-dimensional channel structure consisting of twelve-membered oxygen rings which form straight elliptical cylinders (7.0 × 5.8 Å). Unlike benzene in the liquid state, the translational and rotational broadenings of benzene in the zeolite are very different in magnitude so that the treatment of data is facilitated.

13.6.3.1a. The geometry of the localized rotational motions can be deduced from the EISF. The experimental points obtained at 300 K for the smallest loading (0.05 cm^3 g^{-1}) are represented in Figure 13.15 as open circles. This experimental EISF is independent of the temperature of loading, which means that the rotational motion is the same.

The nature of this rotation can be deduced from a comparison with theoretical curves. In Figure 13.15 the isotropic rotational diffusion model (dashed line), the uniaxial rotational model with jumps of 60° (full line) and 120° (dotted line) are shown. In all cases, the radius of gyration is the same, and is equal to 2.5 Å. It appears that the uniaxial model with jumps of 60° gives the best fit, which indicates that benzene is interacting with the sodium cations via its π electrons, in agreement with quantum mechanical calculations.[66]

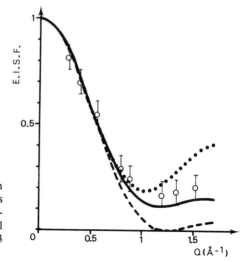

Figure 13.15. EISF of benzene adsorbed in sodium-mordenite: experimental points (O), theoretical EISF for isotropic rotational diffusion (broken line), for uniaxial rotation of $2\pi/6$ (solid line) and of $2\pi/3$ (dotted line).[65]

The scattering law [equation (13.31)] for rotational jumps of 60° is of the form[62]

$$S_{\text{inc}}^{\text{rot}}(Q, \omega) = A_0(Q)\delta(\omega) + (1/\pi) \sum_{l=1}^{5} A_l(Q)\tau_l/(1 + \omega^2\tau_l^2) \qquad (13.34)$$

where

$$A_l(Q) = (1/6) \sum_{j=1}^{6} j_0(2Qr \sin \pi j/6) \cos (2\pi lj/6) \qquad (13.35)$$

and j_0 is the spherical Bessel function of order zero. The width of the Lorentzian functions, τ_l^{-1}, is related to the mean time between successive jumps τ:

$$\tau_l = \tau/2 \sin^2(\pi l/6) \qquad (13.36)$$

By fitting this model to the experimental profiles, the mean time between successive jumps can be calculated to be about 1.5×10^{-12} s at 300 K, with the activation energy for this rotational motion being 4.5 kJ mol^{-1}.

13.6.3.1b. The translational motion along the channels was also measured at low loading. The diffusion coefficient is about 2×10^{-6} cm^2 s^{-1} at 300 K, which is of the same order of magnitude as the value obtained from NMR results for similar systems.

13.6.3.1c. The rotational motion of benzene in another zeolite, ZSM-5, has been characterized recently by QENS at two loadings and in the temperature range 90–400 K.[67] The framework of ZSM-5 contains two types of channels consisting of ten-membered oxygen rings (free diameter ≈ 5.4 Å). There are straight channels interconnected by sinusoidal channels and there are four channel intersections per unit cell. For this system, the EISF was found to be loading- and temperature-dependent. A model of uniaxial rotation in an N-fold cosine potential was used to model the EISF variation. With decreasing temperature, a progressive blocking of the molecules is observed for both loadings but the amplitude of motion is more restricted at higher loading at the same temperature, indicating benzene–benzene interactions. Isotropic rotation is not observed, even if the molecules have enough space available at the channel intersections, in agreement with NMR[68] and theoretical results. The blocking of the benzene molecules is therefore more pronounced in ZSM-5 than in mordenite, which can be related to the smaller channel dimensions.

13.6.3.2. Methanol Adsorbed in H-ZSM-5

The adsorption sites and diffusion mechanism of CH_3OH in HZSM-5 have been studied by QENS.[69]

On a short time scale ($\approx 10^{-11}$ s), the QENS results indicate that there are two different methanol species: adsorbed molecules interacting with the zeolite via hydrogen bonds and mobile molecules diffusing within a volume restricted by the channel dimensions, the proportion of mobile molecules increasing with temperature. Correlation times for the two species, as well as their activation energies, were determined by fitting the quasi-elastic profiles after convoluting the whole scattering law with instrumental resolution; an example of the fit is shown in Figure 13.16.

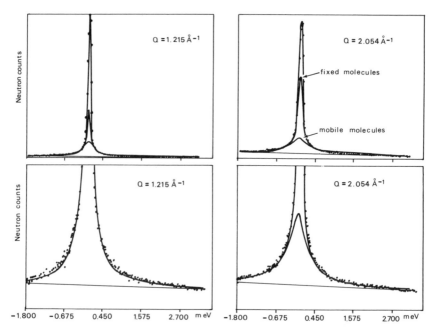

Figure 13.16. Comparison of experimental and calculated QENS profiles for CH_3OH in HZSM-5 at 335 K (there are two different ordinate expansions).[69]

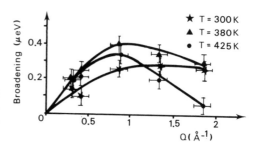

Figure 13.17. Broadening of the elastic peak (in μeV) as a function of the neutron momentum transfer Q, for CH_3OH in HZSM-5, at different temperatures.[69]

On a longer time scale (10^{-9}–10^{-10} s), the diffusion of the molecules along the channels could be determined from the broadening of the elastic peak as a function of the neutron momentum transfer Q (Figure 13.17).

It was found that the molecules jump from one site to another (jump distance ≈ 5 Å) and diffuse along the channels, after a large number of jumps, with a diffusion coefficient of 2×10^{-7} cm^2 s^{-1} at 300 K. This value is about one order of magnitude lower than recent data obtained by NMR in another sample of ZSM-5,[70] but this could be due to the different origin of the samples and to different Si/Al ratios.

13.7. CONCLUSIONS

The number of applications of neutron scattering to surfaces or to catalytic systems is still limited. This is due to restricted access to the neutron facilities, to experimental difficulties, and to the fact that treatment of the data is often laborious. It has been shown, however, in this chapter that this technique is useful in certain areas of surface chemistry as one can study the structure and dynamics of adsorbed phases on different instruments. New spectrometers with better resolution and better signal-to-noise ratios are still being built so that it eventually should be possible to study more complicated systems.

REFERENCES

1. W. Marshall and S. Lovesey, *Theory of Thermal Neutron Scattering,* Oxford University Press, Oxford (1971).
2. ILL Internal Report: Neutron Beam Facilities at the HFR Available for Users (1986).
3. H. J. Lauter and H. Jobic, *Chem. Phys. Lett.* **108**, 393 (1984).
4. J. Penfold and J. Tomkinson, Rutherford Appleton Lab. Rep. RAL-86-019 (1986).
5. W. B. Nelligan, D. J. Lepoire, C. K. Loong, T. O. Brun, and S. H. Chen, *Nucl. Instrum. Meth. Phys. Res.* **A254**, 536 (1987).
6. H. Jobic, *Chem. Phys. Lett.* **106**, 321 (1984).
7. R. K. Thomas, *Prog. Solid State Chem.* **14**, 1 (1982).
8. J. P. Beaufils, Y. Barbaux and B. Saubat, *J. Chem. Soc., Chem. Comm.,* 1212 (1982).
9. R. Kahn, E. Cohen de Lara, P. Thorel, and J.L. Ginoux, *Zeolites* **2**, 260 (1982).

10. A. N. Fitch, H. Jobic, and A. Renouprez, *J. Chem. Soc., Chem. Comm.* **284**, (1985); *J. Phys. Chem.* **90**, 1311 (1986).
11. P. A. Wright, J. M. Thomas, A. K. Cheetham, and A. K. Nowak, *Nature* **318**, 611 (1985).
12. A. Renouprez, H. Jobic, and R. C. Oberthur, *Zeolites* **5**, 222 (1985).
13. A. C. Zemach and R. J. Glauber, *Phys. Rev.* **101**, 118 (1956).
14. H. Jobic, *J. Chem. Phys.* **76**, 2693 (1982).
15. H. Jobic, R. E. Ghosh, and A. Renouprez, *J. Chem. Phys.* **75**, 4025 (1981).
16. A. Griffin and H. Jobic, *J. Chem. Phys.* **75**, 5940 (1981).
17. M. Warner, S. W. Lovesey, and J. Smith, *Z. Phys.* **B51**, 109 (1983).
18. H. Jobic and H. J. Lauter, *J. Chem. Phys.* **88**, 5450 (1988).
19. T. Nakata, *J. Chem. Phys.* **65**, 487 (1976).
20. S. Andersson, *Chem. Phys. Lett.* **55**, 185 (1978).
21. S. Lehwald and H. Ibach, *Surf. Sci.* **89**, 425 (1979).
22. W. Ho, N. J. Dinardo, and E. W. Plummer, *J. Vac. Sci. Technol.* **17**, 134 (1980).
23. H. Ibach and D. Bruchmann, *Phys. Rev. Lett.* **44**, 36 (1980).
24. N. J. Dinarolo and E. W. Plummer, *J. Vac. Sci. Technol.* **20**, 890 (1982).
25. M. Nishijima, S. Masuda, H. Kobayashi, and M. Onchi, *Rev. Sci. Instrum.* **53**, 790 (1982).
26. A. Renouprez, P. Fouilloux, G. Coudurier, D. Toccheti, and R. Stockmeyer, *J. Chem. Soc., Faraday Trans. I* **73**, 1 (1977).
27. R. Stockmeyer, H. Stortnik, I. Natkaniec, and J. Mayer, *J. Ber. Bunsenges. Phys. Chem.* **84**, 79 (1980).
28. (a) R. D. Kelley, J. J. Rush, and T. E. Madey, *Chem. Phys. Lett.* **66**, 159 (1979). (b) R. R. Cavanagh, R. D. Kelley, and J. J. Rush, *J. Chem. Phys.* **77**, 1540 (1982).
29. H. Jobic and A. Renouprez, *J. Chem. Soc., Faraday Trans. I* **80**, 1991 (1984).
30. H. Jobic, G. Clugnet, and A. Renouprez, *J. Electron Spectrosc. Rel. Phenom.* **45**, 281 (1987).
31. C. J. Wright and C. M. Sayers, *Rep. Prog. Phys.* **46**, 773 (1983).
32. D. Graham, J. Howard, and T. C Waddington, *J. Chem. Soc., Faraday Trans. I* **79**, 1281 (1983).
33. J. A. Andrews, U. A. Jayasooriya, I. A. Oxton, B. Powell, N. Sheppard, P. F. Jackson, B. F. G. Johnson, and J. Lewis, *Inorg. Chem.* **19**, 3033 (1980).
34. K. Christmann, R. J. Behm, G. Ertl, M. A. Van Hove, and W. H. Weinberg, *J. Chem. Phys.* **70**, 4168 (1979).
35. T. Engel and K. H. Rieder, *Surf. Sci.* **109**, 140 (1981).
36. J. T. Richardson and T. S. Cale, *J. Catal.* **102**, 419 (1986).
37. J. P. Muscat, *Surf. Sci.* **110**, 85 (1981).
38. G. Piccito, F. Siringo, M. Baldo, and R. Pucci, *Surf. Sci.* **167**, 437 (1986).
39. P. Nordlander, S. Holloway, and J. K. Norskov, *Surf. Sci.* **136**, 59 (1984).
40. C. J. Wright, C. Sampson, D. Fraser, R. B. Moyes, P. B. Wells, and C. Riekel, *J. Chem. Soc., Faraday Trans. I* **76**, 1585 (1980).
41. H. Jobic, J. P. Candy, V. Perrichon, and A. Renouprez, *J. Chem. Soc., Faraday Trans. I* **81**, 1955 (1985).
42. A. Renouprez and H. Jobic, *J. Catal.* **113**, 509 (1988).
43. R. B. Moyes and P. B. Wells, *Adv. Catal.* **23**, 121 (1973).
44. H. Jobic, J. Tomkinson, J. P. Candy, P. Fouilloux, and A. Renouprez, *Surf. Sci.* **95**, 496 (1980).
45. E. B. Wilson Jr., *Phys. Rev.* **45**, 706 (1934).
46. E. B. Wilson Jr., J. C. Decius, and P. C. Cross, *Molecular Vibrations,* McGraw-Hill, New York (1955).
47. J. Favrot, P. Caillet, and M. T. Forel, *J. Chim. Phys.* **71**, 1337 (1974).
48. H. Jobic, A. Renouprez, A. N. Fitch, and H. J. Lauter, *J. Chem. Soc., Faraday Trans. I* **83**, 3199 (1987).
49. H. Jobic and A. Renouprez, *Surf. Sci.* **111**, 53 (1981).

50. J. C. Bertolini, G. Dalmai-Imelic, and J. Rousseau, *Surf. Sci.* **67**, 478 (1977).
51. A. Renouprez, G. Clugnet, and H. Jobic, *J. Catal.* **74**, 296 (1982).
52. J. P. Candy, J. A. Dalmon, P. Fouilloux, and G. A. Martin, *J. Chim. Phys.* **72**, 1075 (1975).
53. H. Jobic and A. Renouprez, *J. Less Common Met.* **129**, 311 (1987).
54. H. Jobic, B. Tardy, and J. C. Bertolini, *J. Electron Spectros. Rel. Phenom.* **38**, 55 (1986).
55. H. Jobic and A. Renouprez, *Proc. ICSS 4 and ECOSS 3, Suppl. Le Vide,* Les couches Minces 201, Vol. 2, (1980), p. 746.
56. R. D. Kelley, R. R. Cavanagh, J. J. Rush, and T. E. Madey, *Surf. Sci.* **155**, 480 (1985).
57. J. P. Candy, H. Jobic, and A. Renouprez, *J. Phys. Chem.* **87**, 1227 (1983).
58. L. Van Hove, *Phys. Rev.* **95**, 249 (1954).
59. C. T. Chudley and R. J. Elliot, *Proc. Phys. Soc.* **77**, 353 (1961).
60. K. S. Singwi and A. Sjolander, *Phys. Rev.* **167**, 152 (1968).
61. P. L. Hall and D. K. Ross, *Mol. Phys.* **42**, 673 (1981).
62. A. J. Dianoux, F. Volino, and H. Hervet, *Mol. Phys.* **30**, 1181 (1975).
63. H. Jobic, M. Bee and A. Renouprez, *Surf. Sci.* **140**, 307 (1984).
64. J. Howard, K. Robson, T. C. Waddington, and Z. A. Kadir, *Zeolites* **2**, 2 (1982).
65. H. Jobic, M. Bee, and A. Renouprez, *Surf. Sci.* **140**, 307 (1984).
66. J. Sauer and D. Deininger, *Zeolites* **2**, 144 (1982).
67. H. Jobic, M. Bee, and A. J. Dianoux, *J. Chem. Soc., Faraday Trans. I* **85**, 2525 (1989).
68. B. Zibrowius, J. Caro, and H. Pfeifer, *J. Chem. Soc., Faraday Trans. I* **84**, 2347 (1988).
69. H. Jobic, A. Renouprez, M. Bee, and C. Poinsignon, *J. Phys. Chem.* **90**, 1059 (1986).
70. J. Caro, M. Bulow, J. Richter-Mendau, J. Karger, M. Junger, D. Freude, and L. V. C. Rees, *J. Chem. Soc., Faraday Trans. I* **83**, 1843 (1987).
71. H. Jobic, M. Bee, and G. J. Kearley, *Zeolites* **9**, 312 (1989).

14

X-RAY ABSORPTION SPECTROSCOPY: EXAFS and XANES

B. Moraweck

14.1. HISTORICAL SURVEY

The discovery of X-rays by Roentgen in 1891 allowed study of crystalline materials on an atomic scale. The X-ray energy range is of the order of 1 keV, which corresponds to wavelengths of 10 Å.

When the X-ray energy is enough to promote a deep electronic level, photon absorption may be observed. This absorption is accompanied by the expulsion of an electron from the absorbing atom, originating from the photoelectric effect. Extended X-ray absorption fine structure (EXAFS) is, from its definition, the fine structure which is developed in the absorption spectrum of the material under observation over an energy range up to 1 keV above the adsorption edge. This phenomenon was first observed in the 1920s by Ficke[1] and Hertz.[2] Hanawalt[3] was the first to study the temperature influence on the EXAFS structure.

The first explanation of this fine structure was tried by Kronig,[4] who related it to long-range order (LRO) theory in solids and to the short-range order (SRO) theory in gases.[5] Petersen[6-8] developed the same idea for molecules by introducing the wave function associated with the photoelectron, the phase shifts of the excited atom, and those of the backscattering atoms. Many years later, Sawada[9] introduced the notion of the mean free path of the electrons in order to take account of the limited lifetime of the photoelectron. In 1961, Smidt[10,11] used the Debye–Waller factor to take account of thermal and static disorder. Thus, all the ingredients of EXAFS theory were identified.

B. Moraweck ● Institut de Recherches sur la Catalyse, CNRS, Villeurbanne, France.

Catalyst Characterization: *Physical Techniques for Solid Materials*, edited by Boris Imelik and Jacques C. Vedrine, Plenum Press, New York (1994).

In 1974, Sterin[12] discovered that the LRO theory was wrong. Apparently, Kronig had never realized that the same basic concepts could explain the EXAFS phenomenon in molecules as well as in solids. According to the SRO theory, EXAFS oscillations are due to the modulation of the final-state wave function of the electron due to its backscattering by the atoms surrounding the absorbing (emitting) atom. Consequently EXAFS technique appears as a local probe which originates in the material itself.

During the 1970s two events promoted X-ray absorption spectroscopy from a laboratory curiosity to a powerful technique for the study of liquids and crystalline and amorphous solids: The first one was the phenomenon of synchrotron sources, which enable the experiment to be performed in minutes instead of days. The second is the simple theoretical formulation given by Sayers et al.,[13] who showed that the Fourier transform of the EXAFS signal gives access to the radial distribution function around the absorbing atom. Thus, it became relatively easy to obtain the distance and the coordination number of the target atom. Moreover, the nature of the atoms surrounding the target atoms can be determined.

14.2. THEORETICAL CONSIDERATIONS

If a material is irradiated by a monochromatic beam of X-ray photons, the decrease in intensity dI is proportional to the intensity I:

$$dI = \mu I dx$$

where dx is an infinitely small thickness in the material and μ a proportionality constant, a linear absorption coefficient, which is a function of energy, $h\nu$. If we integrate this expression over the sample thickness x we obtain Lambert's law:

$$I = I_0 e^{-\mu x}$$

where I_0 is the incident intensity and I the intensity of the beam transmitted through the sample.

Among the different absorption processes (e.g., elastic or Rayleigh scattering and inelastic or Compton scattering) only the photoelectric effect is of interest for EXAFS. In this process the total energy of the photon is transmitted to the promoted electron: a part of this energy is used to extract the electron from a core level and the rest is given in the form of kinetic energy. The linear absorption coefficient μ is proportional to the probability of a photoelectric event, so it can be expressed, within the dipole approximation, as a function of the initial state $|i\rangle$ and the final state $|f\rangle$, at the absorbing atom, following Fermi's golden rule:

$$\mu = (4\pi^2 e^2/c)\nu N |\langle f|\mathbf{E}\cdot\mathbf{R}|i\rangle|^2 \delta(E - E_f - E_0)$$

where e is the electronic charge, c the speed of light in vacuum, N the number of atoms per unit volume, $\mathbf{E}$ the polarization vector of the electric field of the photon, $\mathbf{R}$ the coordinate vector of the electron, and δ the density of the final state.

The major part of the final state is the spherically symmetric wave function of the ejected electron. The kinetic energy of the electron is the difference between the binding energy E_0 and the incident photon's energy E so that the wave vector of the photoelectron is given by

$$k = [(2m/\hbar^2)(E - E_0)]^{1/2}$$

where m is the mass of the electron and $\hbar$ the reduced Planck constant.

The emerging wave will be scattered by neighboring atoms. The final state $|f\rangle$ is a superposition of the wave function of outgoing and backscattering waves. This superposition is constructive or destructive depending on the phase difference between the waves. During its travel through matter the ejected electron experiences several changes: on traversing the potential of the adsorbing atom, hence the potential of the backscattered atoms, and on re-entering the potential of the central atom. The phase factors, specific to both the absorber and the backscatterer, can be accounted for by the global phase factor $2\delta(k) + \theta(k)$, where δ corresponds to the absorber and θ to the backscatterer.

Owing to the construction and destruction of electron waves, the final state $|f\rangle$ is modulated at the absorbing atom and one gets a modulation of the absorption coefficient μ. The amplitude of a spherical wave is proportional to the distance R^{-1} from the origin. During the lifetime of the core hole the photoelectron accomplishes an "*aller-retour*" so that the amplitude of backscattered electrons is proportional to R^{-2}, with a proportionality factor $F(k)/k$ depending on the backscatterer elements. Surrounding atoms generally form a coordination shell enclosing N atoms. As the total variation in μ is a superposition of all interferences caused by all the atoms in different coordination shells, the EXAFS signal is finally given by

$$\chi(k) = - \sum_j (N_j/R_j^2)\sin[2kR_j + 2\delta(k) + \vartheta_j(k)]F_j(k) \qquad (14.1)$$

in which the summation is over all the coordination shells located at the R_j distances from the absorbing (central) atom.

Though equation (14.1) is essentially correct some effects have been neglected, and we have only considered the initial and final states of the photoelectron. We must, moreover, take into account the total initial state of the absorbing atom and the final states of the ionized atom. A different potential will be sensed by the passive electrons which are not directly involved in the absorption process. Consequently the total final state of the atom is modified and the basic EXAFS equation must be corrected by a factor of

$$S_0^2 = \prod_i |\langle p_i | p_i' \rangle|^2 \qquad (14.2)$$

in which $| p_i \rangle$ and $| p_i' \rangle$ are normalized and represent the wave function of the ith electron before and after the X-ray absorption process. It should be noted that S_0^2 is generally less than 1 (about 0.7–0.8).

The losses by inelastic processes limit the lifetime of the core hole and that of the final state $| f \rangle$ and lead to the incorporation of a correction in the EXAFS relationship which is expressed as a function of the mean free path for the electron wave. This mean free path may be approximated by the distance that the electron can travel before it loses its coherence, which is the nearest neighbor distance R_j.

The factor is generally written as $\exp(-2R_j/\lambda)$ but Teo[65] suggested more complicated relationships, in which λ is expressed as a function of k. Due to the thermal vibration and deformation of the lattice, the distance between atoms may be modified. As EXAFS is averaged over all the absorbers around the R_j distance we can take it into account by a Debye–Waller type correction factor, $\exp(-2\sigma_j^2 k^2)$.

It must be noted that this very simple theoretical presentation of the EXAFS phenomenon is true within a series of simple assumptions. First, it is assumed that the sample is unoriented with Gaussian disorder. Systems with non-Gaussian disorder can be studied with cumulant expansions.[73] When the R_j distances in the sample are not randomly oriented (e.g., intercalation compounds), the angle $\mathbf{E} \cdot \mathbf{R}$ must be considered and a factor $-3\cos^2 \theta$ included.

Secondly, the above treatment assumes that in the backscattering process the electron wave is considered far away from its source. This assumption by a plane wave is not valid for low k values and especially for light backscatterers. This limitation can be overcome by curved wave theory, which yields phase and backscattering functions that are dependent on k and on the distance R.

Third, the preceding theory does not consider that at low k (below 2 Å^{-1}) there are multiple scattering effects, including paths involving more than one backscattering atom. In this case, the complete development of equation (14.1) with Green's function theory leads to a series that can be written as[76]:

$$\mu(k) = \mu_0(k)[1 - \chi(k) - k^2\chi'(k) + \cdots] \tag{14.3}$$

The case of single scattering is a particular one for which we retain only the first two terms in equation (14.3). Practically, the existence of multiple scattering means that $\phi(k)$ and $F(k)$ are functions of the global backscattering angle. (For a detailed discussion of this phenomenon and its influence in the EXAFS formulation see, e.g., Reference 76). It can be noted immediately that this phenomenon can occur in some catalytic materials such as the cfc metals. Under the assumption of a single scattering the fourth neighbor of the absorbing atom is completely screened. Due to multiple scattering the first neighbor plays the role of a lens that enhances the influence of the fourth neighbor in the EXAFS spectrum. This phenomenon, known as "forward scattering," corresponds to an important backscattering amplitude for the zero scattering angle.

Enhancements of simple EXAFS theory can be in two directions. The first is to no longer consider the plane wave approximation, which assumes that the atomic radii are much smaller than the interatomic distances. This approximation is generally sufficient for most EXAFS data analysis if we consider only the k range above 4 Å^{-1}. In fact this approximate formalism is only accurate at large kR (high k and/or large R).

The exact curve wave theory was established first by Lee and Beni[66] and Gurman and Pettifer.[67] However, it was recently shown by McKale et al.[59] that the plane wave formalism can be used at lower k (2 Å^{-1} $\approx$ 15 eV) with a correction in the simple EXAFS formula already seen. Taking an R dependence of the backscattering amplitude and phase shift, one can write the total phase shift for a K shell (with p symmetry) as

$$B_j(k, R_j)e^{i\Phi_j(k,R_j)}$$

$$= \frac{kR_j^2}{e^{2ikR_j}} \sum_l \left\{ (2l+1)(-1)^j e^{i\delta_j} \sin \delta_l \left[\frac{l+1}{2l+1} [h_{l+1}^+(kR_j)]^2 + \frac{l}{2l+1} [h_{l-1}^+(kR_j)]^2 \right] \right\}$$

$$(14.4)$$

For an L_{III} shell the final state can have either s or p symmetry. One obtains, then, a more complicated relationship despite the fact that the contribution of s states can be neglected. In these expressions $h^+(kR_j)$ is an outgoing spherical Hankel function. In order to be able to apply such correction formulas one must dispose of the δ_l's. Fortunately, McKale et al.[59] gave the numerical values for B at 2.5 and 4 Å for the backscatterer. They recommend the use of the numerical values tabulated by Teo and Lee[58] for the absorbing atom with a linear extrapolation toward 2.5 Å^{-1}.

The second amelioration concerns a large disorder which cannot be described by a Debye–Waller type factor but only by asymmetric pair distributions. This type of disorder can be formulated by

$$g_n(r) = (1/n!\sigma^{n+1})(r-r_0)^n e^{-(r-r_0)/\sigma} \qquad r \geq r_0$$

$$g_n(r) = 0 \qquad\qquad\qquad\qquad\qquad r < r_0 \qquad (14.5)$$

Crozier and Seary[68] used $g_2(r)$ for the EXAFS analysis of liquid metals, whereas the metallic glasses studied by Teo et al.[69] can be described by $g_1(r)$. It seems that such a disorder can be involved for bimetallic catalysts (Pt-Fe/charcoal) in which the atomic radii of the two metal components are very different.[45]

These distributions are characterized by a mean atomic distance $r_0 + (n+1)\sigma$, but the maximum is found at $r_0 + n\sigma$; r_0 appears here as the minimal approach distance between absorber and backscatterer. The integration of the $g_n(r)$ function in the general expression of EXAFS leads to the introduction of correcting terms into the phase and backscattering amplitudes which depend upon the $g_n(r)$ function chosen. For example, the amplitude correction for $g_0(r)\gamma$

is $[1 + (2k\sigma)^2]^{1/2}$ and the phase correction is $2k\sigma$. More complicated correction factors will be obtained with increasing n.[70]

14.3. PRACTICAL CONSIDERATIONS

In this section we shall briefly describe the different methods of measuring the X-ray absorption coefficient as a function of photon energy and the various problems encountered during the data analysis.

14.3.1. Experimental Arrangements

In order to perform an X-ray absorption experiment we must have an X-ray source. While such experiments can be done with conventional sources (laboratory equipment) they are performed to a great extent from a synchrotron radiation source. The first important part of the setup is the monochromator, which produces a monochromatic beam. The monochromator is generally made of two parallel monocrystals (of the same Miller index) that select a particular photon energy (Bragg's law) out of the white radiation given by the synchrotron source. The usual crystals are diamond-structured such as germanium or silicon for an energy range between 2 and 40 keV. There are two main procedures for measuring the X-ray absorption coefficient. The first one usually measures the intensity of the photon flux before and after the probed sample with two ion chambers. If I_0 and I_1 are the currents measured in the ion chambers, the X-ray absorption of the sample is given by $\mu = \ln(I_0/I_1)$.

This expression is correct to within a few percent if the flux absorbed in the first ion chamber is very small compared with the transmitted flux and if the whole flux transmitted by the sample is completely absorbed by the second ion chamber. On the other hand, corrections must be made. This is a direct method of measurement.

The second group of methods is formed by a set of indirect measurements. In the case of diluted species (e.g., catalysts containing less than 1% of a light metal supported on an absorbing carrier), the fluorescence method is used. The intensity of the fluorescence line is given by

$$I(E) = I_0(E)(\Omega/4\pi)\alpha[\mu_e(E)/\mu_e(E) + \mu_s(E_F) + \mu_M(E)] \qquad (14.6)$$

where α is the fluorescence yield, $\mu_e(E)$ and $\mu_M(E)$ are, respectively, the absorption coefficient of the probed element and the absorption coefficient of the matrix at the working energy E, and $\mu_s(E_F)$ represents the absorption coefficient of the whole sample at the fluorescence energy E_F. If the concentration of the element of interest L is much smaller than the concentration of the other elements in the sample the fluorescence intensity becomes proportional to $\mu_e(E)$. Then $\mu_e(E)$ may be obtained without corrections according to the relationship

$$\mu_e(E) \propto I(E)/I_0(E) \tag{14.7}$$

Fluorescence radiation can be detected with bidimensional proportional chambers filled with a mixture $A + CO_2$ or $A + CH_4$, or with appropriate solid detectors such as Si–Li or scintillation detectors. In order to minimize the background a correctly calibrated filter with the atomic number $Z - 1$ if Z is the probed element can be intercalated between the sample and the detector. Kampers[79] gave some details on this method applied to catalysts. One finds secondary electron detection in this second group of detection methods,[78] as it is based on the fact that some secondary Auger electrons are produced by the irradiated sample surface.

14.3.2. Determination of Parameters

The determination of the EXAFS parameters of interest n_j, R_j, σ_j has been extensively discussed by Teo[71] and by many other authors. We give here only some general principles on the computer use leading to these parameters. Starting from the absorption coefficient measured as indicated in the preceding section, the first step is to compute the absorption background before the edge, which is done by fitting the experimental points over a convenient energy range in a straight line, a Victoreen law or a function $A \exp(-bE)$.[76] This function is then substracted from the original μ, which results in a spectrum that contains all the information concerning the probed element and its neighbors.

The second step is designed to determine the atomic-like absorption background μ_0. This is done by fitting the spectrum from 30 to 50 eV above the edge to various functions. The procedures which are cited in the literature are a Victoreen law, a polynomial of a suitable degree ($2 \leq n \leq 6$), spline functions, or an iterative smoothing technique. In all cases one obtains the oscillatory part of the μ spectrum by normalization:

$$\chi(E) = [\mu(E) - \mu_0(E)]/\mu_0(E) \tag{14.8}$$

The primary characteristic of the basic EXAFS function is the presence of a set of sine functions. Fourier-transforming the $\chi(k)$ function leads to a radial distribution function, the peaks of which correspond to the successive radii of the different coordination spheres around the absorbing (emitting) atom. Before Fourier-transforming, the $\chi(k)$ spectrum was multiplied by a weighting function in order to minimize the parasitic ripples due to an incomplete integration range. The more commonly used windowing function is the Hanning function, and a complete set of windowing functions has been given by Max.[77] The true peak distances are lowered by a quantity which depends mainly on the phase shift function. (This problem will be discussed in the next section.)

As each peak in the Fourier transform (FT) is considered as corresponding to a typical coordination sphere, it is possible to isolate it and to obtain its contribution to the whole spectrum by an inverse FT.

The last step is then to model this contribution. Starting from the phase shift and backscattering amplitude functions (determined as described in the next section) one can determine the EXAFS parameters by a graphic fitting or by an optimization process.

14.3.3. The Phase Shift Problem

14.3.3.1. One-Component Systems

This is the case for monometallic catalysts supported by a carrier. The fundamental EXAFS equation may be considered from two different points of view.

1. The system is of known structure: in this case, the nearest neighbor distance and coordination number are known and for a single shell of neighbors the EXAFS equations may by inverted giving $\theta_{AA}(k)$ and $F(K)$

$$F(k) = (n_j/kR_j^2)A(k)e^{-2\sigma_j^2 k^2}e^{-2R_j/\lambda}$$

$$\Phi(k) = \text{arctg}[I(k)/R(k)] - 2kR_j \tag{14.9}$$

2. The system is of unknown structure: The results obtained in (1) may be used for determining n_j, R_j, and $\Delta\sigma_j$ by fitting the k contribution into k space assuming that the electronic structure is not very different from that of the model. Since this electronic structure may slightly vary for different reasons, the adjustment is made by adding an adjustable parameter E_0 via the relationship:

$$k = [(2m/\hbar^2)(E - E_0)]^{1/2} \tag{14.10}$$

with k the abscissa in reciprocal space. The variation allowed for E_0 takes account of the lack of knowledge of the origin, which can vary from one experiment to another, as well as the uncertainty which exists in the theoretical calculations for describing the potential that is experienced by the ejected photoelectron.

14.3.3.2. Two-Component Systems

In bimetallic systems one can consider that the global EXAFS spectrum is the sum of elementary spectra corresponding to each type of backscatterer. If the solid is made of two metal elements A and B, the EXAFS spectrum at the edge of the A element is given by

$$\chi(k) = \chi_A(k) + \chi_B(k) \tag{14.11}$$

Using the classical EXAFS formula we can develop this expression (for the first coordination sphere) with the following equations:

$$\chi_A(k) = (1/k)(n_{AA}/R_{AA}^2)\sin[2kR_{AA} + \Phi_{AA}(k)]A(k)e^{-2\sigma_{AA}^2 k^2}e^{-2R_{AA}/\lambda}$$

$$- (1/k)(n_{AB}/R_{AB}^2)\sin[2kR_{AB} + \Phi_{AB}(k)]B(k)e^{-2\sigma_{AB}^2 k^2}e^{-2R_{AB}/\lambda} \quad (14.12)$$

We obtain an equivalent relationship $\chi_B(k)$ at the edge of the B element by interchanging the A and B indexes.

From $A(k)$ and/or $B(k)$, we wish to determine n_{AA}, n_{AB}, n_{BA}, n_{BB}, R_{AA}, R_{AB}, R_{BB}, N, and σ_{AA}, σ_{AB}, σ_{BB}. Consequently we must have information regarding the functions of k involved in equation (14.12); these functions may be obtained experimentally or computed theoretically.

14.3.3.2a. Experimental Determination. As for monoatomic materials the F_a, F_b, and Φ_{AA}, Φ_{BB} quantities depend on a unique type of atom. These functions can therefore be extracted from experiments performed with a metal foil of correct thickness.

In the case of a bimetallic system this method would not lead to correct information about $\Phi_{AB}(k)$, but there are three methods that can be used to determine the phase shift of a pair of heteroatoms.

(i) *Intermediate Compounds Method.* This method is based on the additivity rule for phases. It has been used by Sinfelt *et al.*[55] for the study of Ru–Cu/SiO$_2$ catalysts in which the two metals are not miscible. In this case we introduce two compounds AC and BC containing both A and B atoms. Thus we can write

$$\Phi_{AB} = 2\delta_A + \vartheta_B = 2\delta_A + \vartheta_C + 2\delta_B + \vartheta_B - (2\delta_B + \vartheta_C) = \Phi_{AC} + \Phi_{BB} - \Phi_{BC}$$

$$\Phi_{BA} = 2\delta_B + \vartheta_A = 2\delta_B + \vartheta_C + 2\delta_A + \vartheta_A - (2\delta_A + \vartheta_C) = \Phi_{BC} + \Phi_{AA} - \Phi_{AC}$$

$$(14.13)$$

This method can be considered to be correct when one assumes that the change in the electronic configuration for metal atoms is weak. Hence this change may be taken into account by a small adjustment for E_0.

(ii) *Model Compound Method.* This method has never been used for determining the phase shifts of a heteroatomic pair. It needs a compound with a known ordered structure, and this compound must be such that in the first coordination sphere of the A element, one finds only a type B atom. This standard, therefore must have long-range order which can be evidenced by X-ray diffraction superstructure lines or by electrical resistivity measurements. For instance, if the two metals belong to the cfc system and give rise to solid solutions of the same structure, one may find two compositions AB$_3$ and A$_3$B which are ordered.

For these two ordered solid solutions each element A (or B) is surrounded by 12 B (or A) atoms. The AB$_3$ compound leads to Φ_{AB} A$_3$B to Φ_{BA}. This type of experimental determination is limited to structures of high symmetry. It must therefore be noted that a relatively important electronic transfer may be observed between the elements and hence the measured phase shift does not correspond exactly to the sum of elemental phase shifts.

14.3.3.2b. Theoretical Calculation. If none of the above methods is available, one can used theoretical results already published. From the recent work by McKale[59] the phase shifts (and backscattering amplitude) for the backscatterer atoms must be computed from a relationship taking account of the curvature of the backscatterer wave. The result of this work leads to a better approximation of the phase shifts in the low-energy region and consequently allows the use of experimental results between 15 and 50 eV (2.0 and 4.0 Å^{-1}). This is particularly useful for light elements for which the backscattering amplitude becomes weak above 6 Å^{-1}.

Another point must be noted. In the case of heavy elements it is difficult to aproximate the influence of the relativistic effects upon the computed backscattering amplitude and the phase shift. Their role has been exemplified in the case of tungsten,[58] but no extension has been made to other elements. It seems that the problem resides in the solution of the Dirac equation instead of the Schrödinger one, which gives the potential through which the photoelectron travels after it is ejected from a core level. Moreover, it must be stated that there is no general rule to take account of the relaxation of the other electrons after the X-ray photon has been absorbed.

14.3.3.2c. Composite Method.[55,57,58] In this method one calls on model compounds for the homoatom pair and theoretical data for heteroatom pairs. Once more, from the addition rule of phase shifts, we can write

$$\Phi_{AB} + \Phi_{BB} = \Phi_{AB} + \Phi_{BA} \qquad (14.14)$$

where $\Phi_{AB} + \Phi_{BA}$ can be obtained from experiments on pure metals. If we compute the theoretical value from Φ_{AB}, then Φ_{BA} is known and depends only upon E_0^A, an adjustable parameter which is the origin of the abscissa in reciprocal space. The originality of the method proposed by Meitzner *et al.*[56] has its starting point in the fact $R_{AB} = R_{BA}$. For each chosen value of E_0^A, we can compute a set of Φ_{AB} and consequently a corresponding set of Φ_{BA} from equation (14.14). The set retained for the description of the bimetallic catalyst is that which corresponds to $R_{AB} = R_{BA}$.

14.4. SELECTED CATALYTIC SYSTEMS

14.4.1. Hydrodesulfurization Catalysts

The use of hydrodesulfurization (HDS) catalysts is related to the need for up-grading of crude oil fractions or coal-derived liquid. They usually consist of

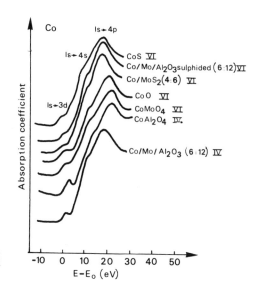

Figure 14.1. Co K edge in cobalt oxides, sulfide and Co–Mo catalysts (adapted from Sankar et al.[21]).

molybdenum (or tungsten) supported on large-surface-area alumina with nickel or cobalt added as a promoter.

In order to describe the active state of these catalysts, three main different models have been proposed: the *monolayer model*, in which cobalt is associated with the alumina; the *contact synergy* model, where cobalt is assumed to be present in the state Co_9, S_8, and the *intercalation model*, in which cobalt is

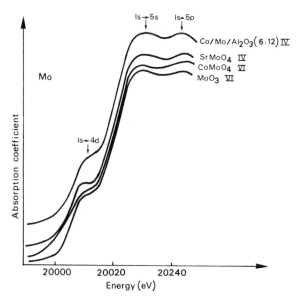

Figure 14.2. Mo K edge in Mo compounds and catalysts (adapted from Sankar[21]).

considered as being intercalated into the large structure of MoS_2. It has been shown by Wivel et al.[14] that a series of $Co–Mo/Al_2O_3$ catalysts with the same content exhibit a typical synergetic promotion behavior as the promoter loading varies.

The catalytic activity reaches a maximum for a Co/Mo ratio of about 1. From Mössbauer emission spectroscopy (MES) it is seen that at low cobalt loading the cobalt is present mainly as Co–Mo–S. When the cobalt loading is increased, Co_9S_8 (the thermodynamically stable cobalt compound at the reaction temperature) is formed and becomes the dominant cobalt phase at high content.

The analysis by X-ray absorption spectroscopy (XAS) is relatively easy because of the important concentration of metal elements (3–5% Co, 12–15% Mo) on the carrier. For these types of catalysts, the qualitative surrounding of cobalt and molybdenum (oxygen or sulfur) can be determined by XANES as shown by the following example.[18]

The principles of such a study are based upon the comparison of XANES spectra of model compounds and catalysts. It is well known that transitions of $1s$ electrons to unoccupied levels of d, s, p symmetry contribute to the formation of the fine structure of the K edge of transition metals. The shape of the absorption coefficient is determined both by the degree of filling of the $3d$, $4s$, and $4p$ bands (or $4d$, $5s$, and $5p$ bands of the second-row elements), which have a different

TABLE 14.1. Model Compound Data and XAS Results at the Co K Edge[a]

Sample	Ligand	N	R (Å)
$CoMoO_4$	O	6	2.09
$CoAl_2O_4$	O	4	1.95
CoO	O	6	2.12
CoO	S	6	2.39
Co–MoS_2 4:6	S	6	2.30
Co–Mo/Al_2O_3 6:1	O	4	1.99

[a] From Sankar et al.[21]

TABLE 14.2. Model Compound Data and XAS Results at the Mo K Edge

Sample	Ligand	N	R (Å)
$CoMoO_4$	O	4	1.98
MoO_3	O	6	2.12
$SrMoO_4$	O	4	1.76
MoS_2	S	6^a	2.41
Co:Mo:S_2 4:6	S	6	2.31
Co–Mo/Al_2O_3	O	6	1.81
After sulfidization	S	6	2.29

[a] Mo atom is at the center of a triangle-based prism.

density of states, and by the degree of hybridization of the d, s, and p orbitals, which determines the probability of the corresponding transition.[61-63]

As shown in Figures 14.1 (Co edge) and 14.2 (Mo edge), it is clear from the model compounds $CoAl_2O_4$ or CoS, CoO, and $CoMoO_4$ that one can differentiate the tetrahedral (IV) from the octahedral (VI) surrounding of the cobalt atom in these compounds. The same remarks can be made in the case of the molybdenum surrounding in $SrMo^{IV}O_4$ and $Mo^{VI}O_3$. We note that the transition $1s \rightarrow 3d$ (or $4d$) is more intense in the case of a tetrahedral surrounding of the probed element (Co or Mo). It thus becomes easy to determine types of metal sites in the catalysts. The results obtained by Sankar et al.[21] are summarized in Table 14.1, which gives XANES and EXAFS results, as well as showing the model compounds that allowed the interpretation of EXAFS spectra and the corresponding metal ligand type and distance.

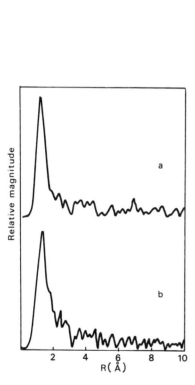

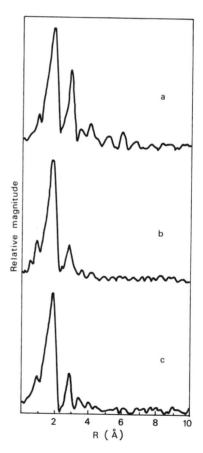

Figure 14.3. Fourier transform of $\chi(k)$ in (a) Mo/Al$_2$O$_3$ catalyst and (b) Co–Mo/Al$_2$O$_3$ catalysts (adapted from Clausen et al.[15]).

Figure 14.4. Fourier transform of $\chi(k)$ in (a) MoS$_2$, (b) sulfided Mo/Al$_2$O$_3$ catalyst, (c) sulfided Co–Mo catalyst.

It can be seen clearly that XAS results (Table 14.2) lead to the conclusion that the sulfidization of catalyst precursors results in a deep modification of the metal sites. This modification is first a lengthening of the metal ligand distance (1.89 to 2.32 Å and 1.81 to 2.29 Å). Secondly, in the case of the cobalt atom, this lengthening is accompanied by a drastic change in the site symmetry, which changes from tetrahedral to octahedral. Co_9S_8 has been shown to have very little catalytic activity and it does not have a significant promoting effect on activity in an HDS reaction.

From EXAFS *in situ* experiments Clausen *et al.*[15] have shown that out of the first coordination shell no distinct peak can be seen in the FT of the catalyst, whereas the peak corresponding to the second coordination sphere is clearly visible (Figure 14.3). This excludes the presence of a well-crystallized MoO_3 phase in the oxidized catalyst. The oxidized phase appears as isolated ions, chains, or very small distorted domains. The Mo—O bond length is about 1.73 Å, which is close to the Mo–O bond length in a tetrahedral coordination such as $Fe_2(MoO_4)_3$.

As the catalysts Mo/Al_2O_3 and Co—Mo/Al_2O_3 are sulfidized, the analysis of EXAFS spectra compared to that of crystallized MoS_2 shows the existence of a second peak in the FT which clearly corresponds to the second shell in MoS_2 (Figure 14.4). In these catalysts molybdenum has the same surrounding of six sulfurs as in MoS_2. The second peak may be analyzed in the same way as the first

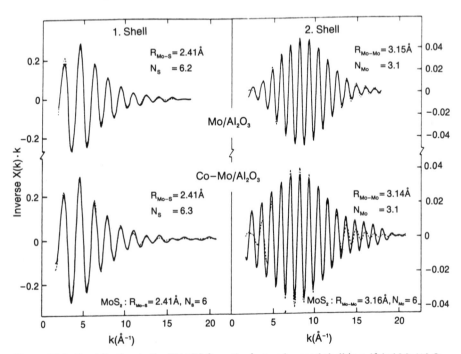

Figure 14.5. Contribution to the EXAFS from the first and second shell in sulfided Mo/Al_2O_3 catalysts.[16]

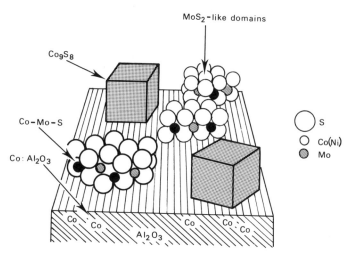

Figure 14.6. Schematic representation of the different phases present in a Co–Mo/alumina catalyst.

one. Clearly the second coordination sphere is due to the same scatterer as in MoS_2, i.e., molybdenum. Moreover this second peak is in the same position. Consequently, in the sulfidized state, a MoS_2-like phase is present in the Mo/Al_2O_3 catalysts. As the amplitude is about a third of that in MoS_2, it seems that the MoS_2 phase is present as very small particles (≈ 10 Å).

In a more recent study[16] the molybdenum environment has been reinvestigated in both the Mo and Mo—Co/Al_2O_3 catalysts. It is clear (Figure 14.5) that there is disagreement between the experimental and model contribution of the second coordination shell. This is obviously due to a modification by cobalt of the second coordination sphere of the molybdenum atom. As there is practically no variation in the number of molybdenum atoms in this sphere, it can be concluded that the size of MoS_2 particles is the same as in the Mo/AlO_3 catalyst and their structure is not changed. The cobalt atoms are consequently considered as surface atoms as shown in the next scheme (Figure 14.6).

Among the three different models proposed the "intercalation" model can be rejected. From EXAFS experiments at the Co K edge it has been shown that the Co—S bond distance (2.19 Å) in Co—Mo—S is significantly lower than that of cobalt in an intercalation site. In the XRD pattern of Co—Mo—S samples exhibiting a three-dimensional MoS_2 structure there is no lattice spacing that corresponds to cobalt intercalation. Moreover, as the cobalt atoms have a very low coordination in sulfur it is probable that they are in edge positions. This agrees with the observation that the cobalt atoms are affected by oxygen exposure and accessible to NO (as shown by IR measurements).

Sankar *et al.*[18] have reached a different conclusion in their EXAFS study at the Co K edge of a similar system. Using a multishell analysis they found that each cobalt is surrounded by six sulfur atoms at 2.33 Å and four oxygen atoms at

2.03 Å. Atmospheric contamination can be ruled out because there is no decrease in the Mo—S coordination. The Co—O distance is too short compared with the Co—O distance in CoO (2.14 Å), and too large for a cobalt atom in a tetrahedral site as in $CoAl_2O_4$. On the other hand, it is very close to that of a low-spin Co^{2+} in an octahedral site (2.05 Å). The partial octahedral coordination of cobalt (four) would be completed to six by sulfur atoms. There appear to be two types of cobalt species: the first one corresponds to a coordination of six as in CoS; the second one with four oxygen atoms is assumed to play a role in anchoring the particle to the Al_2O_3 surface.

Bouwens et al.[19] reinvestigated the role of nickel and cobalt as promoters in carbon-supported cobalt and Co—Mo sulfide catalysts. It appears from the imaginary part of the FT that the coordination number of cobalt increases, in the order Co_9S_8. Co/C, Co—Mo/C. Consequently the Co^{2+} ions in the catalysts have a higher sulfur coordination than in Co_9S_8, and this effect is enhanced in the catalyst, which has a Co—Mo—S structure. Furthermore a XANES study of the $1s \rightarrow 3d$ transition (Figure 14.7) confirms this trend. The most intense peak is obtained for Co_9S_8, which involves 89% tetrahedral and 11% octahedral sites, the lower for CoS_2 in which cobalt is uniquely surrounded by an oxygen octahedron. It may be concluded that the Co/C catalyst contains more octahedral sites than Co_9S_8; thus this effect is more pronounced for the Co—Mo/C catalyst, indicating that the Co^{2+} ions have an octahedral-like sulfur coordination.

This result confirms the proposal of Ledoux[20] that the active sites are those cobalt atoms in octahedral coordination. The other conclusion of this work is that for the Co/C catalyst, cobalt sites can be modified by the carbon carrier.

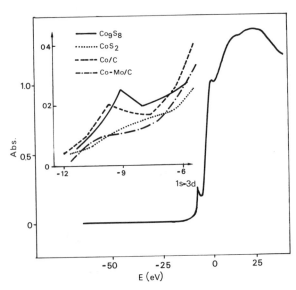

Figure 14.7. XANES spectrum of Co_9S_8. The expanded inset shows $1_s \rightarrow 3d$ transition of Co_9S_8, CoS_2, Co/C and Co–Mo/C.[19]

TABLE 14.3. EXAFS Data of Calculated Mo/Al$_2$O$_3$ Catalyst[a]

Mo (wt.%)	Coordination	Shell radius/Å
1.7	3.9	1.77
2.5	3.6	1.77
3.4	3.7	1.77
6.0	2.4	1.77–1.66
7.2	2.1	1.76–1.66
	1.8	1.91–2.25
21.4	1.1	1.78–1.64
	4.9	1.96–2.2
Na$_2$MoO$_4$, 2H$_2$O	4.0	1.76
MoO$_3$	5.3	1.65–1.93–2.24

[a] From Mensch et al.[23]

14.4.2. Supported Oxides

14.4.2.1. Molybdenum Oxide

Molybdenum oxide on silica and alumina has been characterized by X-ray absorption.[22-26] Only the case of molybdena is described below.

The work on MoO$_3$/Al$_2$O$_3$ is concerned with the preparation of Mo/Al$_2$O$_3$ by equilibrium adsorption of ammonium heptamolybdate (AHM). It is clear from the adsorption isotherm of AHM over alumina that there are two regions that can be explained by the following scheme[24]:

$$Al_s - OH \rightarrow Al_s^+ + OH^-$$

$$Mo_7O_{24}^{6-} + 8OH \rightarrow 7MoO_4^{2-} + 4H_2O$$

$$2Al_s^+ + MoO_4^{2-} \rightarrow (Al_s)_6MoO_4$$

$$6Al_s^+ + Mo_7O_{24}^{6-} \rightarrow (Al_s)_6Mo_7O_{24} \quad (14.15)$$

This scheme requires that there be two types of adsorbed molybdenum anions. The first exists up to 3.8 wt.% Mo and corresponds to a tetrahedrally coordinated molybdenum atom, whereas at greater molybdenum loading octahedrally coordinated Mo^{VI} is involved.

With X-ray absorption both types of molybdenum sites can be identified. A qualitative characterization of Mo^{IV} may be obtained from the pre-edge peak which corresponds to the $1s \rightarrow 4d$ bound states transition, and an intermediate spectrum is observed for MoO$_3$/Al$_2$O$_3$. From the EXAFS experiments it is possible to corroborate this qualitative result quantitatively. Table 14.3 lists the measured coordination numbers (Mo—O bond) and the corresponding shell radii as functions of molybdenum content.

For the samples in which molybdenum loading is lower than 3.8 wt.% almost all the molybdenum atoms are in the Mo^{IV} state. EXAFS gives a mean coordination number of about four oxygen atoms at 1.77 Å, which is very close to the distance measured in Na_2MoO_4 from X-ray diffraction. As the molybdenum content is increased, the initial ratio Mo^{IV} (Mo total) of 0.95 decreases to 0.2 for the 21% sample. Simultaneously several Mo—O distances are detected which partly correspond to Mo^{VI} as in MoO_3. From the shape of the adsorption isotherm and the EXAFS results, it has been concluded that for low loadings molybdenum is preferentially adsorbed on acidic sites. Other sites which correspond to the subsequent formation of Mo^{VI} are assumed to be the coordinately unsaturated Al^{3+} sites. For the 21% sample the radial distribution function differs from those of all other samples because of the appearance of a peak at 3 Å, which is normally attributed to a polymeric species of unknown structure.

14.4.2.2. Vanadium Oxide

Wong et al.[27] have studied the pre-edge and near-edge structure of a whole set of vanadium oxides from VO to V_2O_5. The crystallographic structures and spatial groups of VO to V_2O_5 are known. As the vanadium site symmetry decreases from O_h to C_s and the formal valence of vanadium increases, the shoulder, initially at 3 eV above the zero (before the main edge), is transformed into a strong peak located at 5.6 eV (Figure 14.8). Simultaneously a strong peak remains at about 20 eV, which is attributed to $1s \rightarrow 4p$ transitions. The lower-energy shoulder is assigned as the $1s \rightarrow 4p$ shake-down transition and the main prepeak as the threshold to the forbidden transition $1s \rightarrow 3d$. In the near-edge region there are features whose positions and intensities vary as the formal valence of vanadium is changed. Thus we have a qualitative way of determining the vanadium site symmetry in any solid, especially in catalysts. As in the case of molybdenum oxides, this qualitative technique may be associated with an EXAFS measurement, which gives the V—O distance and the mean coordination number.

14.4.2.2a. V_2O_5 *on Titania.* The first study[25] concerned the behavior of such catalysts for oxidation of o-xylene into phthalic anhydride. The V_2O_5/anatase system is a classic example of support enhancement of the active phase, since V_2O_5, either alone or supported by other substrates, is a rather indifferent catalyst for this reaction.

From the EXAFS results it is clear that in the freshly prepared catalyst (70 °C) and in one calcined at 350 °C structural changes are evident. The method of preparation precludes the absorption of more than one monolayer of V_2O_5 on TiO_2 surface. The resulting EXAFS analysis indicated that the first shell was split with two short and two long V—O bonds of 1.65 Å(2) and 1.90 Å(2), respectively. By using the empirical formula of Brown and Wu,[29] the charge distribution (bond valences) can be calculated. The sum of the bond valence is

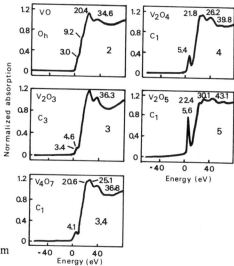

Figure 14.8. Near-edge spectra of vanadium oxides.[27]

4.5 ± 0.5, compared to the expected value of 5.0. The two short bonds are characteristic of terminal double bonds and are comparable to the average value 1.64 Å found in chain vanadates, whereas the longer ones are characteristic of bridging bonds. Because of an abnormally large static disorder, it is difficult to see the fifth V—O bond. The structure of V—O cluster is tentatively described as two double-bonded oxygen ligands attached to vanadium which is anchored to the surface via two V—O—Ti bridges. This model is partially supported by the comparison between the XANES spectrum of the monolayer V_2O_5 with that of $Mn_{1-xx}V_{2-2x}Mo_{2x}O_6$ ($x = 0.36$, represents a vacancy). Both spectra present the same features.

Above 600 °C it is known by X-ray diffraction that the large-area support sinters. During this process the vanadium oxide is freed and crystallizes to V_2O_5. The similarity of its EXAFS spectrum with that of pure V_2O_5 confirms the sintering of vanadium oxide. Furthermore, at 650 °C in the presence of V_2O_5 the anatase transforms into a rutile phase with vanadium occupying the lattice substitutionally as V^{4+} ions, leading to the formation of $Ti_{1-x}V_xO_4$. The comparison of the spectrum of the catalyst heated at 650 °C with that of TiO_2 confirms that this change has taken place.

14.4.2.2b. Hydrated V_2O_5 on Alumina and Silica. Two types of V_2O_5 catalysts were studied by Tanaka *et al.*[26] The first was obtained by impregnating the support with an aqueous solution of NH_4VO_4 and the second with a CH_2Cl_2 solution of $VO(acac)_2$. From Figure 14.9, it is clear that the structure of the catalyst depends on the carrier and not on the precursor. $VO(acac)_2$ and V_2O_5 are compounds in which vanadium atoms are in square-pyramidal coordination, whereas in Na_3VO_4 and NH_4VO_3 the vanadium atoms are in tetrahedral coordination. The common point between the two types of solids is

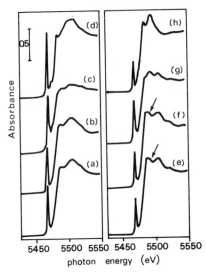

Figure 14.9. Normalized vanadium K-edge XANES spectra of reference compounds and catalysts, supported by Al_2O_3 or SiO_2 (adapted from Tanaka *et al.*[26]): (a) NH_4Vo_3; (b) $VO(acac)_2/Al_2O_3$; (c) NH_4VO_3; (d) Na_3VO_4; (e) NH_4VO_3/Al_2O_3; (f) $VO(acac)_2/SiO_2$; (g) V_2O_5; (h) $VO(acac)_2$.

that they present an intense dipole-allowed pre-edge peak, attributed to the mixing of the $2p$ orbitals of the oxygen atoms with the $3d$ orbitals of vanadium ions. By an *ab-initio* SCF calculation of $V=O(OH)_3$ cluster it has been shown by Koyabashi *et al.*[30] that degenerate LUMO's consist of V_{3d} and O_{2p} atomic orbitals forming π-like antibonding orbitals of $V=O$. Consequently, all the catalysts possess $V=O$ bonds.

The two types of catalysts, however, are different in the surrounding of vanadium atoms. The similarity between XANES spectra of NH_4VO_3 and those of the alumina catalysts demonstrated that vanadium atoms are tetrahedrally coordinated. Moreover, from the sharpness of the first-derivative spectra it may

TABLE 14.4. EXAFS Data for Vanadium-Supported Catalysts[a]

Catalyst	Coordination		Bond length (Å)
1. $Vo(acac)_2/Al_2O_3$	1.3		1.67
(VAD)	2.6	3.9	1.77
2. $NH_4(VO_3)/Al_2O_3$	1.1		1.67
3. $VO(acac)_2/SiO_2$	0.9		1.69
(VSD)	1.5	4.2	1.81
	1.8		1.98
4. NH_4VO_3/SiO_2	0.4		1.60
(VSM)	0.5	2.4	1.76
	0.9		1.89
	0.6		2.03

[a] From Tanaka *et al.*[26]

TABLE 14.6. Modeling of Vanadium Oxide Supported Catalysts[28] a

	Vanadium/SiO$_2$		Vanadium/Al$_2$O$_3$	
	R_{exp}	R_{calc}	R_{exp}	R_{calc}
V=O	1.61	1.35	1.37	1.58
V—O	1.77	1.76	1.77	1.93

^{a} From Shulman *et al.*[28] with model mono- or di-oxo (Figures 14.10 and 14.11).

decomposition of the precursor after impregnation of α-alumina by palladium acetylacetonate solution [Pd(acac)$_2$]. It has been shown by Lesage-Rosenberg[80] that the impregnation of the support takes place as follows:

$$\text{\Big|}-OH + Pd(acac)_2 \longrightarrow \text{\Big|}-O-Pd(acac) + acac\ H_{ads}$$

EXAFS experiments showed that after drying at 100 °C the mean coordination number of palladium, which is 4 oxygen atoms at 1.96 Å changed to 5 oxygen atoms at 1.98 Å. The authors have interpreted this result as an intercalation of Pd(acac) in the aluminum octahedrally vacant sites at the alumina surface. Calcination at higher temperatures resulted in the loss of the second acetylacetone molecule.

The growing of the oxide phase is evidenced by the peak of the second coordination shell around palladium, which corresponds to Pd—Pd distance. The number of nearest neighbors (O atoms) decreases from 5 to 4 at 700 °C with a Pd—O distance of 20 Å (which is close to that in bulk oxide 2.02 Å).

At the same temperature of calcination (700 °C) the catalyst prepared from Pd(NO$_3$)$_2$ shows that in regard to the examined catalyst the amplitude of the Pd—O pair is of the same magnitude as in the bulk oxide, whereas it is only two-thirds for the Pd(acac) catalyst. This shows that the oxide phase obtained from Pd(acac)$_2$ is more dispersed than that obtained from Pd(NO$_3$)$_2$. The existence of an oxide phase is enhanced if the Pd loading is increased.

From this brief study it is concluded that the EXAFS analysis of different shells allows description of the formation of the metal phase during the preparation of catalysts.

14.4.3.2. Rhodium-Based Catalysts

Among the transition metals that are used as catalysts rhodium has been the subject of extensive work by Koningsberger and his colleagues. In the first study van't Blik *et al.*[32] investigated the effect of hydrogen and carbon monoxide chemisorption on 0.57 wt.% Rh/Al$_2$O$_3$. The measurements resulted in a H/Rh ratio of a 1.7 and CO/Rh ratio of 1.9, which indicate a highly dispersed system. After CO chemisorption the IR spectrum exhibits two bonds located at 2095 and 2027 cm^{-1} which are assigned, respectively, to the symmetrical and antisymmetrical stretching frequencies of the $Rh(CO)_2$ species. A careful analysis[33] of the EXAFS spectrum of the catalysts reduced at 773 K revealed that the Rh—Rh distance is 2.65 Å (vs. 2.69 Å in bulk metal) and the coordination number is 5. Consequently, this catalyst is highly dispersed and the metal particles consist of 15–20 rhodium atoms (cf. in a 13 atom cubo-octahedron the mean coordination number is 5.54).

After CO chemisorption (100 kPa) on the evacuated catalysts the EXAFS spectrum changes drastically, due to a desorption of small rhodium crystallites into the $Rh(CO)_2$ species present on the support, but further analysis shows that each rhodium atom is surrounded by two CO molecules (geminal dicarbonyl) and three oxygen atoms belonging to the support. CO adsorption leads to a change in the oxidation state of the rhodium from 0 to $+1$, explained in terms of an oxidative CO chemisorption via CO dissociation. This conclusion is different from that observed with the Pt/Al_2O_3 system on which CO adsorption caused only electronic effects.[34]

Several other rhodium-based catalysts were also studied by the same authors, and it was seen that they vary with the metal loading and the support.[32–33,35] For all these samples it is clear that rhodium is mainly in the reduced state (Rh^0), whereas it is noted that the metal particles present a peculiar interaction with the carrier. Based on a "difference" technique[36] through which one can get minor components in the EXAFS data analysis, it is possible to locate some abnormally large metal–oxygen bonds, whatever the reduction temperature and support. The bond length is of about the same extent as the metal–metal length inside the particles.

Moreover, some catalysts show some Rh–O distances of about the same extent as in bulk Rh_2O_3 (≈ 2.00 Å). The presence of a large Rh–O distance in all these catalysts is interpreted as the formation of an ion-induced dipole bond ($O^2 \rightarrow Rh$). Such a bond is weak and its influence is only felt at very short range.[37] This observation is not in agreement with the work by Vlaic et al.[38] who found, on the contrary, that ruthenium particles are anchored on the carrier via a Ru^{n+}—O bond.

In the case of the 2.85 wt.% Rh/TiO_2 catalyst,[35] after calcination at 623 K and subsequent reduction at 673 K, a Rh—Ti bond (3.42 Å), was found which in the opinion of authors is proof of a structural reorganization in the support in the neighborhood of rhodium metal particles. The formation of a Rh—Ti bond, which is responsible for the SMSI effect is attributed to a surface reduction of TiO_2 into Ti_4O_7. This conclusion is supported by a comparison with a catalyst Rh/Al_2O_3 with the same rhodium content in which a reoxidation is possible at 100 K, which is not the case with a Rh/TiO_2 catalyst.[39] The Rh—O bond length at 2.05 Å is attributed (for a reduction temperature of 473 K) to incomplete reduction of the precursor.

Interesting work[40] was also done which tentatively illustrates the power of EXAFS to calibrate hydrogen chemisorption. It is well known that hydrogen-to-metal ratios exceeding unity have been measured for highly dispersed platinum,

TABLE 14.7. Hydrogen Chemisorption and
EXAFS Results for Pt, Ir, and Rh Dispersed
on Alumina or Silica[a]

wt.%	H/M	N
0.47	1.98	3.8
0.57	0.7	5.1
1.04	1.65	5.8
2.00	1.2	6.3
2.40	1.2	6.6

[a] From Kip et al.[40]

iridium, and rhodium catalysts supported on alumina or silica. For these catalysts the coordination of hydrogen to surface atoms is unknown and consequently the metal surface area cannot be calculated from the chemisorption values, and we must resort to assumptions (e.g., nature of the chemisorption sites and the number of chemisorbed atoms) in order to determine the dispersion of the metal particles.

Whatever the preparation methods (ion exchange or incipient wetness), the metal loadings (0.47 to 2.4 wt.%) or the reduction temperatures (473 to 773 K), the H/M ratio determined by hydrogen chemisorption is an approximately linear function of the average metal coordination number determined by EXAFS, keeping in mind that the hydrogen chemisorption measurements were made at 298 K (Table 14.7).

14.4.3.3. Platinum/Alumina

This system has been studied extensively, undoubtedly because at low precious metal content (0.1–1 wt.%) it is widely used for reforming crude oils in petrochemistry. The successive steps of impregnation, drying, calcination, and activation have been extensively studied by XAS in order to understand the mechanism of formation of small platinum particles over the carrier. These steps must also be considered as an early stage of work necessary for an understanding of the Pt–Re/Al$_2$O$_3$ system. Moreover, Pt/Al$_2$O$_3$ catalysts are often obtained by impregnation of the carrier by H$_2$PtCl$_6$, and chlorine present in the precursor is known to have a decisive influence on the final dispersion of the active metal phase.

In this field, work by Lagarde et al.[86] and Bazin[84] has shown that in the drying step (110 °C) platinum is surrounded by chlorine. This qualitative conclusion was confirmed by Berdala et al.[83] who showed from EXAFS experiments that the anchoring of platinum on the carrier is made via the formation of a mononuclear complex PtCl$_4$(O$_2$). Its existence appears to be dependent on the metal content (0.3%–1 wt.%) and on the chlorine concentration (in the impregnation solution) under the HCl form (1.1 or 2.2 wt.%). After calcination of the precursor at 530 °C the number of oxygen atoms surrounding the platinum atom increases from about 2.0 to 5.5, whereas the coordination in chlorine decreases from about 4.5 to 1.7. The fact that there is no Pt—Pt bond formation indicates the nucleation of platinum particles. Thus, the role of chlorine is to stabilize the anchoring of platinum at the surface of the carrier. The reduction by flowing hydrogen leads to the formation of platinum particles, as indicated by the formation of Pt—Pt bonds (2.74–2.76 Å, coordination number from 3 to 7). For the catalysts of low platinum content (0.37 wt.%) it has been observed that some Pt—O bonds are present (1.6 oxygen atoms at 2.02 Å). This may be an indication of bidimensional particles. On the other hand, an initial concentration of 1.1% chlorine instead of 2.2% leads to a coordination number of 7 around platinum. A high chlorine content favors the dispersion of platinum particles on the carrier.

A similar study was done by Berdala et al.[85] starting from a different precursor, Pt(acac)$_2$, loaded to a catalyst whose platinum content was 0.7 wt.%. At the dried stage (110 °C) the lack of Pt—Pt neighbors shows that the metal stays in the form of isolated mononuclear complexes. The Pt—O bond is intermediate in distance between Pt(acac)$_2$ and PtO$_2$. During impregnation one acetylacetone molecule is freed. From considerations about the shape of the white line and the EXAFS analysis it was concluded that the complex is fixed by two oxygen atoms pertaining to the support and that the two other Pt—O bonds are due to the remaining acetylacetone molecule.

During calcination at 200 °C the complex loses the second molecule of acetylacetone, leading to fixation of the platinum atoms by two or three Pt—O bonds. At 300 and 500 °C the coordination number in oxygen remains unchanged but some Pt—Pt bonds appear. As the coordination number in platinum does not vary, it can be considered that the particles coalescence in bidimensional form without the formation of oxides. This type of growth suggests a strong interaction between the complex and the support. After calcination at 300 °C, the reduction at 400 °C is complete and leads to small platinum particles in which the coordination is 5.4 platinum atoms at 2.75 Å. In the absence of Pt—O detectable bonds it is concluded that the platinum particles are roughly spherical.

14.4.4. Platinum-Based Catalysts

Such catalysts are generally used for the reforming of the naphtha fraction in oil-refining. They require a proper balance between the acidic and metallic function to give an adequate performance. The acidic function is provided by alumina with its acidity promoted by chlorine. Otherwise the metallic function is given by platinum, which is generally promoted by a second metal such as rhenium, tin, or iridium. Great interest on the part of the petroleum industry has led to extensive study of these solids by XAS in order to determine the relative organization of the two metal atoms in the bimetallic phase and to find a correlation between the structure of the catalysts and their activity or selectivity.

TABLE 14.8. EXAFS Studies of Pt–Sn on Silica and Alumina with Pt/Sn $\approx$ 1[a]

Central atom	Neighboring atom	Pt–Sn/SiO$_2$		Pt–Sn/Al$_2$O$_3$	
		N	R	N	R
Pt	Pt	6.5	2.80	3.2	2.75
Pt	Sn	4.7	2.75	< 0.5	2.61
Pt	O	—	—	0.7	2.06
Pt	Cl	—	—	< 0.5	2.34
Sn	Pt	6.8	2.79	< 0.5	2.62
Sn	Sn	0.9	2.85	< 0.5	3.04
Sn	Sn	1.7	3.59		
Sn	O	< 0.5	2.04	1.9	2.04
Sn	Cl	< 0.5	2.27	0.5	2.28

[a] (1.1 wt.% Pt and 0.6 wt.% Sn, from Meitzner et al.[88]).

14.4.4.1. Pt–Sn System

In a relatively recent work Meitzner et al.[88] studied XAS using both EXAFS and absorption edge as complementary techniques. Two kinds of catalysts were examined containing 1.1 wt.% Pt and 0.6 wt.% Sn leading to a Pt/Sn atom ratio close to 1, the bimetallic phase being supported on alumina or silica.

The EXAFS experiments were performed at both Pt L_{III} and Sn K edges. Starting from a set of reference materials it has been possible to extract modified amplitude functions and phase shifts for the following pairs M–M, M–M′, M–O, and M–Cl in which M and M′ represent either platinum or tin atoms. Table 14.8 lists the measured coordination numbers and interatomic distances from complete EXAFS analyses.

14.4.4.1a. Pt–Sn/SiO$_2$. The total coordination number of the platinum atom (11.2) shows clearly that bimetallic particles are larger than 50 Å. This is also demonstrated by the powder diffraction pattern of this sample, which exhibits diffraction lines corresponding approximately to a Pt–Sn phase. The lack of interaction between tin and oxygen or chlorine (as deduced from the coordination numbers) indicates that most of the tin is in the zero oxidation state. This conclusion is supported by the L_{III} absorption edge experiment, which shows that the absorption threshold line is less intense than in a Pt/SiO$_2$ reference material of the same loading in platinum. The presence of tin has the effect of decreasing the number of unfilled d states associated with the platinum atoms.

14.4.4.1b. Pt–Sn/Al$_2$O$_3$. In this catalyst the situation is more complicated. First the coordination numbers of platinum are lower than for the Pt–Sn/SiO$_2$ catalyst, indicating a greater dispersion of the metallic phase. The coordination number of platinum with tin is very low, which leads to the conclusion that most of the tin is not interacting with platinum. Moreover, interaction of platinum or tin with oxygen or chlorine is evident, showing that an important fraction of these atoms are in an oxidation state different from zero. Absorption threshold experiments at Pt L_{III} edge showed that absorption resonance is more intense that in the Pt/SiO_2 catalyst. By comparing this line with the absorption threshold of $[Ph_3PMe]_3[Pt(SnCl_3)_5]$ the authors concluded that the electronic state of platinum seems close to that of this complex. The catalyst appears to be made of platinum clusters dispersed on alumina containing Sn^{2+} at the surface.

As a general consequence the nature of the interactions of the metal atoms between themselves and with the support in the platinum–tin system appears to be strongly dependent on the carrier (silica or alumina) and also on the preparation method of the catalytic material.

14.4.4.2. Pt–Re/Al$_2$O$_3$

An important study has been done on this system, as well as on Pt–Rh/Al$_2$O$_3$ by Bazin et al.[87] The whole can be divided in two parts: the first is concerned with the first stages of preparation (as for Pt/Al$_2$O$_3$) and the second with the study of the catalyst itself after reduction by hydrogen.

14.4.4.2a. Preliminary Steps. The drying step is essentially characterized by the fact that the surrounding of platinum is quite different from that obtained with the monometallic catalyst. The coordination in chlorine is about 2 and in oxygen about 5, independent of the Re/Pt ratios (0.5 to 2). The total coordination of platinum is consequently close to 8 which is obtained only in the case of monometallic catalyst and also for the Pt–Rh with high metal loading. In order to explain this high platinum coordination it is assumed that the platinum atom has partially lost its ionic character and is partly covalent. In this case it can accommodate a higher coordination number than in the starting presursor, which accounts for a weakening of the platinum white line intensity of the completely ionic PtO_2. At the same time the EXAFS analysis at the L_{III} edge of rhenium shows that the rhenium is surrounded by four oxygen atoms at 1.90 Å.

The calcination step slightly modifies the platinum surrounding. The total coordination remains about 8 to 9 and the coordination in chlorine and oxygen shows very little change. Consequently platinum in a dried stage has configuration close to that of the calcined monometallic catalyst. Simultaneously the calcination lowers the coordination number of rhenium by one oxygen atom.

14.4.4.2b. Reduction Step. Whatever the reduction temperatures under flowing hydrogen (300 or 500 °C), platinum appears in reduced state, whereas rhenium is not reduced. As platinum and rhenium are neighbors in the Periodic Table EXAFS is not able to distinguish one from the other in the same coordination sphere because of the similarity of shape of the backscattering amplitude and phase shift. Each platinum atom is surrounded by 3 or 4.5 metal atoms, with a small contribution of oxygen atoms (0.5 to 0.6). On the other hand the rhenium appears to be surrounded essentially by oxygen atoms.

Two conclusions can be drawn from these results:

1. The structure of the bimetallic catalyst is such as one finally obtains with small platinum aggregates suported by rhenium partially oxidized by its anchoring on alumina.
2. The presence of rhenium is undoubtedly to stabilize the small size of the platinum particles, which partly explains the resistance to sintering and consequently to the aging of the catalysts.

14.5. ABSORPTION RESONANCE STUDIES: WHITE LINES

14.5.1. Introduction

Correlations between electronic and chemical properties have always been sought in catalysis, in order to establish a relationship between the d character of the metal catalyst and its catalytic activity. Metals of the third transition series exhibit white lines (L_{III} and L_{II} edges), which do not exist for gold (full d band). It has been concluded that they are mainly due to $2p \rightarrow 5d$ transitions.

In their study of L_{III} adsorption edge for platinum, iridium, and gold, Lytle et al.[41] reported that some compounds of these elements exhibit stronger L_{III} white lines than those of the pure metal. Metal cluster or oscillator strength calculations made by Horsley[43] indicate that the L-edge spectra for the metal and its compounds can be compared on the same basis. The difference in the L-edge resonance area between the metal and its compounds can be used to measure charge transfer.

14.5.2. Theoretical Summary

The $L_{II,III}$ X-ray adsorption edges in the X-ray adsorption spectrum correspond to the process in which an X-ray photon is absorbed by the promotion of an electron from a $2p$ core state. The L_{II} X-ray adsorption edge arises from the $2p_{1/2}$ core state, while the L_{III} X-ray adsorption edge arises from the $2p_{3/2}$ core state and is at a lower energy. One of the characteristics of these edges in that they often exhibit a white line corresponding to an enhanced X-ray absorption. For instance, this white line can be observed for the L_{III} absorption edge for platinum, but not for the L_{II} and L_{I} edges.

For the L_{I} edge the initial core states have s symmetry. Owing to the dipole selection rule the transition probability to d states is insignificant, and a white line cannot be observed. At the L_{II} and L_{III} edges the initial core states have p symmetry. For a free platinum atom the empty states in the $5d$ band have a $J = \frac{5}{2}$ value and thus the transition probability is significant ($\Delta J = 0 \pm 1$). One would therefore expect a transition from the L_{III} edge, where the initial state has a $J = \frac{3}{2}$ value, but not from the l_{II} edge, where the initial state has $J = \frac{1}{2}$. Consequently, we should observe a difference in the platinum L_{II} and L_{III} near-edge structure. In fact we must take into account the spin–orbit coupling. If this coupling exists (as for Pt) the empty d states in the metal, which would be a mixture of atomic states with $J = \frac{3}{2}$ and $J = \frac{5}{2}$, and both edges would give lines of comparable strength. Using the tight binding approximation, we can show that, for platinum, the final states are predominantly those corresponding to $J = \frac{5}{2}$, and contribute 14 times more to the final states than those corresponding to a total angular momentum number $J = \frac{3}{2}$. Thus, we have a method giving the number of unoccupied d electron states in compounds of platinum, especially in catalysts.[47,48]

14.5.3. Monometallic Catalysts

The first approach of this type was made by Lytle et al.[41] in a study of gold clusters supported on silica and alumina. They claimed that the peak intensity at the L_{III} edge is proportional to the electron vacancies. By comparing the difference in area between the normalized spectra of the catalysts (1% Ir or Pt) and the pure metals Short et al.[47] concluded that the concentration of d states is greater in the catalysts than in the pure metal. This fact seems to be due to the small particle size of the metal (dispersion was assumed to be 0.7–1), but one cannot exclude the possibility that platinum or iridium are electron donors to the carrier. A similar conclusion was obtained for osmium catalysts supported on silica.[46] When oxygen is chemisorbed on these small metal particles the concentration of d electron vacancies becomes greater. This behavior is consistent with other work[47-50] in which platinum is supported on silica or Y zeolite. The observed effect is the same but lower than that found for PtO_2 or IrO_2 when compared with metallic platinum or iridium. This effect could be attributed to a lower ionic character of the metal–oxygen bonds in the catalysts covered with oxygen than the metal–oxygen bonds in the bulk oxides.

14.5.4. Bimetallic Catalysts

Such catalysts have been extensively studied by Sinfelt and co-workers, and they reviewed[50] the main results of their studies of the $Os-Cu/SiO_2$ system. The presence of copper in the environment of osmium, as shown by EXAFS experiments, decreases the intensity difference between the spectra of bulk osmium and osmium catalysts. This suggests an electronic interaction between osmium and copper resulting in a decrease in the number of unfilled d states associated with osmium. Such a modification of the electronic structure of osmium would affect the Cu K edge, but there are no details on this kind of experiment. More important is the lack of correlation with catalytic activity.

This aspect has been examined by Moraweck et al.[45] in their study of catalysts of Pt–Fe supported on charcoal. These catalysts are known to be selective in the hydrogenation of cinnamic aldehyde into cinnamic alcohol. The EXAFS measurements at the L_{III} edge of platinum clearly showed that there is an interaction between platinum and iron in the small particles of bimetallic catalysts (50 Å as shown by TEM). A total number of nearest neighbors of about twelve confirms this trend. The ratio between atom numbers of iron and platinum corresponds to the chemical composition. It is only for catalysts containing 20 and 30 at.% Fe that the distance of platinum to its neighbors is the same.

For the two catalysts containing 10 and 55 wt.% Fe, respectively, a rather strong contraction of the Pt–Fe distance was measured, which may be due to an asymmetric distribution.

By evaluating the area difference between L_{III} and L_{II} edges of platinum for a series of Pt–Fe catalysts with different relative concentrations in both metals it has been possible to determine the number of holes in the $5d^{5/2}$ band. From these results it is evident that the number of vacancies in the $5d$ band of platinum is lowered with respect to bulk platinum. The number of vacancies is at a maximum for the more active and selective catalyts (20% Fe).

This behavior may be due to a modification of the relative absorption strengths of the $C=C$ double bond and of the $C=O$ bond induced by the electron transfer from iron toward platinum. The evaluation of the difference in area between the K edge of bimetallic catalysts and pure iron leads to the same conclusion, and constitutes a complementary measurement of those made at the platinum edges. However, it is difficult to give a quantitative interpretation of these results.

14.6. AN EXAMPLE OF CHEMISTRY OF THE SOLID STATE: PREPARATION AND CHARACTERIZATION OF A Pd–Cr/SiO$_2$ CATALYST

Among several methods that may be used to prepare a catalyst of known composition, we can start from a stoichiometric compound that can lead to a bimetallic catalyst with the same composition as the precursor. Here, the chosen method is (i) to prepare the precursor, and (ii) to reduce it by hydrogen after impregnation of the carrier.

14.6.1. Preparation and Characterization of the Precursor

It has been shown by Astier et al.[72] that $NiMoO_4$ allows one to obtain intermetallic Ni_4Mo, after reduction as shown by X-ray diffraction. A type of compound interesting in this way is $[Pd(NH_3)_4]Cr_2O_7$, which may lead to intermetallic $PdCr_2$ after reduction.[51]

With a mixture of solutions $[Pd(NH_3)_4]Cl_2$, H_2O, and $(NH_3)_4Cr_2O_7$, one obtains a precipitate of very small microcrystalline yellow–orange needles, the chemical analysis of which is Pd/N = 0.25 and Pd/Cr = 0.5. The X-ray diffraction pattern shows a new phase for which the parameters are: $a = 9.2065$ Å, $b = 15.0764$ Å, $c = 8.1706$ Å, $\alpha = \beta = 94.492°$, and $\gamma = 93.883°$, corresponding to a monoclinic unit cell[52] containing four molecules of a new compound, which was assumed to be $[(PdNH_3)_4)]Cr_2O_7$. An X-ray adsorption study was performed at both Cr and Pd K edges in order to determine the surrounding of

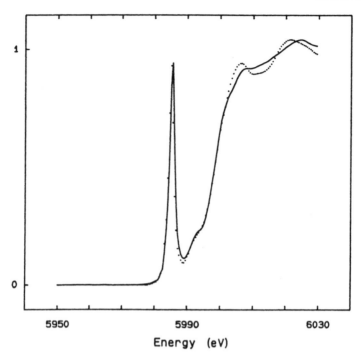

Figure 14.12. XANES spectra of $(NH_4)_2Cr_2O_7$ (dotted line) and $[Pd(NH_3)_4]$ Cr_2O_7 (solid line) at the Cr K edge.

both elements. Figure 14.12 compares the X-ray absorption edges of the new compound and of $(NH_3)_4Cr_2O_7$. The similarity of the two spectra allows one to conclude that the $Cr_2O_7^{2-}$ ion remained during the precipitation reaction.

The intense prepeak is characteristic of a distorted tetrahedron of oxygen atoms around chromium.[74] Moreoever, the EXAFS data at the Cr K edge lead to a determination of the four Cr—O distances, which are, respectively: 1.52, 1.56, 1.81, and 1.90 Å all ± 0.01 Å.

From these values we can see that the tetrahedron is more distorted than in $(NH_4)_2Cr_2O_7$ compound, which explains the difference between the two XANES spectra shown in Figure 14.12. An EXAFS experiment at the Pd K edge gives the surrounding of the palladium atom, which was found to be four nitrogen atoms located at 2.06 Å instead of at 2.05 Å in $[Pd(NH_3)_4]Cl_2, H_2O$.

As a conclusion a compound of known composition and structure has been obtained which has been used as a catalyst precursor.

14.6.2. Preparation and Characterization of the Catalysts

The low solubility of $[Pd(NH_3)_4]Cr_2O_7$ explains the small loading obtained after impregnation of silica (Degussa aerosil — 300 m^2g^{-1}). After reduction overnight by flowing hydrogen at 600 °C, the catalyst was studied by EXAFS at both the Cr and Pd K edges to determine the surrounding of chromium and palladium atoms. The mean particle diameter has been measured by SAXS and TEM experiments and has been found to be equal to 18 Å.

14.6.2.1. Pd K Edge

The contribution of the first coordination sphere was isolated on the FT as shown in Figure 14.13 (triangles). A first modeling was done using only palladium atoms (Figure 14.13a) using an experimental phase shift and amplitude extracted from a similar experiment on a palladium 15 μm foil. It may be observed that a strong disagreement is noted in the 100–200 eV energy range. It seems that another type of atom exists in the first coordination sphere of palladium. Using theoretical phase shifts backscattering amplitude from Teo and Lee,[58] one obtains a better fit (Figure 14.13b) showing that a small quantity of chromium (not oxygen) is interacting with palladium. The results are summarized in Table 14.9.

It is clear from this result that the catalyst is not made of bimetallic particles having the mean composition $PdCr_2$. From the above results, it can be concluded that this composition is about $Pd_{92}Cr_8$. The following questions arise: where is chromium? under what form it is supported on silica? Answers may be obtained by an EXAFS experiment at the Cr K edge in the fluorescence mode.

14.6.2.2. Cr K Edge

The FT of the catalyst's EXAFS spectrum is shown in Figure 14.14 together with the FT of Cr_2O_3. From this comparison it is clear that the first peak may be attributed to Cr—O bonds. It may be noted that the X-ray diffraction study of the reduced unsupported precursor led to the conclusion that two phases are present: a metal phase (Pd or Pd–Cr phase) characterized by weak and very broad lines and an oxide phase identified as being Cr_2O_3 (intense and narrow lines).

From the crystallographic structure of Cr_2O_3[75] the second peak in Figure 14.14 corresponds to Cr—Cr distances. However, this peak may also be attributed to a Cr—Pd distance in the bimetallic phase. This distance may vary between 2.50 Å (bulk chromium) and 2.75 Å (bulk palladium). This leads us to suppose that this second peak may be due to Cr—Cr and/or Cr—Pd distances. The inverse FT of this peak is shown in Figure 14.15. When a model is made by considering only chromium atoms in the second coordination sphere it is evident that such a model is wrong (Figure 14.15a). Consequently, the chromium present in the catalyst is not only in the form of Cr_2O_3. Figure 14.15b shows the fit obtained with chromium and palladium in the neighborhood of chromium atoms. In this case the model becomes very good and it can be concluded that some chromium atoms are interacting with palladium. A global fit may be realized from the inverse FT of the ensemble of both main peaks of the catalyst spectrum.

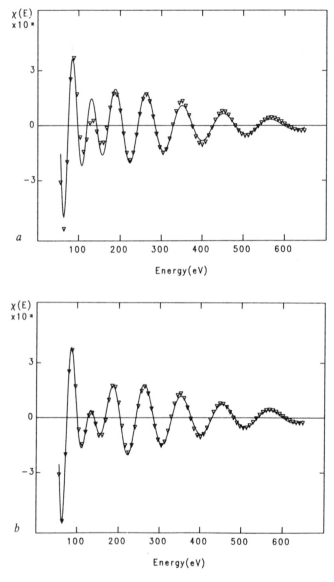

Figure 14.13. EXAFS at the Pd K edge: (a) fit without Cr atoms; (b) fit with Cr atoms introduced in the first shell of Pd.

This model leads to the fit shown in Figure 14.16 and to the results summarized in Table 14.10. By comparing the fit values (coordination numbers and distances) one can conclude that the catalyst contains two types of chromium atoms, the first one consisting of metallic atoms interacting with palladium. It may be noted that the optimization process leads in this case to the

TABLE 14.9. EXAFS Data for a Pd–Cr Catalyst
Reduced at 600 °C at Cr K Edge[a]

Bond	Coordination number	R
Pd—Pd[b]	8.00	2.75
Pd—Pd[c]	8.97	2.78
Pd—Cr	0.7	2.73

[a] From Borgna et al.[53]
[b] See Figure 14.13a.
[c] See Figure 14.13b.

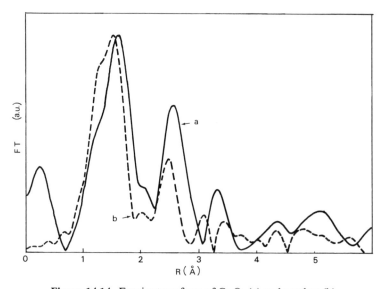

Figure 14.14. Fourier transform of Cr_2O_3 (a) and catalyst (b).

same distance (2.73 Å) as in the case of the model at the Pd K edge. Otherwise, from the comparison between the Cr—O and Cr—Cr distances in the catalyst and in bulk Cr_2O_3, a second type of chromium atom exists. Consequently, the catalyst may be described as a bimetallic phase Pd—Cr and another phase which can be thought as Cr_2O_3.

14.6.2.3. Structure of the Pd–Cr/SiO₂

The chemical analysis as well the STEM experiment in scanning mode on a large area ($\approx 10 \ \mu m^2$) shows that the catalyst has the relative atomic composition Pd/Cr = 0.5. However EXAFS at the Pd K edge leads to a relative

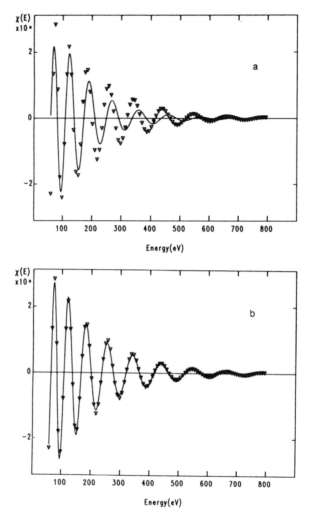

Figure 14.15. Fit of the inverse Fourier transform of the second peak (Figure 14.14) using (a) only Cr atoms, (b) Cr and Pd atoms.

composition $Pd/Cr = 11.5$. In the spot mode, i.e., when the STEM analysis area is about 200 Å^2, the particle composition is found to be $Pd/Cr = 3.5$.

As one does not detect bulk Cr_2O_3 particles by TEM, it becomes clear that a large amount of Cr_2O_3 is finely dispersed at the surface carrier. This hypothesis is confirmed by the fact that the coordination numbers in the oxide phase are lower than in bulk oxide (Table 14.10). Consequently we can give a tentative description of the catalyst. In this model, the bimetallic particles are partially covered by an oxide phase which is presumably Cr_2O_3 and a very dispersed phase CrO_x on the surface of the support.[53]

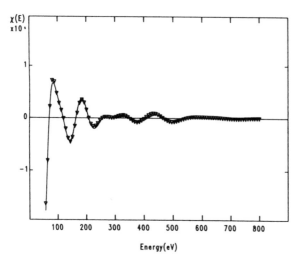

Figure 14.16. Fit of the inverse Fourier transform of the main two peaks of Figure 14.14 (curve b) giving the coordination numbers and distances listed in Table 14.10.

TABLE 14.10. EXAFS Data for a Pd–Cr Catalyst Reduced at 600 °C at Cr K Edge[a]

Bond	Coordination number	R (Å)
Cr—O	4.5	2.03
Cr—Cr	1.4	2.80
Cr—Pd	1.0	2.73

[a] From Borgna *et al.*[(53)]

14.7. CONCLUSIONS

Through the preceding examples it is clear that X-ray adsorption spectroscopy may bring some light in the vast field of electronic and geometrical structure of catalysts. This is not limited to catalyst science but is a general trend in numerous fields, including solid states physics and chemistry, amorphous alloys, biochemistry, and pharmacology. Whatever the studied phenomenon (SEXAFS, XANES, etc.) the trend is always the same: the problem is to get some information on the surrounding of a target element in order to elucidate the physical or chemical properties of catalysts.

However it appears that if the EXAFS technique has become almost routine, some new developments are expected in XANES, especially for theoretical work. However, it actually seems that the building of more powerful synchrotron rings with a weak emittance and a large brillance (for instance in Europe, ESRF) requires an improvement in photon detection. It is not very useful to have large numbers of photons if they cannot be fully detected and used to improve the signal-to-noise ratio and the speed of experiments.

The last example given in this chapter clearly demonstrates that X-ray adsorption spectroscopy is not generally to be considered as a unique characterization tool. In many cases other chemical and physical techniques are necessary to obtain a convenient and precise image of the structure of a heterogeneous catalyst. For example, with recent progress in photon detection for anomalous scattering it is possible to remove the support contribution which is always present in X-ray absorption experiments.[60] X-ray diffraction may also be useful.[64] In other cases, EXAFS is a complementary but necessary tool to explain tentatively some catalytic properties. In all cases, its use remains decisive. An idea of its value may be seen in the numerous proposals submitted to various synchrotron radiation centers.

The recent evolution of X-ray absorption experiments in the field of catalysts and most generally in chemistry shows two tendencies: (i) development of *in situ* experiments on real systems under reaction conditions[54] after an initial study on samples prepared in the laboratory; and (ii) development of kinetic studies under controlled atmospheres and pressures.

Such studies are now possible because synchrotron laboratories now have dispersive EXAFS and special devices such as helium or nitrogen cryostats and furnaces that enable such experiments to be performed under reaction conditions.

REFERENCES

1. H. Ficke, *Phys. Rev.* **16**, 2 (1920).
2. G. Hertz, *Zeit. Phys.* **3**, 19 (1920).
3. J. D. Hanawalt, *Phys. Rev.* **70**, 20 (1931).
4. R. de L. Kronig, *Zeit. Phys.* **70**, 317 (1931).
5. R. de L. Kronig, *Zeit. Phys.* **75**, 468 (1932).
6. H. Petersen, *Zeit. Phys.* **76**, 768 (1932).
7. H. Petersen, *Zeit. Phys.* **80**, 528 (1933).
8. H. Petersen, *Zeit. Phys.* **98**, 569 (1936).
9. M. Sawada, *Rep. Sci. Works,* OSAKA University (1979).
10. V. V. Smidt, *Bull. Acad. Sci. USSR, Ser. Phys.* **25**, 998 (1961).
11. V. V. Smith, *Bull. Acad. Sci. USSR, Ser. Phys.* **27**, 998 (1961).
12. E. A. Stern, *Phys. Rev.* **B10**, 3027 (1974).
13. D. E. Sayers, E. A. Stern, and F. W. Lytle, *Phys. Rev. Lett.* **27**, 1204 (1971).
14. C. Wivel, R. Candia, B. S. Clausen, S. Morups, and H. Topsoe, *J. Catal.* **68**, 453 (1981).
15. B. S. Clausen, H. Topsoe, R. Candia, J. Vailladsen, B. Lengeler, J. Als-Nielsen, and F. Christensen, *J. Phys. Chem.* **85**, 3868 (1981).
16. R. Candia, B. S. Clausen, J. Bartholdy, H. Y. Topsoe, B. Lengeler, and H. Topsoe, *Proc. 8th Inter. Congress Catal* (1984), p. II-375.
17. B. S. Clausen, B. Lengeler, R. Candia, J. Als-Nielsen, and H. Topsoe, *Bull. Soc. Chim. Belg.* **90**, 1249 (1981).
18. G. Sankar, S. Vasudevan, and C. N. R. Rao, *J. Phys. Chem.* **91** (1987).
19. S. M. A. A. Bouwens, D. C. Koningsberger, V. H. S. de Beer, and R. Prins, *Catal. Lett.* **1**, 55 (1988).
20. M. J. Ledoux, J. C. S. *Faraday Trans. I* **83**, 2172 (1987).

21. G. Sankar, P. R. Sarode, A. Srinavasan, C. N. R. Rao, S. Vasudevan, and J. M. Thomas, *Proc. Indian Acad. Sci.* **93**, 321 (1984).
22. N. Kakuta, K. Tohji, and Y. Ugadawa, *J. Phys. Chem.* **92**, 2853 (1988).
23. C. T. J. Mensch, J. A. R. van Veen, van Wingerden, and M. P. van Dijk, *J. Phys. Chem.* **92**, 4961 (1988).
24. J. A. R. van Veen and P. A. J. M. Hendricks, *Polyhedron* **5**, 75 (1986).
25. R. Kozlowski, R. F. Pettifer, and T. M. Thomas, *J. Phys. Chem.* **87**, 5176 (1983).
26. T. Tanaka, H. Yamashita, R. Tsuchitani, T. Funabiki, and S. Yoshida, *J. Chem. Soc., Faraday Trans. I* **84**, 2987 (1988).
27. J. Wong, F. W. Lytle, R. P. Messmer, and D. H. Maylotte, *Phys. Rev.* **B30**, 5569 (1984) and references cited herein.
28. R. G. Shulman, Y. Yafet, P. Eisenberger, and W. E. Blumberg, *Proc. Nat. Acad. Sci. USA* **73**, 1384 (1986).
29. I. D. Brown and U. U. Wu, *Acta Crystallogr.* **B32**, 1957 (1976).
30. Y. Kobayashi, M. Yamagichi, T. Tanaka, and S. Yoshida, *J. Chem. Soc., Faraday Trans. I* **81**, 1513 (1985).
31. S. Yoshida, T. Matsuzaki, T. Kashiwasaki, M. Mori, and K. Tamara, *Bull. Soc. Chim. Japan* **47**, 1564 (1974).
32. H. F. J. van't Blik, J. B. A. D. van Zon, T. Huibinga, J. C. Vis, D. C. Koningsberger, and R. Prins, *J. Phys. Chem.* **83**, 2264 (1983).
33. H. F. J. van't Blik, J. B. A. D. van Zon, J. C. Vis, D. C. Koningsberger, and R. Prins, *J. Mol. Catal.* **25**, 379 (1984).
34. T. Fukushima and J. R. Katzer, *Proc. 7th Inter. Congress on Catalysis* (1980).
35. D. C. Koningsberger, J. H. A. Martens, R. Prins, D. R. Short, and D. E. Sayers, *J. Phys. Chem.* **90**, 3049 (1986).
36. B. K. Teo, *EXAFS: Basic Principles and Data Analysis,* Springer-Verlag, Berlin (1986), p. 139.
37. J. E. Huheey, *Principles of Structure and Reactivity,* Harper and Row, New York (1972), p. 193.
38. G. Vlaic, J. C. J. Bart, W. Cavigliolo, A. Furesi, V. Ragaini, M. G. Cattania-Sabbadini, and E. Burattini, *J. Catal.* **107**, 263 (1987).
39. J. H. A. Martens, R. Prins, H. Zandberger, and D. C. Koningsberger, *J. Phys. Chem.* **92**, 1903 (1988).
40. B. J. Kip, R. Prins, F. B. M. Duivenvoorden, and D. C. Koningsberger, *J. Catal.* **105**, 26 (1987).
41. F. W. Lytle, P. S. W. Wei, R. B. Greegor, G. H. Via, and J. H. Sinfelt, *J. Chem. Phys.* **70**, 4849 (1979).
42. T. K. Sham, *Phys. Rev.* **B31**, 1888 (1985).
43. J. A. Horsley, *J. Chem. Phys.* **76**, 1451 (1982).
44. F. W. Lytle, *J. Catal.* **43**, 376 (1976).
45. B. Moraweck, P. Bondot, D. Goupil, P. Fouilloux, and A. J. Renouprez, *J. Phys. Fr. Coll.* **C8 39**, 263 (1986).
46. J. H. Sinfelt, G. H. Via, F. W. Lytle, and R. B. Greegor, *J. Chem. Phys.* **75**, 5527 (1981).
47. D. R. Short, A. N. Mansour, Jr., J. W. Cook, D. E. Sayers, and J. R. Katzer, *J. Catal.* **82**, 299 (1983).
48. A. N. Mansour, Jr., J. W. Cook, and D. E. Sayers, *J. Phys. Chem.* **88**, 2330 (1984).
49. N. W. Smith, G. K. Wertheim, S. Huefer, and M. M. M. Traum, *Phys. Rev.* **B10**, 3197 (1974).
50. J. H. Sinfelt, G. H. Via, and F. W. Lytle, *Catal. Rev. Sci. Eng.* **26**, 81 (1984).
51. A. Borgna, B. Moraweck, and A. J. Renouprez, *J. Chem. Phys. Biol.* **86**, 1719 (1989).
52. A. Borgna, B. Moraweck, and P. Fessler, *Powder Diffraction* **4**, 217 (1989).
53. A. Borgna, B. Moraweck, J. Massardier, and A. J. Renouprez, *J. Catal.* **128**, 99 (1991).

54. N. Guyot-Sionnest, D. Bazin, J. Lynch, J. P. Bournonville, and H. Dexpert, *Physica* **B158**, 211 (1989).
55. J. H. Sinfelt, G. H. Via, and F. W. Lytle, *J. Chem. Phys.* **72**, 4832 (1980).
56. G. Meitzner, G. H. Via, F. W. Lytle, and J. H. Sinfelt, *J. Chem. Phys.* **78**, 882 (1980).
57. G. Meitzner, G. H. Via, F. W. Lytle, and J. H. Sinfelt, *J. Chem. Phys.* **78**, 2553 (1980).
58. B. K. Teo and P. A. Lee, *J. Am. Chem. Soc.* **101**, 2815 (1979).
59. A. G. McKale, B. W. Veal, A. P. Paulikas, S. K. Chan, and G. S. Knapp, *J. Am. Chem. Soc.* **110**, 3763 (1988).
60. J. M. Tonnerre, Thesis, Université Paris X, Orsay (1989).
61. W. W. Beeman and W. Friedman, *Phys. Rev.* **56**, 392 (1939).
62. S. A. Nemnonov and M. F. Sorokin, *Bull. Acad. Sci. SSSR, Phys. Ser.* **24**, 462 (1960).
63. L. A. Grunes, *Phys. Rev.* **B27**, 2111 (1983).
64. G. Bergeret and P. Gallezot, this volume Chapter 15.
65. B. K. Teo, *J. Am. Soc.* **101**, 3990 (1981).
66. P. A. Lee and G. Beni, *Phys. Rev.* **B15**, 2862 (1977).
67. S. J. Gurman and R. F. Pettifer, *Phil. Mag.* **40**, 345 (1979).
68. E. D. Crozier and A. J. Seary, Can. *J. Phys.* **58**, 1388 (1980).
69. B. K. Teo, H. S. Chang, R. Wang, and M. R. Antonio, *J. Non-Cryst. Solids* **58**, 1388 (1983).
70. B. K. Teo, *EXAFS: Basic Principles and Data Analysis,* Springer-Verlag, Berlin (1986), p. 106.
71. Ibid., p. 34.
72. M. Astier, A. Bertrand, and S. J. Teichner, *Can. J. Chem. Eng.,* **60**, 40 (1982).
73. P. Eisenberger and G. S. Brown, *Solid State Comm.* **29**, 481 (1979).
74. F. W. Kutzler, C. R. Natoli, D. K. Miseder, S. Doniach, and K. O. Hodgson, *J. Chem. Phys.* **73**, 3274 (1980).
75. R. E. Newnham and Y. M. de Haan, *Z. Krist* **177**, 235 (1962).
76. Structures fines d'absorption des rayons X en chimie: des données expérimentales à leur analyse. Ecole du CNRS, GARCHY, 19–24 septembre (1988).
77. J. Max, *Traitement du signal,* Dunod, Paris (1976).
78. M. E. Kordesh and R. W. Hoffmann, *Phys. Rev.* **B29**, 491 (1984).
79. F. W. H. Kampers, *EXAFS in Catalysis: Instrumentation and Applications,* Eindhoven (1988).
80. E. Lesage-Rosenberg, Ph. D. Paris (1984), Technip.
81. B. Moraweck and A. J. Renouprez, *Surf. Sci.* **106**, 35 (1981); **81**, L631 (1979).
82. T. Yokoyama, N. Kosum, K. Asakura, Y. Iwasawa, and H. Kurosa, *J. Phys. Fr. Colloque* C8 **47**, 273 (1986).
83. J. Berdala, E. Freund, and J. P. Lynch, *J. Phys. Fr. Coll.* C8 **47**, 269 (1986).
84. D. Bazin, Thesis, Universitè d'Orsay (1985).
85. J. Berdala, E. Freund, and J. P. Lynch, *J. Phys. Fr. Coll.* C8 **47**, 265 (1986).
86. P. Lagarde, F. Murata, G. Vlaic, E. Freund, H. Dexpert, and J. P. Bournonville, *J. Catal.* **84**, 333 (1983).
87. D. Bazin, H. Dexpert, P. Lagarde, and J. P. Bournonville, *J. Catal.* **110**, 209 (1988).
88. G. Meitzner, G. H. Via, F. W. Lytle, S. C. Fung, and J. H. Sinfelt, *J. Phys. Chem.* **92**, 2925 (1988).

DETERMINATION OF THE ATOMIC STRUCTURE OF SOLID CATALYSTS BY X-RAY DIFFRACTION

G. Bergeret and P. Gallezot

15.1. INTRODUCTION

Solid materials used in heterogeneous catalysis are generally composed of a mixture of several phases which cannot be separated without losing the catalytic properties of the material. These phases are often in a state of fine division, i.e., a large fraction of the atoms are on the surface and therefore they do not have the same coordination as those in the deep layers of a bulk material. Moreover, the structure of these phases as often as not exhibits numerous crystal defects of substitution or displacement with respect to a perfect lattice because the active phase often contains several transition elements without well-defined stoichiometry (e.g., mixed oxide and bimetallic catalyst). It is therefore generally not possible to determine the complete atomic structure of the active phases of a catalyst. However, information on the short-range order of the atoms in catalysts can be obtained from the radial electron distribution function, which gives the distribution of all the interatomic distances present in the solids whatever their degree of order.

Nevertheless, there is an important class of solids used as catalysts or supports of catalysts in which the atoms are ordered enough to allow the determination of their atomic structure by the well-known methods of crystal structure analysis. These solids are the zeolites, which are natural or synthetic aluminosilicates with a lattice built on a three-dimensional arrangement of AlO_4

G. Bergeret and P. Gallezot • Institut de Recherches sur la Catalyse, CNRS, Villeurbanne, France.

Catalyst Characterization: *Physical Techniques for Solid Materials*, edited by Boris Imelik and Jacques C. Vedrine, Plenum Press, New York (1994).

and SiO_4 tetrahedra that encloses a porous network. These open spaces can fill up to 50% of the volume of the zeolite.[1]

We shall first examine how the methods of crystal structure determination can be adapted to the structural analysis of zeolites and some examples will be given. The second part will be devoted to the study of the short-range order in amorphous or poorly crystalline catalysts and in very small metallic particles. We shall then develop briefly the theory of the radial electron distribution and we give some examples of applications.

15.2. CRYSTAL STRUCTURE OF ZEOLITES

Several types of structural studies can be considered:

1. When a new synthetic zeolite is prepared, the structure of the alumino-silicate framework needs to be determined. All the zeolites are formed by chains of tetrahedra linked by corners. A silicon or aluminum atom is sited at the center T of these tetrahedral units and four oxygen atoms form the corners. These chains develop in the three dimensions of the space and form a framework the topology of which is specific to each type of zeolite. The determination of the crystal structure enables us to describe the topology of the zeolite framework and of the microporous network inscribed in this framework. The latter governs the molecular sieve properties and the catalytic properties due to the shape selectivity.

2. Often the structure modifications of a known zeolite are studied. These modifications can result from special synthesis conditions capable of modifying the Si/Al ratio or even of substituting other atoms in the tetrahedral sites (e.g., B, P, Ga, Ge). However, changes are more frequently caused by the treatment that the zeolites undergo during catalytic reactions and subsequent regeneration. These conditions create defects in the framework (e.g., substitution and displacement of atoms, glide planes) and can even lead to a framework reconstruction with a new Si/Al ratio.

3. When the topology and the structure modifications are known, the positions of the exchangeable cations and possibly of the protons, which neutralize the negative charges in excess of the zeolite framework due to the AlO_4 tetrahedra, have to be determined.

4. Finally there is the question of the location of the various other species that may be part of the zeolite composition; reduced metal atoms and molecules coordinated to cations, which occupy zeolite pores and cavities, are especially important.

These questions can be answered by the methods of determination of crystal structure. However, some adaptations are needed and the uncertainty inherent in these methods must always be taken into account.

15.2.1. Methods of Structure Determination

Only a brief survey of the general determination of a crystal structure will be given here, as the methods are treated in great detail in many specialist books.[2-4]

The electron density $\rho(xyz)$ at a point with coordinates x, y, z of a cell of a crystal lattice is given by the triple Fourier series:

$$\rho(xyz) = \frac{1}{V} \sum_h \sum_k \sum_l F_{hkl} \exp - 2\pi i(hx + ky + lz) \qquad (15.1)$$

where V is the unit cell volume and F_{hkl} the structure factor of the hkl plane. The amplitude of the structure factor equals the square root of the intensity I_{hkl} of the X-ray beam scattered by the hkl plane. These intensities are usually obtained from measurements in the whole space and correction of the intensities scattered by a rotating single crystal. When a large enough single crystal is not available or when one wants to study the structure of catalysts under conditions similar to those used in physicochemical characterization studies and in catalysis, the structure can be determined in favorable cases from the powder diffraction pattern. Thus the zeolites used in heterogeneous catalysis have very large unit cells, and their powder pattern shows up to about a hundred reflections, the intensity of which can be measured.

When these lines overlap at large Bragg angles, the measured intensity must be deconvoluted in order to obtain the intensity of each involved line, which is relatively easy because the positions and the profile of these lines are known. A splitting must be performed when a line corresponds to several structure factors (hkl planes with different indices but with the same interplanar spacing d_{hkl}). The intensity is therefore split proportionally to the square of the calculated structure factors.

A newer method of structure refinement from powder diffraction patterns consists of a point-by-point comparison of the profiles of the experimental and calculated patterns. Known as the Rietveld method,[5] it was originally developed for structure determination by neutron diffraction, where, owing to the fact that the angular resolution is much weaker than that obtained in X-ray diffraction, peak overlap is very common. Therefore the Rietveld method does not need deconvolution because the refinement is performed at each point i of the pattern by minimizing the factor $R = \Sigma_i [I_i(\text{obs}) - I_i(\text{cal})]^2$, where $I_i(\text{obs})$ and $I_i(\text{cal})$ are the observed and calculated intensities at the point i, whereas in the other method the refinement is done on the deconvoluted integral intensity of each line hkl. The values of $I_i(\text{cal})$ are calculated by taking the line profile and the background intensity into account, for which some analytical functions are needed. The critical point of the method lies in the choice of these functions.[6] Nevertheless, the Rietveld method is marked out for increasing utilization in X-ray diffraction because of the ease of its use[7] (no deconvolution, no integration) and because automation is easy (development of the mini- and micro-

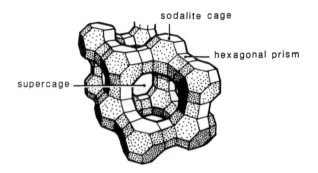

Figure 15.1. Aluminosilicate framework of faujasite-type zeolite (silicon and aluminum atoms of SiO_4 and AlO_4 tetrahedra occupy the corners; the oxygen atoms are at the middle of the edges). The structure is based on an assembly of truncated octahedra (constituted by 24 tetrahedra). The center of the octahedron is occupied by a 6.6.-Å-diameter cage, the sodalite cage, which is accessible by 2.2-Å windows in the hexagonal faces. The truncated octahedra are bound by the hexagonal faces, the small cage between them is the hexagonal prism. The tetrahedral building of the truncated octahedra leaves 12.5-Å-diameter supercages. The crystallographic unit cell of cubic symmetry ($a = 24.7$ Å) contains 192 tetrahedra which constitute 16 hexagonal prisms, 8 sodalite cages, and 8 supercages. The composition of the synthetic zeolite NaY is $Na_{56}Al_{56}Si_{136}O_{384}$, 250 H_2O.

computers). However, the integral method still gives better results in cases where the diffraction pattern shows well-separated lines.

The structure determination with about a hundred independent data is accurate only in the case of high-symmetry cubic zeolites, where only a few independent parameters have to be determined. This is the case with faujasite-type zeolite (synthetic zeolite X and Y), where a single TO_4 tetrahedron (T = Si or Al) is independent in the unit cell, the other 191 being deduced from the first by the symmetry operations of the space group $Fd\bar{3}m$. As the oxygen atoms are sited on symmetry elements, it is enough to determine ten independent atomic coordinates to describe the entire framework of these zeolites. The aluminosilicate framework of faujasite-type zeolite is given in Figure 15.1.

The structure modifications in the course of activation or regeneration treatment and during the catalytic reactions can be studied *in situ* in the case of powder. A sample holder with temperature and atmosphere control fitted on a diffractometer is needed. A Guinier–Lenné focusing camera with film displacement is convenient for structure studies during different treatments. This type of study is easier when the period of time needed for recording the diffraction pattern is not too long compared to the period of time used for the catalyst treatment and for the catalytic activity measure. Therefore it is valuable to employ powerful sources of X-ray (rotating anode, synchrotron radiation) and fast detecting systems, such as the position-sensitive detectors giving the possibility of measuring the diffracted intensities in a large angular 2θ range simultaneously.

15.2.2. Structure of the Zeolite Framework

The structures of a zeolite framework cannot always be determined without ambiguity, because either the zeolite crystals are impure, twinned, and show defects, or an X-ray pattern may correspond to several structures which do not have the same symmetry. Thus, Sherman and Benett[8] have shown that several natural and synthetic zeolites have X-ray patterns similar to that of mordenite although their framework topologies are different. This uncertainty concerning the structure of zeolites similar to mordenite is further increased by the possible presence of stacking faults in the planes perpendicular to the main channels, which can lead to an apparent decrease in the pore diameters.

The zeolite structures are often described with an idealized symmetry. Thus ZSM-5 zeolites are considered orthorhombic although they often show a monoclinic distortion. In addition, structures are more often than not refined assuming that the aluminum and silicon atoms are distributed randomly in the tetrahedral sites; consequently $T-O$ distances are obtained equal to the weighted average of the $Si-O$ (1.61 Å) and $Al-O$ (1.77 Å) distances corresponding to the zeolite composition. Thus the NaA zeolite is often described with a cubic unit cell with a symmetry $Pm3m$ ($a = 12.3$ Å) although Gramlich and Meier[9] have shown, taking into account the superstructure reflections, that the symmetry is $Fm3c$ ($a = 24.6$ Å). This symmetry yields the unambiguous location of silicon and aluminum atoms and the $Si-O$ and $Al-O$ distances. In the same way, Olson[10] has shown that silicon and aluminum atoms are partially ordered in X zeolites. Even in cases where there is no $Si-Al$ long-range order, there is always a short-range order because the $Si-O-Al$ bondings are thermodynamically more likely than the $Al-O-Al$ bondings.

15.2.3. Lattice Defects and Restructuring of the Zeolite Framework

The zeolite lattice shows defects of various types. The lack of long-range order for the silicon and aluminum atoms, the random occupancy of the cationic sites equivalent by symmetry, and the lengthening of the $T-O$ distances when a framework oxygen atom is bound to a cation or to adsorbed molecules are topochemical defects that increase the structure-independent scattering, i.e., the background of the diffraction pattern. The heating of HY zeolites at temperatures higher than 500 °C induces a dehydroxylation, i.e., the removal of a framework oxygen atom in the form of a water molecule, which produces important defects from the displacement of other atoms. Figure 15.2 shows modifications of the diffraction pattern after dehydroxylation of an HY zeolite[3]; it is no longer possible to determine the structure by the methods described in Section 15.2.1. However, the modifications of interatomic distances and of the coordination of atoms can be studied by the radial electron distribution method. This application is described in Section 15.3.2.1a.

The intensive crushing of Y zeolites[11] leads to complete destruction of the lattice, which is shown by the gradual disappearance of the diffraction lines. Defects appear in the course of the ultrastabilization treatments of the HY

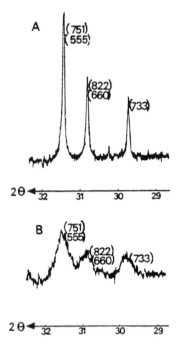

Figure 15.2. Effect of dehydroxylation on the structure of an HY zeolite[3]: (A) HY zeolite heated 15 h at 400 °C; (B) HY zeolite heated 15 h at 550 °C. The diffraction lines broaden and the background increases because of the defects created (microphotometric record of a pattern obtained with a focusing camera, Cu $K\alpha$).

zeolites. For example, under the effect of the water vapor at high temperature, material transport occurs inside the microporous network and this leads to an extraction of the aluminum atoms from the tetrahedral sites.[11] This dealumination leads to temporary vacancies that are filled by silicon atoms migrating from a destroyed part of the solid. The zeolite framework is rebuilt in a more symmetrical way because the aluminum atoms are replaced by silicon atoms, which also gives a greater stability to the lattice. The replacement of the aluminum by silicon was proved directly by Maher *et al.*[12] in the case of a zeolite stabilized by heating with steam and by Gallezot *et al.*[13] in the case of an HY zeolite dealuminated by ethylenediaminetetraacetate (EDTA) complexation. These authors determined the crysal structure of the dealuminated zeolite. It appears that the occupancy factor of the tetrahedral sites remains equal to the unit which involves the filling of the vacancies by silicon atoms. This recrystallization leading to a zeolite richer in silicon is accompanied by a decreasing of the mean distances T—O and, concomitantly, of the parameter a of the unit cell.

15.2.4. Localization of Cations in Zeolites

The exchangeable cations which neutralize the negative charges of the zeolite framework cannot reach a coordination as complete as they usually have in the oxides or in the salts by contact with the walls of the cages or the pores. In a hydrated state, they can be completely solvated by water molecules or bound to the oxygen anions of the cage walls, depending on their nature. During the progressive elimination of the water molecules, cations migrate toward the sites where their coordination is the least unfavorable. The most favorable cation sites are found on the symmetry elements and are generally more numerous than the cations. Therefore both the position of the sites and their mean occupancy factor have to be determined.

When the zeolite contains only cations of a single element, the occupancy factor (averaged in space and time) of a given site is easily obtained from the electron density observed on this type of site and from the atomic number of this element; the number of equivalent sites is known by symmetry, so the complete population occupying these sites can also be deduced. When the zeolite contains cations of two different elements, the attribution of the electron density cannot be done without ambiguity because this is at one and the same time a function of the atomic numbers and of the occupancy factors of the two cations.

The problem becomes still more complicated if there are more than two cations in the zeolite. There is no exact solution for attribution of the electron density observed on the sites to the different cations. Nevertheless, an approximate distribution can be based on the crystallochemical data (bond lengths between the cation and framework oxygen) and on the comparison between the chemical composition (atomic number and loading of the cation species) and the crystal composition. The concentration of localized cations in the unit cell is often smaller than that given by the chemical composition. Indeed, in addition to the uncertainty of the site attribution, it is difficult to localize the cations that occupy sites outside symmetry elements. The occupancy factor of these sites is very low because the number of sites equivalent by symmetry is large. Thus in faujasite-type structures, the occupancy factor of a site on the threefold axis occupied by a single atom is $1/32$ when it is $1/192$ for a site in a general position. At least the accuracy that can be expected in the localization of cations depends heavily on their atomic scattering factor, i.e., the higher their atomic number the easier they are to locate.

It must be emphasized that all these sources of error affect structures determined from single crystals as well as from powder; the ultimate accuracy of the cation distribution depends more on these errors than those introduced by the method of solution. Some examples applied to a Y zeolite exchanged by Cu^{2+} and Ni^{2+} ions are described below. A diagram of the positions of the cations sites typically observed in faujasite-type zeolites is given in Figure 15.3. The structure of two dehydrated zeolites $Cu_{16}Na_{24}Y$ and $Cu_{12}Na_5H_{27}Y$ were determined from their powder diffraction pattern.[14] In the two samples, Cu^{2+} ions were located mainly in SI' sites (11.1 Cu^{2+}/unit cell), where they are coordinated to three framework oxygen atoms. These results were corroborated by Maxwell and de Boer,[15] who, while studying the structure of a single crystal of faujasite exchanged by Cu^{2+} ions, found a cation population of 11.4 Cu^{2+} per unit cell in the SI' sites; therefore there is excellent agreement between the results obtained on single crystals and on powder.

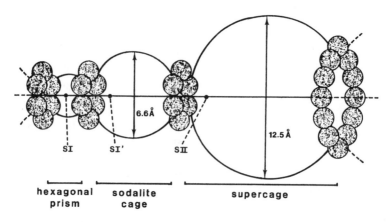

Figure 15.3. Schematic section through the center of the different cages in a faujasite-type zeolite. After dehydration cations generally occupy the *SI*, *SI'*, and *SII* sites close to the hexagonal apertures formed by six-membered oxygen rings. In these positions cations are interacting with the highest number of oxygen anions. The occupancy factors of these sites depend on the nature of the cation and on the presence of residual water molecules.

Study of the structure of dehydrated zeolites $Ni_{19}Na_{15}Y$ and $Ni_{14}Na_{23}Y$ from powder [16,17] has shown that the Ni^{2+} ions preferentially occupied *SI* sites in the hexagonal prism: 11.3 and 11.7 Ni^{2+}/unit cell, respectively. There is also excellent agreement with the results of Olson,[18] who, with a single crystal of nickel faujasite, found a cation population of 10.6 Ni^{2+}/unit cell in the *SI* sites. These results show that when structures are determined from powder diffraction patterns cations can be located with good accuracy. As far as crystallochemistry is concerned, it follows that transition ions occupy positions where the crystal field of the zeolite framework gives them the least imperfect possible coordination. Thus the Ni^{2+} ions reach an octahedral coordination at the center of the hexagonal prisms (*SI* site). The framework then shows some flexibility as the six oxygen anions coordinated to Ni^{2+} ions are brought closer, up to 2.3 Å from the center of the prism, while the distances are from 2.7 to 2.8 Å when these prisms are unoccupied or occupied by Na^+ ions. Nevertheless, the coordination of Ni^{2+} ions remains irregular in that the Ni—O distances are usually from 2.05 to 2.10 Å.

The *SI* and *SI'* sites occupied by Ni^{2+} and Cu^{2+} ions are inaccessible to the reagents adsorbed in the supercages. Therefore the interaction between reagents and cations is possible only if the ions can migrate toward the supercages. This was first shown in the case of the adsorption of pyridine, naphthalene, and butene on CuY zeolites, where the structure analysis showed that some of the Cu^{2+} ions actually migrate toward the supercages to coordinate these molecules.[14] Conversely, the Ni^{2+} ions remain in the *SI* sites whatever the nature of the adsorbed reagent unless the zeolite contains residual water molecules.[17] Then these molecules can play the rôle of carrier: Ni^{2+} ions are extracted from the *SI* sites and transferred toward the supercages by the water molecules. The mechanism can continue only if the reagent molecules displace the water molecules coordinated to cations, the water molecules then being available to extract other cations. This type of migration has been observed with pyridine and nitric oxide, but it is not effective with weaker ligands such as olefins.

As there is no transport mechanism by water molecules in the NiY zeolites, their catalytic activity can appear only for a loading higher than 12 Ni^{2+}/unit cell; i.e., when all the hexagonal prisms are filled the Ni^{2+} ions exchanged in excess become accessible in the supercages. These predictions were tested in the reaction of cyclotrimerization of acetylene into benzene.[19] In dehydrated NiY zeolites exchanged with 19, 14, and 10 Ni^{2+}/unit cell, the reaction rate really follows the sequence Ni 19 > Ni 14 > Ni 10 $\approx$ 0 and the crystal structure analysis corroborates that there are, respectively, 5.8, 2.3, and 0 Ni^{2+} ions in the supercages of these samples.

With CuY zeolites, the catalytic activity is expected to be proportional to the number of Cu^{2+} ions exchanged as these cations migrate easily. Bandiera *et al.*[20] found that the activity of a CuY zeolite in the isomerization of 1-butene and in the transalkylation of toluene increases as a function of the amount of exchanged copper. However, the rate of these reactions does not increase when the loading reaches 8 Cu^{2+} ions/unit cell; this is in agreement with the crystallographic studies showing that only 8 Cu^{2+} ions/unit cell migrate from *SI'* sites toward the supercages.

These results show that cations sited in the cages inaccessible to reagents can still be involved in the catalytic process if they are able to migrate and to interact with the reagents. If we take into account both the unitial positions and the facility for migration of the cations Cu^{2+},[14] Ni^{2+},[16,17] Pd^{2+},[21] and Co^{2+},[22] the possibility of cation–reagent interaction follows the order Pd > Cu > Ni > Co. These results can be used to understand catalytic activities. In the oxidation of methane and of the para–ortho hydrogen conversion on zeolites exchanged with transition cations, Rudham *et al.* [23,24] found the same activity sequence Pd > Cu > Ni > Co. Lastly, many other studies on the cation positions in zeolites have been done by Mortier, Olson, Seff, Smith, and others and these results were reviewed by Smith[25] and by Mortier.[26]

15.2.5. Position of the Adsorbed Molecules

The molecules able to complete the coordination of the cations sited in the accessible cages generally occupy well-defined positions. These complexes, formed by a transition ion grafted onto the zeolite framework and bound to one or several ligands potentially constitute catalytic systems of the supported homogeneous type; the structure of these grafted complexes can be determined because they follow the periodicity of the zeolite lattice and therefore contribute to the intensity of the X-ray diffraction lines. Seff *et al.* have determined the structure of several organometallic complexes inside the cages of A-type zeolites. Thus the stereochemistry of the complexes between acetylene and Mn^{2+}

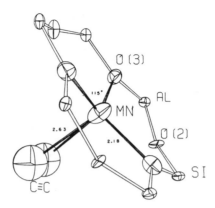

Figure 15.4. Coordination of Mn^{II} ions complexed by acetylene in an A-type zeolite[27] (reprinted with permission from *Inorganic Chemistry*, © 1975, American Chemical Society).

and Co^{2+} ions in the Mn$_{4.5}$Na$_3$A, 4.5 C$_2$H$_2$ and Co$_4$Na$_4$A, 4C$_2$H$_2$ was described precisely.[27] Figure 15.4 shows that the Mn^{2+} ions are coordinated to three $O(3)$ oxygen atoms at 2.18 Å and to a C$_2$H$_2$ molecule with the two carbon atoms equidistant from the Mn^{2+} cation at 2.63 Å. This structure is also found in Co$_4$Na$_4$A, 4 C$_2$H$_4$,[28] where the ethylene molecule is side-on to the cation. These unsaturated molecules interact with the cations in a fairly weak way and can occupy several positions by rotation.

The adsorbed molecules that are not bound to cations or acid sites of the framework, especially the neutral organometallic complexes (e.g., metal carbonyls, and metallocene) can move not only by translation, but also by rotary motion. Therefore it is not possible to determine the precise structure of these species by diffraction methods. However, they can still be located because they are not completely disordered. In fact, the adsorbed molecules occupy certain types of cages only; therefore they still obey the periodicity of the zeolite lattice. In practice, it is assumed that they occupy a sphere of a given radius and the structure is refined using liquid scattering functions which correspond to a model where the scattering matter (the atoms of these molecules) is uniformly distributed inside this sphere.

This method was used for the first time by Simpson and Steinfink[29] to locate nickel complexes in faujasite. With the same method, Gallezot *et al.*[30] showed that the complexes Mo(CO)$_6$ and Ru$_3$(CO)$_{12}$ adsorbed inside an HY zeolite occupy the centers of the apertures of the supercages, while the Re$_2$(CO)$_{10}$ complex is adsorbed in the supercage, with the two rhenium atoms on each side of the center. When Re$_2$(CO)$_{12}$ decomposes partially at 300 °C, rhenium atoms migrate to be bound to the zeolite framework near the apertures of the supercages. Thus the introduction, synthesis, and displacements of organometallic complexes in the cages can be followed by this method. The structure of the molecular clusters synthesized inside the cages can also be determined by radial electron distribution (see Section 15.3.2.2b).

15.2.6. Position of Metallic Atoms

Cations that occupy well-defined crystallographic sites can behave in four different ways upon reduction. They can remain at the same site but at a longer distance from the oxygen atoms (case 1), remain in the same cage but at random sites because their interaction with the zeolite framework is very weak (case 2), migrate out of the cage and agglomerate to form particles encaged in the zeolite (case 3), or form larger particles occluded in the bulk or supported on the external surface (case 4). In the last case, the particles behave as an extraneous phase whose scattering is independent of that of the zeolite lattice; i.e., particles give diffraction lines characteristic of the metallic phase. Case 3 is more complicated to study, but the particles sited in the cages or in the pores have their own scattering, which can be analyzed by the radial electron distribution method (see Section 15.3), which gives the atomic arrangement and the interatomic distances in the particle.

In cases 1 and 2, where the reduction is performed without the migration of atoms, the dispersion states obtained are very unusual because the reduced atoms can be isolated or gathered with at the most two or three in a cage. These atoms are more often partially reduced or electropositive because of electron transfer toward the support.

The AgA zeolites, whose structure was studied by Kim and Seff[31] and the AgF zeolites (F is faujasite with different Si/Al ratios) studied by Gellens et al.[32] are typical examples of case 1. The Ag^+ species still occupy the usual cationic sites, for example, next to the center of the apertures of the sodalite cages, but occupancy of adjacent sites is possible (e.g., occupancy of one SI site and of two SI' sites which are adjacent in a faujasite-type structure). Then the arrangement of the atoms can be described in terms of positively charged clusters. According to Kim and Seff,[31] in the $Ag_{12}A$ zeolite, two-thirds of the sodalite cages are occupied by an octahedral cluster of six Ag^0 atoms in such a way that each Ag^0 atom is found at the center of the face of a cube whose corners are occupied by Ag^+ ions. Conversely Gellens et al.[32] consider that in an $Ag_{6.5}Na_{5.5}A$ zeolite the silver atoms form a linear Ag_3^+ cluster. In the same way in a AgX zeolite, three silver atoms occupying the adjacent sites I', I, I' would form a linear cluster.

The structures of PdY and PtY zeolites activated and reduced at different temperatures were studied by Gallezot et al.[21,33–35] from powder diffraction patterns. Figures 15.5 and 15.6 summarize the different states of dispersion as demonstrated by a combination of methods (determination of crystal structure, small-angle X-ray scattering, electron microscopy). Only dispersion as isolated atoms in the sodalite cages will be consider here. Upon heating at 500–600 °C under oxygen, Pd^{2+} and Pt^{2+} ions occupy the SI' sites in the sodalite cages. After reduction by hydrogen at 25–230 °C in the case of PdY zeolites and at 100–300 °C in the case of PtY zeolites, the initial SI' sites are no longer occupied. Some of the atoms can be found in some new SI' sites, nearer the center of the sodalite cage at longer distances from framework oxygen atoms (2.72 Å instead of 2.01 Å before reduction in the case of PdY zeolites). The

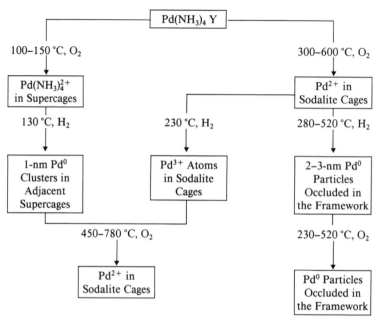

Figure 15.5. States of palladium dispersion in a PdY zeolite[33] (reprinted with permission from the Journal of Physical Chemistry, © 1981, American Chemical Society).

other atoms are no longer located in well-defined sites but remain in the sodalite cages. The structures, refined with liquid scattering functions in order to take into account the X-ray intensity scattered by the reduced palladium atoms, show that these atoms are statistically distributed within a 2.5-Å-radius sphere at the center of the sodalite cages. Indeed there is only one or at most two atoms per sodalite cage because in the PdY and PtY samples there is statistically between one and one and a half cations per cage before reduction and the migration of reduced atoms from one sodalite cage to another is highly unlikely.

Upon heating at temperatures higher than 230–300 °C in the case of PdY zeolites or at 300 °C in the case of PtY zeolites, the atoms isolated in the sodalite cages obtain sufficient energy to migrate through the 2.2-Å apertures of the sodalite cages and agglomerate to form 1.5–2.5-nm particles inside the zeolite framework. This migration is proved both by the disappearance of isolated atoms from the sodalite cages and by the appearance of particles detected and measured by X-ray diffraction and electron microscopy. The atoms isolated in the sodalite cages do not chemisorb hydrogen, probably because of their electropositive character (the binding energies Pt $4f_{7/2}$ and Pd $3d_{5/2}$ are shifted by $+1.3$ and 1.4 eV, respectively, with respect to the metal); the formation of 1.5–2.5-nm particles from the isolated atoms leads to an increase in the hydrogen chemisorption.

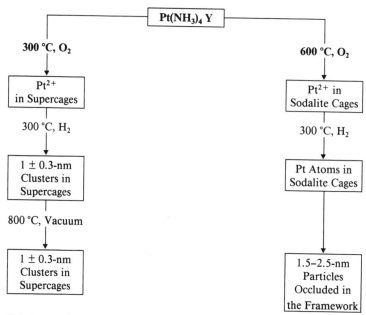

Figure 15.6. States of platinum dispersion in a PtY zeolite[34] (reprinted with permission from the Journal of Catalysis, © 1975, American Chemical Society).

The structures of zeolites can be determined under the conditions of catalytic reaction. Thus Bergeret and Gallezot[35] studied the structure of the PdHY zeolite during the hydrogenation of benzene and ethylene, after activation of the zeolite at 500 °C under oxygen and reduction at 150 °C. The reaction occurs only on the palladium atoms sited outside the sodalite cages; the reaction rate increases when the atoms migrate outside these cages at higher temperatures.

15.3. STRUCTURE OF AMORPHOUS CATALYSTS AND VERY SMALL PARTICLES

Solids that do not exhibit long-range order, i.e., where the atomic arrangement is not recurring in three dimensions, do not give Bragg reflections when they are investigated with X-rays. Such solids are called amorphous, but this term does not mean that their atomic structure is totally disordered because the disorder of the matter is perfect only in the case of diluted monoatomic or inert gases. In fact the so-called amorphous solids always exhibit a short-range order; for example, in vitreous silica and in silica gel, the SiO_4 tetrahedra form regular units and with respect to a reference tetrahedron, the neighboring tetrahedra do not occupy random places because there are characteristic Si—Si distances and Si—O—Si angles. Small particles with sizes below 1.0–1.5 nm constitute a particular category of solids which have no long-range order in that the interatomic distances are at most equal to the diameter of the particles.

X-ray interaction with solids without long-range order gives a weak X-ray scattering spread out on a wide angular range. It is not possible to determine the complete atomic structure of these solids, but a relative distribution of atoms can be obtained from measuring the intensities as a function of the scattering angle. This method, based on the Fourier transform of the intensities, is called radial atom distribution or radial electron distribution.

In Section 15.3.1, the theory of X-ray scattering by randomly oriented matter and the theoretical formula for the radial distribution function is briefly described, followed by the experimental techniques used in the radial distribution method. In Section 15.3.2, we examine some applications including catalyst supports, poorly-crystalline catalysts, small metal aggregates, and supported molecular clusters.

15.3.1. Radial Electron Distribution

15.3.1.1. X-Ray Scattering by Randomly Oriented Matter

Let us consider an object of N atoms. The intensity scattered per object by a collection of these identical objects is given by the Debye equation[36] so long as the orientations between them and with respect to the incident X-ray beam are equally probable:

$$I(s) = \sum_{m=1}^{N} \sum_{n=1}^{N} f_m(s) f_n(s) \sin(sr_{mn})/sr_{mn} \qquad (15.2)$$

where $f_m(s)$ and $f_n(s)$ are the respective atomic scattering factors of the mth and nth atoms; s is equal to $(4\pi/\lambda)\sin\theta$ (θ is the Bragg angle and λ the wavelength), and r_{mn} is the distance between atoms m and n. The double summation is taken over all pairs of atoms in the object.

It must be stressed that equation (15.2) is valid only in the case where the interactomic vectors have all possible orientations in space: rotating crystalline objects, disoriented grains of crystalline (or amorphous) powders, amorphous materials, liquids, and gases. For example, we have applied the Debye formula to calculate the intensity scattered by an increasing number of atoms belonging to the face-centered-cubic lattice.[37] We found that X-ray scattering by the smallest aggregates gives bands next to the theoretical positions corresponding to bulk palladium. Calculations performed on the aggregates containing more and more atoms show the progressive resolution of bands into Bragg lines. It is theoretically possible to determine the structure of small aggregates by computing the intensities scattered from different models and comparing these to the experimental intensities after correction (e.g., for polarization and Compton scattering). This trial-and-error procedure is rarely used in practice because there are too many parameters involved in the computation of the models (number of atoms, structure, shape); moreover information on the structure of small particles or amorphous solids can be obtained directly by the radial electron distribution method.

15.3.1.2. Radial Distribution Functions

The theoretical groundwork for the method is given in books by Guinier,[36] Warren,[38] and Klug and Alexander.[39] The method is based on the Fourier transform of the X-ray intensity scattered by the material. Introducing $\rho(r)$ in equation (15.2), the number of atoms per unit volume at a distance r from a reference atom (atom density) and ρ_0 the value of $\rho(r)$ averaged on the whole sample (average atom density in the sample) gives

$$I(s) = Nf^2(s)\left[1 + \int_0^\infty 4\pi r^2[\rho(r) - \rho_0][\sin(sr)/sr]dr\right] \qquad (15.3)$$

Let $I(s)/Nf^2(s)$ be the interference function and $S(s)$ the structure factor. This term contains all the structural information. Thus equation (15.3) can be written

$$I(s) = Nf^2(s)S(s) \qquad (15.4)$$

where

$$S(s) = 1 + \int_0^\infty 4\pi r^2[\rho(r) - \rho_0][\sin(sr)/sr]dr \qquad (15.5)$$

The Fourier transform of $S(s)$ gives

$$4\pi r^2\rho(r) = 4\pi r^2\rho_0 + (2r/\pi)\int_0^\infty si(s)\sin(rs)ds \qquad (15.6)$$

where

$$i(s) = [I(s)/Nf^2(s)] - 1 = S(s) - 1 \qquad (15.7)$$

The measure of the intensity $I(s)$ scattered by the sample between $s = 0$ and $s = \infty$ gives access to the radial atom distribution function $4\pi r^2\rho(r)$ of the sample, i.e., to the number of atomic centers lying in the volume defined by two spheres of radius r and $r + dr$ around an atom. The radial distribution is often described by the difference function $4\pi r^2[\rho(r) - \rho_0]$.

The radial distribution function given by equation (15.6) applies only to monoatomic solids. However, it can be used for solids consisting of more than one kind of element if the different atomic scattering factors $f(s)$ are assumed to be similarly decreasing as a function of s, which is generally true.

Let a polyatomic material be constituted by the assemblage of N groups of atoms. These groups form the unit of composition (uc) and often correspond to the stoichiometric formula. An average scattering factor of the sample can be expressed (per electron) as

$$f_e(s) = \sum_{uc} x_j f_j(s) \bigg/ \sum_{uc} x_j Z_j \qquad (15.8)$$

with x_j the atomic percentage (or stoichiometric coefficient), $f_j(s)$ the atomic scattering factor, and Z_j the atomic number of the species of type j. The approximation consists of considering each atomic species as an ensemble of K effective electrons (K is often equal to or near to the atomic number). Therefore, the atomic scattering factor of each species is written $f_j(s) = K_j f_e(s)$. Defining $\rho_j^e(r)$ as the number of electrons belonging to the atom j per unit volume at a distance r from a reference atom (partial electron density) and ρ_0^e as the average electron density in the sample, we can now express the radial function as

$$4\pi r^2 \sum_{uc} x_j K_j \rho_j^e(r) = 4\pi r^2 \rho_0^e \left(\sum_{\mu c} x_j K_j \right) + (2r/\pi) \int_0^\infty si(s)\sin(rs)ds \qquad (15.9)$$

where

$$i(s) = \left\{ [I(s)/N] - \sum x_j f_j^2(s) \right\} \bigg/ f_e^2(s) \qquad (15.10)$$

A radial distribution function of electrons in the sample is obtained which represents the superposition (weighted by the atomic percentage x_j and the atomic number Z_j of each species) of a partial radial distribution of electrons $4\pi r^2 \rho_j^e(r)$ belonging to each atomic species j around any atom of the material. Thus, the height of a peak of the electron radial distribution for a given value r is proportional to the number of atom pairs of different atomic species separated by the distance r and to the product of the atomic numbers of these species.

The attribution of a peak to one or several determined atom pairs can be difficult especially in the case of materials with atoms having similar atomic numbers. Thus it is necessary to compare the experimental distribution with the distribution computed from structure models[40,41] or to use the anomalous scattering of X-rays (see Section 15.3.1.3).[42] In addition, significant improvements in the calculation of radial distribution functions have been described by different authors[38,43–45] especially in order to cancel the oscillations which affect the radial distribution function at small r values. These oscillations are due to the errors made in the measurement of weak intensities at wide angles and to the truncation effects in the calculation of Fourier series. This is because the integrals of equation (15.6) and (15.9) are in fact calculated up to a restricted value of s, depending on the experimental conditions. Whatever the improvements in the mathematical treatment of the radial distribution function, the quality of the structure information depends mainly upon the experimental conditions used to measure the scattered intensities.

15.3.1.3. The Anomalous Scattering of X-Ray and Radial Distribution Functions

It is convenient to generalize equation (15.4) for a polyatomic material[42] as

$$I(s) = N \sum_\alpha \sum_\beta x_\alpha f_\alpha(s) \bar{f}_\beta(s) S_{\alpha\beta}(s) \qquad (15.11)$$

where α and β represent the different atomic species present in the material, x the proportion and $f(s)$ the complex atomic scattering factor of each species, $S_{\alpha\beta}(s)$ the partial structure factor $\alpha\beta$. Equation (15.11) contains the structure information concerning the atomic pairs α–β and can be rewritten as

$$S_{\alpha\beta}(s) = 1 + \int_0^\infty 4\pi r^2 [\rho_{\alpha\beta}(r) - \rho_{\beta_0}][\sin(sr)/sr] dr \qquad (15.12)$$

where $\rho_{\alpha\beta}(r)$ is the number of atoms of β type per unit volume at a distance r from an origin atom of α type (partial atomic density) and ρ_{β_0} the value of $\rho_{\alpha\beta}(r)$ averaged on the whole sample (partial average atomic density of β atoms). The Fourier transform of $S_{\alpha\beta}(r)$ gives

$$4\pi r^2 \rho_{\alpha\beta}(r) = 4\pi r^2 \rho_{\beta_0} + (2r/\pi) \int_0^\infty [S_{\alpha\beta}(s) - 1]\sin(rs) ds \qquad (15.13)$$

The function $4\pi r^2 \rho_{\alpha\beta}(r)$ represents the radial distribution function of β atoms around an α atom. Obtaining each partial structure factor $S_{\alpha\beta}(s)$ leads to each partial radial atom distribution after a Fourier transform and consequently to a complete description of the material. These partial atom distributions give a more detailed picture of the material than the single total electron distribution. However, several different experiments are needed to obtain simultaneous equations like equation (15.11) with different $x_\alpha f_\alpha$ coefficients. It is possible to vary the respective proportions x_α of each component of the material and/or the scattering factor f_α either by atomic substitution or by combining neutron scattering (isotopic substitution) with X-ray scattering. However, several samples are needed and there is a risk of structure modification.

With the development of synchrotron radiation, a new way to solve this problem has appeared: the use of anomalous X-ray scattering.[42,46] The atomic scattering factor generally varies smoothly and slowly with the energy of the incident X-ray beam. However, close to the energy of the absorption threshold, the atomic scattering factor varies sharply because of a resonance effect with the incident beam. These variations can be important — as high as 20% at about 100 eV below the absorption threshold of the relevant atom while the scattering factors of the other atoms in the sample do not vary in practice in this energy range because their threshold is found at a very different energy. This is thus a method of contrast variation.

Near the absorption edge the corrections terms f' and f'' must be introduced:

$$f(s) = f_0(s) + f'(E) + if''(E) \qquad (15.14)$$

It is assumed that f' and f'' vary only with the energy of the incident beam and are independent of the scattering angle 2θ. The complex term $f''(E)$ is proportional to the absorption cross section (mass attenuation coefficient) and is directly determined from the studied material by absorption experiments very similar to those performed in extended X-ray absorption fine-structure spectroscopy (EXAFS). Then the real term $f'(E)$ is obtained by numerical integration of a function of $f''(E)$ (dispersion relation). Considering a material with the formula $A_a B_b$ equation (15.11) can be developed:

$$I(s) = a f_A(s) \bar{f}_A(s) S_{AA}(s) + b f_B(s) \bar{f}_B(s) S_{BB}(s)$$

$$+ [a f_A(s) \bar{f}_B(s) + b f_B(s) \bar{f}_A(s)] S_{AB}(s) \qquad (15.15)$$

The difference in scattered intensities $I(s)$ collected at two energies near the absorption threshold of atom A (typically at 100 and 10 eV below the energy of the threshold) enables us to cancel the second term of equation (15.15) because the absorption cross section of atom B does not vary in this region. The Fourier transform of this difference yields a weighted sum of the radial distributions of atomic pairs A–A and A–B, i.e., the local environment of an atom A. Obtaining three separate partial radial atom distributions requires a third experiment which could be carried out at an energy very far from the threshold of atom A (typically at 2–5 keV) or near the threshold of atom B.

Anomalous scattering is beginning to be used to obtain partial radial distributions. Thus, at the Laboratoire pour l'Utilisation du Rayonnement Electromagnétique (LURE), Orsay, France, the structures of amorphous materials like CuY were determined[47] and studies concerning catalysts are in progress.[48] We have studied Pt–Mo bimetallic clusters supported on zeolite[49] at the Stanford Synchrotron Radiation Laboratory (SSRL), Stanford University, USA. It was found that clusters have an atomic structure similar to that of bulk platinum (face centered cubic) with molybdenum atoms deposited epitaxially on the 1-nm platinum clusters. These structural results are in excellent agreement with the reaction mechanism proposed previously[50,51] in order to explain the catalytic properties observed in hydrogenolysis and in the $CO + H_2$ reaction.[51] Pt–Mo bimetallic clusters have shown higher catalytic activity than monometallic clusters of platinum molybdenum. The proposed mechanism needed two kinds of sites: molybdenum atoms were sites for dissociation of hydrocarbons or carbon monoxide, while platinum atoms were sites for dissociation of dihydrogen and hydrogenation of carbon or hydrocarbon fragments.

15.3.1.4. Experimental Requirements of the Radial Distribution Method

The integral in equations (15.6) and (15.9) is computed from the product $s i(s)$ which can have a sizable value at large s, even if the scattered intensity $i(s)$ is small. Therefore it is advisable to measure the intensities scattered at wide angles accurately. In addition, the measurements must be performed in a range of s values as large as possible, therefore with a short wavelength from the smallest to the largest scattering angle (2θ) consistent with the experimental setup. To take these requirements into account, the scattering pattern is usually recorded under the following conditions. The sample is examined by reflection (Bragg–Brentano geometry) between 1 and 140–150° (2θ). In the case where accurate measurements are needed at small angles, a transmission setup must also be used and the data obtained by reflection and transmission have to be scaled. A molybdenum anode is commonly used and the Mo K_α radiation (0.709 Å) is monochromatized as well as possible. The beam is filtered by absorption on a zirconium filter which eliminates the radiation K_β, and the intensities are measured with a scintillation detector connected with a pulse height selector which eliminates the short-wavelength radiations in the continuous spectrum. The solid state detector [Si(Li), Ge(Li) diode, pure Ge] shows much better energy resolution and allows the selection of a better monochromatic radiation without using a zirconium filter owing to the electronic filtering. The best monochromatization is obtained by reflection of the incident or scattered beam on an analyzing monochromator crystal but the intensity is then strongly reduced.

The measurement of the intensities is performed by step-scanning throughout the scattering pattern, recording a number of counts large enough to ensure good statistics. In practice, with a conventional diffractometer equipped with a sealed tube, a week might be required for a complete recording. Time can be saved using a diffractometer with a rotating anode tube. The simultaneous recording of a whole angular region of the pattern with a position-sensitive detector reduces the time needed for the data acquisition to a score of hours.[52] Less than an hour is sufficient if the synchrotron radiation from electron storage rings is used [wiggler lines[49]].

The intensities $I(s)$ of equation (15.7) arise from the experimental intensities after correction and scaling in absolute units (electrons squared).[39] After the polarization and absorption corrections, the experimental intensities are converted to electron units. It is assumed that, at large values of s, the assemblage of atoms no longer has influence on the scattered intensity. This structure-independent intensity is consequently equal to the sum of the coherent scattered intensity (unmodified scattering) and the incoherent intensity scattered with a longer wavelength (Compton modified scattering). Both of these can be computed from the values of the atomic scattering factors and the Compton intensities given in the *Tables of Crystallography*. Once the intensities are in absolute units, $I(s)$ in equation (15.7) can be obtained by subtracting the Compton intensity. The various factors playing rôles in the measure of intensities and in the calculation of the radial distribution functions are reviewed in detail by Ratnasamy and Leonard,[53] especially for catalysts.

15.3.2. Applications of the Radial Distribution

The radial distribution has been used to determine the structure of vitreous solids, polymers, carbons, and amorphous metallic alloys; few studies have dealt with the active phases of heterogeneous catalysts. In the 1960s, the method was developed and applied at Louvain University, Belgium, by Fripiat *et al.* to study the short-range order in solids with a large surface area used as catalysts or as catalyst supports: silica, alumina and silica–alumina.[54–56] Then in the 1970s, work was done on the structure of small metallic aggregates[57] especially at the Institute of Catalysis, Novosibirsk, USSR, by Levitskii, Moroz, and Richter *et al.* to determine the structure and the size of platinum particles or of bimetallic catalysts supported on various supports.[58,59] The method has been extensively applied at Institute de Recherches sur la Catalyse, Villeurbanne, France, to determine the structure of metallic aggregates supported on Y-type zeolite depending on the nature of the adsorbed species.[49,60–70] Other catalysts have also been studied. Some examples are reviewed here.

15.3.2.1. Catalysts and Poorly Crystalline Supports

15.3.2.1a. Structure of a Dehydroxylated Zeolite. A dehydroxylated zeolite has too many defects (see Section 15.2.3), for the structure to be determined by conventional methods. This problem was solved using the radial electron distribution.[71] Figure 15.7 shows the distribution corresponding to an $Na_4H_{52}Y$ zeolite desorbed at 600 K (curve a) and the distribution of this zeolite after dehydroxylation at 920 K under vacuum (curve b). The dehydroxylation produces such a broadening of peaks that the O—O distances at 2.69 Å and the T—O and O—O distances at 3.86 Å are no longer resolved. Curve (a) was subtracted from curve (b) to reveal modifications clearly; the obtained difference curve (c) exhibits positive peaks corresponding to the distances between atoms found in the dehydroxylated zeolite and not in the HY zeolite. Based on the results obtained by Léonard *et al.*,[56] the peaks at 2.88, 3.38, and 4.82 Å can be attributed to Al(VI)—Al(VI), Al(VI)—Al(IV), and Al-(VI)—Al(VI) distances, respectively; the peak at 4.11 Å corresponds to Al—O distances. These results prove that during the dehydroxylation process, the coordination of part of the zeolite aluminum atoms changes from 4 to 6. This conclusion is in agreement with the results obtained by X-ray fluorescence spectroscopy[72] and mass NMR[73] studies. On the other hand, it questions the common assumption of the formation of three-coordinated aluminum atoms in the dehydroxylated zeolites.

15.3.2.1b. Structure of an Oxidation Catalyst. The vandadium–phosphorus mixed oxides are catalysts for the selective oxidation of butane to maleic anhydride. In spite of many patented proposals, because of the simultaneous presence of well-crystallized and amorphous phases in the catalyst, the real active sites still remain ill defined. The catalytic activity has been tentatively correlated with the nature of various crystallized phases but the rôle of the amorphous phase was neglected, principally because of the lack of a physical technique to characterize the structure of this phase. Radial electron distribution, which gives the distribution of all the interatomic distances in the sample whatever its crystallinity, is a suitable tool for this kind of investigation.

Figure 15.8[74] shows the radial distribution of a VPO catalyst (after catalytic reaction) obtained from the Fourier transform of the X-ray scattering intensity (curve 1), and the model distributions of pure compounds $(VO)_2P_2O_7$ (curve 2), and β-$VOPO_4$ (curve 3) computed directly from the positions of atoms in the unit cell obtained from the literature. The radial distribution of the catalyst (curve 1) is very close to that of $(VO)_2P_2O_7$ (curve 2) for interatomic distances higher than 5 Å. Therefore a crystalline $(VO)_2P_2O_7$ phase is present in the catalyst. Below 5 Å, the two radial distributions disagree, which shows the presence of an amorphous phase entirely disordered above 5 Å, in addition to the crystalline $(VO)_2P_2O_7$ phase. From the peaks at 1.55 Å (P—O and V=O distances), 3.31 Å (mainly V—P distances), and 3.97 Å (distance in β-$VOPO_4$), this amorphous phase can be described as very small groups (possibly a two-dimensional array) of corner-sharing VO_6 octahedra surrounded by corner-linked PO_4 tetrahedra. These groups can be viewed as a precursor of a crystalline β-$VOPO_4$ phase. This structural information coming from a real catalyst is not unimportant for the mechanism of the reaction.[74] Recently, other authors have studied a similar catalyst.[75] Radial distribution was also applied to describe the precise structure of the vanadium pentoxide supported on alumina, which forms a so-called monolayer covering the support.[76]

15.3.2.2. Structure of Small Metallic Aggregates

Ratnasamy et al.[57] have shown that the atom packing in 1.5–2.0-nm platinum particles was similar to that of bulk platinum but that oxygen

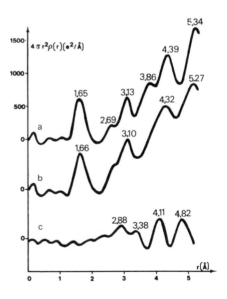

Figure 15.7. Radial electron distribution of HY zeolites[71]: (a) HY zeolite, (b) HY zeolite dehydroxylated at 920 K under vacuum, (c) difference b − a (the positive peaks correspond to distances present in the dehydroxylated zeolite and missing in the HY zeolite) (reprinted with permission from the Journal of Physical Chemistry, © 1985, American Chemical Society).

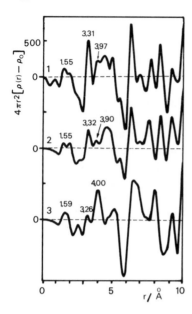

Figure 15.8. Radial electron distribution of a VPO catalyst obtained from X-ray scattering intensity (Curve 1) compared to those of pure model compounds computed from structural data (Curves 2 and 3).[74]

adsorption induced a complete disorder in the external layer of the particles. These results were obtained on catalysts with a very low loading (0.5–1.0 wt.%) in platinum. The resolution of peaks of electron density of platinum with respect to those of the support can be considerably improved by increasing the metal content. Moreover, if the particles are smaller than 1.5 nm, the fraction of atoms exposed at the surface becomes preponderant and the modifications of the particle structure produced by various adsorbates can be precisely probed. This type of study was extended by Gallezot et al.[60–70] to platinum, rhodium, and iridium particles supported in Y-type zeolites, which are suitable materials since 1-nm particles can still be prepared on zeolites with metal concentrations as high as 15 wt.%.

15.3.2.2a. Platinum Aggregates: Effect of Adsorbates. Some results obtained with a zeolite $Pt_{11}Ce_1Na_{19}H_{12}Y$ ($Y = Al_{56}Si_{136}O_{384}$) are summarized here. The adsorption treatments are performed *in situ* in a cell that can be mounted on the diffractometer. Figure 15.9 shows the radial distribution corresponding to the platinum-free NaY support (curve a) and that corresponding to the PtY zeolite (curve b).[63] The latter exhibits large peaks which can all be attributed to platinum. The subtraction of distributions b − a giving curve c, where only the Pt–Pt distances appear, is correct in that the interatomic vectors between the atoms of the particles and the framework atoms are quite random.

Figure 15.10 shows the radial distributions of a Y-type zeolite containing 1-nm platinum particles covered by various adsorbates. These distributions were obtained after subtracting the radial distribution of a platinum-free NaY zeolite. Curve 2 shows the distribution corresponding to the particles covered with hydrogen. It can be seen that the sequence of the coordination shells around an atom agrees with the fcc structure and that the distances are those of bulk platinum. Curve 1 represents the distribution for particles outgassed at 400 °C in vacuum, which have a naked surface. All the interatomic distances are less than those of bulk platinum or those observed on curve 2. The contraction is not uniform for all the distances: the fcc structure is distorted. The contraction of distances is in agreement with the theoretical predictions on the structure of small particles. Hydrogen dissociately adsorbed on these particles completes the coordination of the surface atoms and allows a complete relaxation toward the normal fcc packing. The adsorption of carbon monoxide (curve 3) on the outgassed zeolite also leads to a relaxation of distances, which become equal to those of bulk platinum but the height of the Pt–Pt peaks is reduced by more than 50% against those of hydrogen covered particles (curve 2). This has been explained by a displacement of the platinum atoms bonded to carbon monoxide

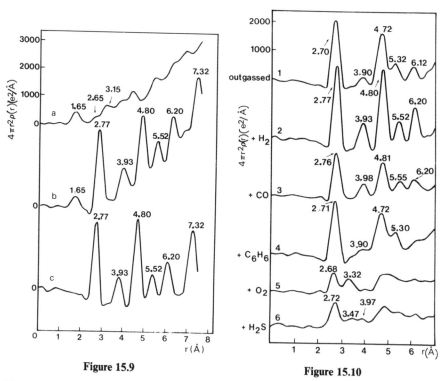

Figure 15.9

Figure 15.9. Radical electron distribution: (a) Pt-free NaY zeolite, (b) Pt_{11}NaY zeolite containing 10-Å platinum particles, (c) difference b − a (the zeolite peaks disappear).[63]

Figure 15.10. Radial electron distribution of a PtY zeolite containing 1-nm platinum particles (the distribution of the support has been subtracted): (1) naked particles, (2) adsorption of hydrogen, (3) adsorption of carbon monoxide, (4) adsorption of benzene, (5) adsorption of oxygen, (6) adsorption of hydrogen sulfide.

in such a way that some distances are elongated and others contracted; the mean distance is little changed. Such a disorder of displacement around a mean position must actually involve a decrease in the area of the Pt–Pt peaks without a change in the distances. The adsorption of oxygen (curve 5) or of hydrogen sulfide (curve 6) leads to similar distributions. In both cases, the adsorbed atoms of oxygen and sulfur strongly perturb the positions of the particle atoms and the structure is disordered. Nevertheless, new peaks appear at 3.32 and 3.47 Å respectively, which correspond to the distances between platinum atoms bridged by an adsorbed oxygen or sulfur atom. This is thus the start of structure reconstruction.

The effects of a catalytic reaction on the structure of platinum particles were also studied.[62,66] Figure 15.11[62] gives the radial distribution of particles covered with hydrogen (curve 1), with benzene (curve 3), and in the course of the catalytic reaction of benzene hydrogenation with the stoichiometric reagent mixture ($H_2/C_6H_6 = 3$) (curve 2). The benzene adsorption (curve 3) induces a weak relaxation of distances compared to the distances in the naked particles (Figure 15.9, curve 1); therefore the interaction of platinum atoms with the

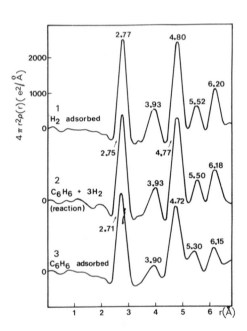

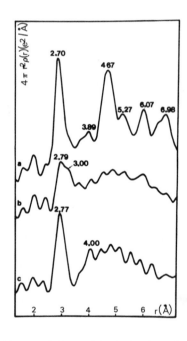

Figure 15.11. Radial electron distribution of a PtY zeolite containing 1-nm platinum particles: (1) particles covered with hydrogen, (2) radial distribution obtained in the course of the catalytic reaction $C_6H_6 + 3H_2$ C_6H_{12} with a stoichiometric mixture or an excess of benzene, (3) particles covered with benzene.[62]

Figure 15.12. Radial distribution of a RhY zeolite containing rhodium aggregates smaller than 1 nm: (a) aggregates covered with hydrogen, (b) adsorption of carbon monoxide, (c) adsorption of $CO + H_2O$.[68]

benzene molecules is weak. The radial distribution obtained when the reaction is carried out with an excess of benzene ($H_2/C_6H_6 = 1$) is similar to curve 2 (stoichiometric mixture). These distributions are intermediate between that of the particles covered with hydrogen (curve 1) and that of the particles covered with benzene (curve 3). It can therefore be concluded that during the catalytic reaction, the surfaces of the aggregates are covered both by benzene and by hydrogen, and that the fraction of the surface covered by each of these reagents does not change even in the presence of a large excess of benzene in the gas phase.

These different results show that the atomic arrangement in the small platinum particles should not be viewed as rigid, because atoms can be displaced from their equilibrium positions according to the nature of the adsorbate. The radial electron distribution method is well suited to following these structure modifications.

15.3.2.2b. Effect of Carbon Monoxide on Rhodium Aggregates. Rhodium aggregates supported in Y-type zeolite were studied by radial electron distribution.[67,68] The peak definition is less accurate than in the case of platinum aggregates because the peak area is proportional to the square of atomic number. Curve a in Figure 15.12 corresponds to aggregates smaller than 1 nm and covered by hydrogen; distances are close to those of bulk rhodium. After CO adsorption (curve b), the Rh–Rh peaks disappear and after coadsorption of CO and H_2O (curve c), a new structure appears with a well-defined peak only at 2.77 Å. Such a distance corresponds to Rh–Rh bonds characteristic of rhodium carbonyl species and especially to $Rh_6(CO)_{16}$. From these results combined with those of an IR spectroscopy study, it can be concluded that the adsorption of CO produces an oxidative disintegration of rhodium aggregates to give monomeric $Rh^I(CO)_2$ species (curve b), which, in presence of water, migrate and condense into polynuclear $Rh_6(CO)_{16}$ carbonyl. *In situ* synthesis of iridium carbonyl clusters encaged in zeolite was also studied by radial distribution.[69] These examples demonstrate that the radial electron distribution is a very suitable method for the study of nuclearity modifications of metallic aggregates and of supported molecular clusters.[70]

15.4. CONCLUSIONS

X-ray diffraction is a basic tool for the determination of the atomic structure of solid phases in heterogeneous catalysis. Determination of the structure of a catalyst does not mean only the identification of the bulk solid phases present in the catalyst, but also the determination of the short-range local order of the surface atoms which constitute the catalytic sites. For example, in the case of a zeolite catalyst, the position of all atoms constituting the internal surface can be known with an error smaller than 0.1 Å with the methods of determination of crystal structure.

The method of radial electron distribution is of obvious use for characterizing those heterogeneous catalysts which consist of a poorly crystallized oxide with a large specific surface area either used just as it is or as a support for small metallic particles. The nature and the structure of the oxide phase and the metal phase can be identified by this method, including such difficult cases as the precise identification of a transition alumina, the characterization of bonds between metal atoms in a bimetallic catalyst, or the characterization of supported polynuclear molecular clusters. In addition, when the fraction of surface atoms becomes preponderant, as in the case of supports with a very large specific surface area (500–1000 m^2/g) or in the case of very small metallic particles (< 2 nm), the radial distribution becomes a method of studying the surfaces. The most interesting case is that of small metal aggregates where modifications of structure and morphology dependent on thermal treatments and on the nature of the absorbate can be determined.[70] The sample can be in contact with reagents in a static or a dynamic mode while measurements are made. Thus the information on surface structures is obtained under real conditions of catalytic reactions.

REFERENCES

1. D. W. Breck, *Zeolite Molecular Sieves,* John Wiley and Sons, New York (1974).
2. G. H. Stout and L. H. Jensen, *X-ray Structure Determination,* MacMillan, New York (1968).
3. G. B. Carpenter, *Principles of Crystal Structure Determination,* W. A. Benjamin, New York (1969).
4. L. H. Schwartz and J. B. Cohen, *Diffraction from Materials,* Academic, New York (1977).
5. H. M. Rietveld, *Acta Cryst.* **22**, 151 (1967); *J. Appl. Cryst.* **2**, 65 (1969).
6. R. A. Young and D. B. Wiles, *J. Appl. Cryst.* **15**, 430 (1982).
7. R. J. Hill and I. C. Madsen, *J. Appl. Cryst.* **17**, 297 (1984).
8. J. D. Sherman and J. M. Benett. in: *Molecular Sieves* (W. M. Meier and J. B. Uytterhoeven eds.), Advances in Chemistry, Vol. 121, American Chemical Society, Washington (1973), p. 52.
9. V. Gramlich and W. M. Meier, *Z. Kristallogr.* **133**, 134 (1971).
10. D. H. Olson, *J. Phys. Chem.* **74**, 2758 (1970).
11. B. Moraweck, P. Gallezot, A. Renouprez, and B. Imelik, *J. Phys. Chem.* **78**, 1959 (1974).
12. P. K. Maher, F. D. Hunder, and J. Scherzer, *Advances in Chemistry, Vol. 101,* American Chemical Society, Washington (1971), p. 266.
13. P. Gallezot, R. Beaumont, and D. Barthomeuf, *J. Phys. Chem.* **78**, 1550 (1974).
14. P. Gallezot, Y. Ben Taarit, and B. Imelik, *J. Catal.* **26**, 295 (1972).
15. I. E. Maxwell and J. J. de Boer, *J. Phys. Chem.* **79**, 1874 (1975).
16. P. Gallezot and B. Imelik, *J. Phys. Chem.* **77**, 652 (1973).
17. P. Gallezot, Y. Ben Taarit, and B. Imelik, *J. Phys. Chem.* **77**, 2556 (1973).
18. D. H. Olson, *J. Phys. Chem.* **72**, 4366 (1968).
19. P. Pichat, J. C. Vedrine, P. Gallezot, and B. Imelik, *J. Catal.* **32**, 190 (1974).
20. J. Bandiera, C. Naccache, and B. Imelik, *C. R. Acad. Sci. Paris.* **282**, 81 (1976).
21. P. Gallezot and B. Imelik, in: *Molecular Sieves* (W. M. Meier and J. B. Uytterhoeven, eds.). Advances in Chemistry, Vol. 121, American Chemical Society, Washington (1973). p. 66.
22. P. Gallezot and B. Imelik, *J. Chim. Phys.* **71**, 155 (1974).
23. R. Rudham and M. R. Sanders, *J. Catal.* **27**, 287 (1972).
24. R. Rudham, A. D. Tullett, and K. D. Wagstaff, *J. Catal.* **38**, 488 (1975).

25. J. V. Smith, in: *Zeolite Chemistry and Catalysis* (J. A. Rabo, ed.), *ACS Monograph,* Vol. 171, American Chemical Society, Washington (1976), p. 1.
26. W. J. Mortier, *Compilation of Extra-Framework Sites in Zeolites,* Structure Commission of the International Zeolite Association, Butterworth, Guildford (1982).
27. D. E. Riley and K. Seff, *Inorg, Chem.* **14**, 714 (1975).
28. D. E. Riley, K. B. Kunz, and K. Seff, *J. Am. Chem. Soc.* **97**, 537 (1975).
29. H. D. Simpson and H. Steinfink, *J. Am. Chem. Soc.* **91**, 6225 (1969).
30. P. Gallezot, G. Coudurier, M. Primet, and B. Imelik, in: *Molecular Sieves II* (J. R. Katzer, ed.) ACS Symposium Series, Vol. 40, American Chemical Society, Washington (1977), p. 144.
31. Y. Kim and K. Seff, *J. Am. Chem. Soc.* **99**, 7055 (1977).
32. L. R. Gellens, W. J. Mortier, and J. B. Uytterhoeven, *Zeolites,* **1**, 11 (1981).
33. G. Bergeret, P. Gallezot, and B. Imelik, *J. Phys. Chem.* **85**, 44 (1981).
34. P. Gallezot, A. Alarcon-Diaz, J. A. Dalmon, A. J. Renouprez, and B. Imelik, *J. Catal.* **39**, 334 (1975).
35. G. Bergeret and P. Gallezot, *J. Phys. Chem.* **87**, 1160 (1983).
36. A. Guinier, *Théorie et Technique de la Radiocristallographie,* Dunod, Paris (1964).
37. P. Gallezot, M. Avalos-Borja, H. Poppa, and K. Heinemann, *Langmuir* **1**, 342 (1985).
38. B. E. Warren, *X-Ray Diffraction.* Addison-Wesley, Reading, MA (1969).
39. H. P. Klug and L. E. Alexander, *X-Ray Diffraction Procedures for Polycrystalline and Amorphous Materials,* John Wiley and Sons, New York (1974), p. 791.
40. R. D. Shannon, G. Bergeret, and P. Gallezot, *Nature* **316**, 736 (1985).
41. G. Bergeret, J. P. Broyer, M. David, P. Gallezot, J. C. Volta, and G. Hecquet, *J. Chem. Soc. Chem. Comm.* 825 (1986).
42. Y. Waseda, *Novel Application of Anomalous (Resonance) X-Ray Scattering for Structural Characterization of Disordered Materials,* Lecture Notes in Physics, Vol. 204, Springer-Verlag, Berlin (1984).
43. R. Kaplow, S. L. Strong, and B. L. Averbach, *Phys. Rev.* **A138**, 1336 (1965).
44. J. H. Konnert and J. Karle, *Acta Cryst.* **A29**, 702 (1973).
45. L. Gatineau, *J. Appl. Cryst.* **5**, 255 (1972).
46. P. H. Fuoss, P. Eisenberger, W. K. Warburton, and A. Bienenstock, *Phys. Rev. Lett.* **46**, 1537 (1981).
47. M. Laridjani, P. Leboucher, D. Raoux, and J. F. Sadoc, *J. Phys. Fr. Coll. C8* **46**, 157 (1985).
48. J. M. Tonnerre, Thesis, Université de Paris-Sud, Centre d'Orsay (1989).
49. M. G. Samant, G. Bergeret, G. Meitzner, and M. Boudart, *J. Phys. Chem.* **92**, 3542 (1988); M. G. Samant, G. Bergeret, G. Meitzner, P. Gallezot, and M. Boudart, *J. Phys. Chem.* **92**, 3547 (1988).
50. T. M. Tri, J. Massardier, P. Gallezot, B. Imelik, *J. Catal.* **85**, 244 (1984); *J. Mol. Catal.* **25**, 151 (1984).
51. F. Børg, P. Gallezot, J. Massardier, and V. Perrichon, *Cl. Mol. Chem.* **1**, 397 (1986).
52. P. Lecante, A. Mosset, J. Galy, *J. Appl. Cryst.* **18**, 214 (1985).
53. P. Ratnasamy and A. J. Léonard, *Catal. Rev.* **6**, 293 (1972).
54. J. J. Fripiat, A. J. Léonard, and N. Baraké, *Bull. Soc. Chim. Fr.* (1963), p. 122.
55. A. J. Léonard, F. Van Cauwelaert, and J. J. Fripiat, *J. Phys. Chem.* **71**, 695 (1967).
56. A. J. Léonard, P. Ratnasamy, F. D. Declerck, and J. J. Fripiat, *Disc. Faraday Soc.* **52**, 98 (1971).
57. P. Ratnasamy, A. J. Léonard, L. Rodrigue, and J. J. Fripiat, *J. Catal.* **29**, 374 (1975).
58. K. Richter, E. A. Levitskii, V. Kolomiichuk, and E. M. Moroz, *Kinet. i Catal.* **16**, 1578 (1975).
59. E. M. Moroz, S. V. Bogdanov, V. A. Ushakov, and E. A. Levitskii, *Kinet. i Catal.* **20**, 138 (1979).
60. P. Gallezot, A. Bienenstock, and M. Boudart, *Nouv. J. Chim.* **2**, 263 (1978).

61. P. Gallezot, *Proc. 5th Int. Conf. on Zeolites* (L. V. C. Rees ed.), Heyden, London (1980), p. 364.
62. P. Gallezot, and G. Bergeret, *J. Catal.* **72**, 294 (1981).
63. P. Gallezot, *Zeolites* **2**, 103 (1982).
64. P. Gallezot, *J. Chim. Phys.* **78**, 881 (1981).
65. G. Bergeret and P. Gallezot, *J. Catal.* **87**, 86 (1964).
66. G. Bergeret and P. Gallezot, *Proc. 8th Int. Congr. on Catalysis,* Dechema, Frankfurt am Main. Vol. V (1985), p. 659, Vol. VI (1985), p. 353.
67. G. Bergeret, P. Gallezot, P. Gelin, Y. Ben Taarit, F. Lefebvre, and R. D. Shannon, *Zeolites* **6**, 392 (1986).
68. G. Bergeret, P. Gallezot, P. Gelin, Y. Ben Taarit, F. Lefebvre, C. Naccache, and R. D. Shannon, *J. Catal.* **104**, 279 (1987).
69. G. Bergeret, P. Gallezot, and F. Lefebvre, *Proc. 7th Int. Congress on Zeolites* (Y. Murakami, A. Iijima and J. W. Ward eds.), Kodansha, Tokyo (1986), p. 401.
70. G. Bergeret and P. Gallezot, *Z. Phys. D. Atoms, Molecules and Clusters* **12**, 591 (1989).
71. R. D. Shannon, K. H. Gardner, R. H. Staley, G. Bergeret, P. Gallezot, and A. Auroux, *J. Phys. Chem.* **89**, 4778 (1985).
72. G. H. Kuhl, *J. Phys. Chem. Sol.* **38**, 1259 (1977).
73. V. Bosacek, D. Freude, T. Frohlich, H. Pfeifer, and H. Schmiedel, *J. Coll. Interf. Sci.* **85**, 502 (1982).
74. G. Bergeret, M. David, J. P. Broyer, J. C. Volta, and G. Hecquet, *Catal. Today,* **1**, 37 (1987).
75. J. B. Parise, H. S. Horowitz, T. Egami, W. Dmowski, and A. W. Sleight, *J. Catal.* **118**, 494 (1989).
76. G. Bergeret, P. Gallezot, K. V. R. Chary, B. Rama Rao, and V. S. Subrahmanyam, *Appl. Catal.* **40**, 191 (1988).

SMALL-ANGLE X-RAY SCATTERING

A. J. Renouprez

16.1. INTRODUCTION

Small-angle X-ray scattering (SAXS) was discovered in 1938 by A. Guinier.[1] It is now a powerful method for characterizing catalysts (particle size, surface area) and disordered materials such as gels, sols, defective alloys, porous oxides or carbons, polymers. Like diffraction, SAXS is a coherent scattering phenomenon, but instead of being produced by the interference of waves scattered by atoms ordered inside a unit cell smaller than 20 Å, SAXS originates from the interference between larger blocks of uniform matter whose diameters are typically 20 to 1000 Å. Then, since direct and reciprocal space are related by Fourier transforms, the normal diffraction is observed at large angles ($\sin 2\theta/\lambda = 0.1$ to $1.5\ \text{Å}^{-1}$), SAXS is limited to a narrow cone near the origin ($< 0.1\ \text{Å}^{-1}$). It is easily understood, from the very nature of this diffraction phenomenon, that its interpretation will be straightforward for a two-phase system, with uniform density in each phase, but much more intricate for multicomponent catalysts. In this case complementary information (electron microscopy, gas adsorption) will be indispensable for interpretation of the scattering data.

16.2. BASIC THEORY

16.2.1. Scattering Cross Section of a Volume of Matter

Let us consider an elementary scattering particle M — nucleus or electron — placed at the origin 0 and struck by a narrow and parallel X-ray beam of

A. J. Renouprez ● Institut de Recherches sur la Catalyse, CNRS, Villeurbanne, France.

Catalyst Characterization: *Physical Techniques for Solid Materials*, edited by Boris Imelik and Jacques C. Vedrine, Plenum Press, New York (1994).

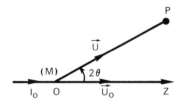
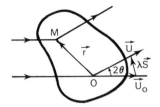

Figure 16.1. Scattering by two electrons located at 0 and M. U_0 and U are the unit vectors in the incident and scattered directions.

Figure 16.2. Scattering by a finite volume of matter, V.

intensity I_0 (Figure 16.1). The intensity measured at the distance L in the θ direction is

$$I(L, \theta) = \sigma(\theta)I_0/L^2 \tag{16.1}$$

where $\sigma(\theta)$ is the cross section of the particle. In the case of X-rays, only electrons have an appreciable cross section, given by the Thomson formula:

$$\sigma_e(\theta) = r_e^2(1 + \cos^2 2\theta)/2$$

where r_e is the radius of the electron. In the small-angle domain $\sigma(\theta)$ is constant and equal to 7.95×10^{-26} cm². If one considers now a finite volume of matter, V, the scattered intensity will be produced by the interference between waves scattered by all the electrons of V (Figure 16.2). The phase shift between the waves scattered by 0 and M, r apart, is given by

$$\Phi = 2\pi \mathbf{r} \cdot \mathbf{s} \tag{16.2}$$

with $|\mathbf{s}| = 2 \sin \theta/\lambda$, where λ is the wavelength.

Hence, the amplitude of the wave scattered by an elementary volume dV, containing $\rho(r)dr$ electrons of electronic density $\rho(r)$, is

$$dA(\mathbf{r}, \mathbf{s}) = (r_e/L)\rho(\mathbf{r})dV \exp(-2 i\pi \mathbf{r} \cdot \mathbf{s}) \tag{16.3}$$

For the total volume V

$$A(\mathbf{s}) = (r_e/L)T_3\rho(\mathbf{r}) \tag{16.4}$$

The scattering amplitude is thus a three-dimensional Fourier transform of the electron density distribution within the object. The coherent scattering cross section is proportional to the absolute square of $A(\mathbf{s})$:

$$\sigma(\mathbf{s}) = L^2|A(\mathbf{s})|^2 = L^2 A(\mathbf{s}) \cdot A(\mathbf{s})^* \tag{16.5}$$

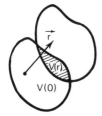

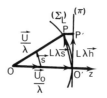

Figure 16.3. Physical representation of the $V(r)$ and $\mathbf{r}$ functions.

Figure 16.4. Representation of the scattering in reciprocal space.

where $A(\mathbf{s})^*$ is the complex conjugate of $A(\mathbf{s})$. Substituting equation (16.4) in (16.5) we set

$$\sigma(\mathbf{s}) = r^2 |T_3\rho(\mathbf{r})| \, |T_3\rho(-\mathbf{r})| = r^2 T_3 |\rho(\mathbf{r}) * \rho(-\mathbf{r})| \qquad (16.6)$$

where $\rho(\mathbf{r}) * \rho(-\mathbf{r})$ is the autocorrelation product of the electronic density. This function $P(\mathbf{r})$ is called the Patterson function in crystallography. Its physical meaning is illustrated in Figure 16.3. A particle of volume V and uniform density ρ is translated by $\mathbf{r}$:

$$P\rho(\mathbf{r}) = \rho^2 V(\mathbf{r}) = \rho^2 V(0)\gamma(\mathbf{r}) \qquad (16.7)$$

The term $V(\mathbf{r})$ is the common volume between V and its shadow translated by $\mathbf{r}$, and $\gamma(\mathbf{r})$, called the correlation function, describes the geometry of the particle; it decreases from 1 to 0 when r varies from 0 to the diameter of the particle. As the scattered intensity is proportional to the Fourier transform of $\gamma(\mathbf{r})$ and because of the reciprocity between ordinary and reciprocal space, large particles will produce a steeply decreasing $I(\mathbf{s})$ at small s values — a Dirac peak near the origin — whereas for small particles, the intensity will decrease slowly, up to relatively large s values (Figure 16.3).

16.2.2. The Scattering Laws

16.2.2.1. Single Scattering

Instead of using the electron density $\rho(\mathbf{r})$, it is more convenient to use the specific mass $\mu(\mathbf{r})$ with $\rho(\mathbf{r})_{\text{el}} \, \text{A}^{-3} = 0.301\mu(\mathbf{r}) \, \text{g cm}^{-3}$, and only density fluctuations $\delta(\mathbf{r})$, around the mean density play a role in the scattered intensity:

$$\delta \, |\mathbf{r}\rangle = \mu \, |\mathbf{r}\rangle - \mu \qquad (16.8)$$

Now, the pinpoint primary beam has an intensity I_0 and a cross section Σ; the observed intensity is

$$I_D(s) = I_0 \Sigma/L^2 \lambda^2 [mvi(s)] \tag{16.9}$$

where L is the radius of the sphere centered on the sample and tangent to the observation plane, m is the mass of sample in g cm^{-3} and $v = 7.16 \times 10^{-3} \lambda^2$, where λ is the wavelength. The total scattered flux is

$$A_D = mv_{A_0} \int i(s)d\mathbf{s} \tag{16.10}$$

The scattering function $i(\mathbf{s})$ is dependent only on the geometry of the sample. Practically, the intensity is not measured in space but in the observation plane π and one has to consider the projected vectors on π: if $\mathbf{h}$ is the projection of $\mathbf{s}$, and $\mathbf{t}$ the projection of $\mathbf{r}$, then $i(\mathbf{h})$ the projection of $i(\mathbf{s})$ is related to $q(\mathbf{t})$ the projection of $P(\mathbf{r})$ (Figure 16.4):

$$i(\mathbf{h}) = T_2 q(\mathbf{t}) \tag{16.11}$$

where T_2 is a two-dimensional Fourier transform and as in (16.10)

$$A_0 = mvA_0 \int i(\mathbf{h})d\mathbf{h} \tag{16.12}$$

If the direct beam, $I_0(h)$, does not have a narrow distribution, equation (16.9) takes the form

$$I_D(\mathbf{h}) = mv_i(\mathbf{h}) * I_0(\mathbf{h}) \tag{16.13}$$

This convolution with the distribution of intensity of the direct beam results in a broadening of the observed phenomenon by the instrumental factor, as in any spectroscopic method. However, a second difficulty arises in SAXS: at very small angles, a fraction of the incident beam, $I_{0'}(\mathbf{h})$, attenuated only by absorption, is superimposed on $I_D(\mathbf{h})$, the intensity scattered by the Thomson effect. Thus the transmitted intensity $I_t(\mathbf{h})$ has the following expression:

$$I_t(\mathbf{h}) = I_0'(\mathbf{h}) + I_D(\mathbf{h}) \tag{16.14}$$

where

$$I_0'(\mathbf{h}) = [1 - m(E + D)]I_0(\mathbf{h}) \tag{16.15}$$

and

$$I_D(\mathbf{h}) = mvA_0 i(\mathbf{h}) \tag{16.16}$$

In equation (16.15), $D = (A_D/A_0)/m$ is the ratio between the flux scattered by the Thomson effect and the incident flux and $E = (1 - A_t/A_0)/m$. Formally equation (16.16) can be written

$$I_t(\mathbf{h}) = I_0(h) * w(\mathbf{h}) \tag{16.17}$$

with

$$w(\mathbf{h}) = [1 - m(D + E)] * \delta(0) + mvi(\mathbf{h}) \tag{16.18}$$

If we take the properties of the Fourier transform into account, equation (16.17) becomes

$$T_2I_t(\mathbf{h}) = T_2I_0(\mathbf{h}) \times [1 - mE - mD + mvq(\mathbf{t})] \tag{16.19}$$

Thus $q(\mathbf{t})$, the projection of $P(\mathbf{r})$, can be calculated, provided the constant D is known, i.e., that the undeviated fraction of the direct beam can be measured. Equation (16.19) is the expression of the single scattering law.

16.2.2.2. Multiple Scattering

Equations (16.12) and (16.13) implicitly assume that the sample is thin and that there is only a single scattering event. Ruland[5] has analyzed the various reasons for the appearance of multiple scattering: Thick samples (large m), strongly scattering elements (large D), or large particles will produce an important multiple scattering and equations (16.15) and (16.17) are no longer valid. Donati[6] and Ruland[5] have shown that equation (16.19) then has to be replaced by

$$\log \frac{T_2I_t(\mathbf{h})}{T_2I_0(\mathbf{h})} = mE + mv[q\,|\,\mathbf{t}\rangle - q\,|\,0\rangle] \tag{16.20}$$

where

$$T_2I_t(\mathbf{h}) = 2\pi \int_0^\infty \mathbf{h}I(\mathbf{h})J_0(2\pi\mathbf{t}\mathbf{h})d\mathbf{h} \tag{16.21}$$

is a Bessel–Fourier transformation of $I(\mathbf{h})$. Guinier and Fournet[7] have shown that $P(\mathbf{r})$ can be determined numerically from $q(\mathbf{t})$:

$$P(\mathbf{r}) = (1/\pi) \int q'(\mathbf{t})[d\mathbf{t}/(\mathbf{t}^2 - \mathbf{r}^2)^{1/2}] \tag{16.22}$$

Finally the determination of $P(\mathbf{r})$, i.e., of $q(\mathbf{t})$, needs a precise evaluation of $I_D(\mathbf{h})$ and thus of $I_0'(\mathbf{h})$, the undeviated residue of the direct beam. This operation can be easily achieved graphically for samples containing small heterogeneities. In the case of the scattering produced by large particles or if significant multiple scattering occurs, one has to use a trial-and-error method; indeed, equation (16.20) leads to the determination of $q(\mathbf{t}) - q(0)$, where $q(0)$ is related to the residue of the direct beam A'_0:

$$q(0) = (1/m)\log(A_t/A_0) \qquad (16.23)$$

Each trial value of A'_0 leads to a $q(0)$ value and thus to a calculated $q_1(t)$; the calculated $q_1(t) - q(0)$ is then compared with the experimental $q(t) - q(0)$. This procedure can be repeated until satisfactory agreement is obtained.

16.3. DETERMINATION OF THE STRUCTURE OF MATERIALS FROM SMALL-ANGLE X-RAY SCATTERING

This section deals with the possibilities of structure determination from the distribution of scattered intensity $I(s)$. Actually all the information needed to describe the morphology of a material is contained in the $P(\mathbf{r})$ function described above. However, one should keep in mind that the determination of the shape of a collection of anisotropic particles (rods, cylinders) and of the distribution of their characteristic dimensions will be extremely difficult with a single SAXS experiment.

16.3.1. The Specific Area

Porod[8] has shown that the surface area Σ between two phases of constant density ρ_1 and ρ_2 is proportional to the first derivative of $P(\mathbf{r})$ for $r = 0$:

$$\Sigma = P'(0)/(\rho_1 - \rho_2)^2 \qquad (16.24)$$

On the other hand, $P(0)$ can be evaluated simply: a Laplace transformation of $rP(r)$ leads to $sI(s)$:

$$sI(s) = P'(0)/2\pi^3 s^3 - P'''(0)/8\pi^5 s^5 + \cdots + s \to \infty \qquad (16.25)$$

Experimentally, $P'(0)$ is determined from the asymptotic value of the $s^4 I(s)$ product at large s values or, if multiple scattering is present, by a numerical derivation of $P(r)$ (Figure 16.5).

Figure 16.5. Scattering in the detector plane.

A determination of Σ, expressed in $m^2 g^{-1}$, needs a knowledge of the mass of the sample per cross-section unit of the flux of the transmitted beam A_t and of the wavelength. The area thus determined takes into account all the heterogeneities, whatever their shape and size (from 10 to 500 Å), the upper size limit being fixed by the instrumental resolution.

16.3.2. The Radius of Gyration: Guinier's Law

In the case of single scattering, equation (16.11) can be written

$$i(s) = 2/s \int_0^\infty rP(r)\sin(2\pi sr)dr \qquad (16.26)$$

Expanding into a power series, this expression becomes

$$i(s) = 2 \sum_0^m (-1)^p \frac{2\pi^{2p-1}}{(2p+1)!} s^{2p} \int_0^\infty r^{(2p+2)}P(r)dr \qquad (16.27)$$

and keeping the first two terms,

$$i(s) = \left[4\pi \int_0^\infty r^2P(r)dr \right](1 - 2\pi^2s^2/3)\frac{\int_0^\infty r^4P(r)dr}{\int_0^\infty r^2P(r)dr} \qquad (16.28)$$

which can be written

$$i(s) = 4\pi \int_0^\infty r^2P(r)dr \exp[(-4\pi^2s^2/3)R_G^2] \qquad (16.29)$$

for $s \to 0$, with

$$R_G^2 = \frac{1}{2} \frac{\int_0^\infty r^4P(r)dr}{\int_0^\infty r^2P(r)dr} \qquad (16.30)$$

Guinier has given an interpretation for R_G, the ratio of the fourth moment of $P(r)$ to the second moment, in terms of the electronic radius of gyration. It is similar to a radius of gyration, where the role of the mass is played by the electrons; for spherical particles, their radius is $R = \sqrt{5/3}R_G$. A plot of $\log I(s)$ vs. s^2 shows a linear descent with a negative slope and $R_G = K \tan \alpha^{1/2}$.

The simple analysis of the data is valid only if three conditions are fulfilled: (1) the approximation holds only for diluted systems; (2) a linear plot is observed only for a collection of particles with a uniform diameter; and (3) for a deviation from the spherical shape, R_G is related in a complex way to the characteristic dimensions, e.g., for a prism with edge lengths A, B, C,

$$R_G^2 = (A^2 + B^2 + C^2)/12 \qquad (16.31)$$

16.3.3. Porosity

One of the advantages of SAXS is the possibility of determining the porous volume, including the internal porosity, which is complementary to information obtained from gas adsorption methods.

The reciprocal form of equation (16.26) can be written

$$P(r) = 2/r \int_0^\infty sI(s)\sin(2\pi sr)ds \qquad (16.32)$$

Moreover, $P(0) = \mu\alpha(1 - \alpha)$, where α is the porosity expressed in volume unit per volume unit. Thus f, the porosity expressed in volume per unit of mass, is equal to

$$f = P(0)/\mu^2 \qquad (16.33)$$

It should noted that precision in the determination of f is related to information regarding the scattered intensity, especially for low values of s; this means that the undeviated part of the direct beam also has to be evaluated precisely.

16.3.4. The Particle Size Distribution

When particles have a well-defined shape, SAXS is able to provide a distribution of their characteristic lengths. However, for systems where the particles are not well separated (nondiluted systems), owing to the Babinet principle, the SAXS curve will also contain information on the distances between particles. Moreover, if the distribution of characteristic distances is broad, the information concerning the shape of the particles will be lost.

In simple cases where particles (or voids) have an isotropic shape (e.g., spheres), the SAS scattering curve leads easily to the distribution of their diameters. In this case, the $\gamma(r)$ function defined in equation (16.7) has the form

$$\gamma(r) = 1 - 3r/2D + \tfrac{1}{2}(r/D)^3 \qquad (16.34)$$

where D is the sphere diameter.

Experimentally, by combination of equations (16.7) and (16.22), one obtains a mean $\gamma(r)$ related to the diameter distribution $f(D)$ by

$$\overline{\gamma|r\rangle} = \int_r^{r_0} \gamma|r\rangle f(D)dD \qquad (16.35)$$

Also

$$\overline{\gamma(r)} = \frac{P(r) - P(r_0)}{P(0) - P(r_0)} \qquad (16.36)$$

On the other hand, it was shown by Donati[6] that equation (16.35) can be

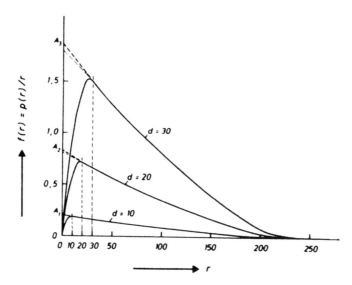

Figure 16.6. A plot of r vs. $f(r)$ for lamellar particles, with basal planes 10×10 nm and thickness 1, 2 and, 3 nm.

reversed and that

$$f_V(D) = -\frac{D^3}{3}\frac{d}{dD}\frac{\gamma''D}{D} = \frac{-D}{3}[\gamma''(D) - \gamma'''(D)] \qquad (16.37)$$

For nonspherical particles, the procedure is more complex, since an analytical inversion of equation (16.35) is generally not possible; thus one usually compares the experimental $\gamma(r)$ with model $\gamma(r)$ to deduce the shape and mean characteristic lengths of the particles. For example, Figure 16.6 shows the shape of $r\gamma(r)$ for flat particles with thickness 10, 20, or 30 Å. It is clear that a maximum is visible on these graphs at an abscissa corresponding to their thickness.

16.3.5. Application to Solid State Materials

As noted earlier, all the expressions given above have a very clear physical meaning for diluted systems, such as a colloidal suspension at low concentration or a porous system of low porosity. When this is not the case, e.g., a porous material such as an active carbon, where the volume fraction of the voids is nearly equal to that of the solid, the interpretation of the data is more difficult. Indeed a distribution of the pore diameters is obscured by the distribution of matter grains which have the same dimension as the pores.

A second difficulty arises in the analysis of inhomogeneous materials like polymers. All the theory presented in Section 16.2 assumes that each phase has a uniform (electronic) density and that the transition between two phases is abrupt. When this is not the case, characteristic parameters such as the mean diameter can still be obtained. Mering and Tchoubar[9] have developed a complete theory covering the various possibilities.

16.3.6. Neutron Small-Angle Scattering

All the principles of structure determination using X-rays can be applied to neutron scattering; the unique difference is that the scattered intensity is proportional to the square of the difference of scattering lengths instead of to the densities. With X-rays, the intensity scattered by materials like polymers or proteins, containing low-Z elements, is very weak, whereas with neutrons it has the same order of magnitude as other elements. For example, this cross section is 5.4 for deuterium, 5.5 for carbon, and 11 for nitrogen, and only 7.8 for copper and 11.2 for platinum. Moreover, several elements have isotopes with a negative scattering length, e.g., hydrogen, lithium, and titanium; consequently, a suitable mixture of isotopes can have a null cross section. Very good sensitivity can thus be achieved in the study of organic molecules either in suspension or in solution in a mixture of heavy and light water, whose scattering lengths are, respectively, -0.56×10^{10} and $+6.3 \times 10^{10}$ cm/cm^3. This principle has been extended by Sturhmann[10] into a more general method known as contrast variation. Indeed the scattering length determining the intensity is given by

$$\rho(r) = \rho(r)_{solute} - \rho(r)_{solvent} \qquad (16.38)$$

This difference can be varied by modifying the proportion of hydrogen and deuterium in the solvent. If we take the example of globular macromolecules in solution in this solvent, at high contrast the structure of the particles no longer contributes to the scattering. Conversely, at vanishing contrast, only the fluctuations of the intramolecular density around the mean scattering of the solute are visible.

The main limitation in the use of neutrons is the presence of inelastic scattering producing, in the case of elements which have a high incoherent cross section, a high level of background which has to be subtracted. A second difficulty is the relatively low flux of neutron sources compared to modern photon sources like synchrotrons. This necessitates the use of a large mass of sample, typically 3 to 10 g.

16.4. EXPERIMENTAL SETUP

All small-angle scattering cameras consist of an optical system providing a thin and monochromatic X-ray beam, a sample holder frequently equipped with an oven or a cryostat and a detection system connected to a computer. One must, however, distinguish between laboratory instruments in which the photon

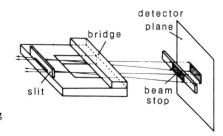

Figure 16.7. Drawing of the Kratky collimating system.

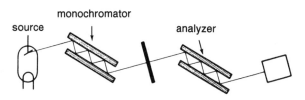

Figure 16.8. The Bonse–Hart collimating system: (2) and (4) are the monochromator and analyzer channel-cut crystals.

source is an X-ray tube (conventional or rotating anode) delivering a diverging beam with an intermediate flux and those located at the synchrotron facilities, where the beam entering the optic elements has a low vertical divergence and a flux two to four orders of magnitude higher.

16.4.1. Laboratory Instruments

Because of the beam divergence, all the instruments have to make a compromise between flux and resolution. The resolution is related to the minimum accessible angle and to the wavelength $R = \lambda/2\theta_{min}$ of the order of 1000 Å for most of the classical instruments.

16.4.1.1. Collimation and Monochromatization

One classical instrument is the Kratky camera[11] shown in Figure 16.7. The divergent beam delivered by a flat graphite crystal enters a fine slit and is collimated by a bridge that eliminates the parasitic scattering in one-half of the detector plane. Its resolution is a direct function of the opening of the entrance slit which limits the flux. Also, owing to the presence of the bridge collimator, the profile of the direct beam is asymmetric. The resolution of this camera reaches 5000 Å.

The Bonse–Hart[12] camera (Figure 16.8) is based on a completely different principle. The collimation and monochromatization is achieved by a channel-cut crystal, on whose walls the beam undergoes six reflections and has a

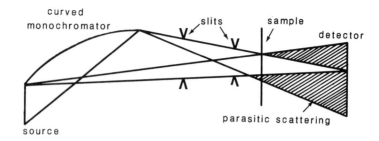

Figure 16.9. The collimating system using an asymmetric Johanson monochromator, focusing in the detector plane.

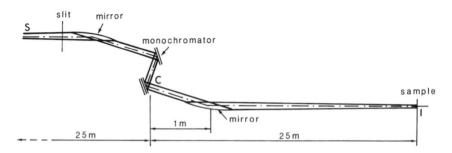

Figure 16.10. The system proposed by Damaschun using the total reflection on a glass mirror combined with the bridge system of the Kratky camera.

divergence of a few seconds of arc. A second similar crystal is placed after the sample and its rotation allows the recording of the intensity scattered by the sample. This collimation is efficient enough to reach a resolution of 50,000 Å; however, if a resolution of 1000 Å is sufficient, the Bonse–Hart camera yields an intensity 300 times lower than the block camera described above.

Another design, proposed by Renouprez et al.,[13] is based on the use of a curved, asymmetric monochromator focusing the beam into a thin rectangular spot in the detector plane. Owing to the asymmetry of the crystal, a large solid angle is admitted on the monochromator and the flux on the sample is thus increased. The parasitic scattering is reduced by slits but the resolution is limited to 2000 Å (Figure 16.9). These optics only deliver a partly monochromatic beam and the high-order harmonics ($\lambda/3$) of the $K\alpha$ radiation is transmitted to the sample.

In the alternative solution proposed by Damaschun,[14] the total reflection on a polished glass surface is used. The position of the energy cut depends on the incident angle (5 to 10 mrad) but 90% of the intensity is retained (Figure 16.10).

All the monochromatization–collimation systems described above are focused in one direction only. The beam in the sample plane therefore has a rectangular shape, which complicates the treatment of data and is not adapted to samples producing anisotropic scattering. Obtaining a point beam requires double focusing by the combined use of a curved monochromator and a bent mirror that focuses the beam in both directions.

16.4.1.2. Detection and Recording

In most conventional systems the recording of the scattered intensity is achieved by moving step by step, with a precise goniometer, a fine analysis slit followed by a detector.

The angular range is limited to 100 mrad but the scattered intensity generally varies by three or four orders of magnitude within the first 5 mrad. Thus very fine steps (0.1 mrad) are needed in the central part of the curve while the outer part is well described by steps of 2–5 mrad. Also as noted above, the high count rate observed at very low angles can be a severe limitation; indeed the gas proportional counters are limited to 10^5 counts s^{-1} and photomultipliers to 10^6 counts s^{-1}, and this rate is not enough even with conventional X-ray tubes. When a rotating anode delivering 10 kW is used, attenuation of the beam with a calibrated filter is needed; this is always necessary when a quantitative determination of the whole transmitted scattered curve is needed. If knowledge of the total flux A_T is not required, the use of standard samples of known scattering power is sufficient and the most commonly used is composed of platelets of pure polyethylene.

An alternative to the goniometer, which appeared in the early 1970s, was the position-sensitive proportional counter. This detector uses a delay line to encode the spatial position of a scattering event and measures the number of impinging photons. It suffers from the same count-rate limitation as an ordinary gas counter in that a beam stop has to be used to register the outer part of the scattering curve and the central part is measured with a filter.

16.4.2. Synchrotron Radiation Instruments

Synchrotron radiation has many advantages over X-ray tubes: (1) The wavelength can be tuned continuously in the full range from 5 to 30 keV; (2) the source has a small cross section (0.2 × 0.5 mm) and a low vertical divergence, of the order of 0.1 mrad; (3) the flux is higher than on conventional sources, typically 10^{14} photons/s/horizontal mrad/0.1% bandwidth. It will be increased by three orders of magnitude on insertion devices of the European facility. Obviously the impact of the source brightness and of the low divergence of the beam is favorable for small scattering experiments, especially if only small samples are available. The increase in intensity is already 10^2 compared to a 3-kW rotating anode tube and will reach 10^4 at the European facility.

A further advantage of using this radiation is the possibility of performing anomalous scattering, which is also a contrast variation method. The scattering factor of a given element varies by 20 to 40% when the photon wavelength is close to the absorption edge of this element. The combined analysis of two scattering curves recorded at two wavelengths determines the surrounding of this element.

One of the instruments located at LURE, D24, on the DCI ring, has the following characteristics: (A) Monochromator-17 cm Ge(111) bent crystal; (B) Wavelength-1 to 2 Å; (C) Sample-detector distance-0.3 to 2m; (D) Measurement range $2\theta/\lambda$-0.0005 to 0.1 Å^{-1}; and (E) Detectors-18-cm linear position-sensitive and 20×20-cm two-dimensional position-sensitive.

On this instrument, a horizontal focusing of the beam is achieved by the bent germanium crystal but the beam is not focused vertically, which leads to a loss of 80% of the photons when the slits are closed to limit the size of the focal spot to 1×1 mm^2. It is planned to install a curved horizontal mirror which would have the advantage, apart from focusing, of eliminating the harmonics transmitted by the crystal.

Actually, all small-angle scattering instruments suffer from a limited detection capacity. The small linear position-sensitive counter is limited to 10^5 photons s^{-1} and the large two-dimensional counter constructed by A. Gabriel to 10^6 photons s^{-1}, if the intensity is spread over the whole surface of the counter. The appearance of second generation sources with flux increases of 10^2 to 10^3 in turn requires the development of new detectors.

A promising innovation will be the development of multistrip (200 strips) germanium or silicon detectors; each strip of 100–250 μm can measure up to 10^5 photons s^{-1} and the energy resolution will be of the order of 5% at 10 keV. If the patterns are obtained on instruments installed on the insertion devices (wiggler, undulators) only this type of counter would allow their recording and make possible kinetic measurements of processes such as association–dissociation of proteins occurring during thermal treatment.

16.5. THE APPLICATIONS OF SMALL-ANGLE SCATTERING

Small-angle scattering is now applied to the study of a large variety of materials including polymers (in both the liquid and solid state), proteins, metallic alloys, porous or finely divided solids. Our purpose here is to describe some of these applications in the field of inorganic substances, especially those concerning catalysts.

16.5.1. Porous and Finely Divided Solids

This is the simplest type of system that can be examined by SAXS, as it is composed of two phases with constant density ρ_1 for the solid and ρ_0 for the air contained in the pores or surrounding the particles. However, a certain number of difficulties have to be overcome: for high-Z elements, ρ_1 has a large value and D, the absorption coefficient, may be much larger than unity, producing a large amount of multiple scattering. To limit this effect, one has to reduce m, the mass per unit area of sample; also, sample thickness must be uniform, which is not easy to achieve with a mass lower than 10 mg cm². An alternative solution is to use a larger mass and perform the study with a short wavelength like Mo K radiation.

A second problem arises from the range of particle sizes to be examined; usually, SAXS can analyze material containing scattering elements whose diameters range from 10 to 500 A; in order to observe the Porod invariant on the scattering curve produced by solids containing the smallest pores (or particles), one has to extend the experimental range up to large s values, i.e., typically 0.1 Å^{-1}.

On the other hand, the possibility of detecting large particles is directly related to the resolution of the instrument, as defined in Section 16.4. Indeed, if particles of 500–2000 Å are to be characterized, one has to use an instrument located at a synchrotron facility, using a highly collimated beam, with a large (1 m) sample detector distance.

Finally, all the parameters discussed in Section 16.3 can be determined on these solids.

16.5.1.1. Specific Area

As shown previously, the specific area is evaluated from the value of the Porod invariant, the asymptotic value of $I(s)s^4$ for large s values. One of the advantages of SAXS is that it gives both internal and external surfaces, compared to gas adsorption methods, so the X-ray value is generally larger. Also, since the density of the second phase has to be precisely defined, a careful outgassing of the solid is necessary. Precise determinations of area have been performed from 0.1 up to 1500 $\text{m}^2 \text{g}^{-1}$, on light materials.

16.5.1.2. Porosity

This parameter is calculated by computing $P(0)/\mu^2$; as shown above, $P(0)$ is related to A_0', the undeviated part of the direct beam. In the case of large pores producing an important amount of scattering at low angles, the precision of this separation is generally poor and is dependent on the instrument resolution.

16.5.1.3. Diameter Distribution

The $P(r)$ function provides diameter distribution curves, assuming that the particles have a spherical shape. It is known that the BJH method, using nitrogen adsorption, assumes a cylindrical pore shape; moreover, this second method can only detect pores with a section larger than that of the probe molecule, 14 Å^2 for nitrogen. As shown on Figure 16.11, the distributions measured by these two methods can be compared.[15] In this case, an active carbon, all the pores are accessible to nitrogen, as the surface areas measured by the two methods are close (1080 and 980 $\text{m}^2 \text{g}^{-1}$).

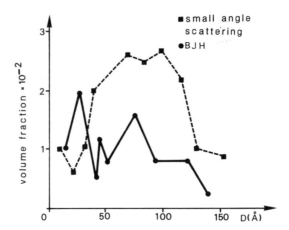

Figure 16.11. Comparison of the pore diameter distributions obtained by (a) SAXS and (b) nitrogen adsorption.

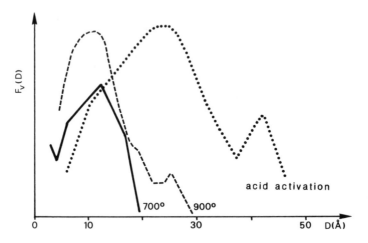

Figure 16.12. The progressive increase of pore diameter as a function of the temperature of the activation treatment of a charcoal.

This is not the case before activation of the carbon. As shown on Figure 16.12, the air-treated vegetal carbons have a closed microporosity in the 10–15 Å range, since the BET area is 30 times lower than the X-ray value. After an acidic (and oxidizing) treatment, larger, open pores appear in the 30 Å range; the X-ray and BET surfaces are then close to each other, indicating that the pores are open (Table 16.1).

TABLE 16.1. Comparison of BET and SAXS Surface Area
as a Function of the Treatment

Treatment	BET area $m^2\,g^{-1}$	SAXS area $m^2\,g^{-1}$
Air at 600 °C	23	380
700 °C	60	390
800 °C	86	450
900 °C	70	420
H_2SO_4/H_3PO_4	875	970

Owing to the Babinet principle, it can be difficult to discriminate between pores and particles on SAXS distribution curves. It is thus necessary to compare electron microscopy measurements, which provide the particle diameter and gas adsorption leading to the pore diameters with the X-ray data.

16.5.2. Supported Metallic Catalysis

In this case, difficulty arises from the presence of three phases, the air, the support, and the metal; it is thus necessary to separate the various contributions to the scattering. The overall sensitivity depends on (i) the amount of metal, and (ii) its density.

16.5.2.1. Model Catalysts with High Metal Concentrations

One example of this group[16] is the study of platinum aggregates occluded in the network of a Y-zeolite. The zeolite produces only a limited amount of scattering, as the pores are crystallographically ordered. Moreover, the density contrast between the silicoaluminate framework and platinum is very high.

Two types of samples were studied: Sample I was prepared by heating the zeolite containing $Pt(NH3)_4^{2+}$ ions first in oxygen at 300 °C and then in hydrogen at 300 °C. As shown on Figure 16.13, particles from 1 to 1.2 nm in size are formed, which are probably located in the large cages of the zeolite. This situation is confirmed by hydrogen chemisorption which indicates a dispersion of 80%. Further heating up to 800 °C does not modify the particle size. At 900 °C, their diameter has grown to 3 nm and there is severe damage in the framework of the support.

Sample II was first treated at 600 °C in oxygen before reduction. Curve a in Figure 16.14 indicates that the particle diameter is also 1 nm; however, the amount of chemisorbed hydrogen is four times lower than in the previous case. Thus, part of the metal is not taken into account by the distribution curve. Actually, a Guinier plot of log I vs. s^2 leads to a radius of gyration of 0.25 nm, corresponding to aggregates composed of three atoms, probably located in the

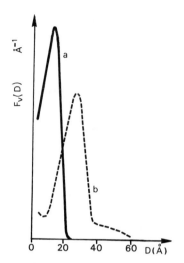

Figure 16.13. The diameter distribution of platinum aggregates in Y-zeolite upon activation at 300 °C in oxygen and reduction: (a) below 800 °C and (b) at 900 °C in hydrogen.

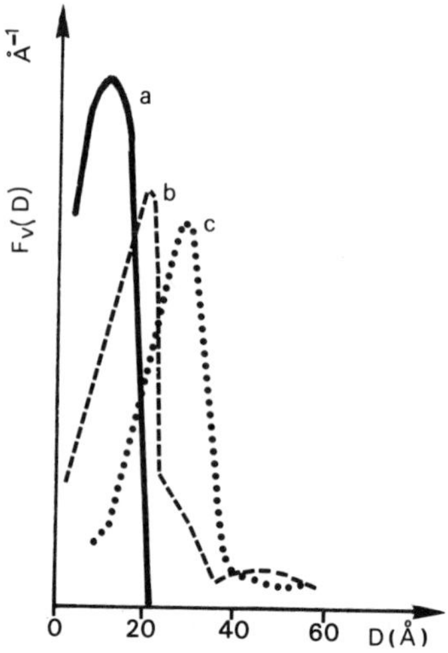

Figure 16.14. Platinum in Y-zeolite: activation at 600 °C and reduction: (a) at 300, (b) at 600, and (c) at 900 °C.

sodalite cages and unable to dissociate hydrogen. A progressive migration of these small clusters toward the supercages is observed during the thermal treatment (Figure 16.14, curves b and c) at higher temperatures, where they agglomerate to larger 1-nm clusters. During this process the hydrogen chemisorption progressively increases to reach a value in agreement with the measured diameter.

16.5.2.2. Reforming Catalysts

When the metal loading is of the order of 0.5 to 1%, especially with low-Z elements such as the transition metals of the first row, the scattering produced by the support can be as high as 95% of the total measured intensity. One thus has to separate these two contributions. The first method to be employed successfully consists of using a contrast modification of the support: in the study of alumina-supported platinum, CH_2I_2 (density 3) was introduced into the pores to eliminate support scattering.[17]

The result is shown in Figure 16.15 for a 0.6% metal concentration. The comparison with electron microscopy shows that the pores no longer contribute to the distribution. Figure 16.16 shows the sintering of a 3.5% Pt/alumina catalyst upon heating at 800 and 900 °C, with the progressive formation of particles 6 to 10 nm in size.

Another method[18] has recently been developed to avoid the tedious procedure of pore masking which involves a balanced subtraction of the support

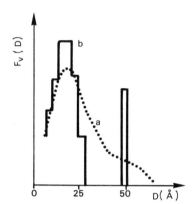

Figure 16.15. A 0.6% Pt/Al$_2$O$_3$ catalyst, when the pores of the alumina are filled with CHI$_3$: (Curve a) SAXS, and (Curve b) electron microscopy, yield similar particle diameter distributions.

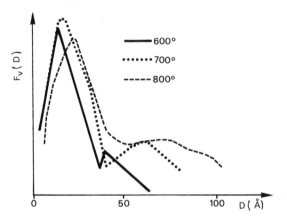

Figure 16.16. The progressive increase in pore diameter as a function of the temperature of the activation treatment of charcoal.

scattering. An experiment is first performed on the support alone, leading to a γ_2 characteristics function. The measurement on the catalyst leads to γ; the $\tilde{\gamma}(r)$ function related to the metal is given by

$$\tilde{\gamma}(r) = \frac{P_{33}(r) - \phi_3^2}{\phi_3(1 - \phi_3)}$$

$$P_{33}(r) - \phi_3 = \frac{1}{n_3^2}(\gamma_3 - 1) \cdot D + \frac{\phi_1\phi_2}{\phi_2 + \phi_3}(1 - \gamma_2)[\phi_3(n_3 - n_1)^2 + \phi_2 n_1^2 - \phi_3 n_3^2]$$

where ϕ_1, ϕ_2, and ϕ_3 are, respectively, the volume fractions of the support, of the void, and of the metal and n_1, n_2, and n_3 their electronic densities.

Curve c in Figure 16.17 illustrates the distribution of particle diameters measured on 3% Pt catalyst. The excellent agreement observed in the compar-

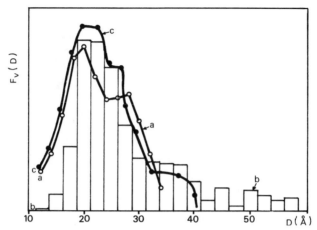

Figure 16.17. Illustration of the balanced subtraction methods used to eliminate the contribution of the pores from the SAXS distribution: (Curve a) simplified analytical procedure, (Curve b) particle diameter distribution obtained from electron microscopy, and (Curve c) correct analytical subtraction.

ison with the electron microscopy histogram (curve b) is proof of a correct elimination of the contribution of the pores to the distribution curve (curve a).

16.6. SMALL-ANGLE NEUTRON SCATTERING

As noted above, one of the main advantages of neutrons with respect to X-rays is the possibility of contrast variation. This was widely used in studying macromolecules or biological molecules in solution in H_2O–D_2O mixtures. Also, the fact that the coherent cross sections do not vary regularly with Z, as

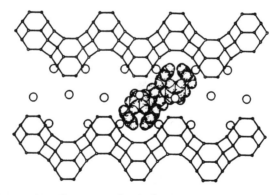

Figure 16.18. Aggregation of benzene molecules in the supercage of a Y-zeolite as measured by neutron small-angle scattering.

those for X-rays, can be an advantage for studying elements close to each other in the Periodic Table. For example, the ratio of the scattering factors for copper and zinc is only 0.96 for X-rays and 0.55 for neutrons.

Finally it is well known that neutrons are also well adapted for the study of light elements that have cross sections comparable to those of heavier elements. A SANS study of the formation of benzene clusters in the network of a Na-Y-zeolite was performed successfully.[19] For a coverage of 1 molecule per super-cage, neutron diffraction shows that each molecule is bonded to a cationic site, and only a very weak small-angle scattering intensity is detected. Conversely, for a coverage of 2.5 molecules per cage, SANS shows that half of these molecules agglomerate into 10–12-Å clusters. A representation of this aggregation process is shown in Figure 16.18.

REFERENCES

1. A. Guinier, *Ann. Phys.* **12**, 161 (1939).
2. J. L. Soule, *P. Phys. Rad.* **18**, 90 A (1958).
3. O. Glatter and O. Kratky, *Small-Angle X-Ray Scattering,* Academic, London (1982).
4. V. Luzzatti, *Acta Cryst.* **10**, 643 (1957).
5. W. Ruland and H. Tompa, *J. Appl. Cryst.* **5**, 1 (1972).
6. J. R. Donati, B. Pascal, and A. J. Renouprez, *Bull. Soc. Fr. Miner. Cristallogr.* 532 (1972).
7. A. Guinier and G. Fournet, in: *Small-Angle Scattering of X-Rays,* John Wiley and Sons, New York (1955).
8. G. Porod, in: *Small-Angle X-Ray Scattering,* Academic, London (1982), p. 29.
9. J. Mering and D. Tchoubar, *J. Appl. Cryst.* **1**, 153 (1968).
10. H. B. Stuhrmann, *J. Appl. Cryst.* **7**, 173 (1974).
11. O. Kratky, *Z. Elektrochem.* **62**, 66 (1958).
12. U. Bonse and M. Hart, *Appl. Phys. Lett.* **6**, 155 (1965).
13. A. Renouprez, H. Bottazzi, D. Weigel, and B. Imelik, *J. Chim. Phys.* **62**, 131 (1965).
14. G. Damaschun, *Naturwiss.* **51**, 378 (1964).
15. A. Renouprez and J. Avom, in: *Characterization of Porous Solids,* Elsevier, Amsterdam **39** (1988).
16. P. Gallezot, A. Alarcondiaz, J. A. Dalmon, and A. Renouprez, *J. Catal.* **39**, 334 (1975).
17. A. Renouprez, C. Van Hoang, and A. Compagnon, *J. Catal.* **31**, 411 (1974).
18. D. Espinat, B. Moraweck, J. F. Larue, and A. Renouprez, *J. Appl. Cryst.* **17**, 269 (1984).
19. A. Renouprez, H. Jobic, and R. C. Oberthur, *Zeolites* **5**, 222 (1985).

17

PHOTOELECTRON SPECTROSCOPIES: XPS AND UPS

J. C. Vedrine

17.1. INTRODUCTION

Photoelectron spectroscopy has been developed in the past 25 years and attracted wide interest among specialists in catalysis, particularly because of its sensitivity to the top first layers of solid materials[1] and to the chemical state of all elements, namely the oxidation state and covalency or ionic types of bondings. The technique suffers from its own physical principle as it must be carried out in a high vacuum. This feature requires cumbersome and expensive high-vacuum equipment, and obviously requires experimental conditions far from those of catalytic reactions.

The principle of the technique is simple.[1,2] A beam of photons of X (X-ray photoelectron spectroscopy) or UV (UV photoelectron spectroscopy) radiation hits the sample, which then emits electrons which are analyzed in terms of numbers and energy. Usually $K\alpha$ emissions from aluminum ($h\nu = 1486.6$ eV) or magnesium ($h\nu = 1253.6$ eV) cathodes are used as X-ray sources and emissions from a helium lamp ($He_I = 21.2$ and $He_{II} = 40.8$ eV) as the UV source. The energy conversion principle allows one to write

$$h\nu = E_k + E_b + \phi_{sp} \tag{17.1}$$

where $h\nu$ is the incident photon energy; E_k the kinetic energy of the electrons emitted analyzed with an appropriate detector; E_b the binding energy of the electrons in their orbital level, i.e., the energy necessary for them to reach the

J. C. Vedrine ● Institut de Recherches sur la Catalyse, IRC, CNRS, Villeurbanne, France.

Catalyst Characterization: Physical Techniques for Solid Materials, edited by Boris Imelik and Jacques C. Vedrine, Plenum Press, New York (1994).

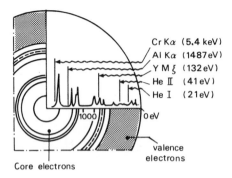

Cr Kα (5.4 keV)
Al Kα (1487eV)
Y M ζ (132eV)
He II (41eV)
He I   (21eV)

0 eV

1000

valence
electrons

Core electrons

Figure 17.1. Principle of the photoemission of electrons as a function of the inner orbitals where they originate.

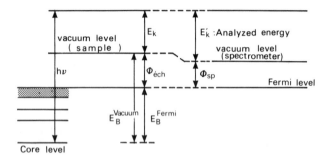

Figure 17.2. Basic principle governing the energy of the electrons emitted under photon impact.

Fermi level; and ϕ_{sp} the work function of the spectrometer, i.e., the energy necessary for the electrons to be analyzed in the spectrometer.

The principle of the technique and the principle governing the energy of the electrons emitted are given in Figures 17.1 and 17.2. It is easy to understand that the depth of the electron levels excited depends upon the energy of the incident beam photons. In UPS only valence band electrons are excited while in XPS both core and valence band electrons are excited (Figure 17.1).

The XPS or UPS spectrum corresponds to the plot of the variations in the numbers of emitted electrons vs. their kinetic energy values, i.e., their binding energy values, the spectrometer work function having been calculated in advance. A typical spectrum is shown in Figure 17.3.

17.2. EXPERIMENTAL APPARATUS

17.2.1. Spectrometer

As the emitted electrons are readily absorbed by any material, the technique requires operation in a high vacuum (10^{-8}–10^{-11} Torr or 10^{-6}–10^{-9} Pa). Spectrometers are usually designed as shown in Figure 17.4.

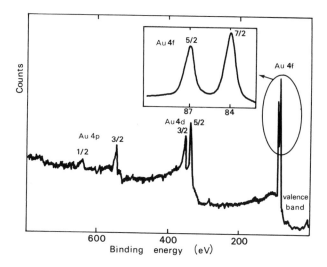

Figure 17.3. Typical XPS spectrum of gold.

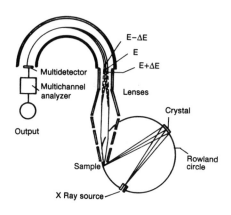

Figure 17.4. Scheme of an XPS spectrometer.

1. A preparation chamber to treat the sample first under given conditions of, e.g., heating and controlled atmosphere, and then to outgas it before passing it to the next chamber.
2. A chamber for analysis under high vacuum which has several components: (a) a source of X rays and a source of UV photons; (b) an analyzer or detector of electrons in number and in energy; (c) an Ar^+ or He^+ ion source for a SIMS or ISS accessory with a mass spectrometry analyzer (the Ar^+ ion source for etching is usually placed in the preparation chamber); (d) an electron beam source for supplementary Auger electron

spectroscopy using the same electron detector and for LEED in single-crysal studies.

3. A computerized system for automatic scanning of energy in several domains chosen for analyzing the peak of the elements to be characterized and for data accumulation.

4. A computer for the analysis of the peaks; e.g., integration, subtraction, deconvolution.

The spectrometer can be restricted to XPS analysis only for routine types of analyses or extended to more sophisticated equipment for basic work, particularly on single crystals. In the latter case accessories as listed above may be added as ISS, SIMS, STM, AFM, AES, and LEED. As such analysis is complex, much time is needed for one sample and thus the technique cannot be used for routine analysis.

17.2.2. Preparation of the Samples

Single-crystal samples are attached to a sample holder that can be rotated in the all spatial directions. This allows an angular type of analysis for which the incident and emitted beam angles can be varied. Examples will be described in Chapter 18. The spectrometer obviously needs a specific spatial manipulator and a specific analysis chamber where the detector can be displaced along a plane.

Powder samples are pressed either on an indium foil fixed on the sample (indium is a soft metal that allows the powder to be retained) or on a grid fixed on the sample holder in order to retain the powder. The latter case is useful when heat treatment is needed since indium melts at a low temperature (423 K). The grid, however, has the disadvantage that the powder pellet may fall during transfer or any manipulation. Modern reaction chambers allow the treatment of a sample under any conditions, even under high pressure or high temperature. A carousel sample holder allows several samples to be treated under the same conditions at the same time.

17.3. XPS PARAMETERS AND BASIC FEATURES[3-10]

A classical XPS spectrum (see Figure 17.3) is composed of peaks superimposed on a background corresponding to electronic emission phenomenon. Unlike in techniques using bombardment by electrons, such as Auger electron spectroscopy, the background is not very intense and the spectrum can be analyzed easily.

17.3.1. Chemical Features

It was shown above that the fundamental relationship may be written as

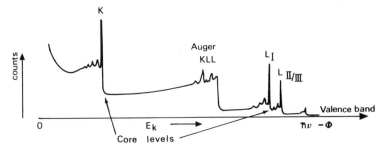

Figure 17.5. Characteristic and theoretical XPS spectrum.

follows:

$$E_b = h\nu - E_k - \phi_{\text{sp}} \qquad (17.2)$$

The value of E_b calculated from the experimental spectrum and equation (17.2) has to be compared with energy values of the electronic levels given in reference tables for all the elements. One can then unambiguously identify all the elements constituting a given material through their different electronic level energy values as described in Figure 17.5. As all the elements, except hydrogen and deuterium, are detectable, the XPS technique is also designated electron spectroscopy for chemical analysis (ESCA).

17.3.1.1. Binding Energy and Chemical Shift[2]

The absolute binding energy value of an electron is defined as the difference in total energy value of the system between its initial state with N electrons, E_i, and its final state, E_f, where a vacancy has been created by the departure of one photoelectron ($N - 1$ electrons). This can be expressed as $E_b = E_f^{N-1} - E_i^N$.

The binding energy value relates to the Fermi level assuming a good electric contact between the sample and the sample holder and between the latter and the spectrometer (equlibrium of all Fermi levels). For heterogeneous catalysts one is often dealing with insulators, and the sample rapidly becomes positively charged because of the ejection of photoelectrons. The charge compensation may be ensured by nearby low-energy electrons. However, depending on the spectrometer design, such a charge compensation may be insufficient. Either peak-broadening and/or a big shift in peak position could occur. This phenomenon is known as a charging effect. To avoid it, one can send additional low-energy electrons from an electron gun accessory. Unfortunately, this procedure is rather arbitrary and results in a strong limitation in validity of chemical shift value measurement (*vide infra*).

The photoemission phenomenon thus appears as a transition between a neutral initial state and an ionized final state. Such transitions are dipolar with $\Delta L = \pm 1$, where L is the orbital momentum. In a monoelectronic transition, the energy conservation principle leads to the basic relationship as seen above:

$$E_b^{exp}(k) = hv - E_k(k) \tag{17.3}$$

where k designates the level of the electron analyzed.

For a chemist it is important to determine how the bonding of a given element and its neighboring atoms modifies the energy value of this element with respect to its free state. One then considers the energy difference between the free state of the element and the same element as part of a given compound and writes

$$\Delta E_b = \Delta E_i - \Delta E_f \tag{17.4}$$

Such a chemical shift is dependent upon the way the element is chemically bound to its environment, i.e., its oxidation state, the number of ligands, and the ionic or covalent character or its bondings to ligands.

17.3.1.1a. Initial State Effects. When an atom loses electrons due to chemical bonding, the interaction of the nucleus with the other electrons remains unchanged but the repulsive forces have decreased. It therefore follows that there is a decrease in binding energy for all electronic levels and consequently a positive shift in E_b. The reverse phenomenon occurs if the atom becomes electron-enriched upon bonding. This crude discussion gives qualitative information concerning electronic transfers. A well-known example is that of ethyltrifluoroacetate ($CF_3COOC_2H_5$), which gives rise to four C_{1s} peaks with chemical shifts equal to 8.2, 4.8, 1.9, and 0.0 eV corresponding to a regular decrease in the positive charge on the carbon atom from CF_3 to CH_3 due to the inductive effect of ligands.

In a simple model, one considers classical electrostatic laws in approximating the element studied to a sphere with radius r. When the oxidation state of the element increases by one unit, this corresponds to the ejection of one electron toward infinity, which requires an energy equal to $-e^2/r$. However, in any compound the charge really borne by an element is different from its oxidation state value because of the more-or-less strongly covalent character of the bondings. Moreover the electron is not ejected toward infinity but rather toward its neighbors. Therefore one usually writes the energy change due to a one unit oxidation state change as

$$\Delta E_i = (e_r^2/r)\Delta V \tag{17.5}$$

where e_r represents the actual change in charge and ΔV the change in the potential created by neighboring atoms at the expense of the atom considered. ΔV can be approximated using the Madelung potential.

An important conclusion that follows is that as the chemical shift is directly related to the charge borne by an element, it is dependent upon the oxidation state of the element and on the nature and number of ligands and chemical

TABLE 17.1. Chemical Shifts Observed as a Function of Oxidation State

Element	Electronic level	Compounds	Chemical shifts (eV)
Al	$2p$	Al^0–Al_2O_3	2.7
Si	$2p$	Si^0–SiO_2	4.0
Co	$2p_{3/2}$	Co^0–CoO	2.1
		Co^0–Co_3O_4	1.8
Ti	$2p_{3/2}$	Ti^0–TiO	0.9
		Ti^0–Ti_2O_3	3.7
		Ti^0–TiO_2	5.1
W	$4f_{7/2}$	W^0–WO_2	1.2
		W^0–$CrWO_4$	2.6
		W^0–WO_3	4.2

TABLE 17.2. Binding Energy Values for Ni^{2+} Cations Engaged in Different Organometallic Compounds or Complexes[a]

Compound	E^b (eV)		
	MP ($2p_{3/2}$)	SP ($2p^{3/2}$)	$E_{3/2}$
$NiSO_46H_2O$	856.5	861.7	5.2
$NiSO_4$	856.3	860.6	4.3
NiO	854.0	860.1	6.1
NiF_2	857.2	863.1	5.9
$NiCl_2$	855.3	860.7	5.4
$NiBr_2$	854.3	859.5	5.2
NiI_2	852.9	857.1	4.2
$[(CH_3)_4N][NiCl_3]$	854.8	859.9	5.1
$[(CH_3)_4N][Ni(NCS)_6]$	855.4	861.8	6.4
$Ni(OAc)_24H_2O$	855.5	860.0	4.5
$[Ni(diars)_2Cl_2]Cl$	853.9	860.9	7.0
$NiCO_3$	854.7	860.5	5.8
$Ni(dipy)_2Cl_2$	854.9	861.2	6.3
$[Ni(phen)_3]Cl_210H_2O$	855.0	862.0	7.0
$[Ni(en)_3]Cl_22H_2O$	854.6	860.9	6.3
$Ni(en)_2Cl_22H_2O$	854.2	860.2	6.0
$Ni(en)_2Br_22H_2O$	854.2	860.2	6.0

[a] From Matienzo et al.[11]
[b] Ni $2p_{3/2}$ main peak (MP) and its satellite (shake up, SP) binding energy values are expressed in eV.

bonds. Some typical shift values are given in Table 17.1, while binding energy values for Ni^{2+} cations engaged in different organometallic complexes are given in Table 17.2 in order to illustrate the foregoing statements. Modification of the chemical environment of a given element results in a spatial rearrangement of valence electrons and thus in a crystal field difference. If A and B are two compounds (e.g., Al and Al_2O_3) with one common element C (Al), the binding energy difference for this element is given by

$$E^C(A, B) = K_C(q^A - q^B) + V^A - V^B \qquad (17.6)$$

The first term in equation (17.6) describes the electron–electron interaction with q^A and q^B the valence charge of element C in A and B, and K_C the overlapping integral between core and valence electrons. This charge q may be treated in a classical way as a point charge:

$$K_C q = q/r_i \qquad (17.7)$$

with r_i being the average radius of the valence band orbital element C. The second term describes the Madelung potential

$$V = \sum_j (q_j/R_j) \qquad (17.8)$$

with R_j the distance between the element studied C and atom j bearing a charge q_j, with summation over all j atoms neighboring element C. Equation (17.8) is known as the potential model, and has been successfuly applied in interpreting chemical shifts in many compounds. The charges q_j are usually derived from the bond ionicity (electronegativity tables from Pauling or from Sanderson).

17.3.1.1b. Final State Effects. When an electron is ejected by the impact of a photon, there is electronic rearrangement within orbitals. A first-order approximation makes the assumption that the atom does not relax immediately, i.e., it is as if the orbitals were frozen (quenching), which is known as the sudden approximation that yields to the Koopmans theorem. This theorem states that if relaxation does not occur, the binding energy of a given orbital i equals $-E_i$, where E_i is the eigenvalue of the orbital i wave function. The difference between $-E_i$ and E^i is the relaxation energy E_R^i left by the ejected electron and corresponds to the relaxation phenomenon. This gives: $E_b^i = -E^i + E_R^i$ and, thus, $\Delta E_b^i = -\Delta E_i + \Delta E_R^i$, where ΔE_i is the contribution of the initial state as described above while E_R represents the energy difference between a neutral atom and the one considered; the values of E_b^i are often large (10 to 20 eV), but the variations in ΔE_R^i are small. Intra-atomic relaxation relative to an atom and extra-atomic relaxation relative to the neighbors action can be distinguished.

In a first-order approximation, valid in heterogeneous catalysis due to the inaccuracy in the experimental data (e.g., broad peaks and a charging effect), one may neglect the influence of intra- and extra-atomic relaxations, the former being independent of the oxidation state. One may then conclude that the chemical shift ΔE_b^i is really a characteristic of the element analyzed. It can even be predicted from an electrostatic model that the chemical shift varies as $1/r$,

i.e., it increases when considering the Periodic Table from left to right and from bottom to top (Z number decreasing).

However, relaxation phenomenon may not be neglected in all catalytic examples. For instance, intuitively it is reasonable to expect that extra-atomic relaxation is important for small clusters and, even more so, isolated atoms in zeolitic cavities or in small pores. This can result in nonnegligible positive chemical shifts, obviously not related to such chemical features as oxidation state or the nature of the bonding with ligands. These shifts have been observed for very tiny platinum particles or platinum atoms entrapped in Y-type zeolite.[12]

From these general considerations, it is worth keeping in mind the main conclusions: To a first approximation the binding energy value of an atom will be dependent on its positive charge. Further, measurement of the binding energy value gives qualitative information on the degree of covalent character of the binding of the atom with its nearby ligands.

Let us recall here that the charging effect, described above, may lead to an erroneous chemical shift. It is therefore important to avoid any charging by using a flood gun accessory and to find a standard level to refer to. The latter aspect is difficult. The C_{1s} line from pollution has sometimes been chosen as a reference since it is almost always present with a binding energy value equal to 284.5 eV. However the C_{1s} line is known to have quite a large chemical shift (several eV) depending on the nature of the species (e.g., graphite, hydrocarbons, or carbonates). It is therefore not a very reliable reference. Evaporation of small drops of gold is a more reliable technique with a peak $Au_{4f_{7/2}}$ at 83.8 eV. However this assumes that the gold drops have taken on the exact charge of the sample and do not experience extra-atomic relaxation because the particles are so tiny. If too large an amount of gold is evaporated, a thin film may be obtained that will take on the charge of the sample holder rather than that of the sample.

17.3.1.2. Multielectronic Effects

Up to now only monoelectronic transitions have been considered. However, many multielectronic processes occur as a result of the electron ejection. Relaxation has already been taken into consideration: other phenomena, such as plasmons, hole coupling, collective excitations as phonons, configuration interaction, shake-up, shake-off, also occur. This results in a broadening of the peaks (phonons), the occurrence of additional peaks (e.g., plasmons, shake-up, configuration interaction), or the presence of a broad background. Only two of these phenomena will be described below.

17.3.1.2a. Shake-Up Phenomenon.[9] This arises from the excitation of a valence electron toward a bound state of the higher-energy system, simultaneously with the photoionization process. This produces a new peak, designated a satellite or shake-up peak, toward lower kinetic energy values (i.e., higher apparent binding energy values) corresponding to the electron having lost part of its kinetic energy in this excitation valence electron process. The difference in energy between the principal peak and its satellite corresponds to the difference between the ground state of the atom and different excited states of the ion after photoemission. If the electron is excited not toward a bound state but toward the continuum, a wide range of energy values is observed, i.e., a broad band, which is designated as the "shake-off process."

For transition metal oxides, the shake-up peaks have been attributed to charge transfer from the ligand toward the metal. This corresponds to np(ligand) $\rightarrow 3d$ (metal) type transitions. Such shake-up lines are usually observed for paramagnetic transition metal ions such as Ni^{2+} and Cu^{2+}. It has also been shown[3] that for organometallic complexes of transition metal ions the energy split between the main and the satellite peaks and the ratio of their respective intensities decreases with an increasing covalent character of the metal–ligand bonds.

Although the actual interpretation of shake-up transitions is still under discussion, their presence is usually interpreted as being due to the presence of paramagnetic ions. This feature has been used to determine the oxidation state. For instance, in the case of cobalt cations, one has a large shake-up peak, 6 eV from the main peak, in CoO and a very small peak for Co_3O_4. This allows easy characterization of cobalt oxides, pure or supported, in catalysts.

17.3.1.2b. Configuration Interaction. The initial state corresponds to given orbital and spin momenta L and S. After photoemission the ion corresponds to momenta L' and S' such as: $L - 1 < L' < L + 1$ and $S' = S \pm \frac{1}{2}$ (dipolar transition rule).

For a closed-shell system, i.e., full orbital, one has $L = S = 0$ and therefore $L' = 1$ and $S = \frac{1}{2}$, i.e., only one final state. For an open shell, i.e., incompletely filled orbitals as in transition metal ions, at least one of the momenta is not nil, which results in several final states. Let us take Mn^{2+} as a example: Mn^{2+}: Ar $3s^23p^63d^5$. This cation has five unpaired electrons with ^6S initial state. The ejection of one electron leads to Mn^{3+} (Mn^{3+}: Ar $3s^13p^63d^5$) with six unpaired electrons. One has two states ^7S and ^5S depending on the spin orientation in the upper case, the $3s$ Mn peak with an energy splitting $E(^7S - {}^5S)$ which represents the exchange integral $G(s, d)$, between $3s$ and $3d$ electrons:

$$\Delta E(^7S - {}^5S) = [(2S + 1)/(2L + 1)]G(s, d) \qquad (17.9)$$

The intensity ratio is equal to the multiplicity ratio, i.e.,

$$I(^7S)/I(^5S) = 7/5 \qquad (17.10)$$

The experimental spectrum is in fact much more complex. For instance, MnF_2 exists in three final states and not one for the ^5S state due to configuration interaction. The barycenter of these three components corresponds to the value expected from Koopmans' approximation as seen in Figures 17.6, 17.7, and

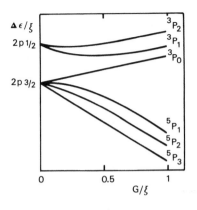

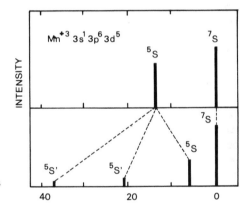

Figure 17.6. Splitting of $2p$ levels by $3d$ electrons as a function of the spin–orbit coupling constant and of the exchange integral $G(2p, 3d)$.

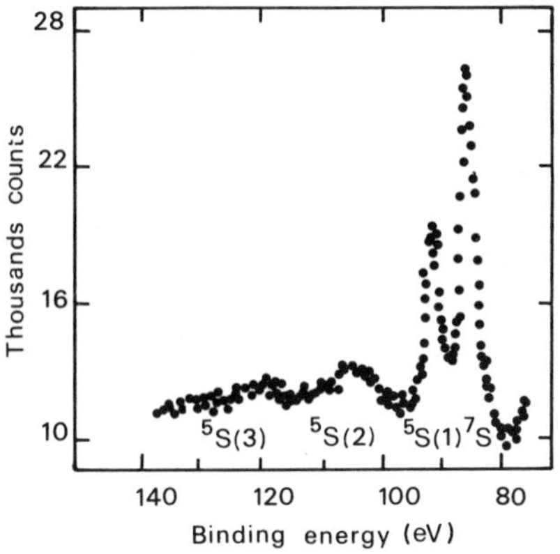

Figure 17.7. Configuration interaction effect on Mn_{3s} XPS spectrum of MnF_2.[10]

Figure 17.8. Splitting in three final states by configuration interaction.[10]

17.8. These different peaks correspond to different transitions and therefore they all have to be taken into account when performing the quantitative measurements described below.

17.3.2. Peak Intensity and Semiquantitative Features

This is certainly the most important parameter with wide applications in heterogeneous catalysis because of its surface sensitivity. The absorption of the emitted electrons by matter follows the well-known Beer–Lambert law, which may be written as

$$I/I_0 = \exp[-x/(\lambda \sin \theta)] \qquad (17.11)$$

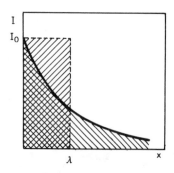

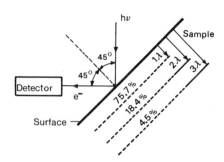

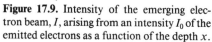

Figure 17.9. Intensity of the emerging electron beam, I, arising from an intensity I_0 of the
emitted electrons as a function of the depth x.

Figure 17.10. Part of the emitted electron intensity analyzed after absorption along different depths of 1, 2, and 3λ.

where λ is the electron mean free path, x the depth from the surface that the
emitted electrons are coming from, and θ the angle of emergence of the electrons
analyzed; I is the intensity of the emitted electron beam as analyzed while I_0
represents the beam intensity as emitted, i.e., before any absorption. Equation
(17.11) indicates that for an infinitely thick sample, the emerging beam intensity
equals that for a normal analysis ($\theta = \pi/2$):

$$I = \int_0^\infty I_0 \exp(x/\lambda)dx = \lambda I_0[\exp(-x/\lambda)]_0^\infty = \lambda I_0 \qquad (17.12)$$

It follows that this intensity is equivalent to the total emitted beam intensity
for a depth equal to λ, i.e., it may be considered that the total beam comes
only from a depth equal to λ without being absorbed. In fact equation (17.11)
shows that 75.7% of the electrons detected do come from the first λ depth,
18.4% from the second λ depth, etc., and almost the whole intensity analyzed
comes from the first 3λ depths. This is shown schematically in Figures 17.9
and 17.10.

The value of the mean free path λ is greatly dependent on the kinetic
energy of the emitted electrons. A universal curve giving λ values vs.
kinetic energy values has been proposed and widely accepted (Figure
17.11). It appears that in the average range of energies used in XPS the λ
values vary from about 0.5 up to 4 nm. This means that XPS is not really
a top-surface technique but rather analyzes several top layers, say to a depth
of 5 nm.

It is worthwhile noting that profiles of concentration may be obtained with
the XPS technique. If single crystals or films are studied, a profile may be
obtained by modifying the emergence angle, as the $1/(\lambda \sin \theta)$ term is involved,
i.e., for grazing emergence only the top surface layers will be analyzed. Moreover, etching of the surface of the sample by argon ion bombardment may also
give information about concentration profiles. However, etching may have
different efficiencies, depending on the elements. It is necessary to be aware of

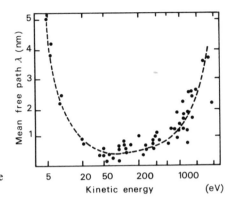

Figure 17.11. Mean-free path values of the electrons *vs.* their kinetic energy values.

this phenomenon and the usual way to allow for it is to take a known and related sample as a standard for calibration.

17.3.2.1. Quantitative Aspects[12]

Quantitative determination of the concentration of different elements in a material is important enough to justify further discussion. The number of photoelectrons detected that came from a depth x from the surface is given by

$$dI(\theta) = \phi \left(N \frac{A_0}{\sin \theta} \right) \left(\frac{d\sigma}{d\Omega} \Omega \right) T \exp - \left(\frac{x}{\lambda \sin \theta} \right) dx \qquad (17.13)$$

where ϕ is the flux of incident photons; N the concentration in atoms per cm³ of the element considered; $A_0/\sin \theta$ the projection of the entrance slit of the analyzer on the sample surface; θ the emergence angle of the analyzed beam with respect to the surface; σ the cross section of photoionization of orbital (n, l) electrons; Ω the solid angle of analysis; and T the transmission factor of the spectrometer, which depends on E_k in a way depending on the spectrometer (data given by the manufacturer).

For an infinite depth one has

$$I_\infty(\theta) = \int_0^\infty \phi N (A_0/\sin \theta)(d\sigma/d\Omega)\Omega T \exp - (x/\lambda \sin \theta)dx \qquad (17.14)$$

$$I_\infty(\theta) = A_0 N (d\sigma/d\Omega)\Omega T \lambda \qquad (17.15)$$

This relationship contains several unknown parameters: A_0, Ω, ϕ, and T. In general to overcome such a problem, the ratios of intensities are taken into consideration. For two elements A and B:

$$N_A/N_B = (I_A/I_B)(\sigma_B/\sigma_A)(T_B/T_A)(\lambda_B/\lambda_A) \qquad (17.16)$$

In equation (17.16) I_A and I_B are the areas of peaks A and B. The T_B/T_A ratio is usually unknown and taken equal to 1, which is valid if the XPS peaks for elements A and B are chosen to be close in kinetic energy. Cross sections σ have been calculated theoretically, especially by Scofield[13] and Wagner,[14] and tabulated. They appear to be sufficiently reliable for most applications. The probability $d\sigma/d\Omega$ of photoelectron ejection depends on the angle α between the incident X-ray beam and the analyzed electron beam and is given by

$$(d\sigma/d\Omega)(\alpha, h\nu) = (\sigma/4\pi)[1 + \tfrac{1}{2}\beta(\tfrac{3}{2}\sin^2\alpha - 1)] \qquad (17.17)$$

where β is an asymmetry parameter varying from -1 to $+2$ depending on the orbital involved. For s orbitals $\beta = 2$ and does not vary with the energy. For p electrons β varies with the energy. Photoionization of a p electron (initial state) may result in final states s and/or d (dipolar selection rules). The variation of β with kinetic energy reflects changes in interference between s and d waves with energy.[15]

The average free mean path of electrons depends both on their energy and on the material considered, as shown in Figure 17.11. Theoretical calculations based upon electron–electron interaction have been performed[16] for free electron materials. Data are given in Table 17.4. However, the model is valid for energies above 200 eV and even for elements which are neither transition metals nor noble metals. One writes

$$\lambda = E_k/a\{[\ln(E_k) - b]\} \qquad (17.18)$$

where a and b are constants depending on the electron concentration in the material, calculated by Penn.[16] In some cases the value of λ can be determined experimentally, but this is rather rare. In practice if two elements A and B have photoelectron peaks of closely similar energies, the same value is taken for λ. If not, then other relations between λ and E_k must be taken. The most widely used relation uses the right part of the universal curve in Figure 17.11, which resembles a parabola. One then writes:

$$\lambda \propto E_k^{1/2}$$

Equation (17.5) may then be written as

$$\frac{N_A}{N_B} = \frac{I_A}{I_B}\frac{\sigma_B(E_K^B)^{1/2}}{\sigma_A(E_K^A)^{1/2}}\frac{T_B}{T_A} \qquad (17.19)$$

This relationship is the one most commonly used in XPS for applications in catalysis with σ values taken from the Scofield[13] or Wagner[14] tables. The T_B/T_A ratio may either be equal to 1 or depend on the kinetic energy value, depending on the spectrometer, which may render equation (17.19) more complicated, e.g., with the ratio $(E_K^A/E_K^B)^{1/2}$ rather than its reverse. Seah and Denck[17] have proposed an empirical relationship stemming from experimental data:

$$\lambda = 2170\,E_k^{-2} + 0.72(dE_k)^{1/2} \quad \text{for monolayers}$$

or

$$\lambda = 2470(dE_k^{-2}) + 0.23(d^3 E_k)^{1/2} \quad \text{for bulk layers}$$

with d the depth of the layers analyzed, determined from $d^3 = 10^{24}/Nn$ and expressed in Å, while E_k is expressed in eV, where N is the molecular density and n the number of atoms in each compound molecule. This formula holds true for inorganic compounds.

17.3.2.2. Limitations of Quantitative Analysis[18]

One aspect in the quantitative determination of relative concentrations is often neglected although it is of great importance. As a matter of fact, contamination of the surface often occurs due to pump grease or even finger grease. In such a case if a well-spread film of contaminant does exist with a thickness d', the Beer–Lambert law should be applied to take account of the absorption of emitted electrons by the contamination layers, and one writes

$$I' = I \exp(-d'/\lambda \sin \theta) \qquad (17.20)$$

and a similar approach to that of the next section on supported catalysts may be taken.

It should also be noted that the whole quantitative determination described above assumes that the sample surface is planar. This does not hold true for solids catalysts. Factors such as rugosity may be very important as they can create shadowed areas for incident and/or emitted photons.[19]

All these factors (e.g., rugosity, contamination, and morphology of the grains) indicate that quantitative determination of relative concentrations of elements in the top layer cannot be very accurate. We estimate that a $\pm 10\%$ accuracy is the most reliable value that can reasonably be expected for solid catalysts.

17.3.2.3. Supported Catalysts[19-24]

As XPS is sensitive to the few first top layers of a solid material, it is reasonable to expect that the technique is particularly suitable to the study of supported catalysts. This holds true especially for e.g., oxides, sulfides, and carbides, deposited on an oxide support, as there is no reliable method, e.g., a chemical technique such as titration, to differentiate the supported phase from its support.

The way the supported phase lies on the surface is important for any theoretical calculation. It may consist, for instance, of well-spread filmlike layers or of more-or-less thick islands with different sizes and thickness distributions.

Let us suppose that the supported phase consists of uniform cubic-type islands of thickness d_p located between layers of thickness d_s of a layered support. Kerkhof et al.[19] have proposed the following relationship:

$$\left(\frac{I_p}{I_s}\right)_{\text{exp}} = \left(\frac{N_p}{N_s}\right)_b \frac{\sigma_p}{\sigma_s} \frac{1 + \exp(-d_s/2\lambda)}{1 - \exp(-d_s/2\lambda)} \frac{1 - \exp(-d_p/\lambda)}{d_p/\lambda} \tag{17.21}$$

where p, s, and b refer to the supported phase, the support, and the chemical (bulk) elemental composition, respectively. The electron mean free path is assumed to be the the the same for the supported phase and the support.

If the supported-phase crystallites are deposited on a semi-infinite support of thickness d_s equation (17.21) may be written as[20]

$$I_p/I_s = (N_p/N_s)(\sigma_p/\sigma_s)(d_s/2\lambda) \tag{17.22}$$

For comparison with the Kerkhof et al. relation [equation (17.21)] for a monolayer $(d_p < \lambda)$, one has

$$[1 - \exp(-d_p/\lambda)]/(d_p/\lambda) = 1 \tag{17.23}$$

Therefore the difference between equations (17.21) and (17.22) is equal to

$$[1 + \exp(-d_s/2\lambda)]/[1 - \exp(-d_s/2\lambda)]$$

For large surface area silica (e.g., $S = 350$ m^2 g^{-1}) as $d_s = 2/(\rho_s S)$, with ρ being the silica density, the correction factor is 3 with $d_s/2\lambda = 0.7$ for Si$_{2p}$ peaks. If $d_s/2\lambda = 1$ the correction factor is only 1.1.

Fung[23] has proposed a more general relationship expressing the shape of the crystallites through a function $F(d, \lambda_p)$

$$I_p/I_s = k[(W/\rho_s Sd)](\lambda_p C_p/\lambda_s C_s)F(d, \lambda_p) \tag{17.24}$$

where k is a constant that depends on the crystallite size distribution, W the weight of the supporting phase of density ρ_p, S the support surface not covered by the phase, and d the thickness of the supported phase. The function F is given by

$$F(d, \lambda_p) = \tfrac{3}{2}\{1 - (2\lambda_p^2/d^2)[1 - \exp(-d/\lambda_p)] + (2\lambda_p/d)\exp(-d/\lambda_p)\} \tag{17.25}$$

for spherical crystallites of diameter d, by

$$F(d, \lambda_p) = 3\{[1 - (8\lambda_p^2/d^2)][1 - \exp(-d/2\lambda_p)] + (4\lambda_p/d)\exp(-d/2\lambda_p)\} \tag{17.26}$$

for hemispherical crystallites of diameter d; and

$$F(d, \lambda_p) = 1 - \exp(-d/\lambda_p) \tag{17.27}$$

for cubic crystallites or planar deposits of thickness d.

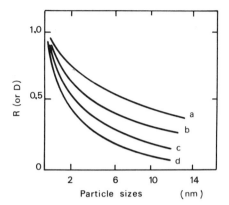

Figure 17.12. Variations of R ratio (or dispersion D) values *vs.* the size of crystallites for various shapes of crystallites. R is calculated on a monolayer basis. The crystallite shapes are: hemispherical (a), spherical (b), cubic (c), and monolayers dispersed (d) (from Ng and Hercules[24]).

The ratio R, defined by

$$R = (I_p/I_s)_{\text{cryst}}/(I_p/I_s)_{\text{monolayer}} \qquad (17.28)$$

is plotted on Figure 17.12 as a function of crystallite size.[24] It represents the intensity ratio of the supported phase to the support assuming that the supported phase lies as a monolayer on the support. It is a representation of the state of dispersion.

The above models are valid if the crystallite size distribution is narrow, i.e., if factor k is constant. This can be checked by plotting the variations of I_p/I_s *vs.* the amount of the dispersed phase. A straight line should be obtained. This was, for instance, observed for $WO_3/\alpha\text{-}Al_2O_3$[24] and $Ni\text{-}W/Al_2O_3$[25] catalysts after reduction and sulfiding.

17.4. APPLICATIONS IN CATALYSIS

Catalysis is by definition a surface phenomenon, so it is surface-sensitive. Any surface characterization technique, despite any limitations, is of much interest in catalysis. XPS turns out to be the most widely used surface technique because it identifies almost all elements and, although operating under high vacuum, it is relatively easy to use and interpret, particularly for quantitative determinations.

17.4.1. Surface Enrichment in Heterogeneous Catalysis[25-29]

This is an important aspect because most of the physical techniques, except when dealing with probe adsorption on surfaces, give mainly bulk analyses. It is, however, important to known whether, during any treatment such as calcination of a precursor giving the catalyst, activation procedure, and even catalytic reactions, some of the elements constituting the catalyst are migrating toward or even away from the surface. Moreover, even if the chemical composition and structural atomic arrangements in the bulk have been characterized, for instance, by chemical and X-ray diffraction analyses, respectively, the actual surface chemical composition and atomic arrangement (surface crystallography) usually remain unknown.

17.4.1.1. Mixed and Complex Oxides

Complex oxides based on bismuth molybdates are excellent partial oxidation catalysts for olefins and are known as multicomponent catalysts. It is important to determine the role of each component in catalytic properties. There are three well-known pure bismuth molybdates, all of which are good catalysts, namely the α-(Bi_2O_3:$3MoO_3$), β-(Bi_2O_3:MoO_3), and γ-(Bi_2O_3:MoO_3) phases. The relative intensity ratio of $Bi4f_{7/2}$ and $Mo3d_{5/2}$ peaks is reported in Figure 17.13 as a function of the treatment. It is observed that reducing treatment (b) results in molybdenum surface enrichment, particular for the α-phase. The authors[28] assigned this result to the deformation structure with MoO_3 migration and evaporation of bismuth species from the surface. Under catalytic reaction conditions the Bi/Mo intensity ratios are close to the bulk ones. This is an important result because it has been suggested[29] that due to their close catalytic properties the three bismuth molybdates phases (α, β, and γ) could have a similar top layer surface composition and structure. This is ruled out by XPS measurements even if the technique is sensitive to several top layers rather than to the uppermost only. This characterization was extended to multicomponent catalysts to find any peculiar surface composition due to a cherry-like particle structure, which could be responsible for unexpected catalytic behavior.

Another important conclusion was drawn from XPS data.[30] It is known that mixtures of several phases are more efficient catalysts than any single

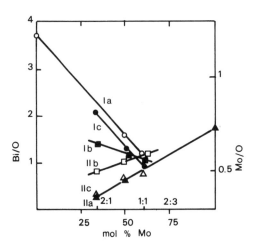

Figure 17.13. Relative intensities of XPS peaks of $Bi_{4f_{7/2}}$ (I) and $Mo_{3d_{5/2}}$ (II) with respect to O_{1s} peak in bismuth molybdates (taken from Grzybowska[28]): (a) starting samples, (b) samples outgassed at 470 °C for 10 h, and (c) samples taken after catalytic reaction at 440 °C with C_3H_6:O_2:$N_2 = 24$:21:55.

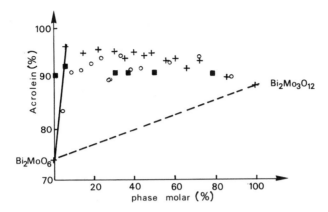

Figure 17.14. Selectivity in acrolein formation in a propene partial oxidation reaction at 380 °C as a function of the amount of α phase added to "pure" phase (from Jamal *et al.*[30]). The symbols refer to mechanical mixtures and calcination of $\alpha + \gamma$ phases: (+) impregnation and calcination of α phase by ammonium molybdate, (■) to α- and γ-MoO_3 mechanical mixtures, and (O) calcination, respectively. Calcination temperature of 480 °C.

component. This phenomenon is known as a synergy effect. It was observed that "pure" α phase was a less efficient catalyst than a mixture of $\gamma + \alpha$ phases whatever the amount of α phase, as shown in Figure 17.14. The results were interpreted by noting that the "pure" γ phase was actually rich in bismuth, i.e., contained a Bi_2O_2 top layer, known to yield total oxidation for propene oxidation. This layer could react with MoO_3 or with α phase (richer in molybdenum than the γ phase) according to the reactions below and give a much better catalyst:

$$MoO_3 + Bi_2O_3 \rightarrow Bi_2MoO_6$$

$$Bi_2O_3 + Bi_2MoO_3O_{12} \rightarrow 2Bi_2MoO_6$$

In other words the synergy effect was actually a kind of cleaning of the bismuth-rich surface corresponding to bulk γ phase composition, which had much better selectivity. In a more recent work[69] on multicomponent catalysts (α-$Bi_2Mo_3O_{12}$ on mixed $Co(Fe)MoO_4$), XPS was the determining technique which demonstrated that during oxidation of propene to acrolein the bismuth molybdate compound spreads over the mixed $Co(Fe)MoO_4$ support, which then constitutes the actual catalyst.

17.4.1.2. Zeolite and Molecular Sieve Catalysts

Zeolitic materials are widely used in catalysis. New developments in zeolite catalysis have arisen from framework substitution of aluminum by other elements such as boron, iron, vanadium titanium, gallium, chromium, and zinc, while a new molecular sieve type family of (AlO_2, PO_2)-based structures was found.

In all these zeolitic-type materials it is important to determine whether the additional element is actually substituting for silicon atoms in the framework and if this element is in the external or the inner layers of the zeolite crystal. The second aspect is of prime importance for shape-selective acid reactions: acidic sites must be located in the inner rather than in the external layers, as an unselective isomerization reaction on the surface may partly cancel out the shape selectivity properties of the interior, e.g., in toluene methylation reactions.

Using thin sections of the sample and the EDX-STEM analysis described in Chapter 19, it was possible to determine whether the additional element was regularly distributed across a crystallite.[31] The XPS technique was also applied to compare the concentration in the additional element with that of the bulk chemical composition. If the surface is enriched in the additional element, this indicates that at least part of it is not incorporated in a framework position. If not, XPS alone cannot give the answer as the additional element may well be in a cationic inner site location, in small encaged oxide particles or in framework positions. Other technique(s) such as X-ray diffraction, EDX-STEM, MAS-NMR, ESR, or IR then have to be used. Surface aluminum concentration was deter- mined successfully by XPS for Y-type[32] or ZSM-5[31,33,34] samples.

When such a zeolitic material is impregnated with a salt and then calcined at 550 °C, it is also important to determine whether the impregnated compound remains on the surface of the crystallite or enters the channel pore system to neutralize inner acidic sites. The XPS technique may then give the answer. For instance, a phosphorus[35] or boron[36] compound has been impregnated on Al-ZSM-5 sample. XPS clearly showed that both compounds had entered the zeolite channels while IR spectroscopy indicated that acid sites (acid OH groups with a band near 3605 cm^{-1}) were partly neutralized. In catalysis the shape selectivity of such impregnated samples was observed to increase. This was attributed to a narrowing of the channels by the added compound in a kind of inner pore coating effect. Similarly, after cationic exchange it was of interest to determine whether the top layer sites were exchanged first or if a more homogeneous exchange had occurred. We studied Rh^{3+} exchanged zeolites,[37] and observed: (1) that homogeneous exchange did occur for large pore zeolites such as Y-type or mordenite, (2) that surface layers were favored with respect to the bulk for ZSM-34, ZSM-5, and ZSM-11 zeolites (medium pore size) as well as for small pore zeolites such as zeolite A, and (3) that Rh^{3+} ions were deposited only on the surface for ZK-5 zeolite (small pore size).

17.4.1.3. Multicomponent Catalysts

These correspond to many industrial catalysts where various additional elements are introduced to basic catalysts. One question that obviously arises is: are such additional elements well distributed all across a particle or are they located more in the top layers of the crystallites or even do they form different phases in different crystallites? The XPS technique is obviously well suited for such a characterization. For example, in Figure 17.15 the changes in surface composition in different elements due to catalytic reaction are given for an

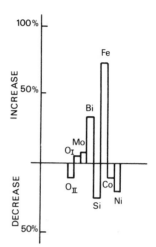

Figure 17.15. Changes in atomic percent of the top layer concentration in different elements measured by XPS for a multicomponent bismuth molybdate-based catalyst for an ammoxidation reaction (from Grzybowska *et al*.[38]).

TABLE 17.3. Effect of Calcination Temperature on the Surface Layer Composition of Exhaust Catalysts as Determined by XPS[39]

Calcination temperature (°C)	Atomic ratios		
	Cu/Al	Cr/Al	Ca/Al
600	29	10	0.8
1000	26	8	2.0

industrial bismuth molybdate-based multicomponent catalyst used for propene ammoxidation.[38]

Another example may be found in an exhaust catalyst $CuO–Cr_2O_3/Al_2O_3$[39] stabilized by calcium oxide. The XPS data are given in Table 17.3 for the catalyst after calcination at 600 or 1000 °C. It is clear that calcium was segregated toward the surface at the highest temperature, which is related to the enhanced stability of the catalyst.

17.4.1.4. SbSnO and SnMoO Systems

Sb–Sn–O mixed oxide catalysts are known to exhibit interesting properties for partial oxidation of propene to acrolein, particularly when the samples are calcined at high temperatures, as shown in Figure 17.16.[40–46]

It was observed that calcination results in extensive sintering of the crystallites from 5–10 nm to 80–150 nm as shown by electron microscopy.[46]

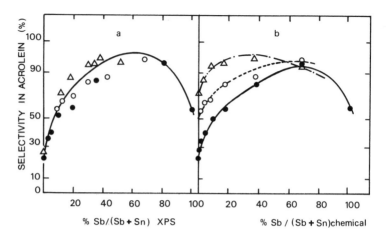

Figure 17.16. Variations of the selectivity in acrolein in propene oxidation reaction at 400 °C as a function of antimony content and calcination temperatures: 500 °C (●), 750 °C (O), and 950 °C (△). In Part a, the experimental points are plotted *vs.* chemical composition determined by XPS and in Part b *vs.* chemical composition determined by chemical analysis.

Some big Sb_2O_4 particles ($\phi \geq 1$ μm) were also detected by IR and X-ray and electron diffractions studies. The presence of Sb^{5+} in solid solution in SnO_2 was detected by electrical conductivity[43] and Mossbaüer[44,45] measurements, while XPS showed that calcined samples at high temperatures have their top layers enriched in antimony.[40,41,46] The whole study[40–46] using a wide variety of physical techniques allowed us to identify the actual catalyst as SnO_2 doped with few Sb^{5+} ions and exhibiting surface antimony enrichment, presumably as an antimony oxide (e.g., Sb_6O_{13} or Sb_2O_4) in a thin film or as islands. Big crystallites of Sb_2O_4 as clearly shown by X-ray and electron diffractions were observed to be rather inactive due to their small surface area. This shows how the XPS technique was important in gaining a better understanding of the actual catalyst composed of an antimony-rich surface (presumably a thin antimony oxide layer) on SnO_2 slightly doped by Sb^{5+} cations. The migration of one element toward the surface upon calcination was often demonstrated by XPS, which obviously modified our thinking concerning the nature of active sites. For instance, such a migration was observed for V/TiO_2 (e.g., 0.5 to 2 V at.%)[47], for Mo/TiO_2 (0.5 to 2 Mo at.%)[48]; Fe_2O_3–Sb_2O_3.[49] In contrast, other techniques such as ESR and UV-visible spectroscopies showed that the foreign elements were incorporated into the lattice of TiO_2.

It is often observed that catalytic properties greatly depend on the preparation procedure. For instance, a TiO_2/Al_2O_3 catalyst with Ti/Al = 0.74 was prepared using either an ammonia solution or urea added to the mixed $Ti(SO_4)_2 + Al_2(SO_4)$ solution. It was observed[50] that the samples containing NH_3 exhibit similar bulk and surface compositions while when urea was used more aluminum was detected in the top layers. On SnO_2–MoO_3 catalysts, Okamoto et al.[51] have shown using the XPS technique that for molybdenum contents less than 75% one has molybdenum-enriched surface samples while for higher contents the surface is tin-enriched. From catalytic studies in oxidative dehydrogenation and dehydration of alcohols reactions the authors have identified two kinds of catalysts depending on whether the molybdenum content was below or above 50%. By XPS analysis using an attached glove box to avoid any contact with air after catalytic reactions took place, the authors have found that in the first group molybdenum ions are in a $+5$ oxidation state while they are in $+4$ for the second group. It was observed by X-ray diffraction that crystallinity of SnO_2 increased with increasing molybdenum content while the crystallite size decreased from 15 to 4 nm. The $Mo{=}O$ IR bands were of low intensity except at high molybdenum content, as if MoO_3 was amorphous or Mo^{5+} cations were dissolved in SnO_2.[52] In the latter case the charge equilibrium was held by electrons delocalized in the conduction band as Mo (V), Sn (III), or Sn (II) cations were not found by XPS, ESR, or Mössbauer spectroscopies. This shows that intimate combinations of SnO_2 and MoO_3 modify lattice oxygen anions and thus the reducibility of molybdenum cations and so catalytic properties.

17.4.2. Active Phase-Support Interactions

This is an important aspect of heterogeneous catalysis since a support is often used for several reasons, namely to:

1. Enhance the surface area of the active phase by its dispersion on the support.
2. Favor mechanical and thermal properties such as attrition resistance and heat transfer in exothermal reactions.
3. Induce bifunctional-type catalytic reaction properties, the support being then active with, e.g., acidic or basic sites.
4. Modify the catalytic properties of the active phase.

17.4.2.1. Hydrodesulfuration (HDS) and Hydrodenitration (HDN) Catalysts

For CoMo- and NiMo-based catalysts used in these reactions both the chemical shift in binding energy values and quantitative analysis have been used in investigations. For MoO_3 deposited on SiO_2 an increase in binding energy value of Mo_{3d} and O_{1s} was observed[53] with increasing molybdenum loading but when cobalt was also present no increase was observed whatever the molybdenum content. Moreover, Mo/Si and Co/Si ratios were observed to remain constant up to 12% of the supported CoMo phase as if a compound with SiO_2 was formed first. The ratio increased sharply above 12%. This overall result was much weakened when cobalt was present with formation of the $CoMoO_4$ phase. If such samples were reduced the mixed $CoMoO_4$ phase was formed because of the weakened MoO_3–SiO_2 interaction rather than a two-dimensional CoMo phase precursor of the active phase. With γ-Al_2O_3 as a support the MoO_3-support interaction was stronger and a two-dimensional layer was formed.[54] However, this interpretation has led to much controversy. Okamoto et al.[55] have proposed that the difference in binding energy value was due instead to a differential charging effect between MoO_3 crystallites and SiO_2 both of which are insulators. To support their conclusion the authors dissolved excess MoO_3 crystallites in water for 10 days and observed that the binding energy value of molybdenum was still enhanced. I am not completely convinced by this argument since it is quite possible that a mixed MoO_3–SiO_2 composition was formed[53,54] and exhibited shifted Mo_{3d} peaks. This discussion shows how difficult the interpretation of chemical shifts in XPS in heterogeneous catalysis can be and that caution is always necessary.

When reducing a $CoMoO_4$ catalyst in a sulfiding atmosphere, the oxidation states of cobalt, molybdenum, and sulfur can be identified easily because all three elements show large chemical shifts (several eV) with oxidation state changes.[56,57] Cobalt atoms are sulfided by an H_2S/H_2 mixture resulting in Co^0 and Co^{2+} identified by their binding energy values and the absence and presence of shake-up peaks, respectively. Sulfate ions are also very easily identified. Moreover the variations in the intensity ratios for S_{2p}, Co_{2p}, Mo_{3d}, and O_{1s} can also be followed by XPS.[58]

When pure oxides as MoO_3, CoO, NiO are present on an Al_2O_3 support they cannot be differentiated by their XPS parameters from mixed oxides such as $NiAl_2O_4$.[59,60] Other physical techniques have to be used, e.g., XRD, Raman, or IR spectroscopies. Mo/Al, Co/Al, and Ni/Al intensity ratios were determined, and it was observed that the presence of molybdenum favors the dispersion of nickel or cobalt on the support. These ratios varied linearly with molybdenum, cobalt, and nickel contents up to the formation of a monolayer.[61] Note also that in some cases the difference between Td and Oh environments of Ni^{2+} or Co^{2+} cations could be established from their binding energy values.[59,62]

Finally from the large amount of work dealing with NiMo-based catalysts, despite some controversy, it may be concluded that the XPS technique was particularly useful, even unique, in showing that the presence of two components (Ni and Mo or Co and Mo) results in a better dispersion of the supported phase. The latter is present in a structure which depends on the interaction of MoO_3 with the support and is catalytically efficient when it is in a two-dimensional layer form.

17.4.2.2. Complexed Oxide Catalysts

The way Co^{2+} and Ni^{2+} cations are held in tetrahedral or octahedral environments depends on the oxide that is formed. By studying the main peak (MP) to the satellite peak (SP) splitting of $2p$ lines and their intensity ratios it was possible[63,64] to determine their environmental symmetry and then to

TABLE 17.4. Binding Energy Values of $Ni_{2p_{3/2}}$ Peaks as a Function of the Support[a]

Support	S $m^2 g^{-1}$	Ni wt.%	Reduction at 500 °C (%)	E_b (eV ± 0.1)
NiO	—	100	100	854.0 855.7
TiO$_2$	58	13.8	97	854.5
SiO$_2$	324	20.0	90	856.3
SiO$_2$–Al$_2$O$_3$ (87–13%)	600	12.9	92.5	856.6
γ-Al$_2$O$_3$	261	14.0	83	856.6
MgO	46	12.9	47	856.7

[a] The reducibility of the samples at 500 °C is given for comparison (from Vedrine *et al.*[63]).

confirm these findings with other techniques. It was shown[63] for nickel on different supports (TiO$_2$, SiO$_2$, SiO$_2$–Al$_2$O$_3$, MgO) with about 20 wt.% Ni that, although the XRD technique showed only the presence of NiO crystallites, the binding energy values of nickel $2p$ peaks were significantly dependent on the support (Table 17.4).

The changes in XPS parameters (E_b values, ΔE_{MP-SP}, MP:SP intensity ratio) were interpreted[64] as changes in the Mott insulator properties of NiO due to the interaction with the support (the $3d$ band near the Fermi level was broadened due to the unpaired electron delocalization toward oxygen ligand ions). It is worthwhile noting here that the samples were exhibiting rather different reducibility properties, which were presumably due to active phase-support interaction.

17.4.2.3. Mono- and Bimetallic Catalysts

It is known that when a metal such as platinum is well dispersed as very small particles ($\phi \approx 1$ nm) on supports such as silica, alumina, and zeolites, its catalytic properties are modified with respect to those of unsupported large crystallites of platinum. It therefore appears that some platinum–support interaction does occur. An XPS study of platinum or iridium on alumina[65] and on Y-type zeolite[12] gave a positive chemical shift (+ 0.7 eV for zeolite) of $4f$ peaks. This shift could be due to changes in the electronic state of platinum (a kind of electron transfer from the metal to the support) or changes in the relaxation energy due to the small size of the particles ($\phi \approx 1.0$ to 2.5 nm). However, such a shift was not observed for the SiO$_2$ support for the same platinum particle size. Moreover, an ESR study of electron donor and electron acceptor properties via the formation of charge transfer complexes has shown that electron donor properties of the zeolite support were sharply enhanced (× 10 to 20) with respect to the support alone. A shift of the threshold X-ray absorption (white line) was also observed supporting the above conclusion. When molybdenum was added to platinum to create a bimetallic system, there was no such shift.[66]

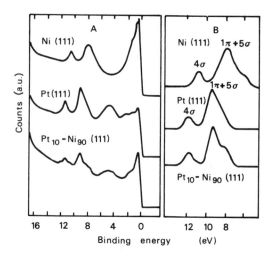

Figure 17.17. UPS spectra of 4σ, 1π, and 5σ molecular orbital bands for CO adsorbed on $Pt_{10}Ni_{90}$ (111) single crystals and on pure (111) monometallic single crystals (from Jugnet *et al.*[67]).

17.4.3. Bimetallic Catalysts

When a bimetallic catalyst is synthesized it is very important to find out if there is an alloy or two juxtaposed metallic phases, or even an elemental segregation of one of the two elements in the surface layers. Quantitative analysis of XPS peaks for both elements is necessary.

Studies on Pt_xNi_{1-x} (111) single crystals have been carried out using both XPS and UPS techniques. By XPS core levels such as Pt_{4f} and Ni_{2p} or Ni_{3p} peaks have been analyzed. It was observed[67] that a large platinum enrichment exists in the surface layers. For $x = 0.1$ the XPS surface layers actually contained 40% Pt and 60% Ni. For $x = 0.5$ only platinum was detected on the surface. These results were supported by a UPS study of CO adsorbed on the same materials (Figure 17.17). It was observed that the molecular orbitals of CO adsorbed on platinum and nickel kept their individuality, even in the alloy, which allowed us to calculate the separate contributions of CO adsorbed on platinum and nickel atoms.

The results were further supported by LEED, AES, and ISS data. It is worth noting that if the crystals are oxidized, nickel atoms migrate toward the surface and return to the bulk upon reduction. A limit to the use of XPS in this way should be noted. If the particles are too small ($\phi \leq 10$ to 5 nm) the technique will not be able to detect any segregation as the electron mean free path is then not negligible with respect to the particle size.

17.4.4. Organometallic Complexes[11]

Previous sections have described how the XPS technique may be very useful in determining the cation oxidation state, the number of ligands, and the extent of covalent binding between the cation and its ligands. If such complexes are deposited on a support, for instance to try to heterogenize homogeneous processes, the XPS technique could be used with benefit not only to determine the number of remaining ligands but also to find any change in the charge borne by the cations due to interaction with the support. One has to be aware of the impossibility of characterizing organic ligands as cation contamination always occurs, and C_{1s} peak electrons exhibit a very small cross section; however, the XPS technique may be used in addition to other techniques such as NMR or IR spectroscopies. Note also that it is very difficult, even impossible, to characterize complexes sensitive to air or oxygen-containing gas mixtures.

17.4.5. Adsorbed Species

The XPS technique requires a high vacuum and is thus not really suitable for the study of adsorbed species. However, when adsorption is strong enough the technique may be applied. For instance the adsorption of pyridine on acidic zeolites has been studied.[68] The presence of Brönsted (H^+) and Lewis (Al) sites was confirmed as a chemical shift for the N_{1s} peak greater than 2 eV was observed for pyridine adsorbed on both sites. The advantage of XPS over IR spectroscopy in this application is that the former can be quantitative.

Another application was a study of CO adsorption on Fischer–Tropsch types of catalysts, where it was important to determine whether the reaction occurs via CO dissociation or not. It was thus necessary to determine the C_{1s} binding energy precisely as it varies greatly among CO, CO_2 or C species. The disadvantage of XPS is, as noted earlier, related to the presence of carbon from contamination. Note also that Auger electron spectroscopy is very useful for characterizing carbonaceous species while UPS allows one to characterize molecular orbitals and thus to detect any dissociation of CO molecules.

17.5. UPS AND APPLICATIONS

In this technique derived from XPS, the photon source is UV light arising from a synchrotron source (see Chapter 18) of variable energy and monochromatized beam or from a helium lamp, He_I and He_{II} photons at 21.2 and 40.8 eV; only valence band electrons will be ejected under such a beam. The technique is thus very suitable for valence band analyses, such as the density of states near the Fermi level, and for molecular orbital characterization of adsorbed molecules. For instance, the 4σ, 1π, and 5σ molecular orbitals of CO are detectable and their energies with respect to those of the gaseous CO phase are a function of chemical bonding to the surface. If dissociation of CO has occurred these orbitals are obviously not detectable.

17.6. CONCLUSIONS

In this chapter we have presented a general overview of the photoelectron-spectroscopy technique, and have shown that three main observations are possible from the analyses:

1. The concentrations of the different elements of a solid material within its few top layers (1 to 5 nm in depth). This is particularly useful if bulk (chemical) and surface concentrations are different owing to elemental segregation.
2. The oxidation states of the elements and covalency character of metal–ligand bonds.
3. The number of ligands associated with a given element (carbon excepted).

The technique is therefore particularly suitable for the study of heterogeneous catalysts. However, one has to be aware of the limitations, namely that (a) a high vacuum is needed, which corresponds to conditions rather far from catalytic conditions!: (b) the X-ray beam may modify the material, particularly in reducing certain elements; (c) the charging effect for insulators may result in misleading chemical shifts; and (d) surface layers of crystallites are analyzed, not the inner part of pores or cavities present in many catalysts, such as zeolites or porous materials.

The actual spectrometers are rather expensive 3.5 M FF or 0.5 M. ECUS in 1993 and many accessories may be added for complementary studies, particularly ISS, SIMS, AES, or LEED. Some efforts have been devoted to decreasing the size of the zone analyzed, and at present it is measured in the μm^2 zone instead of a few mm^2 as it was several years ago. This allows one to determine mapping of chemical composition.

REFERENCES

1. J. C. Vedrine, *Microsc, et Spectrosc. Electron* **8**, 129 (1979). C. R. Brundle and A. D. Baker, *Electron Spectroscopy: Techniques and Applications,* Academic, London, Vol. 1 (1977), Vol. 2 (1978), Vol. 3 (1979), and Vol 4 (1980); J. C. Vedrine, *Surface Properties and Catalysis by Nonmetals* (J. B. Bonnelle *et al.,* eds.) Reidel, Amsterdam (1983), p. 159.
2. K. Siegbahn *et al., ESCA,* Almquist and Wisksell, Stockholm (1979).
3. L. Hedin and G. Johansson, *J. Phys.* **B2**, 1336 (1969).
4. D. A. Shirley, *Chem. Phys. Lett.* **16**, 220 (1973).
5. J. C. Slater, *Quantum Theory of Atomic Structure* 2, McGraw-Hill, New York (1960), p. 287.
6. D. W. Davis and D. A. Shirley, *Chem. Phys. Lett.* **15**, 185 (1972). L. Ley, S. P. Kowalczyk, F. R. McFeeley, F. R. Pollak, and D. A. Shirley, *Phys. Rev.* **B8**, 2392 (1973).
7. T. Robert, *Chem. Phys.* **8**, 123 (1975).
8. S. P. Kowalczyk, L. Ley, R. A. Pollack, F. R. Feely, and F. R. Shirley, *Phys. Rev.* **B7**, 4009.
9. P. S. Bagus, A. J. Freeman, and F. Sasaki, *Phys. Rev. Lett.* **30**, 850 (1973).
10. S. T. Manson, *J. Electron Spectrosc.* **1**, 413 (1972, 1973).
11. L. J. Matienzo, L. I. Yin, S. O. Grim, and W. E. Swartz, *Inorg. Chem.* **12**, 2762 (1973).

12. J. C. Vedrine, M. Dufaux, M. Naccache, and B. Imelik, *J. Chem. Soc. Faraday Trans. I* **74**, 440 (1978).
13. J. H. Scofield, *J. Electron Spectros,* **8**, 129 (1976).
14. C. D. Wagner, *Anal. Chem.* **49**, 1282 (1977).
15. R. L. Reilman, A. Useane, and S. Y. Manson, *J. Electron Spectrosc.* **8**, 398 (1976).
16. D. R. Penn, *J. Electron Spectrosc.* **8**, 29 (1976).
17. M. P. Seah and W. A. Denck, *Surf. Interface Anal.* **1**, 2 (1979).
18. C. S. Fadley, R. J. Baird, W. Sickhaus, T. Novakov, and S. A. L. Bertstrom, *J. Electron Spectrosc.* **4**, 93 (1974).
19. F. P. F. Kerkhof and J. A. Mouljin, *J. Phys. Chem.* **83**, 1612 (1979).
20. P. J. Angevine, J. C. Vartuli, and W. N. Deglass, *Proc. VIth Inter. Congr. on Catalysis, Vol. 2* (G. C. Bond and P. B. Tompkins, eds.), The Chemical Society, London (1977), p. 611.
21. R. B. Shalvoy and P. L. Reucroft, *J. Electron Spectrosc.* **12**, 351 (1977).
22. C. Defosse, P. Canesson, P. G. Rouxhet, and B. Delmon, *J. Catal.* **51**, 269 (1978).
23. S. C. Fung, *J. Catal.* **58**, 454 (1979).
24. T. K. Ng and D. M. Hercules, *J. Phys. Chem.* **80**, 2094 (1976).
25. P. G. Menon and T. S. R. Rao Prasada, *Catal. Rev.* **20**, 97 (1979).
26. P. H. Emmett and S. Brunauer, *J. Am. Chem. Soc.* **59**, 310 (1937).
27. S. Brunauer and P. H. Emmett, *J. Am. Chem. Soc.* **88**, 1732 (1940).
28. B. Grzybowska, J. Haber, W. Marczewaski, and L. Ungier, *J. Catal.* **42**, 327 (1976).
29. A. G. Michell, P. M. Lyne, K. F. Scott, and J. Phillips, *J. Chem. Soc., Faraday Trans. I* **77**, 2417 (1981).
30. M. El Jamal, M. Forissier, G. Coudurier, and J. C. Vedrine, *Proc. 9th Intern. Congr. on Catalysis 4* (M. J. Philipps and M. Ternan, eds.), The Chemical Institute of Canada, Ottawa (1988), p. 1617.
31. A. Auroux, M. Dexpert, C. Leclerq, and J. C. Vedrine, *Appl. Catal.* **6**, 95 (1983). G. Coudurier and J. C. Vedrine, *Pure Appl. Chem.* **58**, 1389 (1986).
32. J. F. Tempere, D. Delafosse and J. P. Contour, *Molecular Sieves II, Vol. 40* (J. Katzer, ed.) American Chemical Society, Washington (1977), p. 16.
33. E. G. Derouane, S. Detremmerie, Z. Gabelica, and N. Blom, *Appl. Catal.* **6**, 95 (1983).
34. A. E. Hugues, K. G. Wilshier, B. A. Sexton, and P. Smart, *J. Catal.* **80**, 221 (1983).
35. J. C. Vedrine, A. Auroux, P. Dejaifve, V. Ducarme, H. Hoser, and S. B. Shou, *J. Catal.* **73**, 147 (1982).
36. M. B. Sayed and J. C. Vedrine, *J. Catal.* **101**, 43 (1986). M. B. Sayed, A. Auroux, and J. C. Vedrine, *Appl. Catal.* **23**, 49 (1986).
37. J. C. Vedrine, *Zeolite Chemistry and Catalysis* (P. A. Jacobs *et al.,* eds.), Elsevier, Amsterdam (1991), p. 25.
38. T. B. R. Rao Prasada and P. G. Menon, *J. Catal.* **51**, 64 (1978).
39. J. S. Brinen, *J. Electron Spectrosc.* **51**, 377 (1974).
40. Y. Boudeville, F. Figueras, M. Forissier, J. L. Portefaix, and J. C. Vedrine, *J. Catal.* **58**, 52 (1979).
41. J. C. Vedrine, *Analysis* **9**, 199 (1981). D. R. Pyke and Y. M. Cross, *J. Catal.* **58**, 61 (1979).
42. J. C. Volta, B. Benaichouba, I. Mutin, and J. C. Vedrine, *Appl. Catal.* **8**, 215 (1983).
43. J. M. Herrmann and J. L. Portefaix, *React Kinet. Catal. Lett.* **12**, 51 (1979). B. Benaichouba and J. M. Herrmann, *React Kinet. Catal. Lett.* **22**, 209 (1983).
44. J. L. Portefaix, P. Bussière, F. Figueras, M. Forissier, J. M. Friedt, F. Theobald, and J. P. Sanchez, *J. Chem. Soc., Faraday Trans. I* **76**, 1652 (1980).
45. J. C. Volta, G. Coudurier, I. Mutin, and J. C. Vedrine, *J. Chem. Soc. Chem. Commun.,* 1044 (1982).
46. J. C. Volta, P. Bussière, G. Coudurier, J. M. Herrmann, and J. C. Vedrine, *Appl. Catal.* **16**, 315 (1985).
47. P. Mériaudeau and J. C. Vedrine, *Nouv. J. Chim.* **2**, 133 (1978).

48. J. C. Vedrine, H. Praliaud, P. Mériaudeau, and M. Che, *Surf. Sci.* **8**, 101 (1979).
49. I. Aso, T. Amamoto, N. Yamazoe, and T. Seiyama, *Chem. Soc. Japan, Chem. Lett.* 365 (1980).
50. E. Rodenas, H. Hattori, and I. Toyoshima, *Chem. Soc. Japan. Chem. Lett.* (1980).
51. Y. Okamoto, T. Yashimoto, T. Imanaka, and S. Teranishi, *Chem. Lett.* 1035 (1978). Y. Okamoto, K. Oh-Hiraki, T. Imanaka, and S. Teranishi, *J. Catal.* **71**, 99 (1981).
52. Ph. de Montgolfier, P. Mériaudeau, Y. Boudeville, and M. Che, *Phys. Rev.* **B14**, 1788 (1976).
53. P. Gajardo, C. Defosse, and B. Delmon, *J. Catal.* **63**, 201 (1980).
54. T. A. Patterson, J. C. Carver, D. E. Leyden, and D. M. Hercules, *J. Phys. Chem.* **80**, 1700 (1976).
55. Y. Okamoto, T. Imanaka, and S. Teranishi, *J. Phys. Chem.* **85**, 3798 (1981).
56. J. S. Brienen and W. D. Armstrong, *J. Catal.* **54**, 57 (1978).
57. M. Breysse, B. A. Bennett, and D. Chadwick, *J. Catal.* **71**, 430 (1981).
58. B. A. Bennett, D. Chadwick, A. R. Jawahery, M. Breysse, and M. Vrinat, *Proc. Symposium on Structure and Activity of Sulfided Hydroprocessing Catalysts.* ACS Meeting, Kansas City, September (1982).
59. P. Dufresne, E. Payen, J. Grimblot, and J. P. Bonnelle, *J. Phys. Chem.* **85**, 2344 (1981).
60. H. Knozinger, H. Jeriorowski, and E. Taglauer, *Proceed. 7th Int. Congr. on Catalysis* (T. Seiyama and K. Tanabe, eds.) Elsevier, Amsterdam (1981), p. 604.
61. J. Bachelier, J. C. Duchet, and D. Cornet, *Bull. Soc. Chim. Fr.* (1978), p. 112.
62. E. Payen, J. Barbillat, J. Grimblot, and J. P. Bonnelle, *Spectrosc. Lett.* **11**, 997 (1978).
63. J. C. Vedrine, G. Hollinger, and Tran Minh Duc, *J. Phys. Chem.* **82**, 1515 (1978).
64. A. D'Huysser, B. Lerebourg-Hannoyer, M. Lenglet, and J. P. Bonnelle, *J. Sol. Stat. Chem.* **39**, 246 (1981).
65. J. Escard, B. Pontvianne, and J. P. Contour, *J. Electron Spectrosc.* **6**, 17 (1975).
66. T. M. Tri, J. P. Candy, P. Gallezot, J. Massardier, M. Primet, J. C. Vedrine, and B. Imelik, *J. Catal.* **79**, 396 (1983).
67. Y. Jugnet, J. C. Bertolini, J. Masardier, B. Tardy, Tran Minh Duc, and J. C. Vedrine, *Surf. Sci. Lett.* **107**, L320 (1981).
68. C. Defosse, B. Delmon, P. Canesson, and P. G. Rouxet, *J. Catal.* **51**, 268 (1978).
69. H. Ponceblanc, J. M. M. Millet, G. Coudurier, J. M. Herrmann, and J. C. Vedrine, *J. Catal.* **142**, 381 (1993).

PHOTOELECTRON DIFFRACTION AND SURFACE CRYSTALLOGRAPHY

Y. Jugnet

18.1. INTRODUCTION

This chapter is a short introduction to the technique of photoelectron diffraction, and the reader can refer to general articles[1–4] for more information. Here, we are dealing with a newly developed technique which until now has been applied mainly to simple systems. No real catalytic application has yet been investigated, but as a surface technique, its future development is expected to give new insights to catalysis.

In Chapter 17, we were mostly concerned with angle integrated photoemission experiments. However, depending on the value of the analyzer acceptance angle $\Delta\Omega$, we can proceed to an integrated experiment ($\Delta\Omega \geq 5\text{–}10°$) or to an angle resolved experiment ($\Delta\Omega \leq 5\text{–}10°$). In the latter case, not only is the magnitude of the photoelectron wave vector k known (as was the case for integrated experiments), but also the direction, which gives a new dimension to the photoemission technique.

As with integrated photoemission, both core levels and valence levels can be measured. In order to study the surface crystallography of monocrystalline samples with or without adsorbates, angle resolved photoemission exploits the full benefit of this newly defined parameter (photoelectron emission direction). Schematically, in an integrated experiment, core level measurements lead to the well-known ESCA (Electron Spectroscopy for Chemical Analysis) method, which describes the chemical environment of an atom through the chemical shift,

Y. Jugnet ● Institut de Physique Nucléaire de Lyon, CNRS and Université Claude Bernard, Villeurbanne, France. Present address: Institut de Recherches sur la Catalyse CNRS, F. 69626, Villeurbanne, France.

Catalyst Characterization: *Physical Techniques for Solid Materials*, edited by Boris Imelik and Jacques C. Vedrine, Plenum Press, New York (1994).

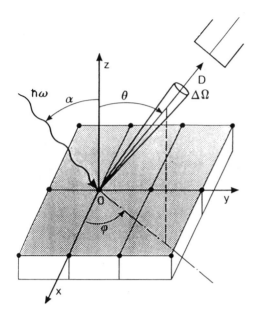

Figure 18.1. Schematic representation of an angular photoemission experiment.

while the valence level measurement leads to the DOS (Density of Occupied States). In an angle resolved experiment the measurement of the valence level energy and its dispersion as a function of the photoelectron wave vector leads to the band structure determination of any monocrystalline sample. Looking now to the core level peak intensity instead of its energy, we can probe the surface crystallography through the photoelectron diffraction process, which is the aspect we shall describe now.

18.2. ANGULAR PHOTOEMISSION GEOMETRY

In an angle resolved experiment, we define α, the light incidence angle relative to the normal to the surface OZ (Figure 18.1). The emitted photoelectrons are analyzed along a direction defined through the polar angle Θ relative to OZ and the azimuthal angle Φ, which corresponds to the angle between the projection of the analysis direction in the surface plane and the OX axis.

The components K_x, K_y, and K_z of the **K** wave vector are thus defined as

$$K_x = \sqrt{(2mE/\hbar^2)}\sin\Theta\cos\Phi$$

$$K_y = \sqrt{(2mE/\hbar^2)}\sin\Theta\sin\Phi$$

$$K_z = \sqrt{(2mE/\hbar^2)}\cos\Theta \qquad (18.1)$$

with E corresponding to the photoelectron kinetic energy.

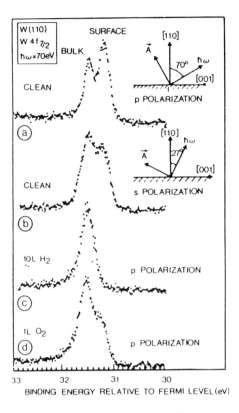

Figure 18.2. W(110) $4f_{7/2}$ core level photoelectron spectra[5]: (a) clean surface (p polarization), (b) clean surface (s polarization), (c) 10 L H_2 on clean surface (p polarization), (d) 1 L H_2 on clean surface (p polarization). (1 L or 1 Langmuir, defines an exposure to 10^{-6} Torr for 1 s.)

18.3. SURFACE SENSITIVITY

In photoemission, the value $3\lambda \cos \Theta$, where λ is the electron mean free path and Θ the polar angle, is generally agreed to be an estimate of the analyzed depth. As the mean free path λ varies with the kinetic energy E_{kin} roughly as $\sqrt{E_{kin}}$, the low kinetic energy range, i.e., $30 \text{ eV} \leq E_{kin} \leq 200 \text{ eV}$ appears to be a good choice for surface enhancement. By use of synchrotron radiation with tunable photon energy, it is possible to choose the appropriate photon energy leading to the lowest photoelectron kinetic energy and thus the highest surface sensitivity. Such an experiment is described in Figure 18.2,[5] which shows different $4f_{7/2}$ core level photoelectron spectra of clean W(110) using two different polarizations of the light (Figure 18.2a and b) and adsorbate covered W(110) (Figure 18.2c and d). This experiment has been done for a photon energy of 70 eV and detection along the normal to the surface. On the clean surface, two structures appear clearly at binding energies of 31.2 and 31.5 eV;

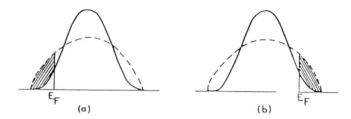

Figure 18.3. Scheme of surface and bulk density of states. Hatched areas display (a) the lack or (b) the excess of electrons.[7]

the one at the lower energy being enhanced when using p polarized light (A the electric field vector of the light is almost perpendicular to the analyzed surface). Hydrogen or oxygen adsorption on the W(110) surface leads to a strong decrease of the low binding energy component intensity. The behavior of this low binding energy structure (high intensity with p polarized light, high sensitivity to chemisorption) allows us to identify it as a surface (first atomic layer) specific component, the other structure being representative of the bulk.

In this example, W(110), photoelectrons coming from the surface top layer have a binding energy which is lower than the one coming from the bulk. This is not true for tantalum, for instance, where the opposite is observed.[6] Indeed, this variant behavior can be understood in the following way[7]: owing to the broken bonds in the surface plane, the density of d states DOS (in the case of $3d$ transition metals) is less in the surface as compared to the bulk (resulting in a narrower DOS band). This phenomenon is more pronounced for lower coordination numbers (Figure 18.3). Depending on the Fermi level location, the surface band will be more-or-less populated as compared to the bulk. For a filling of less than five electrons, the surface band will show a lack of electrons (Figure 18.3a); on the other hand, for a filling of more than five electrons, the bulk band will display a lack of electrons (Figure 18.3b). To preserve the charge neutrality, the surface band has to be shifted toward higher binding energy values for a d-band filling of less than five electrons, or toward lower binding energy for a d-band filling greater than five electrons. Calculations have shown that the core levels display a similar shift although it is slightly larger.[7] Such a surface core level shift measurable on narrow core levels with a photon source with high-energy resolution and a tunable energy (synchrotron radiation) is very interesting since we now have two probes, one for the surface top layer and the other characteristic of the bulk.

18.4. PHOTOELECTRON DIFFRACTION

Core level photoelectrons are emitted from localized and incoherent atomic sources (Figure 18.4). The detected intensity can be pictured as the coherent sum of the direct photoelectron wave (I_0) propagating from the emitting atom toward the detector, and all of the other waves (I_1 and I_2) emitted in other directions and scattered by the atomic sites surrounding the emitter toward the detector. The interference between these different coherent waves leads to the photoelectron diffraction.

Experimentally, a photoelectron diffraction experiment consists of measuring a core level peak intensity as a function of the **K** wave vector. Changes of the photoelectron **K** vector value can be achieved in two ways: either by varying the photoelectron kinetic energy E in a fixed detection geometry or by varying the detection geometry (Θ, Φ) at a fixed kinetic energy. In the first case, the experiment evidently requires synchrotron radiation as a photon source with tunable photon energy. In that case the photoionization cross section is varying as well as the analyzed depth (through the electron mean free path). In the second case, the experiment can be run using any angle resolved photoemission apparatus, (1) by varying the polar angle Θ (each Θ value implying its own analyzed depth), or (2) by varying the azimuthal angle Φ, which from a theoretical point of view is much simpler as cross section and analyzed depth are constant for a given curve.

The photoemission process includes three steps: the photoionization process (photon absorption and photoelectron emission), the transfer of the emitted electron to the surface (damping plus scattering events), and lastly the escape through the sample surface into the vacuum. All of these steps will contribute to the measured peak intensity, which can be schematically expressed as a summation over all the intermediate states:

$$I(E, \Theta, \Phi)$$

$$= \sum_{\substack{\text{intermediate} \\ \text{states}}} \left| \left\langle z \left| \begin{matrix} \text{surface} \\ \text{escape} \end{matrix} \right| y \right\rangle \langle y | \text{diffraction} | x \rangle \langle x | \text{photoion.} | n, l \rangle \right|^2 \quad (18.2)$$

The photon absorption results in a nonisotropic electron emission, which is a function of incident light characteristics (photon energy, degree of polarization, and incidence angle), as well as photoemitted electron characteristics (binding energy and quantum numbers). This anisotropy is depicted by the β

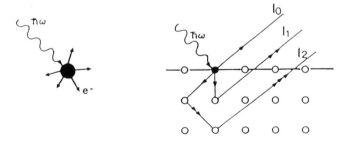

Figure 18.4. Schematic representation of photoelectron diffraction effects. The detected photocurrent appears as the interference between the directly emitted wave, I_0, and the waves emitted after scattering on one atom, I_1, or several atoms, I_2.

asymmetry parameter, which takes into account tion, and incidence angle), as well as photoemitted electron characteristics (binding energy and quantum numbers). This anisotropy is depicted by the β interference between the two outgoing $(l + 1)$ and $(l - 1)$ partial waves allowed by the dipolar selection rules applied to the l initial state. This asymmetry parameter is directly connected to the differential photoionization cross section $d\sigma/d\omega$:

$$d\sigma/d\omega = (\sigma/4\pi)\{[1 - \beta(E)/4](3\cos^2\Omega - 1)\} \qquad (18.3)$$

for unpolarized light, Ω being the angle between the incident light and the photoejection direction, and

$$d\sigma/d\omega = \sigma/4\pi\{[1 + \beta(E)/4](3\cos^2\Omega - 1)\} \qquad (18.4)$$

for polarized light, Ω now being the angle between the polarization vector $\mathbf{A}$ and the photoejection direction. The β parameter shows rapid changes near the threshold but then varies slowly to reach its asymptotic value.

The escape through the surface is commonly treated as a simple refraction effect. The photoelectron external kinetic energy E_{ext} decreases from its value inside the crystal E_{int} by an amount equal to the inner potential V_0 ($V_0 = 10$–15 eV):

$$E_{ext} = E_{int} - V_0 \qquad (18.5)$$

The momentum conservation laws ($k_{int} = k_{ext}$) give rise to the refraction relation

$$\sin\Theta_{ext} = [(E + V_0)/E]^{-1/2}\sin\Theta_{int} \qquad (18.6)$$

where E is still the measured kinetic energy, Θ_{ext} and Θ_{int} are, respectively, the polar angles before and after the electron escapes through the surface. The smallest refraction effects are found at the highest kinetic energy.

The last step describes the scattering and diffraction processes. Depending on the photoelectron kinetic energy, two scattering kinetic energy regimes can be distinguished. Although no fixed value can be established as the boundary between the regimes, we shall put it at 200 eV, for instance, in the case of $3d$ transition metals. For kinetic energies greater than or equal to 200 eV, photoelectrons will have scattered once before detection — this is the kinematical approach. For lower kinetic energies, the electrons might have been scattered several times before crossing the surface and being detected; this is the multiple scattering well known from LEED experiments. Furthermore, forward scattering strongly dominates the backscattering in the photoelectron diffraction process at high kinetic energy, and this simplifies the interpretation of the data considerably. In contrast to the low kinetic energy regime, where diffraction patterns are more sensitive to physical parameters such as electron energy and crystal potential, high kinetic energy diffraction patterns reflect the crystal surface geometry more directly.

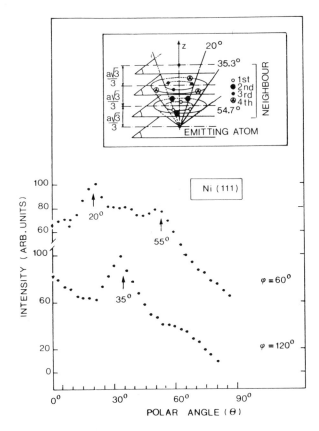

Figure 18.5. Experimental polar intensity distribution for $Ni_{2p_{3/2}}$ from a clean Ni(111) single crystal at $\Phi = 60°$ and $\Phi = 120°$. The fcc(111) internuclear axes are shown in the insert.[3]

18.5. APPLICATIONS

Most photoelectron diffraction work* during the last decade has been done on carefully chosen surfaces, e.g., clean crystals, overlayer covered surfaces, epitaxial growth, and adsorbate molecules in order to determine their orientation), in the high kinetic energy range. This can be related to the fact that it is easier to handle the data when high (≥ 200 eV) kinetic energy photoelectrons are analyzed. Scattering then takes place along the forward direction, and as a result leads to an intensity enhancement (constructive interference) along internuclear axes. This is illustrated in Figure 18.5, for the case of the Ni(111) face of pure nickel, where we report the $Ni_{2p_{3/2}}$ intensity as a function of the polar angle Θ for two different azimuthal angle $\Phi = 60°$ and $\Phi = 120°$.[8,9] In the case of amorphous nickel, these curves would have been similar whatever the angle,

* The state of the art can be found in references in Grenet *et al.*[3]

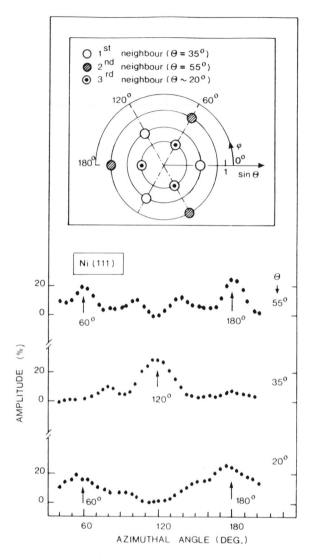

Figure 18.6. Experimental azimuthal percent intensity distribution for $Ni_{2p_{3/2}}$ from a clean Ni(111) single crystal at $\Theta = 20$, 35, and 55°. The fcc(111) projective view is shown in the insert.[3]

and with a $\cos \Omega$ shape describing the polar cross-section variation. The observed sharp modulations in the case of a monocrystalline Ni(111) sample are attributed to diffraction effects. The observed peaks appear at $\Theta = 20°$, 35°, and 55°, depending on the azimuthal angle value, and correspond to the internuclear axes between the emitting atom and the third, first, and second neighbors, respectively. The azimuthal plots (see Figure 18.6) measured at $\Theta = 20°$, 35°, and 55° (i.e., along the internuclear directions) give a direct indication of the

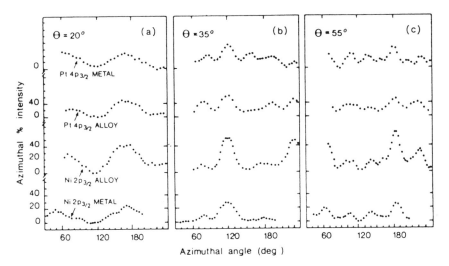

Figure 18.7. Experimental azimuthal percent intensity distribution for $Pt_{4p_{3/2}}$ of pure platinum and alloy and for $Ni_{2p_{3/2}}$ of pure nickel and alloy: (a) at $\Theta = 20°$, (b) at $\Theta = 35°$, and (c) at $\Theta = 55°$.[9]

symmetry in a plane that is defined by the chosen Θ angle. At $\Theta = 20°$, two peaks appear at $\Phi = 60°$ and $120°$ showing a threefold symmetry. When measuring the same curve at $\Theta = 35°$, one main peak is observed at $\Phi = 120°$. Looking now along $\Theta = 55°$, we again find two main peaks at $\Phi = 60°$ and $120°$. All these peaks indicate a threefold symmetry with second and third neighbors rotated by $60°$ as compared to first neighbors.

This direct real-space probe has led different research groups to study the CO molecule orientation on several transition metals[10-12] but little work has been done with other molecules. Exceptions are the recent work on the structure of formate and methoxy species adsorbed on copper surfaces,[13-15] and on the C_2H_2, C_2H_4 orientation on platinum.[15] Indeed, as previously noted, this technique has not yet been really applied in catalysis. One good reason is that the photoelectron diffraction technique is evidently strictly limited to mono-crystalline samples; a second reason is found in the fact that light elements, which constitute most of the organic compounds used in catalysis (e.g., C, N, O, S, . . .) are not good candidates for photoelectron diffraction studies owing to their low scattering efficiency. However, future use of synchrotron radiation, with its high brilliance and the possibility of choosing the energy range, may lead to new investigations of low-Z elements.

As a last example, we shall see how the rate of segregation in an alloy can be investigated through the magnitude of azimuthal anisotropy. Figure 18.7 displays the intensity of $Ni_{2p_{3/2}}$ and Pt_{4p} core levels measured on monocrystalline $Pt_{50}Ni_{50}(111)$ and on pure $Pt(111)$ and $Ni(111)$, as a function of the azimuthal angle for the three previously described polar angles, $\Theta = 20°$, $35°$, and $55°$

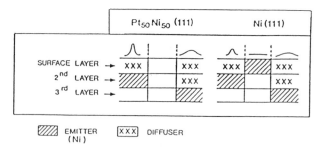

Figure 18.8. Scheme of the emission diffraction processes for the case of $Pt_{50}Ni_{50}(111)$ and Ni(111).

characteristic of fcc(111) faces. A comparison between pure metals and alloys indicates that the Ni_{2p} diffracts more strongly (amplitude 50–60%) in the alloy than in the pure metal (about 30%), that the Pt_{4p} diffracts more strongly (about 40%) in the pure metal than in the alloy (about 30%) and that whatever the polar angle, $Ni_{2p_{3/2}}$ plots display much stronger diffraction than Pt_{4p} plots as far as the alloy is concerned. These results are in good agreement with a platinum segregation at the top layer as illustrated in Figure 18.8. Here, we show a scheme of the different diffraction processes expected in the case of pure metals (Pt and Ni) as well as the alloy, depending on the photoionized (or emitter) atom.

We restrict our discussion to the contribution from the first three layers of the crystal. Let us consider first, the pure metals. The emitter can be located in any one of the three layers. When the emitter is in the first plane, it cannot scatter on any atom on top of it in a model of forward scattering. This emitter, if located in the second plane, will have one layer on top of it where electrons will scatter. This induces a sharp diffraction peak. Lastly, if the emitter is in the third plane, the electrons emitted will scatter first in the second layer, then in the first surface layer. These successive scatterings will be averaged and contribute to a smaller diffraction effect. For the alloy, the layer-by-layer composition is assumed to be 100% Pt in the first layer, 100% Ni in the second layer, and 50% Pt and 50% Ni in the third one.[16] The fact that the strongest diffraction amplitude is seen when the emitter is in the second layer explains why nickel diffracts strongly whereas the diffraction effects are quite small when the emitter is a platinum atom.

18.6. CONCLUSIONS

The study of directional effects appears as a logical evolution in the development of the ESCA technique. Photoelectron diffraction adds a new dimension to the ESCA technique as it combines sample crystallographic properties with the physico chemical identification obtained by conventional ESCA. It appears as a real-space scanning of the sample especially at high kinetic energies. By going to the highest kinetic energies we get the most direct structural information. At low kinetic energies (less than about 200 eV), the electronic structure becomes more important compared with the electron energy, and the surface (top layer) crystallographic information is more difficult to extract although the surface sensitivity (mean free path variation as a function of kinetic energy) is higher. Thus, depending on the kinetic energy range, photoelectron diffraction is related to low-energy electron diffraction in the case of low kinetic energy or to extended X-ray Absorption Fine Structure spectroscopy in the case of high kinetic energy, as far as local order is concerned. Although still in its early days, this technique appears to be a promising tool for a range of surface studies including adsorbate molecule orientation with respect to the substrate top layer, segregation investigation, top layer characterization, epitaxial growth, and surface relaxation or reconstruction.

REFERENCES

1. C. S. Fadley, *Progress in Surface Science, Vol. 16* (S. G. Davison, ed.), Pergamon, Oxford (1984), p. 275.
2. M. Sarguton, E. L. Bullock, and C. S. Fadley, *Surf. Sci.* **182**, 287 (1987).
3. G. Grenet, Y. Jugnet, S. Holmberg, H. C. Poon, and Tran Minh Duc, *Surf. Interface Anal.* **14**, 367 (1989).
4. J. J. Barton, C. C. Bahr, S. W. Robey, Z. Hussain, E. Umbach, and D. A. Shirley, *Phys. Rev.* **B34**, 3807 (1986).
5. Tran Minh Duc, C. Guillot, Y. Lassailly, J. Lecante, Y. Jugnet, and J. C. Vedrine, *Phys. Rev. Lett.* **43**, 789 (1979).
6. Y. Jugnet, G. Grenet and Tran Minh Duc, *Handbook on Synchrotron Radiation, Vol. 2,* (G. V. Marr, ed.), North Holland, Amsterdam (1987).
7. M. C. Desjonquères, D. Spanjaard, Y. Lassailly, and C. Guillot, *Solid State Comm.* **34**, 807 (1980).
8. Y. Jugnet, G. Grenet, N. S. Prakash, Tran Minh Duc, and H. C. Poon, *Surf. Sci.* **189/190**, 649 (1987).
9. Y. Jugnet, G. Grenet, N. S. Prakash, Tran Minh Duc, and H. C. Poon. *Phys. Rev.* **B38**, 5281 (1988).
10. D. A. Wesner, G. Pirug, F. P. Coenen, and H. P. Bonzel, *Surf. Sci.* **178**, 608 (1986).
11. D. A. Wesner, F. P. Coenen, and H. P. Bonzel, *Phys. Rev.* **B33**, 8837 (1986).
12. K. A. Thompson and C. S. Fadley, *J. Electron Spectrosc. Rel. Phenon.* **33**, 29 (1984).
13. D. P. Woodruff, C. F. McConville, A. L. D. Kilcoyne, Th. Lindner, J. Somers, M. Surman, G. Paolucci, and A. M. Bradshaw, *Surf. Sci.* **201**, 228 (1988).
14. Th. Lindner, J. Somers, A. M. Bradshaw, A. L. D. Kilcoyne, and D. P. Woodruff, *Surf. Sci.* **203**, 333 (1988).
15. D. A. Wesner, F. P. Coenen, and H. P. Bonzel, *J. Vac. Sci. Technol.* **A5**, 927 (1987).
16. Y. Gauthier, Y. Joly, R. Baudoing, and J. Rundgren, *Phys. Rev.* **B31**, 6216 (1985).

CHARACTERIZATION OF CATALYSTS BY CONVENTIONAL AND ANALYTICAL ELECTRON MICROSCOPY

P. Gallezot and C. Leclercq

19.1. INTRODUCTION

Heterogeneous catalysts usually consist of highly divided solid phases that are closely interconnected and thus difficult to characterize. Conventional transmission electron microscopy (CTEM) offers the unique advantages of allowing the direct observation of catalyst morphology with a resolution tunable in the range 10^{-4}–10^{-10} m and of obtaining structural information by lattice imaging and microdiffraction techniques. Moreover, scanning transmission electron microscopes (STEM) equipped with X-ray analyzers can be used to determine the local composition of catalysts with a spatial resolution as good as 1 nm in the case of field emission gun STEM. This is why electron microscopy is now in widespread use for catalyst characterization.

There are many textbooks on electron microscopy.[1-8] A few review articles have been devoted to electron microscopy as applied to catalyst characterization,[9-11] but the recent, rapid development of high-resolution analytical microscopy has not been properly covered. The present chapter is intended to give only a brief account of CTEM instrumentation and principles, but application to catalyst morphology and structure characterization will be developed. Emphasis will be laid on STEM and its associated analytical capabilities: energy

P. Gallezot and C. Leclercq ● Institute de Recherches sur la Catalyse, CNRS, Villeurbanne, France.

Catalyst Characterization: Physical Techniques for Solid Materials, edited by Boris Imelik and Jacques C. Vedrine, Plenum Press, New York (1994).

dispersive X-ray emission spectroscopy (EDX), electron energy loss spectroscopy (EELS), and diffraction on nanodomains (nanodiffraction). More specifically, dedicated STEM, working only in scanning mode, will be considered.

19.2. CONVENTIONAL AND ANALYTICAL MICROSCOPES

19.2.1. Electron Interactions with Atoms

19.2.1.1. Electron-Induced Signals

Electrons have an associated wavelength given by $\lambda[\text{Å}] = (150/V)^{1/2}$, where V is the accelerating potential in volts. Thus, while they can be used in diffraction experiments, they have much shorter wavelengths than X-rays (for 100 kV, $\lambda = 0.387$ Å) and they interact more strongly with atoms. Thus, the mean free path for 100-kV electrons is only a few thousand angstroms. These interactions lead to various kinds of signals that can be used for analytical purposes.

Figure 19.1 gives the most common signals produced when an electron beam interacts with a thin specimen. There are three types of transmitted electrons, namely: (1) unscattered electrons, (2) elastically scattered (or diffracted electrons, and (3) inelastically scattered electrons. Potentially, they can

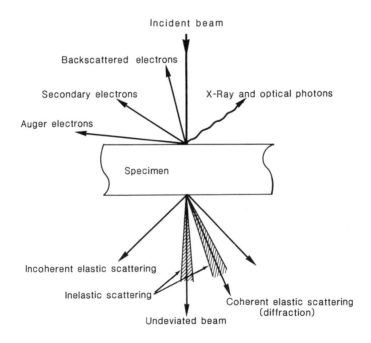

Figure 19.1. Interactions between electrons and the specimen.

be used to make an image of the specimen, the most commonly used being Type 1 (bright-field image) and type-2 (dark-field image). Inelastic electrons can be used in a STEM equipped with an electron spectrometer to make energy-filtered images and electron spectra.

There are also three types of electronic signals emitted from the surface of the specimen exposed to the incident beam: (1) backscattered electrons which have an energy close to that of the incident electrons, (2) secondary electrons of low energy (50 eV) which have undergone random multienergy loss processes, and (3) Auger electrons produced by the de-excitation of atoms. All these signals can be used for imaging purposes. Signals 1 and 2 are used in the scanning electron microscope (SEM) and Signal 3 is used in the scanning Auger microscope to make energy-filtered images, i.e., images of the solid phases containing a given element. The de-excitation of atoms also produces a photon within an emission domain ranging from the X-ray to the visible. X-ray emission spectroscopy is the most useful tool for the qualitative and quantitative analysis of the elements present in the volume scanned by the electron beam.

19.2.1.2. Elastically Scattered Electrons

Electrons are scattered coherently (diffracted) by the lattice of solid phases. Figure 19.2 gives the Ewalds construction showing that a diffracted beam is

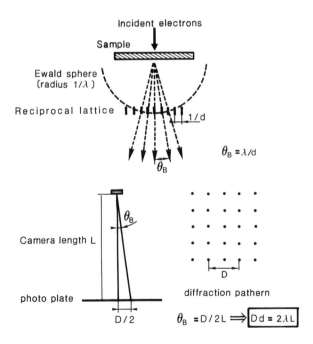

Figure 19.2. Principles of electron diffraction: relation between interatomic distances in the direct and in the reciprocal lattice.

generated when a node $h^*k^*l^*$ of the reciprocal lattice is located on the surface of the Ewald sphere of radius $1/\lambda$. Because of the very short wavelength associated with the electrons, the radius is very large with respect to the reciprocal lattice parameter. Therefore the diffraction pattern can be assimilated to a section of the reciprocal lattice by a plane perpendicular to the incident beam. In the case of a collection of very small crystallites oriented at random, all the lattice modes are statistically present on the Ewald sphere. This leads to a Bragg pattern of concentric rings which are more or less broken depending upon the number of crystallites reflecting the incident beam. Amorphous solids or very small metal clusters give broad scattering bands.[12] The kinematical theory of diffraction discussed so far is valid for X-ray diffraction and is a good approximation for electron diffraction on thin crystals. However, as the thickness increases, the diffracted beam can be diffracted again and the resulting multiple scattering is described by the dynamical theory of diffraction.

19.2.1.3. Inelastically Scattered Electrons

Electrons can lose energy by interacting either with loosely bound electrons (conduction or valence band electrons) or with core electrons $(K, L, M, \ldots)$. These interactions lead to low-energy-loss peaks (plasmons) and high-energy-loss peaks corresponding to the characteristic internal transitions in atoms. This is the basis of electron energy loss spectroscopy (EELS) discussed in Section 19.2.4.3. The de-excitation of atoms occurs by emission of either X-ray photons or Auger electrons (Figure 19.3). This is the basis for X-ray emission spectroscopy which will be examined in Section 19.2.4.2.

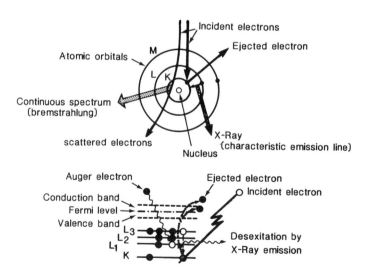

Figure 19.3. Interactions between electrons and atoms: X-ray emission.

19.2.2. The High-Resolution Transmission Microscope

19.2.2.1. Principles and Resolution

The formation of images in a conventional TEM is very similar to that in an optical microscope (Figure 19.4). The TEM column indicates: (1) the electron gun, which is either a tungsten filament heated at 2500 °C or a LaB_6 cathode heated at 1600 °C giving an intensity ten times higher. The acceleration voltage is commonly 100–200 kV but higher voltages are available on commercial microscopes. (2) The system of condenser lenses gives a demagnified image of the source on the specimen. (3) The objective lens gives a magnified image (typically $\times$ 100) of the specimen on the image plane. (4) A system of projection lenses gives 10^3–10^4 magnification of the intermediate image on a fluorescent screen. The total magnification is between 10^4–10^6.

The specimen should be thin enough (e.g., less than 100 nm for a 100-kV beam) to transmit the electron beam. The limit of resolution in the absence of lens aberration is $0.61\lambda/\alpha$, where α is the aperture angle of the beam determined by the objective aperture. Lens astigmatism can be corrected by additional coils. The most limiting aberration is spherical aberration, which causes the electron beam to focus in different image planes for different aperture angles α. The aberration is proportional to $C_s\alpha^3$, where C_s is the spherical aberration coeffi-

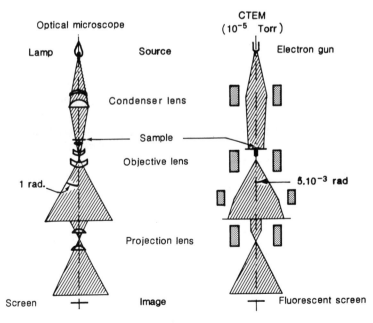

Figure 19.4. Comparison between the optical microscope and the conventional transmission electron microscope.

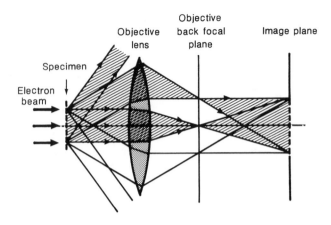

Figure 19.5. Formation of diffraction pattern and image by the objective lens.

cient, and the resolution limit is given by $C_s\alpha^3 + 0.61\lambda/\alpha$. For an optimal aperture angle, this sum is at the minimum and the resolution limit is $\tau_{min} = C_s^{1/4}\lambda^{3/4}$. With $C_s = 1$ mm, $\lambda = 0.037$ Å (100 kV), $\alpha = 5.5$ mrad, and $\tau_{min} = 4.7$ Å. This value is too conservative; taking into account the radial distribution of intensity, τ_{min} is typically between 2 and 3 Å.

19.2.2.2. Fundamentals of Imaging and Contrast

Figure 19.5 gives the principles of image formation with the objective lens. As the specimen is under parallel illumination, the unscattered electrons are focused on the lens axis in the back focal plane of the lens. Scattered electrons are also brought into focus in the back focal plane; therefore the complete diffraction pattern is formed in this plane. The image of the diffraction pattern on the fluorescent screen can be obtained by focusing the projection lenses on the back focal plane. After passing through the back focal plane, the scattered and unscattered waves combine to form an image in the image plane. The image can be produced on the screen by focusing the projection lenses on the image plane. It is interesting to note that two successive Fourier transforms take place in the electron microscope: object (real space) → diffraction pattern (reciprocal space) → image (real space).

Little of the contrast of images in the transmission electron microscope is due to absorption contrast as in optical microscopy but rather it depends on the intensity scattered by the specimen and on the fraction of that intensity collected through the objective aperture in the back focal plane. Let us first take the example of an amorphous particle which is a weak electron scatterer. If all the scattering waves are admitted through the objective aperture there will be no contrast. Figure 19.6a shows that image contrast can be obtained by introducing a small objective aperture (or contrast aperture) which eliminates most of the scattered waves but keeps the direct beam and the small-angle scattering for image formation. The particle appears darker than the background; this is the so-called bright-field image. The larger the scattering, the higher the contrast. If the particle is a small crystallite, the contrast will not be very different from that of an amorphous particle if the atomic planes are out of a Bragg position. However, if the crystallite gives a diffracted beam and if the diffracted intensity is stopped by the objective aperture as shown in Figure 19.6b, the image of the particle will be highly contrasted.

Another type of image is obtained by centering the objective aperture on the diffracted beam, thus excluding all the unscattered waves (Figure 19.6c). Then the image of the particle is bright against a dark background (dark-field image). Lattice imaging is obtained by combining one or several diffracted beams with the direct beam. This is done by letting one or several beams pass through the objective aperture. Let us suppose that the direct beam plus one *hkl* reflection are selected in the back focal plane. These beams will combine to give interference fringes on the image plane which are the images of the (*hkl*) lattice planes.

When the image of the specimen is formed on the screen, one may want to take the diffraction pattern of a specific area. This selection is made by inserting the selected area aperture that encompasses the zone of interest into the image plane. Since the magnification of the objective aperture is about 100, a 10-μm-diameter aperture selects a 100-nm circular zone from which the diffraction pattern is obtained by focusing the projection lens on the back focal plane.

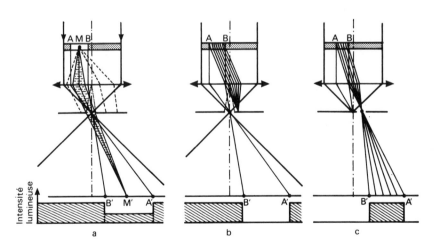

Figure 19.6. Interpretation of image contrast: (a) scattering by an amorphous particle, (b) scattering by a crystallite, bright-field image, and (c) scattering by a crystallite, dark-field image.

19.2.3. The Scanning Transmission Electron Microscope

The scanning transmission electron microscope (STEM) combines imaging and analytical capabilities such as X-ray emission spectroscopy, electron energy loss spectroscopy and Auger spectroscopy. Most of the TEM microscopes are optionally equipped with scanning coils which allow them to work in the STEM mode. However, we will concentrate our attention on dedicated-STEMs working only in the scanning mode. These have the best analytical performance in terms of spatial resolution and intensity and thus they are the best suited for use in catalyst characterization.

19.2.3.1. Principles of the STEM

A scheme of the STEM HB501 manufactured by Vacuum Generators is given in Figure 19.7. The field emission gun is located at the bottom of the microscope column. It consists of a tungsten tip with a curvature radius of a few tens of nanometers from which electrons are extracted at room temperature by a voltage of from 3 to 5 kV provided the vacuum is lower than 5×10^{-9} Pa. The electron beam accelerated under 100 kV is demagnified and focused on the specimen by two condenser lenses and the objective lens. The incident probe

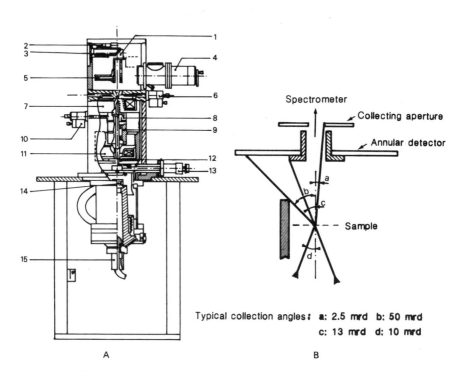

Typical collection angles : a: 2.5 mrd   b: 50 mrd
c: 13 mrd   d: 10 mrd

A   B

Figure 19.7. Scheme of the optical column of the VG HB5 STEM.

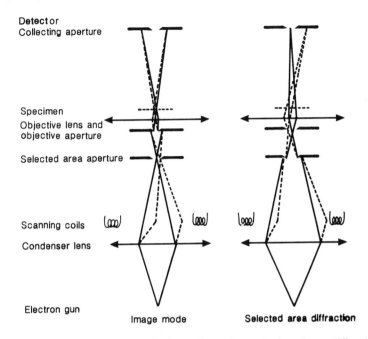

Figure 19.8. Electron pathways for image formation and selected area diffraction.

can be as small as 0.5 nm and as intense at 10^{-9} A. The beam is scanned over the specimen by two scanning coils and transmitted electrons are recorded by two separate detectors: (i) a central detector collects the unscattered and the inelastically scattered electrons at low angles (bright-field detector) (an electron spectrometer can be installed in front of the bright-field detector to select electrons of a given energy); (ii) an annular detector, which allows the unscattered electrons to pass through a hole drilled through its center, collects the scattered (diffracted) electrons (dark-field detector).

The bright-field and dark-field images available simultaneously are formed sequentially: for each position of the beam on the specimen the corresponding unscattered and scattered electrons are recorded by each detector and then displayed on two cathode-ray tubes scanned synchronously with the scanning coils in the microscope column. The magnification of the image is given by the ratio of the CRT screen area to the area scanned on the specimen). The bright- and dark-field images can be combined in different ways (linear combination, division). Note that the two images are available during EDX measurements and that the dark-field image is available during EELS measurements.

19.2.3.2. Image Formation and Contrasts

Figure 19.8 gives the principles of image formation in the STEM. The point-to-point resolution is limited by the size of the electron probe (0.3–

0.5 nm). Lattice planes can be imaged with a 0.2-nm resolution, but that is not as good as that of a conventional microscope. Very small clusters and even isolated heavy atoms can be imaged on very thin supporting materials with the annular dark-field detector which collects all the electrons scattered by the atoms very efficiently.[13] This type of contrast can be improved still further by dividing the dark-field signal by the bright-field signal obtained by selecting the plasmon peak with the electron spectrometer (the intensity of this is proportional to the thickness of the supporting material) (see Section 19.2.4.3). This technique, known as Z-contrast, is useful for visualizing heavy atoms on an amorphous support or on a crystalline support oriented outside the Bragg reflection.

19.2.4. Analytical Tools Associated with the STEM

Different types of signals (See Section 19.2.1.1) are potentially useful for analytical purposes. We will consider only diffracted electrons (nanodiffraction), X-ray emission (EDX), and inelastic electrons (EELS).

19.2.4.1. Nanodiffraction

Selected-area microdiffraction (see Section 19.2.1.2) in the conventional microscope is useful for taking the electron diffraction pattern of crystallites larger than a few tens of nanometers or for taking the ring pattern of a collection of small particles (see Section 19.3.4.3). The spatial selectivity and the beam intensity must be much higher to record the diffraction pattern of individual catalyst particles, which are often smaller than 5 nm. This is possible with the small electron probe of the FEG-STEM, which can be stopped or scanned over the area of interest. The diffraction pattern can be visualized on a fluorescent screen placed in the microscope column above the specimen. The screen is observed or photographed with an inclined miror.[14] A small objective aperture should be used to minimize the beam divergence thus reducing the size of the diffraction spots. Even so, it is difficult to measure spot positions accurately and thus determine precise lattice parameters. However, information on the lattice symmetry and on the orientation of the particles is readily obtained for particles as small as 2 nm (see Section 19.3.3.4).

19.2.4.2. Energy Dispersive X-Ray Emission (EDX)

X-ray emission spectra are recorded with a Si–Li solid state detector installed in the microscope column close to the specimen grid. In most cases, these detectors have a beryllium window which absorbs X-ray photons of energies smaller than 1 keV and thus does not allow the analysis of elements lighter than sodium. The detection of lighter elements (C, N, O) is possible with windowless detectors.

Quantitative EDX analysis of the local composition of thin specimens is well documented. [15,16] The solid phases should be as thin as possible to minimize absorption corrections. Since there are no reliable reference compounds on a scale of a few nanometers, standardless analysis is performed. The ratio C_A/C_B of the concentration of two elements is given by $C_A/C_B = (K_A/K_B)(I_A/I_B)$, where I_A and I_B are the intensities of K or L emission lines after background subtraction, K_A and K_B are calculated from the theoretical values of the ionization cross section, the fluorescent yield, the relative intensities of emission lines, and the detector yield for elements A and B.

With a FEG-STEM the electron probe can be as small as 0.5 nm so that the spatial resolution of EDX analysis is close to 1 nm on thin samples where the broadening due to electron and X-ray scattering in the solid is minimal. Moreover as a very intense beam is available, typically 1 to 1.5 nm, particles containing about 50 atoms can be determined after counting for a few minutes.[17]

19.2.4.3. Electron Energy Loss Spectroscopy (EELS)

Electrons interacting with atoms suffer an energy loss corresponding to the excitation process, which is measured with an electron spectrometer collecting the transmitted electrons that have been dispersed in energy by a magnetic prism. The electron spectrum thus obtained is scanned sequentially across a small slit placed in front of the electron detector. Figure 19.9 shows a typical EELS spectrum. The large peak at zero energy loss is the elastic peak corresponding to electrons transmitted without interaction with atoms. The plasmon peak is due to the excitation of valence electrons or free electrons in metals. The sample thickness can be determined from the intensity ratio of the plasmon peak to the elastic peak. The intensity of the plasmon peak is high enough to be used for the formation of video images. As its energy position depends upon the oxidation state, energy-filtered images can be obtained, e.g., to visualize an oxide layer around a metal core.

At higher energy loss, i.e., from a few tens to a few thousand electron volts, the EELS peaks correspond to electron transitions from core to outer atomic levels. These transitions are similar to the edges observed in X-ray absorption spectroscopy. The energy domain between 0 and 50 eV from the onset of the peak is known as the energy loss near edge structure (ELNES), similar to XANES in X-ray spectroscopy. These structures provide information on the oxidation state and on the coordination of the excited atoms, but further theoretical development is needed to draw quantitative information from ELNES. The energy domain between fifty and a few hundred electron volts is the extended energy loss fine structure (EXELFS), the equivalent to EXAFS.

Analytical and structural information can be obtained from EELS data.[15-18] The intensity of an EELS peak is proportional to the number of atoms present in the scanned area of the specimen. This is the basis for a quantitative analysis of specimen composition. The ratio of the concentration between two elements is given by $C_A/C_B = (S_A/S_B)(\sigma_A/\sigma_B)$, where σ_A and σ_B are the ionization cross sections obtained from theoretical calculations. This technique is useful for light elements such as carbon and oxygen which have large σ values and

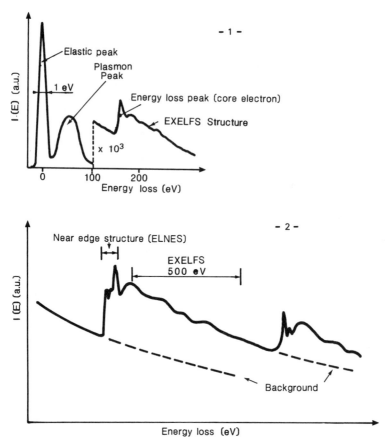

Figure 19.9. Electron energy loss spectroscopy (EELS): (1) scheme of an EELS spectrum; (2) details in the region of core electron energy loss peaks.

cannot be analyzed by EDX. The difficulty lies in the correction for background noise, which is very high with respect to EELS peaks and decreases rapidly with energy. Another critical factor is sample thickness, which should be as small as possible to obtain good signal-to-noise ratios and energy resolution. Typically, for thicknesses in the range 20–30 nm and a collecting aperture of 0.5 mrad, the energy resolution given by the width at half-height of the elastic peak is 1 to 2 eV. The sequential acquisition of EELS data is also a limiting factor because during the period of time needed to record a spectrum (at least 10 min) the specimen may drift too much or may be modified by the electron beam. New types of detectors based on diode arrays (PEELS) will be available soon, and they should reduce the period of time for data acquisition by a factor of fifty.

Structural information can be obtained from the EXELFS and from the ELNES. Indeed EXELFS data can be treated like EXAFS, i.e., by Fourier transform or curve fitting of the oscillations of the spectra. However, the signal-to-noise ratio is usually much too low with present electron spectrometers so that development of the technique must await the advent of detection in an energy-dispersive mode. On the other hand, the signal in the ELNES domain is sufficiently large but quantitative information on the local environment cannot be extracted in the present state of the theory. In the meantime qualitative structural information can be obtained by comparing EELS spectra to those of reference compounds of known electronic and atomic structure.

Finally, it should be noted that EELS could be used to produce energy-filtered images. The intensities of EELS peaks are too weak (except the plasmon peak) to give video images. However, by controlling the scanning of the electron beam in the microscope with a computer it is possible to record digital images that are stored in the computer memory. As in the case of quantitative EELS analysis, the background has to be subtracted from the EELS peak of the element of interest.

19.3. APPLICATION OF ELECTRON MICROSCOPY TO THE STUDY OF HETEROGENEOUS CATALYSTS

19.3.1. Introduction

Conventional and analytical electron microscopy are unique tools for the exploration of the local composition, structure, and texture of catalysts. Figure 19.10 shows a scheme of the solid phases (grain, micrograin, flakes, metal particles), present in a Pt/Al_2O_3 reforming catalyst. The compositions of the millimeter-size grain is determined on a macroscopic scale by chemical analysis of X-ray fluorescence analysis and the size is controlled by sieving. Micrometer-sized micrograins are imaged by SEM and analyzed by EDX (electron microprobe). The characterization of the texture of grains is very important for industrial catalysts as mechanical properties (e.g., attrition resistance) and the

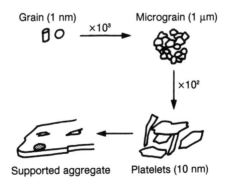

Figure 19.10. Scheme of the texture of a heterogeneous catalyst.

transport of heat and matter are dependent on how the micrograins are assembled. However, we will focus our attention rather on the characterization of nanodomains on catalysts.

19.3.2. Preparation of Samples for TEM Examination

Specimen preparation is a critical step in electron microscopy because the quality of the image and the significance of the analyses are highly dependent on how the different solid phases are dispersed on the microscope grid and how thin they are. The thickness should be less than 50 to 100 nm to allow transmittance through the sample — the thinner the specimen, the better will be the resolution and contrast of the image. Thus chromatic aberration due to the presence of a spectrum of electron energy will be minimized on a thin specimen as electron energy loss due to interaction with the atoms of the specimen will be less frequent. Another important factor is the stability of the specimen preparation. Thus to keep a 0.3-nm resolution, the specimen drift should be negligible with respect to 0.3 nm during the period of time needed to take the photograph (typically 1 to 3 s at high magnification).

Only the most often used techniques of specimen preparation useful for highly divided solids will be examined in this section. A more complete description of preparative techniques in electron microscopy can be found in specialized texts.[19]

19.3.2.1. Preparation of Microscope Grids

Specimens have to be deposited on a support for TEM examination. The standard supports are grids 2 or 3 mm in diameter covered with a film. The grid usually consists of a 100 to 400 mesh copper sieve. Plastic or carbon films are used: the latter are better suited because their thermal and electrical conductivities limit heating and charging effects under the electron beam. Figure 19.11 shows the different steps in the preparation of a carbon-covered grid with the final step being the dissolution of the collodion by isoamyl acetate. For studies requiring high resolution and high contrast, the carbon film may add unwanted background. Thus the use of holey carbon films is recommended, i.e., carbon films punctured by many holes, usually in the range 0.2–10 μm. TEM observations are then carried out with the solid across the holes. Holey carbon films are commercially available or can be made using a technique described by Fukami.[20]

19.3.2.2. Specimen Preparation

Several preparation techniques can be used according to the nature and thickness of the specimen.

19.3.2.2a. Preparation for Direct Observation. The most common technique is to disperse the catalyst powder so that the particles are well separated

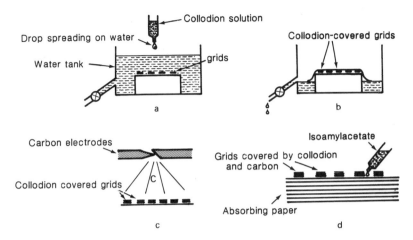

Figure 19.11. Preparation of specimen grids.

on the specimen grid. The particles should preferably be less than 50–100 nm especially if smaller particles carried on these primary particles have to be detected. Figure 19.12 shows the influence of support thickness on the image of a supported metal catalyst. The easiest way is to disperse the catalyst powder in a liquid (water or alcohol). A few milligrams of the powder are ultrasonically dispersed in a few milliliters of the liquid. A drop of the suspension is then deposited on the specimen grid, which can be examined in the microscope after evaporation of the liquid.

If it is not possible to disperse the powder in a liquid, or if the sample composition is modified in the liquid phase (e.g., by dissolution of a component), a dry dispersion should be carried out as described in Figure 19.13. An aerosol of the powder is produced by a shock wave in the syringe, then the aerosol sediments on a grid. Unfortunately, the adherence of the dry-deposited particles is usually poor and this leads to specimen instability during TEM observation.

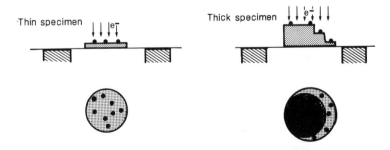

Figure 19.12. Influence of specimen thickness on image contrast.

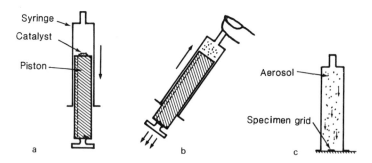

Figure 19.13. Dry dispersion: formation of an aerosol in a syringe and deposition on a specimen grid.

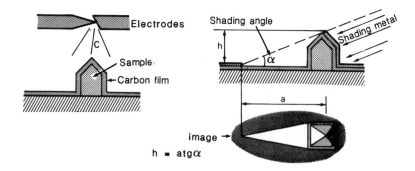

Figure 19.14. Preparation of a shadowed replica.

Since TEM gives a two-dimensional image it is difficult to visualize the shape of the particles. Information on the morphology can be obtained by using a shadowed replica as shown in Figure 19.14. A carbon film is evaporated on the powder dispersed on a clean surface such as freshly-cleaved mica. The sample is then dissolved in an appropriate solvent leaving a carbon replica, and the replica is shadowed by sputtering a heavy metal (Pd, Pt), along a known incidence angle. Shadowing can also be done before sample dissolution or even before the carbon film deposition. An example of a shadowed replica is given in Figure 19.18b. This technique is seldom used now because the morphology of large particles is more easily observed by scanning microscopy.

19.3.2.2b. Preparation of Replicas and Thin Sections. Whenever particles are too thick for direct observation, either extractive replicas or thin sections have to be made. Extractive replicas are useful for visualizing small particles deposited on the surface of a support which scatters or adsorbs the electron

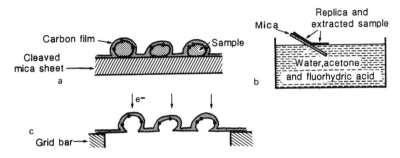

Figure 19.15. Preparation of an extractive replica.

beam too much. The catalyst powder is ultrasonically dispersed in a liquid and a drop of the suspension is deposited on freshly cleaved mica. After drying, the dispersed powder is covered by a carbon film. The mica is then plunged into a solution containing a mixture of water, acetone, and hydrofluoric acid. The acid concentration is chosen to dissolve the support without dissolving the metal particles. They remain stuck to the carbon film (see Figure 19.15), which is collected on a copper grid.

Extractive replicas are needed whenever supports are not transparent to electrons, and even if the support is thin enough, they enable a better contrast. Thus replicas are recommended for visualizing the smallest particles. Electron diffraction patterns of the extracted particles are more easily interpreted in the absence of extraneous diffraction spots or background from supporting materials.[21] However, the technique cannot be used if the metal dispersion is modified by the dissolving bath. In this respect, it is necessary to check that there is no sample modification by comparing the image of metal particles on the replica and on the support whenever this is possible. The replica technique has been used successfully for noble metals supported on silica, transition aluminas, zeolites, and transition or rare-earth metal oxides (CeO_2). However it cannot be applied on easily oxidizable metals of the first transition row (Fe, Ni, Co) which are dissolved at the same time as the support. Figure 19.16 gives the image of an extractive replica obtained from a Pt/Al_2O_3 catalyst.

When the catalyst grains are too thick for direct observation they can be cut into thinner sections by ultramicrotomy. This technique, initially used for biological objects, has been adapted for solids. The powder is embedded in a polymeric resin (e.g., Polarbed kit from Biorad) and cut into a section as small as 20 nm with an ultramicrotome as shown in Figure 19.17. A diamond knife is required to cut most of the solid phases except α-alumina, which is too hard. This technique is painstaking and requires skill, but this is the only way to image the internal morphology and metal dispersion in the microporous grains of zeolites or active charcoals. Figures 19.18 and 19.19 illustrate TEM views taken on zeolites with different specimen preparations.[22-23]

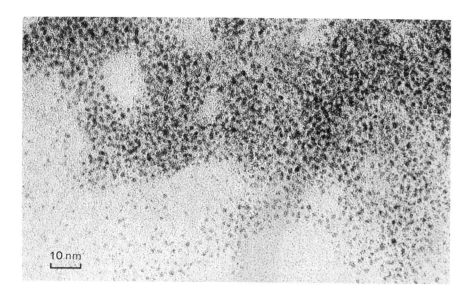

Figure 19.16. Extractive replica of a Pt/Al₂O₃ catalyst.

As there is a very small amount of catalyst on a specimen grid and only a small fraction of the grid is examined, the question arises, are the photographs representative of whole catalyst sample? Ideally, several grids should be prepared with different portions of the catalyst and many zones on the grids should be examined. However, this is so time-consuming that it cannot be used practically. Fortunately, catalysts usually fall into two classes: (1) those with a homogeneous morphology and composition from which a good description can be obtained within a reasonable period of time (e.g., one working day); and (2) those having a heterogeneous composition which cannot be characterized much further because too many views and analyses would be needed to make a statistical evaluation.

19.3.3. Morphology and Structure of Catalysts

The primary goal of electron microscopy is observation of the shape and dimensions of solid phases, their porosity, and their interfaces. In addition to a merely qualitative description of the TEM views, an effort is being made to develop more quantitative treatments of images, e.g., to measure the distribution of particle sizes (granulometry). As TEM images are projected views of objects on a plane perpendicular to the direction of the electron beam, a complete description of the morphology of small particles (< 50 nm) is hard to obtain. The specific techniques used for the study of catalyst morphology are discussed in this section.

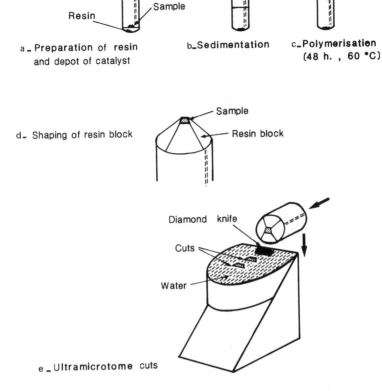

Gelatin capsule

Resin

Sample

a_ Preparation of resin and depot of catalyst

b_Sedimentation

c_Polymerisation (48 h. , 60 °C)

d_ Shaping of resin block

Sample

Resin block

Diamond knife

Cuts

Water

e_Ultramicrotome cuts

Figure 19.17. Ultramicrotomy of resin-embedded samples.

19.3.3.1. Imaging of Catalyst Particles

19.3.3.1a. Unsupported Catalysts. Unsupported catalysts are usually monophasic solids of various types: oxides, sulfides, zeolites, and metal powders. Electron microscopy is used to monitor the homogeneity of the preparation and to give a good description of the catalyst morphology.

Let us take first the example of the characterization of a series of ZSM-11 zeolites prepared with different particles sizes.[23] Figure 19.20a gives a view of a shadowed replica of the preparation showing spherulitic grains of $6 \pm 2 \mu m$ with a rough surface. Small microcrystals (20–30 nm) are present in the core of the zeolite as shown by a view taken through a thin section (Figure 19.20c). From the images of the lattice planes and from microdiffraction experiments it is possible to conclude that these microcrystals are mutually oriented. The external surface of the spherulites is covered by well-crystallized needles (Figure 19.20d) which are connected to the core of the grain. A subsequent STEM–EDX analysis indicates that Si/Al ratios are two to four times smaller near the core of

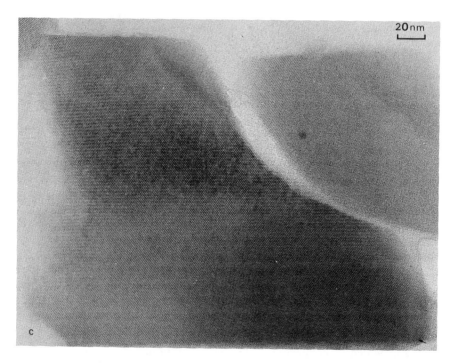

Figure 19.18. ZSM-5 zeolite: (a) direct view, (b) shadowed replica, and (c) view through ultramicrotome cut.

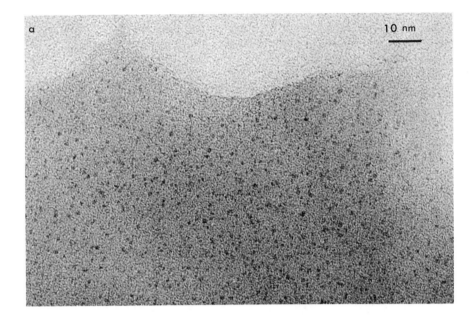

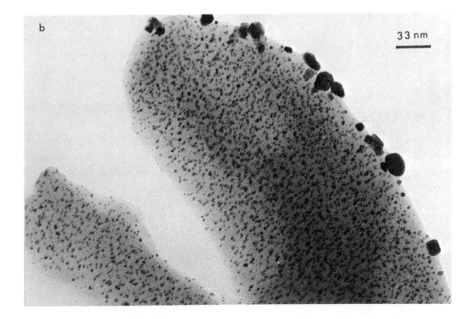

Figure 19.19. Platinum supported on Y-type zeolite — views through cuts: (a) sample with encaged 1-nm particles, (b) sample with occluded platinum particles of 2–5-nm and with particles of 5–20-nm on the external surface.

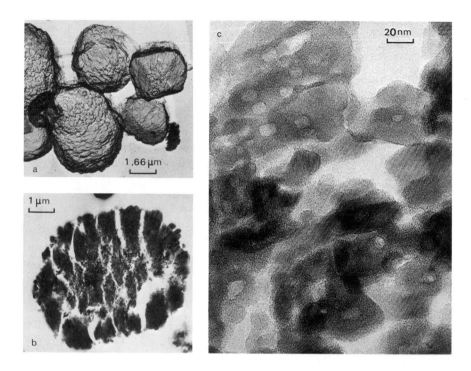

Figure 19.20. ZSM-11 zeolite: (a) replica, (b) but of a zeolite grain at low magnification, (c) view through a cut near the center of the grain, and (d) view near the grain edge.

the zeolite grain than near the external surface. These results are useful in interpreting catalytic results. Indeed catalytic reactions in zeolites depend upon the diffusion of reactants and products in the micropores and thus on the size of the crystals. For instance, the distribution of hydrocarbons obtained from methanol conversion was found to be different in the zeolites described above and in another ZSM-11 zeolite sample synthesized with ten times smaller spherical grains.

Electron microscopy is also very useful for following the modification of catalyst morphology occurring after various treatments of reactions. The TEM photographs given in Figure 19.21 show nickel oxide grains attacked by different reducing treatments.[24-26] It turns out that the etching figures produced by reduction depend upon treatment conditions. Thus, hydrogen reduction leads to random etching (Figures 19.21b, c, d), whereas the attack starts from the external surface upon NH_3 reduction (Figure 19.21e). Carbon deposits can be observed on Figure 19.21f after incomplete reduction by n-butane. Textural modifications of zeolites can be followed by TEM observations of thin sections, and Figure 19.22a is a transmission view through a Y-type zeolite. The images of the (111) lattice planes are clearly observed throughout the crystal. After hydrothermal aging at high temperature, simulating the treatment condition of an fcc catalyst in the regenerator unit, there are many low-contrast zones (Figure 19.22b) corresponding to collapsed portions of the zeolite lattice, either amorphous zones or mesopores filled with various amounts of amorphous alumina. More severe hydrothermal treatment leads to the complete disappearance of the lattice plane images.[27,28]

19.3.3.1b. Supported Catalysts. In the case of supported catalysts, the aim of the TEM study is to establish the size of the supported particles and their distribution on (or in) the support. These studies are particularly useful for characterizing metal-supported catalysts. The best contrasts are obtained when metals of high atomic number are deposited on a thin support containing light elements. When direct observation is not possible because of the lack of contrast between the two phases, preparation techniques described in the previous section, namely extractive replica and ultramicrotomy, can be used.

Particle sizes and particle distribution on the support depend upon the preparation methods as well as on the nature of the metal and the support. Figure 19.23a shows a TEM view of platinum on a nondehydroxylated alumina, the particles spread on the support forming a continuous deposit of low specific surface area. On the other hand, on a support dehydroxylated by high-temperature calcination, the platinum is in the form of homogeneously dispersed particles (Figure 19.23b).

In most cases the supported metal particles are viewed from above so that it is not possible to know the morphology of the particle at the interface between particle and support. However, when the support is spherical, it is possible to observe the particles edge-on. Thus, in Figures 19.24a and b corresponding to Pt/Al_2O_3 and Pd/Al_2O_3 catalysts, respectively, the particles are clearly hemi-

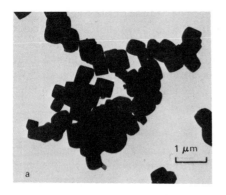
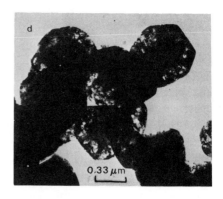
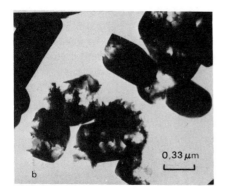
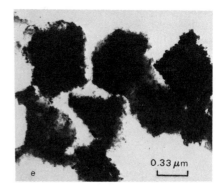
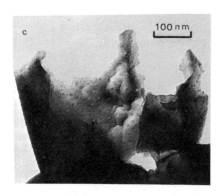
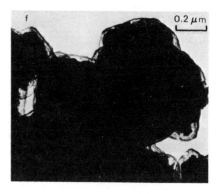

Figure 19.21. TEM study of NiO reduction (metal was dissolved in bromine-methanol solutions): (a) initial NiO, (b–d) corrosion figures after hydrogen reduction, (e) reduction by NH_3, and (f) carbon deposits.

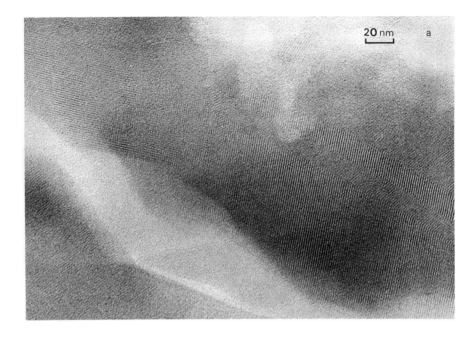

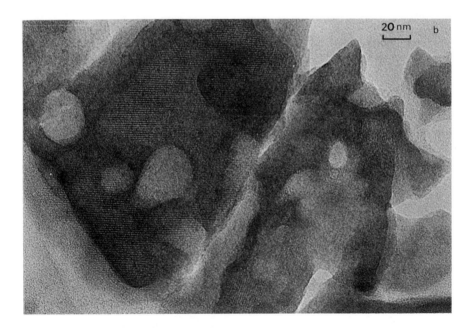

Figure 19.22. Modification of the texture and struture of a Y-type zeolite: (a) before hydro-thermal aging, and (b) after hydrothermal aging.

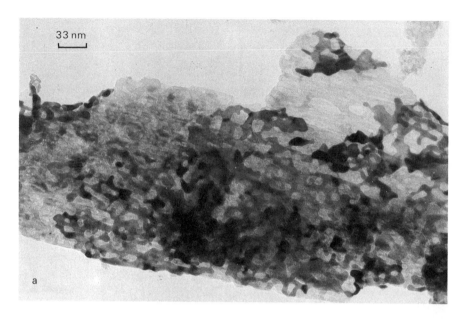

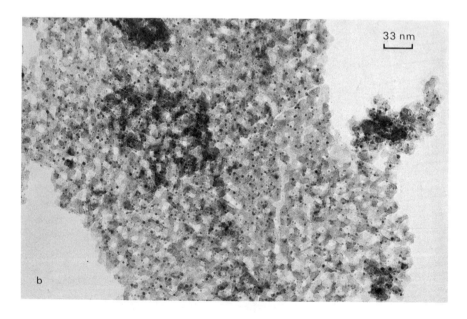

Figure 19.23. Alumina-supported platinum catalyst (10 wt.%Pt): (a) prepared by impregnation of the alumina with surface hydroxyl groups; and (b) prepared by impregnation of a dehydroxylated alumina.

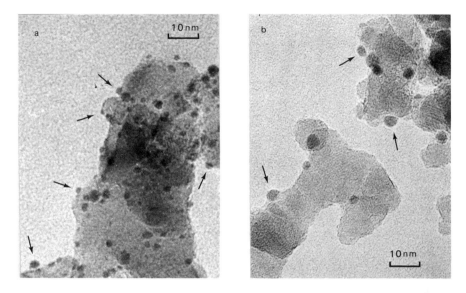

Figure 19.24. Hemispherical shape of particles on a support: (a) Pt/Al$_2$O$_3$, and (b) Pt/Al$_2$O$_3$.

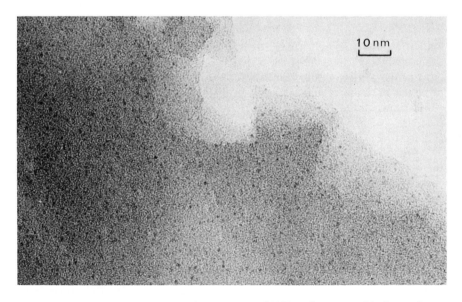

Figure 19.25. View through an ultramicrotome cut of RhY zeolite prepared by ion exchange. The 1-nm rhodium particles are uniformly distributed in the zeolite.

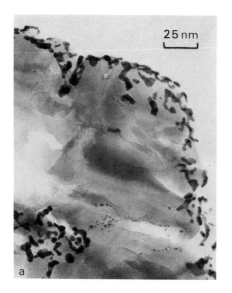 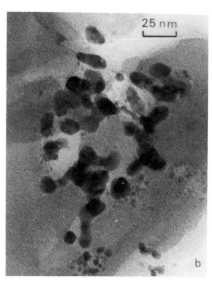

Figure 19.26. View through an ultramicrotome cut of RhY zeolite prepared by decomposition of $Rh_6(CO)_{16}$. The metal is concentrated near the edge (a) or in mesopores (b).

spherical with the flat surface in contact with the support. By tilting the specimen stage, it is possible to confirm that many other particles are hemispherical.

The TEM views given in Figures 19.25 and 19.26 correspond to rhodium-loaded zeolite prepared under two different conditions. When the zeolite is prepared by ion exchange of $Rh(NH_3)_4Cl_2$ and specific activation and reduction treatments, a view through an ultramicrotome cut (Figure 19.25) shows the presence of 1-nm rhodium particles distributed throughout the zeolite. In contrast when the zeolite is loaded by adsorption of $Rh_6(CO)_{16}$, the metal particles are large and located essentially outside the microporous lattice (Figure 19.26).

A good knowledge of the theory of image formation is required to choose the best conditions for specimen observation and to interpret image contrasts correctly, especially those associated with electron diffraction (e.g., lattice imaging and thickness fringes). Thus the contrast of the image is a function of the orientation of the particles with respect to the electron beam. For instance, Figure 19.27 gives two TEM views taken after tilting the specimen by a few degrees. It is conspicuous that the contrast of the particles marked by an arrow is completely inverted after the tilt: particles become darker when a set of lattice planes is in Bragg orientation with respect to the beam. This shows clearly that the contrast is due mainly to diffraction rather than to the thickness of the particles. Supported phases other than metals can also be imaged although the contrast with the support is usually much less favorable.

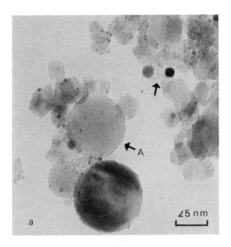
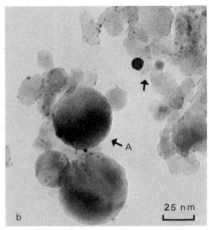

Figure 19.27. Influence of the particle orientation with respect to the beam (Pt/Al$_2$O$_3$): (a) image taken with specimen stage at 0°, and (b) image taken after a 5° tilt. Particle contrast is inverted because the alumina grain A is no longer in the Bragg position.

Figures 19.28a and b show lattice fringes corresponding to two samples of mixed nickel–tungsten sulfides supported on alumina. The sulfide layers ($d_{002} = 6.18$ Å) are imaged provided that they are oriented at $\pm 10°$ from the electron beam. Relationships have been established between the catalytic properties of two different preparations of a sulfide and the number of stacked planes which is larger in the sulfide observed in Figure 19.28a than in the other preparation (Figure 19.28b).

19.3.3.2. Particle Size Measurement

Electron microscopy has been widely used to determine the average size and the size distribution of particles. Particle diameters are measured manually (or with a simple mechanical counter) or automatically with a computer after digitization and storing of the image. However, automatic image treatment and particle counting is possible only in favorable cases, e.g., when the background is smooth enough and the particles well separated from each other. The first step in a granulometric calculation is to define a suitable size increment Δd so that particles will be counted in successive intervals $0-\Delta d$, $\Delta d-2\Delta D$, ..., the average diameter d_i for each interval being $\Delta d/2$, $3\Delta d/2$, ... Let n_i be the number of particles in a given interval d_i. Several types of distribution can be calculated and plotted. The number distribution is obtained by plotting n_i as a function of d_i. The corresponding mean diameter is then $\Sigma n_i d_i / \Sigma n_i$. The area distribution is obtained by plotting $n_i d_i^2$ as a function of d_i, with a mean surface diameter $d_{sp} = \Sigma n_i d_i^3 / \Sigma n_i d_i^2$. Table 19.1 gives an example of particle size measurement on 1000 particles ranging from 0 to 10 nm.

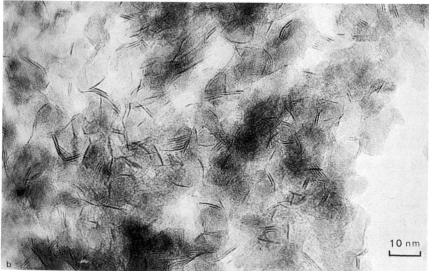

Figure 19.28. Mixed sulfides (Ni–W) S_2 supported on alumina. The number of stacked (002) planes in each sulfide sheet is smaller in (a) than in (b).

The distribution in number, I, and in surface, II, are given in Figure 19.29, the corresponding mean diameters being 4.1 and 5.6 nm, respectively. Distribution I shows a maximum between 2 and 3 nm which accounts for 30% of the total number of particles. However, distribution II shows that these particles account for only 9% of the total surface area whereas particles between 6 and

TABLE 19.1. Distribution of Particle Size: Sample of Data Obtained by Counting Particles on a TEM Micrograph

Interval (nm)	Mean diameter d_i (nm)	Number of particles n_i	$n_i d_i$	$n_i d_i^2$	$n_i d_i^3$
0–1	0.5	0	0	0	0
1–2	1.5	100	150	225	337.5
2–3	2.5	300	750	1875	4687.5
3–4	3.5	150	525	1837.5	6431.25
4–5	4.5	100	450	2025	9112.5
5–6	5.5	100	550	3025	16637.5
6–7	6.5	200	1300	8750	54925
7–8	7.5	50	375	2812.5	21093.75
8–9	8.5	0	0	0	0
9–10	9.5	0	0	0	0
		$\Sigma n_i = 1000$	$\Sigma n_i d_i = 4100$	$\Sigma n_i d_i^2 = 20{,}250$	$\Sigma n_i d_i^3 = 113{,}225$

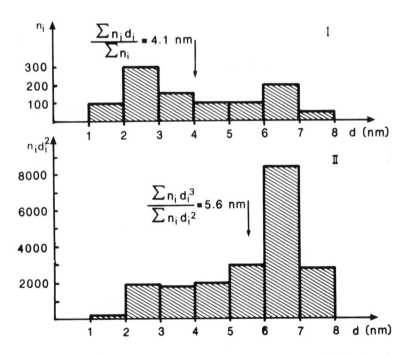

Figure 19.29. Number distribution and area distribution calculated with the data given in Table 19.1.

7 nm account for 42%. It is much more useful to determine area distribution rather than number distribution, as catalytic properties are directly related to surface area. Furthermore, the mean surface diameter can be readily compared with mean diameters calculated from gas adsorption measurements or from small-angle X-ray scattering. If granulometric data obtained by electron microscopy have to be compared with techniques sensitive to the volume of particles (e.g., granulometry by magnetic measurement) the mean volume diameter $\Sigma n_i d_i^4 / \Sigma n_i d_i^3$ should be used.

Particle size measurements should ideally be performed on a large number of particles on several TEM views taken on different zones of a specimen grid and on several grids corresponding to different samplings of the catalyst batch. Particle size distribution can be established by counting about 1000 particles provided the distribution, either monomodal or bimodal, is narrow enough. The best strategy is to choose at random different zones containing a few tens of particles and to measure all the particles in the different zones. If the sizes are too disperse, a prohibitive number of measurements would have to be performed to get a representative distribution.

The agreement with particle size measurements performed by other techniques, especially gas chemisorption, is usually good. Discrepancies are usually due to the limitation of one or the other of the techniques (e.g., particles too small or not sufficiently contrasted to be measured by TEM). The palladium dispersion in a Pd/MgO catalyst was studied by both TEM and hydrogen-adsorption measurements,[36] the mean surface diameters, 3.5 and 2.5 nm, respectively, were not in good agreement. The 2.5 nm value was calculated by assuming equal proportions of the three denser atomic planes: (111), (100), and (110). This discrepancy was understood in view of electron diffraction patterns showing that only (111) planes are exposed on the surface. The particle diameter recalculated from the chemisorption data was then 3.2 nm, in good agreement with the size measured by TEM.

19.3.3.3. Morphology and Orientation of Particles

Particle size measurement by TEM gives the dimensions of the projected area of the particle on a plane perpendicular to the beam, but a three-dimensional description of the particle morphology can be obtained by a combination of electron microscopy and electron diffraction.

A useful technique called "weak-beam, dark-field" was developed by Yacaman[30] to establish the morphology of metal particles. It is based on the presence of equal-thickness fringes produced by a dynamical combination of diffracted beams. By choosing the appropriate conditions of specimen tilt, the spacings between the fringes are small enough so that the shape of particles larger than 5 nm can be mapped by elevation lines. From the distances between the fringes it is possible to calculate the angles between the base plane and the other faces and thus to deduce the nature of these faces. Figure 19.30a gives bright-field image of a particle with a hexagonal

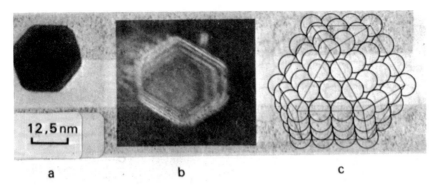

Figure 19.30. Cubo-octahedral particle: (a) bright-field; (b) image in weak-beam, dark-field; and (c) corresponding scheme of atom packing.

contour. From the weak-beam dark-field image given in Figure 19.30b it can be concluded that the particle is a truncated octahedron similar to the scheme given in Figure 19.30c.

A complete study of the morphology and structure of iron oxide has been carried out by an electron diffraction study and by the weak-beam technique. Figure 19.31 is a TEM view of Fe_2O_3 showing the presence of faceted particles with homogeneous sizes and shapes. Electron diffraction patterns and weak-beam images taken under different orientations show that these particles are cubo-octahedra exposing (100) and (111) faces. The relative development of the exposed faces as a function of particle size has been correlated with the catalytic properties.[31-34]

According to the nature of the support and the preparation conditions, different exposed planes can develop on the surface of metal particles. Electron diffraction is very useful in characterizing the nature and orientation of the exposed metal faces. Thus silica-supported nickel catalysts obtained by reduction of nickel antigorite have been studied to determine the type of atomic planes exposed on nickel crystallites.[35] Figure 19.32a is a TEM view showing a layer of silica support covered with nickel particles. The electron pattern given in Figure 19.32b indicates that a large number of nickel particles are oriented with the (111) plane parallel to the support. This is confirmed by the dark-field image given in Figure 19.33 showing only the oriented particles. The comparison of the number of particles detected in Figure 19.33 with the total number of particles detected in the bright-field image (Figure 19.32a) gives the fraction of particles oriented with the (111) plane parallel to the support. This is a useful parameter to interpret catalytic properties for structure-sensitive reactions. A similar study was carried out on a Pd/MgO catalyst.[36] The combination of TEM (Figure 19.34) and electron diffraction studies indicates that the (111) plane of palladium is parallel to the (111) plane of MgO and the Pd[110] direction is aligned with the MgO[110] direction. The particles are octahedra or cubo-octahedra exposing (111) and (100) planes.[36]

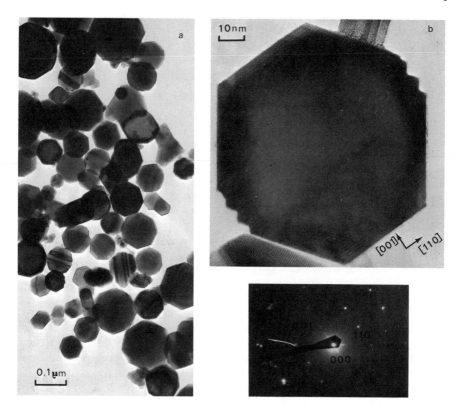

Figure 19.31. Iron oxide (γ-Fe$_2$O$_3$): (a) general view, and (b) γ-Fe$_2$O$_3$ grain viewed along [110] axis and diffraction pattern.

Metal particles are usually much smaller than the smallest zone that can be probed by selected-area diffraction with a TEM so that it is not possible to record the pattern of individual metal particles. On the other hand, in the STEM mode the electron beam can be focused on particles, and the smaller the beam, the better the spatial resolution. With a FEG–STEM, the electron beam can be smaller than 1 nm so that nanodiffraction experiments can be carried out on particles or on support areas as small as 1–2 nm.

The morphology of platinum particles supported on γ alumina has been studied, before and after catalytic methane combustion, by combined nanodiffraction and high-resolution TEM studies.[37] Figure 19.35a is a TEM view of a γ-alumina-supported platinum catalyst. A nanodiffraction pattern was taken with a FEG–STEM on the platinum particle marked with an arrow (Figure 19.35b). A second pattern was taken on the alumina support in the vicinity of the particle (Figure 19.35c), and comparison of the two patterns shows that the (110) plane of platinum is oriented parallel to the (110) plane of alumina; furthermore the [111] axes of the platinum and alumina lattices are parallel.

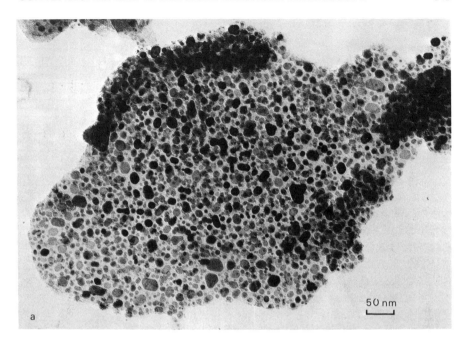

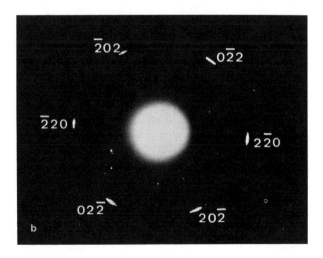

Figure 19.32. Ni/SiO$_2$ catalyst obtained by reduction of nickel antigorite: (a) bright-field image, and (b) diffraction pattern.

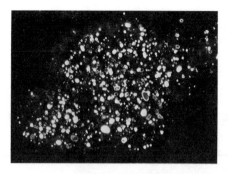

Figure 19.33. Dark-field image of the Ni/SiO$_2$ catalyst obtained with a 220 reflection.

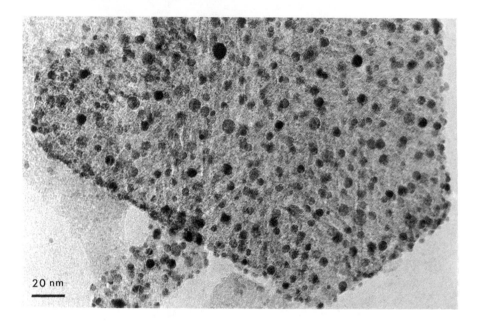

20 nm

Figure 19.34. Bright-field view of a Pd/MgO catalyst.

This perfect epitaxy is due to the fact that both lattices are fcc and the unit-cell parameter of alumina is exactly twice that of platinum. Another example of an epitaxial relation between metal particle and support (graphite) is given in the next section. Nanodiffraction patterns are also useful in understanding the structure of multitwinned particles.

19.3.3.4. Structure of Catalysts

The structure of catalysts can be characterized by electron diffraction, lattice imaging, and EELS. A number of combined TEM and diffraction studies

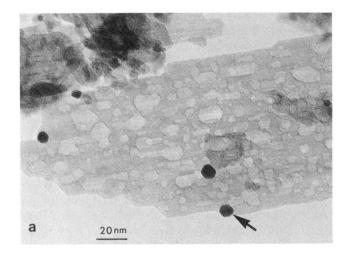

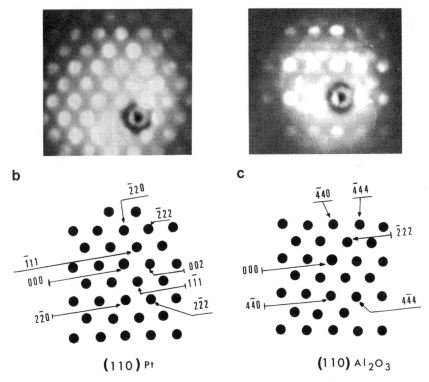

Figure 19.35. (a) Bright-field image of a Pt/Al₂O₃ catalyst, (b) nanodiffraction pattern taken with a FEG-STEM on the platinum particle, and (c) nanodiffraction pattern taken on alumina (from Briot *et al*.[37]).

have been carried out on supported metal aggregates to detect the presence of particles with fivefold symmetry. These particles form icosahedra or decahedra, the structure of which can be described as multitwinned distorted tetrahedra. A precise description of these studies is given in Gillet,[38] Yacaman et al.,[39] Dominguez and Yacaman,[41] and Gallezot et al.[42] Lattice distortions in small metal particles can occur as a result of an epitaxial interaction with the support. Thus graphite-supported 2-nm platinum particles have been studied by nanodiffraction in a STEM.[42] It has been shown that most of the particles are oriented with their (110) plane parallel to the graphite basal plane (001). Furthermore, because of a strong epitaxy, the angles between the Pt(111) planes are distorted to get closer to 60°, the angle between the carbon atom rows in the basal plane. In contrast, diffraction patterns taken on 2-nm platinum particles supported on an amorphous active charcoal show that there is no distortion of the lattice (Figure 19.36).

Lattice imaging is a useful technique for detecting changes in lattice spacings and lattice defects and characterizing the structure of interfaces. Figure 19.37 is a view of a 15-nm multitwinned metal particle taken with a high-voltage microscope. The particle structure can be described as an association of a decahedron and an icosahedron sharing two tetrahedra. Figure 19.38 is a view through an ultramicrotome cut of a faujasite-type zeolite containing 1- to 3-nm occluded particles. The lattice planes Pt(111) $(d = 2.26 \text{ Å})$ and Pt(100) $(d = 1.96 \text{ Å})$ are imaged clearly. The lattice spacings are the same as in bulk platinum. However, change in lattice spacing between the edge and the core of platinum particles has been reported.[43,44]

Electron energy loss spectroscopy (EELS) can provide information on the local coordination of atoms (Section 19.2.4.3). The EXELFS data can in principle be treated like EXAFS data to obtain the distance between the absorbing atom and its neighbors. However, the signal-to-noise ratio is not good enough to be processed quantitatively. The ELNES signal is stronger but a quantitative treatment of the data is not yet possible because the theory accounting for ELNES is still being developed. However, useful structural information can be obtained by comparison with EELS spectra of reference compounds. Figure 19.39a gives the spectra of graphite, amorphous carbon, coronene, and pentacene recorded with an electron spectrometer attached to a STEM, VG HB 501. The spectra were taken on very thin zones of the carbon compounds to minimize plural scattering. The rapidly decreasing background was subtracted and the different spectra were normalized.

These reference spectra were compared to EELS spectra taken on coke deposits formed on the external surface of zeolites[45] or on reforming catalysts.[46] Figures 19.39b and c give TEM views of external coke deposits on US-HY and HZSM-5 to compare with their EELS spectra. The coke deposited on US-HY forms filaments as small as 1.5 nm which emerge from the zeolite surface. The EELS spectra show that their structure is neither

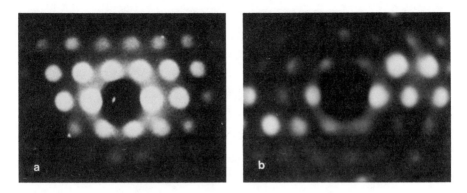

Figure 19.36. Nanodiffraction pattern of 2-nm platinum particles: (a) supported on active charcoal, and (b) supported on graphite.

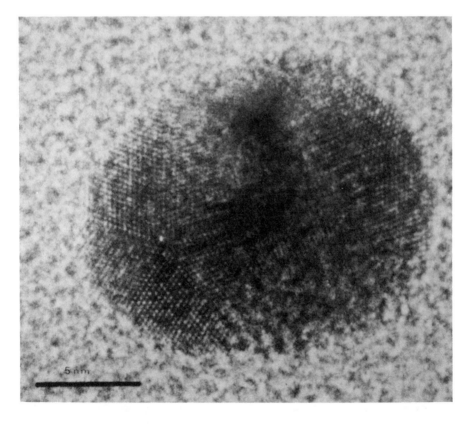

Figure 19.37. Lattice-imaging of a multitwinned particle.

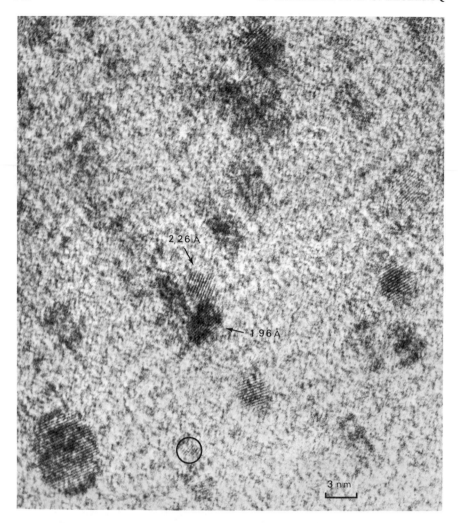

Figure 19.38. Lattice-imaging of platinum particles in Y-type zeolite.

pregraphitic nor amorphous but rather similar to that of a linear polyaromatic molecule such as pentacene, which can form and diffuse in micro-pores. In contrast, the coke formed on HZSM-5 is clearly graphitic. The EELS study of coke deposits on a Pt/Al$_2$O$_3$ catalyst has shown that the coke produced by cyclopentane cracking is deposited on the metal particle and on the support at a limited distance (20 nm) from a given particle. Coke deposits can also be detected by electron diffraction patterns or by lattice imaging when the coke is graphitic.

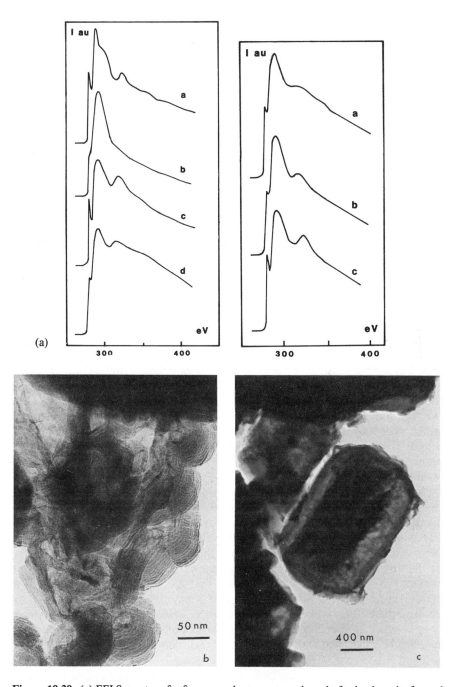

Figure 19.39. (a) EELS spectra of reference carbon compounds and of coke deposits formed on the external surface of zeolite crystallites [left field: (a) graphite, (b) amorphous carbon, (c) coronene, (d) pentacene; right field: (a) carbon on HY, (b) carbon on alumina, (c) carbon on ZSM-5]. (b) TEM view of carbon deposit on USH-Y zeolite. (c) TEM view of a coked ZSM-5 zeolite (from Gallezot *et al.*[(45)]).

19.3.4. Composition of Catalysts

Heterogeneous catalysts can easily be characterized by energy dispersive X-ray emission analysis (EDX, see Section 19.2.4.2). Because the catalyst phases are usually thin enough, quantitative analysis without standards is possible and absorption corrections are seldom required. EDX analysis is usually performed with a conventional microscope working in the STEM mode. However, a dedicated STEM equipped with a field-emission gun is better suited as the intense and sharp beam allows a fast analysis with high spatial resolution. The use of a virtual objective aperture reduces the parasitic electron and X-ray scatterings considerably. However, a few limitations have to be considered. The analyzed zone can be contaminated by carbon deposits formed by the cracking of organic molecules present as impurities in the microscope column, on the specimen grid, and on the catalyst itself. These deposits can reduce the X-ray signal considerably. In FEG–STEM operating under high vacuum, the main source of contamination is the specimen. It should be outgassed before it is introduced into the microscope column, preferably in the airlock as then there will be no further passage in air.

Specimen drift can be compensated as the image of the analyzed zone is available during analysis in FEG–STEM. Specimen instability under the electron beam is the main cause of limitation. Damages depend upon the electron dose but the composition of the specimen is of primary importance. There is always a correlation between the thermal stability of the solid and its resistance to irradiation.

The composition of the specimen can be modified under the electron beam, which may lead to errors in EDX measurements. This problem is well illustrated in the case of a silica-supported palladium catalyst promoted with potassium.[47] This element is introduced in the form of KNO_3, which should normally spread over the surface of silica upon heating at 650 K. Indeed, EDX measurements with an analysis window larger than 10 μm^2 shows that the K/Si ratio is constant whatever the area analyzed and close to the nominal K/Si ratio given by chemical analysis. Therefore at least on micrometer-sized zones potassium is well distributed, probably in the form of a superficial oxide compound. However, Figure 19.40 giving the K/Si ratio as a function of the area of the analysis window (scanned area) shows that the smaller the area, the smaller the K/Si ratio. It was also seen that a second analysis on the same area gives a smaller ratio than the first. Clearly there is a loss of potassium because the diffusion of the oxide is enhanced when the beam is concentrated on too small an area. At higher temperatures of calcination, the loss is less marked because a stable superficial silicate compound with potassium is formed.

Metal particles can be unstable under the beam depending upon the nature of the metal (melting point), the size of the particles, and any interaction with the support. Irradiation can displace or even vaporize the particles. Figure 19.41 shows an example of cation reduction into metal under the electron beam. View 1 corresponds to a Pd/Al_2O_3 precursor where the metal is not yet reduced.

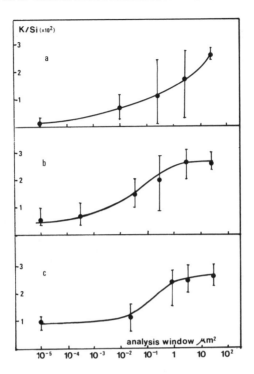

Figure 19.40. EDX data taken on a palladium catalyst supported on silica and promoted with potassium. The K/Si ratios are given as a function of the size of the analysis window; curves a, b, and c correspond to samples heated at 300, 400, and 600 °C, respectively (points correspond to the average of at least ten measurements performed with a given analysis window on different zones).[47]

The framed zone has been analyzed for 100 s, a weak but significant Pd($L\alpha$) signal is detected (spectrum 1). The image of the same zone after analysis (view 2) shows dark contrast corresponding to a palladium particle (spectrum 2). Therefore ions have been reduced to palladium atoms, which migrate and agglomerate to form metal particles. Interestingly, this experiment indicates that even if an element present in the sample cannot be detected in images it can still be localized by EDX analysis. The sensitivity is such that a few tens of atoms can be detected by EDX with a FEG–STEM while the spatial resolution of analysis is better than 2 nm. Several examples of STEM–EDX analyses on different types of heterogeneous catalysts follow.

19.3.4.1. Bimetallic Catalysts

The usual problem is to establish the composition of individual metal particles, to know in what proportion the two metals are associated and how it compares to the overall composition. With a FEG–STEM this is possible on

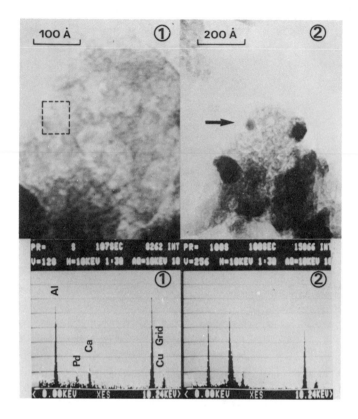

Figure 19.41. Modification of the local composition of a palladium-catalyst under an electron beam.

particles as small as 1–1.5 nm, and for particles larger than 5 nm a composition profile can even be determined to find the distribution of the two metals in a given particle (e.g., to detect segregation).

Figure 19.42a is a transmission view through a charcoal-supported, Pt–Fe catalyst prepared by coimpregnation. Most of the particles are in the size range 2–4 nm. The STEM–EDX analysis shows that the compositions of the individual metal particles are within 5% of the nominal Pt:Fe = 80:20 composition. A typical spectrum taken on a 2-nm particle is given in Figure 19.42b. It was seen that after use in the liquid phase hydrogenation of cinnamaldehyde there was no change of composition.[48] Another interesting application of STEM–EDX is given by the study of metal particles prepared by decomposition of heteropoly-nuclear molecular clusters of known structure.[49] It was shown that the metal particles retain the composition of the molecular precursor although the nuclearity of the bimetallic particles is at least a hundred times larger than that of the bimetallic molecular cluster.

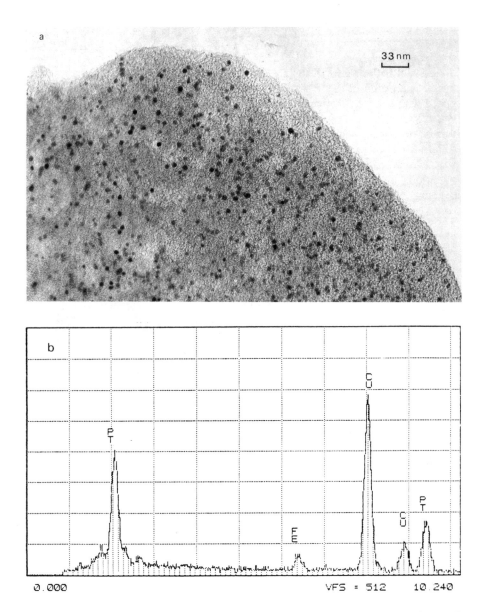

Figure 19.42. Bimetallic $Pt_{80}Fe_{20}$ catalyst support on active charcoal: (a) TEM view, and (b) EDX analysis on a 2-nm platinum particle.

19.3.4.2. Zeolites

A frequent application of STEM–EDX analysis to zeolites is to check on a nanometer-sized scale the variation of composition across the zeolite microcrystals of framework elements (Si, Al, Fe, Ga, ...) or of extraframework species (cations, host molecules). These studies should be carried out preferably on ultramicrotome cuts of the zeolite. The hydrothermal aging of a LaY zeolite has been studied in the presence of vanadium.[50] The V($K\alpha$) peak is superimposed with the La($L\beta$) peak. However, since the La($L\alpha$)/La($L\beta$) intensity ratio is known from previous analysis without vanadium, the contribution of V($K\alpha$) can be determined accurately. Figure 19.43 gives three EDX spectra taken at different places with a 30-nm^2 analysis window (scanned area). The upper left spectrum taken in the core of the zeolite crystal shows that the concentration of vanadium is very low. On the other hand, analysis of the dark particles on the external surface or isolated from the zeolite grain gives La/V = 1. The most likely compound is LaVO$_4$ and this was confirmed by electron diffraction. It was concluded that vanadium enhances the destruction of the zeolite structure because it combines with La$^+$ ions which stabilize the structure of the zeolite.

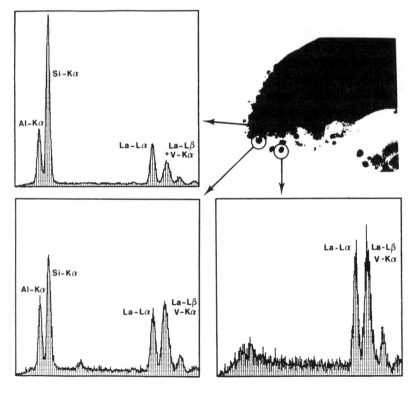

Figure 19.43. EDX spectra taken at different places on a hydrothermally aged LaY zeolite in presence of vanadium.

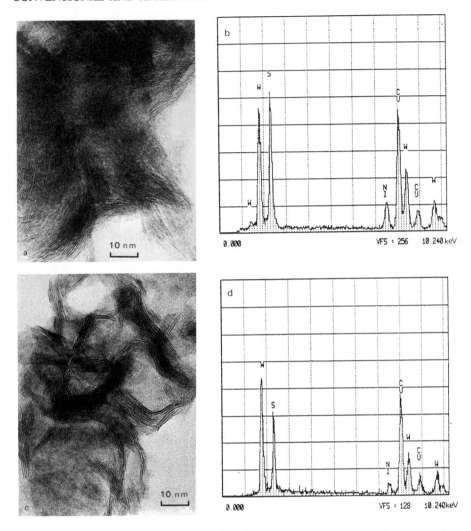

Figure 19.44. Mixed Ni–W sulfide obtained by decomposition of a heteronuclear thiometallate: (a) TEM view before catalytic reaction, (b) corresponding EDX spectrum, (c) view after reaction, and (d) corresponding spectrum.

19.3.4.3. Mixed Oxides and Sulfides

Care should be taken to analyze the zones which are thin enough, because oxides and sulfides of transition metals are usually strongly absorbing materials. Mixed oxides of the type SnSbO and SnSbFeO have variable Sb/Sn ratios depending upon the calcination temperature.[51] At 500 °C, a homogeneous solid solution was detected. At higher temperatures (750–900 °C) antimony segregates in the form of Sb_2O_4.

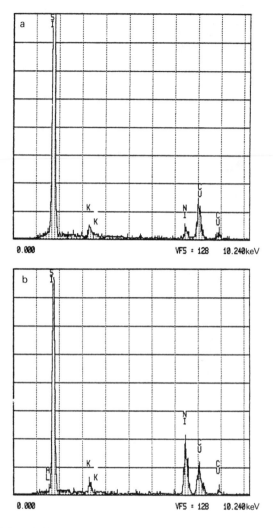

Figure 19.45. EDX analysis on a Ni/SiO$_2$ catalyst promoted with potassium: (a) spectrum of the support, and (b) spectrum of a 3-nm nickel particle.

Figure 19.44 gives a TEM view of a Ni–W mixed sulfide obtained by decomposition and sulfiding of a thiometallate.[52] The WS$_2$ lattice planes are highly disordered (Figure 19.44a) because of the presence of heteroatoms. STEM–EDX analysis indicates a very homogeneous local composition with a ratio (Ni/Ni + W = 0.36). After biphenyl hydrogenation on this catalyst, segregation occurs with the formation of two phases, detected by STEM–EDX analysis: a pure nickel sulfide Ni$_3$S$_2$ and a WS$_2$ phase containing much less nickel than previously (Ni/Ni + W = 0.2). Accordingly, the structure is less disordered because there are fewer heteroatoms present (Figure 19.44c).

19.3.4.4. Distribution of Promoters and Poisons

In addition to active phases, industrial catalysts usually contain various promoters, added to increase their activity or stability. Therefore it is essential to know the distribution of the promoter on the catalyst (e.g., on the support or on the metal). For the same reasons the distribution of poisons is an important parameter in understanding catalytic properties.

The distribution of potassium added on a Ni/SiO_2 catalyst to modify the selectivity has been studied by analyzing the concentration of potassium on the support and on the nickel particles (Figure 19.45). Most of the potassium is uniformly distributed over the silica. It forms a surface compound which is stable under the electron beam, unlike potassium on a Pd/SiO_2 catalyst (see Figure 19.40).

REFERENCES

1. P. B. Hirsh, A. Howie, R. B. Nicholson, D. W. Pashley, M. J. W. Whelan, *Electron Microscopy of Thin Crystals*, Krieger, Malabar, Florida (1977).
2. D. Kay, *Techniques for Electron Microscopy*, Blackwell, Oxford (1967).
3. J. M. Cowley, *Diffraction Physics*, North Holland, Amsterdam (1981).
4. S. Amelinckx, *Diffraction and Imaging Techniques in Materials Sciences*, Elsevier, Amsterdam (1978).
5. G. Thomas and M. J. Goring, *Transmission Electron Microscopy of Materials*, John Wiley, and Sons, New York (1979).
6. J. C. H. Spence, *Experimental High Resolution Electron Microscopy*, Clarendon, Oxford (1981).
7. B. Jouffrey, A. Bourret, and C. Colliex, *Microscopie Elecronique en Science des Matériaux*, Editions du CNRS, Paris (1983).
8. J. P. Eberhart, *Méthodes physiques d'étude des minéraux et des matériaux solids*, Doin, Paris (1976).
9. T. Baird, *Catalysis* **5**, 172 (1981).
10. A. Howie, in: *Characterization of Catalysts* (J. M. Thomas and R. M. Lambert, eds.), John Wiley and Sons, New York (1980), p. 89.
11. J. V. Sanders, *Catalysis Science and Technology, Vol. 7* (J. R. Anderson and M. Boudart, eds.), Springer-Verlag, Berlin (1985), p. 51.
12. P. Gallezot, M. Avalos-Borja, H. Poppa, and K. Heinemann, *Langmuir* **1**, 342 (1985).
13. A. V. Crewe, J. Wall, and J. Langmore, *Science* **168**, 1338 (1970).
14. J. P. Lynch, H. F. F. Dexpert, and E. Freund, *Electron Microscopy and Analysis*, Institute of Physics Conference Series Vol. 61 (M. J. Goringe, ed.), Institute of Physics, London (1981), p. 67.
15. J. J. Hren, J. I. Goldstein, and D. C. Joy, eds., *Introduction to Analytical Electron Microscopy*, Plenum, New York (1979).
16. F. Mavroce, L. Meuy, and R. Toxver (eds.), *Microanalyse et Microscopie Electronique à Balayage*, Les Editions de physique, Orsay (1979).
17. H. Dexpert, E. Freund, J. P. Lynch, S. J. Pennycook, *Electron Microscopy and Analysis*, Institute of Physics Conferences Series, Vol. 61 (M. J. Goringe, ed.), Institute of Physics, London (1981), p. 209.
18. R. F. Egerton, *Electron Energy-Loss Spectroscopy in the Electron Microscope*, Plenum, New York (1986).
19. D. H. Kay, *Techniques for Electron Microscopy*, Blackwell, Oxford (1965).

558 P. GALLEZOT AND C. LECLERCQ

20. A. Fukami and K. Adachi, *J. Electr. Microsc.* **14**, 112 (1965).
21. G. Dalmai-Imelik, C. Leclercq, and I. Mutin, *J. Microsc.* **20**, 123 (1974).
22. P. Gallezot, I. Mutin, and G. Dalamai-Imelik, *J. Microsc. Spectrosc. Electron,* **1**, 1 (1976).
23. A. Auroux, H. Dexpert, C. Leclercq, and C. J. Vedrine, *Appl. Catal.* **6**, 95 (1983).
24. H. Charcosset, G. Dalamai, R. Frety, and C. Leclercq, *C. R. Acad. Sc.* **264**, serie C, 151 (1967).
25. G. Labbe, R. Frety, H. Charcosset, and Y. Trambouze, *J. Chem. Phys.* **70**, 1721 (1973).
26. R. Frety, *Ann. Chem.* **4**, 453 (1969).
27. F. Mauge, A. Auroux, J. C. Courcelle, Ph. Engelhard, P. Gallezot, and J. G. Grosmangin, *Catalysis by Acids and Bases* (B. Imelik *et al.,* eds.), Elsevier, Amsterdam (1985), p. 91.
28. F. Mauge, J. C. Courcelle, Ph. Engelhard, P. Gallezot, J. Grosmangin, P. Primet, and B. Trusson, *Zeolites: Synthesis, Structure Technology and Applications* (B. Drzag *et al.,* eds.), Elsevier, Amsterdam (1985), p. 401.
29. M. Breysee *et al., Symposium on Advances in Hydrotreating,* ACS Denver meeting, 5–10 April (1987).
30. M. J. Yacaman and T. Ocana, *Phys. Stat. Sol.* **42**, 571 (1977).
31. H. Batis, C. Leclercq, and P. Vergnon, *J. Microsc. Spectrosc, Electron* **7**, 149 (1982).
32. C. Leclercq, H. Batis, and M. Boudeulle, *J. Microsc. Spectrosc. Electron* **8**, 243 (1983).
33. M. Boudeulle, H. Baris, C. Leclercq, and P. Vergnon, *J. Sol. Stat. Chem.* **48**, 21 (1983).
34. P. Vergnon and H. Batis, *Bull. Soc. Chem., 9–10,* Part 1 (1984), p. 265.
35. G. Dalmai-Imelik, C. Leclercq, and A. Maubert-Muguet, *J. Sol. State. Chem.* **16**, 129 (1976).
36. F. Figueras, S. Fuentes, and C. Leclercq, *Growth and Properties of Metal Clusters* (J. Bourdon, ed.), Elsevier, Amsterdam (1980), p. 525.
37. P. Briot, G. Gallezot, C. Leclercq, and M. Primet, *Microsc. Microanal. Microstruct.* **1**, 149 (1990).
38. M. Gillet, *Surf. Sci.* **67**, 139 (1977).
39. M. J. Yacaman, K. Heinemann, C. Y. Yang, and H. Poppa, *J. Cryst. Growth* **47**, 187 (1979).
40. C. Y. Yang, *J. Cryst. Growth* **47**, 274 (1979).
41. J. M. Dominguez and M. J. Yacaman, *J. Catal.* **64**, 223 (1980).
42. P. Gallezot, C. Leclercq, I. Mutin, C. Nicot, and D. Richard, *J. Microsc. Spectrosc. Electron.* **10**, 479 (1985).
43. D. J. Smith and L. D. Marks, *Phil. Mag.* **44**, 735 (1981).
44. J. Turkevich, L. L., Ban, and J. H. Wall, *Perspectives in Catalysis* (R. Larsson ed.), CWK, Gleerup (1981), p. 59.
45. P. Gallezot, C. Leclercq, M. Guisnet, and P. Magnoux, *J. Catal.* **114**, 100 (1988).
46. P. Gallezot, C. Leclercq, J. Barbier, and P. Marecot, *J. Catal.* **116**, 164 (1989).
47. V. Pitchon, P. Gallezot, C. Nicot, and H. Praliaud, *Appl. Catal.* **47**, 357 (1989).
48. D. Goupil, Thèse de Doctorat no. 9086, Lyon (1986).
49. A. Choplin, L. Huang, A. Theolier, P. Gallezot, J. M. Basset, U. Siriwardane, S. G. Shore, and R. Mathieu, *J. Am. Chem. Soc.* **108**, 4224 (1986).
50 F. Mauge, J. C. Courcelle, Ph. Engelhard, P. Gallezot, and J. Grosmangin, *New Developments in Zeolite Science and Technology* (Y. Murakami, A. Iijima, and J. M. Ward, eds.), Elsevier, Amsterdam (1986), p. 803.
51. J. C. Volta, B. Benaichouba, I. Mutin, and J. C. Vedrine, *Appl. Catal.* **8**, 215 (1983).
52. W. Eltzner, M. Breysse, M. Lacroix, C. Leclercq, M. Vrinat, M. A. Muller, E. Diemann, *Polyhedron* **7**, 2405 (1988).

APPLICATIONS OF ELECTRICAL CONDUCTIVITY MEASUREMENTS IN HETEROGENEOUS CATALYSIS

J.-M. Herrmann

20.1. INTRODUCTION

The use of electrical conductivity measurements in catalysis was developed in the 1950s when the Electronic Theory of Catalysis appeared. As the catalytic process is a succession of chemical reactions with the rupture of chemical bonds and the creation of new ones, it was natural to look at the electronic properties of the different partners — the reactants and the catalyst. It was expected that by controlling the electronic properties of the solid, its reactivity with respect to a given catalytic reaction could be controlled. However, as electronic processes are very fast phenomena, they do not constitute the rate-limiting step of a catalytic reaction. Thus the electronic theory of catalysis was found not to apply in numerous examples. By contrast, electical conductivity appeared as a fruitful technique to characterize many catalytic systems, as exemplified in this chapter.

20.2. THEORY AND DEFINITIONS

In a solid, the charge carriers are electrons e^- and positive holes p^+, the latter corresponding to electronic vacancies. In a single crystal, when an electric field $\mathbf{E}$ (in volt m^{-1}) is applied, the electrons move in the opposite direction to the field and the holes in the same direction with respective velocities $\mathbf{v}_n$ and $\mathbf{v}_p$,

J.-M. Herrmann • Laboratoire de Photocatalyse, Catalyse et Environnement, URA CNRS, Ecole Centrale de Lyon, Ecully, France.

Catalyst Characterization: *Physical Techniques for Solid Materials*, edited by Boris Imelik and Jacques C. Vedrine, Plenum Press, New York (1994).

which are proportional to this field:

$$\mathbf{v}_n = \mu_n \mathbf{E} \qquad \mathbf{v}_p = \mu_p \mathbf{E} \tag{20.1}$$

where μ_n and μ_p are the mobility tensors of the electrons and the holes, respectively. For an elementary volume dV of section S, one can define a current density vector $\mathbf{j} = i/S$, where i is the current intensity across S. Since $|i| = dQ/dt = q([e^-] + [p^+])dV/dt$, with q the charge of the electron and $[e^-]$ and $[p^+]$ the electron and hole concentrations, we can write

$$\mathbf{j} = \{|q|([e^-] + [p^+])Sdl\}/Sdt = |q|([e^-]\mathbf{v}_n + [p^+]\mathbf{v}_p) \tag{20.2}$$

Thus,

$$\mathbf{j} = ([e^-]\mu_n + [p^+]\mu_p)q\mathbf{E} = \sigma\mathbf{E} \tag{20.3}$$

The factor σ is called electrical conductivity. In a single crystal σ is an anisotropic tensorial characteristic of the solid, which means that it varies with orientation in the lattice. In a metal whose conduction is purely electronic ($[p^+] = 0$), σ simplied as

$$\sigma = nq\mu \tag{20.4}$$

with $n = [e^-]$.

20.2.1. Electrical Conductivity of Powder Catalysts

Laboratory catalysts are generally powders whose surface area varies from a few to several hundreds $m^2 g^{-1}$. If the contact points between the particles are considered good electrical contacts, the powder sample can be regarded as a conductor whose conductivity is given by

$$\mathbf{j} = i/S = \sigma E = -\sigma \operatorname{grad} \mathbf{V} \tag{20.5}$$

$$|i|/S = V/h \tag{20.6}$$

where h is the thickness of the powder. Thus

$$V = (1/\sigma)(h/S) = Ri \tag{20.7}$$

which is Ohm's law. Consequently from the experimental value of the electrical resistance R, the electrical conductivity can be deduced:

$$\sigma = 1/\rho = (1/R)(h/S) \tag{20.8}$$

where R is the experimental electrical resistance, h the thickness of the powder sample, S the section area of the electrodes of the cell, and ρ the resistivity.

The experimental value of σ only has a relative meaning. It depends on the number of contact points, on the shape and the size of the grains (texture), and on the compression of the powder. However, the relative variations of σ as a function of a given parameter are indicative of the electronic properties of the solid with respect to this parameter and can provide information identical to that found on single crystals.

The electrical conductivity of a powder sample will be given by

$$\sigma = Ac \tag{20.9}$$

where c is the concentration of the majority charge carrier and A is a constant which includes the charge of the electron, the mobility of the charge carriers, and a factor linked to the number and the quality of the contact points.

20.2.2. Definitions

1. Semiconductor: A compound whose electrical conductivity varies exponentially with temperature according to an Arrhenius type law:

$$\sigma = \sigma_0 \exp(-E_c/RT) \tag{20.10}$$

   Generally: $10^{-12} \leq \sigma \leq 10^{-2}$ ohm^{-1} cm^{-1} and $20 \leq E_c \leq 200$ kJ mol^{-1}.
2. n-type semiconductor: A semiconductor whose charge carriers are electrons. For a semiconductor oxide, the conductivity σ decreases when the oxygen pressure P_{O_2} increases ($\delta\sigma/\delta P_{O_2} < 0$).
3. p-type semiconductor: A semiconductor whose charge carriers are positive holes (electronic vacancies). For an oxide, the conductivity σ increases with the oxygen pressure ($\partial\sigma/\partial P_{O_2} > 0$).
4. Intrinsic semiconductor: A semiconductor whose electrons rise from the valence band into the conduction band. If the charge carriers arise from defects in the solid, the compound is called an extrinsic semiconductor.
5. Quasi-metallic conductor: A solid that behaves as a metal (i.e., with an activation energy of conduction equal to zero), but with a conductivity value within a domain equidistant from the metal and the semiconductors.
6. Metallic conductor: A solid of high conductivity ($\sigma > 10^6$ ohm^{-1} cm^{-1}) whose electrical resistance varies linearly with temperature $R = R_0(1 + \alpha T)$.
7. Orders of magnitude: The conductivity value of the different types of conductors are presented along the axis given in Scheme I.

20.2.3. Apparatus

The electrical conductivity cells are either of the static type to study the electronic interactions between a solid and the gas phase or of the dynamic type to measure *in situ* the conductivity during a catalytic reaction in an attempt to identify the reaction (charged) intermediates. Figure 20.1 represents a static model,[1] where the powder sample is slightly compressed between two platinum electrodes under a constant pressure of about 10^5 Pa. Figure 20.2 repre-

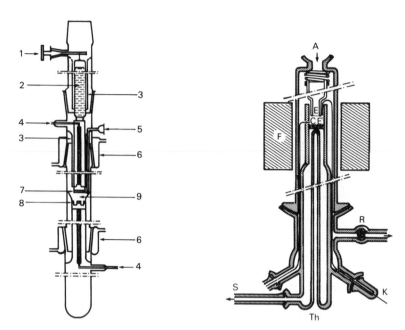

Figure 20.1. Static-type electrical conductivity cell: (1) lifting key; (2) mercury containing sealable bulb; (3) piston-guide; (4) thermocouple; (5) gas inlet; (6) watercooled ground joint; (7, 8) electrodes; and (9) sample holder.

Figure 20.2. Dynamic flow reactor cell: (A) gas inlet; (E, E') stainless steel electrodes; (F) furnace; (R) tap for reactor purging; (Th) thermocouple well; (K) Kovar passage; (S) gas outlet.

sents a dynamic flow reactor cell,[2] where the catalyst is placed between two grid electrodes E and E' made of stainless steel through which the gaseous reactants are able to flow. A constant pressure is applied to the powder by a spring.

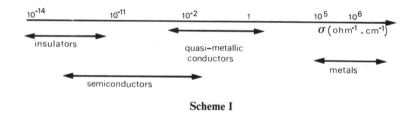

Scheme I

20.3. STRUCTURE DEFECTS IN PURE OXIDE CATALYSTS

20.3.1. The n-Type Semiconductors

Conduction electrons are generated by two kinds of solid defects: interstitials or anionic vacancies.

20.3.1.1. Interstitials

Interstitial cations can be formed in the case of ceria[3] according to the equilibrium

$$CeO_2 \rightleftarrows Ce_{int}^{4+} + O_2(g) + 4e^-$$ (20.11)

The mass action law

$$K_{11} = [Ce_{int}^{4+}]P_{O_2}[e^-]^4$$ (20.12)

combined with the electroneutrality condition of the crystal, $[e^-] = 4[Ce_{int}^{4+}]$, yields the equation

$$[e^-] = (K_{11}/4)^{1/5}P_{O_2}^{-1/5}$$ (20.13)

The electrical conductivity σ, which is proportional to $[e^-]$ $(\sigma = A[e^-])$, will be given by

$$\sigma = A[e^-] = A[(K_0)_{11}]^{1/5}\exp -(\Delta H_{11}/5RT)P_{O_2}^{-1/5}$$ (20.14)

If the conductivity of an oxide MO_2 is found to vary as $P_{O_2}^{-1/5}$ it means that, as in ceria, the main defects are tetravalent M^{4+} interstitial ions. The enthalpy ΔH_{11} of formation of these interstitials is equal to five times E_c, the experimental activation energy of conduction.

20.3.1.2. Anionic Vacancies

Anionic vacancies in oxides are produced either by reduction or by increasing the temperature under low pressure. They are in equilibrium with oxygen of the gas phase according to

$$O^{2-} \rightleftarrows \tfrac{1}{2}O_2(g) + V_{O^{2-}}$$ (20.15)

where $V_{O^{2-}}$ represents an anionic vacancy with two trapped electrons (neutral entity with respect to the lattice). They can become singly or doubly ionized by losing successively their first and second electron:

$$V_{O^{2-}} \rightleftarrows V_{O^{2-}}^+ + e^-$$ (20.16)

$$V_{O^{2-}}^+ \rightleftarrows V_{O^{2-}}^{2+} + e^-$$ (20.17)

The condition of electroneutrality of the solid

$$[e^-] = [V_{O^{2-}}^+] + 2[V_{O^{2-}}^{2+}]$$ (20.18)

combined with the mass action law applied to equations (20.15)–(20.17) yields

$$[e^-]^3 = K_{15}K_{16}P_{O_2}^{-1/2}([e^-] + 2K_{17}) \qquad (20.19)$$

There are two limiting cases:

1. $K_{17} \ll [e^-]$: The ionization of the anionic vacancies is limited to that of the first electron. Equation (20.19) then becomes

$$\sigma = A[e^-] = A[(K_{15})_0(K_{16})_0]^{1/2} \exp -[(\Delta H_{15} + \Delta H_{16})/2RT]P_{O_2}^{-1/4} \quad (20.20)$$

Consequently, the electrical conductivity will vary as $P_{O_2}^{-1/4}$ if the main defects are singly ionized vacancies. This has been observed for titania[4] (Figure 20.3) and tin oxide.[5] Moreover, the slopes $\partial\sigma/\partial P_{O_2}^{-1/4}$ of the linear transforms $\sigma = f(P_{O_2}^{-1/4})$ obey the Arrhenius law. This is shown in Figure 20.4. The slope of the straight line is equal to $(\Delta H_{15} + \Delta H_{16})/2R$. Generally, the ionization energy ΔH_{16} of the first electron is small and can be neglected with respect to the enthalpy ΔH_{15} of formation of an anionic vacancy. Consequently one can obtain this energy directly, and it was found to be equal to 343 and 221 kJ/mol^{-1} for TiO_2 and SnO_2, respectively.

2. $K_{17} \gg [e^-]$: The ionization of the anionic vacancies involves both electrons, and equation (20.19) becomes

$$\sigma = A[e^-] = A(2K_{15}K_{16}K_{17})^{1/3}P_{O_2}^{-1/6} \qquad (20.21)$$

Consequently, if the main defects are doubly ionized anionic vacancies, the electrical conductivity will vary as $P_{O_2}^{-1/6}$. The variations of σ under constant oxygen pressure as a function of temperature give an activation energy of conduction which is equal to

$$E_c = (\Delta H_{15} + \Delta H_{16} + \Delta H_{17})/3 \qquad (20.22)$$

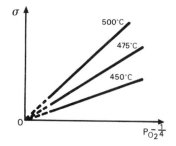

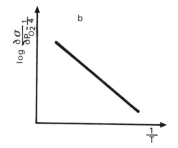

Figure 20.3. Variations of the electrical conductivity σ of TiO_2 as a function of $P_{O_2}^{-1/4}$.

Figure 20.4. Arrhenius plot of the slopes of the lines in Figure 20.3: $\log(\partial\sigma/\partial P_{O_2}^{-1/4}) = f(1/T)$.

Such variations ($\sigma = f(P_{O_2}^{-1/6})$) have been observed for vanadium-doped titania[6] with an activation energy of 142 kJ mol^{-1}. Consequently it was deduced that

$$\Delta H_{15} + \Delta H_{16} + \Delta H_{17} = 3E_c = 427 \text{ kJ/mol}^{-1} \tag{20.23}$$

As $\Delta H_{15} + \Delta H_{17} = 335$ kJ/mol^{-1} for pure TiO_2 had been obtained previously, it followed that the ionization of the second electron required 92 kJ mol^{-1} (≈ 1 eV), which was in excellent agreement with results in the literature obtained with single crystals.

20.3.2. The p-Type Semiconductors

The main defects are cationic vacancies. For example in NiO,[7] the appearance of positive holes p^+ will occur on double ionization of these cationic vacancies:

$$\tfrac{1}{2}O_2(g) \rightleftarrows O^{2-} + V_{Ni^{2+}} \tag{20.24}$$

$$V_{Ni^{2+}} \rightleftarrows V_{Ni^{2+}}^- + p^+ \tag{20.25}$$

$$V_{Ni^{2+}}^- \rightleftarrows V_{Ni^{2+}}^{2-} + p^+ \tag{20.26}$$

The symbol $V_{Ni^{2+}}$ represents cationic vacancy in nickel with the two positive charges localized around the defect. This entity is neutral with respect to the lattice. The mass action law applied to equations (20.24)–(20.26) yields

$$K_{24} = [V_{Ni^{2+}}]/P_{O_2}^{1/2} \tag{20.27}$$

$$K_{25} = [V_{Ni^{2+}}^-][p^+]/[V_{Ni^{2+}}] \tag{20.28}$$

$$K_{26} = [V_{Ni^{2+}}^{2-}][e^-]/[V_{Ni^{2+}}^-] \tag{20.29}$$

The condition of electroneutrality for the whole crystal requires

$$[p^+] = [V_{Ni^{2+}}^-] + 2[V_{Ni^{2+}}^{2-}] \tag{20.30}$$

From equations (20.27)–(20.30) one gets

$$[p^+]^3 = K_{24}K_{25}P_{O_2}^{1/2}([p^+] + 2K_{26}) \tag{20.31}$$

If the ionization of the cationic vacancies is limited to the first stage, $[p^+] \gg 2K_{26}$, and thus,

$$[p^+] = (K_{24}K_{25})^{1/2}P_{O_2}^{+1/4} \tag{20.32}$$

If the ionization of the cationic vacancies concerns the second level ($K_{26} \gg [p^+]$), the electrical conductivity σ will vary as

$$\sigma = A[p^+] = A(2K_{24}K_{25}K_{26})^{1/3}P_{O_2}^{+1/6} \tag{20.33}$$

Consequently, depending on the exponent of the oxygen pressure in the isotherms $\sigma = f(P_{O_2}^{1/n})$, it will be possible to determine the nature of the defect and its ionization state.

20.4. ELECTRONIC STRUCTURE OF DOPED OXIDES

The doping of an oxide consists of dissolving a heterocation in the lattice positions of this host oxide. The content is generally of a few percent.

20.4.1. Valence Induction

The semiconductivity of a solid is obtained by valence induction when the charge carriers exist independently of the existence of structure defects and directly from the presence of heterovalent doping agents. The required conditions are the following.

1. The cation of the host oxide must have several valence states.
2. The added heterocation must differ by at least one valence unit with respect to the main cation.
3. It must form a solid solution with the host oxide. This will be easily obtained if the cations have ionic radii close to each other.

The electrical conductivity of a p-type oxide is increased by doping it with a lower valence heterocation (for instance Li^+ in NiO). For an n-type semiconductor, the electrical conductivity σ is increased by doping with a higher valence heterocation (e.g., Nb^{5+} in TiO_2). In contrast, the electrical conductivity will be decreased by doping a p-type semiconductor with a higher valence heterocation (e.g., La^{3+} in NiO) and by doping an n-type semiconductor with a lower valence heterocation (e.g., Ga^{3+} in TiO_2). In Figure 20.5, the valence induction is illustrated for an n-type semiconductor (TiO_2) doped with Nb^{5+} and Ga^{3+}, respectively (ionic radii $r_{Ga^{3+}} = 0.69$ Å; $r_{Nb^{5+}} = 0.69$ Å; $r_{Ti^{4+}} = 0.68$ Å).

In the case of doping TiO_2 with Nb^{5+}, the dissolved pentavalent niobium ion shares four electrons with the four neighboring oxygen anions. The fifth electron, which cannot enter into a Nb—O bond, is delocalized around Nb^{5+}. A very small energy is then sufficient to ionize this electron into the conduction band. Generally, at room temperature all the donor centers are ionized and the concentration of free electrons is equal to that of the doping agent. In the case of Ga^{3+}-doped titania, each trivalent gallium ion shares three electrons with the

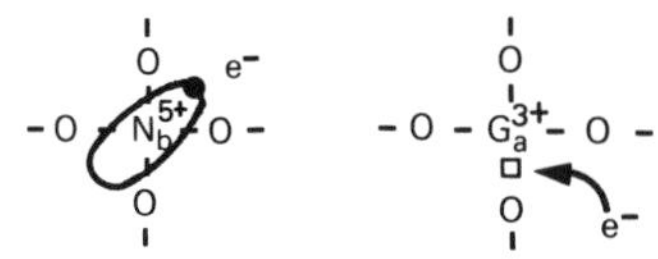

Figure 20.5. Dissolution of Nb^{5+} and Ga^{3+} in the lattice of TiO_2 (doping effect).

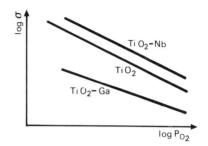

Figure 20.6. A plot of log $\sigma = f(\log P_{O_2})$ isotherms for pure, Ga^{3+}-doped, and Nb^{5+}-doped titania.

four neighboring oxygen anions. One of the four $Ga^{3+}-O^{2-}$ bonds will lack one electron and thus will constitute a trap for a conduction electron originating from an anionic vacancy. The doping of TiO_2 by Ga^{3+} ions creates acceptor centers, which explains the decrease in electrical conductivity predicted by the induction valence law.

The electrical conductivity of TiO_2 and Ga- and Nb-doped titania as a function of oxygen pressure is given in Figure 20.6. For pure TiO_2 at 575 °C, σ varies as $P_{O_2}^{-1/4}$, which implies the existence of singly ionized vacancies as des- cribed in Section 20.3.1.2. The isotherm for niobium-doped titania is parallel to that for TiO_2 (slope $-1/4$) and located above it. This means that electrons originate from two sources: (1) the anionic vacancies (in the singly ionized state ($\partial \log \sigma / \partial \log P_{O_2} = -1/4$) and (2) the n-type doping induced by the dissolution of Nb^{5+} in the lattice position of Ti^{4+} (Figure 20.5). In contrast, the isotherm relative to Ga-doped titania is located below that of TiO_2 with a slope different from $-1/4$ and equal to $-1/5.7$. This illustrates that the presence of Ga^{3+} ions has induced the formation of acceptor centers, which decrease the conductivity, i.e., the concentration of conduction elec- trons. To account for the fractional slope ($-1/5.7$) of the isotherm relative to Ga-doped titania, a model including the simultaneous presence of singly and doubly ionized vacancies has been developed.[4] The electrical conductivity is consequently given by the equation

$$\sigma^3 = BP_{O_2}^{-1/2}\sigma + CP_{O_2}^{-1/2} \qquad (20.34)$$

where $B = A^2 K_{15} K_{16}$ and $C = 2A^2 K_{15} K_{16} K_{17}$. A is a constant of proportionality between σ and $[e^-]$ ($\sigma = A[e^-]$). Coefficients B and C obey the Arrhenius law (Figure 20.7). Consequently enthalpy changes can be calculated:

$$-\partial \ln B/\partial(1/T) = (\Delta H_{15} + \Delta H_{16})/R = \Delta H_{15}/R = 339 \text{ kJ mol}^{-1} \quad (20.35)$$

$$-\partial \ln C/\partial(1/T) = (\Delta H_{15} + \Delta H_{16} + \Delta H_{17})/R = 443 \text{ kJ mol}^{-1} \quad (20.36)$$

For the three samples (TiO_2, $Ga^{3+}-TiO_2$, and $Nb^{5+}-TiO_2$), the same energies ΔH_{15} and ΔH_{17} were found. This shows that doping has not affected the thermo- dynamic properties of titania. The heat of formation of singly ionized vacancies

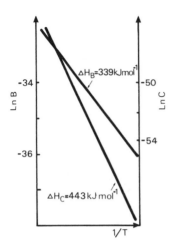

Figure 20.7. Arrhenius diagrams of coefficients B and C.

is equal to 339 kJ mol^{-1}, whereas the heat of formation of doubly ionized vacancies is equal to 443 kJ mol^{-1}. The difference between the two values gives the enthalpy of ionization of the second electron of the vacancies ($\Delta H_{17} = 104$ kJ mol^{-1}). These energies, obtained with powder samples, are equal to those obtained at relatively low temperatures ($\approx 500\,°C$), whereas measurements on single crystals require temperatures ranging between 900 and 1200 °C.

20.4.2. Consequence for Catalysis

The existence of anionic vacancies constitutes dissociative sites for oxygen chemisorption. After pretreatment of TiO_2 under vacuum or in hydrogen, the catalytic oxidation of CO in a static reactor exhibits a partial order with respect to oxygen equal to 1/2, confirming the dissociation of oxygen to form active species. Such dissociated active species are also found in CO oxidation in a flow system as demonstrated in Section 20.7.1.

20.5. STRUCTURE OF MIXED OXIDES: CASE STUDY OF Sn–Sb–O

20.5.1. Electrical Conductivity of Homogeneous Solids Calcined at 500 °C

The structure of mixed oxides can be elucidated with the help of electrical conductivity as the dissolution of a minor cation in the lattice of the main oxide can induce the formation of charge carriers by valence induction. This is exemplified by the case of Sn–Sb mixed oxides.[5] The electrical conductivity of Sn–Sb mixed oxides, found in SnO_2 and Sb_2O_4, is represented in Figure 20.8. Qualitatively, the progressive incorporation of antimony in SnO_2 increases the conductivity by three orders of magnitude for Sb ≤ 7 at.%. Then σ decreases moderately for $7\% \leq$ Sb $\leq 40\%$ before decreasing abruptly for $40\% \leq$ Sb $\leq 100\%$. Experimentally, for all the catalysts, the electrical conductivity can be expressed

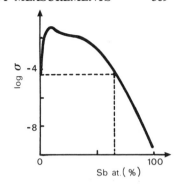

Figure 20.8. Variations of the electrical conductivity of Sn–Sb–O mixed oxides as a function of the mole fraction in antimony.

as a function of oxygen pressure and temperature according to the general equation

$$\sigma = kP_{O_2}^{-1/n}\exp(-E_c/RT) \tag{20.37}$$

P_{O_2} varies from 10^{-2} to 160 Torr and T from 623 to 708 K, a temperature range corresponding to the domain where all these solids are active in the mild oxidation of propene into acrolein. The exponent n has an integral value (equal to 4) only in the case of pure SnO_2, which corresponds to the existence in this oxide of singly ionized anionic vacancies. For all the other samples, n is equal to infinity, which means that these solids have a conductivity independent of the oxygen pressure.

The activation energy of conduction is equal to 111 kJ mol^{-1} for pure SnO_2 and decreases to zero when the antimony content varies between 1 and 40 at.%. For Sb = 100 at.% (Sb_2O_4), E_c is equal to 134 kJ mol^{-1}. Because of their nil activation energy of conduction and of the value of σ, Sn–Sb mixed oxides can be considered as quasi-metallic conductors. However, Sb_2O_4 can be considered as an intrinsic semiconductor since (1) it has a low conductivity, and (2) σ is independent of the oxygen pressure, which means that there are no defects in equilibrium with gaseous oxygen. Its activation energy of conduction yields a band gap energy E_G ($E_G = 2E_c$) equal to 268 kJ mol^{-1} (i.e., 2.8 eV) in good agreement with the value of 2.95 eV obtained by UV spectroscopy.

The conductivity spectrum of Figure 20.8 can be interpreted as follows. The addition of 1.5 at.% Sb increases the conductivity of SnO_2 by three orders of magnitude. This implies a doping effect of Sb^{5+} ions in the lattice of SnO_2 in agreement with the valence induction illustrated in Figure 20.5. Each antimony ion dissolved yields one free electron of conduction. This is confirmed in Figure 20.9, where it is shown that for Sb ≤ 6 at.% the electrical conductivity is proportional to [Sb]. The valence induction requires that antimony be essentially present as Sb^{5+}, which was later confirmed by Mössbauer spectroscopy. The maximum of σ corresponds to 6.1. at.%, i.e., 1.8×10^{21} Sb^{5+} atoms cm^{-3}.

Above 6.1 at.% in Sb^{5+}, the electrical conductivity decreases although the dissolution of antimony still occurs. This has to be accounted for by the existence of an electron-consuming reaction. In particular, it can be explained

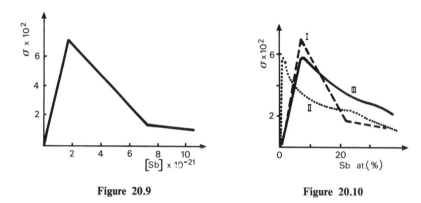

Figure 20.9   **Figure 20.10**

Figure 20.9. Variations of the electrical conductivity of Sn–Sb–O mixed oxides as a function of the concentration in antimony (in atoms cm^{-3}).

Figure 20.10. Variations of the electrical conductivity of Sn–Sb–O mixed oxides as a function of Sb at.%: (curve I) samples calcined at 500 °C; (curve II) samples calcined at 950 °C; (curve III) samples calcined at 950 °C, but with the surface content in antimony determined by XPS.

by formation of Sb^{3+} ions according to the reaction

$$Sb^{5+} + 2e^- \rightleftarrows Sb^{3+} \tag{20.38}$$

The appearance of Sb^{3+} ions was demonstrated by Mössbauer spectroscopy. The progressive introduction of antimony induces the formation of more Sb^{3+} responsible for the decrease in σ down to a level corresponding to Sb $\approx$ 20 at.%, where a plateau develops in the 20–40 Sb at.% range. It was deduced that the dissolution had reached its maximum at 20 Sb at.%. At this level of antimony the probability of finding Sb^{5+} and Sb^{3+} ions in close proximity made possible the formation of a new Sb$_2$O$_4$ phase. This was confirmed by X-ray analysis, which showed the presence of a supported Sb$_2$O$_4$ phase for antimony ranging between 20 and 40 at.%.

For high Sb%, the conductivity decreases strongly to the low level of Sb$_2$O$_4$. However, as the main constituent is almost an insulator, the conductivity should be lower than the values found in Figure 20.8. The reason is a doping effect of Sn^{4+} dissolved in the trivalent sites of Sb$_2$O$_4$.

20.5.2. Catalytic Properties of Sn–Sb Mixed Oxides

In the catalytic oxidation of propene, the selectivity in acrolein increases with the antimony content and becomes a maximum in the 20–40 Sb at.% region. There is no direct relationship between the catalytic properties and the optimum solid solution of Sb^{5+} in SnO$_2$ at 6.1 at.%. However, it can be inferred that the most selective solids are those in which the solid solution supports an Sb$_2$O$_4$ phase in the range $20 \leq$ Sb ≤ 40 at.%.

20.5.3. Influence of Calcination at 950 °C

The calcination of Sn–Sb mixed oxides increases the selectivity in acrolein at a given antimony percentage. The electrical conductivity of mixed oxides (Sb < 40%) calcined at 950 °C are presented in Figure 20.10 (curve II). It can be observed that curve II is shifted to the y axis as compared to curve I which is relative to samples calcined at 500 °C. It appears that the solids calcined at 950 °C have an antimony content higher than its nominal value. In fact, it was shown by XPS that these solids were enriched in antimony in their surface region. Since the solids are heterogeneous, the conductivity mode has changed and has become a surface conductivity. If σ is now plotted as a function of the actual surface antimony content determined by XPS, curve II will be shifted to curve III which has the same profiles as curve I determined for homogeneous solids calcined at 500 °C. This means that the surface region which controls the conductivity behaves as a homogeneous solid. This explains why, for a given nominal content of antimony, an increase in the calcination temperature improved the selectivity in acrolein, which requires the simultaneous presence of the solid solution of antimony in SnO_2 and of Sb_2O_4.

20.5.4. Structural Study of Ternary Fe–Sn–Sb Oxides

Industrial Sn–Sb mixed oxide catalysts contain iron as an additive. A series of Sn–Sb oxides was prepared with the same iron content of 5 at.%. The electrical conductivity was studied at room temperature under a pressure of 1 atm air (Figure 20.11). Two observations can be made: (1) the maximum

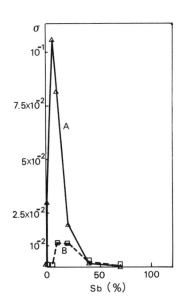

Figure 20.11. Room temperature electrical conductivity of Fe–Sn–Sb–O mixed oxides as a function of antimony content.

conductivity is decreased by one order of magnitude when compared to that of binary Sn–Sb oxides, and (2) it is shifted from 6 to 11 Sb at.%.

The decrease in σ for Fe–Sn–Sb oxides is due to a doping effect of Fe^{3+} in the lattice of SnO_2. Fe^{3+} ions in substitutional lattice positions create acceptor centers which trap the free electrons arising from the anionic vacancies and from the n-type doping of Sb^{5+} simultaneously dissolved in Sn^{4+} positions. As long as $[Sb^{5+}]$ remains lower than $[Fe^{3+}]$ (5 at.%), the conductivity is very low. When $[Sb^{5+}]$ exceeds $[Fe^{3+}]$, the n-type doping by Sb^{5+} becomes preponderant and σ increases. This can be symbolized as follows:

$$[(Sb^{5+})e^-] \rightleftarrows (Sb^{5+}) + e^-$$

$$D \rightleftarrows D^+ + e^-$$

$$(Fe^{3+}) + e^- \rightleftarrows [(Fe^{3+})e^-]$$

$$A + e^- \rightleftarrows A^- \tag{20.39}$$

where D and A are donor and acceptor centers, respectively.

There is a titration of A centers by D centers:

$$A + D \rightleftarrows A^- + D^+ \tag{20.40}$$

which explains why the maximum conductivity is shifted from 6 to 11% in Sb when 5 at.% Fe is added into the solids. From the catalytic point of view, added Fe^{3+} ions have no influence upon activity or selectivity. They are known as a stabilizer of the mixed oxides, preventing the migration of antimony to the surface. Such a property can be partly explained by the above equations, where the presence of acceptor Fe^{3+} centers can stabilize the dissolution of Sb^{5+} ions by electrostatic interaction.

20.6. STRUCTURAL STUDY OF CATALYSTS BY TEMPERATURE PROGRAMMED CONDUCTIVITY

It is possible to determine accurately the temperature of transition between two allotropic forms of the same compound by measuring its electrical conductivity as a function of programmed temperature. This was done with cobalt molybdate as well as with iron molybdate, for which the differential thermal analysis is often not sensitive enough. $CoMoO_4$ exists in two forms (α and β). The β form is violet and can exist at room temperature as a metastable phase. It can be transformed into the green α form by merely grinding at room temperature.

With a programmed rise in temperature, the kinetic curve $\sigma(\alpha) = f(t)$ describes the typical exponential law of a semiconductor:

$$\sigma = \sigma_0 \exp(-E_c/RT) \tag{20.41}$$

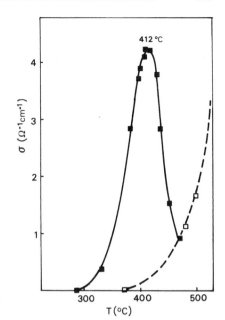

Figure 20.12. Evidence by temperature programmed electrical conductivity measurements for the temperature of transition between the α and β phases of $CoMoO_4$.

with T proportional to time t (Figure 20.12). When α-$CoMoO_4$ reaches its transition temperature, the electrical conductivity of the sample becomes progressively more governed by that of β-$CoMoO_4$, which is much less conductive. The curve $\sigma(t)$ exhibits a decrease which corresponds to the crossover period between the two exponential curves. When the time is long enough for a total conversion of the α form into β, $\sigma(t)$ describes the exponential variations of $\sigma(\beta)$. The temperature at the peak is considered to be the temperature of transition. This method is very accurate and sensitive and enables the determination of transition temperatures of very low thermicity processes in conditions of catalysis.

20.7. DETERMINATION OF REACTION MECHANISMS AND IDENTIFICATION OF CHARGED ACTIVE INTERMEDIATES BY *IN SITU* CONDUCTIVITY MEASUREMENTS DURING CATALYSIS

20.7.1. Oxidation of CO on TiO_2

The oxidation of CO was carried out in the cell reactor described in Figure 20.2, which allowed simultaneous measurement of the electrical conductivity of the catalyst and of the reaction rate in a differential flow regime. The catalyst was slightly compressed between two grids which act as electrodes and allowed the gaseous reactants to flow through the solid. The electrical conductivity σ and the catalytic activity a were studied as functions of temperature ($420 \leq T \leq 480\,°C$) and of the partial pressures of the reactants. In the stationary state, σ and a varied with temperature as

$$\sigma = C\exp(-E'_c RT) \qquad E'_c = 144.4\text{ kJ mol}^{-1}$$

$$a = A\exp(-E_a/RT) \quad E_a = 102.5\text{ kJ mol}^{-1}$$

At constant temperature, when P_{CO} and P_{O_2} were independently varied between 50 and 400 Torr and the total pressure of a stoichiometric mixture between 100 and 400 Torr, σ and a were found to fit the following power forms:

$$\sigma = K_c P_{O_2}^{-0.51} P_{CO}^{0.33} \tag{20.42}$$

$$a = K_a P_{O_2}^{0.12} P_{CO}^{0.72} \tag{20.43}$$

As both catalytic activity and electrical conductivity vary simultaneously as functions of various parameters, it can be inferred that one of the active species is electrically charged. Because of the high temperature region, let us assume the existence of O^- as an active species involved in a Langmuir–Hinshelwood mechanism, where S is an adsorption site for CO:

$$CO(g) + S \rightleftarrows CO(ads) \qquad (I)$$

$$CO(ads) + O^- \rightleftarrows CO_2(ads) + e^- \qquad (II)$$

$$CO_2(ads) \rightleftarrows CO_2(g) + S \qquad (III)$$

$$O_2(g) + 2e^- \rightleftarrows 2O^-(ads) \qquad (IV)$$

The rate of the overall reaction corresponds to the rate in the adsorbed phase: $r = r_{II}\theta_{CO}[O^-(ads)]$. At steady state, the concentration $[O^-(ads)]$ must be constant:

$$d[O^-(ads)]/dt = 2k_{IV}P_{O_2}[e^-]^2 - k_{II}\theta_{CO}[O^-_{(ads)}] = 0 \tag{20.44}$$

Thus,

$$[e^-] = \frac{\{k_{II}\theta_{CO}[O^-(ads)]\}^{-1/2}}{(2k_{IV})^{1/2}P_{O_2}^{1/2}} = \frac{r^{1/2}}{(2k_{IV})^{1/2}P_{O_2}^{1/2}} \tag{20.45}$$

Consequently, at a given temperature, the electrical conductivity should vary as the square root of the ratio r/P_{O_2}. This is confirmed by the diagram $\sigma P_{O_2}^{0.5} = f(r)$ of Figure 20.13 (slope $+\frac{1}{2}$). The concentration in free electrons $[e^-]$ depends on two factors: (1) the semiconductor character of TiO_2 with $[e^-] = e_0\exp(-E_c/RT)$, where E_c is the activation energy of conduction without reaction ($E_c = 77.4$

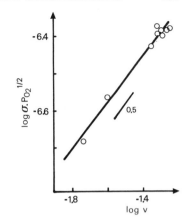

Figure 20.13. Linear relationship between the electrical conductivity of TiO_2 (measured *in situ* during CO oxidation), the oxygen pressure, and the reaction rate $r[\sigma P_{O_2}^{0.5} = f(r)]$.

kJ mol^{-1} obtained under nitrogen as a carrier gas); and (2) the catalytic reaction according to equation (20.45).

It follows that

$$\sigma = A[e^-] = C\exp(-E_c/RT)\{r^{1/2}/[(2k_{IV})^{1/2}P_{O_2}^{1/2}]\} \qquad (2.46)$$

Equation (20.46) enables one to interpret all the experimental results obtained. The reaction rate varies as

$$r = kP_{O_2}^{0.12}P^{0.72} \qquad (20.47)$$

The partial order of oxygen is small. This means that O^- species are observed almost at their saturation level whereas θ_{CO} varies as $P^{0.72}$. If the kinetic partial orders are introduced into equation (20.46), conductivity partial orders can be calculated with respect to P_{O_2} and P_{CO}. These orders are, respectively, equal to

$$n_{O_2} = 0.12/2 + (-0.5) = -0.44$$
$$n_{CO} = 0.72/2 = 0.36$$

These calculated values are in close agreement with the experimental values $(-0.51$ and $0.33)$. In addition, equation (20.46) accounts for the relationship among the various activation energies. From all the factors in equation (20.46) which vary exponentially with temperature, we derive

$$E_c' = E_c + E_a/2 - E_{IV}/2 \qquad (20.48)$$

where E_{IV} is the activation energy of reaction IV. Since the ionosorption of oxygen as O^- species is a very fast process, independent of temperature, E_{IV} can be considered to be zero. The calculated activation energy of conduction E_c during catalysis is equal to $77.4 + 102.5/2 = 129$ kJ mol^{-1} in reasonable agreement with the experimental value (144 kJ mol^{-1}).

In conclusion, the reaction mechanism (Reactions I to IV), based on the existence of ionosorbed O^- species as the oxygen-active intermediate enables one to account for all the experimental results concerning the kinetic and conductivity partial orders as well as the different activation energies of reaction and conduction. Moreover, it is possible to reject other mechanisms which are described below, because they do not give equation (20.45). For instance, active oxygen species cannot be O^{2-} as that would cause σ to vary as $P_{O_2}^{-1/4}$ ($\sigma \propto r^{1/2}/P_{O_2}^{1/4}$). They also cannot be O_2^- species, since in that case σ would vary as $\sigma \propto r/P_{O_2}$, which is not consistent with the experimental results.

20.7.2. Photocatalytic Oxidation of Isobutane on TiO_2

Various semiconductor oxides such as TiO_2, ZnO, ZrO_2, SnO_2, WO_3, CeO_2, Sb_2O_4, and Fe_2O_3 become photoconductors and active in photocatalysis when they are illuminated with photons whose energy is equal to or greater than their band gap energy.[8]

$$SC + h\nu \rightarrow e^- + p^+ \qquad (h\nu \geq E_g) \qquad (20.49)$$

where e^- and p^+ are the photoproduced electrons and holes.

20.7.2.1. Identification of Photoadsorbed Oxygen Species

At room temperature, associatively and dissociatively chemisorbed oxygen can ionosorb by capturing photoelectrons:

$$O_2(g) \rightleftarrows O_2(ads) \qquad (20.50)$$

$$O_2(g) \rightleftarrows 2O(ads) \qquad (20.51)$$

$$O_2(ads) + e^- \rightarrow O_2(ads) \qquad (20.52)$$

$$O(ads) + e^- \rightarrow O^-(ads) \qquad (20.53)$$

A stationary state is obtained when photoadsorption is compensated by photodesorption, occurring via the neutralization by positive photoholes p^+ of negatively charged oxygen species:

$$O_x^- + p^+ \rightarrow (x/2)O_2(g) \qquad (x = 1 \text{ or } 2)$$

Consequently for the photoconductivity σ measured at equilibrium, the isotherm is a function of P_{O_2} of the form

$$\sigma = \sigma_0 + \alpha(1 - R)\Phi/(k_{52}K_{50}P_{O_2} + k_{53}K_{51}^{1/2}P_{O_2}^{1/2}) \qquad (20.54)$$

where σ_0 is the conductivity in the dark, α the adsorption coefficient, R the reflectance, and Φ the light flux. Then σ varies as $P_{O_2}^{-1}$ if oxygen is the major photoadsorbed species or as $P_{O_2}^{-1/2}$ if oxygen photosorbs dissociatively. The results obtained for various oxides are given in Table 20.1.

TABLE 20.1. Photoadsorbed Oxygen Species on Various Oxide Photocatalysts

Oxide	$\dfrac{\partial \log(\sigma - \sigma_0)}{\partial \log P_{O_2}}$	Photoadsorbed oxygen species	Oxidative photocatalytic activity
TiO_2	$-1; -1/2$	O_2^- [a]; O^-	Very high
ZrO_2	$-1/2$	O^-	High
ZnO	$-1/2$	O^-	Medium
CeO_2 $\Big\}$ Sb_2O_4	$O(P_{O_2} < 0.5 \text{ Torr})$ $\Big\}$ $-1/2(P_{O_2} > 0.5 \text{ Torr})$	O^-	Small
SnO_2	$-1/2$	O^-	Very small
WO_3	$-1/2$	O^-	Very small
V_2O_5	0	—	None

[a] Also detected by EPR.

20.7.2.2. Nature of the Photoactive Oxygen Species on TiO_2

To determine the nature of the oxygen photoactive species as well as that of its precursor (O_2 or O^-), a simultaneous measurement of photoactivity a and photoconductivity σ was performed in a photocell reactor of the following dynamic type.[9] The reaction chosen was the photocatalytic oxidation of isobutane into acetone. As for the oxidation of CO on TiO_2 (see Section 20.6.1), simultaneous variations of σ and a were obtained as a function of the pressures of the reactants:

$$\sigma = k P_{iso}^0 P_{O_2}^{-1}$$
$$a = k_a P_{iso}^{0.35} P_{O_2}^0 \tag{20.55}$$

The kinetic order of isobutane indicates a Langmuir-type isotherm with a coverage θ_{iso} varying as

$$\theta_{iso} = K P_{iso}/(1 + K P_{iso}) \approx K' P_{iso}^{0.35} \tag{20.56}$$

The exponent 0.35 was verified by photogravimetry measurements. The zero order of isobutane with respect to conductivity indicates that this reactant is not part of a (photo)-electronic interaction with the solid. For oxygen, its partial order relative to σ is equal to -1, which means that under the conditions used, oxygen interacts with the solid as the O_2^- photoadsorbed species. However, the partial kinetic order of oxygen is nil, which means that the O_2^- species, which controls the photoelectronic interaction between gaseous O_2 and TiO_2, is not involved in the photocatalytic process. The zero kinetic order of oxygen implies that the photoactive species and its precursor are in the saturation state. This is the case of O^- species, which can produce photoactive oxygen according to the reaction:

$$O^-(ads) + p^+ \to O^*(ads) \tag{20.57}$$

The activation O* species is very active and unstable because of its excess energy and its electrophilic character, and it will react spontaneously with a substrate molecule M:

$$M(ads) + O^*(ads) \to [MO(ads)]^* \begin{array}{c} \longrightarrow MO(g) \\ \underset{O_2}{\longleftarrow} \\ \longrightarrow \text{Final products} \end{array} \tag{20.58}$$

The activated complex [MO*] can lose its excess energy either by desorbing or by subsequently reacting with adsorbed oxygen to produce more oxygenated final products. The absence of a molecular photoactive oxygen species, shown by simultaneous σ and a measurements, has been corroborated by several experimental facts: (1) the absence of hydroperoxides as intermediates; (2) the absence of O_2^- photoadsorbed oxygen species on all the photocatalysts other than TiO_2 (see Table 20.1); and (3) the triatomic mechanism of oxygen isotopic exchange:

$$^{18}O - {}^{18}O + {}^{16}O_S \to \quad \overset{\displaystyle {}^{18}O{-}{-}{-}{}^{18}O}{\underset{\displaystyle {}^{16}O_S}{\diagdown\!\diagup}} \quad \to {}^{16}O - {}^{18}O + {}^{18}O_S \tag{20.59}$$

20.7.2.3. Consequences for other Photocatalytic Oxidation Reactions

The photocatalytic process based on equations (20.57) and (20.58) has been tested on various oxides listed in Table 20.1 with many oxidizable substrates M, both inorganic (CO, I^-, Br^-, NH_3, SO_3^{2-}) and organic [isobutane, propene, cyclohexane, alkyltoluenes $r - \varphi - CH_3$ (with $R = CH_3$; C_2H_5; $i - C_3H_7$; $t - C_4H_9$; $O - CH_3$; Cl, ...), aliphatic alcohols C_1–C_4; oxalic acid...]. The best catalysts are those which exhibit the best photon absorption combined with the best adsorption capacity for the reactants. Titania was always found to be the best catalyst in all the reactions tested.

From a fundamental point of view, the photoconductivity measurements allow the definition of three criteria to determine a priori if a solid can be photocatalytically active in oxidation reactions.

1. The solid must exhibit an increase in conductivity when illuminated under vacuum (ability to produce photoelectrons and photoholes).
2. The solid must exhibit an increase in conductivity when illuminated in oxygen (ability to photosorb oxygen).
3. The photoconductivity must vary as $P_{O_2}^{-1/2}$ (ability to photosorb oxygen in the form of O^-, precursor of the active species O*).

The existence of O* is corroborated by using NO instead of O_2 as the oxidizing agent. NO photosorbs as NO^- (detected by photoconductivity as $\partial \log \sigma / \partial \log P_{NO} = -1$) and then decomposes after neutralization by photoholes:

$$
\begin{aligned}
NO(g) &\rightleftarrows NO(ads) \\
NO(ads) + e^- &\rightarrow NO^-(ads) \\
NO^- + p^+ &\rightarrow NO^* \\
NO^* &\rightarrow O_s + N^\cdot
\end{aligned}
$$

$$
\begin{aligned}
&\xrightarrow{+N} N_2 \quad (\text{Low } P_{NO}) \\
&\xrightarrow{+NO} N_2O \quad (\text{High } P_{NO})
\end{aligned}
\qquad (20.60)
$$

The atomic oxygen formed can oxidize various substrates such as aliphatic alcohols into the corresponding aldehydes or ketones.

20.8. SUPPORTED METAL CATALYSTS

Generally, metal catalysts are present in the form of small crystallites regularly distributed over all the particles of the support. Their content is often ≤ 1 wt.%, i.e., much smaller than the "percolation threshold," which indicates the level where a continuous percolation path constituted by the conducting component can exist between the two electrodes.[10] Consequently, the conductivity of supported metal catalysts is essentially that of the support and by comparing the results with those obtained on the bare support, some information can be deduced on the metal–support interactions.

20.8.1. Evidence for Hydrogen Spillover

The migration of atomic hydrogen from a metal crystallite onto the support is now a well-known phenomenon called "hydrogen spillover."[11] As well as other techniques (e.g., IR or NMR spectroscopies), electrical conductivity is a useful method since it responds only to the support, i.e., the solid onto which atomic hydrogen has spilt over from the metal.

After a standard reducing treatment (hydrogen at 200 °C, vacuum at 400 °C, cooling to room temperature), the electrical conductivity of a supported metal catalyst increases with the hydrogen pressure. This was observed for Pt, Rh, Ni/TiO_2, Pd/ZnO, and Ni/CeO_2. An electrical conductivity isotherm (Figure 20.14, curve A) gives a linear transform (Figure 20.14, curve B) by the equation

$$
\sigma = a + b P_{H_2}^{1/2} \qquad (20.61)
$$

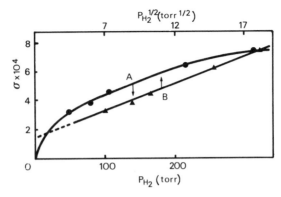

Figure 20.14. Pt/TiO$_2$: electrical conductivity isotherm $\sigma = f(P_{H_2})$ (curve A), and its linear transform $\sigma = a + bP^{1/2}$ (curve B).

This relationship can be explained by the following processes:

(a) Dissociative chemisorption of hydrogen on a surface metal atom M$_s$:

$$\tfrac{1}{2}H_2 + M_s \rightleftarrows M_s - H \qquad (20.62)$$

(b) Migration of atomic hydrogen on the surface of the oxide (hydrogen "spillover") via its protonation in contact with O^{2-} anions and release of a free electron of conduction:

$$M_s - H + O^{2-} \rightleftarrows M_s + OH^- + e^- \qquad (20.63)$$

(c) Alignment of the Fermi levels of the metal and the support with a supplementary migration of electrons from the support to the metal:

$$e^- + M \rightleftarrows e_M^- \qquad (20.64)$$

At a stationary state, the electrical conductivity remains constant, so we have

$$d\sigma/dt \propto d[e^-]/dt = r_{63} + r_{-64} - r_{-63} - r_{64} = 0 \qquad (20.65)$$

By substituting the rates in equation (20.65), we obtain an equation for the electrical conductivity:

$$\sigma \propto [e^-] = \frac{k_{-64}[e_M^-]}{k_{64}[M] + k_{-63}[M_s][OH^-]} + \frac{k_{63}K_{62}[M_s][O^{2-}]}{k_{64}[M] + k_{-63}[M_s][OH^-]} P_{H_2}^{1/2} \qquad (20.66)$$

Equation (20.66) is formally identical to equation (20.61) and accounts semi-quantitatively for the results obtained. The ordinate at the origin a contains a term $[e_M^-]/[M]$ that relates to the electron enrichment of the metal, which substantially increases when the solid is in the strong metal-support interaction (SMSI) state (see Section 20.8.2). The slope b, which contains the ratio $[M_s]/[M]$, shows that the extent of hydrogen spillover will be greater, the greater the dispersion of the metal.

Spilt over hydrogen is present as a proton. Kinetic measurements of conductivity $\sigma = f(t)$ have shown that hydrogen spillover is more rapid on fully hydroxylated or hydrated surfaces because of a higher mobility of the protons. From the catalytic point of view, it was shown that hydrogen spillover was not affected by UV illumination. As a consequence, it was established that all the photocatalytic reactions involving hydrogen (alcohol dehydrogenation, deuterium–alkane isotopic exchange) occurred on the support and that H_2 (or HD) was involved in the gas phase from the metal crystallites after a reverse spillover of H (or D) atom at the surface of titania.

20.8.2. Electronic Origin of Strong Metal–Support Interactions

Ten years ago, workers at Exxon[12] showed that when a metal is deposited on a reducible support such as titania, it acquires peculiar properties. If the sample is reduced in hydrogen at 200 °C, its adorption capacity is normal, i.e., the dispersion and the metallic area determined by hydrogen chemisorption are in agreement with those determined by electron microscopy. On the other hand, if the reduction temperature is increased to 500 °C, the adsorption capacity for H_2 or CO falls sharply without any apparent modification of the particle size. In parallel, the catalytic activity in certain sensitive reactions such as hydrogenolysis is also strongly decreased. However, if the catalyst is reoxidized by oxygen (or oxygen-containing molecules such as H_2O), it recovers its normal adsorptive properties after a new reduction at low temperature. This effect was called strong metal-support interaction (SMSI). As the conductivity of a deposited metal catalyst refers to that of the support, it is possible, by comparison with a bare support, to discover possible electronic interactions between the two phases.

20.8.2.1. Reduction at Low Temperature (200 °C)

The variation of the electrical conductivity of Pt, Rh, and Ni/TiO₂ as a function of various sequences (reduction in hydrogen at 200 °C; desorption of hydrogen at 400 °C; cooling to room temperature; adsorption of hydrogen at room temperature) are given in Figure 20.15 and can be interpreted as follows: In hydrogen at 200 °C, the conductivity of titania supporting a metal is 1 to 2 orders of magnitude greater than that of bare titania. The presence of a metal favors the reduction of titania, which occurs with the formation of singly ionized anionic vacancies:

$$O^{2-} + H_2 \rightleftarrows H_2O + V_{O^{2-}}^+ + e^- \tag{20.67}$$

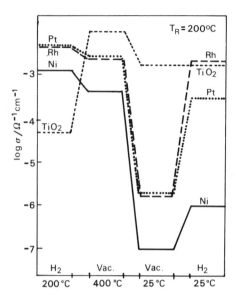

Figure 20.15. Pt, Rh, and Ni/TiO₂ vs. bare TiO₂: electrical conductivity variations during various sequences after reduction in hydrogen (250 Torr) at 200 °C.

The existence of $V_{O^{2-}}^+$ has been confirmed by the introduction of oxygen, which gives variations of σ as $P_{O_2}^{-1/4}$ as described in Section 20.3.1.2. The subsequent evacuation at 400 °C, a temperature high enough for desorbing all the hydrogen present, causes an increase of $\sigma(TiO_2)$ and a decrease of $\sigma(M/TiO_2)$, which has become lower than $\sigma(TiO_2)$. The increase of $\sigma(TiO_2)$ is due to the formation of new anionic vacancies by increasing T from 200 to 400 °C. The same phenomenon occurs for M/TiO₂ but a simultaneous stronger effect — the migration of free electrons from the support to the metal — makes $\sigma(M/TiO_2)$:

$$e^- + M \rightleftarrows e_M^- \tag{20.68}$$

Reaction (20.68) occurs in agreement with the alignment of the Fermi levels of M and TiO₂ according to the values of their work functions.

When the solid is cooled down to room temperature, $\sigma(TiO_2)$ varies little: TiO₂ has become a degenerate semiconductor (activation energy of conduction is zero), i.e., a quasi-metallic conductor. On the other hand, because of the trapping at the metal particles [equation (20.68)]. M/TiO₂ has remained a semiconductor with an activation energy of conduction equal to 29, 31, and 42 kJ mol⁻¹ for platinum, rhodium, and nickel, respectively. Finally when hydrogen is introduced at room temperature (Figure 20.15, final sequence), there is an increase in $\sigma(M/TiO_2)$, unlike $\sigma(TiO_2)$, which is due to hydrogen spillover ($\sigma = a + bP_{H_2}^{1/2}$), discussed in Section 20.7.1.

20.8.2.2. *Reduction at High Temperature (500 °C)*

The same sequence in Figure 20.15 was repeated for a reduction temperature of 500 °C. In this case, the electrical conductivities of M/TiO_2 and TiO_2 were two orders of magnitude greater at 500 °C than after reduction at 200 °C. Both solids had become quasi-metallic conductors. The higher degree of reduction of titania makes the electron transfer from the semiconductor to the metal [equation (20.68)] more important. Consequently, one can correlate the SMSI state to an excess of electrons within the metal crystallites. The inhibition of hydrogen chemisorption after reduction at 500 °C could be accounted for in two different ways: First, the excess of electrons fills the d bands of the metal which makes them tend to a d^{10} configuration, at least for the outer atoms of the particle. Nickel, rhodium, and platinum would progressively resemble copper, silver, and gold, known as poor hydrogen adsorbents and poor catalysts for reactions involving hydrogen. Secondly, the excess of electrons within the metal can counteract hydrogen (or CO) chemisorption if this gas chemisorbs by forming dipoles at the surface with a donor character to the metal. The dipoles $M^{\delta-}$–$H^{\delta+}$ will be less easily formed if the metal is already negatively charged. This observation could explain the decrease from 92 to 75 kJ mol^{-1} of the initial heat of adsorption of hydrogen on Pt/TiO_2 when this solid is put in the SMSI state.[13] This electronic interpretation does not necessarily exclude the alternative view of decoration of the metal by TiO_x suboxide species as underlined by White.[14] All these considerations have been exhaustively discussed in a recent review by Resasco and Haller.[15]

20.9. GENERAL CONCLUSIONS

In this chapter, different applications of electrical conductivity measurements in the field of heterogeneous catalysis have been presented. This technique appears to be a valuable tool in various areas such as: (i) the existence of solid solutions and solubility limits; (ii) the evidence for doping effects; (iii) the determination of temperatures of phase transition; (iv) the identification of electrically charged active species; (v) the photosensitivity of certain catalysts; and (vi) the electronic interactions between a metal and its support.

Many other potential applications exist and have yet to be explored. Most of the favorable cases deal with chalcogenides (e.g., oxides and sulfides, in the form of pure, doped, or mixed polyphasic samples). A single limitation can be noted: clearly the study of catalysts supported on insulators is not possible, although electrical conductivity is sometimes induced by the action of light or reducing agents. Moreover semiconductor oxides such as TiO_2 are nowadays often used as supports, which makes possible the utilization of electrical conductivity as an investigation technique. Finally it is worth emphasizing that this technique can easily be used *in situ* during a catalytic reaction.

REFERENCES

1. B. M. Arghiropoulos and S. J. Teichner, *J. Catal.* **3**, 477 (1964).
2. M. Breysse, B. Claudel, M. Guénin, H. Latreille, and J. Véron, *J. Catal.* **27**, 275 (1972).
3. R. N. Blumenthal, P. W. Lee, and R. J. Panlener, *J. Electrochem. Soc.* **118**, 123 (1971).
4. J. M. Herrmann, *J. Chim. Phys.* **73**, 474 (1976); *J. Chim. Phys.* **73**, 479 (1976).
5. J. M. Herrmann, J. L. Portefaix, M. Forissier, F. Figueras, and P. Pichat, *J. Chem. Soc., Faraday Trans. I* **75**, 1346 (1979).
6. J. M. Herrmann, P. Vergnon, and S. J. Teichner, *Bull. Soc. Chim. Fr.* (1976), p. 1056.
7. C. M. Osburn and R. W. West. *J. Chem. Solids* **32**, 1331 (1971).
8. J. M. Herrmann, J. Disdier, and P. Pichat, *J. Chem. Soc., Faraday Trans. I* **77**, 2815 (1980).
9. J. M. Herrmann, J. Disdier, M. N. Mozzanega, and P. Pichat, *J. Catal.* **60**, 369 (1979).
10. A. Ovenston and J. R. Walls, *J. Phys.* **18**, 1859 (1985).
11. G. M. Pajonk, S. J. Teichner and J. E. Germain (ed.), *Spillover of Adsorbed Species, Studies in Surface Science and Catalysis,* Vol. 17, Elsevier, Amsterdam (1983).
12. S. J. Tauster and S. G. Fung, *J. Catal.* **55**, 29 (1978).
13. J. M. Herrmann, *J. Catal.* **118**, 43 (1989).
14. D. M. Belton, Y. M. Sun, and J. M. White, *J. Phys. Chem.* **88**, 5172 (1984).
15. D. E. Resasco and G. L. Haller, *Adv. Catal.* **36**, 173–235 (1989).

Literature

E. Spenke, *Semiconducteurs électroniques,* Dunod, Paris (1959).
A. Many, Y. Golstein, and N. B. Grover, *Semiconductor Surfaces,* North Holland, Amsterdam (1971).
N. B. Hannay, *Solid-State Chemistry,* Prentice-Hall, Englewood Cliffs, NJ (1967).
J. P. Suchet, *Chimie Physique des Semiconducteurs,* Monographie Dunod, Paris (1972).
A. Rose, *Photoconductivité: Modèles et problèmes annexes,* Monographie Dunod, Paris (1966).
R. H. Bube, *Photoconductivity of Solids,* John Wiley and Sons, New York (1967).
S. R. Morisson, *The Chemical Physics of Surfaces,* Plenum, New York (1977).
J. Mort and D. M. Pai, *Photoconductivity and Related Phenomena,* Elsevier, Amsterdam (1976).

MAGNETIC MEASUREMENTS AND CATALYSIS

J.-A. Dalmon

21.1. INTRODUCTION

Among the various theories dealing with catalysis, many try to link catalytic and electronic properties. As electronic properties are often related to the magnetic state of the active phase, it is important to study catalysts and catalysis in relation to magnetism.

Catalysts very often present paramagnetic or ferromagnetic properties. There have long been susceptibility and magnetization measurement techniques available to obtain information concerning the physicochemical state of the active phase and its adsorption properties. These studies, initiated by P. W. Selwood,[1,2] have been applied most often to para- and ferromagnetic catalysts. After a description of the main features of various magnetic states, the following topics are covered in this chapter: (1) a brief description of the experimental methods, (2) catalyst characterization, and (3) a study of adsorbed phases and their relation to catalysis.

21.2. DIFFERENT MAGNETIC STATES

This section provides a simple description of some aspects of magnetism that can be related to catalysis. Readers interested in more details and theory can refer to specialized works.[3-6] Definitions and symbols are given in the Appendix.

J.-A. Dalmon ● Institut de Recherches sur la Catalyse, CNRS, Villeurbanne, France.

Catalyst Characterization: Physical Techniques for Solid Materials, edited by Boris Imelik and Jacques C. Vedrine, Plenum Press, New York (1994).

21.2.1. Paramagnetism

21.2.1.1. Paramagnetism due to Ions

This is the magnetism of ions of rare-earth and transition elements. Magnetic moments usually range from 1.7 to 10 Bohr magnetons (BM) and susceptibilities are of the order of 10^{-4} emucgs.

The first description of paramagnetism was given by Langevin in terms of classical mechanics. In this model the magnetic moments, which can take any direction, are subjected to the orientating effect of the applied field and to thermal agitation. Boltzmann-type statistics lead to the Langevin law, which gives the magnetization M of an ensemble of N paramagnetic moments μ vs. temperature T and field H:

$$M = M\mu[\coth(\mu H/kT) - (kT/\mu H)] \qquad (21.1)$$

$N\mu$ (all the moments in the direction of the field) is the saturation magnetization M_s. When experimental conditions are far from saturation ($\mu H \ll kT$), which is usually the case, equation (21.1) can be written

$$M = N\mu(\mu H/3kT) \qquad (21.2)$$

and the magnetic susceptibility $\chi = dM/dH$ is given by

$$\chi = N\mu^2/3kT = C/T \qquad (21.3)$$

where C is a constant. Equation (21.3) is the Curie law, characteristic of paramagnetic compounds, which is only valid when there is no interaction between magnetic moments, as is the case for high dilutions. However, these interactions often exist and can be accounted for by assuming that there is an additional field, dependent on M. Equation (21.3) can then be written

$$\chi = C/(T - \Theta) \qquad (21.4)$$

In this Curie–Weiss law, Θ, the Weiss constant, represents the interaction between the magnetic moments. Combining equations (21.4) and (21.3) leads to

$$\mu = 2.84[\chi(T - \Theta)]^{1/2} \qquad (21.5)$$

This equation, where μ is expressed in BM and χ in emucgs, allows calculation of the magnetic moment of ions, μ, from experimental data (χ, T, Θ).

This classical model was revised by Hund and Van Vleck on the basis of quantum mechanics: μ is a quantum concept and is created by the applied field through changes in energy levels. The magnetization is now linked to H and T

by a Brillouin-type function:

$$M = (N\beta g/2)\{(2J + 1)\coth[\beta g(2J + 1)H/2kT] - \coth(\beta gH/2kT)\} \qquad (21.6)$$

where β is the Bohr magneton, g the Landé splitting factor, and J the total quantum number. Equation (21.6) tends to (21.1) when $J \rightarrow \infty$, in good agreement with Langevin's intial hypothesis.

For low values of H/T

$$M = N[J(J + 1)/3kT]\beta^2 g^2 H \qquad (21.7)$$

Combining equations (21.2) and (21.7) leads to

$$\mu = g[J(J + 1)]^{1/2} \qquad (21.8)$$

For ions of transition elements only the spin component is important and equation (21.8) may be written as

$$\mu = 2[S(S + 1)]^{1/2} \qquad (21.9)$$

where S is the spin quantum number equal to half the number of unpaired spins in the atom.

21.2.1.2. Paramagnetism due to Metals: Pauli Spin Paramagnetism

This is the magnetism of various metals frequently used in catalysis such as Pt, Pd, Ru, Rh, Re, Os, and Ir. Magnetic susceptibilities are smaller than for ions and are weakly temperature-dependent.

The transition from atomic to metallic state involves a complete rearrangement of the electronic levels leading to the so-called band structure. In this model, when no field is applied, the spin-up and spin-down bands are equally filled; in the presence of an external field some electrons are shifted from one to the other half-band which creates the magnetization. The Pauli magnetic susceptibility is therefore proportional to the density of states at the Fermi level.

As the potential energy induced by the magnetic field is small with respect to the maximum energy of the occupied states, the Pauli susceptibility is small (in comparison with ionic susceptibilities); for the same reason the effect of the temperature is limited.

21.2.2. Ferromagnetism and Superparamagnetism

The analogies between magnetization curves and variations of the density $vs.$ pressure relationships (paramagnetism = gas, ferromagnetism = liquid) led P. Curie to propose positive interactions between magnetic moments in the ferromagnetic state. To account for this observation Weiss represented this interaction by a magnetic field, the molecular field H_m, which is proportional to the actual magnetization and adds its effects to the experimental field H_e. Equation (21.2), which is valid at high temperatures, can be written as

$$M = N\mu[\mu(H_e + H_m)/3kT] \tag{21.10}$$

If we let $K = N\mu^2/3k$ and set $H_m = wM$, as is assumed in the Weiss hypothesis,

$$M = KH/(T - T_c) \tag{21.11}$$

with $T_c = wK$. Equation (21.11), which is valid at high temperatures, shows that when T decreases down to T_c, $\chi = M/H$ becomes infinite and M thus has a finite value in a zero field, which is a characteristic of the ferromagnetic state. T_c, the Curie point, is the temperature of transition between the para- and ferromagnetic states. Equation (21.11) predicts a spontaneous magnetization (i.e., a magnetization in the absence of an applied field) for ferromagnetic materials, which is, however, not always the case. Weiss therefore proposed that this spontaneous magnetization is restricted to small domains dividing the solid; as these domains are magnetically randomly oriented, the overall effect is zero. The applied magnetic field induces shifts in the walls dividing the domains, which produces the magnetization.

As for paramagnetism, the classical model for ferromagnetism was revised on the basis of quantum mechanics and the positive interaction of the magnetic moments originates in an exchange energy between itinerant electrons. In the band model spontaneous magnetization exists if the two half-bands are unequally filled in the absence of an applied field, a state which depends on the energy balance between the kinetic and exchange terms. Stoner showed that ferromagnetism will occur if

$$UN(E_F) > 1 \tag{21.12}$$

where U is the intra-atomic exchange energy and $N(E_F)$ the density of state at the Fermi level. The calculation shows that $UN(E_F)$ is generally near 1, which indicates that small perturbations (surface, adsorption) could modify the magnetic state of the atom.

For ferromagnetic metals the Fermi level may fall outside one of the half-bands (Ni) or span both half-bands (Fe).

When the ferromagnetic material is finely divided, the small particles can be single magnetic domains. This state is generally observed when the particle size is under about 20 nm, which is the case for most dispersed catalysts. As all the atomic moments μ_0 of the particle are parallel, the particle has a giant moment μ proportional to n, the number of atomic moments; μ therefore depends on the particle size:

$$\mu = n\mu_0 \tag{21.13}$$

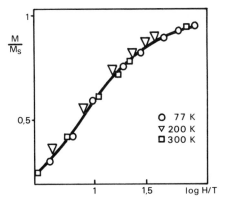

Figure 21.1. Superparamagnetic behavior on a NiCu/SiO$_2$ catalyst (M: magnetization, M_s: saturation magnetization, H: magnetic field, T: temperature).[25]

The state in which these giant moments and the applied field are in thermodynamic equilibrium is called superparamagnetism. Owing to thermal agitation, in the absence of a field, these moments will be randomly oriented. If a field H is applied, each moment undergoes a couple μH. These were Langevin's starting hypotheses in his description of paramagnetism. The magnetization M of an ensemble of N superparamagnetic particles whose magnetic moments are equal to μ will thus follow a Langevin-type function:

$$M = N\mu[\coth(\mu H/kT) - (kT/\mu H)] \qquad (21.14)$$

with M_s (saturation magnetization) equal to $N\mu$.

The experimental determination of superparamagnetism is based on equation (21.14): M/M_s vs. H/T plots should give the same curve (Figure 21.1). Equation (21.14) can be written as

$$M = M_s(\mu H/3kT) \qquad (21.15)$$

for low fields (or low H/T ratios) and as

$$M = M_s(1 - kT/\mu H) \qquad (21.16)$$

for high fields (or high H/T ratios).

Equation (21.16) shows that it is possible to obtain M_s by extrapolating a graph of $M(1/H)$ to $1/H = 0$ (for applications, see the Appendix).

21.3. SHORT DESCRIPTION OF EXPERIMENTAL METHODS

Experimental techniques can be divided into (i) classical methods, especially designed for magnetic measurements and described here, and (ii) methods that originated in other physical measurements and are described in other chapters: ferromagnetic resonance (Chapter 7), Mössbauer resonance (Chapter 8), and neutron diffraction (Chapter 13).

21.3.1. Static Methods

Static methods are often based on the Faraday method[7-9] and are used regularly for measuring paramagnetic susceptibilities. The principle of this method is the following: When a paramagnetic compound is placed in a non-uniform magnetic field H having a gradient in a direction x, a force F is applied to the compound such that

$$F = m\chi H(dH/dx) \tag{21.17}$$

Most experimental devices use an electromagnet with asymmetrical pole pieces that produces a large field gradient. F can be measured in various ways, e.g., with a torsion balance or an induction coil attached to the sample holder. Apart from these methods based on equation (21.17), other systems such as permeameters, which use the change of inductance in a coil that results from the presence of a magnetic sample, have been developed.[8,10]

21.3.2. Dynamic Methods

Most dynamic methods make use of the variation of flux induced in a coil placed in a magnetic field when moving the magnetic sample (Weiss extraction method) or when changing the magnetization of the sample.[11,12] These methods are often devoted to measuring the magnetization curves of well-dispersed ferromagnetic samples and thus need high magnetic fields, which can be provided by electromagnets or superconducting coils. These systems have been improved significantly through the use of integrators and calculators and, more recently, of SQUID (superconducting quantum interference device based on the Josephson effect).

The sample is normally placed in a glass or metallic cell which can be connected to other systems for adsorption or catalytic activity measurements. With the use of different devices (a specially designed cryostat or oven) the magnetic measurements can be performed over a large range of temperatures.

21.4. MAGNETIC CHARACTERIZATION OF CATALYSTS

21.4.1. Nonmetallic Species

Most of this information is obtained through paramagnetic susceptibility measurements. Equation (21.5) allows us to calculate μ, the atomic magnetic moment, from experimental data (χ, T, Θ).

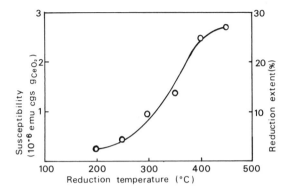

Figure 21.2. Ce^{4+} (CeO_2) reduction followed by susceptibility measurements.[16]

For ions of transition elements [equation (21.9)] μ is directly related to the number n of unpaired electrons (spin-only contribution). In the case of Ni^{2+} in its ground state the electronic configuration has two unpaired electrons ($3d^8$, $S = 1$), which corresponds to $\mu = 2.83$ BM, in good agreement with experimental observations for many Ni^{2+} salts. For iron, several oxidation states exist and susceptibility measurements can indicate whether Fe^{2+} ($3d^6$; $\mu = 4.9$ BM) or Fe^{3+} ($3d^5$; $\mu = 5.9$ BM) species are present: in the course of a study dealing with bismuth molybdovanadates promoted by iron, magnetic measurements showed the presence of Fe^{3+} ions.[13] On unsupported cobalt-promoted molybdenum sulfide, magnetic susceptibility measurements gave evidence of the presence of low-spin, sulfur-complexed cobalt.[14] Recently, the oxidation states of iron species in zeolites have been determined through magnetic measurements.[15]

Another recent application of the use of susceptibility data to determine the oxidation state of elements concerns the study of the reduction of Ce O_2 by hydrogen.[16] The reduction of Ce^{4+} to Ce^{3+} (Figure 21.2) has been followed in a magnetic balance (Ce^{4+} has no unpaired spin and is therefore diamagnetic whereas Ce^{3+} is paramagnetic with a magnetic moment of 2.5 BM).

As indicated above, the Weiss constant Θ [equation (21.3)] gives information on the interaction between magnetic moments. Selwood[17] studied a series of alumina-supported chromium oxides and measured μ and Θ for different chromium loadings. When μ remains almost constant for chromium loadings ranging from 1 to 35 wt.%, Θ decreases rapidly under 5 wt.%. This decrease was attributed to a weakening of the interaction between moments due to the formation of small chromium oxide islands dispersed on the alumina carrier. An increase in activity (n-heptane dehydrocyclization) parallels this decrease in Θ, in good agreement with better access of the gaseous phase to the active surface due to the formation of the small chromium patches. More recent studies[17,18] made use of similar measurements.

21.4.2. Metallic Species

Most of the studies of metallic species deal with ferromagnetic metals or alloys by means of magnetization measurements. Information concerning the amount of metal in the metallic state (or ferromagnetic metal titration), the extent of alloy formation, and the particle size distribution can be obtained. These results are deduced from saturation magnetization, Curie points, or magnetization curve (M vs. H) determinations.

21.4.2.1. Ferromagnetic Metal Titration

Due to the collective effect of ferromagnetism, small amounts of metal in the zero state can easily be detected via saturation magnetization (M_s) determination: plots of M vs. $1/H$ lead to M_s by extrapolating m to $1/H = 0$ [see equation (21.16)], Section 21.2.2 and the Appendix). The amount of ferromagnetic metal is obtained by dividing M_s by the specific magnetization of the metal at the experimental temperature. It needs to be emphasized that when the metal is very finely divided accurate determination of M_s can be difficult (see below).

This ferromagnetic metal titration has been used extensively for measuring the extent of reduction of nickel-based catalysts as a function of the reduction (activation) conditions.[19] In another study[20] the reducibility of supported nickel was shown to depend largely on the nature of the support due to the degree of interaction between unreduced Ni^{2+} species and the support (Figure 21.3). For cobalt-based catalysts the effects of different promotors (Cu, Zn, Na) on the reduction of cobalt have also been studied and compared with TPR data.[21]

Apart from these reduction studies, magnetic titrations of ferromagnetic metals have also been used to follow deactivation phenomena, especially when the deactivation processes involve a loss of ferromagnetic metal (as, e.g., during CO dismutation or hydrogenation on nickel-based catalysts which leads to the formation of inactive, nonferromagnetic, nickel carbides).[22] Magnetic titrations associated with TPH (temperature programmed hydrogenation) experi-

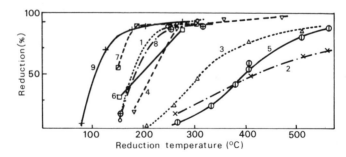

Figure 21.3. Nickel reduction as a function of the support: (1) ○ unsupported Ni, (2) ✕ Ni/MgO, (3) △ Ni/Al$_2$O$_3$, (4) ▽ Ni/Cr$_2$O$_3$, (5) ◐ Ni/SiO$_2$, (6) □ Ni/TiO$_2$, (7) ▨ Ni/ThO$_2$, (8) ⊕ Ni/ZrO$_2$, and (9) + Ni/CeO$_2$.[20]

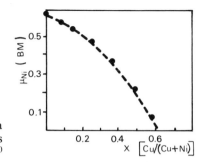

Figure 21.4. Nickel atomic magnetic moment as a function of the copper content in NiCu/SiO$_2$ catalysts (the dashed line corresponds to bulk Ni–Cu alloys).[25]

ments on carbon deposits can discriminate between carbon linked to the metal and fouling carbon.[22] For iron-based catalysts, carbides and oxides as well as the metal can have ferromagnetic properties, so the measurements of Curie temperatures can provided information concerning the various species that are present before or after reaction.[23]

These methods also gave results that indicated the formation of a strong Ni–Si (alloy-type) interaction in the so-called SMSI state for Ni/SiO$_2$ catalysts.[24]

21.4.2.2. Alloy Formation

The magnetic properties of ferromagnetic metals are heavily modified upon alloying.[8] M_s and Curie point determination can thus contribute to the study of the metal-A–metal-B interaction, in an AB bimetallic catalyst. Silica-supported NiCu catalysts have been characterized by these means.[25] Figure 21.4 shows that after reduction at high temperature, the specific magnetization of the supported metallic phase agrees well with the change in magnetization of bulk, homogeneous alloys. Moreover Curie point measurements[8] of these catalysts showed that experimental values can be determined easily (linear extrapolation) and they are very near those reported in the literature, suggesting good homogeneity of the alloyed phase.[25]

21.4.2.3. Particle Size Distribution of Superparamagnetic Samples

Let us consider an ensemble of superparamagnetic particles of various sizes. Each of them bears a magnetic moment μ which depends directly on the particle size (see Section 21.2.2). When no field is applied, all these magnetic moments are randomly orientated because of thermal agitation, the resulting effect being zero (Figure 21.5). In the presence of a moderate field H_1, the largest particles (with large magnetic moments μ) will respond to strong couples (proportional to μH_1) and will then orientate themselves in the field. Small particles, on the other hand, have small moments and the thermal agitation will be high enough to maintain a random distribution of their magnetic moments (Figure 21.5). High magnetic fields (or low temperatures) will then be necessary

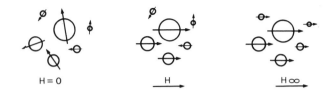

Figure 21.5. Magnetization as a function of different particle sizes.

to orientate the sample completely (Figure 21.5), that is, to obtain magnetic saturation M_s). The experimental curve M vs. H then reflects mainly the orientation of the large particles of the sample in its low-field part and that of the small particles in its high-field part.

It should be emphasized that this is only a qualitative description of the Langevin function[12]: Figure 21.6 shows the theoretical curves for different particle sizes and for a real catalyst.

Let us assume that the volume distribution $f_v(D)$ of the particle diameters of this real sample is such that

$$\int_0^\infty f_v(D)dD = 1 \tag{21.18}$$

where $f_v(D)dD$ is the volume fraction of the particles with diameters ranging from D to $D + dD$. The magnetic moment μ of a particle of diameter D is

$$\mu = (\pi\rho\sigma_s/6)D^3 \tag{21.19}$$

where ρ is the mass per unit volume and σ_s the saturation magnetization per unit mass).

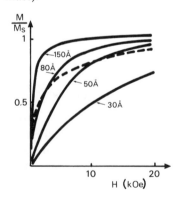

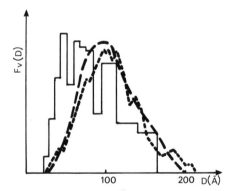

Figure 21.6. Magnetization curves as a function of particle sizes (T = 300 K). The dashed line corresponds to the experimental curve obtained on a heterodispersed sample.

Figure 21.7. Particle size distribution of a Ni/SiO$_2$ sample: (dashed line) magnetic distribution, (solid line) electron microscopy, and (dotted line) small-angle X-ray scattering.[26]

The specific magnetization σ of this ensemble of particles is linked to the applied field H and to the temperature T by a Langevin function:

$$\sigma = \sigma_s \int_0^\infty f_v(D)[\coth(\pi\rho\sigma_s HD^3/6kT) - (6kT/\pi\rho\sigma_s HD^3)]dD \quad (21.20)$$

This equation can be approximated for low H/T ratios:

$$\sigma = \sigma_s^2(\pi\rho H/18kT)\int_0^\infty f_v(D)D^3 dD \quad (21.21)$$

and for high H/T ratios:

$$\sigma = \sigma_s\left[1 - (6kT/\pi\rho\sigma_s H)\int_0^\infty f_v(D)D^{-3}dD\right] \quad (21.22)$$

These two equations can be used to calculate average diameters D_1 and D_2 for low and high H/T ratios, respectively:

$$D_1 = [\mathcal{M}_3]^{1/3} \quad (21.23)$$
$$D_2 = [\mathcal{M}_{-3}]^{-1/3} \quad (21.24)$$

with

$$\mathcal{M}_n = \int_0^\infty D^n f_v(D)dD \quad (n\text{th order moment}) \quad (21.25)$$

In the low-field part of the curve (low H/T ratios), the average diameter D_1 relates mainly to the large particles of the sample:

$$D_1 = [(\sigma/H)(18kT/\pi\rho\sigma_s^2)]^{1/3} \quad (21.26)$$

For the high-field part of the curve, the average diameter D_2 relates mainly to the small particles of the sample:

$$D_2 = [6kT/(\sigma_s - \sigma)\pi\rho H]^{1/3} \quad (21.27)$$

An application of the D_1 and D_2 calculations is given in the Appendix.

It should be noted that the surface average diameter D_s, which is linked to the specific metallic area, is expressed as

$$D_s = [\mathcal{M}_{-1}]^{-1} \quad (21.28)$$

Then the relation

$$[\mathcal{M}_{-3}]^{-1/3} < [\mathcal{M}_{-1}]^{-1} < [\mathcal{M}_3]^{1/3} \quad (21.27)$$

can be used to calculate the surface diameter D_s conveniently:

$$D_s = (D_1 + D_2)/2 \quad (21.30)$$

The difference between D_1 and D_2 gives an estimate of the particle size heterogeneity. One can then characterize the distribution width by the parameter $(D_1 - D_2)/D_s$.

The complete distribution, $f_v(D)$ vs. D, can be obtained by computer processing of the $M(H)$ curve.[26] Figure 21.7 shows that the results are in good agreement with those obtained by other physical methods.

In concluding this section, let us emphasize the importance of the use of high magnetic fields to saturate the sample when the metallic phase includes very small particles,[27] especially because these particles contribute a large part of the exposed active area of the sample.

21.4.2.4. Particle Size Distribution of Nonsuperparamagnetic Samples

For nonsuperparamagnetic conditions, the magnetic moments of some particles are not in thermodynamic equilibrium with the applied field. Such behavior is represented in Figure 21.8. In state 1 there is no applied field and the magnetic moments of the particles are randomly distributed along the axes of easy magnetization (see the Appendix). State 2 corresponds to magnetic saturation. After going back to zero field (state 3), the small particles, as a result of thermal agitation kT, have their magnetic moments randomly redistributed along the axes of easy magnetization, and their global effect on the sample magnetization is zero. In fact, the energy required for returning the magnetic moment along the axis of easy magnetization depends on the energy of magnetic anisotropy E_a, which is directly proportional to the volume v of the particle:

$$E_a = Kv \qquad (21.31)$$

For small particles $kT > Kv$, and they do not contribute to the magnetization in state 3 (these particles are superparamagnetic). Conversely, large particles have $Kv > kT$ and thermal agitation does not lead to their moments being random along the axes of easy magnetization. These moments will thus retain the direction of the magnetic field and have a preferential orientation nearest to the applied field. The moments of these large particles thus create a remanent magnetization M_r. For $T = 0$ (no thermal agitation) all the particles will have

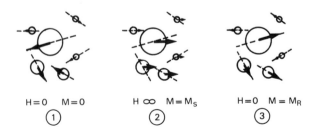

$$H = 0 \quad M = 0 \qquad\qquad H \propto \quad M = M_S \qquad\qquad H = 0 \quad M = M_R$$
$$(1) \qquad\qquad\qquad (2) \qquad\qquad\qquad (3)$$

Figure 21.8. Magnetization of a nonsuperparamagnetic sample as a function of the field (the dashed line represents the direction of easy magnetization).

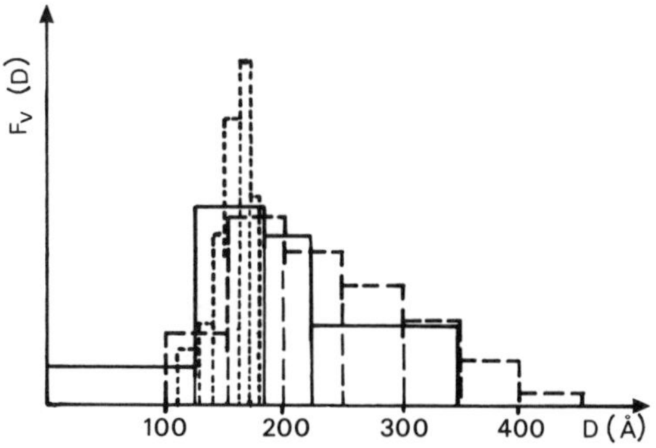

Figure 21.9. Particle size distribution of a Co/SiO₂ sample (nonsuperparamagnetic sample): (dashed line) magnetic (remanence method) distribution, (dotted line) magnetic (alternative field method) distribution, and (solid line) electron microscopy.

their moments blocked in this way, and the calculation shows that for particles with uniaxial anisotropy:

$$M_r/M_s = 0.5 \qquad (21.32)$$

For temperatures above 0 K, the thermal agitation will redistribute the magnetic moments of the smallest particles and the ratio M_r/M_s becomes smaller than 0.5 so that $0.5 - M_r/M_s$ is proportional to the fraction of those particles whose moments become randomly oriented. Besides the temperature T, the time t plays a role and Néel drew up the relation linking t, T, and the critical volume V above which the magnetic moments will contribute to M_r:

$$1/t = f_0 \exp(-V_c I_s H_c/2kT) \qquad (21.33)$$

where f_0 is a frequency factor of the order of 10^9 s^{-1}, I_s the magnetization per unit volume, and H_c the coercive field as measured at low temperatures ($V_c I_s H_c$ represents the energy of anisotropy E_a).

Values of M_r vs. T then allow calculation of the volume fraction of the particles of diameters greater than the critical diameter given by equation (21.33). This method of particle size determination is known as the remanence or Weil method.[28] Equation (21.33) also shows that parameter t can be used instead of T. If the sample is placed, for instance, in an alternative magnetic field (whose frequency will determine the relaxation time t), only particles of diameter greater than the critical diameter will contribute to the magnetic susceptibility.[29] Figure 21.9 shows results obtained by both methods.

21.5. STUDY OF ADSORBED PHASES AND THEIR RELATION TO CATALYSIS

21.5.1. Surface and Magnetism

All the magnetic data discussed above concern bulk properties. New questions arise when studying surface phenomena: e.g., what are the magnetic properties at the surface of the particle? are they affected by the size of the particle or by the nature of the exposed planes? Theoretical calculations have been developed recently on the basis of different considerations and models. In the case of metals, s–d rearrangements, exchange interactions, and changes in the density of states due to the surface have been used.[30,31] Another scheme[32] based upon a cross between the molecular clusters approach and tight-binding theory showed that the surface magnetic moment of nickel was little changed when compared to bulk values: increases of about 5% and 2% are observed for the $\langle 100 \rangle$ and $\langle 111 \rangle$ planes, respectively.

Experimental data have been obtained for both Pauli and ferromagnetic metals. On palladium, it was shown[33] that when a correction is made for ferromagnetic impurities, the susceptibility is partly affected by the dispersion of the supported palladium particles. The susceptibility of surface atoms ranges between 0.7 and 0.5 times that of bulk atoms.[33,34] Platinum susceptibility[35] and NMR[36] measurements indicate that the magnetic moment of surface atoms is also clearly lower than that of bulk atoms. All these results suggest that surface atoms of Pauli metals have a density of states at the Fermi level, $N(E_F)$, lower than that of bulk atoms.

More results are available for ferromagnetic metals. For nickel, saturation magnetization and Curie points of well-dispersed supported powders[37] or films[38] are in good agreement with bulk values (provided that the preparation keeps the samples free of Ni^{2+} species). Ferromagnetic resonance on films[39] and PLEED (polarized low-energy electron diffraction) on single crystals[40] also indicate that the nickel surface atom is ferromagnetic. All these results suggest that $N(E_F)$ is little modified at the surface, in good agreement with theoretical calculations.[31,32] Mössbauer spectroscopy experiments on iron and cobalt films[41] also indicate that the surface magnetic properties are similar to those of the bulk, which is supported by theoretical calculations of the surface local density of states.[32]

21.5.2. Study of Adsorbed Species

As surface atoms have magnetic properties, adsorption of gases on the surface may induce some modifications of these properties. The magnetic study of adsorption has been developed by different groups along the line of the pioneering work of Selwood in the 1950s.[2] Low-field techniques, e.g., permeameters[42–45] or high-field techniques using electromagnets[46–51] have been used. The two methods have been compared[52] and the high-field method appears to be better, especially if large nonsuperparamagnetic particles are present.

21.5.2.1. Pauli Paramagnetic Metals

On palladium, the adsorption of hydrogen induces a linear decrease of the susceptibility[33,34] This decrease corresponds approximately to the magnetic contribution of one surface palladium atom per adsorbed hydrogen atom, which

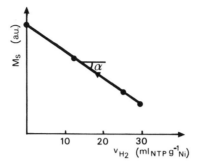

Figure 21.10. Magnetic effect [α (BM) per adsorbed molecule] of hydrogen adsorption on a Ni/SiO$_2$ sample: (circles) adsorption, and (triangle) desorption.

suggests that the surface palladium atom linked to hydrogen loses its paramagnetism. As Pauli susceptibilities are proportional to $N(E_F)$, the magnetic effect of hydrogen adsorption suggests a decrease in $N(E_F)$, which is in good agreement with UPS data related to hydrogen adsorption on the $\langle 111 \rangle$ plane of palladium.[53]

Apart from hydrogen, adsorptions of CO and C$_2$ hydrocarbons on supported palladium catalysts have also been studied by means of susceptibility measurements.[34] These adsorptions induce decreases (per adsorbed molecule) in the susceptibility greater than those associated with hydrogen. These effects have been interpreted on the basis of the model developed for nickel (see Section 21.5.3).

21.5.2.2. Ferromagnetic Metals

Nickel is by far the most studied of the ferromagnetic metals. The first experiments were performed by Selwood, who observed that adsorption on nickel induces a decrease in magnetization that depends on the nature of the gas.[1]

21.5.2.2a. Hydrogen. On supported nickel powders, the magnetic isotherm (saturation magnetization M_s vs. coverage) is linear and reversible (Figure 21.10), which suggests that the gas–metal interaction is coverage-independent. The slope of the isotherm, α_{H_2}, expressed in Bohr magnetons (BM) per adsorbed molecule is about 1.4.[2] As μ_{Ni}, the atomic magnetic moment of nickel, is equal to 0.6 BM, we can write

$$\alpha_{H_2} \approx 2\mu_{Ni} \qquad (21.34)$$

This relation suggests that the dissociative adsorption of hydrogen results in the disappearance of the ferromagnetism of two nickel surface atoms linked to hydrogen. The decrease of magnetization upon hydrogen adsorption has also

TABLE 21.1. Values of α_{H_2} for Three Ferromagnetic Methods

Metal	α_{H_2} (BM)
Ni	1.4
Co	0.3, 0.7
Fe	0

been observed on nickel single crystals[54] and films.[55,56] As in the case of palladium, dissolution of hydrogen in nickel results in a decrease of magnetization which is close to that due to adsorption.

Fewer studies have been devoted to the magnetic effects of hydrogen adsorption on cobalt or iron surfaces. Two states were detected (at 20 and 200 °C) on silica-supported cobalt by means of magnetization measurements[57]; in contrast to nickel, the slope of the isotherm indicates that the H–Co interaction does not induce complete magnetic disappearance of the surface atom. On metallic iron powders, hydrogen adsorption has no effect on the magnetic properties[58] (note that in the presence of unreduced Fe^{x+} species a redox equilibrium can activate hydrogen via a mechanism that is different from chemisorption on the metal[58]).

The values of α_{H_2} for the three ferromagnetic metals are listed in Table 21.1.

21.5.2.2b. The Magnetic Bond Number. Equation (21.34) has been tentatively generalized by Selwood to the adsorptions of various gases on nickel. For a given gas x equation (21.34) leads to

$$\alpha_x = n_x \mu_{Ni} \qquad (21.35)$$

where α_x is the magnetic effect (in BM) of the adsorption of one molecule of x. The term n_x, which represents the number of surface nickel atoms that are magnetically decoupled by the adsorption of the molecule x, has been called the magnetic bond number.[2] However, this concept appears to be suitable only in the case of nickel. Experimentally, hydrogen chemisorption on cobalt or iron does not induce magnetic effects that are whole multiples of the atomic magnetic moments.

21.5.2.2c. Carbon Monoxide on Nickel. Equation (21.35) received some experimental verification in the course of a combined IR-magnetic study of CO adsorption on silica-supported nickel and NiCu catalysts.[48,49] On pure nickel, CO adsorption leads to $n = 1.8$ and gives IR bands at 2050 and 1950 cm^{-1}. Upon increasing the copper content in the NiCu alloy the 1950 cm^{-1} band gradually disappears and the magnetic bond number n tends to 1. The 2050 cm^{-1} band ($n = 1$) corresponds to the so-called linear species where CO is linked on the top of one nickel surface atom. The comparison between n

and the ratio of the intensities of the two bands for different samples shows that the 1950 cm^{-1} band can be assigned to a bridged species linked to two surface atoms ($n = 2$). The good agreement between the IR and magnetic results verifies equation (21.35) and suggests that the so-called magnetic bond number n could correspond to the stoichiometry of the chemical bond between the adsorbed molecule and the surface. Magnetic measurements have therefore been used to obtain n for various adsorptions and to propose models for the corresponding adspecies.

21.5.2.2d. Hydrocarbons on Nickel. The adsorption of numerous hydrocarbons has been studied by magnetic methods.[2,45,46] Thermomagnetic curves corresponding to the plots of the parameter α of the hydrocarbon *vs.* the temperature of the hydrocarbon–catalyst system have been determined.[46,59] Figure 21.11 shows a typical curve. At low temperatures [the first part of the curve (1), $T < 0\,°C$] only physisorption occurs, which does not produce a magnetic effect. Upon increasing the temperature, chemisorption takes place (2) and different thermostable adspecies may appear, a plateau corresponding to one type of chemisorption. At higher temperatures [(3), $T > 100\,°C$] there is no drastic change as the adsorbed molecules are probably cracked (suggested by single-crystal results). Let us now consider this third state, which corresponds to the following transformation:

$$C_n H_m \;-\;-\;-\;\rightarrow\; nCH_x + (m - nx)H$$

Let us assume that the global magnetic effect observed in this state, $\alpha_{(3)}$, results from the addition of the magnetic effects of the cracking products:

$$\alpha_{(3)} = n\alpha_{CH_x} + (m - nx)\alpha_H$$

The solution of this equation has been carried out[50] on the basis of $\alpha_{(3)}$ data obtained for various hydrocarbons (e.g., alkanes, olefins, and aromatics) and

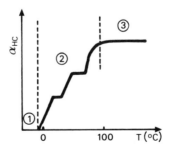

Figure 21.11. Typical thermomagnetic curve of hydrocarbon adsorption on nickel [(1) physisorption, (2) thermostable species, (3) complete cracking].

TABLE 21.2. Influence of Impurities on Magnetic Perturbations

Element	Adsorbed	Dissolved
H	0.7	0.6
C	1.8	3.5

leads to

$$x = 0$$

$$\alpha_H = 0.7 \text{ BM}$$

$$\alpha_C = 1.8 \text{ BM}$$

These results indicate that the carbonaceous species coming from the hydrocarbon is completely dehydrogenated and has a magnetic effect corresponding to a magnetic bond number of 3 [equation (21.35)]. This surface carbon (C_s) interacts with nickel through three surface atoms leading to the global formula Ni_3C_s. The hydrogen atom from the hydrocarbon has the same magnetic effect as one from molecular hydrogen.

Finally it should be noted that the magnetic perturbations on nickel as well as on palladium depend directly on the valence of the impurity adsorbed or dissolved (Table 21.2).

21.5.3. Interpretation of the Magnetic Effects

The analogies between the magnetic effects of adsorption and absorption led to similar interpretations for each.[60] Let us consider the presence of an impurity T at the surface (adsorption) or in the bulk (absorption) of a metal M; Z, the difference in electronic charge between T and M, will create a rearrangement of the M electrons around T to screen this charge Z. This screening corresponds to about one atomic distance and concerns d electrons.

For ferromagnetic metals, the change in magnetic properties due to the presence of T can be accounted for by the variations in the number of states $Z(+)$ and $Z(-)$ of the two half-bands. Then

$$\alpha = Z(+) - Z(-)$$

The total number of states that have been displaced should equal Z[60]:

$$Z = Z(+) + Z(-)$$

Let us consider the adsorption of one hydrogen atom on nickel. For this example $Z = 1$ and $\alpha = 0.7$, which leads to

$$Z(+) = 0.15 \text{ BM}$$
$$Z(-) = 0.85 \text{ BM}$$

The screening occurs mainly in the $(-)$ half-band and the number of states displaced in the $(+)$ half-band is low. In the case of iron no magnetic effect is observed (see experimental results) and then

$$Z(+) + Z(-) = 0.5 \text{ BM}$$

The screening is equally distributed in the two half-bands. This difference between nickel and iron is probably related to the electronic structures of the two metals (see Section 21.2.2). In the case of nickel, the Fermi level falls outside one of the half-bands and thus the screening is strongly asymmetrical; conversely in iron the Fermi level spans both half-bands leading to symmetrical screening.

The chemisorption of CO on small nickel clusters has also been studied from the theoretical viewpoint.[61] Interactions with Ni–CO antibonding orbitals lead to a destabilization of Ni(d) levels, which occurs predominantly in one of the half-bands. Support effects may affect this change in the magnetic properties.

Let us now consider the concept of magnetic bond number. As shown in Section 21.5.2, this concept is only valid for adsorption on nickel. Adsorption on nickel seems to "demetallize" the surface atoms,[62] which suggests that it leads to the formation of a surface compound, independent of the bulk metallic nickel, and the magnetic bond number will then give the stoichiometry of the surface compound. However, further studies[63,64] led to a better description of the phenomenon: the surface atom linked to the adsorbate remains in the metallic state but loses its magnetic component. This magnetic decoupling is very likely a localized effect as it corresponds to the perturbation of one surface atom. This local disappearance of ferromagnetism is probably related to the Stoner criterion of ferromagnetism [see equation (21.12)]:

$$UN(E_F) > 1 \tag{21.36}$$

Experimentally, UPS results on hydrogen adsorption on a Ni $\langle 111 \rangle$ surface indicate a decrease in $N(E_F)$ upon adsorption,[53] which could create a local defect of the ferromagnetic state, in good agreement with theoretical calculations.[65] The magnetic bond number thus corresponds to the number of nickel surface atoms for which the density of states at the Fermi level is sufficiently decreased during adsorption to break Stoner's rule. Further, UPS does not indicate any change in $N(E_F)$ during adsorption of hydrogen on iron, in good agreement with the nonrelevance of the magnetic bond number in the case of iron.

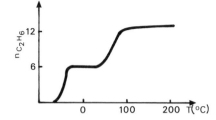

Figure 21.12. Thermomagnetic curve of ethane adsorption on a Ni/SiO$_2$ catalyst (magnetic bond number as a function of the holding temperature).[46]

21.5.4. Relation to Catalysis

Few magnetic measurements have been made in the course of a catalytic reaction.[45,66] Most of the results on catalytic mechanisms have been deduced from adsorption studies of the partners of the reaction: alkane hydrogenolysis,[67] D$_2$–hydrocarbon exchange,[68] CO,[69] and C$_6$H$_6$[70] hydrogenation.

One example of these studies is the case of ethane hydrogenolysis.[67] Figure 21.12 shows the thermomagnetic curve (Section 21.5.2) of C$_2$H$_6$ adsorption on a Ni/SiO$_2$ catalyst. According to Martin and Imelik,[46] the plateau observed near room temperature (magnetic bond number = 6) can be attributed to

$$
\begin{array}{cc}
\text{H} \qquad\qquad \text{H} & \text{H} \\
\diagdown \qquad\qquad \diagup & | \\
\text{C}\!=\!\text{C} \qquad + 4 & | \\
\diagup \qquad\qquad \diagdown & | \\
\text{Ni} \qquad\qquad \text{Ni} & \text{Ni}
\end{array}
\qquad\qquad (A)
$$

Above 70 °C there is a second plateau that corresponds to $n = 12$. According to the discussion on hydrocarbon adsorption (Section 21.5.2) this state corresponds to a completely cracked and dehydrogenated species:

$$
\begin{array}{cc}
\text{C} & \text{H} \\
2 \quad \triangle & + 6 \quad | \\
\text{Ni Ni Ni} & \text{Ni}
\end{array}
\qquad\qquad (B)
$$

corresponding to 12 surface atoms of nickel per ethane molecule. Only the B species can be hydrogenated in CH$_4$ by adding hydrogen to these two adspecies (the A species with $n = 6$ leads to C$_2$H$_6$ when reacting with hydrogen). The B species (linked to 12 nickel atoms) was then proposed as an intermediate in the hydrogenolysis of ethane.[59]

The reaction was studied dynamically over nickel-based catalysts. On supported NiCu alloys (with known surface composition[25]), the measured activity (per metallic unit area) $A_{(x)}$ of an alloy with a copper content x

($x = $ Cu/(Cu + Ni) atom ratio) is linked to x by[67]:

$$A_{(x)} = A_{(x=0)}(1-x)^N \qquad (21.37)$$

The experimental points fit well with the value $N = 12 \pm 2$. With the assumption of a random distribution of nickel and copper atoms,[25] $(1-x)^N$ is proportional to the probability of finding an ensemble of N neighboring nickel atoms in the surface of the alloy. Thus the activity decreases as the probability of the existence of an ensemble of about 12 adjacent nickel atoms decreases, in good agreement with the intermediate deduced from magnetic measurements (species B with a magnetic bond number of 12).

APPENDIX

Definitions and Symbols

1. Magnetization M: With the assumption that a magnetic material can be represented by an ensemble of magnetic moments μ randomly oriented, the magnetization M is the sum of the projections of μ in the direction of the experimental magnetic field H. When H increases, the couple μH tends to orientate the moment in the direction of H, and M increases. $M(H)$ is the magnetization curve.
2. Saturation magnetization M_s: The (maximum) magnetization obtained when all the moments μ are parallel to H.
3. Remanent magnetization M_r: The magnetization remaining after removal of the applied field.
4. Magnetic susceptibility χ: The slope of the magnetization curve $M(H)$: $\chi = \delta M / \delta H$.
5. Magnetic anisotropy: The magnetization of a solid can depend on the direction of the applied field. The magnetic anisotropy can have different causes: e.g., crystal structure, shape, surface, and constraint. There can be one or several axes of easy magnetization.
6. Bohr magneton BM: Magnetic moment corresponding to the electron spin: 1 BM $= eh/4\pi mc = 0.93 \times 10^{-20}$ emucgs.
7. emucgs: Electromagnetic unit in the cgs system. For magnetic susceptibilities emucgs values have to be multiplied by 12.56×10^{-3} to obtain SI units (m^3 kg^{-1}).

Example of Application: Characterization of a Ni/SiO$_2$ Catalyst

The catalyst was prepared by the hexammine method.[51] The nickel loading was 20.8 wt.% and a 55 mg sample was reduced at 600 °C for 15 h. Figure 21.13 gives the magnetization curve of the sample as measured at room temperature using the extraction method (see Section 21.3.1) in an electromagnet giving fields up to 21 kOe.

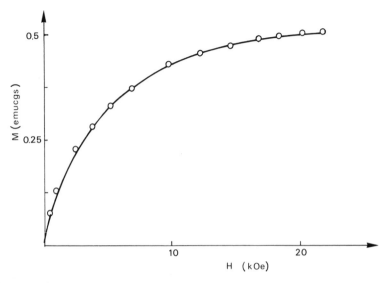

Figure 21.13. Magnetization curve of the Ni/SiO$_2$ sample ($T = 300$ K, $m = 55$ mg, Ni loading = 20.8 wt.%).

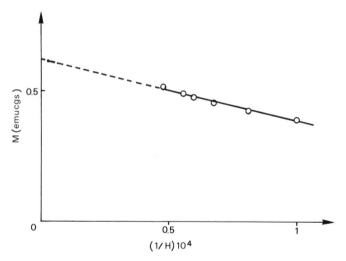

Figure 21.14. Saturation magnetization determination: extrapolation at $1/H = 0$ of the magnetization M of the Ni/SiO$_2$ sample ($T = 300$ K, $m = 55$ mg, Ni loading = 20.8 wt.%).

The saturation magnetization M_s is obtained (Figure 21.14) by extrapolating plots of M vs. $1/H$ at $1/H = 0$ [equation (21.14)]. The extrapolation gives $M_s = 0.618$ emucgs, which (accounting for a specific magnetization of metallic nickel $\sigma_s = 54.4$ emucgs at 300 K) corresponds to a Ni$^\circ$ amount of

$$m\,\text{Ni}^\circ = 0.618/54.4 = 11.4 \qquad 10^{-3}\,\text{g}$$

The extent of reduction r is thus

$$r = [11.4/(55 \times 0.208)] \times 100 = 99.7\%$$

Equations (21.26) and (21.27) can be used to calculate the surface average diameter D_s (see Section 21.4.2). D_1 (low H/T ratio) is obtained from equation (21.26) using the transformed form (D_1 in nm):

$$D_1 = 44.6[(T/\sigma_s)(1/M_s)(M/H)_{H \to 0}]^{1/3}$$

As H tends to 0 the ratio M/H tends (from data on Figure 21.13) to about 1.25×10^{-4} emucgs and then

$$D_1 = 4.6 \text{ nm}$$

Equation (21.27), corresponding to the high-field part of the curve, can be written as (D_2 in nm)

$$D_2 = 1.44\{(T/\sigma_s)[M_s/(M_s - M')]\}^{1/3}$$

where M' is the magnetization for $H = 10$ kOe, which leads to

$$D_2 = 3.8 \text{ nm}$$

and then

$$D_s = 4.2 \text{ nm}$$

REFERENCES

1. P. W. Selwood: *Magnetochemistry, Interscience,* New York (1956).
2. P. W. Selwood: *Chemisorption and Magnetization,* Academic Press, New York (1975).
3. A. Herpin: *Théorie du Magnétisme,* P.U.F. (1968).
4. D. H. Martin: *Magnetism in Solids,* MIT Press, Cambridge (1967).
5. M. Cyrot (ed.): *Magnetism of Metals and Alloys,* North-Holland, Amsterdam (1982).
6. E. P. Wohlfarth (ed.): *Ferromagnetic Materials,* North-Holland, Amsterdam (1980).
7. M. Faraday: *Experimental Researches III,* Taylor and Francis, London (1855).
8. R. M. Bozorth: *Ferromagnetism,* Van Nostrand, New York (1953).
9. B. N. Figgis and J. Lewis: *Techniques of Inorganic Chemistry,* Vol. IV, Interscience, New York (1965), p. 137.
10. W. Heukelom, J. J. Broeder, and L. L. van Reijen: *J. Chim. Phys.* **51**, 474 (1954).
11. R. Pauthenet, *Ann. Phys.* **7**, 710 (1952).
12. Y. J. Chang and C. F. Ng: *Rev. Sci. Instrum.* **59**, 342 (1988).
13. D. Cordischi, M. Lo Jacono, G. Minelli, and P. Porta: *J. Catal.* **102**, 1 (1989).
14. J. T. Richardson, *J. Catal.* **112**, 313 (1988)
15. J. Patarin, H. Kessler, and J. L. Guth, *Zeolites* **10**, 674 (1990).
16. A. Laachir et al.: *J. Chem. Soc. Faraday Trans.* **87**, 1601 (1991).
17. R. P. Eischens and P. W. Selwood: *J. Am. Chem. Soc.* **70**, 2271 (1948).

J.-A. DALMON

608

J.-A. DALMON

18. P. Brant, A. N. Speca, and D. C. Johnston: *J. Catal.* **113**, 250 (1988).
19. G. A. Martin, C. Mirodatos, and H. Praliaud, *Appl. Catal.* **1**, 367 (1981).
20. P. Turlier, H. Praliaud, P. Moral, G. A. Martin, and J.-A. Dalmon: *Appl. Catal.* **19**, 287 (1985).
21. J.-A. Dalmon, P. Chaumette, and C. Mirodatos, *Catal. Today,* **15**, 101 (1992).
22. C. Mirodatos, J.-A. Dalmon, and G. A. Martin: *Catalysis on the Energy Scene,* Studies in Surface Science and Catalysis, Vol. 19, Elsevier, Amsterdam (1984), p. 505.
23. G. Le Caer, J. M. Dubois, M. Pijolat, V. Perrichon, and P. Bussière, *J. Phys. Chem.* **68**, 4799 (1982).
24. H. Praliaud, and G. A. Martin: *J. Catal.* **72**, 394 (1981).
25. J-A. Dalmon: *J. Catal.* **60**, 325 (1979).
26. P. de Montgolfier, G. A. Martin, and J-A. Dalmon: *J. Phys. Chem. Solids* **34**, 801 (1973).
27. J. T. Richardson, and P. Desai: *J. Catal.* **42**, 294 (1976).
28. L. Weil, *J. Chim. Phys.* **51**, 715 (1954).
29. G. A. Martin and B. Imelik: *C. R. Acad. Sci.* **270**, 127 (1970).
30. M. Cyrot, in: *Les agrégats,* (F. Cyrot-Lackmann, ed.) (1982), p. 269.
31. M. C. Desjonquères and F. Cyrot-Lackman: *J. Phys.* **35**, 245 (1975).
32. F. Liu, M. R. Press, S. N. Khanna, and P. Jena: *Phys. Rev.* **B 38**, 5760 (1988).
33. S. Laddas, R. A. Dalla Betta, and M. Boudart: *J. Catal.* **53**, 356 (1978).
34. J.-P. Candy and V. Perrichon: *J. Catal.* **89**, 93 (1984).
35. R. F. Markze, W. Glaunsinger, and M. Bayard: *Solid State Comm.* **18**, 1025 (1976).
36. C. P. Slichter: *Surf. Sci.* **106**, 382 (1981).
37. E. G. Derouane, A. Simoens, C. Colin, G. A. Martin, J.-A. Dalmon, and J. C. Vedrine: *J. Catal.* **52**, 50 (1978).
38. J. K. Blumand, and W. Gopel: *Thin Solid Films* **42**, 7 (1977).
39. J. K. Blumand, and W. Gopel: *J. Magnetism and Magnetic Mat.* **6**, 186 (1975).
40. M. Campagna and M. Landolt: *Phys. Rev. Lett.* **38**, 663 (1977).
41. T. Shingo, S. Hine, and T. Takada: *Proc. 7th Int. Vacuum Congr.* Vol. III (R. Dobrozewsky *et al.,* eds.), (1977), p. 2655.
42. J. J. Broeder, L. L. van Reijen, W. M. H. Sachtler, and G. C. A., Schuit: *Z. Elektrochem.* **60**, 838 (1956).
43. J. Geus, A. P. P. Nobel, and P. Zwietering: *J. Catal.* **1**, 8 (1962).
44. G. A. Martin, N. Ceaphalan, P. de Montgolfier, and B. Imelik: *J. Chem. Phys.* **10**, 1422 (1973).
45. R. Z. C. van Meerten, T. F. M. de Graaf, and J. W. E. Coenen: *J. Catal.* **46**, 1 (1977).
46. G. A. Martin and B. Imelik: *Surf. Sci.* **42**, 157 (1974).
47. J.-A. Dalmon, G. A. Martin, and B. Imelik: *Surf. Sci.* **41**, 587 (1974).
48. J.-A. Dalmon, M. Primet, G. A. Martin, and B. Imelik: *Surf. Sci.* **50**, 95 (1975).
49. M. Primet, J.-A. Dalmon, and G. A. Martin: *J. Catal.* **46**, 25 (1977).
50. G. A. Martin, J.-A. Dalmon, and C. Mirodatos: *Bull. Soc. Chim. Belg.* **88**, 559 (1979).
51. G. Marcellin and J. E. Lester: *J. Catal.* **93**, 270 (1985).
52. G. A. Martin, P. de Montgolfier, and B. Imelik: *Surf. Sci.* **3**, 675 (1973).
53. H. Conrad, G. Ertl, J. Kuppers, and E. E. Latta: *Surf. Sci.* **58**, 578 (1976).
54. M. Campagna and M. Landolt: *Phys. Rev. Lett.* **39**, 568 (1977).
55. S. Morup, B. S. Clausen, and H. Topsoe: *J. Phys. Coll.* **40**, 78 (1979).
56. J. K. Blumand, and W. Gopel: *Proc. 7th Int. Vacuum Congr.* Vol. III (R. Dobrozewsky *et al.,* eds.) (1977), p. 2655.
57. J.-A. Dalmon, G. A. Martin, and B. Imelik: Colloques Internationaux du CNRS, *Thermo-chimie,* Vol. 201 (1972), p. 593.
58. R. Dutartre, P. Bussière, J.-A. Dalmon, and G. A. Martin: *J. Catal.* **59**, 382 (1979).
59. J.-A. Dalmon, J.-P. Candy, and G. A. Martin: *Proc. 6th Int. Cong. on Catalysis* (G. C. Bond, P. B. Wells, and F. C. Tompkins, eds.) The Chemical Society, London (1977), p. 903.

60. J. Friedel: *Can. J. Phys.* **34**, 1190 (1956).
61. F. Raatz and D. R. Salahub: *Surf. Sci.* **156**, 982 (1985).
62. W. M. H. Sachtler and P. Van Der Planck: *Surf. Sci.* **18**, 62 (1969).
63. J.-A. Dalmon, G. A. Martin, and B. Imelik, *Jap. J. Appl. Phys.* **2**, 264 (1974).
64. P. W. Selwood, *J. Catal.* **42**, 148 (1976).
65. M. C. Desjonquères and F. Cyrot-Lackmann: *Surf. Sci.* **80**, 208 (1979).
66. P. W. Selwood: *Adsorption and Collective Paramagnetism,* Academic Press, New York (1962).
67. J.-A. Dalmon, and G. A. Martin: *J. Catal.* **66**, 214 (1980).
68. H. F. Leach, C. Mirodatos, and D. A. Whan: *J. Catal.* **63**, 138 (1980).
69. J.-A. Dalmon and G. A. Martin: *J. Catal.* **84**, 456 (1983).
70. C. Mirodatos, J.-A. Dalmon, and G. A. Martin: *J. Catal.* **105**, 405 (1987).

THERMAL METHODS:
CALORIMETRY, DIFFERENTIAL
THERMAL ANALYSIS, AND
THERMOGRAVIMETRY

A. Auroux

22.1. INTRODUCTION

Calorimetry and the simpler method of differential thermal analysis both contribute to the study of internal transformations of a sample, or reactions of a sample with the exterior, and mass transfers associated with liberation or absorption of energy. Thermogravimetry, which is a process of weighing is specific to reactions with a mass transfer.

22.2. CALORIMETRY

22.2.1. Introduction

Calorimetry is the direct measurement of heat and gives access to the energies of transformation and combination. Its application is universal because only rarely is a system altered without modification of its energetic state. The method is not specific: it is as appropriate for a simple system, such as a gas or an ionized solution, as for a heterogeneous system like a metallic structure. It is used in two ways: most often for its fundamental purpose (to obtain an energy balance sheet), but also more and more frequently as an analytical method or as a criterion of progress of a reaction (Table 22.1).

A. Auroux ● Institut de Recherches sur la Catalyse, CNRS, Villeurbanne, France.

Catalyst Characterization: Physical Techniques for Solid Materials, edited by Boris Imelik and Jacques C. Vedrine, Plenum Press, New York (1994).

TABLE 22.1. Various Techniques of Thermal
Analysis

DSC	Differential Scanning Calorimetry
DTA	Differential Thermal Analysis
DTG	Differential Thermogravimetry
TG	Thermogravimetry
TGA	Thermogravimetric Analysis

The thermodynamic quantities obtained by direct measurement are enthalpy, specific heat, and heat capacity, which give access to other values such as entropy and internal energy.

The fundamental principles of calorimetry and its uses have been the subject of many reviews,[1-7] and its applications are extremely varied:

1. Heats of hydration, dilution, reaction, combustion, determination of equilibrium constants.
2. Measurement of heat capacities, enthalpies of fusion and vaporization.
3. Study of the bidimensional changes of phase, characterization of micropores, evaluation of specific areas, demonstration and characterization of superficial heterogeneities.
4. Study of liquid–solid and gas–solid interactions which lead to a transformation of the sample (oxidation, reduction) or to an adsorption (physical and chemical).

The use of calorimetry in gaseous adsorption is very common in catalysis and has been the subject of many articles.[8-15] The definition of the various heats of adsorption and their relationship with thermodynamic quantities directly obtainable by calorimetry are given in Table 22.2.[12-15]

22.2.2. Principles of Calorimetry

A calorimeter consists basically of an experimental vessel in which the thermal phenomena occur. In general this vessel is placed in a cavity whose walls have either a constant or a variable temperature. The wall of the calorimetric container constitutes the internal vessel and the wall of the cavity in which it is placed constitutes the eternal vessel. According to the quantities of heat exchanged between the two vessels, three main types of apparatus can be identified.

22.2.2.1. Adiabatic Calorimeters

In adiabatic calorimeters (Figure 22.1),[16] the thermal insulation between the internal and external vessels is supposed to be perfect. In this case, the rise in temperature is measured within the internal vessel. The quantity of heat

TABLE 22.2. Definition of the Various Heats of Adsorption[a]

Name and symbol	Unit	Definition	Method of measurement
Integral heat of adsorption: Q_{int}	J	Quantity of heat received by the system when n_a mol are adsorbed at constant temperature on an adsorbent initially *in vacuo*, without a change in the exterior volume of the cell: $Q_{int} = n_a(U_a - U_g)$	Isotherm calorimetry
Molar integral heat: $Q_{int} = Q_{int}/n_a$	J mol^{-1}	Integral heat reduced to an adsorbed mole	Isotherm calorimetry (finished variation at constant T, V, A)
Differential heat of adsorption: q_{diff}	J mol^{-1}	Defined from integral heat by $q_{diff} = (\delta Q_{int}/\delta n_a)_{T,A}$	Isotherm calorimetry (infinitesimal variation of n or P at constant A and T)
Isothermal heat of adsorption: q_{th}	J mol^{-1}	Quantity of heat received by the cell for a reversibly adsorbed mole at constant T and A. It is not a characteristic parameter of the adsorption phenomenon alone: $q_{th} = q_{diff} - RT - V_g(\delta P/\delta n_a)_{T,A}$	Isothermal calorimetry
Adiabatic heat of adsorption: q_{ad}	J mol^{-1}	$q_{ad} = -C_{tot}(dT/dn_a)$	Adiabatic calorimetry infinitesimal variation of P at constant S, A, n
Isosteric heat: q_{st}	J mol^{-1}	Heat of adsorption calculated from isotherms: $q_{st} = -RT^2(\delta \ln P/\Delta T)_{n_a}$ (Clapeyron)	Isotherms of adsorption

[a] a is the adsorbed phase, A the surface of the adsorbent, T the temperature, V the volume, P the Pressure, g the gaseous phase, n the number of moles, U the internal energy, S the entropy, and C_{tot} heat capacity of the system (container–gas–solid–adsorbed phases).

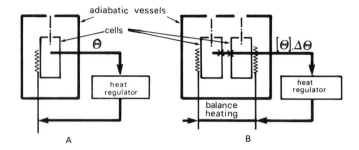

Figure 22.1. Scheme of adiabatic calorimeters: (A) mono, and (B) differential.

evolved or adsorbed can be calculated from this rise in temperature if the heat capacities of the calorimeter and of the reactants present are known.

22.2.2.2. Isothermal Calorimeters

In isothermal calorimeters heat must not accumulate within the internal vessel, and the measurement made is then of the thermal flux. An isothermal calorimeter is characterized by a vessel whose temperature must remain strictly constant. To achieve this, the evolved or adsorbed heat of the studied phenomenon must be compensated for as perfectly and as quickly as possible by bringing in heat or by pumping it out.[17] The addition of heat is generally accomplished by Joule effect heating, and dissipation is achieved by means of Peltier cooling which operates on the principle of heat-pumps.

22.2.2.3. Tian–Calvet Calorimeters

Conduction or heat-flow calorimeters of the Tian–Calvet type represent an intermediate solution between the adiabatic and isothermal models. In this case, thermocouples are used which simultaneously perform thermal conduction and measure the heat flow. Experience has shown that this type of apparatus has a high sensitivity and it is the most suitable for following the kinetics of a reaction.

Calvet[2] increased the thermal conductivity between the internal calorimetric vessel and its surrounding considerably and measured the thermal transfer by conduction directly. The system so created becomes noticeably isothermal due to stabilization of the surrounding temperature by a precisely thermostated calorimetric block, which, owing to its high conductivity, keeps the internal vessel at this constant temperature. The differential assembly of two identical vessels in the same calorimetric block results in very high stability: the variations in temperature are usually around 10^{-6} °C and, under favorable experimental conditions, can be as small as 10^{-7} °C. The principle and the theory of Tian–Calvet microcalorimeters are as follows.

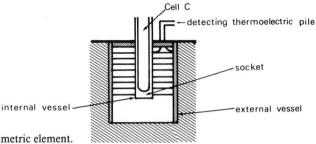

Figure 22.2. Microcalorimetric element.

22.2.2.3a. Principle. The Tian–Calvet microcalorimeter records, as a function of time, the total heat flow exchanged by the system in the form of a thermoelectric potential which is always strictly proportional.

As described by Souchon,[18] the main part of the apparatus is the microcalorimetric element (Figure 22.2) made up of a thermopile and a cylindrical cell in which the phenomenon to be studied is observed. The cell, surrounded by a silver socket, is the internal vessel of the microcalorimeter, and it is connected to the block of the apparatus or external vessel by the thermopile.

The thermopile consists of thermocouples that allow controllable and measurable thermal conduction between the internal and external vessels (Figure 22.3). The external vessel contains the heat insulator, which insulates the thermostat surrounding the calorimeter. The thermostat is made up of concentric aluminum jackets, the outer jacket containing the electrical heating components and the detecting circuit controlling the temperature regulator. The battery of vessels uniformly regulates thermal changes induced by the exterior. The inner part of the block holds two or four thermopiles symmetrically set and connected in opposition. A symmetrical assembly is preferable. If the temperature of the calorimetric block changes very slightly, the cell follows this temperature.

The heat that is transmitted by the fluxmeter induces a parasitic signal. To eliminate such a drift, an identical cell is set in the block and connected in opposition. To obtain very good stability, it is desirable to match the thermal capacity of the compensation cell accurately. The extended distance between the cell and the entrance reduces any perturbing effects of the external temperature to a minimum.

22.2.2.3b. Theory.[2,18] Part of the heat produced in the cell increases the temperature of the internal vessel by $d\theta$ during the time dt. If μ is the calorimetric capacity of the internal vessel, the resulting power is equal to $\mu d\theta/dt$. This variation in temperature is small enough that it does not modify the isothermal operating conditions of the reaction. The other part of the evolved heat is evacuated toward the external vessel by the thermopile. The electric potential produced by this pile is proportional to the thermal flux, which can be formulated as follows:

$$\sum_{i=1}^{i=n} \alpha\Delta\varphi_i = \alpha \sum_{i=1}^{i=n} \Delta\varphi_i = \alpha\Phi \qquad (22.1)$$

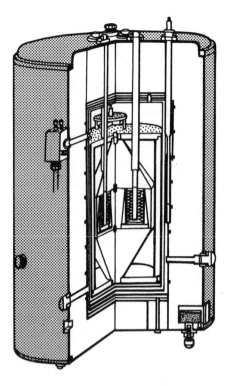

Figure 22.3. Calvet microcalorimeter.

where $\Delta\varphi_i$ is the heat flow transmitted by a surface element, n the number of surface elements, and α the fraction of the surface element within the welds.

Let P be the thermal conduction coefficient, i.e., the total heat flow between the two vessels as a result of a temperature change of 1 °C. P is measured in J s^{-1}°C^{-1}. Then $\Phi = P(\theta_i - \theta_e)$ can be written, with θ_i and θ_e as the temperatures of the internal and external cavities. If a thermal evaluation is made, the thermal power evolved in the cell at an instant t can be written

$$w = P\theta + \mu d\theta/dt \qquad (22.2)$$

with $\theta = \theta_i - \theta_e$.

The deviation Δ of the recorder is proportional to the electric potential Φ of the pile, and therefore to θ, so that

$$\Delta = g\theta$$

Hence,

$$w = (P/g)\Delta + (\mu/g)(d\Delta/dt) = (P/g)[\Delta + (\mu/P)(d\Delta/dt)] \qquad (22.3)$$

In the hypothesis, μ is not a function of the variation of the internal vessel temperature; $\mu/P = \tau$ is a constant which has the dimensions of time, and it characterizes the inertia of the apparatus. Then equation (22.3) can be written as

$$w = (P/g)(\Delta + \tau d\Delta/dt) \tag{22.4}$$

which is the Tian equation. This relation shows that w is not proportional to Δ and that the recorded signal follows behind the signal produced in the internal vessel.

22.2.2.3c. Calibration of the Microcalorimeter. In order to use the microcalorimeter for quantitative measurements, it is necessary to know the ratios P/g and μ/g that appear in the Tian equation. The determination of the constants P/g and τ make up the static and dynamic calibration of the calorimeter. The responses of the apparatus to an instantaneous liberation of energy (Figure 22.4) and to a power step (Figure 22.5) are as follows: P/g can be obtained easily from a constant Joule heating ($w = RI^2$) produced in the internal vessel which acts as a reactor. As a matter of fact, we have seen that at steady state,

$$d\Delta/dt = d\theta/dt = 0 \quad \text{and} \quad w = RI^2 = (P/g)\Delta \tag{22.5}$$

i.e., $P/g = RI^2/\Delta$.

In the same way the curve of the response (Figure 22.6b) corresponding to the passage $w = RI^2$ to $w = 0$ will allow us to determine the value of the constants τ_i. When the heat production stops abruptly, the return to the base line is described in a first approximation by a law of exponential decreases. A better description is obtained by a more complex equation involving several time constants.

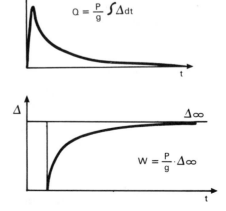

Figure 22.4. Instantaneous response of the calorimeter.

Figure 22.5. Power step response of the calorimeter.

The Camia method[19] is generally used to evaluate the time constants τ_i of the microcalorimeter and they are determined from the cooling curve of the cell and its contents after a Joule heating. This curve is, in fact, an exponential sum such that

$$\Delta(t) = \alpha_1 e^{-(t/\tau_1)} + \alpha_2 e^{(-t/\tau_2)} + \cdots + \alpha_n e^{(-t/\tau_n)} \qquad (22.6)$$

i.e.,

$$\Delta(t) = \Delta_1(t) + \Delta_2(t) + \cdots + \Delta_n(t) \qquad (22.7)$$

For a long enough time t, all the terms can be neglected except $\Delta_1(t)$, which enables the calculation of α_1 and τ_1 from two points sufficiently distant on the curve. The first time constant can thus easily be obtained through studying the CD curve (Figure 22.6). If the elongation at instant t is Δ and the maximum elongation is Δ_{max}:

$$\Delta = \Delta_{max} e^{(-t/\tau)} \qquad (22.8)$$

or

$$\ln(\Delta/\Delta_{max}) = -(t/\tau_i) \qquad (22.9)$$

The time $t_{1/2}$ taken by the recorder to reach a half-deviation $\Delta = \Delta_{max}/2$ is given by $t_{1/2}/\tau_1 = \ln 2$ or $\tau_1 = t_{1/2}/\ln 2$.

22.2.2.3d. Advantages of Flowmeter Calorimeters. The measured quantity is not the temperature of the studied system but the quantity of heat it exchanges with the exterior while changing at a constant temperature. It is the heat flow passing from one vessel into the other which is evaluated. The variations in temperature of the calorimeter are very small. This temperature constancy is particularly suitable in the case of measurements of differential heats of adsorption. The almost complete integration of the heat flow emanating from the cell (due to the large number of thermocouples used) eliminates the difficulty of an imperfect transfer of heat between the mass of the sample and the thermometer.

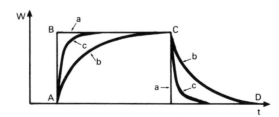

Figure 22.6. Signals caused by Joule heating: (a) theoretical signal, (b) uncorrected signal, and (c) signal corrected with the help of a degree of analog correction.

There are two advantages to this method:

1. The measurement determines the heat of reaction directly (enthalpy if the experiment is performed at a constant volume) without requiring a knowledge of the heat capacity of the reagents present.
2. The method can be applied to fast reactions as well as to slow transformations; there is a direct relation between the thermal power generated by the reaction and the measurement of transferred power, whatever the level of power and however long the experiment takes.

This is not the case with an adiabatic apparatus which is only suitable for short time measurements, and introduces distortions in the long term owing to parasitical thermal exchanges.

A large number of different calorimeters that operate according to the same principle are now on the market: the thermometer block transfers its own temperature to the sample and the flowmeter measures the heat transfers between the sample and the block.

22.2.2.4. Differential Scanning Calorimeters (DSC)

The International Confederation of Thermal Analysis (ICTA) gives the following description of the DSC method (differential scanning calorimetry): a technique in which the differences in energetic behavior between a substance and reference material are measured as a function of the temperature when the substance and the reference material undergo a controlled temperature program.

The differential scanning calorimeter can be: (1) either flowmatic, i.e., derived from the Tian–Calvet calorimeter[16,20] or (2) power compensation DSC, in which each of the two cells, reference and sample, has a specific heating element; the heat supplied to one cell only is measured, according to the principle of power compensation; a signal, proportional to the difference in the heat supplied to the sample and to the reference, dH/dt, is recorded as a function of the temperature T.[21]

A DSC can measure not only enthalpy changes, but also the rates of reaction which lead, in principle, to kinetic parameters and to reaction mechanisms.[21]

22.2.3. Use of Calorimeters

As there are technological differences among calorimeters, there are different modes of use (temperature, programming rate, effective volume, sensitivity), different types of applications (specific heat, pressure, mixture, gas–solid interaction), and various possibilities of coupling with other methods of analysis.

22.2.3.1. Temperature Range

The temperature range can extend from -200 to $1500\,^\circ\mathrm{C}$ in certain specific cases. Experiments at very low temperatures include the fusion and solidification of organic liquids, measurements of vitreous transitions of elastomers and silicones, and physisorption. At high temperatures, typical studies are the changes of states or of structure of inorganic products, decompostion of organic products, gas–solid interactions (oxidation, reduction), specific heats, and phase transitions.

22.2.3.2. Measurement Under Pressure

The cells have to be specially adapted for this work. Depending on the experiment, the pressure in the cell can correspond to the vapor pressure of the sample or to its decomposition vapors, and it can be controlled during the test. It can also be preset before the experiment. For example, a DSC was used in a study of catalytic reactions under a pressure that simulated the rehydrogenation of a solvent during coal[22] hydroliquefaction.

22.2.3.3. Programming Rate

In each case, a calorimetric measurement can be performed in either an isothermal or a temperature-scanning (DSC) mode. The choice of programming rate is an important factor as it not only determines the duration of the experiment but also the accuracy of the measurement. These two factors are generally in opposition to one another and it is usually difficult to make precise heat measurements while the sample is being heated very quickly. In this case, the sample is never in thermal equilibrium with its environment. A high programming rate is useful for studying the behavior of a sample over a broad temperature range. However, if its specific heat or the enthalpy associated with the detected transformations are wanted with accuracy, a low programming rate is recommended.

22.2.3.4. Effective Volume

The choice of the amount of material to be analyzed depends upon several criteria, including diffusion phenomena, molar mass, and reactivity. At a high programming rate the analysis must be made on a small mass of sample to avoid thermal gradients.

22.2.3.5. Sensitivity

Within the temperature range, sensitivity is an essential parameter. It can be considered under two aspects: absolute sensitivity and relative sensitivity per unit of volume.

22.2.3.6. Data Acquisition

The wide variety of applications of calorimetry requires suitable systems for processing the data. All calorimeters are thus connected to calculators to process the calorimetric signals. The calorimeter gives an analog signal which is amplified before it is passed on to the recorder. The amplified analog signal is converted into a digital signal by means of a digital voltmeter. A digital clock, most commonly built into the calculator, gives the necessary time basis for a mathematical treatment of the data. A digital thermometer receives and produces a signal corresponding to the temperature of the detector.

22.2.3.7. Coupling with Other Methods of Analysis

22.2.3.7a. Coupled Calorimetry–Conductimetry. The changing electric conductivity of the sample during the reaction is recorded simultaneously as a function of time and the reaction conversion rate. This is also applied to the thermal flow.[23] In the study of surfaces with gas–solid interactions the following techniques are the ones used most often.

22.2.3.7b. Coupled Calorimetry Mass Spectrometry. The analysis of the gaseous phase of the sample can cover all or only certain gases, selected in advance.

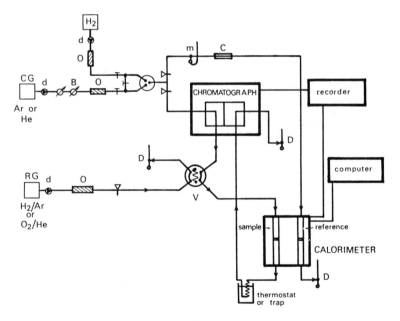

Figure 22.7. Scheme of apparatus for coupled calorimetry–chromatography: (B) flow regulator, (C) capillary, (d) double expansion valve manometer, (D) bubble flowmeter, (GR) reactant gas, (GV) carrier gas, (m) manometer, (O) oxosorb traps, and (v) injection valve.

Example. Decomposition of ammonia-loaded zeolites. This coupling allows characterization of the different stages of thermodesorption studied by DSC.[24,25]

22.2.3.7c. Coupled Calorimetry–Thermogravimetry. A symmetrical microbalance measures the fluctuations in weight of a suspended sample heated in the calorimeter.[16,150,151] The applications of this type of apparatus are developed in Section 22.4.2.3d on thermogravimetric analysis.

22.2.3.7d. Coupling Calorimetry–Chromatography in the Gaseous Phase. This type of coupling is very useful in the analysis of the gaseous phase of the sample: decomposition, adsorption, and desorption of gas. The reactor, either flow or pulse, containing the catalyst is placed in the calorimeter (Figure 22.7).

When this system is used with a pulse reactor, it allows the measurement of adsorption heats of a gaseous reactant on a solid or interaction heats between a gaseous reactant and preadsorbed species. With a flow reactor it allows the kinetic study of catalytic reactions as well as the study of the activation or the aging of the catalyst.[26]

Examples. (1) Study of the bond energy of oxygen with silver-supported catalysts.[27] (2) Study of the adsorption of hydrogen on a platinum catalyst (SiO_2 support).[24] The catalyst is placed on a glass frit in a gas circulation cell in the calorimeter. The hydrogen is injected at regular intervals into an inert carrier gas and is adsorbed on the catalyst. The chromatograph detects the nonadsorbed hydrogen on the sample (Figure 22.8). In the same way the desorption of hydrogen can be observed by temperature programming (Figure 22.9).

22.2.3.7e. Coupled Calorimetry–Volumetry. A system that links microcalorimetry to the volumetric measurement of quantities of adsorbed reactants

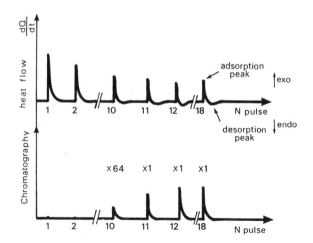

Figure 22.8. Adsorption of hydrogen on Pt/SiO_2 at 20 °C from 7% H_2/93% Ar.

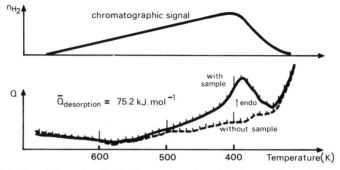

Figure 22.9. Coupled studies DSC–chromatography of hydrogen desorption on a Pt/SiO$_2$ catalyst; $V = 10$ K min^{-1}.

enables the study of gas–solid interactions and catalytic reactions. The admission of gases to the calorimeter can be performed either in a discontinuous way (by successive doses) by means of a valve, or in a continuous manner by means of a capillary. The pulse technique is used more often in the case of the gas–solid systems (Figure 22.10).[8,9,28]

The volumetric determination of the adsorbed amounts of gas is performed in a constant-volume vessel, linked to a vacuum pump which allows residual pressures of about 0.13 m Pa to be obtained. The apparatus consists of two parts: the measuring element (between stopcocks R0 and R1) on which there is a differential gauge, and the cell section (between R0 and R2) which includes the cells placed in the calorimeter (a sample cell in which the adsorbent solid is set and a reference cell) (Figure 22.10). The volume of this vessel is determined by the expansion of a known quantity of gas, contained in the measuring part of the assembly, into the cell section previously evacuated. This determination must be made with the same gas and under the same temperature conditions as the proposed study.

The experiment consists of sending successive increments of the reactant gas onto the catalyst and waiting for thermal equilibrium after each increment.

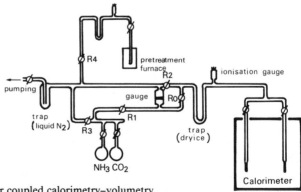

Figure 22.10. Apparatus for coupled calorimetry–volumetry.

The aim is to minimize the volume of each increment in order to follow as closely as possible any changes as a function of the coverage. The adsorption is considered completed when, for a significant increase in pressure, there is no detectable heat evolution or gas adsorption.

The desorption is obtained by a succession of expansions of the gaseous phase in equilibrium with the solid in a known volume, followed by a direct evacuation of the sample when the equilibrium pressure is so low that the desorbed quantity cannot be measured accurately by means of a simple expansion. A second adsorption–desorption cycle gives the measurement of the reversible fraction.

For every dose or expansion, the equilibrium data, such as the pressure P_i, the adsorbed amount Δna_i, and the integral evolved heat ΔQ_{int_i}, are measured. From these results we can also get the equilibrium pressure P, the adsorbed (or desorbed) amount up to the dose i ($\Sigma \Delta na_i = n_a$), and the corresponding evolved heat ($\Sigma \Delta Q_{int_i} = Q_{int}$). Kinetic results for each dose are also measured, such as the heat flow and the evolution of the gas pressure as a function of time. The processing of data and examples of the application of this method are developed in Section 22.2.4.3b on gas–solid interactions.

The method, which allows a continuous introduction of the adsorbate on the solid, is also particularly useful for studies on physical adsorption phenomena. The quasi-equilibrium pressure and the thermal power dQ/dt are recorded in a continuous manner as a function of time.[14,29]

22.2.4. Applications: Data Processing

22.2.4.1. Specific Heat Measurements[16]

Three methods of measurement of specific heat are used: (1) continuous temperature scanning, (2) discontinuous temperature scanning (increments), and (3) dropping samples between T_1 and T_2.

22.2.4.1a. Continuous Temperature Scanning.[16,30] Equation (22.10) shows the principle of the measurement:

$$C_p = (dH/dT)_P$$

$$C_p = (dH/dT)_P/(dT/dt)_P \qquad (22.10)$$

where C_p = Calorimetric signal/Scanning rate.

Two tests must be performed under identical experimental conditions. In the first, the two cells are empty. In the second, the sample is placed in the measuring cell, and the reference cell remains empty. The measured variation between the two calorimetric plots characterizes the specific heat of the sample.

For liquid samples, accurate measurement of the specific heat requires some precautions. The main source of error is insufficient information about the vapor phase above the sample during the temperature scan. To remedy this, open-tube cells are used.

22.2.4.1b. Discontinuous Scanning Temperature. Equation (22.11) shows the principle of the measurement:

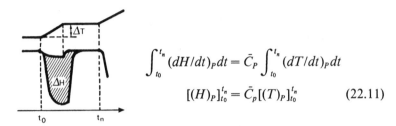

$$\int_{t_0}^{t_n} (dH/dt)_P \, dt = \bar{C}_P \int_{t_0}^{t_n} (dT/dt)_P \, dt$$

$$[(H)_P]_{t_0}^{t_n} = \bar{C}_P [(T)_P]_{t_0}^{t_n} \qquad (22.11)$$

with $\bar{C}_p$ = Area of peak ΔH/Variation of temperature ΔT.

A temperature increment is programmed between two temperatures T_1 and T_2 and the rise in temperature of the sample for this increase of temperature ΔT is measured. As before, two tests are necessary to measure the specific heat with accuracy. The method is applicable to the measurement of specific heats of solids and liquids.

22.2.4.1c. Drop Method. The principle of the method consists of taking the sample from a temperature T_1 to a temperature T_2 by means of a fall within the calorimeter. The heat associated with the rise in temperature of the sample is measured and compared to that of a reference sample (alumina, platinum) analyzed under identical conditions. The use of this method requires an accurate knowledge of temperatures T_1 and T_2.

22.2.4.2. Measurement of Reaction Heats

Heterogeneous reactions: The catalyst or solid reactant is placed inside the calorimetric cell, and the gaseous reactants are preheated and brought into contact with the solid. Simultaneous measurements are made of the heat evolved and the state of progress of the reaction, e.g., by a linkup to a gas chromatograph or a mass spectrometer. This allows the calculation of the heat of reaction associated with an end product or a consumed reactant for a reaction in a steady state. In the case of complex reactions when more than one product is formed, variation of the kinetic parameters of the reaction can give a system of n equations with n unknowns. The heat of conversion of the reactants to each of the end products can be calculated from these equations.

Numerous examples of reactions of the solid–liquid mixture can be given, e.g., dissolution,[31] hydration and wetting.[32]

22.2.4.3. Measurement of Heats of Interaction

22.2.4.3a. Liquid–Solid Interaction.[14,33] In liquid–solid adsorption, the solid is allowed to contact solutions of adsorbent in increasing concentrations. However, it is necessary to use a new sample of the solid for every test.

(i) *Measurements of Surface Area* (immersion heat). It is possible to determine the surface area of a powder from the thermal effect caused by the displacement of the solid–gas interface by means of immersion in water. The powder must be previously wetted and the solid must be covered with no more than two layers of water molecules (which is enough to mask the superficial chemical interactions). The specific area A of the powder is then calculated from the enthalpy of immersion ΔH_{imm} according to:

$$|\Delta H_{imm}| = A[\gamma^{l,g} - T(\partial\gamma^{l,g}/\partial T)] \tag{22.12}$$

is which $\gamma^{l,g}$ is the surface tension of the immersion liquid at temperature T. This method is adaptable to the measurement of specific areas between 1 and $50\ m^2\ g^{-1}$ with pore diameters of at least 10 nm.[13]

(ii) *Thermoporometry* (study of the porosity of adsorbents and their state of division). Through calorimetric analysis, this technique uses the liquid–solid transformation of a condensate (water–benzene) maintained in a porous solid.[13,38,39,40] Based on the fact that in a porous medium, the freezing point of any condensed liquid is lowered, it is an analytical method giving the actual size of the cavities and to a certain extent, their shape.

Figure 22.11 shows the type of calorimetric curves obtained by Brun *et al.*,[39,40] either during solidification (curve 1 shows solidification of the excess of condensate, and curve 2 solidification in the pores) or during melting (curves 3 and 4). The width and the height of peak 2 can be linked to the distribution of pore radii, either following the theory (based on the Laplace and Clapeyron equations) or through experiment (measurement of the lowering of the freezing point and of the melting energy as a function of the porous radii).

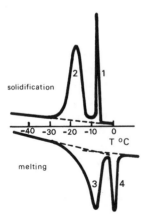

Figure 22.11. Calorimetric detection of the solidification and the melting of an adsorbed phase in a porous medium, after Brun *et al.*[39,40]

(iii) *Adsorption of Liquids by Displacement*. A certain amount of solution is put into contact with a suspension of the solid adsorbent in a given solvent.[14] Adsorption of pyridine on alumina in an alkane solution is an example.[41]

22.2.4.3b. Gas–Solid Interactions.

These involve a wide range of energies according to the nature of the reactants. The phenomena of physical adsorption (physisorption characterized by low energies) and chemical adsorption (chemisorption generally characterized by higher energies, specific to the established bond type) can be distinguished in this way.

The classical technique of adsorption calorimetry by doses is the most appropriate way to measure the energy between the adsorbed species and the catalyst (cf. Section 22.2.3.7e). If the surface can be considered *a priori* as heterogeneous, the adsorption heat, the amount adsorbed, and the kinetics of adsorption must be measured by very small successive doses of the adsorbate so as to obtain information on the variation of these quantities as a function of the coverage. Nevertheless, the method is not precise enough for a direct determination of the distribution spectrum of the energetic sites.[10]

(i) *Data Processing*. As we have seen (Section 22.2.3.7e, Figure 22.10) the volumetric apparatus can give the adsorbed quantity and the equilibrium pressure of the gas (and thus the adsorption isotherm can be plotted) while the variations in the thermal signal indicate the amount of heat evolved.

The *equilibrium data*, obtained directly from the calorimetric measurements, pressure, integral heat, and amount adsorbed, can be used in several ways depending on the gas–solid system under study (Figure 22.12).[9] These data can be expressed in five ways:

1. The volumetric isotherms (n_a, P) for a cycle of adsorption I, desorption, adsorption II [Figure 22.12(1)].

2. The corresponding calorimetric isotherms (Q_{int}, P) [Figure 22.12(2)].

From these two types of isotherms (calorimetric and volumetric) the energy and the molar entropy of adsorption [equations (1) and (2) in Figure 22.12] can be evaluated as functions of the degree of coverage, using the pairs of values (n_a, Q_{int}) measured at the same equilibrium pressure (Figure 22.12).

When there are several possible adsorption mechanisms (e.g., reversible and irreversible adsorption), these isotherms can be used to evaluate the sequence and the importance (number of active sites and bond energy) of different processes. Three limiting cases are illustrated: (α) Irreversible chemisorption which occurs without a detectable equilibrium pressure; this adsorption, on reaching saturation, is followed by a reversible adsorption, dependent on the pressure: adsorption I (open circles) and desorption I (black dots), which coincide. (β) At low coverage, an irreversible chemisorption takes place; at higher coverage, irreversible and reversible processes occur simultaneously, and then at an even higher-pressure region, only reversible adsorption occurs. Thus there is a pressure region in which adsorption I and desorption I do not coincide. (γ) The reversible and irreversible processes occur simultaneously

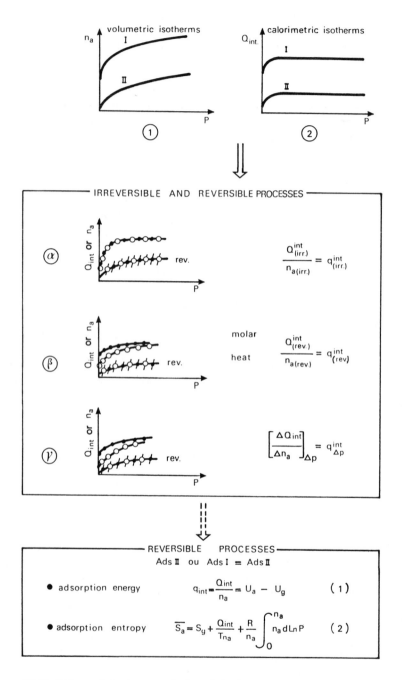

Figure 22.12. Volumetric isotherms, calorimetric isotherms and exploitation of the equilibrium data obtained from calorimetric measurements.[9]

throughout the whole pressure range investigated. Adsorption I and desorption I do not coincide in any part of the pressure-dependent isotherm. Adsorption II and desorption II which represent only the reversible adsorption coincide on the same curve in the three cases. From the calorimetric and volumetric isotherms, one can evaluate the integral molar heat related to the irreversible process $Q_{int(irr)}$ and to the reversible process $Q_{int(rev)}$ in all the cases investigated.

3. Another representation of calorimetric data is that of integral heats as a function of the adsorbed quantities (Q_{int}). This representation leads to the

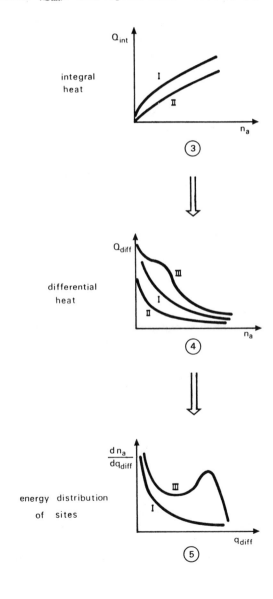

Figure 22.12 (*continued*)

detection of coverage ranges with constant heat of adsorption, those for which the evolved heat is a linear function of the coverage [Figure 22.12(3)].

4. Complementary information is given by the variation in the differential heats (molar heat of each dose of adsorbate),

$$q_{\text{diff}} = \Delta Q_{\text{int}} / \Delta n_{a_i} \qquad (22.13)$$

as a function of n_a [Figure 22.12(4)]. The ratio of the amount of heat evolved for each increment to the number of moles adsorbed in the same period is equal to the average value of the differential enthalpy of adsorption in the interval of the adsorbed quantity considered. The curve showing the differential heat variations in relation to the adsorbed amount is traditionally represented by histograms.[42] However, for simplification, the histogram steps are often replaced by a continuous curve connecting the centers of the histograms.[43] The decrease in the differential heat values as a function of the coverage is generally attributed to heterogeneity of the adsorption sites on the surface of the solids or to repulsive interactions between adsorbed molecules. On the surface of metals and sometimes of semiconductors, an induced heterogeneity is thought to exist, due to changes in the properties of the surface during adsorption.[9]

5. Finally, it is possible to obtain a distribution of the energies of the adsorption sites by expressing dn_a/dq_{diff} as a function of q_{diff} which is represented either by a continuous curve (numeric derivative) or by histograms (incremental ratio). The area under the curve is representative of the number of molecules that are adsorbed with a given evolved heat. This type of representation is, nevertheless, a little less accurate than the previous one [Figure 22.12(5)].[69]

The *kinetic data* necessary to detect mechanisms of reaction are often distorted by the inertia of the calorimeter (2) and calculations to correct for inertia are necessary in order to arrive at the actual thermokinetics of the phenomenon under investigation. It is theoretically possible to deconvolve any curve of the variation of thermal signal as a function of time to recover the heat flow at every instant (Figure 22.6a).[44,45] This operation enables several successive stages (e.g., physiosorption and chemisorption) to be distinguished. Figure 22.13 illustrates some of the information that can be deduced from direct kinetic data: notably the evolution of pressure on the adsorbent during the adsorption of a dose and the peaks of evolved heat.

On a logarithmic scale [Figure 22.13(1)] a distinction can be made between very fast and very slow phenomena, and it is possible to calculate the half-reaction times corresponding to each increment. When the final decrease in the peak becomes linear with time, it is usually an indication that the release of heat heat has stopped. However, in certain cases, a very quick evolution of heat, and then an exponential decrease, can be followed or accompanied by slow phenomena, only detetcable in the tail of the calorimetric curve [Figure 22.13(2)].

Another and closer representation of the adsorption kinetics (n_a, t) is given by the evolution of the integral heat as a function of time ($Q_{\text{int}}\%$, t) during the

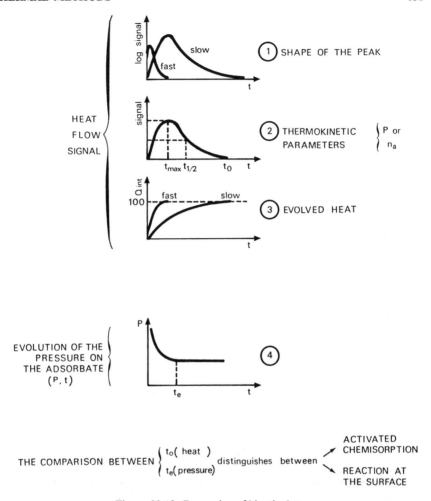

Figure 22.13. Processing of kinetic data.

adsorption of each injection [Figure 22.13(3)]. If the evolution of heat produced by adsorption is slow when compared to the response time of the calorimeter, it is possible to obtain kinetic data.[8]

The evolution of gas pressure is related to the kinetics of the adsorption process [Figure 22.13(4)]. If there is a correlation in the case of a slow phenomenon between the kinetic and thermokinetic results, i.e., if the time necessary to reach the equilibrium of adsorption in the gaseous phase (t_e) coincides with the time it takes for the calorimetric peak to return to baseline (t_0), it can be inferred that the kinetically prevailing process is an activated chemisorption. However, if the value of t_e is much lower than the value of t_0 the slow phenomenon detected through thermokinetics must be attributed to surface reactions following rapid adsorption.[9]

Having described the different types of information that can be obtained from adsorption calorimetry, we can now examine a few examples of gas–solid interactions.

(ii) *Physisorption*.[14,33,46] Calorimetry is a precise technique for characterizing porous textures.[13]

Example 1. In the adsorption of inert gases (N_2, Kr, Ar) at low temperatures on more or less structured solids, the combination of the measurement of adsorption and the differential heats can demonstrate singular discontinuities. These indicate changes of phase of the adsorbate (passage from two-dimensional fluid states to two-dimensional solidified states). These phase transitions can take place under pressure conditions depending on the structure of the solid investigated.[29]

Example 2. The reversible adsorption, at room temperature, of water vapor on a transition alumina with a completely rehydrated surface consists first of the localized adsorption of a monolayer of water by the formation of hydrogen bonds with surface hydroxyl groups. This is followed by the adsorption of other layers and finally by capillary condensation.[9] The heats of adsorption related to the first process decrease exponentially with the coverage, which, according to the authors, confirms the heterogeneous nature of the surface of transition alumina. The heats related to the next processes are strictly constant from the second layer until the capillary condensation is complete. The distribution of pore radii can be inferred from calorimetric isotherms.[9]

(iii) *Chemisorption*. The variations in energy measured during strong chemisorptions are specific to the type of bonds formed. For the same adsorbate these variations are a function of the nature of the active site. These strong interactions are often used to study the surface state of metallic catalysts, either supported or pure, by means of adsorption of hydrogen, carbon monoxide, or oxgyen. In the same way the adsorption of acids or bases on basic or acidic catalysts has also been used to characterize the strength and number of surface sites on such solids as a function of various pretreatments or modifications.

Example 1. Characterization of metallic surfaces by adsorption of hydrogen, oxygen, or carbon monoxide[26,47–62]: The strength of the metal–hydrogen bond depends on the nature of the metal, and for the same metal it is sensitive to the immediate environment of the atom that bonds the hydrogen. The coordination of a surface atom and the electron density around such an atom, determines — at least partly — the interaction energy of the atom with an adsorbate.

These two parameters, in the first instance, depend upon the size of the particle. Particular geometrical shapes can be favored depending on the number of atoms making up the particle: e.g., icosahedra, ubo-octahedra. Therefore the nature of the exposed faces (pentagons, squares, hexagons...) will vary, which entails a variation in the coordination of the surface atoms. Furthermore, the electron density around a given atom will depend upon the number of atoms making up the particle: it will thus change with size. It has been demonstrated that the heat of adsorption of hydrogen on atomically dispersed platinum on a carbon filament (according to the authors) is about 150 kJ mol^{-1} H_2.[47] In contrast, platinum black produces heats that vary between 42 and 80 kJ mol^{-1}.[47]

Nevertheless, it is found that heat decreases with the degree of coverage. This phenomenon could be explained either by the distribution of the sizes of particles or by a heterogeneity of the sites within the same particle. The continuous modification of the energy of sites (all initially identical) owing to the adsorption of hydrogen (collective effects) is another possible explanation. Basset et al.[50] observed an average value of 75 kJ mol^{-1} H$_2$ for platinum/Al$_2$O$_3$ and complementary measurements of chemisorption of oxygen sequentially with hydrogen enabled them to verify the titration equations proposed by Benson and Boudart[63]:

$$Pt_s + \tfrac{1}{2}H_2(g) \Leftrightarrow Pt_s\text{-}H \qquad \Delta H = -75 \text{ kJ mol}^{-1} \text{ H}_2$$

$$Pt_s + \tfrac{1}{2}O_2(g) \Leftrightarrow Pt_s\text{-}O \qquad \Delta H = -242 \text{ kJ mol}^{-1} \text{ O}_2$$

$$Pt_s\text{-}O + \tfrac{3}{2}H_2(g) \Leftrightarrow Pt_s\text{-}H + H_2O \qquad \Delta H = -226 \text{ kJ mol}^{-1} \text{ H}_2$$

$$Pt_s\text{-}H + \tfrac{3}{4}O_2(g) \Leftrightarrow Pt_s\text{-}O + \tfrac{1}{2}H_2O \qquad \Delta H = -250 \text{ kJ mol}^{-1} \text{ O}_2$$

Platinum on silica was similarly characterized by the heat of hydrogen adsorption varying from 150 kJ mol^{-1} ($\theta = 0$) to 46 kJ mol^{-1} ($\theta = 0.9$).[47,49,52]

When the disparity among the results in the literature is considered, there is little evidence for a heterogeneity of the sites, whether due to the distribution of sites within particles of a similar size, or to the distribution particle size, or to some other factor. In most of the studies made by various authors several parameters vary simultaneously (e.g., degree of coverage, size of particles, distribution of sizes, nature of the support, temperature of adsorption), which makes any comparison difficult. Thus it is necessary to control the parameters suspected of influencing the energy of interaction if any of these effects (e.g., size, support, or additives) are to be observed. Moreover, the platinum surfaces may be variously contaminated before the calorimetric measurements, depending upon the experimental and instrumental conditions used by the different authors. Thus a minimal contamination of the surface by oxygen or other reactants would be enough for the observed heats to be noticeably higher than the expected values. Most studies concerning the heats of adsorption of hydrogen, carbon dioxide, and oxygen on transition metal surfaces have been reviewed by Toyashima and Somorjai,[48] Cerny et al.,[47] Vannice et al.,[49] and Wedler et al.[53,54]

Others factors can influence the bonding energy M–H which is so important in evaluating numerous catalytic reactions involving hydrocarbons. While it has been impossible to investigate them through this method, modification of the properties of metals by alloying has been studied through the adsorption of CO. The heat of adsorption of CO on pure nickel and copper–nickel alloys on the same support was measured for various degrees of coverage with CO and for different ratios of Ni/Cu (Figure 22.14).[62] The authors observed that the heat of chemisorption increases with the amount of copper. They attributed this to the progressive filling of the d band of nickel by the 4s electron of copper, which strengthens the M—C bond as the amount of copper increases.[62] Furthermore,

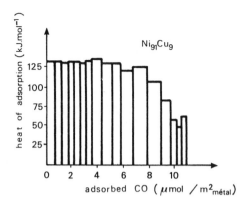

Figure 22.14. Adsorption heat of CO adsorbed on $Ni_{91}Cu_9$ alloy.

it was noted that the heat of chemisorption increases linearly with the ratio bridged-CO/linear-CO, which therefore also seems dependent on the amount of copper.[62]

Example 2. Acid-base interactions: Adsorption of basic molecules (ammonia, pyridine, n-butylamine...) or acidic molecules (CO_2, SO_2, ...) on silicas, aluminas, zeolites, and other oxides has been investigated extensively by means of the calorimetry of adsorption.[64-130] The work of Tsutsumi *et al.*,[86-99] Klyachko *et al.*,[100-113] Auroux *et al.*,[64-85] Stone and Whalley,[117] and Cartraud *et al.*[118-120] is especially notable.

Knowledge of the strength and number of acidic or basic sites in a catalyst is very important but measurement in the solid phase is very difficult. IR spectrometry can certainly be used for the determination (e.g., by adsorption of pyridine) of Brönsted acids and Lewis acids. However, quantitative aspects and the acid strength distribution are only partially accessible through this method. The strength of sites and their distribution are more precisely and more directly measured by the determination of the differential heats of adsorption of a basic (NH_3, pyridine, n-butylamine) or acidic (CO_2, CH_3COOH) probe molecule, depending on the solid.

It is worth noting that the nature of the sites, e.g., whether Lewis or Brönsted, cannot be determined by such a measurement alone. A complete study at various pretreatment temperatures and a complementary investigation by IR spectroscopy are necessary.[67,73,82]

For example, the zeolite HZM-5, whose main catalytic interest is the spatial selectivity of reactions such as the conversion of methanol or the alkylation of aromatics (benzene, toluene), has been shown to be very acidic compared to other zeolites, mordenite, or faujasite. Pretreatment at 400 °C eliminates molecular water; pretreatment at 800 °C generates isolated Lewis sites by the elimination of hydroxyl groups from the surface (Figure 22.15).[64] The strength of the strongest sites increased as the ratio Si/Al increased, reaching a maximum for a ratio of about 35, then decreased.[67] The number of acidic sites decreased linearly with this ratio, which is as expected. Moreover after use and coking of the catalyst, acidic sites were neutralized but strong sites remained accessible to reactants.[68]

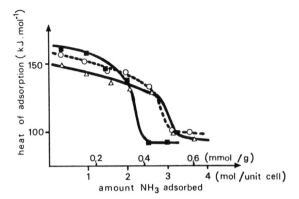

Figure 22.15. Differential heats of adsorption on H-ZSM-5 zeolite pretreated at (△) 743 K, (○) 923 K and (■) 1073 K; calorimeter temperature 416 K.

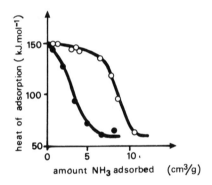

Figure 22.16. Differential heats of adsorption of NH_3 on HZSM-5 before (○) and after (●) conversion of methanol at 643 K; pretreatment temperature 673 K; calorimeter temperature 423 K.

Thus from different kinds of information provided by the calorimetry of adsorption, it is possible to obtain a distribution of the acidic sites on the surface and even some indication of the location of these sites.[69] Finally this method can be used to follow the modifications of acidity resulting from various pretreatments, hydration, or rehydration, additives such as phosphorus[70] or boron,[77,79,80] poisoning,[68] substitution,[76,81] and dealumination.[74,83,85] Figures 22.16 and 22.17 show the differential heats of adsorption of NH_3 and the corresponding isotherms of a ZSM-5 zeolite and of the same coked sample, as a function of the coverage. $[Q_{diff} = F(\theta)]$.[68]

The acidity spectrum of the zeolite can also be represented by the curve $d\theta/dQ$ or dn_a/dQ as a function of Q showing the number of NH_3 molecules that are adsorbed with the production of a given heat and therefore the number of sites corresponding to a given acid strength (Figure 22.18).[69]

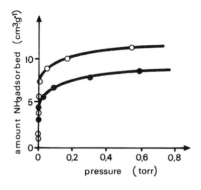

Figure 22.17. Adsorption isotherms of NH3 on HZSM-5 before (○) and after (●) conversion of methanol; calorimeter temperature 423 K.

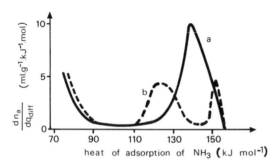

Figure 22.18. Acidity spectrum from ammonia adsorption of a HZSM-5 zeolite evacuated at 660 K (a), and of the same sample evacuated at 1070 K, treated by H_2O vapor at 300 K and then evacuated again at 670 K (b); calorimeter temperature 515 K.

We have given here a nonexhaustive list of examples using adsorption calorimetry in catalysis. This method can be used equally well for the study of other solids such as sulfurs,[131] oxides.[132,133]

22.3. DIFFERENTIAL THERMAL ANALYSIS

22.3.1. Introduction

Very frequently, changes in the physicochemical state of substances under changing conditions of temperature are accompanied by exothermal or endothermal effects that can be measured quantitatively by differential thermal analysis (DTA). In this technique the temperature difference between a substance and a reference material is measured as function of the temperature when the substance and the reference material are subjected to the same temperature control program. DTA gives information about energies relating to the sample and so this is fundamentally different from gravimetric information. It gives the energies of the reactions of the sample with the surrounding medium and also of its internal structural transformations. It can also be a method of control or analysis.

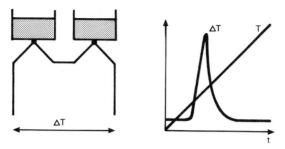

Figure 22.19. ATD principle.

22.3.2. Principles[134–142]

In DTA the active sample and a reference sample (generally inert and symmetrically placed) are heated in a vessel with a programmed rise in temperature (Figure 22.19). A system with thermocouples measures the difference in temperature between the two samples. In the absence of reactions or transformations, the difference in temperature is low and constant: this is the base line. When a change in the active sample takes place, it involves a quantity of energy and the sample temperature then deviates from that of the reference sample run on. The differential temperature ΔT is recorded in the shape of a peak or a series of peaks as a function of time t. The temperature of the sample T is simultaneously recorded (Figure 22.19).[16] When DTA is conducted under well-defined conditions (type of crucible, nature of the atmosphere), the differential temperature signal accurately corresponds to a thermal exchange value between the active crucible and the atmosphere. The deviation is proportional to the power; the area of the peak is proportional to the total amount of heat involved (analogy with the DSC method).

There are many examples of the use of DTA,[134–142] including the study of polymerization, the control of purity, the study of structure or state changes in a solid, determination of a phase diagram, and studies of thermodesorption. An example is the reduction of a copper chromium oxide catalyst by hydrogen (Figure 22.20).[16]

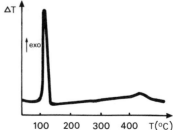

Figure 22.20. Reduction of an oxide CuCr–O by hydrogen: 21.6 mg of catalyst; 3 °C min^{-1}; 20 cm^3 min^{-1} H$_2$.

22.4. THERMOGRAVIMETRIC ANALYSIS

Thermogravimetric analysis is the measurement of the gain or loss in weight of a material as a function of the composition of the atmosphere and the temperature.

22.4.1. Introduction

Thermogravimetry is a very useful technique for the study of solid–gas systems. Its use in the chemistry of solids to characterize phase transformations and, if necessary, to dose them is well known.[143-147] Most physical, chemical, or physicochemical phenomena are characterized by variations in the masses of reactive samples when these samples are subjected to various environments, whether fixed or changing, such as temperature, atmosphere, vacuum, magnetic field, pressure, and irradiation. A recorded, continuous and precise weighing can follow the changes in a substance under these environmental conditions, which are also recorded continuously. As the main parameter is often the temperature, thus the common name of thermogravimetry. The measuring instrument is therefore a weighing device. It has to remembered that changes in the mass of the sample clearly imply evolution or uptake of matter by the sample. In weighing the sample it is assumed that it can be connected to the weighing device. The method only works well if the sample is in compact form. Generally this is a solid or a powder, sometimes a liquid placed in an open crucible or even fixed on a solid phase.

In practice, the sample can be: (a) a liquid or a homogeneous solid (solid solutions), (b) adsorbed phases at the surface of a solid or a liquid in pores, (c) distinct solid phases in granular solid (metals), (d) layers on a surface (oxide films), or (e) heterogeneous liquids (emulsions). It is always advantageous to complement thermogravimetry with other methods of analysis.

22.4.2. Principles and Use[16,143-148]

Figure 22.21 represents the process: the initial mass M (weight P) becomes $M - \Delta M$ (weight $P - \Delta P$) through a transfer of mass ΔM to the environment. The reverse is also possible: the surrounding phase, generally gaseous, can react with the sample to give a mass increase. The more usual case is the loss of weight

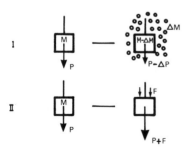

Figure 22.21. TGA principle.

as a function of temperature or time, in an atmosphere with a given composition and pressure:

1. Evaporation, sublimation, slow decomposition.
2. Structural or chemical change with or without stoichiometric loss of one or several constituting elements.
3. Corrosion with formation of a volatile compound.

reaction of an element from the surrounding atmosphere with the sample is frequently investigated:

1. Oxidation (corrosion), chlorination, fluorination, etc.
2. Adsorption (specific surfaces), diameter and surface of pores, bond energy.
3. Formation of a defined compound (hydration, carbonation).

Thus we can appreciate the importance of the sample environment, not only its temperature, which is generally also that of the sample, but also its composition. A gravimetric thermoanalyzer must therefore enable precise control of the sample atmosphere.

22.4.2.1. Instrumentation

A gravimetric thermoanalyzer consists of essentially: (a) an automatic magnetic re-equilibrating balance with continuous recording of the signal; (b) a furnace in which the required temperatures can be generated; (c) a gas-flow circuit for controlling the gaseous environment (e.g., a primary vacuum, a secondary vacuum, pressure gauges, and manometers).

This weighing and conditioning equipment is completed with the means to allow coupled meaurements: derivative gravimetry DTG, DTA, and gas analysis. There are also facilities for on-line data acquisition and processing: numeric converters and interfaces, calculators, printer, and plotter.

22.4.2.2. The Balance

The balance consists of a blade or ribbbon-mounted beam. The variable mass of sample is hung at one end. The other end supports the automatic re-equilibrating device, which consists essentially of a permanent magnet with one pole located in the core of a fixed surrounding electromagnetic coil. A flag is attached to the beam, and depending on the position of the beam, it obstructs the incoming light, triggering a photoresistant cell. The signal from the photocell controls an amplifier that energizes the coil. The system is thus stabilized, the current intensity in the coil increasing as the beam moves from its initial position. The weight measurement is obtained from the value of the electric current passing through the coil (Figure 22.22).

At present, many balances have a symmetrical beam suspended by a torsion ribbon. Automatic electromagnetic re-equilibration compensates for the weight variations of the sample and generates a linear electric signal proportional to the

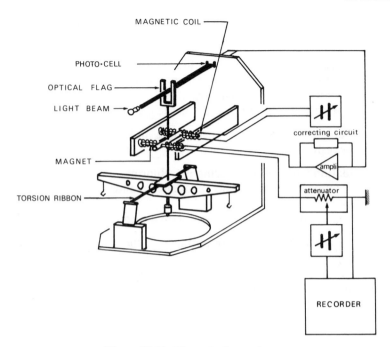

Figure 22.22. Thermobalance scheme.

mass exchanges between the sample and the gaseous phase. The balance, which is placed in a sealed vessel, is generally swept downward around each suspension (sample or counterbalance) by a sealed gas-flow system. The whole balance can thus be set under a vacuum or operated under controlled gas flows.

22.4.2.3. Simultaneous Methods

It is often advantageous to associate thermogravimetry with other analytical methods. This allows a more precise investigation of the process, including the formation of intermediate phases, surface structure, and evolved gaseous species. Thermogravimetry is often only considered the complement of more specific methods — it has the advantages of being a precise quantitative method but remains a nonspecific one.

There are three kinds of combined methods:

1. Simultaneous methods: Only one sample is used. Simultaneity is sometimes an obstacle to the optimization of the two respective methods. It has the advantage of the identity of experimental conditions and of strict coincidence of the various sources of information.
2. Combined methods: Two similar samples are placed in the same environment, each connected to a different detector.

3. Separated methods: The experiments are performed in different apparatus on different samples. The correlation takes place at the stage of processing the results.

Four distinct methods are often linked with gravimetry.

22.4.2.3a. Differential Thermogravimetry (DTG). This method is simultaneous by definition. The signal is derived through filtering and is a derivative of the main signal. The treatment is performed by analog or digital methods.

22.4.2.3b. Evolved Gas Analysis (EGA). In EGA by *chromatography* the chromatograph connected to the thermoanalyzer comprises a gas injector and two switchable columns. The switching provides considerable flexibility of adaptation to suit the diluted gas. Injections are automatically repeated rapidly so that successive chromatograms give an account of the gases evolved during a decomposition.

Alternatively, in EGA by *mass spectrometry*, for certain applications, the chromatograph is replaced by a mass spectrometer, generally of the quadrupole type: this analysis is faster, compatible with low concentrations, and applies to a wider range of compounds. On the other hand, its is more complex owing to the multiplicity of ionized radicals and identical radicals which can originate from different species. Examples of the application of this coupling are described by Courtault.[149]

22.4.2.3c. Differential Thermal Analysis DTA. The sample is set is the DTA detector, suspended on a balance. Differential thermal analysis gives information on the energy changes in the sample. As the sample changes, its mass variations are recorded simultaneously with the corresponding thermal effects. DTA also detects transitions not associated with mass variations. The coupling TG/DTA/DTG is being used more and more (Example 3 in Section 22.4.3)[164] and its evolution leads to the TG/DTG/DSC system.

22.4.2.3d. Differential Scanning Calorimetry (DSC). A symmetrical microbalance measures the weight variations of samples suspended and heated in a calorimeter.[150,151] The crucibles suspended on the beam contain the sample at one end and the inert reference at the other. The calorimetric measurement is made with fluxmeters which surround the tube in which the cells enter without contact. The gas-tight vessel enables work to be done under vacuum or under a controlled gas flow. The coupling *in situ* of a differential scanning calorimeter and a thermobalance allows simultaneous control of heat and matter transfer; hence it finds application in temperature programmed thermodesorption (Example 4 in Section 22.4.3).

22.4.3. Applications of Thermogravimetry and Associated Techniques

The range of application of these methods of analysis is extremely broad. Among the possible applications some of the most notable are the following:

1. Catalysis: Preparation of catalysts, measurement of their specific area, change of phases, plotting of the adsorption isotherms of various gases or vapors and calculation of the corresponding isosteric heats,[152–160] study of the reduction or reoxidation of oxide catalysts,[161] and temperature-programmed thermodesorption.[167–172]
2. Chemistry: Thermal decomposition with either solid–solid or gas–solid reaction.
3. Measurement of specific surfaces of powders, adsorption of gases, determination of humidity ratio, rate of evaporation, and rate of drying.
4. Metals and alloys: Calcination, oxidation, reduction, and corrosion.

Example 1. Hydrogen–oxygen titrations on $Pt-Re/Al_2O_3$ and $Pt-Ir/Al_2O_3$ catalysts in a microbalance under vacuum (Figure 22.23).[162]

The activated sample of catalyst was put into contact with the adsorbate in a microbalance. Hydrogen was the first gas introduced at a given pressure and the total gain in mass was then determined at equilibrium. A further evacuation gave the amount of irreversibly adsorbed hydrogen. The subsequent introduction of oxygen increased the weight further: this increase corresponded to the total amount of oxygen combined with the chemisorbed or physisorbed hydrogen previously adsorbed. Evacuation then allowed the separation of the chemisorbed oxygen from the reversibly adsorbed oxygen.

A back titration with hydrogen showed that the water formed remained adsorbed on the catalyst, verified the adsorption stoichiometries, and determined the dispersion. Comparison with electron microscopy confirmed the hypothesis related to the postulated stoichiometries.

Figure 22.23 shows the weight variations of the sample (0.3% Ir/Al_2O_3) as a function of time during hydrogen chemisorption and H_2-O_2 titrations.

Example 2. Figure 22.24 illustrates the thermoprogrammed reduction of a V_2O_5 catalyst studied by combined TG–DSC under various gas flows.[161]

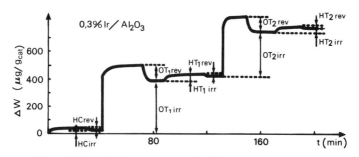

Figure 22.23. Hydrogen–oxygen titration on an Ir/Al_2O_3 catalyst.

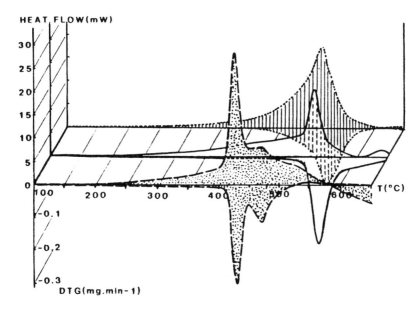

Figure 22.24. Heats of reduction of V_2O_5 and associated derivatives of the thermogravimetric curves as a function of the temperature, for various reducing agents: (solid lines) hydrogen, (dashed lines) ethane, (dots) ethylene.

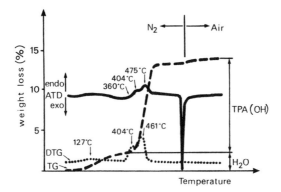

Figure 22.25. TG/DTA/DTG diagrams typical of the activation of a zeolite under N_2/air (Na, TPA)–ZSM-5 (Si/Al = 240).[163,164]

Example 3. Some authors (e.g., Gabelica *et al.*,[163–165] Choudhary and Pataskar[166]) have investigated the synthesis, structure, catalytic reactivity, and deactivation of ZSM-5 zeolites through the methods of simultaneous thermal analysis (TG/DTA/DTG). They emphasize the possibility of studying various catalytic reactions quantitatively (among them, synthesis of hydrocarbons from methanol) performed on these catalysts *in situ* in the thermobalance. Thus it is possible to control catalyst deactivation by quantitatively analyzing coking with

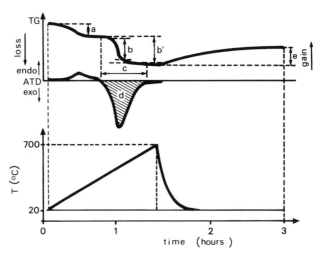

Figure 22.26. Schematic curve of a TG/DTA diagram corresponding to the activation of (TPA)–ZSM-5 under air. Definition of the various parameters which can characterize a quantitative decomposition.[163,164]

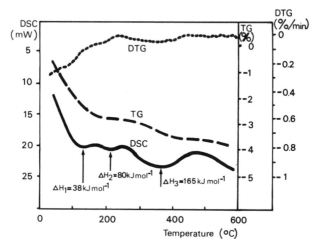

Figure 22.27. DSC, TG, and DTG curves during the heating under vacuum of NH4–ZSM-5 zeolite (initial mass 62.9 mg).

TG/DTA.[163–165] TG/DTA/DTG diagrams, which characterize the decomposition of the precursor of a ZSM-5 zeolite [the latter having been synthesized from corresponding tetrapropylammonium salts (TPA) under nitrogen] are shown in Figure 22.25. It is noticeable that the process passes through the successive stages:

1. A slow dehydration between 20 and 300 °C (DTA peak at 127 °C) corresponding to the loss of water molecules.

2. A more important loss, which takes place in two steps, between 350 and 500 °C, corresponding to the decomposition of the TPA ions.

The introduction of air at 500 °C causes the immediate oxidation of these residues (which is shown by an intense exothermic peak in DTA and a loss of weight).

Figure 22.26 schematically redefines the various parameters that characterize the decomposition of a precursor of ZSM-5: (a) loss of weight due to dehydration, (b or b') loss of weight due to TPA decomposition, (c) interval of temperature characterizing this decomposition, (d) area of the corresponding exothermal DTA peak, and (e) gain of weight due to the adsorption of dry nitrogen by the activated material.

Example 4. Study of the temperature-programmed desorption of ammonia (TPDA) from zeolites (ZSM) with the methods of simultaneous analysis TG/DTG/DSC (Figure 22.27)[24,150,151,172]: In this experiment, the sample, initially in its ammoniated form, was heated under vacuum at a rate of 10 K min^{-1} up to 590 °C, and the weight loss was determined. Ammonia was then adsorbed at room temperature and the physisorbed part was evacuated at the same temperature. Finally the sample was heated again up to 600 °C under vacuum at a rate of 10 K min^{-1}. The mass and thermal curves were used to determine the acidity spectrum of this zeolite.[172]

22.5. CONCLUSION

Thermal methods give the thermodynamic variables necessary for a knowledge of chemical transformations. Consequently these techniques enable workers to solve given problems taking into account means and the time allowed for their study.

The simplest technique, DTA, provides among other results the easy detection of thermal transformations. TGA brings more quantitative data as it provides a mass balance. Finally, calorimetry, often linked to other techniques, offers not only the possibility of obtaining the essential thermodynamic variables but also gives access to the surface properties of the material studied. In particular, in heterogeneous catalysts, the strength of surface sites can be defined and the strength distribution evaluated.

REFERENCES

1. W. Hemminger and G. Hone, *Grundlagen der Kalorimetrie,* Verlag Chemie, Weinhein (1979); *Calorimetry — Fundamentals and Practice,* Verlag Chemie, Weinhein (1984).
2. E. Calvet and H. Prat, *Recent Progress in Microcalorimetry* (H. A. Skinner, ed.), Pergamon, Oxford (1963).
3. M. Brun and P. Claudy, *Techniques de l'ingénier* **67**, 1200 (1983).

4. Microcalorimétrie et thermogénèse, *Colloques internationaux du CNRS, No. 156,* Editions du CNRS, Paris (1967).
5. I. Lamprecht and B. Schaarschmidt, eds., *Application of Calorimetry in Life Sciences,* Proc. Int. Conf. on Thermal Analyses, Walter de Cruyter, Berlin (1977).
6. S. Sunner and M. Masson, eds., *Combustion Calorimetry, Experimental Chemical Thermodynamics, Vol. 1,* Pergamon, Oxford (1979).
7. R. S. Porter and J. P. Johnson, eds., *Analytical Calorimetry Vols. 1–4,* Plenum, New York (1984); J. P. Johnson and P. S. Gillin, eds., *Analytical Calorimetry Vol. 5,* Plenum, New York (1984).
8. P. C. Gravelle, *Adv. Catal.* **22,** 191 (1972); *Catal. Rev. Sci. Eng.* **16,** 37 (1977); *J. Therm. Anal.* **14,** 53 (1978); *Thermochim. Acta* **72,** 103 (1984); **83,** 117 (1985); **96,** 365 (1985).
9. B. Fubini, *Rev. Gén. Therm. Fr.* **209,** 297 (1979).
10. V. Bolis, G. Della Gatta, B. Fubini, E. Giamello, L. Stradella, and G. Venturello, *Gaz. Chim. Ital.* **112,** 83 (1982).
11. G. Della Gatta, *Thermochim. Acta* **96,** 349 (1985).
12. C. Letoquard, F. Rouquerol, and J. Rouquerol, *J. Chim. Phys.* **3,** 559 (1973).
13. J. Rouquerol, *Pure and Appl. Chem.* **47,** 315–322 (1976).
14. J. Rouquerol, *Thermochim. Acta* **96,** 377 (1985).
15. P. Cavalotti and S. Carra, *La Chimica e l'Industria* **58,** 5 (1976).
16. *Documentation SETARAM,* 7 rue de l'Observatoire, 69300 Caluire, France.
17. G. I. Berezin, A. V. Kiselev, R. I. Sagatelyan, and M. V. Serdbodov, *Russ. J. Phys. Chem.* **43,** 118 (1969).
18. A. Souchon, *Thèse,* Grenoble (1977).
19. F. M. Camia, J. C. Garriues, and R. Roux, *J. Chem. Phys.* **62,** 1126 (1965); E. Calvet and F. M. Camia, *J. Chim. Phys.* **55,** 818 (1958).
20. J. Mercier, *J. Thermal Anal.* **14,** 161 (1978).
21. J. L. McNaughton and C. T. Mortimer, *Physical Chemistry Series, Vol. 2,* Butterworths, London (1975), p. 10.
22. M. Andrès, H. Charcosset, P. Le Parlouer, and J. Mercier, *C. R. Journées AFCAT* **13,** 64 (1982).
23. E. Karmazsin, *Thèse,* Lyon (1978).
24. A. Auroux and L. Benoist, unpublished results.
25. H. L. Krauss and R. Hopel, *Proc. 2nd European Symp. on Thermal Analysis* ESTA, Aberdeen (1981), p. 175.
26. M. Gruia, M. Jarjoui, and P. C. Gravelle, *J. Chim. Phys.* **73,** 634 (1976).
27. A. Auroux and P. C. Gravelle, *Thermochim, Acta* **47,** 333 (1981).
28. J. P. Reymond and P. C. Gravelle, *Bull. Soc. Chim. Fr.* (1978), 25.
29. Y. Grillet, F. Rouquerol, and J. Rouquerol, *J. Chim. Phys.* **74,** 179 (1977); **74,** 778 (1977).
30. P. Cléchet, C. Martelet, and Nguyen Dinh Cau, *Analysis* **6,** 220 (1978).
31. A. Auroux, G. Perachon, R. D. Joly, and J. M. Letoffe, *J. Chim. Phys.* **69,** 1371 (1972); A. Auroux, J. Bousquet, and J. M. Letoffe, *J. Chim. Phys.* **70,** 1133 (1973).
32. S. Zaini, S. Partyka, and B. Brun, *C. R. Journées AFCAT* **14,** 133 (1983).
33. J. Rouquerol, S. Partyka, and F. Rouquerol, *Adsorption at the Gas–Solid and Liquid–Solid Interface,* Elsevier, Amsterdam (1982).
34. A. C. Zettlemoyer, in: *Interface for Polymer Coatings,* Elsevier, Amsterdam (1969), p. 208.
35. L. Robert, Z. Kessaissia, and G. Tabak, *C. R. Acad. Sci. C* **276,** 451 (1973).
36. R. Tsunoda and Y. Fukazawa, *Bull. Chem. Soc. Jpn* **51,** 1883 (1978).
37. W. Rudzinski, J. Michalek, J. Zajac, and S. Partyka, in: *Proc. of the Engineering Foundation Conferences* (A. L. Myers and G. Belfort, eds.), (1983), p. 513.
38. J. F. Quinson, J. Dumas, and J. Serughetti, *J. Non-Crystalline Sol.* **79,** 397 (1986).
39. M. Brun, A. Lallemand, G. Lorette, J. F. Quinson, M. Richard, L. Eyraud, and C. Eyraud, *J. Chim. Phys.* **70,** 973 (1973).

40. M. Brun, A. Lallemand, J. F. Quinson, and C. Eyraud, *Thermochim. Acta* **21**, 59 (1977).
41. L. Stradella, G. Della Gatta, and G. Venturello, *Z. Phys. Chem. Neue Folge Wiesbaden* **115**, 25 (1979).
42. G. Della Gatta, B. Fubini, and C. Antonione, *J. Chem. Phys.* **72**, 66 (1975).
43. O. M. Dzhigit, A. V. Kiselev, T. A. Rachmanova, and S. P. Zhdanov, *J. Chem. Soc., Faraday Trans. I* **75**, 2662 (1973).
44. J. Navarro, E. Rojas, and V. Torra, *Rev. Gen. Thermoanal,* **11**, 143 (1973).
45. C. Brie, J. L. Petit, and P. C. Gravelle, *J. Chim. Phys.* **70**, 1108 (1973).
46. W. Rudzinski and J. Baszynska, *Z. Phys. Chem. Leipzig* **262**, 533 (1981).
47. S. Cerny, M. Smuteck, and F. Buzek, *J. Catal.* **38**, 245 (1975); **47**, 166 (1977).
48. I. Toyashima and G. A. Somorjai, *Catal. Rev. Sci. Eng.* **19**, 105 (1979).
49. M. A. Vannice, L. C. Hasselbring, and B. Sen. *J. Catal.* **95**, 57 (1985).
50. J. M. Basset, A. Théolier, M. Primet, and M. Prettre, in: *Proc. 5th Int. Congr. on Catalysis* (J. W. Hightower, ed.) Elsevier Amsterdam (1973), p. 915.
51. B. Tardy, H. Charcosset, M. Abon, and G. Bergeret, *Thermochim. Acta* **28**, 71 (1979).
52. J. B. Lantz and R. D. Gonzalez, *J. Catal.* **41**, 293 (1976).
53. G. Wedler, *J. Therm. Anal.* **14**, 15 (1978).
54. G. Wedler, K. P. Geuss, K. G. Colb, and G. McElhinery. *Appl. Surf. Sci.* **1**, 471 (1978).
55. Y. D. Pankrat'ev, E. M. Malyshev, Y. A. Ryndin, V. M. Turkov, B. N. Kuznetsov, and Y. I. Ermakov, *Kinet, i Katal.* **19**, 1543 (1978).
56. E. G. Igranova and V. E. Ostrovskii, *Dokl. Akad. Nauk. SSSR* **221**, 1351 (1975).
57. V. E. Ostrovskii and M. I. Temkin, *Kinet. i Katal.* **7**, 529 (1966).
58. G. D. Zakumbaeva and S. V. Artamonov, *React. Kinet. Catal. Lett.* **10**, 183 (1979).
59. H. Kobel and H. Roberg, *Ber. Buns. Phys. Chem.* **75**, 1101 (1971).
60. H. Topsoe, N. Topsoe and H. Bohlbro, in: *Proc. 7th Int. Congr. on Catalysis* (T. Seiyama and K. Tanabe, eds.) Elsevier, Amsterdam (1980), p. 247.
61. L. Tournayan, A. Auroux, H. Charcosset and R. Szymanski, *Adsorp. Sci. Tech.* **2**, 55 (1986).
62. J. Prinsloo and P. C. Gravelle, *J. Chem. Soc., Faraday Trans. I* **76**, 512 (1980); **76**, 2221 (1980); **78**, 273 (1982).
63. J. E. Benson and M. Boudart, *J. Catal.* **4**, 704 (1965).
64. J. C. Vedrine, A. Auroux, V. Bolis, P. Dejaifve, C. Naccache, P. Wierzchowski, E. G. Derouane, J. B. Nagy, J. P. Gilson, J. C. H. Van Hooff, J. P. Van den Berg, and J. Wolthuizen, *J. Catal.* **59**, 248 (1979).
65. A. Auroux, P. Wierzchowski, and P. C. Gravelle, *Thermochim. Acta* **32**, 165 (1979).
66. A. Auroux, V. Bolis, P. Wierzchowski, P. C. Gravelle, and J. C. Vedrine, *J. Chem. Soc., Faraday Trans. II* **75**, 2544 (1979).
67. A. Auroux, P. C. Gravelle, J. C. Vedrine and M. Rekas, in: *Proc. 5th Int. Zeolite Conf.* (L. V. Rees, ed.) Heyden, London (1980), p. 433.
68. P. Dejaifve, A. Auroux, P. C. Gravelle, J. C. Vedrine, Z. Gabelica, and E. G. Derouane, *J. Catal.* **70**, 123 (1981).
69. A. Auroux, J. C. Vedrine, and P. C. Gravelle, *Stud. Surf. Sci. Catal.* **10**, 305 (1982).
70. J. C. Vedrine, A. Auroux, P. Dejaifve, V. Ducarme, H. Hoser, and S. Zhou, *J. Catal.* **73**, 147 (1982).
71. A. Auroux, H. Dexpert, C. Leclercq, and J. C. Vedrine, *Appl. Catal.* **6**, 95 (1983).
72. J. C. Vedrine, A. Auroux, G. Coudurier, Ph. Engelhard, J. P. Gallez, and G. Szabo, in: *Proc. 6th Int. Zeolite Conf.* (D. Olson and A. Bisio, eds.) Butterworths, London (1983), p. 497.
73. J. C. Vedrine, A. Auroux, and G. Coudurier, in: *ACS Symposium Series 248,* American Chemical Society, Washington (1984), p. 254.
74. F. Maugé, A. Auroux, P. Gallezot, J. C. Courcelle, Ph. Engelhard, and J. Grosmangin, *Stud. Surf. Sci. Catal.* **20**, 91 (1985).

75. A. Auroux and J. C. Vedrine, *Stud. Surf. Sci. Catal.* **20**, 311 (1985).
76. A. Auroux, *Stud. Surf. Sci. Catal.* **37**, 385 (1988).
77. A. Auroux, M. B. Sayed, and J. C. Vedrine, *Thermochim. Acta* **93**, 557 (1985).
78. R. D. Shannon, K. Gardner, R. H. Staley, G. Bergeret, P. Gallezot, and A. Auroux, *J. Phys. Chem.* **89**, 4778 (1985).
79. M. B. Sayed, A. Auroux, and J. C. Vedrine, *Appl. Catal.* **23**, 49 (1986).
80. M. B. Sayed, A. Auroux, and J. C. Vedrine, *J. Catal.* **116**, 1 (1989).
81. G. Coudurier, A. Auroux, J. C. Vedrine, R. D. Farlee, L. Abrams, and R. D. Shannon, *J. Catal.* **108**, 1 (1988).
82. R. D. Shannon, R. H. Staley, and A. Auroux, *Zeolites* **7**, 301 (1987).
83. Y. S. Jin, A. Auroux, and J. C. Vedrine, *Appl. Catal.* **37**, 1 (1988).
84. C. Fernandez, A. Auroux, J. C. Vedrine, and J. Grosmangin, in: *Proc. 7th Int. Zeolite Conf.* (Y. Murakami, A. Iijima and J. W. Ward, eds.) Kodansha, Tokyo (1986), p. 345.
85. Z. C. Shi, A. Auroux, and Y. Ben Taarit, *Can. J. Chem.* **66**, 1013 (1988).
86. T. Masuda, H. Taniguchi, K. Tsutsumi, and H. Takahashi, *Bull. Chem. Soc. Jpn* **51**, 633 (1978).
87. T. Masuda, H. Taniguchi, K. Tsutsumi and H. Takahashi, *J. Japan Petrol. Inst.* **22**, 67 (1979).
88. T. Masuda, K. Tsutsumi and H. Takahashi, in: *Proc. 5th Int. Zeolite Conf.* (L. V. Rees, ed.) Heyden, London (1980), p. 483.
89. K. Tsutsumi, H. Q. Koh, S. Hagiwara, and H. Takahashi, *Bull. Chem. Soc. Jpn* **48**, 3576 (1975).
90. T. Masuda, H. Taniguchi, K. Tsutsumi, and H. Takahashi, *Bull Chem. Soc. Jpn* **51**, 1965 (1978).
91. H. Taniguchi, T. Masuda, K. Tsutsumi, and H. Takahashi, *Bull. Chem. Soc. Jpn* **51**, 1970 (1978).
92. T. Masuda, H. Taniguchi, K. Tsutsumi, and H. Takahashi, *Bull. Chem. Soc. Jpn* **52**, 2849 (1979).
93. H. Taniguchi, T. Masuda, K. Tsutsumi, and H. Takahashi, *Bull. Chem. Soc. Jpn* **53**, 362 (1980).
94. H. Taniguchi, T. Masuda, K. Tsutsumi, and H. Takahashi, *Bull. Chem. Soc. Jpn* **53**, 2463 (1980).
95. Y. Miwa, K. Tsutsumi and H. Takahashi, *Bull. Chem. Soc. Jpn* **53**, 2800 (1980); *Zeolites* **1**, 3 (1981); **1**, 98 (1981).
96. K. Tsutsumi, S. Hagiwara, Y. Mitani, and H. Takahashi, *Bull. Chem. Soc. Jpn* **55**, 2572 (1982).
97. K. Tsutsumi, Y. Mitani and H. Takahashi, *Bull. Chem. Soc. Jpn* **56**, 1912 (1983).
98. Y. Mitani, K. Tsutsumi, and H. Takahashi, *Bull. Chem. Soc. Jpn* **56**, 1917 (1983).
99. Y. Mitani, K. Tsutsumi, and H. Takahashi, *Bull. Chem. Soc. Jpn* **56**, 1921 (1983); *Colloid Polymer Sci.* **263**, 832 (1985); **263**, 838 (1985); **264**, 445 (1985).
100. T. R. Brueva, A. L. Klyachko-Gurvich, and A. M. Rubinshtein, *Izv. Akad. Nauk. SSSR. Ser. Khim.* **12**, 2807 (1972).
101. I. V. Mishin, A. L. Klyachko-Gurvich, T. R. Brueva, and A. M. Rubinshtein, *Dokl. Akad. Nauk. SSSR* **208**, 1100 (1973).
102. T. R. Brueva, A. L. Klyachko-Gurvich, I. V. Mishin, and A. M. Rubinshtein, *Izv. Akad. Nauk. SSSR. Ser. Khim.* **6**, 1254 (1974).
103. T. R. Brueva, A. L. Klyachko, I. V. Mishin, and A. M. Rubinshtein, *Izv. Akad. Nauk. SSSR. Ser. Khim.* **4**, 939 (1975).
104. A. L. Klyachko, T. R. Brueva, I. V. Mishin, G. I. Kapustin, and A. M. Rubinshtein, *Acta Phys. Chem.* **24**, 183 (1978).
105. V. F. Vogt, H. Wolf, H. Bremer, A. M. Rubinshtein, A. L. Klyachko, T. R. Brueva, and I. V. Mishin, *Z. Anorg. Allg. Chem.* **439**, 153 (1978).
106. A. L. Klyachko, *Kinet. i Katal.* **19**, 1218 (1978).

107. T. R. Brueva, A. L. Klyachko, I. V. Mishin, and A. M. Rubinshtein, *Kinet. i Katal.* **20**, 990 (1979).

108. A. L. Klyachko, T. R. Brueva, and A. M. Rubinshtein, *Kinet. i Katal.* **20**, 1256 (1979).

109. A. D. Rukhadze, G. I. Kapustin, T. R. Brueva, A. L. Klyachko, and A. M. Rubinshtein, *Kinet. i Katal.* **22**, 474 (1981).

110. G. I. Kapustin, T. R. Brueva, A. L. Klyachko, and A. M. Rubinshtein, *Kinet. i Katal.* **22**, 1561 (1981).

111. G. I. Kapustin, T. R. Brueva, A. L. Klyachko, A. D. Rukhadze, and A. M. Rubinshtein, *Kinet. i Katal.* **23**, 972 (1982).

112. G. I. Kapustin, L. M. Kustov, G. O. Glonti, T. R. Brueva, V. Y. Borovskov, A. L. Klyachko, A. M. Rubinshtein, and V. B. Kazanskii, *Kinet, i Katal.* **25**, 1129 (1984).

113. A. L. Klyachko, G. I. Kapustin, G. O. Glonti, T. R. Brueva, and A. M. Rubinshtein, *Kinet. i Katal.* **26**, 706 (1985).

114. S. S. Khvoschev, V. E. Skazyvaev, S. P. Zhdanov, and I. V. Karetina, *Izv. Akad. Nauk. SSSR. Ser. Khim.* **1**, 23 (1978).

115. S. S. Khvoshchev, V. E. Skazyvaev and E. A. Vasil'eva, in: *Proc. 5th Int. Zeolite Conf.* (L. V. Rees, ed.) Heyden, London (1980), p. 476.

116. S. S. Khvoshchev and E. A. Vasil'eva, *Izv. Akad. Nauk SSSR. Ser. Khim.* **5**, 973 (1982); **1**, 15 (1983).

117. F. S. Stone and M. Whalley, *J. Catal.* **8**, 173 (1967).

118. P. Cartraud, *Thermochim. Acta* **16**, 197 (1967).

119. P. Cartraud, A. Cointot, and B. Chauveau, in: *Molecular Sieves II. ACS Symposium Series, Vol 40* (J. R. Katzer, ed.), American Chemical Society, Washington (1977), p. 367.

120. P. Cartraud, B. Chauveau, M. Bernard and A. Cointot, *J. Therm. Anal.* **11**, 51 (1977).

121. J. M. Lopez Cuesta, *Thèse, Grenoble* (1985).

122. G. Della Gatta, B. Fubini, and E. Giamello, *Stud. Surf. Sci. Catal.* **10**, 331 (1982).

123. J. C. Petit and H. James, in: *Proc. 2nd Symp. on Thermal Analysis,* ESTA, Aberdeen (1981), p. 178; *Thermochim. Acta* **35**, 111 (1980).

124. H. Thamm, H. Stach, and W. Fiegig, *Zeolites* **3**, 95 (1983).

125. H. Stach, U. Lhose, H. Thamm, and W. Schirmer, *Zeolites* **6**, 74 (1986).

126. V. Kevorkian and R. O. Steiner. *J. Phys. Chem.* **67**, 545 (1963).

127. F. Pequignot, M. Bastick, and J. Bastick, *C. R. Acad. Sci. Ser.* C **279**, 981 (1974).

128. C. Meyer and J. Bastick, *Bull. Soc. Chim. Fr.* (1974), 59.

129. P. Y. Hsieh, *J. Catal.* **2**, 211 (1963).

130. G. Curthoys, V. Y. Davydov, A. V. Kiselev, S. A. Kiselev, and B. V. Kuznetsov, *J. Colloid Interf. Sci.* **48**, 58 (1974).

131. A. Auroux, P. C. Gravelle, and M. Poleski, *Calorim. Anal. Therm.* **14**, 84 (1983).

132. E. Garrone, G. Ghiotti, E. Giamello, and B. Fubini, *J. Chem. Soc., Faraday Trans. I* **77**, 2613 (1981).

133. A. Auroux and A. Gervasini, *J. Phys. Chem.* **94**, 6371 (1990).

134. M. I. Pope and M. D. Judd, *Differential Thermal Analysis: A Guide to the Technique and Its Applications,* Heyden, London (1977).

135. Ph. E. Slade and L. T. Jenkins, *Thermal Analysis, Vol. 1* Marcel Dekker, New York (1970); *Thermal Characterization Techniques,* Marcel Dekker, New York (1970).

136. W. W. Wendlandt, *Thermal Methods of Analysis,* Interscience, New York (1964).

137. W. W. Wendlandt, in: *Chemical Analysis, Vol. 19,* 2nd edition, John Wiley and Sons, New York (1974).

138. R. Sh, Mikkail and E. Robens, *Microstructure and Thermal Analysis of Solid Surfaces,* John Wiley and Sons, New York (1983).

139. W. J. Smothers and Y. Chiang, *Handbook of Differential Thermal Analysis,* Chemical Publishing Co., New York (1966).

140. J. Paulik and F. Paulik, in: *Comprehensive Analytical Chemistry, Vol. 12* (G. Svehla, ed.), Elsevier, Amsterdam (1981).
141. R. F. Schwenker and P. D. Garn, *Thermal Analysis, Vols. 1 and 2*, Academic, New York (1969).
142. R. C. Mackenzie, *Differential Thermal Analysis, Vols. 1 and 2*, Academic, London (1972); R. C. Mackenzie and P. G. Laye, *Chem. in Britain* **22**, 1005 (1986).
143. C. Duval, *Inorganic Thermogravimetric Analysis*, Elsevier, Amsterdam (1963).
144. P. Vallet, *Thermogravimétrie. Etude critique et théorique, utilisation, principaux usages*, Gauthier-Villard, Paris (1972).
145. C. J. Keattch and D. Dollimore, *An Introduction to Thermogravimetry*, Heyden, London (1975).
146. T. Gast and E. Robens, *Progress in Vacuum Microbalance Techniques, Vol. 1*, Heyden, London (1972); S. C. Bevan, S. J. Gregg and N. D. Parkins, *ibid. Vol. 2*, (1973); C. Eyraud and M. Escoubés, *ibid. Vol. 3*, (1975).
147. P. Barret, *l'Actualité Chimique* (1977), 15.
148. J. Mercier, in: *Progress in Vacuum Microbalance Techniques, Vol. 3*, Heyden, London (1975), p. 409.
149. B. Courtault, *C. R. Journées AFCAT* **4**, 31 (1978).
150. P. Le Parlouer, *Themochim. Acta* **92**, 371 (1985).
151. P. Le Parlouer, *Thermochim. Acta* **103**, 21 (1986).
152. P. A. Compagnon, C. Hoang-Van, and S. J. Teichner, *Bull. Soc. Chim Fr.* (1974), 2317.
153. J. R. Kiovsky, W. J. Goyette, and T. M. Notermann, *J. Catal.* **52**, 25 (1978).
154. H. Stach, T. Peinze, K. Fiedler, and W. Schirmer, *Z. Phys. Chem. Leipzig* **259**, 913 (1978).
155. W. Schirmer, H. Stach, K. Fiedler, W. Rudzunski, and J. Jagiello, *Zeolites* **3**, 199 (1983).
156. J. E. Benson, K. Ushiba, and M. Boudart, *J. Catal.* **9**, 91 (1967).
157. M. D. Sefcik and H. K. Yuen, *Thermochim. Acta* **26**, 297 (1978).
158. P. A. Jacobs, H. K. Beyerand, and J. Valyon, *Zeolites* **1**, 161 (1981).
159. R. M. Barrer and R. M. Gibbons, *Trans. Faraday Soc.* **59**, 2569 (1963); **61**, 948 (1965).
160. S. G. T. Bhat, in: *Symposium on Science of Catalysis and Its Application in Industry*, FPDIL, Sindri (1979), p. 25.
161. J. Le Bars, A. Auroux, J. C. Vedrine, B. Pommier, and G. Pajonk, *J. Phys. Chem.* **96**, 2217 (1992).
162. L. Tournayan, H. Charcosset, R. Fréty, C. Leclerc, and P. Turlier, *Thermochim. Acta* **27**, 95 (1978).
163. Z. Gabelica, J. P. Gilson, J. B. Nagy, and E. G. Derouane, *C. R. AFCAT* **14**, 360 (1983).
164. Z. Gabelica, J. P. Gilson, J. B. Nagy, and E. G. Derouane, *Clay Minerals* **19**, 803 (1984).
165. A. Nastro, Z. Gabelica, P. Bodart, and J. B. Nagy, *Calorim. Anal. Therm.* **15**, 206 (1984).
166. V. R. Choudhary and S. G. Pataskar, *Thermochim. Acta* **97**, 1 (1986).
167. C. T. W. Chu C. D. Chang, *J. Phys. Chem.* **89**, 1569 (1985).
168. V. Penchev, Ch. Minchev, V. Kanazirev, O. Pencheva, N. Borisova, L. Kosova, H. Lechert, and H. Kacirek, *Zeolites* **3**, 253 (1983).
169. H. Beyer, P. A. Jacobs, J. B. Uytterhoeven, and F. Till, *J. Chem. Soc., Faraday Trans. I* **73**, 1111 (1977).
170. R. E. Richards and L. V. C. Rees, *Zeolites* **6**, 17 (1986).
171. M. T. Aronson, R. J. Gorte, and W. E. Farneth, *J. Catal.* **98**, 434 (1986).
172. A. Auroux, Y. S. Jin, J. C. Vedrine, and L. Benoist, *J. Catal.* **36**, 323 (1988).

MASS SPECTROMETRY: PRINCIPLES AND APPLICATIONS IN CATALYSIS

C. Mirodatos

23.1. BASIC PRINCIPLES AND SYSTEMS OF MASS SPECTROMETRY

Mass spectrometry (MS) is unique among the molecular methods of analysis such as IR, UV, or NMR spectroscopy as it allows mass discrimination via moleculer ionization. This irreversible process makes mass spectrometry a destructive means of analysis. Basically, ionization proceeds when the first ionization potential of a neutral molecule has been reached, according to

$$M: \rightarrow M^+ + e^- \tag{23.1}$$

The mass spectrometer measures the ratio of mass to charge (m/e) of the molecular ion formed. The energy typically used for ionization (e.g., 70 eV, standard value in mass spectrometry) is generally far in excess of that needed for this process and the ion formed will be prone to dissociation, mainly due to extra vibrational energy; this is called fragmentation:

$$M^+ \rightarrow A^+ + B \quad \text{or} \quad A + B^+ \tag{23.2}$$

A mass spectrum is formed from the pattern of the different molecular and fragment ions obtained for a given molecule. The functions of the mass

C. Mirodatos ● Institut de Recherches sur la Catalyse, CNRS, Villeurbanne, France.

Catalyst Characterization: Physical Techniques for Solid Materials, edited by Boris Imelik and Jacques C. Vedrine, Plenum Press, New York (1994).

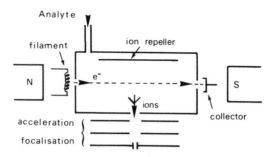

Figure 23.1. Scheme of an ion source using electron impact.

spectrometer are ionization of molecules (ion source), separation according to the mass-to-charge ratio (analyzer), and quantitative measurement of the different ions (detector).

The way to introduce the sample (gas, liquid, or solid) into the mass spectrometer depends on the particular use of this technique, e.g., *in situ* reactions, chemical analysis for species identification, on-line gas chromatography/mass spectrometry (GC/MS) analysis. This point will be treated further on in the different practical examples.

23.1.1. Ion Sources

Ionization is most commonly obtained through electron impact, i.e., bombarding the sample molecules with high-energy electrons (Figure 23.1). The sample is introduced at a pressure of about 10^{-5} Torr, which corresponds to a mean free path of about 5 m, assuming $I = 5 \times 10^{-5}/P$ (l in meters and P in Torr). Electrons are emitted from a heated metal filament and accelerated according to the potential difference between the filament and the source enclosure. The electron beam is focused by a magnetic field, travels across the ion chamber at a constant speed (isopotential), and interacts with the electrons of the sample molecule M, leading to ionization:

$$M + e^- \rightarrow M^+ + 2e^- \qquad (23.3)$$

The electrons moving out of the ionization chamber are trapped by an electrode that regulates the electron beam. For a typical current of 10^{-4} A (around $6 \times 10^{14}\, e^-\, s^{-1}$), 1/1000 molecule will be ionized. For a gaseous volume of 2 ml at 10^{-5} Torr, i.e., 10^{12} molecules, $(10^{12}/10^3) = 10^9$ ions will be formed per second, which gives an ionic current of 10^{-10} A. The ionization efficiency, which depends on the internal energy distribution of ions, is generally optimized for an electronic current of 70 eV. These ions are ejected from the source due to an electric field created between a plate maintained at a positive potential (ion repeller) and the source enclosure. They are then accelerated and focused through high-voltage plates to the analyzer unit.

This ionization method is used almost exclusively in the low-mass range generally required for heterogeneous catalysis (< 200 amu). Analytical problems coming from the high fragmentation of molecular ions can be partially solved by decreasing the ionization energy of the electron beam, although with a detrimental effect on sensitivity. This possibility of changing the ionization energy easily, often ignored by MS manufacturers, should be considered when choosing a model to equip a laboratory.

Chemical ionization is another technique suitable for catalysis. This gives reduced and controlled fragmentation by ionizing the sample not directly with electrons but with ions of another gas maintained at a constant pressure in the source (for instance 1 Torr of methane). Other ionization techniques are thermal ionization (for liquid and solid samples), photoionization (discharge tube in helium), and ionization by field emission.

23.1.2. Analyzers

23.1.2.1. Magnetic Sector

After acceleration (voltage up to 10 kV), ions with a charge e and a mass m have acquired a kinetic energy of

$$eV = \tfrac{1}{2}mv^2 \tag{23.4}$$

The role of the analyzer is to separate the different ions according to their mass-to-charge ratio. If the accelerated ions are subjected to a magnetic force created by a magnetic field H perpendicular to their path, they will travel the field following a circular path of radius R due to the balance between the magnetic force and the centrifugal force:

$$Hev = mv^2/R \tag{23.5}$$

From equations (23.4) and (23.5), one gets

$$m/e = H^2R^2/2V \tag{23.6}$$

Therefore, each ion will have a proper trajectory according to its mass-to-charge ratio (Figure 23.2). By placing a select slot at the magnetic sector outlet, only ions with a given m/e ratio will reach the detector. By scanning the accelerating voltage or the magnetic field, ions of different m/e ratio will be analyzed in turns. Figure 23.3 shows as an example of ionic separation the spectrum of methane with the peak $m/e = 16$ (ion CH_4^+), the peak 17 ($^{13}CH_4^+$ and CH_3D^+, natural isotopes of methane), and the peaks 15, 14, 13, 12, 2, 1 (fragments CH_3^+, CH_2^+, CH^+, C^+, H_2^+, and H^+). In practice, the ions entering the analyzer do not have exactly the same kinetic energy (Boltzmann distribu-

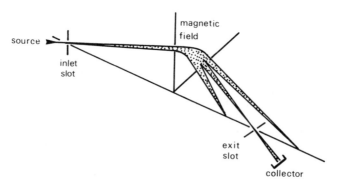

Figure 23.2. Ion deflection in a magnetic analyzer.

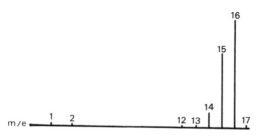

Figure 23.3. Mass spectrum of methane (ionization at 70 eV).

tion of ion energies and source heterogeneity), and this induces a peak spread around a given mass.

The resolving power (ability to distinguish two adjacent peaks) of a magnetic sector analyzer as described above depends on the radius R of the magnetic sector and on the width w of the selecting slots:

$$\delta m/m = 2w/R \qquad (23.7)$$

Values up to 1/1000 are usual for such a configuration. Due to the quasi-logarithmic scale of masses, the resolving power remains constant on the whole range of mass. The resolution can be considerably improved if a first deflection through an electrostatic field (this separates the ions according to their kinetic energy) is placed in front of the magnetic deflection. This is the principle of double-focusing analyzers.

Due to their stability and high resolving power, magnetic analyzers are well adapted to quantitative analysis of gaseous mixtures or of traces. However, their technology is complex and delicate (high voltage and magnetic fields) and so they are rather expensive. Furthermore, scanning a magnetic field remains a relatively slow process, which limits the study of fast phenomena often characteristic of catalytic kinetics.

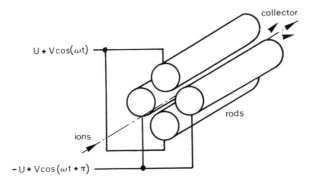

$U + V\cos(\omega t)$

collector

rods

ions

$-U + V\cos(\omega t + \pi)$

Figure 23.4. Scheme of a quadrupole analyzer.

23.1.2.2. Quadrupole Mass Analyzer

Unlike the magnetic analyzer, the quadrupole analyzer requires low-energy ions (accelerated under a few volts only). These ions enter a space delimited by four parallel conducting rods, electrically connected in pairs, as shown in Figure 23.4. One pair is connected with a direct current voltage U and an alternative current voltage at a radio frequency $V_{\cos}(\omega t)$, while the other is connected to the reverse dc voltage $-U$ and the same ac voltage but 180° out of phase. The ions which experience these complex fields move along oscillating trajectories between the rods, and only those having a given initial kinetic energy will reach the detector. By varying dc and ac voltages together (constant intensity ratio), a mass scanning is accomplished, and this can be achieved in a few milliseconds for the full mass range.

Due to a linear scan of voltages, the quadrupole analyzer produces a linear mass scale, which allows much easier computer control than in the magnetic analyzer, as for instance in multiple ion monitoring, i.e., the ability to follow only preselected ions in the mass range. Furthermore, these systems are relatively cheap to build. The combination of these advantages, despite a generally lower resolving power, makes the quadrupole analyzer the most commonly used system in the area of catalysis, especially for mechanistic studies.

23.1.2.3. Time-of-Flight Analyzer

For this kind of analyzer, gas ionization is carried out in a pulsed mode, which produces ion clusters, first trapped between two electrodes, then sent into a linear-tube (about 1 m long) to the ion collector and the detector. Separation is achieved according to the time the ions will "fly" through the analyzer. Such a system allows a spectrum acquisition within a millisecond, a performance well adapted to chemical kinetics. However, some problems dealing with the procedure of synchronization of the pulsed sequences still limit the marketing of this analyzer.

C. MIRODATOS

23.1.3. Detectors

The detector allows the ionic current to be converted either before the analyzer (total ionic current) or after the analyzer (filtered ions) into electronic current for amplification and data treatment.

23.1.3.1. Total Ionic Current

The total ionic current is obtained by means of an electrode intercepting part of the ionic beam coming from the source, producing a signal proportional to the ion concentration. This current reflects any inlet of analyte into the source, like a chromatographic detector at a column outlet. Thus, in the case of a GC/MS system, the total current and the chromatographic detector give similar information.

23.1.3.2. Ion Collector

This collector is simply a Faraday cage that is placed after the ion analyzer, and it includes an electrode that prevents the detections of secondary ions. This detector is convenient for quantitative analysis, but cannot be used for traces.

23.1.3.3. Electron Multiplier

The electron multiplier amplifies the ionic current by several orders of magnitude. Basically, the ions filtered by the analyzer first strike a metallic electrode (known as a dynode), which releases a shower of electrons to a second dynode, then to a third, starting a cascade effect up to a stabilized gain (generally obtained with a decade of dynodes). Different geometries have been tested for assembling the dynodes (e.g., Allen or venetian-blind types), in order to optimize amplification and resistance to poisoning. Some models used on recent systems, such as the channeltron (Bendix Corporation), are made of a circular glass tube coated with a conducting material, and allow a continuous cascading effect based on the cycloidal motion of ions and electrons.

23.2. NATURE AND TREATMENT OF DATA

Data produced by the mass spectrometer (total ion current, distribution of ionic abundancies as a function of the m/e ratios, i.e., mass spectra) are amplified, then computed. The computer monitors the different parameters of the mass spectrometer, acquires the data, and treats them in real time or off time (by, e.g., fragmentometry, concentration calculations, or kinetics).

23.2.1. Analysis of a Multicomponent Gas Mixture

During the analysis of a multicomponent gas mixture, it is most likely that the spectra from the different gases will overlap. It follows that no gas will have even one peak of its spectrum directly proportional to its concentration. A matrix technique can generally overcome this problem.

Let us assume that the gas mixture to be analyzed has p components, the concentration of the jth being c_j. If there is no chemical interaction between the components, then the peak height at mass i can be expressed as

$$h_i = h_{i_{res}} + \sum_j f_{ij} c_j \qquad (23.8)$$

where $h_{i_{res}}$ is the background contribution to the mass i (residual peak height before mixture introduction), $\sum_j f_{ij} c_j$ is the sum of the contributions of the p components to the mass i. The contribution of the jth component is expressed as the product of the actual concentration c_j by the factor of fragmentation f_{ij}. The latter represents the peak height at mass i that would give the jth component if sent pure into the mass spectrometer (at a concentration equal to one). These fragmentation factors are then readily obtained from calibration measurements. The system made of the equations (23.8) can be treated as a matrix system, which allows the calculation of the different c_j values by adjusting them in order to minimize the differences betwen the experimental and the calculated peak heights, and by checking finally to ensure that $\sum_j c_j = 1$. In most case, MS manufacturers now use software systems adapted to this quantitative analysis of gas mixtures.

23.2.2. Analysis of a Mixture of Stable Isotopes

This case is typical of mechanistic studies in catalysis. Whatever the mixture of isotopic components, a first step in the calculation consists of correcting the peak heights for the naturally occurring isotopes.

23.2.2.1. Correction of the Natural Isotope Abundancy

Each element of a given atomic number may have one or several stable isotopes in nearly constant proportion in the natural state. Table 23.1 lists the concentration of stable isotopes for the main elements of organic chemistry.

Let us consider a hydrocarbon molecule of formula $C_n H_m$, of mass $M = 12n + m$. Let us call c the natural concentration of ^{13}C in carbon and h the concentration of ^2H (deuterium) in hydrogen. The probability of having x atoms of ^{13}C among the n atoms of the molecule is

$$C_n^x c^x (1 - c)^{n-x} \qquad (23.9)$$

Thus the probability that all the carbon atoms are ^{12}C is $(1 - c)^n$, the same type

TABLE 23.1. Natural Isotope Abundance

Element	Atomic number	Atomic mass	Relative abundance (%)
Carbon	6	12	98.892
		13	1.108
Hydrogen	1	1	99.985
		2	0.015
Oxygen	8	16	99.759
		17	0.037
		18	0.204
Nitrogen	7	14	99.635
		15	0.365

of formula holding for hydrogen, $(1 - h)^m$. The product of these two probabilities represents the probability of a molecule with only ^{12}C and ^1H, that is to say the parent ion of mass M:

$$P_M = (1 - c)^n (1 - h)^m \qquad (23.10)$$

The probability of having an ion of mass $M + 1$ will correspond to the probability of having either an ion with one ^{13}C and only ^1H atoms or with only ^{12}C atoms and one ^2H.

$$P_{M+1} = C_n^1 c (1 - c)^{n-1} (1 - h)^m + C_m^1 h (1 - h)^{m-1} (1 - c)^n \qquad (23.11)$$

The ratio $P_{M+1}/P_M = nc(1 - c) + mh(1 - h)$ gives the fraction of natural isotopes which contributes to the mass $M + 1$, originating from the molecule of mass M. The same holds for the ratio P_{M+2}/P_M. The correction for the natural isotope abundancy will then be achieved by subtracting from the peak heights $M + 1$ and $M + 2$ the values (height of peak M) P_{M+1}/P_M and (height of peak M) P_{M+2}/P_M, respectively. There is no need generally to correct more than the two upper masses due to the usual uncertainties of the measurements.

For the case frequent in catalysis where the hydrocarbon itself contains isotopic atoms (a labeled molecule), e.g., x ^{13}C and y ^2H, its formula can be written $^{12}C_{n-x} {}^{13}C_x {}^1H_{m-y} {}^2H_y$ and its mass as $M + x + y$. The correction then concerns the masses $M + x + y + 1$ and $M + x + y + 2$, and is carried out, as described earlier by changing n and m by $n - x$ and $m - y$, respectively. Table 23.2 gives an example of correction for the methane molecule and its deuterated isotope CH_3D.

This example of correction on hydrocarbons can be readily extended to more complex molecules, taking into account other natural isotopes such as oxygen and nitrogen. The presence of elements having several natural isotopes

TABLE 23.2. Example of Contribution of the Naturally
Occurring Isotopes

Molecule	Parent ion M	Ion M + 1
CH_4	CH_4^+, $m/e = 16$ 100%	$^{13}CH_4^+ + CH_3D^+$, $m/e = 17$ 1.168%
CH_3D	CH_3D^+ 100%	$^{13}CH_3D^+ + CH_2D_2^+$, $m/e = 18$ 1.153%

TABLE 23.3. Example of Fragmentation Factors of
Methane as a Function of the Ionization Energy

Ionization energy (eV)	f_{1H}	f_{2H}	f_{1D}	f_{2D}
70	0.76	0.06	0.49	0.04
15	0.25	0.003	0.16	0.002

in nonnegligible amounts such as chlorine and sulfur makes the correcting
procedure more delicate.

23.2.2.2. Correction for Fragmentation with a Mixture of Isotopes

This procedure will be illustrated with a mixture of the deuterated isotopes
of methane (CH_4, CH_3D, CH_2D_2, CHD_3, CD_4). Each molecule will undergo
ionization and fragmentation in the mass spectrometer. The ratios between the
concentration of fragments and molecular ions (or parent ions) define the
fragmentation factors (e.g., $CH_3^+/CH_4^+ = f_{1H}$, related to the loss of one hydrogen).
Table 23.3 lists these fragmentation factors related to the loss of one and two
hydrogen or deuterium atoms in a methane molecule for two ionzation energies.

It may be noted in Table 23.3 that a methane molecule will lose a hydrogen
more easily than a deuterium atom, which reflects a primary isotopic effect
based on the energy differences between the C—H and C—D bonds (see
below). Table 23.3 also illustrates the effect of the ionization energy on the
fragmentation level. The simultaneous loss of one hydrogen and one deuterium,
f_{HD}, may be estimated as a statistical combination of the fragmentation factors
f_{2H} and f_{2D}. Thus for CHD_3.

$$f_{HD} = \tfrac{1}{4} f_{2H} + \tfrac{3}{4} f_{2D} \qquad (23.12)$$

23.2.2.3. Correction by Recurrence

As the fragmentation of a molecular ion contributes to lower masses, the
correction of the peak heights has to be done by a recurrent process which starts

TABLE 23.4. Example of Correction for Fragmentation of a Deuterated Methane Mixture

Molecular ions	m/e	Corrected peak height
CD_4	20	$h'_{20} = h_{20}$
CHD_3	19	$h'_{19} = h_{19}$
CH_2D_2	18	$h'_{18} = h_{18} - h'_{20} f_{1D} - \frac{1}{4} h'_{19} f_{1H}$
CH_3D	17	$h'_{17} = h_{17} - \frac{3}{4} h'_{19} f_{1D} - \frac{1}{2} h'_{18} f_{1H}$
CH_4	16	$h'_{16} = h_{16} - h'_{20} f_{2D} - \frac{1}{6} h'_{19} f_{HD} - h'_{18}(\frac{1}{2} f_{1D} + \frac{1}{6} f_{2H}) - \frac{3}{4} h'_{17} f_{1H}$
CH_3	15	$h'_{15} = h_{15} - \cdots = 0$
CH_2	14	$h'_{14} = h_{14} - \cdots = 0$

from the higher masses. Thus, in the present case of a mixture of methanes, the correction will proceed from CD_4 to CH_4. The calculation consists of subtracting from a given peak height (already corrected for background and natural isotopes) the contribution related to the fragmentation of the higher-mass ions. These contributions are equal to the products of peak heights and fragmentation factors. Table 23.4 details the recurrent procedure for the labeled methanes.

The quality of the correction is tested by checking if all the peaks which are due only to fragments (e.g., 15, 14, 13... for the methanes) are properly corrected to zero. After the corrections have been completed, one gets finally the concentration of the different components of the analyzed mixture (for this case, the isotopic distribution of methanes).

If the mixture of isotopes results from an exchange or equilibration reaction (e.g., the catalytic reaction $CH_4 + D_2$), it may be worthwhile, from a mechanistic point of view, to compare the experimental distribution to the one obtained if the same amount of deuterium was randomly distributed among the methane molecules. This random distribution is calculated statistically.

Let us express the experimental distribution:

	CH_4	CH_3D	CH_2D_2	CHD_3	CD_4
%	d_0	d_1	d_2	d_3	d_4

The mean number of deuterium atoms per methane molecule is

$$\Phi = \sum_i [(id_i)/100] \qquad (23.13)$$

The probability of having a C—H bond is $p = 1 - \Phi/4$ and a C—D bond is $\Phi/4$. Then the probability of having a methane molecule CH_iD_j is $C_4^i p^i (1 - p)^j$. These values correspond to the different terms of the developed binomial formula:

$$[p + (1 - p)]^4 \qquad (23.14)$$

Therefore, the random distribution is a binomial distribution.

For the case where more than two isotopes of an element are randomly distributed in a molecule containing n atoms of this element, one has to use the polynomial formula

$$(p_1 + p_2 + \cdots + p_i)^n \qquad (23.15)$$

Further mathematical treatment can be also carried out on the basic mass spectrometer data (e.g., kinetics calculation and modeling). They will be referred to below in the examples.

23.3. SELECTED APPLICATIONS IN CATALYSIS

23.3.1. Homogeneous and Surface Organometallic Catalysis

The mass spectrometry is used in this research area mainly as a tool for analysis — identification and concentration of products. Due to the fact that organic mixtures are often complex, GC/MS and LC/MS techniques are generally required. In all cases, the gaseous, liquid, or solid analyte is sampled from the reactor (often static like a Schlink tube), and then introduced into the analysis system. The type of inlet device depends on the nature of the sample (calibrated volume leaking continuously to the MS or injected into a chromatographic column connected to the MS). The GC/MS coupling can be direct (with a capillary column) or via a separator (with a packed column) which both decreases the inlet pressure and concentrates the sample by suppressing most of the carrier gas (helium). These coupled systems which associate the resolving power of chromatography and the mass detection of mass spectrometry are now available in compact form (quadrupole analyzer) well adapted to the identification of labeled or unlabeled organic and organometallic products in a catalysis laboratory.[5]

23.3.2. Adsorption/Desorption Processes on Solid Surfaces

23.3.2.1. Well-Defined Surface Studies

In surface studies (e.g., single crystals, polycrystalline alloys, and growing oxides), it is quite usual to add mass spectrometry to the assembly of surface spectroscopic techniques. It is generally a quadrupole analyzer, the source of which is directly connected to the ultrahigh vaccum chamber of the system, or via a molecular leak if gases are admitted into the chamber. Its main use is to analyze the gaseous residues produced by a solid during a thermal treatment (e.g., a programmed thermodesorption) or when a catalytic reaction is carried out in a side chamber allowing pressure and temperature.

23.3.2.2. Flash Thermodesorption on Oxides

In order to identify and discriminate among the different adsorption modes of molecules on a solid, a technique of flash thermodesorption under high vacuum has been developed which depends on the time resolution of mass spectrometry. Halpern and Germain have used this technique to study the modes of oxygen adsorption on oxide catalysts.[9] The thermodesorption cell is connected directly to a time-of-flight mass spectrometer which analyzes the oxygen evolving from the catalyst during the flash (from 20 to 600 °C in 30s) continuously. From the experimental spectrum (Figure 23.5), the activation energies for desorption are calculated on the basis of the following equation:

$$E/kT_m = \ln(v/B) + \ln(1/T_m)\ln(v/B) \qquad (23.16)$$

where T_m is the temperature at the peak maximum, B the rate of the temperature increase, and v the frequency factor related to the vibration states of the molecules and their activated complexes. The shapes of the different peaks are also analyzed. Distinct states of adsorption were identified in a series of Group IV oxides. It was shown that the calculated desorption energies followed approximately the general activity pattern of oxidation catalysis, low desorption energies implying high activities.

Thus, the flash desorption technique, which is directly dependent on the mass spectrometer performance, appears to be particularly convenient for discerning, discrete states of gas adsorption.[8] Falconer and Schwarz have described in detail the theoretical approach required to quantify binding energies obtained from temperature programmed thermodesorption curves.[9]

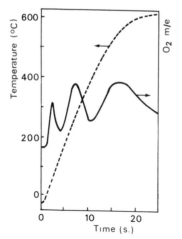

Figure 23.5. Spectrum of oxygen flash thermodesorption on Fe_2O_3.[7]

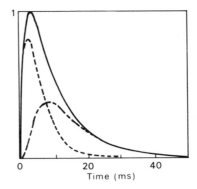

Figure 23.6. TAP response to a gas pulse with its deconvolution in nonadsorbing molecules (---) and adsorbing molecules (– · –).[10]

23.3.2.3. Transient Adsorption/Desorption Processes

In order to minimize the diffusional effects which most often influence a TPD profile and which are not easily controlled by the experimentalist, a specific system called "temporal analysis of products" (TAP from Autoclave Engineers, Inc.) has been designed with the aim of carrying out isothermal experiments at very low pressures where Knudsen flow predominates.[10] Basically, the TAP system consists of a differential microreactor fed by a set of high-speed, molecular-beam-type pulse valves and a quadrupole mass spectrometer. Pulses of reactant are introduced to the reactor in a desired sequence and plugs of reacting components leave the reactor in a differentially pumped vacuum chamber which makes them travel to the mass spectrometer for analysis with a time resolution of less than a millisecond.

For analysis of an adsorption/desorption process (assuming first-order desorption kinetics and Knudsen diffusion through the reactor), a single pulse of a reactant A is used to determine the average residence time of molecules in the reactor. This time t_a which is obtained by analyzing the TAP curves (moments calculation) (Figure 23.6) is related to the adsorption (k_1) and desorption (k_2) rate constants, to the effective diffusivity D_a, and to the reactor length L through the equation

$$t_a = (L^2/2D_a)(1 + k_1/k_2) \qquad (23.17)$$

For a nonadsorbing inert gas added as a tracer to the reactant feed

$$t_i = L^2/2D_i \qquad (23.18)$$

Then D_a can be calculated using the simple relation between the two diffusivities and the related molecular weights,

$$D_a/D_i = (M_i/M_a)^{1/2} \qquad (23.19)$$

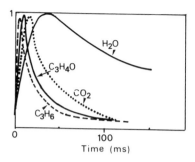

Figure 23.7. TAP response in propene oxidation on bismuth molybdate.[10]

By taking $L^2/2D_a$ as t_{ia}, the average residence time of A molecules which would cross the reactor without an adsorption/desorption delay, one gets

$$k_2/k_1 = t_{ia}/(t_a - t_{ia}) \qquad (23.20)$$

which allows the determination of the adsorption and desorption energies E_a and E_d via the Arrhenius transform:

$$\ln(t_a - t_{ia}) = (E_d - E_a)/RT + \text{constant} \qquad (23.21)$$

This technique has been used, for example, to study the selective oxidation of propene into acrolein[12]:

1. By pulsing mixtures of acrolein and argon at various temperatures on a bismuth molybdate catalyst (Figure 23.7), a desorption energy of 17 kcal/mol has been determined, assuming a nonactivated adsorption ($E_d \gg E_a$). As this value was obtained at temperatures close to those of propene oxidation, it can be compared with other values determined from classical TPD experiments, e.g., by Matsuura[11] and Keulks et al.[12] Moreover, this value is close to the activation energy of the propene oxidation, so it can be deduced that acrolein desorption is a step that controls the rate of acrolein formation from propene.
2. By pulsing a mixture of propene and oxygen in reacting conditions, a CO_2 peak is detected but later than the acrolein peak and moreover much broader than a peak obtained from a CO_2 pulse. This change in the CO_2 desorption mode suggests that CO_2 comes mainly from further oxidation of acrolein rather than from a direct oxidation of propene, in accord with ^{14}C labeling studies carried out by Keulks et al.[12] Other applications of the TAP system dealing more specifically with reaction kinetics are developed in Gleaves et al.[10]

23.3.3. Exchange Reactions

Exchange reactions have been widely developed in catalysis on metals and oxides as a tool for investigating the elementary steps of adsorption and desorption in connection with the properties and surface states of the reacting catalysts.[14-16]

23.3.3.1. Exchange Reactions on Oxides

The propene oxidation, described above for illustrating the determination of desorption energies, may also be referred to for depicting some classical uses of oxygen isotopes in catalysis. For this reaction, Keulks[13] has used a recirculation reactor directly coupled to a quadrupole mass spectrometer, which gives a system convenient for optimizing the isotope consumption. In a first step, the equilibration reaction of an equimolecular mixture of $^{16}O_2$ and $^{18}O_2$ over bismuth molybdate

$$^{16}O_2 + {}^{18}O_2 \Leftrightarrow 2\ {}^{16}O^{18}O$$

has been shown to be negligible, ruling out any fast adsorption/desorption process of oxygen on the oxide, at least those involving a molecular dissociation. When the oxidation reaction is carried out with a $^{18}O_2$/propene mixture, a fast oxygen uptake is noted together with the disappearance of propene but only a very small fraction of labeled oxygen is detected in the oxidation products (Figure 23.8).

From this elegant experiment, completed by other isotopic observations, it is concluded that the oxygen atoms incorporated into the oxidation products are lattice oxygens. These atoms are replaced by bulk atoms which migrate rapidly throughout the oxide structure, being supplied by oxygen adsorption on separate sites (e.g., layer edges).

23.3.3.2. Exchange Reactions on Metals

As catalysis on metals is frequently concerned with hydrocarbon conversion or synthesis, hydrogen/deuterium exchange is naturally considered a relevant approach for mechanistic studies.

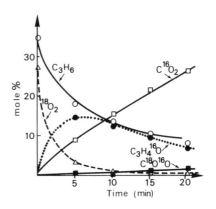

Figure 23.8. Gas phase composition for the reaction of C_3H_6 with $^{18}O_2$ at 425 °C.[13]

23.3.3.2a. Methane/Deuterium Exchange. Kemball[14] and Frennet[17] have studied this reaction on metallic films deposited under low pressure in detail. Dalmon and Mirodatos[18] have shown that the kinetic parameters of exchange on nickel catalysts are related to the changes in the metal properties after alloying (Ni–Cu), alkaline promotion (K) and poisoning (H₂S). For this study, a plug-flow type of microreactor working in a differential kinetic regime was connected to a mass spectrometer (magnetic sector) through a pumped capillary and a calibrated molecular leak. A CH_4/D_2 mixture was admitted on the reduced catalyst and analyzed continuously at the reactor outlet. Comparing the isotopic distributions that were obtained after reaction to the corresponding binomial distributions (see Section 23.2.3) led to the following conclusions:

1. The main isotopic exchange products, CH_3D (single exchange) and CD_4 (multiple exchange) are formed according to different and unrelated kinetics, which suggests distinct mechanisms.
2. The rate limiting step of these processes is related to the adsorption of a methane molecule on a surface mainly covered with hydrogen atoms, the coverage with hydrocarbon species being negligible (as a matter of fact, almost no methane is initially adsorbed on the catalyst when analyzing the transient response of the mass spectrometer after the first gas mixture admission).
3. With the assumption that the rate of formation for each of the deuterated species is proportional to the probability of a light methane molecule colliding with specific surface ensembles, it is proposed that the single exchange proceeds via an associative adsorption on a site including a preadsorbed deuterium atom. In contrast the multiple exchange implies a dissociative adsorption leading to a surface carbide intermediate (Ni_3C), also identified by magnetic methods after methane adsorption (see Chapter 21 for more details).

On the basis of the previous conclusions, it is shown that:

1. The Ni–Cu alloying does not modify the intrinsic kinetic parameters of the exchange and inferentially the electronic state of the nickel. A pure effect of statistical dilution of the nickel-active phase by inert copper atoms is therefore postulated.
2. The addition of potassium induces changes in the kinetic parameters, so it has an electronic effect for nickel atoms. An electronic transfer $K \rightarrow Ni$ able to weaken the Ni—H bonds is proposed in connected with IR observations (ν_{CO} shifts in similar conditions[19]).
3. Poisoning nickel by sulfur simply induces a decrease in the exchange rate proportional to the amount of deposited sulfur. This suggests that the loss of active surface does not proceed by statistical dilution but more likely via wave front or spreading patches in the catalytic bed.

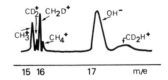

Figure 23.9. High-resolution spectrum for CX_3^+ ions.[21]

15 16   17   m/e

23.3.3.2b. Hydrocarbon/Deuterium Exchange. With reactions involving changes in molecular structure (e.g., hydrogenolysis, hydrogenation), the H/D exchange of the reactant and the products has to be considered. Except for cases where light hydrocarbons (e.g., ethane, propane, and isopentane) or molecules with limited fragmentation (benzene) are involved, the direct isotopic analysis soon becomes impossible and must be replaced by a GC/MS analysis. The gas chromatograph then provides peaks which are isotopic mixtures of separated components and the mass spectrometer has to determine the isotopic distribution during the peak elution (a fast scanning is required since some shift in the different isotopes can occur under the chromatographic peak envelope). This elegant but technically complex procedure is well developed today due to computer facilities. One example is the study of ^{12}CO and ^{13}CO hydrogenation (Fischer–Tropsch synthesis) by Mims and McCandlish.[20]

The place of the labeled atoms in the reaction products can give valuable information for determining a mechanism. Thus, Guczy and Ujszaski[21] studied ethane adsorption on metals by analysis of the incorporation of deuterium with a high-resolution mass spectrometer. Depending on the mode of ethane adsorption,

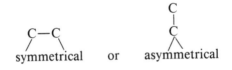

the deuterium atoms will incorporate into the ethane molecule in different ways; unfortunately, the actual process cannot be identified solely from the isotopic distribution of ethanes. To overcome this problem, the authors calculated the fragmentation patterns of ethane that are expected according to the possible modes of adsorption. In order to collect the related experimental data, i.e., the relative abundances of the various ethane fragments, they had to use a high-resolution mass spectrometer able to separate fragments and molecular products with a close m/e ratio [e.g., the ions CD_2, CH_2D, and CH_4 are all close to a mass/charge ratio of 16 (Figure 23.9)]. The comparison between modeled and experimental fragment distributions has shown that on platinum and nickel the model of symmetrical adsorption was correct at low temperatures while asymmetrical adsorption developed at higher temperatures, corresponding to the hydrogenolysis process. However, the use of high-resolution mass spectrometry is neither easy nor frequent in catalysis laboratories. The location of isotopes has to be determined by other techniques such as microwave and ^{1}H and ^{13}C NMR spectroscopies. This obviously marks the limits of mass spectrometry for the present topic.

23.3.3.3. Isotopic Effects

Often neglected in isotopic exchange studies, the isotopic effects which result from the difference of reactivity between a reactant and its labeled equivalent can provide valuable information concerning a reaction mechanism. Let us consider a molecule AH and its labeled homolog AD reacting, respectively, according to the following equilibria:

$$AH + B \leftrightharpoons CH + E \quad K_1 \tag{23.22}$$

$$AD + B \leftrightharpoons CD + E \quad K_2 \tag{23.23}$$

The difference between 1 and the ratio K_1/K_2 defines the isotopic effect, which is equivalent to the equilibrium

$$AH + CD \leftrightharpoons AD + CH \quad K_3 = K_1/K_2 \tag{23.24}$$

If it is assumed that the translation and rotation movements do not interfere significantly in the energetic balance of a chemical reaction, the isotopic effect is generally restricted to vibration movements of the molecules.[22] Therefore, the equilibrium constant K_3 can be written

$$K_3 \prod_i f(v_{i_{AD}} v_{i_{AH}}) \Big/ \prod_j f(v_{j_{CD}} v_{j_{CH}})$$

where $v_{i,j}$ are vibrational mode frequencies of molecules. Similarly, the rate constants of reactions (23.22) and (23.23) can be expressed following the theory of the transition state:

$$AH + B \leftrightharpoons (AHB) \quad k_H \tag{23.25}$$

$$AD + B \leftrightharpoons (ABD) \quad k_D \tag{23.26}$$

As the vibration frequency of an X—H bond is higher than that of an X—D bond, the rate constant is higher for the light activated complex formation than for the heavy one, $k_H/k_D > 1$, but this effect tends to vanish when the temperature is increased.

Reverse isotopic effects may be observed, in particular when reactants such as hydrogen and deuterium are in atomic form, which is often the case in surface catalysis.

23.3.3.3a. Benzene H/D Exchange. On platinum and nickel[(23)] the two exchange reactions

$$C_6H_6 + D_2 \rightarrow C_6H_5D + HD \quad k_{C_6H_6} \qquad (23.27)$$

$$C_6D_6 + H_2 \rightarrow C_6D_5H + HD \quad k_{C_6D_6} \qquad (23.28)$$

show an important isotopic effect $k_{C_6H_6}/k_{C_6D_6} = 6\text{--}7$. For this reaction, two mechanisms have been proposed:

Associative — $C_6H_6 + D \Leftrightarrow (C_6H_6D) \rightarrow C_6H_5D + H \qquad (23.29)$

Dissociative — $C_6H_6 + * + D \Leftrightarrow (C_6H_5*) + H + D \rightarrow C_6H_5D + H + * \quad (23.30)$

The expected isotopic effect related to the energy changes between the benzene molecule and its two possible transition states has been evaluated for comparison with the experimental data:

model	associative	dissociative
calculated effect	0.6–1.8	5.0–12.2

From these data, the dissociative mechanism is proposed for benzene exchange on platinum and nickel. However, the theoretical calculation of the isotopic effect includes a number of assumptions such as, e.g., the vibration frequencies of the transition states. Even a small error in these data can make a significant difference in the final value.

23.3.3.3b. Oxidative Dimerization of Methane. The oxidative dimerization of methane into ethane and ethylene is carried out on basic oxides at high temperatures ($> 600 \,°C$), but reactions in the gas phase are within limits set by thermodynamic equilibria. One of the key points for possible industrial development of this catalytic process concerns the rate limiting step: if it is a heterogeneous catalytic step, then better performance (selectivity and yield of dimers) may be expected from catalyst optimization.

By comparing the reactions

$$2\,CH_4 + \tfrac{1}{2}O_2 \rightarrow C_2H_6 + H_2O \quad k_{CH_4} \qquad (23.31)$$

$$2\,CH_4 + \tfrac{1}{2}O_2 \rightarrow C_2D_6 + D_2O \quad k_{CD_4} \qquad (23.32)$$

an isotopic effect $k_{CH_4}/k_{CD_4} = 1.5\text{--}2$ was found experimentally.[(24)] A calculation based on a limiting step including the C—H bond breaking leads to a similar isotopic effect (a value which corresponds to the ratio of the fragmentation factors f_{1H}/f_{1D} given in Table 23.3). This suggests that the rate limiting step in methane coupling includes the elementary step of the C—H bond activation on a catalytic site.

23.3.3.4. Transient Kinetics

Steady-state kinetics generally give a reaction rate which appears as a product of rate constants by function of intermediate species coverages and reactant pressures:

$$r = k_i f(P_i, O_i) \tag{23.33}$$

If the steady-state regime is abruptly disturbed (e.g., by interrupting the inlet flow of reactant), a transient reponse of the system is induced at the reactor outlet (an exponential decrease of the product concentration in the simplest case). This transient response carries unique information: its time constant, which can reflect the reverse value of the rate constant k_i of equation (23.33). From this value, the coverages Θ_i may be determined, which brings a new and often critical insight to mechanistic interpretation. In the analysis of a transient phenomenon, it is assumed that the rate constants k_i of the steady-state regime are not affected by the removal of reactants, which is not true for numerous cases in heterogeneous catalysis. For instance, in CO/H_2 chemistry, the abrupt removal of CO generally induces as a first step a transient increase in hydrocarbon production before the expected exponential decrease. This indicates that during the transient period, carbon species that accumulate on the catalyst, without participating in the steady-state reaction, start to react when the CO pressure is decreased. Thus the steady-state reaction pathway is changed completely. These unwanted effects are suppressed if the transient phenomenon is induced by an isotopic signal (e.g., replacing a reactant by its labeled equivalent). This allows the observation of the full transient kinetic information without disturbing the overall steady state (to the extent that isotopic scrambling is absent and that the possible isotopic effects are taken into account).

However, the use of the transient kinetic methods is limited by practical features such as[25]:

1. Gas turbulence causing partial mixing of the labeled and unlabeled phases in the isotopic step. This may induce a limit to the time resolution close to 1 s, depending on the catalytic test configuration (e.g., flow rates, void volumes, and manifold sections).
2. The trade-off between response time and sensitivity of the analytical system, i.e., the mass spectrometer. This can be ignored, as most mass spectrometers allow much lower resolution times.
3. The rates of mass transfer between the solid and the gas phase. Unless the transient technique is applied specifically to mass transfer studies (such as diffusion effects in zeolites), the mass transfer limitations have to be minimized by solid size and porosity adjustments. In most flow experiments on powder or agglomerated catalysts, it can be shown that diffusion times may easily be kept below 1 s.

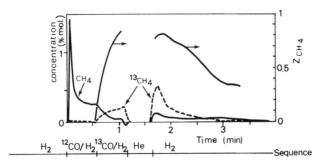

Figure 23.10. Transient concentration of $^{13}CH_4$ and $^{12}CH_4$ formed by ^{12}CO and ^{13}CO hydrogenation on Fe/Al_2O_3 (Z_{CH_4} is the ^{13}C content in CH_4).[29]

It appears therefore that the mean residence time of the active intermediates has to be higher than the resolution time of the system for a correct evaluation; fortunately, this is often the case in catalysis.

Although there has been a continuing interest in the application of transient isotopic tracing since the early studies,[26] new developments have recently appeared,[27] partly due to the improved possibilities of routine mass spectrometry. CO hydrogenation,[20,28-31] benzene hydrogenation,[32] and oxidative dimerization of methane[33] are typical examples.

23.3.3.4a. CO Hydrogenation. Figure 23.10 gives an example of transient responses obtained during CO hydrogenation on Fe/Al_2O_3.[29] Nonisotopic sequences (change from pure H_2 flow to CO/H_2 mixture or the reverse) were combined with isotopic tracing (switch from $^{12}CO/H_2$ to $^{13}CO/H_2$). For the latter, the outlet effluent was sampled during the transient by means of a 16-loop rotating valve; each sample was analyzed by GC to separate the different hydrocarbons, MS analysis giving the related ^{13}C contents. This type of experiment leads to the following conclusions:

1. The first transient spike in methane formation after the H_2/syngas switch denotes that the metallic surface, highly active in its initial state, is rapidly deactivated, indicating a carburization process.
2. The initial slope of the $^{12}CH_4$ and $^{13}CH_4$ transient curves resulting from the switch $^{12}CO/H_2 \rightarrow ^{13}CO/H_2$ gives the mean residence time of the most active intermediates for the methane formation. Therefore the amount of these adspecies can be calculated, as shown above from equation (23.33).
3. The total amount of carbon deposited after a given time on steam is readily obtained by integrating the transient curves of methane formation obtained when switching from the reacting mixture to a pure hydrogen flow (with a brief intermediate helium flush).

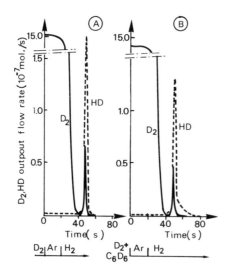

Figure 23.11. Titration of deuterium adsorbed on Ni/SiO_2 by hydrogen, following the sequences: (A) D_2 (steady state) → Ar (20-s flush) → H_2; (B) $D_2 + C_6D_6$ (steady state) → Ar (20-s flush) → H_2.[32]

4. A comparison of the amounts of active and total carbon shows that only a limited fraction of the surface carbonaceous deposits participates in the steady-state reaction (these species are assumed to be mainly CH fragments).

5. Similar transients carried out after various times on stream showed further that the active intermediate species become less and less active as the reaction proceeds, although their concentration remains stable. This suggests that the deactivation does not come from a surface blocking by coke formation but rather by changes in reactivity of the sites.

6. The similarly between the transient curves observed for methane and for higher hydrocarbons suggests that the related kinetic processes are all rapid once initiated and that the concentrations of intermediates for methane or for chain growth are small in most cases. Similar conclusions are proposed in Wolfsberg.[22]

23.3.3.4b. Benzene Hydrogenation. Isotopic transients of the benzene hydrogenation[32] have been carried out by switching C_6H_6/H_2 to C_6D_6/H_2, which minimizes any isotopic effects compared, e.g., with a hydrogen labeling switch. The concentrations of intermediate species leading to cyclohexane formation are calculated from the time constants of the transient curves as described in the previous example. The surface coverage with hydrogen species is measured by switching sequentially from H_2 to D_2 after an intermediate flush of argon, as shown in Figure 23.11. This procedure gives the same quantitative data as classical volumetry for pure hydrogen adsorption, but in addition it has the unique advantage that hydrogen adsorption can be measured in the presence of other adsorbents such as benzene.

TABLE 23.5. Quantitative Evaluation of a Catalytic Surface
(Ni/SiO$_2$) Occupancy during Benzene Hydrogenation[32]

Temperature (°C)	Monolayer		
	Reversible benzene	Hydrogen	Reacting intermediates
25	1.5	0.72	0.002
80	0.1	0.68	0.005

For Ni/SiO$_2$ catalysts, the active surface in the hydrogenation conditions may be depicted as mainly covered with dissociated hydrogen, surrounded with a reservoir of loosely and reversibly adsorbed benzene. The active sites which allow the irreversible adsorption of benzene with the formation of intermediates leading to cyclohexane are shown from the transients to exist in a very low concentration as shown in Table 23.5. Analysis of the effects of temperatures and pressures shows that the concentration of reacting intermediates can be expressed as a function of the hydrogen coverage Θ_H:

$$\text{Reacting intermediates} = (1 - \Theta_H)^X \quad \text{with X = 3-4} \qquad (23.34)$$

This function represents the probability of the adsorption of a benzene molecule on a site formed with X atoms of nickel free from hydrogen. On this experimental basis, completed with other mass spectrometry experiments such as the analysis of the cyclohexane isotopic distribution, an overall mechanism has been proposed, different from the mechanisms deduced from classical steady-state kinetics. Its rate determining step is a bimolecular reaction between a molecule of benzene adsorbed on a nickel site free from hydrogen and a molecular hydrogen species.[32]

Thus steady-state and transient isotopic tracing, based mainly on the use of mass spectrometry, are techniques which can give unique and often crucial information about the mechanisms of heterogeneous catalytic reactions.

23.4. CONCLUSION

Mass spectrometry was formerly restricted to those laboratories which specialized in chemical analysis or isotopic separation and heavy and expensive equipment was often needed. Owing to miniaturization (e.g., the development of the quadrupole analyzer and turbomolecular pumping groups) and computer facilities, mass spectrometry now offers a wide range of applications in research on catalysis. However, combinations with other techniques (chromatography, TGA, *in situ* FTIR and Raman spectroscopy, microwaves...) are required to enlarge its potential for investigation.

REFERENCES

1. J. H. Benon, *Mass Spectrometry and Its Applications to Organic Chemistry,* Elsevier, Amsterdam (1960).
2. F. W. Mc Lafferty, *Mass Spectrometry of Organic Ions.* Academic, New York (1963).
3. P. J. Derrick and K. F. Donchi in: *Comprehensive Chemical Kinetics, Vol 24,* Elsevier, Amsterdam (1963), p. 53.
4. E. Murand, *Spectrometric Techniques,* Vol. IV (G. A. Vanasse ed.), Academic, New York (1985).
5. J. de Graeve, F. Berthou, and M. Prost, *Méthodes Chromatographiques couplées à la spectrométrie de masse,* Masson, Paris (1986).
6. R. Davis and M. Frearson, *Mass Spectrometry,* John Wiley and Sons, New York (1987).
7. B. Halpern and J. E. Germain, *J. Catal.* **37**, 44 (1975).
8. A. Perrard and J. P. Jolly, *Vacuum* **39**, 551 (1989).
9. J. L. Falconer and J. A. Schwarz, *Catal. Rev. Sci. Eng.* **25**, 141 (1983).
10. J. T. Gleaves, J. R. Ebner, and T. C. Kuechler, *Cat. Rev. Sci. Eng.* **30**, 49 (1988).
11. I. Matsuura, *J. Catal.* **33**, 420 (1974).
12. L. D. Krenzke, O. W. Keulks, A. V. Sklyarov, A. A. Frisova, M. Yu. Kutirev, L. Y. Margolis, and O. V. Krylov, *J. Catal.* **52**, 418 (1978); O. W. Keulks, L. D. Krenze, and T. M. Noterman, *Adv. Catal.* **27**, 183 (1978).
13. O. W. Keulks, *J. Catal.* **19**, 235 (1970).
14. C. Kemball, *Adv. Catal.* **11**, 223 (1959).
15. R. L. Burwell, Jr., *Catal. Rev.* **7**, 325 (1972).
16. G. K. Boreskov, *Adv. Catal.* **15**, 285 (1969).
17. A. Frennet, *Catal. Rev. Sci. Eng.* **10**, 37 (1974).
18. J. A. Dalmon and C. Mirodatos, *J. Molec. Catal.* **25**, 161 (1984).
19. H. Praliaud, J. A. Dalmon, C. Mirodatos, and G. A. Martin, *J. Catal.* **97**, 344 (1986).
20. C. A. Mims and L. E. Mc Candlish, *J. Amer. Chem. Soc.* **107**, 696 (1985).
21. L. Guczy and K. Ujszaski, *Reac. Kinet. Catal. Lett.* **8**, 489 (1978).
22. M. Wolfsberg in: *Stable Isotopes* (H. L. Schmidt *et al.,* eds.), Elsevier, Amsterdam (1982), p. 3; L. Melander, *Isotope Effects on Reactions Rates,* Ronald, New York, (1960).
23. R. Z. C. Van Meerten, A. Morales, J. Barbier, and R. Maurel, *J. Catal.* **58**, 43 (1979).
24. N. W. Cant, C. A. Lukey, P. F. Nelson, and R. J. Tyler, *J. Chem. Soc. Chem. Commun.,* 766 (1988); C. Mirodatos, A. Holmen, R. Mariscal and G. A. Martin, *Catalysis Today* **6**, 601 (1990).
25. P. Biloen, *J. Molec. Catal.* **21**, 17 (1983).
26. M. B. Neiman and D. Dal. *The Kinetic Isotope Method and Its Applications,* Elsevier, Amsterdam (1971).
27. C. Mirodatos, *Catalysis Today* **9**, 83 (1991).
28. P. Winslow and A. T. Bell. *J. Catal.* **86**, 158 (1984).
29. D. M. Stockwell, D. Bianchi, and C. O. Bennet, *J. Catal.* **113**, 25 (1988).
30. J. Happel, *Isotopic Assessment of Heterogeneous Catalysis,* Academic, New York (1986).
31. X. Zhang and P. Biloen, *J. Catal.* **98**, 468 (1986).
32. C. Mirodatos, *J. Phys. Chem.* **90**, 481 (1986); C. Mirodatos, J. A. Dalmon and G. A. Martin, *J. Catal.* **105**, 405 (1987).
33. A. Ekstrom and J. A. Lapszewicz, *J. Phys. Chem.* **93**, 5230 (1989); K. P. Peil, J. G. Goodwin, Jr., and G. Marcelin, *J. Phys. Chem.* **93**, 5977 (1989); *J. Am. Chem. Soc.* **112**, 6129 (1990).

SCANNING TUNNELING AND ATOMIC FORCE MICROSCOPIES

P. Gallezot

24.1. INTRODUCTION

The scanning tunneling microscope (STM) developed by Binnig and Rohrer[1-3] at the IBM Zürich research laboratory in the early 1980s was the first example of a new family of instruments based on a concept radically different from that of the optical and electron microscopes. The basic idea is to examine a surface at very close range (near field) with a scanning probe, the position of which can be perfectly controlled in three dimensions to within ± 0.1 nm. The signal generated on each point of the surface by the interaction between the probe and the surface atoms is translated electronically into an image of the surface. The resolution depends essentially on the probe size and on the accuracy of its positioning. There are potentially a wide variety of probes and signals which can be used to image a surface. We will focus our attention on the scanning tunneling microscope (STM) based on the tunneling current generated between the extremity of a metal tip and the surface atoms of a conducting material at a different potential. We will also examine the atomic force microscope (AFM) based on the repulsive force between the atoms of a diamond tip and surface atoms. The AFM, developed more recently,[4,5] has a potentially wider application range since it does not require conducting materials. Finally a few other scanning probe microscopes will be mentioned. They do not give atomic resolution but do not require a high surface flatness. So far applications in surface science have been carried out mainly with the STM.

P. Gallezot • Institut de Recherches sur la Catalyse, CNRS, Villeurbanne, France.

Catalyst Characterization: Physical Techniques for Solid Materials, edited by Boris Imelik and Jacques C. Vedrine, Plenum Press, New York (1994).

24.2. DESIGN AND PRINCIPLES OF SCANNING PROBE MICROSCOPES

24.2.1. Scanning Tunneling Microscope (STM)

According to quantum mechanics there is a finite probability that electrons will pass between two conducting materials at different potentials separated by a small gap. The tunneling current I is proportional to the potential V and decreases exponentially with the distance s between the two metals, i.e., $I = V \exp(-ks)$, where k depends upon the work function of the two metals. Typically the current decreases by an order of magnitude as the gap is widened by 0.1 nm. The probe is a metal tip, usually tungsten, which is mechanically ground and electrochemically etched. The ultimate dimension of the tip end is not known; it can be a cluster of atoms or even consist of only a single atom. The tips are not easy to prepare reproducibly and some variation in instrumental resolution can be observed even with equal tip preparation. The tip is mounted on three piezoelectric ceramics which can control the position of the tip along the X, Y, and Z directions. Piezoelectric Z controls the position of the tip to within 1 nm above the surface with an accuracy of 0.1 nm. Piezoelectrics X and Y move the tip back and forth over the surface in a raster pattern with tracks separated by a fraction of a nanometer. When a small voltage is applied (of the order of 1 V or less) electrons tunnel across the gap. If the probe is maintained at constant height as it scans the surface, the current fluctuates, a high current corresponding to bumps, e.g., to adatoms or steps on a flat surface. Alternatively, the microscope can be operated in the constant current mode: a feedback mechanism senses the

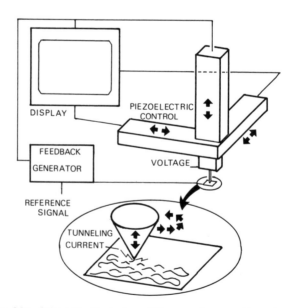

Figure 24.1. Schematic illustration of the scanning tunneling microscope.

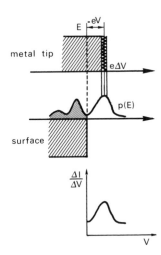

Figure 24.2. Relation between the surface density of states and the signal $\Delta I/\Delta V$.

variation in tunneling current and the Z piezoelectric moves the tip vertically up and down to maintain a constant current. The variation in the voltage applied to the Z piezoelectric is electronically translated into a topographic image of the surface. A scheme of the STM is given in Figure 24.1. The size of newer instruments is comparable to that of a match box (pocket-size STM). They do not need complicated vibration isolation, as did the first design, and can operate over a wide range of temperatures and in various media (vacuum, gas, liquids). The resolution of STM can be as good as 5×10^{-3} nm in the Z direction and better than 0.2 nm in the X and Y directions.

In addition to topographic data on surfaces, STM provides information on the local electronic structure. Indeed, the tunneling current is proportional to the local density of states (LDOS) at the Fermi level energy.[6,7] In the constant current mode, the STM image represents a contour of constant LDOS. In the usual case of an extended Fermi surface, the image gives basically the contour of constant total charge density. Spectroscopic measurements can be performed at different voltages. On a given point of the surface, measurements of the differential conductivity give information on the LDOS (Figure 24.2). A cartography of the LDOS can be obtained at different energies. More simply, the comparison of two conventional images (obtained by scanning at constant current) acquired at two different voltages can be used to distinguish the nature of the different atoms present on the surface. This technique, proposed by Feenstra et al.,[8] provides both analytical and topographical information on surfaces.

24.2.2. Atomic Force Microscope (AFM)

Applications of the STM are limited to conducting materials such as metals and semiconductors. Even these materials can be difficult to study since they are often covered with an insulating layer of oxide. To overcome these limitations, a different type of scanning probe microscope, the atomic force microscope, was

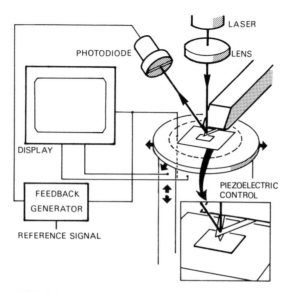

Figure 24.3. Schematic illustration of the atomic force microscope.

designed.[4,5] The probe is a diamond shard mounted on a flexible metal foil which keeps it pressed against the surface. The signal is now the repulsive force (atomic force) generated when the electron cloud of surface atoms overlap with the electron cloud of the atoms at the tip extremity. In a sense, the AFM tip reads the surface like the stylus of a phonograph. In the initial design the vertical deflection of the metal foil supporting the tip was monitored by the tunneling current between the metal foil and a tungsten tip mounted just above it. A feedback mechanism maintained the tunneling current, and thus the deflection, constant, by adjusting the voltage of the Z piezoelectric, moving the sample up and down, while two other piezoelectrics scanned the sample in the X and Y directions. The voltage applied to the Z piezoelectric was translated into a topographic image. In a more recent design,[9] the deflecting of the metal foil is monitored by a laser beam reflected from the top of the foil to a photodiode sensor. A feedback mechanism responds to any change in the beam paths by activating a piezoelectric control which adjusts the height of the sample so that the deflection of the arm remains constant (Figure 24.3).

The resolution of the AFM is high enough to reveal individual atoms on a surface as the STM does. However, as it is not based on a tunneling current it can be used to image the surface of nonconducting materials. Its main drawback comes from the high pressure exerted by the diamond tip on the surface which can modify and damage the structure of the surface. The new design using an optical sensor to detect the metal foil deflection is much more sensitive than the tunneling sensor; therefore, the pressure can be considerably reduced, thus decreasing the risk of damage to the surface.

24.2.3. Other Scanned-Probe Microscopes

The STM and AFM give topographic images of a surface with an atomic resolution provided the sample is large enough and smooth enough so that the tip can follow the surface relief. The study of deep surface relief and of fine powders is not possible. New types of scanned-probe microscopes have been designed with a tip much further from the surface than in the STM. The resolution is then much lower but rough surface relief can be studied. In the laser force microscope (LFM),[10] the sample is scanned by a tungsten or silicon wire as far as 2 to 20 nm from the surface. The signal is a small attractive force, a thousand times smaller than the repulsive force involved in the AFM, which is due to Van der Waals forces and to surface tension of water condensed between the tip and the surface.

The wire is maintained vibrating close to its resonance frequency by a piezoelectric transducer powered by an alternating current. When the vibrating tip is scanned over the surface, the weak attractive force, and thus the vibration amplitude, changes according to the surface topography. These amplitude changes are detected by interferometry. A feedback mechanism responds to any amplitude change by varying the voltage across a Z piezoelectric to keep the vibration amplitude constant, and thus maintain the probe-to-surface distance. These voltage fluctuations are then translated into a surface profile as the sample is scanned. The LFM can detect surface relief as small as 5 nm and because the tip senses topography from a distance it can probe features inside deep, narrow clefts.

There are potentially a variety of scanned-probe microscopes based on various chemical or physical signals. A scanning thermal probe based on a 30-nm-wide thermocouple measuring variations in the heat loss as a function of surface relief has been described.[11] Hansma et al.[9] designed a scanning ion-conductance microscope (SICM), where the probe is a glass micropipette containing a tiny electrode which can be used to study biological samples bathed in electrolytes. Electromagnetic radiations such as visible light or X-rays passing through a pinhole close to the surface can also be used as near-field probes. The resolution does not depend upon the wavelength but only on the probe size. Near-field optical scanning microscopes are expected in the near future to have a resolution of 10 nm, namely a 25-fold improvement over the best conventional optical microscopes.

24.3. APPLICATION OF STM TO SURFACE CHEMISTRY

STM and AFM require samples with large areas of flat surface and therefore cannot be applied to actual catalysts, which consist of very small and irregular particles. Therefore applications have been limited so far to extended surfaces of metals, semiconductors, and pyrolytic graphite. Review papers have been published on early STM work[12,13] and many applications have been reported in the proceedings of the international conferences on STM.[14-16] We

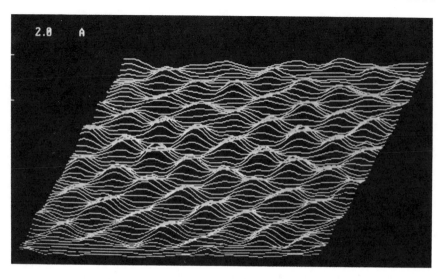

Figure 24.4. STM image of the basal plane of graphite at high resolution (vacuum generator).

will examine briefly some of these applications with emphasis on those concerning surface chemistry.

24.3.1. Arrangements of Atoms on Surfaces

From the start, the capability of STM to image monoatomic steps and individual atoms on surfaces was fully demonstrated.[12] Surface reconstructions were shown for a number of metal and semiconductor surfaces. In addition to topographic information, it was soon discovered that STM can be used to resolve the electronic structure of a surface on an atomic scale. The feasibility of studying surface structure in air, water, and electrolytes has been demonstrated. An example of a STM image of the basal plane of graphite at atomic resolution is given in Figure 24.4. The graphite lattice does not exhibit the expected honeycomb structure but rather an hexagonal arrangement where each carbon atom has six nearest neighbors in the basal plane. In fact, according to Binnig,[12] the STM image reflects the local density of states at the Fermi level and not the position of atoms.

24.3.2. Topography of Surfaces

There have been comparatively few STM studies at lower resolution (e.g., 1 nm) with the aim of determining the surface topography of large sample arrays. This cannot be done with electron microscopy: scanning electron microscopy (SEM) has too low a resolution and transmission electron microscopy (TEM) gives a projected image on a plane perpendicular to the electron beam and thus little information can be obained on the surface relief in the direction of projection.

The capability of STM to reveal a surface profile is well illustrated in a study on the morphological changes in a graphite oxidized by sodium hypochlorite.[17] This treatment was intended to develop both the number of steps and the acidic groups associated with these steps, in order to prepare graphite-supported platinum catalysts by ion-exchange with platinum particles grafted along the steps. A TEM view of the graphite after NaClO treatment gives little information on the etching effect because steps (adlayers) can only be seen if there is a good contrast, i.e., if the step heights are not too small with respect to the total thickness of the graphite slab. The STM image before NaClO oxidation shows that the surface is perfectly smooth over large areas. In contrast the STM image after oxidation (Figure 24.5) shows clearly a high density of etching figures. Two types of heterogeneities can be seen in the STM image: (1) parallel or wedge-shaped steps with heights in the range from one to a few tens of nanometers, and (2) domains of high surface roughness with bumps as high as 5 nm. It can be concluded that graphite layers have been etched by the oxidizing treatment, and that steps and bumps are left as a result of incomplete graphite layer attack.

24.3.3. Image of Supported Clusters

Conventional TEM is a powerful tool for the study of the morphology and structure of metal clusters. However, it is difficult on a projected image to

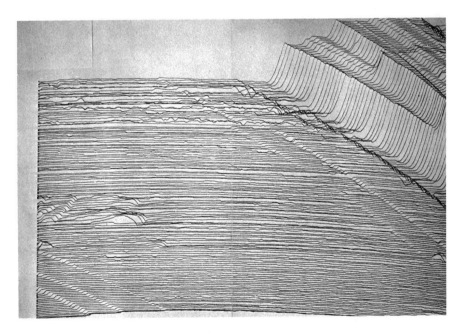

Figure 24.5. STM image of the basal plane of graphite oxidized by NaClO.

differentiate between raft-like and spherical particles. Furthermore, electron bombardment in a TEM can alter the cluster structure. The first attempts to image metallic clusters by STM were made on gold and silver clusters deposited at room temperature on cleaved graphite.[18,19] Isolated atoms and bidimensional islands with a structure different from three-dimensional crystals were observed.

Gold clusters deposited on graphite under UHV conditions were studied by STM without losing vacuum.[20] Figure 24.6 shows a 1–1.5 nm large cluster where both metal atoms and carbon atoms are clearly resolved. The cluster is two layers thick and the atoms are close-packed. There is no evidence for a regular polyhedral arrangement. The interaction with the graphite substrate is strong as the cluster tends to spread on the graphite surface with one edge parallel to an axis of the graphite lattice. Several images were recorded without any change, indicating that the cluster is stable and not perturbed by the STM tip. Humbert *et al.*[21] have also obtained STM images of palladium atoms on graphite. Figure 24.7 shows that they form monolayer islands with interatomic distances as large as 0.4 nm, indicating that there is little interaction among the palladium atoms but a strong interaction with the carbon atoms of the graphite plane.

A study of platinum particles supported on graphite was reported several years ago.[22] The particles were obtained by methods similar to those used in

Figure 24.6. STM image of a gold cluster deposited on graphite.

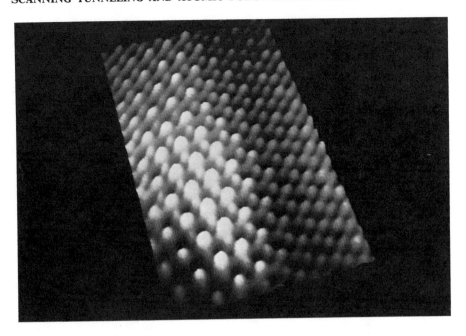

Figure 24.7. STM image of palladium atoms on graphite.

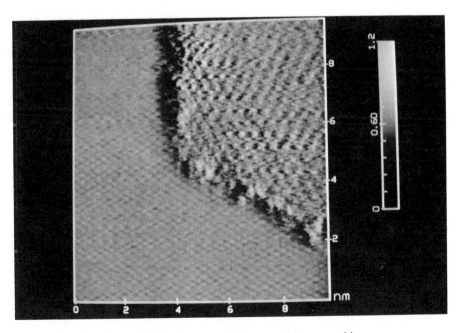

Figure 24.8. STM image of platinum particles on graphite.

catalyst preparation, such as impregnation and ion exchange. Figure 24.8 shows an example of a particle prepared by ion exchange of a platinum tetrammine nitrate solution with functionalized nuclear grade high-purity graphite (NGHPG), subsequently reduced in hydrogen. The image displays at high resolution a corner of a platinum crystallite of about 20 nm showing surface atoms organized in two different structures. The image was obtained under air in the current imaging mode with a Nanoscope II instrument (Digital Instruments). These studies demonstrate that STM is a unique tool for probing the atomic arrangement in clusters and the interaction of clusters with substrate.

Figure 24.9 shows two STM images of palladium aggregates obtained by the atomic deposition technique on pyrolytic graphite.[23] The palladium coverage was $\theta_{Pd} = 5 \times 10^{13}$ atoms cm^{-2}. STM measurements were carried out before (Figure 24.9a), and after (Figure 24.9b) 1–3 butadiene hydrogenation. In both cases observations were made *in situ* on particles after exposure to air. Most of the initial particles before catalysis present a monoatomic height whereas the particles exposed to reactants are higher and narrower in size. These observations indicate that palladium clusters reconstruct during the reactions.

24.3.4. Image of Organic Adsorbates

To image an adsorbed molecule, the surface should be sufficiently flat, there should be little or no molecular mobility, and the tunneling current should be high enough. Thus the surface of gold is flat enough but the mobility of adsorbates is too high. This is the reason that the gold surface can be imaged at atomic resolution in various atmospheres; the molecules are so mobile that they hardly hinder the detection of surface atoms. On the other hand, the bonding between NH$_3$ molecules and silicon atoms on a Si(111)–(7 $\times$ 7) surface[24] has been clearly demonstrated. Different reactivities between NH$_3$ molecules and different silicon sites (terraces, corner, adatoms) have been revealed.

Images at high resolution have been obtained on Rh(111) covered with coadsorbed benzene and carbon monoxide.[25] The image is stable at complete monolayer packing but individual molecules at submonolayer coverage move during the STM scan. Similarly, it is difficult to image single phthalocyanine molecules on silicon[26] but regular arrays of these molecules have been detected on Cu(100). Multilayers of adsorbed organic molecules are difficult to image because the conductivity is then too small and the mobility too large. Similarly, large molecules of biological interest are difficult to image because the tunneling distances are too large.

In conclusion, STM images can reveal the structure of adsorbed molecules provided the interaction substrate–adsorbate is high enough. Close-packed monolayers are easier to study than isolated molecules because of restricted mobility. The physics involved in the imaging process of organic molecules is

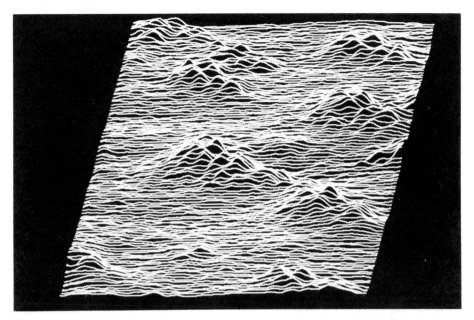

(a)

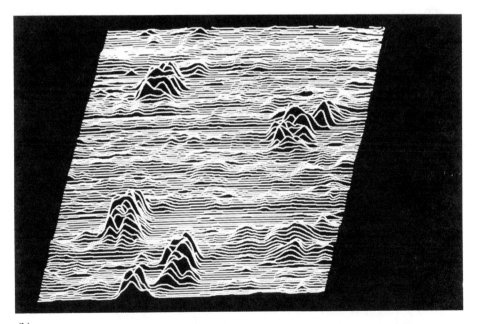

(b)

Figure 24.9. STM image of palladium aggregates vapor-deposited on graphite before (a) and after (b) butadiene hydrogenation at room temperature. The image size corresponds to 20×20 nm.

still the subject of debate, but it is clear that the images represent the charge distribution of free molecular states near the Fermi level.

24.4. CONCLUSION

Although STM has proved to have unique capabilities for revealing the structure of surfaces and adsorbed species, the appeal and impact of STM lie in more than the observation of surface atoms. As in conventional microscopy, studies at the highest resolution need much time and great expertise, so they will be applied only to specific problems, on a limited number of materials. Many other less spectacular aspects can be addressed, e.g., topographical studies on surfaces less smooth than the model surfaces used so far. Many other materials, e.g., those involved in catalyst formulation, can then be studied. This evolution from instrumental development to its application to materials science is made easier by the advent of present-day STM units, which are smaller, simpler, and more affordable, and require less technical expertise.

However, the question arises as to whether one should invest in STM or in the new scanned-probe microscopes which do not require conducting materials (AFM) or which can probe rougher surface relief (e.g., LFM, scanning near-field optical microscope). There have been too few applications of these new instruments to answer this question at the present time.

REFERENCES

1. G. Binnig and H. Rohrer, *Helv. Phys. Acta* **55**, 726 (1982).
2. G. Binnig, H. Rohrer, C. Gerber, and E. Weibel, *Phys. Rev. Lett.* **49**, 57 (1982).
3. G. Binnig, H. Rohrer, C. Gerber, and E. Weibel, *Phys. Rev. Lett.* **50**, 120 (1983).
4. G. Binnig, C. F. Quate, and C. Gerber, *Phys. Rev. Lett.* **56**, 930 (1986).
5. G. Binnig, C. Gerber, E. Stoll, C. F. Albercht, and C. F. Quate, *Europhys. Europhys. Lett.* **3**, 1281 (1987).
6. J. Tersoff and D. R. Hamann, *Phys. Rev. Lett.* **50**, 1998 (1983).
7. J. Tersoff and D. R. Hamann, *Phys. Rev.* **B31**, 805 (1985).
8. R. M. Feenstra, J. A. Stroscio, J. Tersoff, and A. D. Ferri, *Phys. Rev. Lett.* **58**, 1668 (1987).
9. D. K. Hansma, V. B. Blings, and C. E. Braker, *Science* **242**, 209 (1988).
10. Y. Martin, C. C. Williams, and H. K. Wickramasinghe, *Scan. Microsc.* **2**, 3–8 (1988).
11. C. C. Williams and H. K. Wickramasinghe, *Appl. Phys. Lett.* **49**, 1587 (1987).
12. G. Binnig and H. Rohrer, *Anger Chem. Int.* **26**, 606 (1987).
13. P. K. Hansma and J. Tersoff, *J. Appl. Phys.* **91**, R1 (1987).
14. Proc. 1st Int. Conf. on STM, *Surf. Sci* **181** (1987).
15. Proc. 2nd Int. Conf. on STM, *J. Vac. Sci. Technol.* **A6**, 256 (1988).
16. Proc. 3nd Int. Conf. on STM, *J. Microsc.* **152** (1988).
17. L. P. Porte, D. Richard, and D. Gallezot, *J. Micros.* **125**, 515 (1988).
18. E. Ganz, K. Sattler, and J. Clarke, *J. Vac., Sci. Technol.* **A6**, 419 (1988).
19. E. Ganz, K. Sattler, and J. Clarke, *Phys. Rev. Lett.* **60**, 1856 (1988).
20. A. Humbert, P. Pierrisnard, S. Sangay, C. Chapon, C. R. Henry, and C. Claeys, *Europhys. Lett.* **10**, 533 (1989).

21. C. Chapon, C. R. Henry, A. Humbert, M. Dayez, and S. Sangay, Congrès de la SFP Lyon, Septembre (1989).
22. K. L. Yeung and E. E. Wolf, Prof. 5th Int. STM Conference, Philadelphia, 1990.
23. L. Porte, M. Phaner, C. Noupa, B. Tardy, and J. C. Bertolini, *Ultramicroscopy* **42–44**, 1355 (1992).
24. R. Wolkow and Ph. Avouris, *Phys. Rev. Lett.* **60**, 1049 (1988).
25. H. Ohtani, R. J. Wilson, S. Chiang, and R. M. Mate, *Phys. Rev. Lett.* **60**, 2398 (1988).
26. J. K. Gimzewski, E. Stoll, and R. R. Schlittler, *Surf. Sci.* **181**, 267 (1987).

FINAL REMARKS

B. Imelik[†] and J. C. Vedrine

25.1. INTRODUCTION

It is useful to conclude this book with some general remarks to show what physical methods can contribute to catalysis. Physical methods contribute mainly to the description of the catalyst and to a lesser extent to that of the reagent in interaction with the catalyst. This stems from the technical difficulties encountered when adapting the physical methods to the conditions of catalytic reactions.

The description of the catalytic solid involves the determination of its bulk features (composition, structure, morphology, and the bulk properties of its constituent atoms or ions) and of its surface features (exposed crystal planes, superficial composition, and arrangement). Methods described in this book allow the determination of the morphology of crystallites (SEM and TEM), their structure at long range (X-ray diffraction) or at short range (IR, Raman, EXAFS, STM, AFM, and RED). UV–visible, XPS, EPR, NMR, GNR, and EDX–STEM spectroscopies and magnetism give access to the chemical nature of the catalyst's constituent atoms or ions, to the symmetry of their environment, their degree of oxidation, and the degree of covalence of the interatomic bonds. The description of a catalyst depends on its degree of order: more abundant information is obtained from more crystalline solids. Metals and alloys usually yield detailed descriptions. Because of their high degree of order, metallic particles even of a small size (≈ 1 nm) can be studied. In the last two decades much work has been done on the influence of the size of metallic crystallites upon their catalytic activity, on the structure of hyperdivided metals, and on the specificity of crystal faces. The concept of "faults" is often used to interpret catalytic results although it is, in general, difficult to characterize these "faults."

B. Imelik and J. C. Vedrine ● Institut de Recherches sur la Catalyse, CNRS, Villeurbanne, France.

Catalyst Characterization: *Physical Techniques for Solid Materials*, edited by Boris Imelik and Jacques C. Vedrine, Plenum Press, New York (1994).

In some cases LEED, STM, and TEM have established the presence of structural defects such as steps, corners, and ridges in metallic monocrystals.

Unfortunately, good catalytic qualities in, e.g., oxides, sulfides, and carbides are often associated with a high degree of disorder. Nevertheless, for certain oxides and sulfides with a high degree of crystallinity, a specificity of reaction of the crystalline faces has been shown as for metals. For example, with the oxides of molybdenum and antimony certain faces are specifically active in total oxidation and others in selective oxidation of propene, butenes, or methanol. For other solids, the catalytic properties depend on both the degree of order and of disorder, e.g., the VPO systems where the intergrowth of a well crystallized phase of $(VO)_2P_2O_7$ (detected by X-ray diffraction) and a phase of amorphous $VOPO_4$ (detected by X-ray radial electronic distribution or Raman spectroscopy and characterized by ^{31}P MAS-NMR) favors the oxidation of n-butane to maleic anhydride.

The case of zeolites is exceptional since their high degree of crystallinity enables the acquisition of data far in excess of that required for the simple "identity card" of the catalyst. We can now locate cationic sites, and absorbates determine the migration of cations and reveal the occurrence of allotropic transformations during the adsorption of molecules or hydrothermal treatment.

We show several examples which need the use of numerous techniques, each making its contributions by its specificity and by the energy level involved.

25.2. EXAMPLES

25.2.1. Platinum/Y-Zeolite

Supported platinum catalysts are used extensively in industry, in particular in catalytic reforming and hydrocracking. For some applications the classical supports, e.g., aluminas, can be advantageously replaced by zeolites. Let us consider a Pt/Y-zeolite catalyst, prepared by the exchange of the Na^+ ions of the NaY-zeolite by Pt^{2+} ions and subsequent hydrogen reduction at 350 °C. These catalysts are slightly more active than Pt/SiO_2 or Pt/Al_2O_3 in the hydrogenation of benzene, but they are seven times more active in the hydrogenolysis of butane. It is important to characterize these materials in order to explain these differences in catalytic properties. We will now see how to choose the physical techniques as a function of the questions to be answered.

I. What are the size and morphology of the metallic particles?

Four methods can be used:

1. Electron Microscopy: Size, morphology, and distribution of particle sizes.
2. X-ray Diffraction: From line broadening, the average particle size, within a certain range.

3. Classical H_2–O_2 Titration: (Assuming a 1:1 Pt/H and Pt/O stoichiometry) a measure of the metal dispersion.
4. ^{129}Xe-NMR Spectroscopy: (From the mean free path of adsorbed xenon) the size of very small metallic particles and free cavities.

The platinum particles on Y-zeolites have a homogeneous size range of 0.8 to 3 nm.

II. Are the metallic particles at the exterior or the interior of the grains of the zeolite?

Two techniques provide an answer:

1. Electron Microscopy: (Thin-section and replica methods) the platinum particles are homogeneously distributed, probably in the supercages. Thus the 1.3-nm cages have to burst locally in order to accommodate the 2.3-nm particles.
2. XPS-Spectroscopy: Comparison between the Pt/Si ratio in the first superficial layers ($d \approx 2$ nm) with that of the bulk indicates an even distribution of the metallic particles within the zeolite grain.

III. Is there an interaction between the surface and the support?

This question is of major importance but, because of the very low metal content, the answer is incomplete. Most physical techniques, especially those used to characterize metal–ligand bonds, are not sufficiently sensitive.

Four techniques give some information:

1. X-Ray Photoemission Spectroscopy (XPS): For comparable particles sizes, the binding energy of the platinum-core levels ($4f$ lines) is displaced by $+0.7$ eV with respect to that of the metallic bulk platinum. This effect, which disappears on chemisorption of CO or in the presence of molybdenum ions, can be interpreted as an increase in the positive charge carried by platinum in the platinum/zeolite catalyst.
2. X-Ray Absorption: (Analysis of X-ray absorption discontinuity at the threshold level). The results for the platinum/zeolite agree with XPS data, indicating an electronic deficiency of the metal on zeolite.
3. EPR Spectroscopy: With platinum the zeolitic support has ten times more electron donor sites capable of ionizing tetracyanoethylene than in its absence. The number of sites returns to the value for the pure supports when the metal is poisoned (carbon monoxide or molybdenum ions).
4. Infrared Spectroscopy: Shifts in the ν_{CO} vibration of adsorbed carbon monoxide correspond to an electron deficiency in the platinum particle when the latter is supported on the zeolite.

Platinum particles are electron deficient with electron transfer properties to the support.

692

B. IMELIK AND J. C. VEDRINE

IV. Is the metal modified by the zeolite?

As the catalytic properties of metallic platinum vary with the nature of the support there is some modification of the metal.

1. UPS Technique: (Variations in the "density of electronic states" near the Fermi level). Unfortunately, due to the low platinum concentration, contributions from platinum and the support cannot be distinguished.
2. EXAFS and Radial Electronic Distribution of X-Ray Diffusion: The Pt—Pt distance in the platinum-zeolite under vacuum is shorter (0.270 nm) than those in the bulk metal, but return close to the values for the fcc bulk structure (0.277 nm) in the presence of adsorbates (H_2, CO, etc.).

The small platinum particles have a more compact configuration than in the bulk metal but return to normal configuration in the presence of adsorbates.

25.2.2. Antimony/Tin Mixed Oxides

These are catalysts for the oxidation and ammoxidation of propene to acrolein and acrylonitrile, respectively. Selectivities increase with antimony content and especially with calcination temperature. The simple oxides SnO_2, Sb_2O_4, and Sb_6O_{13} have much lower selectivities.

I. After calcination, is the catalyst a solid solution of Sb^{n+} in SnO_2 or a mixture of phases?

1. X-Ray Diffraction: (All the solids are crystalline). The X-ray diffraction diagram of specimens calcined at 500 °C corresponds to SnO_2. The unit cell parameter calculated from the d-spacing ($d = 0.177$ nm) of the (211) plane varies with the antimony content according to Végard's law. In contrast, for specimens calcined at 950 °C, the variations of this unit cell parameter are close to the accuracy of the measurements and for antimony greater than 20%, lines of Sb_2O_4 are supported on those of SnO_2. Thus, the presence of solid solutions in the samples cannot be determined.
2. Infrared Spectroscopy: (Intensities of absorptions of Sb—O and Sn—O vibrations). Quantification of Sb_2O_4 and SnO_2 contents in the sample calcined at 950 °C is possible.

Thus calcination of 500 °C leads to solid solutions of antimony in SnO_2. By heating to 950 °C, antimony migrates out the solid solution to give Sb_2O_4.

II. What is the valency state of antimony in the solid solution: Sb^{5+} or Sb^{3+}?

1. Mössbauer Spectroscopy: Low-temperature (4 K) GNR spectrum of antimony differentiates Sb^{3+} from Sb^{5+} and measures their concentra-

tions. As all Sb^{3+}–Sb^{5+} pairs belong to Sb_2O_4 the remaining Sb^{5+} ions can be attributed to solid solution. The 500 °C calcined SnO_2 lattice can accommodate up to 40% Sb^{5+}. The Sn^{4+} ions of the SnO_2 lattice undergo distortion which is detected by Sn GNR.

2. Electrical Conductivity: (Electrical changes occasioned by the incorporation of Sb^{5+} in SnO_2 or by its extrusion by heating at 950 °C). Up to 6 at.% dissolved Sb^{5+} is found for the solid calcined at 500 °C and the conductivity increases by 10^6. Above 6% some Sb^{3+} cations may be dissolved.

III. Is the Sb_2O_4 extruded from the solid solution the catalytically active and selective phase?

This appears possible even though pure Sb_2O_4 is neither very active nor selective in propene oxidation.

1. EDX-STEM and Electron Diffraction: In samples of high selectivity the antimony is located both in large Sb_2O_4 crystallites and in low concentrations in small SnO_2 crystallites.
2. XPS: Surface of grains of SnO_2 is richer in antimony than the bulk.

Thus it was proposed that the active and selective phase is composed of SnO_2 whose surface is enriched in antimony with a thin peripheral layer of antimony oxide on the antimony-poor particles of the Sb–SnO_2 solid solution. The large Sb_2O_4 crystallites extruded from the solid solution by calcination have a negligible effect in catalysis because of their low surface area.

25.3. GENERAL CONCLUSION

These two examples demonstrate that it is possible to draw up a list of various bulk and surface properties by the judicious application of different physical techniques. In other words, it is possible to draw up what we have called an "identity card" and even to propose a description of the active site.

The description of the active site implies a knowlege of the surface crystallography under the reaction conditions. In general the surface arrangement of the atoms or ions of a solid is deduced from their arrangement in the bulk, as determined by X-ray diffraction or as defined by low-energy electron diffraction (LEED) or angular distribution photoemission spectroscopy (ADES) at temperatures and pressures very different from those used in the reaction. It has been proved that the surface atoms of a metallic oxide or sulfide particle are modified by the influence of adsorbates, and techniques such as IR, Raman, XRD, EXAFS, and XANES, are then very useful when used in catalytic reaction conditions. The description depends upon the number of active sites. This number may be high (e.g., surface atoms or ions), or, on the other hand, very low (e.g., very dilute active phase, faults, particular atomic arrangements, corners, and steps). In the latter case, with the exception of methods like EPR whose sensitivity is less than 1 ppm, the physical techniques are powerless since the results obtained are dominated by the contribution from the much more numerous inactive sites.

The continuous progress achieved in solid state physics — instrumentation and theory — and the market improvements in the spatial resolution and sensitivity of physical techniques, as well as their adaptation to the conditions of catalysis, all point to a bright future for their use for the study of catalysts and catalysis.

It must always be remembered that the comparative study of catalytic reactions is the only way to determine whether a surface site, as defined by physical techniques, is indeed an active site. Such studies are possible for metals and acid catalysts but are more difficult for, e.g., the oxides, sulfides, nitrides, and carbides as the precise relationship between the catalytic properties and the surface sites remain ill-defined.

INDEX